Lecture Notes in Computer Science 4513

Commenced Publication in 1973
Founding and Former Series Editors:
Gerhard Goos, Juris Hartmanis, and Jan van Leeuwen

Matteo Fischetti   David P. Williamson (Eds.)

Integer Programming and Combinatorial Optimization

12th International IPCO Conference
Ithaca, NY, USA, June 25-27, 2007
Proceedings

 Springer

Volume Editors

Matteo Fischetti
University of Padova
DEI, Dipartimento di Ingegneria dell'Informazione
via Gradenigo, 6/A, 35131 Padova, Italy
E-mail: matteo.fischetti@unipd.it

David P. Williamson
Cornell University
School of Operations Research and Information Engineering
Rhodes 236, Ithaca, NY 14853, USA
E-mail: dpw@orie.cornell.edu

Library of Congress Control Number: 2007928347

CR Subject Classification (1998): G.1.6, G.2.1, F.2.2, I.3.5

LNCS Sublibrary: SL 1 – Theoretical Computer Science and General Issues

ISSN	0302-9743
ISBN-10	3-540-72791-4 Springer Berlin Heidelberg New York
ISBN-13	978-3-540-72791-0 Springer Berlin Heidelberg New York

Springer is a part of Springer Science+Business Media

springer.com

© Springer-Verlag Berlin Heidelberg 2007
Printed in Germany

Typesetting: Camera-ready by author, data conversion by Scientific Publishing Services, Chennai, India
Printed on acid-free paper   SPIN: 12069569   06/3180   5 4 3 2 1 0

Preface

This volume contains the papers selected for presentation at IPCO XII, the 12th Conference on Integer Programming and Combinatorial Optimization, held June 25–27, 2007, in Ithaca, New York, USA. Since its inception in 1990, the IPCO conference series has become an important forum for researchers and practitioners working on various aspects of integer programming and combinatorial optimization. The aim of the conference is to present recent developments in theory, computation, and applications in these areas.

IPCO is sponsored by the Mathematical Programming Society, and is held in those years in which no International Symposium on Mathematical Programming takes place. The previous Symposium was held in 2006 in Rio de Janeiro, Brazil, and the previous two IPCOs were held in 2004 and 2005 in New York, USA and Berlin, Germany, respectively.

There were over 120 submissions to the conference. During its meeting in early January of 2007, the Program Committee carefully selected 36 papers for presentation in non-parallel sessions at the conference. Because of the limited number of time slots for presentations, many excellent submissions could not be accepted.

During the selection process, the extended abstracts were refereed according to the standards of refereed conferences. As a result, this volume contains papers describing high-quality research efforts. The page limit for contributions to these proceedings was set to 15. We expect full versions of these papers to appear in scientific journals in the near future.

We gratefully acknowledge IBM Research, ILOG, and the Office of Naval Research for their sponsorship of IPCO 2007. We are grateful for the use of EasyChair (www.easychair.org), which greatly simplified the process of collecting submissions, reviewing papers, and assembling this proceedings volume. We thank Phoebe Sengers and the Culturally Embedded Computing Group at Cornell, whose server was used to host the IPCO 2007 Web site. We thank the members of the Program Committee and the many subreferees who spent untold hours examining all of the submissions. And finally, we especially thank the many authors for submitting their work to the conference.

March 2007

Matteo Fischetti
David P. Williamson

Conference Organization

Program Committee

Dimitris Bertsimas (MIT)
Dan Bienstock (Columbia)
Alberto Caprara (Bologna)
Bill Cook (Georgia Tech)
Gérard Cornuéjols (CMU)
Matteo Fischetti, Chair (Padova)

Bertrand Guenin (Waterloo)
Christoph Helmberg (TU Chemnitz)
Tibor Jordán (ELTE Budapest)
Tom McCormick (UBC)
David P. Williamson (Cornell)
Gerhard Woeginger (Eindhoven)

Local Organization

David P. Williamson (Cornell)

Conference Sponsors

IBM Research

Table of Contents

Session 5

Session 6

Session 7

Session 8

Inequalities from Two Rows of a Simplex Tableau[*]

Kent Andersen[1], Quentin Louveaux[2], Robert Weismantel[3],
and Laurence A. Wolsey[4]

[1] Institute for Mathematical Sciences, University of Copenhagen, Denmark
`kha@math.ku.dk`
[2] CORE and INMA, Université catholique de Louvain, Belgium
`louveaux@core.ucl.ac.be`
[3] Department of Mathematics, Otto-von-Guericke Universität, Magdeburg, Germany
`weismant@mail.math.uni-magdeburg.de`
[4] CORE and INMA, Université catholique de Louvain, Belgium
`wolsey@core.ucl.ac.be`

Abstract. In this paper we explore the geometry of the integer points in a cone rooted at a rational point. This basic geometric object allows us to establish some links between lattice point free bodies and the derivation of inequalities for mixed integer linear programs by considering two rows of a simplex tableau simultaneously.

1 Introduction

Throughout this paper we investigate a mixed integer linear program (MIP) with rational data defined for a set I of integer variables and a set C of continuous variables

(MIP)   max $c^T x$ subject to $Ax = b$, $x \geq 0$, $x_i \in \mathbb{Z}$ for $i \in I$.

Let LP denote the linear programming relaxation of MIP. From the theory of linear programming it follows that a vertex x^* of the LP corresponds to a basic feasible solution of a simplex tableau associated with subsets B and N of basic and nonbasic variables

$$x_i + \sum_{j \in N} \bar{a}_{i,j} x_j = \bar{b}_i \text{ for } i \in B.$$

Any row associated with an index $i \in B \cap I$ such that $\bar{b}_i \notin \mathbb{Z}$ gives rise to a set

$$X(i) := \left\{ x \in \mathbb{R}^{|N|} \mid \bar{b}_i - \sum_{j \in N} \bar{a}_{i,j} x_j \in \mathbb{Z}, \ x_j \geq 0 \text{ for all } j \in N \right\}$$

[*] This work was partly carried out within the framework of ADONET, a European network in Algorithmic Discrete Optimization, contract no. MRTN-CT-2003-504438. The second author is supported by FRS-FNRS. This text presents research results of the Belgian Program on Interuniversity Poles of Attraction initiated by the Belgian State, Prime Minister's Office, Science Policy Programming. The scientific responsibility is assumed by the authors.

whose analysis provides inequalities that are violated by x^*. Indeed, Gomory's mixed integer cuts [4] and mixed integer rounding cuts [6] are derived from such a basic set $X(i)$ using additional information about integrality of some of the variables. Interestingly, unlike in the pure integer case, no finite convergence proof of a cutting plane algorithm is known when Gomory's mixed integer cuts or mixed integer rounding cuts are applied only. More drastically, in [3], an interesting mixed integer program in three variables is presented, and it is shown that applying split cuts iteratively does not suffice to generate the cut that is needed to solve this problem.

Example 1: [3] Consider the mixed integer set

$$t \leq x_1,$$
$$t \leq x_2,$$
$$x_1 + x_2 + t \leq 2,$$
$$x \in \mathbb{Z}^2 \text{ and } t \in \mathbb{R}_+^1.$$

The projection of this set onto the space of x_1 and x_2 variables is given by $\{(x_1, x_2) \in \mathbb{R}_+^2 : x_1 + x_2 \leq 2\}$ and is illustrated in Fig. 1. A simple analysis shows that the inequality $x_1 + x_2 \leq 2$, or equivalently $t \leq 0$, is valid. In [3] it is, however, shown that with the objective function $z = \max t$, a cutting plane algorithm based on split cuts does not converge finitely. $\qquad\square$

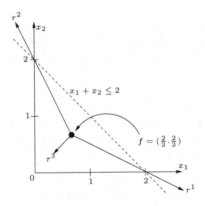

Fig. 1. The Instance in [3]

The analysis given in this paper will allow us to explain the cut $t \leq 0$ of Example 1. To this end we consider two indices $i_1, i_2 \in B \cap I$ simultaneously. It turns out that the underlying basic geometric object is significantly more complex than its one-variable counterpart. The set that we denote by $X(i_1, i_2)$ is described as

$$X(i_1, i_2) := \left\{ x \in \mathbb{R}^{|N|} \mid \bar{b}_i - \sum_{j \in N} \bar{a}_{i,j} x_j \in \mathbb{Z} \text{ for } i = i_1, i_2, \ x_j \geq 0 \text{ for all } j \in N \right\}.$$

Setting

$$f := \left(\bar{b}_{i_1}, \bar{b}_{i_2}\right)^T \in \mathbb{R}^2, \text{ and}$$
$$r^j := \left(\bar{a}_{i_1,j}, \bar{a}_{i_2,j}\right)^T \in \mathbb{R}^2,$$

the set obtained from two rows of a simplex tableau can be represented as

$$P_I := \{(x,s) \in \mathbb{Z}^2 \times \mathbb{R}^n_+ : x = f + \sum_{j \in N} s_j r^j\},$$

where f is fractional and $r^j \in \mathbb{R}^2$ for all $j \in N$. Valid inequalities for the set P_I was studied in [5] by using superadditive functions related to the group problem associated with two rows. In this paper, we give a characterization of the facets of $\text{conv}(P_I)$ based on its geometry.

Example 1 (revisited): For the instance of Example 1, introduce slack variables, s_1, s_2 and y_1 in the three constraints. Then, solving as a linear program, the constraints of the optimal simplex tableau are

$$
\begin{array}{rrrrr}
t & +\frac{1}{3}s_1 & +\frac{1}{3}s_2 & +\frac{1}{3}y_1 & = \frac{2}{3} \\
x_1 & -\frac{2}{3}s_1 & +\frac{1}{3}s_2 & +\frac{1}{3}y_1 & = \frac{2}{3} \\
x_2 & +\frac{1}{3}s_1 & -\frac{2}{3}s_2 & +\frac{1}{3}y_1 & = \frac{2}{3}
\end{array}
$$

Taking the last two rows, and rescaling using $s'_i = s_i/3$ for $i = 1, 2$, we obtain the set P_I

$$
\begin{array}{rrrr}
x_1 & -2s'_1 & +1s'_2 & +\frac{1}{3}y_1 & = +\frac{2}{3} \\
x_2 & +1s'_1 & -2s'_2 & +\frac{1}{3}y_1 & = +\frac{2}{3}
\end{array}
$$
$$x \in \mathbb{Z}^2, s \in \mathbb{R}^2_+, y_1 \in \mathbb{R}^1_+.$$

Letting $f = (\frac{2}{3}, \frac{2}{3})^T$, $r^1 = (2, -1)^T$, $r^2 = (-1, 2)^T$ and $r_3 = (-\frac{1}{3}, -\frac{1}{3})^T$ (see Fig. 1), one can derive a cut for $\text{conv}(P_I)$ of the form

$$x_1 + x_2 + y_1 \geq 2 \text{ or equivalently } t \leq 0,$$

which, when used in a cutting plane algorithm, yields immediate termination.

□

Our main contribution is to characterize geometrically all facets of $\text{conv}(P_I)$. All facets are *intersection cuts* [2], *i.e.*, they can be obtained from a (two-dimensional) convex body that does not contain any integer points in its interior. Our geometric approach is based on two important facts that we prove in this paper

- every facet is derivable from at most four nonbasic variables.
- with every facet F one can associate three or four particular vertices of $\text{conv}(P_I)$. The classification of F depends on how the corresponding $k = 3, 4$ integer points in $\mathbb{Z}^2$ can be partitioned into k sets of cardinality at most two.

More precisely, the facets of $\text{conv}(P_I)$ can be distinguished with respect to the number of sets that contain two integer points. Since $k = 3$ or $k = 4$, the following interesting situations occur

- no sets with cardinality two: all the $k \in \{3,4\}$ sets contain exactly one tight integer point. We call cuts of this type *disection cuts*.
- exactly one set has cardinality two: in this case we show that the inequality can be derived from lifting a cut associated with a two-variable subproblem to k variables. We call these cuts *lifted two-variable cuts*.
- two sets have cardinality two. In this case we show that the corresponding cuts are *split cuts*.

Furthermore, we show that inequalities of the first two families are not split cuts. Our geometric approach allows us to generalize the cut introduced in Example 1. More specifically, the cut of Example 1 is a degenerate case in the sense that it is "almost" a disection cut and "almost" a lifted two-variable cut: by perturbing the vectors r^1, r^2 and r^3 slightly, the cut in Example 1 can become both a disection cut and a lifted two-variable cut.

We review some basic facts about the structure of $\operatorname{conv}(P_I)$ in Section 2. In Section 3 we explore the geometry of all the feasible points that are tight for a given facet of $\operatorname{conv}(P_I)$, explain our main result and presents the classification of all the facets of $\operatorname{conv}(P_I)$.

2   Basic Structure of conv(P_I)

The basic mixed-integer set considered in this paper is

$$P_I := \{(x,s) \in \mathbb{Z}^2 \times \mathbb{R}^n_+ : x = f + \sum_{j \in N} s_j r^j\}, \qquad (1)$$

where $N := \{1,2,\ldots,n\}$, $f \in \mathbb{Q}^2 \setminus \mathbb{Z}^2$ and $r^j \in \mathbb{Q}^2$ for all $j \in N$. The set $P_{LP} := \{(x,s) \in \mathbb{R}^2 \times \mathbb{R}^n_+ : x = f + \sum_{j \in N} s_j r^j\}$ denotes the LP relaxation of P_I. The jth unit vector in $\mathbb{R}^n$ is denoted e_j. In this section, we describe some basic properties of $\operatorname{conv}(P_I)$. The vectors $\{r^j\}_{j \in N}$ are called *rays*, and we assume $r^j \neq 0$ for all $j \in N$.

In the remainder of the paper we assume $P_I \neq \emptyset$. The next lemma gives a characterization of $\operatorname{conv}(P_I)$ in terms of vertices and extreme rays.

Lemma 1.

(i) *The dimension of* $\operatorname{conv}(P_I)$ *is* n.
(ii) *The extreme rays of* $\operatorname{conv}(P_I)$ *are* (r^j, e_j) *for* $j \in N$.
(iii) *The vertices* (x^I, s^I) *of* $\operatorname{conv}(P_I)$ *take the following two forms:*
   (a) $(x^I, s^I) = (x^I, s^I_j e_j)$, *where* $x^I = f + s^I_j r^j \in \mathbb{Z}^2$ *and* $j \in N$
       (an integer point on the ray $\{f + s_j r^j : s_j \geq 0\}$*).*
   (b) $(x^I, s^I) = (x^I, s^I_j e_j + s^I_k e_k)$, *where* $x^I = f + s^I_j r^j + s^I_k r^k \in \mathbb{Z}^2$ *and* $j,k \in N$
       (an integer point in the set $f + \operatorname{cone}(\{r^j, r^k\})$*).*

Using Lemma 1, we now give a simple form for the valid inequalities for $\operatorname{conv}(P_I)$ considered in the remainder of the paper.

Corollary 1. *Every non-trivial valid inequality for P_I that is tight at a point $(\bar{x}, \bar{s}) \in P_I$ can be written in the form*

$$\sum_{j \in N} \alpha_j s_j \geq 1, \tag{2}$$

where $\alpha_j \geq 0$ for all $j \in N$.

For an inequality $\sum_{j \in N} \alpha_j s_j \geq 1$ of the form (2), let $N_\alpha^0 := \{j \in N : \alpha_j = 0\}$ denote the variables with coefficient zero, and let $N_\alpha^{\neq 0} := N \setminus N_\alpha^0$ denote the remainder of the variables. We now introduce an object that is associated with the inequality $\sum_{j \in N} \alpha_j s_j \geq 1$. We will use this object to obtain a two dimensional representation of the facets of conv(P_I).

Lemma 2. *Let $\sum_{j \in N} \alpha_j s_j \geq 1$ be a valid inequality for conv(P_I) of the form (2). Define $v^j := f + \frac{1}{\alpha_j} r^j$ for $j \in N_\alpha^{\neq 0}$. Consider the convex polyhedron in $\mathbb{R}^2$*

$$L_\alpha := \{x \in \mathbb{R}^2 : \text{ there exists } s \in \mathbb{R}_+^n \text{ s.t. } (x, s) \in P_{LP} \text{ and } \sum_{j \in N} \alpha_j s_j \leq 1\}.$$

(i) $L_\alpha = \text{conv}(\{f\} \cup \{v^j\}_{j \in N_\alpha^{\neq 0}}) + \text{cone}(\{r^j\}_{j \in N_\alpha^0})$.
(ii) interior(L_α) does not contain any integer points.
(iii) If $\text{cone}(\{r^j\}_{j \in N}) = \mathbb{R}^2$, then $f \in \text{interior}(L_\alpha)$.

Example 2: Consider the set

$$P_I = \{(x, s) : x = f + \begin{pmatrix} 2 \\ 1 \end{pmatrix} s_1 + \begin{pmatrix} 1 \\ 1 \end{pmatrix} s_2 + \begin{pmatrix} -3 \\ 2 \end{pmatrix} s_3 + \begin{pmatrix} 0 \\ -1 \end{pmatrix} s_4 + \begin{pmatrix} 1 \\ -2 \end{pmatrix} s_5\},$$

where $f = \begin{pmatrix} \frac{1}{4} \\ \frac{1}{2} \end{pmatrix}$, and consider the inequality

$$2s_1 + 2s_2 + 4s_3 + s_4 + \frac{12}{7} s_5 \geq 1. \tag{3}$$

The corresponding set L_α is shown in Fig. 2. As can be seen from the figure, L_α does not contain any integer points in its interior. It follows that (3) is valid for conv(P_I). Note that, conversely, the coefficients α_j for $j = 1, 2, \ldots, 5$ can be obtained from the polygon L_α as follows: α_j is the ratio between the length of r^j and the distance between f and v^j. In particular, if the length of r^j is 1, then α_j is the inverse of the distance from f to v^j.   □

The interior of L_α gives a two-dimensional representation of the points $x \in \mathbb{R}^2$ that are affected by the addition of the inequality $\sum_{j \in N} \alpha_j s_j \geq 1$ to the LP relaxation P_{LP} of P_I. In other words, for any $(x, s) \in P_{LP}$ that satisfies $\sum_{j \in N} \alpha_j s_j < 1$, we have $x \in \text{interior}(L_\alpha)$. Furthermore, for a facet defining inequality $\sum_{j \in N} \alpha_j s_j \geq 1$ of conv(P_I), there exist n affinely independent points

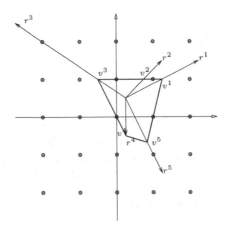

Fig. 2. The set L_α for a valid inequality for $\mathrm{conv}(P_I)$

$(x^i, s^i) \in P_I$, $i = 1, 2, \ldots, n$, such that $\sum_{j \in N} \alpha_j s^i_j = 1$. The integer points $\{x^i\}_{i \in N}$ are on the boundary of L_α, i.e., they belong to the integer set:

$$X_\alpha := \{x \in \mathbb{Z}^2 : \exists s \in \mathbb{R}^n_+ \text{ s.t. } (x, s) \in P_{LP} \text{ and } \sum_{j \in N} \alpha_j s_j = 1\}.$$

We have $X_\alpha = L_\alpha \cap \mathbb{Z}^2$, and $X_\alpha \neq \emptyset$ whenever $\sum_{j \in N} \alpha_j s_j \geq 1$ defines a facet of $\mathrm{conv}(P_I)$. We first characterize the facets of $\mathrm{conv}(P_I)$ that have zero coefficients.

Lemma 3. *Any facet defining inequality $\sum_{j \in N} \alpha_j s_j \geq 1$ for $\mathrm{conv}(P_I)$ of the form (2) that has zero coefficients is a split cut. In other words, if $N^0_\alpha \neq \emptyset$, there exists $(\pi, \pi_0) \in \mathbb{Z}^2 \times \mathbb{Z}$ such that $L_\alpha \subseteq \{(x_1, x_2) : \pi_0 \leq \pi_1 x_1 + \pi_2 x_2 \leq \pi_0 + 1\}$.*

Proof: Let $k \in N^0_\alpha$ be arbitrary. Then the line $\{f + \mu r^k : \mu \in \mathbb{R}\}$ does not contain any integer points. Furthermore, if $j \in N^0_\alpha$, $j \neq k$, is such that r^k and r^j are *not* parallel, then $f + \mathrm{cone}(\{r^k, r^j\})$ contains integer points. It follows that all rays $\{r^j\}_{j \in N^0_\alpha}$ are parallel. By letting $\pi' := (r^k)^\perp = (-r^k_2, r^k_1)^T$ and $\pi'_0 := (\pi')^T f$, we have that $\{f + \mu r^k : \mu \in \mathbb{R}\} = \{x \in \mathbb{R}^2 : \pi'_1 x_1 + \pi'_2 x_2 = \pi'_0\}$. Now define:

$$\pi^1_0 := \max\{\pi'_1 x_1 + \pi'_2 x_2 : \pi'_1 x_1 + \pi'_2 x_2 \leq \pi'_0 \text{ and } x \in \mathbb{Z}^2\}, \text{ and}$$
$$\pi^2_0 := \min\{\pi'_1 x_1 + \pi'_2 x_2 : \pi'_1 x_1 + \pi'_2 x_2 \geq \pi'_0 \text{ and } x \in \mathbb{Z}^2\}.$$

We have $\pi^1_0 < \pi'_0 < \pi^2_0$, and the set $S_\pi := \{x \in \mathbb{R}^2 : \pi^1_0 \leq \pi'_1 x_1 + \pi'_2 x_2 \leq \pi^2_0\}$ does not contain any integer points in its interior. We now show $L_\alpha \subseteq S_\pi$ by showing that every vertex $v^m = f + \frac{1}{\alpha_m} r^m$ of L_α, where $m \in N^{\neq 0}_\alpha$, satisfies $v^m \in S_\pi$. Suppose v^m satisfies $\pi'_1 v^m_1 + \pi'_2 v^m_2 < \pi^1_0$ (the case $\pi'_1 v^m_1 + \pi'_2 v^m_2 > \pi^2_0$ is symmetric). By definition of π^1_0, there exists $x^I \in \mathbb{Z}^2$ such that $\pi'_1 x^I_1 + \pi'_2 x^I_2 = \pi^1_0$, and $x^I = \lambda v^m + (1 - \lambda)(f + \delta r^k)$, where $\lambda \in]0, 1[$, for some $\delta > 0$. We then have

$x^I = f + \frac{\lambda}{\alpha_m} r^m + \delta(1-\lambda) r^k$. Inserting this representation of x^I into the inequality $\sum_{j \in N} \alpha_j s_j \geq 1$ then gives $\alpha_m \frac{\lambda}{\alpha_m} + \alpha_k \delta(1 - \lambda) = \lambda < 1$, which contradicts the validity of $\sum_{j \in N} \alpha_j s_j \geq 1$ for P_I. Hence $L_\alpha \subseteq S_\pi$.

To finish the proof, we show that we may write $S_\pi = \{x \in \mathbb{R}^2 : \pi_0 \leq \pi_1 x_1 + \pi_2 x_2 \leq \pi_0 + 1\}$ for some $(\pi, \pi_0) \in \mathbb{Z}^2 \times \mathbb{Z}$. First observe that we can assume (by scaling) that π', π_0^1 and π_0^2 are integers. Next observe that any common divisor of π_1' and π_2' also divides both π_0^1 and π_0^2 (this follows from the fact that there exists $x^1, x^2 \in \mathbb{Z}^2$ such that $\pi_1' x_1^1 + \pi_2' x_2^1 = \pi_0^1$ and $\pi_1' x_1^2 + \pi_2' x_2^2 = \pi_0^2$). Hence we can assume that π_1' and π_2' are relative prime. Now the Integral Farkas Lemma (see [8]) implies that the set $\{x \in \mathbb{Z}^2 : \pi_1' x_1 + \pi_2' x_2 = 1\}$ is non-empty. It follows that we must have $\pi_0^2 = \pi_0^1 + 1$, since otherwise the point $\bar{y} := x' + x^1 \in \mathbb{Z}^2$, where $x' \in \{x \in \mathbb{Z}^2 : \pi_1' x_1 + \pi_2' x_2 = 1\}$ and $x^1 \in \{x \in \mathbb{Z}^2 : \pi_1' x_1 + \pi_2' x_2 = \pi_0^1\}$, satisfies $\pi_0^1 < \pi_1' \bar{y}_1 + \pi_2' \bar{y}_2 < \pi_0^2$, which contradicts that S_π does not contain any integer points in its interior.   □

3   A Characterization of conv(X_α) and conv(P_I)

As a preliminary step of our analysis, we first characterize the set conv(X_α). We assume $\alpha_j > 0$ for all $j \in N$. Clearly conv(X_α) is a convex polygon with only integer vertices, and since $X_\alpha \subseteq L_\alpha$, conv($X_\alpha$) does not have any integer points in its interior. We first limit the number of vertices of conv(X_α) to four (see [1] and [7] for this and related results).

Lemma 4. *Let $P \subset \mathbb{R}^2$ be a convex polygon with integer vertices that has no integer points in its interior.*

(i) P has at most four vertices
(ii) If P has four vertices, then at least two of its four facets are parallel.
(iii) If P is not a triangle with integer points in the interior of all three facets (see Fig. 3.(c)), then there exists parallel lines $\pi x = \pi_0$ and $\pi x = \pi_0 + 1$, $(\pi, \pi_0) \in \mathbb{Z}^3$, such that P is contained in the corresponding split set, i.e., $P \subseteq \{x \in \mathbb{R}^2 : \pi_0 \leq \pi x \leq \pi_0 + 1\}$.

Lemma 4 shows that the polygons in Fig. 3 include all possible polygons that can be included in the set L_α in the case when L_α is bounded and of dimension 2. The dashed lines in Fig. 3 indicate the possible split sets that include P. We excluded from Fig. 3 the cases when X_α is of dimension 1. We note that Lemma 4.(iii) (existence of split sets) proves that there cannot be any triangles where two facets have interior integer points, and also that no quadrilateral can have more than two facets that have integer points in the interior.

Next, we focus on the set L_α. As before we assume $\alpha_j > 0$ for all $j \in N$. Due to the direct correspondence between the set L_α and a facet defining inequality $\sum_{j \in N} \alpha_j s_j \geq 1$ for conv(P_I), this gives a characterization of the facets

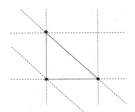
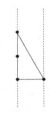

(a) A triangle: no facet has interior integer points

(b) A triangle: one facet has interior integer points

(c) A triangle: all facets have interior integer points

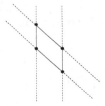

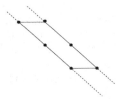

(d) A quadrilateral: no facet has interior integer points

(e) A quadrilateral: one facet has interior integer points

(f) A quadrilateral: two facets have interior integer points

Fig. 3. All integer polygons that do not have interior integer points

of $\text{conv}(P_I)$. The main result in this section is that L_α can have at most four vertices. In other words, we prove

Theorem 1. *Let $\sum_{j\in N} \alpha_j s_j \geq 1$ be a facet defining inequality for $\text{conv}(P_I)$ that satisfies $\alpha_j > 0$ for all $j \in N$. Then L_α is a polygon with at most four vertices.*

Theorem 1 shows that there exists a set $S \subseteq N$ such that $|S| \leq 4$ and $\sum_{j\in S} \alpha_j s_j \geq 1$ is facet defining for $\text{conv}(P_I(S))$, where

$$P_I(S) := \{(x,s) \in \mathbb{Z}^2 \times \mathbb{R}_+^{|S|} : x = f + \sum_{j\in S} s_j r^j\}.$$

Throughout this section we assume that no two rays point in the same direction. If two variables $j_1, j_2 \in N$ are such that $j_1 \neq j_2$ and $r^{j_1} = \delta r^{j_2}$ for some $\delta > 0$, then the halflines $\{x \in \mathbb{R}^2 : x = f + s_{j_1} r^{j_1}, s_{j_1} \geq 0\}$ and $\{x \in \mathbb{R}^2 : x = f + s_{j_2} r^{j_2}, s_{j_2} \geq 0\}$ intersect the boundary of L_α at the same point, and therefore $L_\alpha = \text{conv}(\{f\} \cup \{v^j\}_{j\in N}) = \text{conv}(\{f\} \cup \{v^j\}_{j\in N\setminus\{j_2\}})$, where $v^j := f + \frac{1}{\alpha_j} r^j$ for $j \in N$. This assumption does therefore not affect the validity of Theorem 1.

The proof of Theorem 1 is based on characterizing the vertices $\text{conv}(P_I)$ that are tight for $\sum_{j\in N} \alpha_j s_j \geq 1$. We show that there exists a subset $S \subseteq N$ of variables and a set of $|S|$ affinely independent vertices of $\text{conv}(P_I)$ such that

$|S| \leq 4$ and $\{\alpha_j\}_{j \in S}$ is the unique solution to the equality system of the polar defined by these vertices. The following notation will be used intensively in the remainder of this section.

Notation 1

(i) *The number $k \leq 4$ denotes the number of vertices of* $\mathrm{conv}(X_\alpha)$.

(ii) *The set $\{x^v\}_{v \in K}$ denotes the vertices of* $\mathrm{conv}(X_\alpha)$, *where $K := \{1, 2, \ldots, k\}$.*

Recall that Lemma 1.(iii) demonstrates that for a vertex $(\bar{x}, \bar{s})$ of $\mathrm{conv}(P_I)$, $\bar{s}$ is positive on at most two coordinates $j_1, j_2 \in N$, and in that case $\bar{x} \in f + \mathrm{cone}(\{r^{j_1}, r^{j_2}\})$. If $\bar{s}$ is positive on only one coordinate $j \in N$, then $\bar{x} = f + \bar{s}_j r^j$, and the inequality of the polar corresponding to $(\bar{x}, \bar{s})$ is $\alpha_j \bar{s}_j \geq 1$, which simply states $\alpha_j \geq \frac{1}{\bar{s}_j}$. A point $\bar{x} \in \mathbb{Z}^2$ that satisfies $\bar{x} \in \{x \in \mathbb{R}^2 : x = f + s_j r^j, s_j \geq 0\}$ for some $j \in N$ is called a *ray point* in the remainder of the paper. In order to characterize the tight inequalities of the polar that correspond to vertices x^v of $\mathrm{conv}(X_\alpha)$ that are *not* ray points, we introduce the following concepts.

Definition 1. *Let $\sum_{j \in N} \alpha_j s_j \geq 1$ be valid for* $\mathrm{conv}(P_I)$. *Suppose $\bar{x} \in \mathbb{Z}^2$ is not a ray point, and that $\bar{x} \in f + \mathrm{cone}(\{r^{j_1}, r^{j_2}\})$, where $j_1, j_2 \in N$. This implies $\bar{x} = f + s_{j_1} r^{j_1} + s_{j_2} r^{j_2}$, where $s_{j_1}, s_{j_1} > 0$ are unique.*

(a) *The pair (j_1, j_2) is said to give a* representation of $\bar{x}$.

(b) *If $\alpha_{j_1} s_{j_1} + \alpha_{j_2} s_{j_2} = 1$, (j_1, j_2) is said to give a* tight representation of $\bar{x}$ *wrt.* $\sum_{j \in N} \alpha_j s_j \geq 1$.

(c) *If $(i_1, i_2) \in N \times N$ satisfies $\mathrm{cone}(\{r^{i_1}, r^{i_2}\}) \subseteq \mathrm{cone}(\{r^{j_1}, r^{j_2}\})$, the pair (i_1, i_2) is said to define a* subcone of (j_1, j_2).

Example 2 (continued): Consider again the set

$$P_I = \{(x, s) : x = f + \begin{pmatrix} 2 \\ 1 \end{pmatrix} s_1 + \begin{pmatrix} 1 \\ 1 \end{pmatrix} s_2 + \begin{pmatrix} -3 \\ 2 \end{pmatrix} s_3 + \begin{pmatrix} 0 \\ -1 \end{pmatrix} s_4 + \begin{pmatrix} 1 \\ -2 \end{pmatrix} s_5\},$$

where $f = \begin{pmatrix} \frac{1}{4} \\ \frac{1}{2} \end{pmatrix}$, and the valid inequality $2s_1 + 2s_2 + 4s_3 + s_4 + \frac{12}{7} s_5 \geq 1$ for $\mathrm{conv}(P_I)$. The point $\bar{x} = (1, 1)$ is on the boundary of L_α (see Fig. 2). We have that $\bar{x}$ can be written in any of the following forms

$$\bar{x} = f + \frac{1}{4} r^1 + \frac{1}{4} r^2,$$

$$\bar{x} = f + \frac{3}{7} r^1 \qquad + \frac{1}{28} r^3,$$

$$\bar{x} = f \qquad + \frac{3}{4} r^2 \qquad + \frac{1}{4} r^4.$$

It follows that $(1, 2), (1, 3)$ and $(2, 4)$ all give representations of $\bar{x}$. Note that $(1, 2)$ and $(1, 3)$ give tight representations of $\bar{x}$ wrt. the inequality $2s_1 + 2s_2 +$

$4s_3 + s_4 + \frac{12}{7}s_5 \geq 1$, whereas $(2,4)$ does not. Finally note that $(1,5)$ defines a subcone of $(2,4)$. □

Observe that, for a vertex x^v of $\mathrm{conv}(X_\alpha)$ which is not a ray point, and a tight representation (j_1, j_2) of x^v, the corresponding inequality of the polar satisfies $\alpha_{j_1} t_{j_1} + \alpha_{j_2} t_{j_2} = 1$, where $t_{j_1}, t_{j_2} > 0$. We now characterize the set of tight representations of an integer point $\bar{x} \in \mathbb{Z}^2$, which is not a ray point

$$T_\alpha(\bar{x}) := \{(j_1, j_2) : (j_1, j_2) \text{ gives a tight representation of } \bar{x} \text{ wrt. } \sum_{j \in N} \alpha_j s_j \geq 1\}.$$

We show that $T_\alpha(\bar{x})$ contains a unique maximal representation $(j_1^{\bar{x}}, j_2^{\bar{x}}) \in T_\alpha(\bar{x})$ with the following properties.

Lemma 5. *There exists a unique maximal representation $(j_1^{\bar{x}}, j_2^{\bar{x}}) \in T_\alpha(\bar{x})$ of $\bar{x}$ that satisfies:*

(i) Every subcone (j_1, j_2) of $(j_1^{\bar{x}}, j_2^{\bar{x}})$ that gives a representation of $\bar{x}$ satisfies $(j_1, j_2) \in T_\alpha(\bar{x})$.
(ii) Conversely, every $(j_1, j_2) \in T_\alpha(\bar{x})$ defines a subcone of $(j_1^{\bar{x}}, j_2^{\bar{x}})$.

To prove Lemma 5, there are two cases to consider. For two representations (i_1, i_2) and (j_1, j_2) of $\bar{x}$, either one of the two cones (i_1, i_2) and (j_1, j_2) is contained in the other (Lemma 6), or their intersection defines a subcone of both (i_1, i_2) and (j_1, j_2) (Lemma 7). Note that we cannot have that their intersection is empty, since they both give a representation of $\bar{x}$.

Lemma 6. *Let $\sum_{j \in N} \alpha_j s_j \geq 1$ be a facet defining inequality for $\mathrm{conv}(P_I)$ that satisfies $\alpha_j > 0$ for all $j \in N$, and let $\bar{x} \in \mathbb{Z}^2$. Then $(j_1, j_2) \in T_\alpha(\bar{x})$ implies $(i_1, i_2) \in T_\alpha(\bar{x})$ for every subcone (i_1, i_2) of (j_1, j_2) that gives a representation of $\bar{x}$.*

Proof: Suppose $(j_1, j_2) \in T_\alpha(\bar{x})$. Observe that it suffices to prove the following: for any $j_3 \in N$ such that $r^{j_3} \in \mathrm{cone}(\{r^{j_1}, r^{j_2}\})$ and (j_1, j_3) gives a representation of $\bar{x}$, the representation (j_1, j_3) is tight wrt. $\sum_{j \in N} \alpha_j s_j \geq 1$. The result for all remaining subcones of (j_1, j_2) follows from repeated application of this result. For simplicity we assume $j_1 = 1$, $j_2 = 2$ and $j_3 = 3$.

Since $\bar{x} \in f + \mathrm{cone}(\{r^1, r^2\})$, $\bar{x} \in f + \mathrm{cone}(\{r^1, r^3\})$ and $r^3 \in \mathrm{cone}(\{r^1, r^2\})$, we may write $\bar{x} = f + u_1 r^1 + u_2 r^2$, $\bar{x} = f + v_1 r^1 + v_3 r^3$ and $r^3 = w_1 r^1 + w_2 r^2$, where $u_1, u_2, v_1, v_3, w_1, w_2 \geq 0$. Furthermore, since $(1,2)$ gives a tight representation of $\bar{x}$ wrt. $\sum_{j \in N} \alpha_j s_j \geq 1$, we have $\alpha_1 u_1 + \alpha_2 u_2 = 1$. Finally we have $\alpha_1 v_1 + \alpha_3 v_3 \geq 1$, since $\sum_{j \in N} \alpha_j s_j \geq 1$ is valid for P_I. If also $\alpha_1 v_1 + \alpha_3 v_3 = 1$, we are done, so suppose $\alpha_1 v_1 + \alpha_3 v_3 > 1$.

We first argue that this implies $\alpha_3 > \alpha_1 w_1 + \alpha_2 w_2$. Since $\bar{x} = f + u_1 r^1 + u_2 r^2 = f + v_1 r^1 + v_3 r^3$, it follows that $(u_1 - v_1) r^1 = v_3 r^3 - u_2 r^2$. Now, using the representation $r^3 = w_1 r^1 + w_2 r^2$, we get $(u_1 - v_1 - v_3 w_1) r^1 + (u_2 - v_3 w_2) r^2 = 0$. Since r^1 and r^2 are linearly independent, we obtain:

$$(u_1 - v_1) = v_3 w_1 \text{ and } u_2 = v_3 w_2.$$

Now we have $\alpha_1 v_1 + \alpha_3 v_3 > 1 = \alpha_1 u_1 + \alpha_2 u_2$, which implies $(v_1 - u_1)\alpha_1 - \alpha_2 u_2 + \alpha_3 v_3 > 0$. Using the identities derived above, we get $-v_3 w_1 \alpha_1 - \alpha_2 v_3 w_2 + \alpha_3 v_3 > 0$, or equivalently $v_3(-w_1\alpha_1 - \alpha_2 w_2 + \alpha_3) > 0$. It follows that $\alpha_3 > \alpha_1 w_1 + \alpha_2 w_2$.

We now derive a contradiction to the identity $\alpha_3 > \alpha_1 w_1 + \alpha_2 w_2$. Since $\sum_{j \in N} \alpha_j s_j \geq 1$ defines a facet of $\text{conv}(P_I)$, there must exist $x' \in \mathbb{Z}^2$ and $k \in N$ such that $(3, k)$ gives a tight representation of x' wrt. $\sum_{j \in N} \alpha_j s_j \geq 1$. In other words, there exists $x' \in \mathbb{Z}^2$, $k \in N$ and $\delta_3, \delta_k \geq 0$ such that $x' = f + \delta_3 r^3 + \delta_k r^k$ and $\alpha_3 \delta_3 + \alpha_k \delta_k = 1$. Furthermore, we can choose x', δ_3 and δ_k such that r^3 is used in the representation of x', i.e., we can assume $\delta_3 > 0$.

Now, using the representation $r^3 = w_1 r^1 + w_2 r^2$ then gives $x' = f + \delta_3 r^3 + \delta_k r^k = f + \delta_3 w_1 r^1 + \delta_3 w_2 r^2 + \delta_k r^k$. Since $\sum_{j \in N} \alpha_j s_j \geq 1$ is valid for P_I, we have $\alpha_1 \delta_3 w_1 + \alpha_2 \delta_3 w_2 + \alpha_k \delta_k \geq 1 = \alpha_3 \delta_3 + \alpha_k \delta_k$. This implies $\delta_3(\alpha_3 - \alpha_1 w_1 - \alpha_2 w_2) \leq 0$, and therefore $\alpha_3 \leq \alpha_1 w_1 - \alpha_2 w_2$, which is a contradiction. $\square$

Lemma 7. *Let $\sum_{j \in N} \alpha_j s_j \geq 1$ be a facet defining inequality for $\text{conv}(P_I)$ satisfying $\alpha_j > 0$ for $j \in N$, and suppose $\bar{x} \in \mathbb{Z}^2$ is not a ray point. Also suppose the intersection between the cones $(j_1, j_2), (j_3, j_4) \in T_\alpha(\bar{x})$ is given by the subcone (j_2, j_3) of both (j_1, j_2) and (j_3, j_4). Then $(j_1, j_4) \in T_\alpha(\bar{x})$, i.e., (j_1, j_4) also gives a tight representation of $\bar{x}$.*

Proof: For simplicity assume $j_1 = 1$, $j_2 = 2$, $j_3 = 3$ and $j_4 = 4$. Since the cones $(1, 2)$ and $(3, 4)$ intersect in the subcone $(2, 3)$, we have $r^3 \in \text{cone}(\{r^1, r^2\})$, $r^2 \in \text{cone}(\{r^3, r^4\})$, $r^4 \notin \text{cone}(\{r^1, r^2\})$ and $r^1 \notin \text{cone}(\{r^3, r^4\})$. We first represent $\bar{x}$ in the translated cones in which we have a tight representation of $\bar{x}$. In other words, we can write

$$\bar{x} = f + u_1 r^1 + u_2 r^2, \tag{4}$$

$$\bar{x} = f + v_3 r^3 + v_4 r^4 \text{ and} \tag{5}$$

$$\bar{x} = f + z_2 r^2 + z_3 r^3, \tag{6}$$

where $u_1, u_2, v_3, v_4, z_2, z_3 > 0$. Note that Lemma 6 proves that (6) gives a tight representation of $\bar{x}$. Using (4)-(6), we obtain the relation

$$\begin{pmatrix} T_{1,1} I_2 & T_{1,2} I_2 \\ T_{2,1} I_2 & T_{2,2} I_2 \end{pmatrix} \begin{pmatrix} r^2 \\ r^3 \end{pmatrix} = \begin{pmatrix} u_1 r^1 \\ v_4 r^4 \end{pmatrix}, \tag{7}$$

where T is the 2×2 matrix $T := \begin{pmatrix} T_{1,1} & T_{1,2} \\ T_{2,1} & T_{2,2} \end{pmatrix} = \begin{pmatrix} (z_2 - u_2) & z_3 \\ z_2 & (z_3 - v_3) \end{pmatrix}$ and I_2 is the 2×2 identity matrix. On the other hand, the tightness of the representations (4)-(6) leads to the following identities

$$\alpha_1 u_1 + \alpha_2 u_2 = 1, \tag{8}$$

$$\alpha_3 v_3 + \alpha_4 v_4 = 1 \text{ and} \tag{9}$$

$$\alpha_2 z_2 + \alpha_4 z_3 = 1, \tag{10}$$

where, again, the last identity follows from Lemma 6. Using (8)-(10), we obtain the relation

$$\begin{pmatrix} T_{1,1} & T_{1,2} \\ T_{2,1} & T_{2,2} \end{pmatrix}\begin{pmatrix} \alpha_2 \\ \alpha_3 \end{pmatrix} = \begin{pmatrix} u_1\alpha_1 \\ v_4\alpha_4 \end{pmatrix}. \tag{11}$$

We now argue that T is non-singular. Suppose, for a contradiction, that $T_{1,1}T_{2,2} = T_{1,2}T_{2,1}$. From (5) and (6) we obtain $v_4 r^4 = (z_3 - v_3)r^3 + z_2 r^2$, which implies $z_3 < v_3$, since $r^4 \notin \text{cone}(\{r^1, r^2\}) \supseteq \text{cone}(\{r^2, r^3\})$. Multiplying the first equation of (11) with $T_{2,2}$ gives $T_{2,2}T_{1,1}\alpha_2 + T_{2,2}T_{1,2}\alpha_3 = u_1 T_{2,2}\alpha_1$, which implies $T_{1,2}(T_{2,1}\alpha_2 + T_{2,2}\alpha_3) = u_1 T_{2,2}\alpha_1$. By using the definition of T, this can be rewritten as $z_3(\alpha_2 z_2 + (z_3 - v_3)\alpha_3) = u_1\alpha_1(z_3 - v_3)$. Since $z_2\alpha_2 + z_3\alpha_3 = 1$, this implies $z_3(1 - v_3\alpha_3) = u_1\alpha_1(z_3 - v_3)$. However, from (9) we have $v_3\alpha_3 \in]0, 1[$, so $z_3(1 - v_3\alpha_3) > 0$ and $u_1\alpha_1(z_3 - v_3) < 0$, which is a contradiction. Hence T is non-singular.

We now solve (7) for an expression of r^2 and r^3 in terms of r^1 and r^4. The inverse of the coefficient matrix on the left hand side of (7) is given by $\begin{pmatrix} T_{1,1}^{-1}I_2 & T_{1,2}^{-1}I_2 \\ T_{2,1}^{-1}I_2 & T_{2,2}^{-1}I_2 \end{pmatrix}$, where $T^{-1} := \begin{pmatrix} T_{1,1}^{-1} & T_{1,2}^{-1} \\ T_{2,1}^{-1} & T_{2,2}^{-1} \end{pmatrix}$ denotes the inverse of T. We therefore obtain

$$r^2 = \lambda_1 r^1 + \lambda_4 r^4 \quad \text{and} \tag{12}$$
$$r^3 = \mu_1 r^1 + \mu_4 r^4, \tag{13}$$

where $\lambda_1 := u_1 T_{1,1}^{-1}, \lambda_4 := v_4 T_{1,2}^{-1}, \mu_1 := u_1 T_{2,1}^{-1}$ and $\mu_4 := v_4 T_{2,2}^{-1}$. Similarly, solving (11) to express α_2 and α_3 in terms of α_1 and α_4 gives

$$\alpha_2 = \lambda_1\alpha_1 + \lambda_4\alpha_4 \quad \text{and} \tag{14}$$
$$\alpha_3 = \mu_1\alpha_1 + \mu_4\alpha_4. \tag{15}$$

Now, using for instance (4) and (12), we obtain

$$\bar{x} = f + (u_1 + u_2\lambda_1)r^1 + (u_2\lambda_4)r^4, \quad \text{and:}$$

$$(u_1 + u_2\lambda_1)\alpha_1 + (u_2\lambda_4)\alpha_4 \qquad = \quad \text{(using (8))}$$
$$(1 - u_2\alpha_2) + u_2\lambda_1\alpha_1 + (u_2\lambda_4)\alpha_4 \qquad =$$
$$1 + u_2(\lambda_1\alpha_1 + \lambda_4\alpha_4 - \alpha_2) \qquad = 1. \text{ (using (14))}$$

To finish the proof, we only need to argue that we indeed have $\bar{x} \in f + \text{cone}(\{r^1, r^4\})$, i.e., that $\bar{x} = f + \delta_1 r^1 + \delta_4 r^4$ with $\delta_1 = u_1 + u_2\lambda_1$ and $\delta_4 = u_2\lambda_4$ satisfying $\delta_1, \delta_4 \geq 0$. If $\delta_1 \leq 0$ and $\delta_4 > 0$, we have $\bar{x} = f + \delta_1 r^1 + \delta_4 r^4 = f + u_1 r^1 + u_2 r^2$, which means $\delta_4 r^4 = (u_1 - \delta_1)r^1 + u_2 r^2 \in \text{cone}(\{r^1, r^2\})$, which is a contradiction. Similarly, if $\delta_1 > 0$ and $\delta_4 \leq 0$, we have $\bar{x} = f + \delta_1 r^1 + \delta_4 r^4 = f + v_3 r^3 + v^4 r^4$, which implies $\delta_1 r^1 = v_3 r^3 + (v_4 - \delta_4)r^4 \in \text{cone}(\{r^3, r^4\})$, which is also a contradiction. Hence we can assume $\delta_1, \delta_4 \leq 0$. However, since $\delta_1 = u_1 + u_2\lambda_1$ and $\delta_4 = u_2\lambda_4$, this implies $\lambda_1, \lambda_4 \leq 0$, and this contradicts what was shown above, namely that the representation $\bar{x} = f + \delta_1 r^1 + \delta_4 r^4$ satisfies $\alpha_1\delta_1 + \alpha_4\delta_4 = 1$.   □

It follows that only one tight representation of every point x of $\text{conv}(X_\alpha)$ is needed. We now use Lemma 5 to limit the number of vertices of L_α to four. The following notation is introduced. The set $J^x := \cup_{(j_1,j_2)\in T_\alpha(x)}\{j_1, j_2\}$ denotes the set of variables that are involved in tight representations of x. As above, $(j_1^x, j_2^x) \in T_\alpha(x)$ denotes the unique maximal representation of x. Furthermore, given any $(j_1, j_2) \in T_\alpha(x)$, let $(t_{j_1}^{j_2}(x), t_{j_2}^{j_1}(x))$ satisfy $x = f + t_{j_1}^{j_2}(x)r^{j_1} + t_{j_2}^{j_1}(x)r^{j_2}$. Lemma 5 implies that $r^j \in \text{cone}(r^{j_1^x}, r^{j_2^x})$ for every $j \in J^x$. Let $(w_1^j(x), w_2^j(x))$ satisfy $r^j = w_1^j(x)r^{j_1^x} + w_2^j(x)r^{j_2^x}$, where $w_1^j(x), w_2^j(x) \geq 0$ are unique.

Let $\sum_{j\in N} \alpha_j s_j \geq 1$ be a valid inequality for $\text{conv}(P_I)$ that satisfies $\alpha_j > 0$ for $j \in N$. The inequality $\sum_{j\in N} \alpha_j s_j \geq 1$ is facet defining for $\text{conv}(P_I)$, if and only if the coefficients $\{\alpha_j\}_{j\in N}$ define a vertex of the polar of $\text{conv}(P_I)$. Hence $\sum_{j\in N} \alpha_j s_j \geq 1$ is facet defining for $\text{conv}(P_I)$, if and only if the solution to the system

$$\alpha_{j_1} t_{j_1}^{j_2}(x) + \alpha_{j_2} t_{j_2}^{j_1}(x) = 1, \quad \text{for every } x \in X_\alpha \text{ and } (j_1, j_2) \in T_\alpha(x). \tag{16}$$

is unique. We now rewrite the subsystem of (16) that corresponds to a fixed point $x \in X_\alpha$.

Lemma 8. *Let* $\sum_{j\in N} \alpha_j s_j \geq 1$ *be a facet defining inequality for* $\text{conv}(P_I)$ *that satisfies* $\alpha_j > 0$ *for* $j \in N$. *Suppose* $x \in X_\alpha$ *is not a ray point. The system*

$$\alpha_{j_1} t_{j_1}^{j_2}(x) + \alpha_{j_2} t_{j_2}^{j_1}(x) = 1, \quad \text{for every } (j_1, j_2) \in T_\alpha(x). \tag{17}$$

has the same set of solutions $\{\alpha_j\}_{j\in J^x}$ *as the system*

$$1 = t_{j_1}^{j_2}(x)\alpha_{j_1} + t_{j_2}^{j_1}(x)\alpha_{j_2}, \qquad \text{for } (j_1, j_2) = (j_1^x, j_2^x), \tag{18}$$

$$\alpha_j = w_1^j(x)\alpha_{j_1^x} + w_2^j(x)\alpha_{j_2^x}, \qquad \text{for } j \in J^x \setminus \{j_1^x, j_2^x\}. \tag{19}$$

We next show that it suffices to consider vertices of $\text{conv}(X_\alpha)$ in (16).

Lemma 9. *Let* $\sum_{j\in N} \alpha_j s_j \geq 1$ *be a facet defining inequality for* $\text{conv}(P_I)$ *that satisfies* $\alpha_j > 0$ *for* $j \in N$. *Suppose* $x \in X_\alpha$ *is not a vertex of* $\text{conv}(X_\alpha)$. *Then there exists vertices* y *and* z *of* $\text{conv}(X_\alpha)$ *such that the equalities*

$$\alpha_{j_1} t_{j_1}^{j_2}(y) + \alpha_{j_2} t_{j_2}^{j_1}(y) = 1, \quad \text{for every } (j_1, j_2) \in T_\alpha(y) \text{ and} \tag{20}$$

$$\alpha_{j_1} t_{j_1}^{j_2}(z) + \alpha_{j_2} t_{j_2}^{j_1}(z) = 1, \quad \text{for every } (j_1, j_2) \in T_\alpha(z) \tag{21}$$

imply the equalities:

$$\alpha_{j_1} t_{j_1}^{j_2}(x) + \alpha_{j_2} t_{j_2}^{j_1}(x) = 1, \quad \text{for every } (j_1, j_2) \in T_\alpha(x). \tag{22}$$

By combining Lemma 8 and Lemma 9 we have that, if the solution to (16) is unique, then the solution to the system

$$t_{j_1^x}^{j_2^x}(x)\alpha_{j_1^x} + t_{j_2^x}^{j_1^x}(x)\alpha_{j_2^x} = 1, \qquad \text{for every vertex } x \text{ of } \text{conv}(X_\alpha). \tag{23}$$

is unique. Since (23) involves exactly $k \leq 4$ equalities and has a unique solution, exactly $k \leq 4$ variables are involved in (23) as well. This finishes the proof of Theorem 1.

We note that from an inequality $\sum_{j \in S} \alpha_j s_j \geq 1$ that defines a facet of $\mathrm{conv}(P_I(S))$, where $|S| = k$, the coefficients on the variables $j \in N \setminus S$ can be simultaneously lifted by computing the intersection point between the halfline $\{f + s_j r^j : s_j \geq 0\}$ and the boundary of L_α.

We now use Theorem 2 to categorize the inequalities $\sum_{j \in N} \alpha_j s_j \geq 1$ that define facets of $\mathrm{conv}(P_I)$. For simplicity, we only consider the most general case, namely when none of the vertices of $\mathrm{conv}(X_\alpha)$ are ray points. Furthermore, we only consider $k = 3$ and $k = 4$. When $k = 2$, $\sum_{j \in N} \alpha_j s_j \geq 1$ is a facet defining inequality for a cone defined by two rays. We divide the remaining facets of $\mathrm{conv}(P_I)$ into the following three main categories.

(i) *Disection cuts (Fig. 4.(a) and Fig. 4.(b)):*
    Every vertex of $\mathrm{conv}(X_\alpha)$ belongs to a different facet of L_α.
(ii) *Lifted two-variable cuts (Fig. 4.(c) and Fig. 4.(d)):*
    Exactly one facet of L_α contains two vertices of $\mathrm{conv}(X_\alpha)$. Observe that this implies that there is a set $S \subset N$, $|S| = 2$, such that $\sum_{j \in S} \alpha_j s_j \geq 1$ is facet defining for $\mathrm{conv}(P_I(S))$.
(iii) *Split cuts:*
    Two facets of L_α each contain two vertices of $\mathrm{conv}(X_\alpha)$.

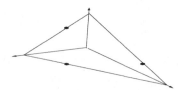

(a) Disection cut from a triangle

(b) Disection cut from a quadrilateral

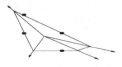

(c) Lifted two-variable cut from quadrilateral

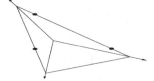

(d) Lifted two-variable cut from triangle

Fig. 4. Disection cuts and lifted two-variable cuts

An example of a cut that is not a split cut was given in [3] (see Fig. 1). This cut is the only cut when $\mathrm{conv}(X_\alpha)$ is the triangle of Fig. 4.(c), and, necessarily, $L_\alpha = \mathrm{conv}(X_\alpha)$ in this case. Hence, *all* three rays that define this triangle are ray points. As mentioned in the introduction, the cut in [3] can be viewed as being on the boundary between disection cuts and lifted two-variable cuts.

Since the cut presented in [3] is not a split cut, and this cut can be viewed as being on the boundary between disection cuts and lifted two-variable cuts, a natural question is whether or not disection cuts and lifted two-variable cuts are split cuts. We finish this section by answering this question.

Lemma 10. *Let* $\sum_{j \in N} \alpha_j s_j \geq 1$ *be a facet defining inequality for* $\mathrm{conv}(P_I)$ *satisfying* $\alpha_j > 0$ *for* $j \in N$*. Also suppose* $\sum_{j \in N} \alpha_j s_j \geq 1$ *is either a disection cut or a lifted two-variable cut. Then* $\sum_{j \in N} \alpha_j s_j \geq 1$ *is not a split cut.*

Proof: Observe that, if $\sum_{j \in N} \alpha_j s_j \geq 1$ is a split cut, then there exists $(\pi, \pi_0) \in \mathbb{Z}^2 \times \mathbb{Z}$ such that L_α is contained in the split set $S_\pi := \{x \in \mathbb{R}^2 : \pi_0 \leq \pi_1 x_1 + \pi_2 x_2 \leq \pi_0 + 1\}$. Furthermore, all points $x \in X_\alpha$ and all vertices of L_α must be either on the line $\pi^T x = \pi_0$, or on the line $\pi^T x = \pi_0 + 1$. However, this implies that there must be two facets of L_α that do not contain any integer points. $\square$

References

1. J. R. Arkinstall. Minimal requirements for Minkowski's theorem in the plane I. *Bulletin of the Australian Mathematical Society*, 22:259–274, 1980.
2. E. Balas. Intersection cuts - a new type of cutting planes for integer programming. *Operations Research*, 19:19–39, 1971.
3. W.J. Cook, R. Kannan, and A. Schrijver. Chvátal closures for mixed integer programming problems. *Mathematical Programming*, 47:155–174, 1990.
4. R.E. Gomory. An algorithm for the mixed integer problem. *Technical Report RM-2597, The Rand Corporation*, 1960a.
5. E. Johnson. On the group problem for mixed integer programming. *Mathematical Programming*, 2:137–179, 1974.
6. G.L. Nemhauser and L.A. Wolsey. Integer and combinatorial optimization. *Wiley*, 1988.
7. S. Rabinowitz. A census of convex lattice polygons with at most one interior lattice point. *Ars Combinatoria*, 28:83–96, 1989.
8. A. Schrijver. Theory of linear and integer programming. *Wiley*, 1986.

Cuts for Conic Mixed-Integer Programming

Alper Atamtürk and Vishnu Narayanan

Department of Industrial Engineering and Operations Research, University of
California, Berkeley, CA 94720-1777 USA
atamturk@berkeley.edu, vishnu@ieor.berkeley.edu

Abstract. A conic integer program is an integer programming problem
with conic constraints. Conic integer programming has important applications in finance, engineering, statistical learning, and probabilistic
integer programming.

Here we study mixed-integer sets defined by second-order conic constraints. We describe general-purpose conic mixed-integer rounding cuts
based on polyhedral conic substructures of second-order conic sets. These
cuts can be readily incorporated in branch-and-bound algorithms that
solve continuous conic programming relaxations at the nodes of the
search tree. Our preliminary computational experiments with the new
cuts show that they are quite effective in reducing the integrality gap of
continuous relaxations of conic mixed-integer programs.

Keywords: Integer programming, conic programming, branch-and-cut.

1 Introduction

In the last two decades there have been major advances in our capability of
solving linear integer programming problems. Strong cutting planes obtained
through polyhedral analysis of problem structure contributed to this success
substantially by strengthening linear programming relaxations of integer programming problems. Powerful cutting planes based on simpler substructures of
problems have become standard features of leading optimization software packages. The use of such structural cuts has improved the performance of the linear
integer programming solvers dramatically.

On another front, since late 1980's we have experienced significant advances
in convex optimization, particularly in conic optimization. Starting with Nesterov and Nemirovski [22, 23, 24] polynomial interior point algorithms that have
earlier been developed for linear programming have been extended to conic optimization problems such as convex quadratically constrained quadratic programs
(QCQP's) and semidefinite programs (SDP's).

Availability of efficient algorithms and publicly available software (CDSP[9],
DSDP[7], SDPA[33], SDPT3[32], SeDuMi[30]) for conic optimization spurred
many optimization and control applications in diverse areas ranging from med-
ical imaging to signal processing, from robust portfolio optimization to truss
design. Commercial software vendors (e.g. ILOG, MOSEK, XPRESS-MP) have

M. Fischetti and D.P. Williamson (Eds.): IPCO 2007, LNCS 4513, pp. 16–29, 2007.

responded to the demand for solving (continuous) conic optimization problems by including stable solvers for second-order cone programming (SOCP) in their recent versions.

Unfortunately, the phenomenal advances in continuous conic programming and linear integer programming have so far not translated to improvements in conic integer programming, i.e., integer programs with conic constraints. Solution methods for conic integer programming are limited to branch-and-bound algorithms that solve continuous conic relaxations at the nodes of the search tree. In terms of development, conic integer programming today is where linear integer programming was before 1980's when solvers relied on pure branch-and-bound algorithms without the use of any cuts for improving the continuous relaxations at the nodes of the search tree.

Here we attempt to improve the solvability of conic integer programs. We develop general purpose cuts that can be incorporated into branch-and-bound solvers for conic integer programs. Toward this end, we describe valid cuts for the second-order conic mixed-integer constraints (defined in Section 2). The choice of second-order conic mixed-integer constraint is based on (i) the existence of many important applications modeled with such constraints, (ii) the availability of efficient and stable solvers for their continuous SOCP relaxations, and (iii) the fact that one can form SOCP relaxations for the more general conic programs, which make the cuts presented here widely applicable to conic integer programming.

1.1   Outline

In Section 2 we introduce conic integer mixed-programming, briefly review the relevant literature and explain our approach for generating valid cuts. In Section 3 we describe conic mixed-integer rounding cuts for second-order conic mixed-integer programming and in Section 4 we summarize our preliminary computational results with the cuts.

2   Conic Integer Programming

A conic integer program (CIP) is an integer program with conic constraints. We limit the presentation here to second-order conic integer programming. However, as one can relax more general conic programs to second-order conic programs [14] our results are indeed applicable more generally.

A *second-order conic mixed-integer program* is an optimization problem of the form

$$\min\ cx + ry$$
$$\text{(SOCMIP)}\quad \text{s.t.}\ \parallel A_i x + G_i y - b_i \parallel\ \leq d_i x + e_i y - h_i,\ \ i = 1, 2, \ldots, k$$
$$x \in \mathbb{Z}^n,\ y \in \mathbb{R}^p\ .$$

Here $\parallel \cdot \parallel$ is the Euclidean norm, A_i, G_i, b are rational matrices with m_i rows, and c, r, d_i, e_i are rational row vectors of appropriate dimension, and h_i is a rational scalar. Each constraint of SOCMIP can be equivalently stated as $(A_i x + G_i y - b_i, d_i x + e_i y - h) \in \mathcal{Q}^{m_i+1}$, where

$$\mathcal{Q}^{m_i+1} := \{(t, t_o) \in \mathbb{R}^{m_i} \times \mathbb{R} : \| t \| \leq t_o\} \ .$$

For $n = 0$, SOCMIP reduces to SOCP, which is a generalization of linear programming as well as convex quadratically constrained quadratic programming. If $G_i = 0$ for all i, then SOCP reduces to linear programming. If $e_i = 0$ for all i, then it reduces to QCQP after squaring the constraints. In addition, convex optimization problems with more general norms, fractional quadratic functions, hyperbolic functions and others can be formulated as an SOCP. We refer the reader to [2, 6, 10, 18, 25] for a detailed exposure to conic optimization and many applications of SOCP.

2.1   Relevant Literature

There has been significant work on deriving conic (in particular SDP) relaxations for (linear) combinatorial optimization problems [1, 13, 19] for obtaining stronger bounds for such problems than the ones given by their natural linear programming relaxations. We refer the reader to Goemans [12] for a survey on this topic. However, our interest here is not to find conic relaxations for linear integer problems, but for conic integer problems.

Clearly any method for general nonlinear integer programming applies to conic integer programming as well. Reformulation-Linearization Technique (RLT) of Sherali and Adams [27] initially developed for linear 0-1 programming has been extended to nonconvex optimization problems [28]. Stubbs and Mehrotra [29] generalize the lift-and-project method [5] of Balas et. al for 0-1 mixed convex programming. See also Balas [4] and Sherali and Shetti [26] on disjunctive programming methods. Kojima and Tunçel [15] give successive semidefinite relaxations converging to the convex hull of a nonconvex set defined by quadratic functions. Lasserre [16] describes a hierarchy of semidefinite relaxations nonlinear 0-1 programs. Common to all of these general approaches is a hierarchy of convex relaxations in higher dimensional spaces whose size grows exponentially with the size of the original formulation. Therefore using such convex relaxations in higher dimensions is impractical except for very small instances. On the other hand, projecting these formulations to the original space of variables is also very difficult except for certain special cases.

Another stream of more practically applicable research is the development of branch-and-bound algorithms for nonlinear integer programming based on linear outer approximations [8, 17, 31]. The advantage of linear approximations is that they can be solved fast; however, the bounds from linear approximations may not be strong. However, in the case of conic programming, and in particular second-order cone programming, existence of efficient algorithms permits the use of continuous conic relaxations at the nodes of the branch-and-bound tree.

The only study that we are aware of on developing valid inequalities for conic integer sets directly is due to Çezik and Iyengar [11]. For a pointed, closed, convex cone $\mathcal{K} \subseteq \mathbb{R}^m$ with nonempty interior, given $S = \{x \in \mathbb{Z}^n : b - Ax \in \mathcal{K}\}$, their approach is to write a linear aggregation

$$\lambda' Ax \le \lambda' b \text{ for some fixed } \lambda \in \mathcal{K}^*, \tag{1}$$

where $\mathcal{K}^*$ is the dual cone of $\mathcal{K}$ and then apply the Chvátal-Gomory (CG) integer rounding cuts [20] to this linear inequality. Hence, the resulting cuts are linear in x as well. For the mixed-integer case as the convex hull feasible points is not polyhedral and has curved boundary (see Figure 2 in Section 3). Therefore, nonlinear inequalities may be more effective for describing or approximating the convex hull of solutions.

2.2   A New Approach

Our approach for deriving valid inequalities for SOCMIP is to decompose the second-order conic constraint into simpler polyhedral sets and analyze each of these sets. Specifically, given a second-order conic constraint

$$\| Ax + Gy - b \| \le dx + ey - h \tag{2}$$

and the corresponding second-order conic mixed-integer set

$$C := \left\{ x \in \mathbb{Z}_+^n, \ y \in \mathbb{R}_+^p : (x, y) \text{ satisfies } (2) \right\},$$

by introducing auxiliary variables $(t, t_o) \in \mathbb{R}^{m+1}$, we reformulate (2) as

$$t_o \le dx + ey - h \tag{3}$$
$$t_i \ge |a_i x + g_i y - b_i|, \ i = 1, \ldots, m \tag{4}$$
$$t_o \ge \| t \|, \tag{5}$$

where a_i and g_i denote the ith rows of matrices A and G, respectively. Observe that each constraint (4) is indeed a second-order conic constraint as $(a_i x + g_i y - b_i, t_i) \in \mathcal{Q}^{1+1}$, yet polyhedral. Consequently, we refer to a constraint of the form (4) as a *polyhedral second-order conic constraint*.

Breaking (2) into polyhedral conic constraints allows us to exploit the implicit polyhedral set for each term in a second-order cone constraint. Cuts obtained for C in this way are linear in (x, y, t); however, they are nonlinear in the original space of (x, y).

Our approach extends the successful polyhedral method for linear integer programming where one studies the facial structure of simpler building blocks to second-order conic integer programming. To the best of our knowledge such an analysis for second-order conic mixed-integer sets has not been done before.

3   Conic Mixed-Integer Rounding

For a mixed integer set $X \subseteq \mathbb{Z}^n \times \mathbb{R}^p$, we use relax$(X)$ to denote its continuous relaxation in $\mathbb{R}^n \times \mathbb{R}^p$ obtained by dropping the integrality restrictions and conv(X) to denote the convex hull of X. In this section we will describe the cuts for conic mixed-integer programming, first on a simple case with a single integer variable. Subsequently we will present the general inequalities.

3.1   The Simple Case

Let us first consider the mixed-integer set

$$S_0 := \{(x, y, w, t) \in \mathbb{Z} \times \mathbb{R}_+^3 : |\, x + y - w - b \,| \le t\} \qquad (6)$$

defined by a simple, yet non-trivial polyhedral second-order conic constraint with one integer variable. The continuous relaxation relax(S_0) has four extreme rays: $(1, 0, 0, 1), (-1, 0, 0, 1), (1, 0, 1, 0)$, and $(-1, 1, 0, 0)$, and one extreme point: $(b, 0, 0, 0)$, which is infeasible for S_0 if $f := b - \lfloor b \rfloor > 0$. It is easy to see that if $f > 0$, conv(S_0) has four extreme points: $(\lfloor b \rfloor, f, 0, 0), (\lfloor b \rfloor, 0, 0, f), (\lceil b \rceil, 0, 1 - f, 0)$ and $(\lceil b \rceil, 0, 0, 1 - f)$. Figure 1 illustrates S_0 for the restriction $y = w = 0$.

Proposition 1. *The simple conic mixed-integer rounding inequality*

$$(1 - 2f)(x - \lfloor b \rfloor) + f \le t + y + w \qquad (7)$$

cuts off all points in relax$(S_0) \setminus$ conv(S_0).

Observe that inequality (7) is satisfied at equality at all extreme points of conv(S_0). Proposition 1 can be proved by simply checking that every intersection of the hyperplanes defining S_0 and (7) is one of the four extreme points of conv(S_0) listed above.

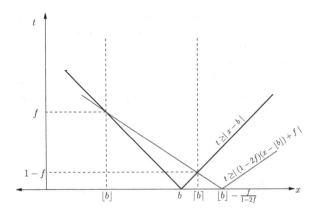

Fig. 1. Simple conic mixed-integer rounding cut

The simple conic mixed-integer rounding inequality (7) can be used to derive nonlinear conic mixed-integer inequalities for nonlinear conic mixed-integer sets. The first observation useful in this direction is that the piecewise-linear conic inequality

$$|(1 - 2f)(x - \lfloor b \rfloor) + f| \leq t + y + w \tag{8}$$

is valid for S_0. See Figure 1 for the restriction $y = w = 0$.

In order to illustrate the nonlinear conic cuts, based on cuts for the polyhedral second-order conic constraints (4), let us now consider the simplest nonlinear second-order conic mixed-integer set

$$T_0 := \left\{ (x, y, t) \in \mathbb{Z} \times \mathbb{R} \times \mathbb{R} \; : \; \sqrt{(x - b)^2 + y^2} \leq t \right\} . \tag{9}$$

The continuous relaxation relax(T_0) has exactly one extreme point $(x, y, t) = (b, 0, 0)$, which is infeasible for T_0 if $b \notin \mathbb{Z}$. Formulating T_0 as

$$t_1 \geq |x - b| \tag{10}$$

$$t \geq \sqrt{t_1^2 + y^2}, \tag{11}$$

we write the piecewise-linear conic inequality (8) for (10). Substituting out the auxiliary variable t_1, we obtain the *simple nonlinear conic mixed-integer rounding inequality*

$$\sqrt{\left((1 - 2f)(x - \lfloor b \rfloor) + f\right)^2 + y^2} \leq t, \tag{12}$$

which is valid for T_0.

Proposition 2. *The simple nonlinear conic mixed-integer rounding inequality* (12) *cuts off all points in* relax(T_0) \ conv(T_0).

Proof. First, observe that for $x = \lfloor b \rfloor - \delta$, the constraint in (9) becomes $t \geq \sqrt{(\delta + f)^2 + y^2}$, and (12) becomes $t \geq \sqrt{(f - (1 - 2f)\delta)^2 + y^2}$. Since $(\delta + f)^2 - (f - (1 - 2f)\delta)^2 = 4f\delta(1 + \delta)(1 - f) \geq 0$ for $\delta \geq 0$ and for $\delta \leq -1$, we see that (12) is dominated by relax(T_0) unless $\lfloor b \rfloor < x < \lceil b \rceil$. When $-1 < \delta < 0$ (i.e., $x \in (\lfloor b \rfloor, \lceil b \rceil)$), $4f\delta(1 + \delta)(1 - f) < 0$, implying that (12) dominates the constraint in (9).

We now show that if $(x_1, y_1, t_1) \in$ relax(T_0) and satisfies (12), then $(x_1, y_1, t_1) \in$ conv(T_0). If $x_1 \notin (\lfloor b \rfloor, \lceil b \rceil)$, it is sufficient to consider $(x_1, y_1, t_1) \in$ relax(T_0) as (12) is dominated by relax(T_0) in this case. Now, the ray $R_1 := \{(b, 0, 0) + \alpha(x_1 - b, y_1, t_1) \; : \; \alpha \in \mathbb{R}_+\} \subseteq$ relax(T_0). Let the intersections of R_1 with the hyperplanes $x = \lfloor x_1 \rfloor$ and $x = \lceil x_1 \rceil$ be $(\lfloor x_1 \rfloor, \bar{y}_1, \bar{t}_1)$, $(\lceil x_1 \rceil, \hat{y}_1, \hat{t}_1)$, which belong to T_0. Then (x_1, y_1, t_1) can be written as a convex combination of points $(\lfloor x_1 \rfloor, \bar{y}_1, \bar{t}_1)$, $(\lceil x_1 \rceil, \hat{y}_1, \hat{t}_1)$; hence $(x_1, y_1, t_1) \in$ conv(T_0).

On the other hand, if $x_1 \in (\lfloor b \rfloor, \lceil b \rceil)$, it is sufficient to consider (x_1, y_1, t_1) that satisfies (12), since (12) dominates the constraint in (9) for $x \in [\lfloor b \rfloor, \lceil b \rceil]$. If $f = 1/2$, (x_1, y_1, t_1) is a convex combination of $(\lfloor b \rfloor, y_1, t_1)$ and $(\lceil b \rceil, y_1, t_1)$. Otherwise, all points on the ray $R_2 := \{(x_0, 0, 0) + \alpha(x_1 - x_0, y_1, t_1) \; : \; \alpha \in \mathbb{R}_+\}$, where $x_0 = \lfloor b \rfloor - \frac{f}{1 - 2f}$, satisfy (12). Let the intersections of R_2 with the hyperplanes $x = \lfloor b \rfloor$ and $x = \lceil b \rceil$ be $(\lfloor b \rfloor, \bar{y}_1, \bar{t}_1)$, $(\lceil b \rceil, \hat{y}_1, \hat{t}_1)$, which belong to T_0. Note that the intersections are nonempty because $x_0 \notin [\lfloor b \rfloor, \lceil b \rceil]$. Then we see that (x_1, y_1, t_1) can be written as a convex combination of $(\lfloor b \rfloor, \bar{y}, \bar{t})$ and $(\lceil b \rceil, \hat{y}, \hat{t})$. Hence, $(x_1, y_1, t_1) \in \mathrm{conv}(T_0)$ in this case as well.   □

Proposition 2 shows that the curved convex hull of T_0 can be described using only two second-order conic constraints. The following example illustrates Proposition 2.

Example 1. Consider the second-order conic set given as

$$T_0 = \left\{ (x, y, t) \in \mathbb{Z} \times \mathbb{R} \times \mathbb{R} \; : \; \sqrt{\left(x - \frac{4}{3}\right)^2 + (y-1)^2} \le t \right\} .$$

The unique extreme point of $\mathrm{relax}(T_0)$ $(\frac{4}{3}, 1, 0)$ is fractional. Here $\lfloor b \rfloor = 1$ and $f = \frac{1}{3}$; therefore,

$$\mathrm{conv}(T_0) = \left\{ (x, y, t) \in \mathbb{R}^3 : \sqrt{\left(x - \frac{4}{3}\right)^2 + (y-1)^2} \le t, \; \sqrt{\frac{1}{9}x^2 + (y-1)^2} \le t \right\} .$$

We show the inequality $\sqrt{\frac{1}{9}x^2 + (y-1)^2} \le t$ and the region it cuts off in Figure 2. Observe that the function values are equal at $x = 1$ and $x = 2$ and the cut eliminates the points between them.

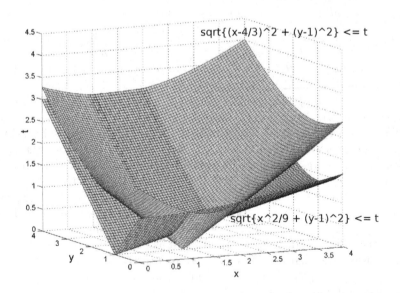

Fig. 2. Nonlinear conic integer rounding cut

3.2   The General Case

In this section we present valid inequalities for the mixed-integer sets defined by general polyhedral second-order conic constraints (4). Toward this end, let

$$S := \{x \in \mathbb{Z}_+^n, y \in \mathbb{R}_+^p, \ t \in \mathbb{R} : \ t \geq |ax + gy - b|\} \ .$$

We refer to the inequalities used in describing S as the trivial inequalities. The following result simplifies the presentation.

Proposition 3. *Any non-trivial facet-defining inequality for* $\mathrm{conv}(S)$ *is of the form* $\gamma x + \pi y \leq \pi_0 + t$. *Moreover, the following statements hold:*

1. $\pi_i < 0$ *for all* $i = 1, \ldots, p$;
2. $\frac{\pi_i}{\pi_j} = \left| \frac{g_i}{g_j} \right|$ *for all* $i, j = 1, \ldots, p$.

Hence it is sufficient to consider the polyhedral second-order conic constraint

$$\left| ax + y^+ - y^- - b \right| \leq t, \tag{13}$$

where all continuous variables with positive coefficients are aggregated into $y^+ \in \mathbb{R}_+$ and those with negative coefficients are aggregated into $y^- \in \mathbb{R}_+$ to represent a general polyhedral conic constraint of the form (4).

Definition 1. *For* $0 \leq f < 1$ *let the conic mixed-integer rounding function* $\varphi_f : \mathbb{R} \to \mathbb{R}$ *be*

$$\varphi_f(v) = \begin{cases} (1 - 2f)n - (v - n), & \text{if } n \leq v < n + f, \\ (1 - 2f)n + (v - n) - 2f, & \text{if } n + f \leq v < n + 1 \ . \end{cases} \quad n \in \mathbb{Z} \tag{14}$$

The conic mixed-integer rounding function φ_f is piecewise linear and continuous. Figure 3 illustrates φ_f.

Lemma 1. *The conic mixed-integer rounding function* φ_f *is superadditive on* $\mathbb{R}$.

Theorem 1. *For any* $\alpha \neq 0$ *the conic mixed-integer rounding inequality*

$$\sum_{j=1}^n \varphi_{f_\alpha}(a_j/\alpha)x_j - \varphi_{f_\alpha}(b/\alpha) \leq (t + y^+ + y^-)/|\alpha|, \tag{15}$$

where $f_\alpha = b/\alpha - \lfloor b/\alpha \rfloor$, *is valid for* S. *Moreover, if* $\alpha = a_j$ *and* $b/a_j > 0$ *for some* $j \in \{1, \ldots, n\}$, *then* (15) *is facet-defining for* $\mathrm{conv}(S)$.

Proof. (Sketch) It can be shown that $\varphi_{f_{a_j}}$ is the lifting function of inequality

$$(1 - 2f)(x - \lfloor b \rfloor) + f \leq (t + y^+ + y^-)/|a_j| \tag{16}$$

for the restriction

$$\left| a_j x_j + y^+ - y^- - b \right| \leq t$$

of (13) with $x_i = 0$ for $i \neq j$. Then the validity as well as the facet claim follows from superadditive lifting [3] of (16) with x_i for $i \neq j$. For $\alpha \neq 0$ validity follows by introducing an auxiliary integer variable x_o with coefficient α and lifting inequality

$$\left| \alpha x_o + y^+ - y^- - b \right| \leq t$$

with all x_i, $i = 1, \ldots, n$ and then setting $x_o = 0$.   □

Remark 1. The continuous relaxation relax(S) has at most n fractional extreme points $(x^j, 0, 0, 0)$ of the form $x^j_j = b/a_j > 0$, and $x^j_i = 0$ for all $i \neq j$, which are infeasible if $b/a_j \notin \mathbb{Z}$. It is easy to check that conic mixed-integer rounding inequalities with $\alpha = a_j$ are sufficient to cut off all fractional extreme points $(x^j, 0, 0, 0)$ of relax(S) as for $x^j_i = 0$ inequality (15) reduces to (7).

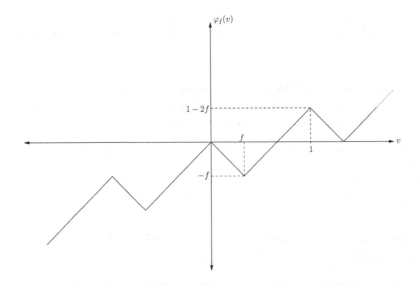

Fig. 3. Conic mixed-integer rounding function.

Next we show that mixed-integer rounding (MIR) inequalities [21, 20] for linear mixed-integer programming can be obtained as conic MIR inequalities. Consider a linear mixed-integer set

$$ax - y \leq b, \ x \geq 0, \ y \geq 0, \ x \in \mathbb{Z}^n, \ y \in \mathbb{R} \tag{17}$$

and the corresponding valid MIR inequality

$$\sum_j \left(\lfloor a_j \rfloor + \frac{(f_j - f)^+}{1 - f} \right) x_j - \frac{1}{1 - f} y \leq \lfloor b \rfloor, \tag{18}$$

where $f_j := a_j - \lfloor a_j \rfloor$ for $j = 1, \ldots, n$ and $f := b - \lfloor b \rfloor$.

Proposition 4. *Every MIR inequality is a conic MIR inequality.*

Proof. We first rewrite inequalities $ax - y \leq b$ and $y \geq 0$, in the conic form

$$-ax + 2y + b \geq |ax - b|$$

and then split the terms involving integer variables x on the right hand side into their integral and fractional parts as

$$-ax + 2y + b \geq \left| \left(\sum_{f_j \leq f} \lfloor a_j \rfloor x_j + \sum_{f_j > f} \lceil a_j \rceil x_j \right) + \sum_{f_j \leq f} f_j x_j - \sum_{f_j > f} (1 - f_j) x_j - b \right| .$$

Then, since $z = \sum_{f_j \leq f} \lfloor a_j \rfloor x_j + \sum_{f_j > f} \lceil a_j \rceil x_j$ is integer and $y^+ = \sum_{f_j \leq f} f_j x_j \in \mathbb{R}_+$ and $y^- = \sum_{f_j > f} (1 - f_j) x_j \in \mathbb{R}_+$, we write the simple conic MIR inequality (8)

$$-ax + 2y + b + \sum_{f_j \leq f} f_j x_j + \sum_{f_j > f} (1 - f_j) x_j$$

$$\geq (1 - 2f) \left(\sum_{f_j \leq f} \lfloor a_j \rfloor x_j + \sum_{f_j > f} \lceil a_j \rceil x_j - \lfloor b \rfloor \right) + f .$$

After rearranging this inequality as

$$2y + 2(1 - f)\lfloor b \rfloor \geq \sum_{f_j \leq f} ((1 - 2f)\lfloor a_j \rfloor - f_j + a_j) x_j + \sum_{f_j > f} ((1 - 2f)\lceil a_j \rceil - (1 - f_j) + a_j) x_j$$

and dividing it by $2(1 - f)$ we obtain the MIR inequality (18). □

Example 2. In this example we illustrate that conic mixed-integer rounding cuts can be used to generate valid inequalities that are difficult to obtain by Chvátal-Gomory (CG) integer rounding in the case of pure integer programming. It is well-known that CG rank of the polytope given by inequalities

$$-kx_1 + x_2 \leq 1, \ kx_1 + x_2 \leq k + 1, \ x_1 \leq 1, \ x_1, x_2 \geq 0$$

for a positive integer k equals exactly k [20]. Below we show that the non-trivial facet $x_2 \leq 1$ of the convex hull of integer points can be obtained by a single application of the conic MIR cut.

Writing constraints $-kx_1 + x_2 \leq 1$ and $kx_1 + x_2 \leq k + 1$ in conic form, we obtain

$$\left| kx_1 - \frac{k}{2} \right| \leq \frac{k}{2} + 1 - x_2 . \tag{19}$$

Dividing the conic constraint (19) by k and treating $1/2 + 1/k - x_2/k$ as a continuous variable, we obtain the conic MIR cut

$$\frac{1}{2} \leq \frac{1}{2} + \frac{1}{k} - \frac{x_2}{k}$$

which is equivalent to $x_2 \leq 1$.

Conic Aggregation

We can generate other cuts for the second order conic mixed integer set C by aggregating constraints (4) in conic form: for $\lambda, \mu \in \mathbb{R}^m_+$, we have $\lambda' t \geq \lambda'(Ax + Gy - b)$, and $\mu' t \geq \mu'(-Ax - Gy + b)$. Writing these two inequalities in conic form, we obtain

$$
\begin{aligned}
&\left(\frac{\lambda + \mu}{2}\right)' t + \left(\frac{\mu - \lambda}{2}\right)' (Ax + Gy) + \left(\frac{\lambda - \mu}{2}\right)' b \\
&\geq \left| \left(\frac{\mu - \lambda}{2}\right)' t + \left(\frac{\lambda + \mu}{2}\right)' (Ax + Gy) - \left(\frac{\lambda + \mu}{2}\right)' b \right| .
\end{aligned}
\tag{20}
$$

Then we can write the corresponding conic MIR inequalities for (20) by treating the left-hand-side of inequality (20) as a single continuous variable. Constraint (20) allows us to utilize multiple polyhedral conic constraints (4) simultaneously.

4   Preliminary Computational Results

In this section we report our preliminary computational results with the conic mixed-integer rounding inequalities. We tested the effectiveness of the cuts on SOCMIP instances with cones $\mathcal{Q}^2$, $\mathcal{Q}^{25}$, and $\mathcal{Q}^{50}$. The coefficients of A, G, and b were uniformly generated from the interval [0,3]. All experiments were performed on a 3.2 GHz Pentium 4 Linux workstation with 1GB main memory using CPLEX[1] (Version 10.1) second-order conic MIP solver. CPLEX uses a barrier algorithm to solve SOCPs at the nodes of a branch-and-bound algorithm.

Conic MIR cuts (15) were added only at the root node using a simple separation heuristic. We performed a simple version of conic aggregation (20) on pairs of constraints using only $0 - 1$ valued multipliers λ and μ, and checked for violation of conic MIR cut (15) for each integer variable x_j with fractional value for the continuous relaxation.

In Table 1 we report the size of the cone (m), number (n) of integer variables in the formulation, the number of cuts, the integrality gap (the percentage gap between the optimal solution and the continuous relaxation), the number of nodes explored in the search tree, and CPU time (in seconds) with and without adding the conic mixed-integer rounding cuts (15). Each row of the table represents the averages for five instances. We have used the default settings of CPLEX except that the primal heuristics were turned off. CPLEX added a small number of MIR cuts (18) to the formulations in a few instances.

We see in Table 1 the conic MIR cuts have been very effective in closing the integrality gap. Most of the instances had 0% gap at the root node after adding the cuts and were solved without branching. The remaining ones were solved within only a few nodes. These preliminary computational results are quite encouraging on the positive impact of conic MIR cuts on solving conic mixed-integer programs.

---

[1] CPLEX is a registered trademark of ILOG, Inc.

Table 1. Effectiveness of conic MIR cuts (15)

		without cuts			with cuts			
m	n	% gap	nodes	time	cuts	% gap	nodes	time
	100	95.8	19	0	87	0.4	1	0
	200	90.8	29	0	192	0.6	1	0
2	300	90.3	38	0	248	0.6	1	0
	400	85.2	62	0	322	0.0	0	0
	500	86.4	71	0	349	0.7	1	0
	100	8.6	10	0	35	2.6	2	0
	200	41.2	80	2	101	4.5	12	1
25	300	46.1	112	4	20	0.0	0	2
	400	68.3	5951	295	99	17.8	63	12
	500	74.6	505	24	116	3.4	6	3
	100	24.5	7	1	42	0.0	0	1
	200	51.3	67	6	44	0.0	0	1
50	300	52.6	105	13	51	3.2	3	2
	400	55.6	158	20	49	5.4	7	5
	500	66.9	233	43	62	1.3	2	3

References

[1] F. Alizadeh. Interior point methods in semidefinite programming and applications to combinatorial optimization. *SIAM Journal on Optimization*, 5:13–51, 1995.

[2] F. Alizadeh and D. Goldfarb. Second-order cone programming. *Mathematical Programming*, 95:3–51, 2003.

[3] A. Atamtürk. Sequence independent lifting for mixed–integer programming. *Operations Research*, 52:487–490, 2004.

[4] E. Balas. Disjunctive programming. *Annals of Discrete Mathematics*, 5:3–51, 1979.

[5] E. Balas, S. Ceria, and G. Cornuéjols. A lift-and-project cutting plane algorithm for mixed 0-1 programs. *Mathematical Programming*, 58:295–324, 1993.

[6] A. Ben-Tal and A. Nemirovski. *Lectures on Modern Convex Optimization: Analysis, Algorithms, and Engineering Applications*. SIAM, 2001.

[7] S. J. Benson and Y. Ye. DSDP5: Software for semidefinite programming. Technical Report ANL/MCS-P1289-0905, Mathematics and Computer Science Division, Argonne National Laboratory, September 2005.

[8] P. Bonami, L. T. Biegler, A. R. Conn, G. Cornuéjols, Grossmann I. E, C. D. Laird, J. Lee, A. Lodi, F. Margot, N. Sawaya, and A. Wächter. An algorithmic framework for convex mixed integer nonlinear programs. Technical Report RC23771, IBM, November 2005.

[9] B. Borchers. CDSP, a C library for semidefinite programing. *Optimization Methods and Software*, 11:613–623, 1999.

[10] S. Boyd and L. Vandenberghe. *Convex Optimization*. Cambridge University Press, Cambridge, 2004.

[11] M. T. Çezik and G. Iyengar. Cuts for mixed 0-1 conic programming. *Mathematical Programming*, 104:179–202, 2005.

[12] M. X. Goemans. Semidefinite programming in combinatorial optimization. *Mathematical Programming*, 79:143–161, 1997.

[13] M. X. Goemans and D. P. Williamson. Improved approximation algorithms for maximum cut and satisfyibility problems using semidefinite programming. *Journal of the ACM*, 42:1115–1145, 1995.

[14] S. Kim, M. Kojima, and M. Yamashita. Second order cone programming relaxation of a positive semidefinite constraint. *Optimization Methods and Software*, 18:535–451, 2003.

[15] M. Kojima and L. Tuncel. Cones of matrices and successive convex relaxations of nonconvex sets. *SIAM Journal on Optimization*, 10:750–778, 2000.

[16] J. B. Lasserre. An explicit exact SDP relaxation for nonlinear 0-1 programs. In K. Aardal and A. M. H. Gerards, editors, *Lecture Notes in Computer Science*, volume 2081, pages 293–303. 2001.

[17] J. Linderoth. A simplical branch-and-bound algorithm for solving quadratically constrained quadratic programs. *Mathematical Programming*, 103:251–282, 2005.

[18] M. Lobo, L. Vandenberghe, S. Boyd, and H. Lebret. Applications of second-order cone programming. *Linear Algebra and its Applications*, 284:193–228, 1998.

[19] L. Lovász and A. Schrijver. Cones of matrices and set-functions and 0-1 optimization. *SIAM Journal on Optimization*, 1:166–190, 1991.

[20] G. L. Nemhauser and L. A. Wolsey. *Integer and Combinatorial Optimization*. John Wiley and Sons, New York, 1988.

[21] G. L. Nemhauser and L. A. Wolsey. A recursive procedure for generating all cuts for 0-1 mixed integer programs. *Mathematical Programming*, 46:379–390, 1990.

[22] Y. Nesterov and A. Nemirovski. A general approach to polynomial-time algorithm design for convex programming. Technical report, Center. Econ. & Math. Inst, USSR Acad. Sci., Moskow, USSR, 1988.

[23] Y. Nesterov and A. Nemirovski. Self-concordant functions and polynomial time methods in convex programming. Technical report, Center. Econ. & Math. Inst, USSR Acad. Sci., Moskow, USSR, 1990.

[24] Y. Nesterov and A. Nemirovski. Conic formulation of a convex programming problem and duality. Technical report, Center. Econ. & Math. Inst, USSR Acad. Sci., Moskow, USSR, 1991.

[25] Y. Nesterov and A. Nemirovski. *Interior-point polynomial algorithms for convex programming*. SIAM, Philedelphia, 1993.

[26] H. D. Sherali and C. Shetti. *Optimization with disjunctive constraints*, volume 181 of *Lectures on Econ. Math. Systems*. Springer Verlag, Berlin, Heidelberg, New York, 1980.

[27] H. D. Sherali and C. H. Tunçbilek. A hierarchy of relaxations between continuous and convex hull representations for zero-one programming problems. *SIAM Journal on Discrete Mathematics*, 3:411–430, 1990.

[28] H. D. Sherali and C. H. Tunçbilek. A reformulation-convexification approach for solving nonconvex quadratic programming problems. *Journal of Global Optimization*, 7:1–31, 1995.

[29] R. Stubbs and S. Mehrotra. A branch-and-cut methods for 0-1 mixed convex programming. *Mathematical Programming*, 86:515–532, 1999.

[30] J. Sturm. Using SeDuMi 1.02, a MATLAB toolbox for ptimization over symmetric cones. *Optimization Methods and Software*, 11:625–653, 1999.

[31] M. Tawarmalani and N. V. Sahinidis. Global optimization of mixed-integer non-linear programs: A theoretical and computational study. *Mathematical Programming*, 99:563–591, 2004.

[32] K.C. Toh, M. J. Todd, and R. H. Tütüncü. SDPT3 – a Matlab software package for semidefinite programming. *Optimization Methods and Software*, 11/12:545–581, 1999.

[33] M. Yamashita, K. Fujisawa, and M. Kojima. Implementation and evaluation of SDPA 6.0 (SemiDefinite Programming Algorithm 6.0). *Optimization Methods and Software*, 18:491–505, 2003.

Sequential-Merge Facets for Two-Dimensional Group Problems*

Santanu S. Dey and Jean-Philippe P. Richard

School of Industrial Engineering, Purdue University,
315 N. Grant Street, West Lafayette, IN 47906-2023

Abstract. In this paper, we show how to generate strong cuts for unstructured mixed integer programs through the study of high-dimensional group problems. We present a new operation that generates facet-defining inequalities for two-dimensional group problems by combining two facet-defining inequalities of one-dimensional group problems. Because the procedure allows the use of a large variety of one-dimensional constituent inequalities, it yields large families of new cutting planes for MIPs that we call *sequential-merge inequalities*. We show that sequential-merge inequalities can be used to generate inequalities whose continuous variable coefficients are stronger than those of one-dimensional cuts and can be used to derive the three-gradient facet-defining inequality introduced by Dey and Richard [4].

1 Introduction

Over the last decade, a vast amount of research has been directed towards generating strong general purpose cutting planes for unstructured integer programs; see Marchand et al. [13] and Johnson et al. [12]. One approach to generate strong cutting planes is to use constraints of the problems one at a time. This approach has proven to be successful in many cases and cuts generated from single constraint relaxations of MIPs are currently used in all commercial MIP solvers. It seems however that an option to generate stronger cutting planes is to use information from multiple constraints concurrently. In this paper, we show how to generate such strong cuts through the study of two-dimensional group relaxations.

In a series of papers Gomory [6], Gomory and Johnson [7, 8, 9], Gomory et al. [10], and Johnson [11] showed how to use group relaxations to generate cutting planes for general Integer Programs. Although their theory applies to problems with multiple constraints, most research has considered only one-dimensional group relaxations; see Gomory and Johnson [7, 8, 9], Gomory et al. [10], Aráoz et al. [2], Miller et al. [14], Richard et al. [15], and Dash and Günlük [3]. There are only a few papers that focus on the systematic study of group problems with multiple constraints. In [11], Johnson presents general theoretical results for group relaxations of Mixed Integer Programs with multiple constraints. Recently, Dey and Richard [4] introduced tools to study two-dimensional infinite

---

* This research was supported by NSF Grant DMI-03-48611.

M. Fischetti and D.P. Williamson (Eds.): IPCO 2007, LNCS 4513, pp. 30–42, 2007.

group problems and introduced two families of facet-defining inequalities for two-dimensional group relaxations. We note however that very few families of facet-defining inequalities are known for two-dimensional group problems. Further, Gomory and Johnson [9] recently write about strong inequalities of two-dimensional group problems that

> "There are reasons to think that such inequalities would be stronger since they deal with the properties of two rows, not one. They can also much more accurately reflect the structure of the continuous variables."

Similarly, in a recent review of non-traditional approaches to Mixed Integer Programming, Aardal, Weismantel and Wolsey [1] mention that:

> "Given the recent computational interest in using Gomorys fractional cuts, mixed integer rounding inequalities and Gomorys mixed integer cuts, this reopens questions about the possible use of alternative subadditive functions to generate practically effective cutting planes. It is also natural to ask whether interesting higher dimensional functions can be found and put to use..."

In this paper, we present a general procedure for generating large families of facet-defining inequalities for two-dimensional infinite group problems. This procedure in turn yields a large number of new cutting planes for general MIPs. Although Dey and Richard [4] already showed that a specific aggregation scheme yields facet-defining inequalities for two-dimensional group problems from facet-defining inequalities of the one-dimensional group problem, the procedure presented in this paper shows different, richer relations between facets of one-dimensional and two-dimensional group problems.

In Sect. 2 we introduce and present fundamental results and concepts about the group problem. We also describe its relation to lifting. In Sect. 3, we present a sequential-merge procedure that generates inequalities for the two-dimensional group problem by combining inequalities for the one-dimensional group problem in a specific fashion. We also show that the procedure shares some relationship with the two-step MIR procedure of Dash and Günlük [3] and can be used to derive the family of three-gradient facet of Dey and Richard [4]. In Sect. 4 we show that, under mild conditions, the procedure presented in Sect. 3 generates facets for the two-dimensional infinite group problem. We conclude in Sect. 5 with directions of future research.

2   Group Problem and Lifting-Space

In this section, we present important results about group problems that were introduced and proven by Gomory and Johnson [9]. We then introduce the notion of valid and facet-defining inequalities for group problems and discuss how such inequalities can be derived from certain lifting functions. We denote by $\mathbb{I}^m$ the group of real m-dimensional vectors where the group operation is performed as addition modulo 1 componentwise, i.e., $\mathbb{I}^m = \{(x_1, x_2...x_m) \mid 0 \leq x_i < 1 \ \forall 1 \leq$

$i \leq m$}. In particular, the symbol $+$ is used to denote both the addition in $\mathbb{R}^m$ and in $\mathbb{I}^m$. We refer to the vector $(0, 0, ..., 0) \in \mathbb{I}^m$ as o. Next we give a formal definition of the group problem.

Definition 1 ([11]). *For $r \in \mathbb{I}^m$ with $r \neq o$, the group problem $PI(r, m)$ is the set of functions $t : \mathbb{I}^m \to \mathbb{R}$ such that*

1. *t has a finite support, i.e., $t(u) > 0$ for a finite subset of $\mathbb{I}^m$.*
2. *$t(u)$ is a non-negative integer for all $u \in \mathbb{I}^m$,*
3. *$\sum_{u \in \mathbb{I}^m} ut(u) = r$.*

Next we define the concept of a valid inequality for the group problem.

Definition 2 ([11]). *A function $\phi : \mathbb{I}^m \to \mathbb{R}_+$ is said to define a valid inequality for $PI(r,m)$ if $\phi(o) = 0$, $\phi(r) = 1$ and $\sum_{u \in \mathbb{I}^m} \phi(u)t(u) \geq 1$, $\forall t \in PI(r, m)$.*

In the remainder of this paper, we will use the terms valid function and valid inequality interchangeably. For a vector $a \in \mathbb{R}^m$, define $\mathbb{P}(a) = (a_1(mod1),$ $...a_m(mod1))$.

It can be verified that given m rows of the simplex tableau $\sum_{i=1}^n a_i x_i = b$ of an integer program P, the inequality $\sum_{i=1}^n \phi(\mathbb{P}(a_i))x_i \geq 1$ is valid for P, if ϕ is valid for $PI(r, m)$, and $\mathbb{P}(b) = r$; see Gomory and Johnson [9]. We next describe necessary conditions for valid inequalities ϕ to be strong.

Definition 3 ([7]). *A valid inequality ϕ for $PI(r, m)$ is said to be subadditive if $\phi(u) + \phi(v) \geq \phi(u + v)$, $\forall u, v \in \mathbb{I}^m$.*

Gomory and Johnson [7] prove that all valid functions of $PI(r, m)$ that are not subadditive are dominated by valid subadditive functions of $PI(r, m)$. Therefore it is sufficient to study the valid subadditive functions of $PI(r, m)$. Next we introduce a definition to characterize strong inequalities.

Definition 4 ([11]). *A valid inequality ϕ is minimal for $PI(r, m)$ if there does not exist a valid function ϕ^* for $PI(r, m)$ different from ϕ such that $\phi^*(u) \leq \phi(u)$ $\forall u \in \mathbb{I}^m$.*

We next present a result characterizing minimal functions. This result is proven in Gomory and Johnson [7] and Johnson [11].

Theorem 1 ([7]). *If ϕ is a valid function for $PI(r, m)$ and $\phi(u) + \phi(r - u) = 1$ $\forall u \in \mathbb{I}^m$ then ϕ is minimal.* □

Minimal inequalities for $PI(r, m)$ are strong because they are not dominated by any single valid inequality. However, there is a stronger class of valid inequalities that Gomory and Johnson refer to as facet-defining inequalities. Next, we present the definition of facet-defining inequality in the context of $PI(r, 2)$.

Definition 5 (Facet). *Let $P(\phi) = \{t \in PI(r, 2)| \sum_{u \in \mathbb{I}^2, t(u) > 0} \phi(u)t(u) = 1\}$. We say that an inequality ϕ is facet-defining for $PI(r, 2)$ if there does not exist a valid function ϕ^* such that $P(\phi^*) \supsetneq P(\phi)$.*

Gomory and Johnson [8] proved that all facet-defining inequalities are minimal inequalities. To prove that a function is facet-defining, Gomory and Johnson [9] introduced a tool that they refer to as Facet Theorem. We describe the Facet Theorem in Theorem 2 and introduce the necessary definitions next.

Definition 6 (Equality Set, [9]). *For each point* $u \in \mathbb{I}^2$, *we define* $g(u)$ *to be the variable corresponding to the point* u. *We define the set of equalities of* ϕ *to be the system of equations* $g(u) + g(v) = g(u + v)$ *for all* $u, v \in \mathbb{I}^2$ *such that* $\phi(u) + \phi(v) = \phi(u + v)$. *We denote this set as* $E(\phi)$.

Theorem 2 (Facet Theorem, [9]). *If* ϕ *is minimal and subadditive, and if* ϕ *is the unique solution of* $E(\phi)$ *then* ϕ *is facet-defining.* □

Currently all known facets for infinite group problems are piecewise linear functions. A function ϕ is defined to be piecewise linear, if $\mathbb{I}^2$ can be divided into polytopes such that the function ϕ is linear over each polytope; see Gomory and Johnson [9] and Dey and Richard [4]. Further, Gomory and Johnson [9] conjectured that all facets of infinite group problems are piecewise linear. Therefore, when introducing tools to prove that inequalities are facet-defining, it is usual to assume that the inequality under study is piecewise linear. Next we present in Theorem 4 a result regarding the continuity of functions of $PI(r, 2)$ that is used in the proof of the Sequential-Merge Theorem of Sect. 4. Theorem 4 is proven using the following preliminary result.

Theorem 3 ([5]). *If a valid function* ϕ *for* $PI(r, m)$ *satisfies the following conditions*

1. $\phi(x) + \phi(y) \geq \phi(x + y) \; \forall x, y \in \mathbb{I}^m$,
2. $\lim_{h \downarrow 0} \frac{\phi(hd)}{h}$ *exists for any* $d \in \mathbb{R}^m$,

then ϕ *is continuous.* □

Theorem 4. *Let* ϕ *be a minimal piecewise linear and continuous function for* $PI(r, 2)$. *If* ψ *is a valid function for* $PI(r, 2)$ *such that* $E(\phi) \subseteq E(\psi)$ *then* ψ *is continuous.* □

Generating strong inequalities for group problems is often difficult. Richard et al. [15] showed that lifting can be used to derive valid and facet-defining inequalities for one-dimensional group problems. The family of facet-defining inequalities we present here is also easier to derive using lifting functions. In the remainder of this section, given any $x \in \mathbb{I}^m$, we denote $\tilde{x}$ as the element of $\mathbb{R}^m$ with the same numerical value as x. Similarly, for $x \in \mathbb{R}^m$ such that $0 \leq x_i < 1 \; \forall 1 \leq i \leq m$, we denote $\dot{x}$ to be the element of $\mathbb{I}^m$ such that $\tilde{\dot{x}} = x$.

Definition 7 (Lifting-Space Representation). *Given a valid inequality* ϕ *for* $PI(r, m)$, *we define the lifting-space representation of* ϕ *as* $[\phi]_r : \mathbb{R}^m \to \mathbb{R}$ *where*

$$[\phi]_r(x) = \sum_{i=1}^{m} x_i - \sum_{i=1}^{m} \tilde{r}_i \phi(\mathbb{P}(x)).$$

To illustrate the idea that motivates this definition, we discuss the case where $m = 1$. Consider a row of the simplex tableau $\sum_{i=1}^{n} a_i x_i = a_0$ of an integer program, where $a_i \in \mathbb{R}$, the fractional part of a_0 is $r \neq 0$, and x_i are nonnegative integer variables. If ϕ is a valid function for $PI(r, 1)$ we have that $\sum_{i=1}^{n} \phi(a_i) x_i \geq 1$ is a valid cut for the original IP. Multiplying this cut with r and then subtracting it from the original row we obtain $\sum_{i=1}^{n} [\phi]_r(a_i) x_i \leq [\phi]_r(a_0)$. One well-known example of the relation between the group-space and the lifting-space representation of an inequality is that of Gomory Mixed Integer Cut (GMIC) and the Mixed Integer Rounding (MIR) inequality. It can be easily verified that the form in which MIR is presented is $[GMIC]_r$. Thus, intuitively, the construction of the lifting-space representation given in Definition 7 is a generalization of the relation that GMIC shares with MIR to other group cuts of one- and higher-dimensions.

Propositions 1 and 2 are generalizations of results from Richard et al. [15].

Proposition 1. *If ϕ is valid function for $PI(r, m)$,*

1. *$[\phi]_r(x + e_i) = [\phi]_r(x) + 1$, where e_i is the ith unit vector of $\mathbb{R}^m$. We say that $[\phi]_r$ is pseudo-symmetric.*
2. *$[\phi]_r$ is superadditive iff ϕ is subadditive.* $\qquad\qquad\square$

Motivated by Definition 7, we define next the inverse operation to $[\phi]_r(x)$.

Definition 8 (Group-Space Representation). *Given a superadditive function $\psi : \mathbb{R}^m \to \mathbb{R}$ which is pseudo-symmetric, we define the group-space representation of ψ as $[\psi]_r^{-1} : \mathbb{I}^m \to \mathbb{R}$ where $[\psi]_r^{-1}(\dot{x}) = \frac{\sum_{i=1}^{m} x_i - \psi(x)}{\sum_{i=1}^{m} r_i}$.*

In Fig. 1, a three-gradient facet [4] of the two-dimensional group problem is shown in its group-space and lifting-space representation.

Proposition 2. *A valid group-space function $g : \mathbb{I}^m \to \mathbb{R}$ is minimal iff $[g]_r$ is superadditive and $[g]_r(x) + [g]_r(r - x) = 0$.* $\qquad\qquad\square$

3   Sequential-Merge Inequalities for Two-Dimensional Group Problems

In this section, we introduce an operation that produces valid inequalities for $PI(r, 2)$ from valid inequalities for $PI(r', 1)$. To simplify the notation, we denote $\tilde{x}$ and $\dot{x}$ by x since the symbol is clear from the context.

Definition 9 (Sequential-merge inequality). *Assume that g and h are valid functions for $PI(r_1, 1)$ and $PI(r_2, 1)$ respectively. We define the sequential-merge of g and h as the function $g \lozenge h : \mathbb{I}^2 \to \mathbb{R}_+$ where*

$$g \lozenge h(x_1, x_2) = [[g]_{r_1}(x_1 + [h]_{r_2}(x_2))]_r^{-1}(x_1, x_2) \qquad (1)$$

and $r = (r_1, r_2)$. In this construction, we refer to g as the outer function and to h as the inner function.

Figure 2 gives an example of a sequential-merge inequality that is facet-defining of $PI((r_1, r_2), 2)$.

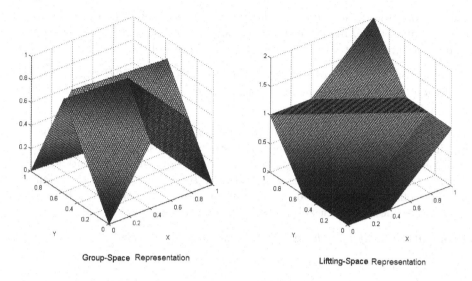

Fig. 1. Group-space and lifting-space representations of a three-gradient facet of PI(r,2)

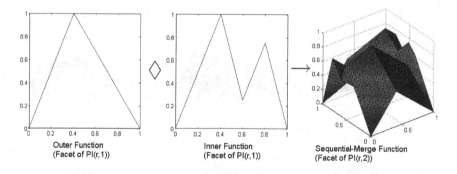

Fig. 2. Examples of sequential-merge operation

We first observe that there is an intuitive interpretation to the construction presented in Definition 9. Given two rows of a simplex tableau, we first generate a cutting plane in the lifting-space of the first row. This cutting plane is added to the second row of the tableau to generate a combined inequality. Finally, a one-dimensional cutting plane is generated from the combined inequality. Proposition 4 states that the group-space representation of inequalities generated using this procedure are valid inequalities for $PI(r, 2)$ under the condition that the outer function is nondecreasing in the lifting-space.

Before we present this result, we give a formula for the sequential-merge inequality in terms of the inner and outer functions in their group-space representations.

Proposition 3. $g \Diamond h(x_1, x_2) = \frac{r_2 h(x_2) + r_1 g(\mathbb{P}(x_1 + x_2 - r_2 h(x_2)))}{r_1 + r_2}.$ $\qquad \square$

Using Proposition 3 it is easy to verify that the sequential-merge $\Diamond$ operator is non-commutative. In the next proposition, we record that sequential-merge inequalities are valid for the two-dimensional group problem.

Proposition 4. *If g, h are valid functions for $PI(r_1, 1)$ and $PI(r_2, 1)$ respectively, and $[g]_{r_1}$ is nondecreasing then $g \Diamond h$ is a valid function for $PI(r, 2)$ where $r \equiv (r_1, r_2)$.* □

In Fig. 3 we illustrate all the different types of valid inequalities that can be obtained using GMIC, a two-step MIR and a three-slope facet of the one-dimensional group problem as inner and outer functions in the sequential-merge construction. These inequalities are valid for $PI(r, 2)$ since all the three building functions used have non-decreasing lifting-space representations. It can be proven that the inequalities obtained in this way are strong.

Proposition 5. *If g and h are minimal and $[g]_{r_1}$ is nondecreasing, then $g \Diamond h$ is minimal.* □

We next give two examples of well-known valid inequalities for group problems that can be obtained using the sequential-merge procedure.

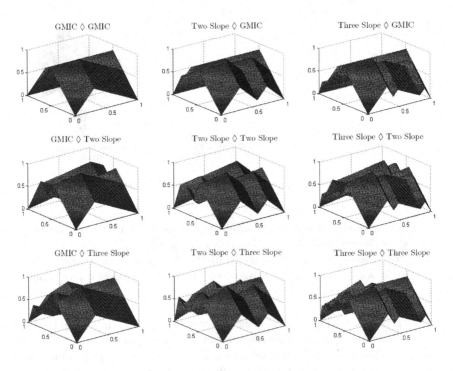

Fig. 3. Examples of sequential-merge inequalities for $PI(r, 2)$

Proposition 6. *Consider* $\kappa(x) = [[\xi]_r(x + [\xi]_r(x))]^{-1}_{(r,r)}(x,x)$, *where* ξ *is the GMIC, i.e.,* $\kappa(x)$ *is the sequential-merge inequality obtained using the same constraint twice and using GMIC as both the inner and outer function. Then* $\kappa(x)$ *is a two-step MIR function from Dash and Günlük [3].* □

We observe that sequential-merge procedure shares some relations with the two-step MIR procedure of Dash and Günlük [3]. An important difference however is that the sequential-merge procedure uses in general two different rows of a simplex tableau. Also the two-step MIR procedure only uses MIR inequalities as constituent functions.

We describe in the next proposition another family of facets for the two-dimensional group problem that can be obtained using the sequential-merge procedure.

Proposition 7. *Consider* $\rho(x,y) = [[\xi]_r(x + [\xi]_r(y))]^{-1}_{(r_1,r_2)}(x,y)$, *where* ξ *is the GMIC, i.e.* $\rho(x,y)$ *is the sequential-merge inequality obtained using GMIC as both the inner and outer function. This inequality is the three-gradient facet-defining inequality for* $P((r_1,r_2),2)$ *presented in Dey and Richard [4].* □

4 Facet-Defining Sequential-Merge Inequalities

In this section, we derive conditions under which sequential-merge inequalities are facet-defining for the two-dimensional group problem $PI(r,2)$. We begin by studying some geometric properties of $g\Diamond h$.

Definition 10. *We define the set of points* $\{(x,y) \mid x = (-y + r_2h(y))(mod\,1)\}$ *as the support of the function* $g\Diamond h$. *We denote the support of* $g\Diamond h$ *as* $\mathcal{S}(g\Diamond h)$.

It is easy to verify that given a value of y, there is an unique value of x such that $(x,y) \in \mathcal{S}(g\Diamond h)$.

In Fig. 4 we illustrate the support of a function $g\Diamond h$ for the case where the inner function is the three-slope facet defining inequality of Gomory and Johnson [9] with right-hand-side of 0.2. The support of $g\Diamond h$ is important because

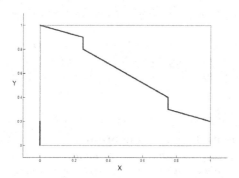

Fig. 4. Example of $\mathcal{S}(g\Diamond h)$ where h is the three-slope facet of Gomory and Johnson [9]

it contains all the equalities that h satisfies. In particular, the next proposition states that for every equality that h satisfies, there exists a related equality that $g \Diamond h$ satisfies, which only involves points of its support.

Proposition 8. *Let g and h be valid subadditive inequalities and let $[g]_{r_1}$ be nondecreasing. If $v_1, v_2 \in \mathbb{I}^1$ are such that $h(v_1) + h(v_2) = h(v_1 + v_2)$ and $(u_1, v_1), (u_2, v_2) \in \mathcal{S}(g \Diamond h)$ then*

1. $(u_1 + u_2, v_1 + v_2) \in \mathcal{S}(g \Diamond h)$.
2. $g \Diamond h(u_1, v_1) + g \Diamond h(u_2, v_2) = g \Diamond h(u_1 + u_2, v_1 + v_2)$ $\square$

Definition 11. *Let ϕ be a valid continuous function for $PI(r, 1)$. We say $E(\phi)$ is unique up to scaling if for any other continuous function $\phi' : \mathbb{I}^1 \rightarrow \mathbb{R}_+$, $E(\phi') \supseteq E(\phi)$ implies that $\phi' = c\phi$ for $c \in \mathbb{R}$.*

Intuitively, because the function $g \Diamond h$ has the equalities of h on its support, $E(g \Diamond h)$ will have an unique solution on its support up to scaling whenever $E(h)$ has a unique solution up to to scaling. This key result is used in the proof of the Sequential-Merge Theorem 5 to show that $E(g \Diamond h)$ is unique and therefore show that $g \Diamond h$ is facet-defining.

Proposition 9. *Let g, h be piecewise linear and continuous valid inequalities for $PI(r_1, 1)$ and $PI(r_2, 1)$ respectively and assume that $E(h)$ has an unique solution up to scaling. Let ψ be a valid function for $PI(r, 2)$ such that $E(\psi) \supseteq E(g \Diamond h)$, then the value of $\psi(u_1, u_2) = cg \Diamond h(u_1, u_2) = \frac{cr_2}{r_1 + r_2} h(u_2) \; \forall (u_1, u_2) \in \mathcal{S}(g \Diamond h)$ where c is a nonnegative real number.* $\square$

Although Proposition 9 establishes that $E(g \Diamond h)$ has an unique solution up to scaling on its support, it falls short of proving that $E(g \Diamond h)$ has an unique solution over $\mathbb{I}^2$. Therefore, we identify in Propositions 10 and 11 some equalities that $g \Diamond h$ satisfies that help in extending the result of Proposition 9 to $\mathbb{I}^2$.

Proposition 10. *Let g and h be valid functions for $PI(r_1, 1)$ and $PI(r_2, 1)$ respectively such that $[g]_{r_1}$ is nondecreasing, then $g \Diamond h(x_1, y_1) + g \Diamond h(\delta, 0) = g \Diamond h(x_1 + \delta, y_1) \; \forall \delta \in \mathbb{I}^1$ and $\forall (x_1, y_1) \in \mathcal{S}(g \Diamond h)$.* $\square$

Proposition 11. *Let g and h be valid functions for $PI(r_1, 1)$ and $PI(r_2, 1)$ respectively and assume that $[g]_{r_1}$ and $[h]_{r_2}$ are nondecreasing functions. Then $g \Diamond h(x_1, x_2) = \frac{x_1 + x_2}{r_1 + r_2}$ for $0 \leq x_1 \leq r_1$, and $0 \leq x_2 \leq r_2$. Furthermore $g \Diamond h(u_1, v_1) + g \Diamond h(u_2, v_2) = g \Diamond h(u_1 + u_2, v_1 + v_2)$ whenever $u_1, u_2, u_1 + u_2 \leq r_1$ and $v_1, v_2, v_1 + v_2 \leq r_2$.* $\square$

Theorem 5 (Sequential-Merge Theorem). *Assume that g and h are continuous, piecewise linear, facet-defining inequalities of $PI(r_1, 1)$ and $PI(r_2, 1)$ respectively. Assume also that $E(g)$ and $E(h)$ are unique up to scaling and $[g]_{r_1}$ and $[h]_{r_2}$ are nondecreasing. Then $g \Diamond h$ is a facet-defining inequality for $PI((r_1, r_2), 2)$.* $\square$

We briefly present the outline of the proof of the above theorem. We first assume by contradiction that $g\Diamond h$ is not facet-defining. Then using Theorem 2 we conclude that there exists a function ϕ' that is different from $g\Diamond h$ and satisfies all the equalities of $g\Diamond h$. Using Theorem 4 we can prove that ϕ' is continuous. Using Proposition 9 we show that the function ϕ' is a scalar multiple times $g\Diamond h$ over $\mathcal{S}(g\Diamond h)$. Finally, we use Proposition 10 and Proposition 11 and the fact that $E(g)$ is unique up to scaling to show that the value of this scalar is 1 and that $\phi'(u) = g\Diamond h(u) \; \forall u \in \mathbb{I}^2$, which is the required contradiction.

In Theorem 5, we assumed the technical condition that $E(h)$ and $E(g)$ are unique up to scaling. This assumption is not very restrictive as it is satisfied by all known facet-defining inequalities for $PI(r,1)$. The condition that $[g]_{r_1}$ and $[h]_{r_2}$ are nondecreasing on the other hand is more restrictive since there exists facet-defining inequalities of $PI(r,1)$ that do not satisfy this condition. Finally note that, Theorem 5 implies that all the functions illustrated in Fig. 3 are facet-defining for the two-dimensional group problem.

We now extend the family of inequalities obtained in Theorem 5 to the mixed integer case. To this end we use a result from Johnson [11] which states that the coefficient of a continuous variable in a minimal group cut ϕ can be found as $\mu_\phi(u) = lim_{h\to0+} \frac{\phi(\mathbb{P}(hu))}{h}$ where $u \in \mathbb{R}^2$ is the column vector of coefficients of this continuous variable in the simplex tableau.

The following proposition describes how the sequential-merge facets obtained for $PI(r,2)$ can be extended into two-dimensional mixed integer group cuts.

Proposition 12. *Let* $c_g^+ = lim_{\epsilon\to0+}\frac{g(\epsilon)}{\epsilon} = \frac{1}{r_1}$, $c_g^- = lim_{\epsilon\to0+}\frac{g(1-\epsilon)}{\epsilon}$, $c_h^+ = lim_{\epsilon\to0+}\frac{h(\epsilon)}{\epsilon} = \frac{1}{r_2}$ *and* $c_h^- = lim_{\epsilon\to0+}\frac{h(1-\epsilon)}{\epsilon}$. *The coefficients of the continuous variables for* $g\Diamond h$ *are given by*

$$\mu_{g\Diamond h}(u_1,u_2) = \begin{cases} \frac{u_1+u_2}{r_1+r_2} & u_1 \geq 0, u_2 \geq 0 \\[2mm] \frac{u_2-r_1c_g^-(u_1)}{r_1+r_2} & u_1 < 0, u_2 \geq 0 \\[2mm] \frac{-r_2c_h^-u_2+r_1c_g^-(-u_1-u_2-r_2c_h^-u_2)}{r_1+r_2} & u_1 \leq 0, u_2 \leq 0 \\[2mm] \frac{u_1+u_2}{r_1+r_2} & \begin{cases} u_1 > 0, u_2 < 0 \\ u_1+u_2+r_2c_h^-u_2 > 0 \end{cases} \\[2mm] \frac{-r_2c_h^-u_2+r_1c_g^-(-u_1-u_2-r_2c_h^-u_2)}{r_1+r_2} & \begin{cases} u_1 > 0, u_2 < 0 \\ u_1+u_2+r_2c_h^-u_2 \leq 0 \end{cases} \end{cases} \qquad (2)$$

□

Next we illustrate on an example how the sequential-merge procedure can be applied to mixed integer programs.

Example 1. Consider the following mixed integer set

$$\frac{7}{17}x - \frac{7}{17}y \leq \frac{154}{85} \qquad (3)$$

$$\frac{7}{17}x + \frac{10}{17}y \leq \frac{359}{170} \tag{4}$$

$$x, y \in \mathbb{Z}_+. \tag{5}$$

We introduce non-negative slack variables s_1 and s_2 and perform a few simplex iterations to obtain

$$x + 1.4286s_1 + s_2 = 4.7$$

$$y - s_1 + s_2 = 0.3.$$

Using Proposition 12 and using GMIC as both the inner and outer functions we obtain the sequential-merge cut $0.4286s_1 + 2s_2 \geq 1$ which is equivalent to $x + y \leq 4$. It can easily be verified that this inequality is facet-defining for the convex hull of solutions to (3), (4) and (5).

Moreover the two GMICs generated from the individual rows are $2.048s_1 + 1.4286s_2 \geq 1$ and $1.4286s_1 + 3.3333s_2 \geq 1$ which are equivalent to $x \leq 4$ and $10x + 7y \leq 44$. It can be verified that these inequalities are not facet-defining for the convex hull of solutions to (3), (4) and (5).

It can be seen from Proposition 12 that the sequential-merge inequalities yield very diverse coefficients for continuous variables. To understand the strength of the continuous coefficients in sequential-merge inequalities we consider the following general example.

Example 2. Consider a continuous variable with $u_1 > 0$, $u_2 < 0$, $u_1 + u_2 + r_2c_h^- u_2 = 0$. The coefficient of this continuous variable in $g \Diamond h$ is $\frac{r_2}{r_1+r_2}(-u_2c_h^-)$. If the group cut h was used to generate a cut from the second constraint alone, the coefficient of the continuous variable would have been $-u_2c_h^- > \frac{r_2}{r_1+r_2}(-u_2c_h^-)$. Similarly, if the group cut g was derived using the first constraint alone, the coefficient of the continuous variable would have been $\frac{1}{r_1}u_1$. Since $u_1 + u_2 + r_2c_h^- u_2 = 0$ the coefficient of the continuous variable using $g \Diamond h$, is $\frac{r_2}{r_1+r_2}(-u_2c_h^-) = \frac{u_1+u_2}{r_1+r_2} < \frac{1}{r_1}u_1$ as $u_2 < 0$. Therefore in this case the continuous coefficients generated using the two different cuts g and h individually will be strictly weaker than those generated using $g \Diamond h$.

We conclude from Example 2 that if both the inner and outer functions used in the sequential-merge procedure are GMICs then the coefficient generated for the continuous variable is stronger than the coefficient generated using each of the individual group cuts when the column corresponding to the continuous variable is (u_1, u_2) with $u_1 > 0$, $u_2 < 0$, $u_1 + u_2 + r_2c_h^- u_2 = 0$ (i.e., the coefficients of the sequential-merge inequalities are not dominated by the GMIC). This result is significant because it can be proven that GMIC generates the strongest possible coefficients for continuous variables among all facets of one-dimensional group problems. We note that this result was numerically observed for the three-gradient facet in Dey and Richard [4].

Note also that although the above discussion was based on the specific case where $u_1 > 0$, $u_2 < 0$ and $u_1 + u_2 + r_2c_h^- u_2 = 0$, there exists a large range of continuous variables coefficient for which the sequential-merge procedure yields inequalities whose coefficients are not dominated by the continuous coefficient of the one-dimensional group cuts derived from individual rows.

5 Conclusion

In this paper we presented a general procedure that produces a wide array of facet-defining inequalities for two-dimensional group problems. We showed that, under very general conditions, these inequalities are facet-defining. These cuts illustrate that strong coefficients for continuous variables can be found by considering group relaxations with multiple constraints. In particular, it is possible to obtain inequalities that are not dominated by group cuts generated from individual constraints. Sequential-merge inequalities are also interesting because they show strong relations between facet-defining inequalities of one-dimensional and two-dimensional group problems.

A few important theoretical and practical questions arise from this paper. First we observe that all the known facet-defining inequalities for the two-dimensional group problem obtained to date are derived either using aggregation [4] or using the sequential-merge procedure. This is an interesting characterization of a subset of facets of the two-dimensional group problem. However this implies that all known facet-defining inequalities of the two-dimensional problem are tightly related to facet-defining inequalities of the one-dimensional group problem. An interesting open question is that of finding a family of group cuts for the two-dimensional infinite group problem that cannot be easily obtained using one-dimensional group cuts.

Second because the sequential merge approach can be applied using the same constraint twice instead of using two different constraints, one interesting question is that of determining when the sequential-merge procedure generates strong inequalities for one-dimensional group problems. The question is particularly interesting since we have shown in Sect. 3 that some two-step MIRs can be derived in this way.

Finally, a large numerical experimentation is needed to determine how and when to use multi-dimensional group cuts to solve MIPs. In particular, numerical determination of how much two-dimensional group cuts improve on one-dimensional group cuts is an important direction of research.

References

[1] K. Aardal, R. Weismantel, and L. A. Wolsey. Non-standard approaches to integer programming. *Discrete Applied Mathematics*, 123:5–74, 2002.
[2] J. Aráoz, L. Evans, R. E. Gomory, and E. L. Johnson. Cyclic groups and knapsack facets. *Mathematical Programming*, 96:377–408, 2003.
[3] S. Dash and O. Günlük. Valid inequalities based on simple mixed integer set. *Mathematical Programming*, 106:29–53, 2006.

[4] S. S. Dey and J.-P. P. Richard. Facets of two-dimensional infinite group problems. *Under Review*, 2006.
     http://www.optimization-online.org/DB_HTML/2006/01/1280.html.
[5] S. S. Dey, J.-P. P. Richard, L. A. Miller, and Y. Li. Extreme inequalities for infinite group problems. 2006.
     http://www.optimization-online.org/DB_HTML/2006/04/1356.html.
[6] R. E. Gomory. Some polyhedra related to combinatorial problems. *Journal of Linear Algebra and Applications*, 2:341–375, 1969.
[7] R. E. Gomory and E. L. Johnson. Some continuous functions related to corner polyhedra, part I. *Mathematical Programming*, 3:23–85, 1972.
[8] R. E. Gomory and E. L. Johnson. Some continuous functions related to corner polyhedra, part II. *Mathematical Programming*, 3:359–389, 1972.
[9] R. E. Gomory and E. L. Johnson. T-space and cutting planes. *Mathematical Programming*, 96:341–375, 2003.
[10] R. E. Gomory, E. L. Johnson, and L. Evans. Corner polyhedra and their connection with cutting planes. *Mathematical Programming*, 96:321–339, 2003.
[11] E. L. Johnson. On the group problem for mixed integer programming. *Mathematical Programming Study*, 2:137–179, 1974.
[12] E. L. Johnson, G. L. Nemhauser, and M. W. P. Savelsbergh. Progress in linear programming-based algorithms for integer programming: an exposition. *INFORMS Journal of Computing*, 12:2–23, 2000.
[13] H. Marchand, A. Martin, R. Weismantel, and L. A. Wolsey. Cutting planes in integer and mixed integer programming. *Discrete Applied Mathematics*, 123:397–446, 2002.
[14] L. Miller, Y. Li, and J.-P. P. Richard. New facets for finite and infinite group problems from approximate lifting. Technical Report MN-ISYE-TR-06-004, University of Minnesota Graduate Program in Industrial and Systems Engineering, 2006.
[15] J.-P. P. Richard, Y. Li, and L. A. Miller. Strong valid inequalities for mips and group polyhedra from approximate lifting. *Under Review*, 2006.

Triangle-Free Simple 2-Matchings in Subcubic Graphs (Extended Abstract)

David Hartvigsen[1] and Yanjun Li[2]

[1] Mendoza College of Business, University of Notre Dame
Notre Dame, IN 46556, USA
dhartvig@nd.edu
[2] Krannert School of Management, Purdue University
West Lafayette, IN 47907, USA
li14@purdue.edu

Abstract. A *simple 2-matching* in an edge-weighted graph is a subgraph all of whose vertices have degree 1 or 2. We consider the problem of finding a maximum weight simple 2-matching that contains no triangles, which is closely related to a class of relaxations of the TSP. Our main results are, for graphs with maximum degree 3, a complete description of the convex hull of incidence vectors of triangle-free simple 2-matchings and a strongly polynomial time algorithm for the above problem. Our system requires the use of a type of comb inequality (introduced by Grötschel and Padberg for the TSP polytope) that has {0,1,2}-coefficients and hence is more general than the well-known blossom inequality used in Edmonds' characterization of the simple 2-matching polytope.

1   Introduction

We consider undirected graphs $G = (V, E)$ with no parallel edges or loops. With every edge $e \in E$ we associate a real weight w_e. A *simple 2-matching* in a graph $G = (V, E)$ is a subgraph of G all of whose vertices have degree 1 or 2. Hence the connected components of a simple 2-matching are simple paths or cycles, each with at least one edge. (For the sake of brevity, we henceforth drop the adjective "simple.") A *2-factor* in G is a special type of 2-matching that contains all the vertices of G and all of whose components are cycles. The problems of finding a maximum weight 2-matching and a maximum weight 2-factor in a graph are well studied. Polynomial time algorithms (see Johnson [19]) and polyhedral characterizations (see Edmonds [10]) are known for both, as well as many other results (see Schrijver [25] for a thorough survey). The key type of inequalities used for the polyhedron is typically called *blossom inequalities* (they are different from the blossoms used by Edmonds [9] for the classical matching problem).

In this paper we focus on a variation of these problems, which we next define. A 2-matching or 2-factor is called C_k-*free* if it contains no cycles of length $\leq k$, for k a positive integer. The C_k-*free 2-matching (2-factor) problem* is to find a C_k-*free* 2-matching (2-factor) with maximum weight.

M. Fischetti and D.P. Williamson (Eds.): IPCO 2007, LNCS 4513, pp. 43–52, 2007.

These problems were originally studied in the context of the travelling sales-man problem by Fisher, Nemhauser, and Wolsey [11] and by Cornuéjols and Pulleyblank [7]. It is shown in [11] how solutions to the C_k-free 2-factor prob-lem in a complete graph, for increasing values of k, yield increasingly accurate appoximations of the optimal value of a travelling salesman tour. Observe that for $n/2 \leq k \leq n - 1$ (where n is the number of vertices in the graph), solving the C_k-free 2-factor problem is equivalent to the TSP.

Other results are known for these problems, which lead to some open ques-tions. For $k \geq 5$, the C_k-*free* 2-factor problem with weights all $0, 1$ was shown to be NP-hard by Papadimitriou (the proof appears in [7]). This implies that, for $k \geq 5$, the C_k-*free* 2-matching problem with general weights is NP-hard. (See also Hell et al [16], where similar complexity results are presented.) Vornberger [29] showed that the C_4-free 2-factor problem (with general weights) is NP-hard. For the case that the edge weights are all $0, 1$, an algorithm solving the C_3-free 2-matching problem (hence the C_3-free 2-factor problem) appears in [14]. (It is quite complex.) So two obvious open problems (discussed in [29] and [7]) are to find the complexity of the C_3-*free* 2-factor (2-matching) problem, with general weights, and the C_4-*free* 2-factor (2-matching) problem, with weights all $0, 1$. Another obvious open problem is to describe the polytope associated with the C_3-*free* 2-factor (2-matching) problem (assuming it is polynomial).

One way of approaching the open problems mentioned above has been to con-sider them on special classes of graphs. For example, Hartvigsen [15] presented, for bipartite graphs, a polynomial time algorithm for the C_4-*free* 2-matching problem with weights all $0, 1$ (hence for the corresponding 2-factor problem as well). Nam [22] presented a polynomial time algorithm for the C_4-*free* 2-factor problem for general graphs with the property that no two squares share a vertex. (The algorithm is quite complex.) In this paper we consider such problems on the *cubic* (*subcubic*) graphs; that is, those graphs for which every vertex has degree 3 (at most 3). Some work in this regard has been done. For cubic graphs, Vornberger [29] presented a polynomial time algorithm for the C_3-*free* 2-factor problem. Furthermore, his proof that the C_4-free 2-factor problem (with general weights) is NP-hard was done for cubic graphs. He also showed that the C_5-*free* 2-factor problem with weights all $0, 1$ is NP-hard for cubic graphs. Some partial polyhedral results have also been obtained by Cunningham and Wang [8], who presented a class of valid inequalities for the polytope associated with the C_k-*free* 2-factor problem and studied the special structure of these inequalities for the C_3-*free* 2-factor problem.

Our main results are, for subcubic graphs, a complete description of the convex hull of C_3-free 2-matchings and a strongly polynomial time algorithm for finding a maximum weight C_3-free 2-matching. (This polytope immediately yields the polytope for C_3-free 2-factors in subcubic graphs.) An interesting property of this polytope is that it requires the use of two types of inequalities not needed for the 2-matching polytope. One type is straightforward and simply requires that, for each triangle in the graph, the sum of the associated edge variables is at most 2. The second type is a class, we call them *tri-combs*, that properly falls between the *blossom inequalities* (as introduced by Edmonds for the 2-matching polytope [10]) and the more general *comb inequalities* (introduced by Grötschel and Padberg [13] for the travelling salesman polytope; another class, the Chvátal combs [4], also properly falls between the blossom and comb inequalities, but is different from our tri-combs). Results of Cunningham and Wang [8] show that the tri-combs are not sufficient to describe the C_3-free 2-factor polytope on general graphs. They show that additional needed inequalities include, but are not limited to, the so-called bipartition inequalities.

A tri-comb is a comb such that every tooth has at most 3 vertices and every tooth has exactly one node not in the handle. A blossom is a comb such that every tooth has exactly two vertices. Another property of our tri-comb inequalities that distinquishes them from blossoms is that they are not $\{0,1\}$-inequalities; i.e., the variables have coefficients in $\{0,1,2\}$. We also show, somewhat surprisingly, that only the $\{0,1\}$-inequalities in this class (i.e., the blossoms) are needed to describe the polytope for C_3-free 2-factors in subcubic graphs (which is a face of the polytope for C_3-free 2-matchings). Thus we see that C_3-free 2-matchings and C_3-free 2-factors, in this domain, have significantly different polyhedral descriptions. To the best of our knowledge, this is the only matching example known to have this property.

Our main polyhedral result is proved using the algorithm, which is primal-dual in the style used by Edmonds [9] for the classical matching problem. Polyhedral and algorithmic results for the 2-matching problem (and more general problems) are often proved in the literature by using a reduction to classical matchings due to Tutte [28] (see Schrijver [25]). Algorithms that avoid such a reduction are typically more efficient (but more complex) and have also been studied (e.g., see Johnson [19] and Gabow [12]). We have been unable to find such a reduction for the C_3-free 2-matching problem in subcubic graphs, hence our algorithm works directly on the original graph. However, for the restricted case of C_3-free 2-factors, a reduction was discovered by Vornberger [29]. It yields a polynomial algorithm for finding C_3-free 2-factors in cubic graphs (which can be extended, in a straightforward manner, to finding C_3-free 2-factors in subcubic graphs). We show that this same reduction idea, combined with the 3-cut polyhedral reduction idea for the TSP problem developed by Cornuéjols, Naddef, and Pulleyblank [6], yields the C_3-free 2-factor polytope for subcubic graphs.

Let us remark on a topic one might expect to accompany work of this type. In Edmonds' primal-dual algorithm for classical matchings, the primal stage is essentially an algorithm for finding a maximum $(0, 1)$-weight matching, which is applied to a special subgraph of the original graph. Hence, as a by-product, one obtains a simpler algorithm for finding a maximum $(0, 1)$-weight matching and one can prove, directly from the algorithm, theorems such as Tutte's characterization of the graphs with a perfect matching [27] and the min-max Tutte-Berge theorem [1]. The algorithm we have d eveloped does not appear to have these nice by-products, at least in a direct way. Hence we plan to address the special case of $(0, 1)$-weights elsewhere.

This extended abstract is organized as follows. In Section 2 we state our two main polyhedral results. Section 3 contains an outline of our non-algorithmic proof of the second, simpler result. The final section contains an overview of the primal-dual algorithm for C_3-free 2-matchings. The details of the proofs and algorithm will appear elsewhere.

We close this section by referencing some related research. A $\{0,1,2\}$-matching is an assignment of 0,1, or 2 to each edge in a graph so that the sum of the values on the edges incident with each node is at most 2. (In this terminology, we consider $\{0,1\}$-matchings in this paper.) Cornuéjols and Pulleyblank in [7] completely characterize the convex hull of C_3-free $\{0,1,2\}$-matchings and present a polynomial time algorithm for finding maximum weight C_3-free $\{0,1,2\}$-matchings. Their algorithm is similar in style to the one presented in this paper: it is primal-dual (as in [9]) and provides a proof of the polyhedral result.

Finally, we note that there is a fairly extensive literature involving the study of matching problems in regular graphs, particularly in cubic graphs. Here are some of the highlights:

- Kaiser and Skrekovski [20] showed that every bridgeless cubic graph has a 2-factor that intersects all edge cuts of size 3 or 4. This result implies that every bridgeless cubic graph has a C_3-free 2-factor.
- Petersen [23] showed that every bridgeless cubic graph can be decomposed into a 2-factor and a perfect matching.
- Tait [26] showed that every planar bridgeless cubic graph can be decomposed into three perfect matchings iff the 4-color conjecture holds (which, of course, is now a theorem). And Petersen [24] showed that this is not true for non-planar bridgeless cubic graphs by exhibiting what we now call the Petersen graph. Holyer [18] showed that it's NP-complete to decide if a cubic graph can be decomposed into three matchings.
- Bertram and Horak [2] showed that there is a polynomial time algorithm to decompose any 4-regular graph into two C_3-free 2-factors, if such a decomposition exists.
- The study of Hamilton cycles in cubic graphs is fairly extensive. A key result is that the problem of deciding if a planar, bipartite, subcubic graph has a Hamiltonian cycle is NP-complete (see [21]). Additional results can be found in [17].

2   Main Results

In this section we present our main polyhedral results: complete descriptions of the convex hulls of C_3-free 2-matchings and 2-factors for subcubic graphs. We begin with some definitions.

Let $G = (V, E)$ be a graph. For $V' \subseteq V$, let $\delta(V')$ denote the set of edges of G with exactly one vertex in V'; and let $\gamma(V')$ denote the set of edges of G with both vertices in V'. For $S \subseteq E$ and $x \in \mathbb{R}^E$, let $x(S) = \sum_{e \in S} x_e$. If G' is a subgraph of G, then a vector $x \in \{0, 1\}^E$ is called the 0-1 *incidence vector* for

G' if $x_e = 1$ if and only if e is in G'. A *triangle* of G is a set of three edges of G that form a cycle. A C_3-free 2-matching is also called a *triangle-free 2-matching*.

Let $P^M(G)$ denote the convex hull of incidence vectors of triangle-free 2-matchings in G. Let $P^F(G)$ denote the convex hull of incidence vectors of triangle-free 2-factors in G.

A *tri-comb* of G is a set $\{H, T_1, \ldots T_{2k+1}\}$ of subsets of V, where $k \geq 1$, that satisfy the following conditions:

1. $T_1, \ldots T_{2k+1}$ are pairwise disjoint;
2. For each i, $\gamma(T_i)$ is either a single edge or a triangle;
3. Each T_i has exactly one vertex not in H.

We call H the *handle* and $T_1, \ldots T_{2k+1}$ the *teeth* of the tri-comb. A tooth with two vertices is called an *edge-tooth* and a tooth with three edges is called a *triangle-tooth*. Observe that every triangle-tooth has exactly one edge in common with $\gamma(H)$, which we call a *common edge*. See Fig. 1, which contains a tri-comb with two triangle-teeth, one edge-tooth, and two common edges.

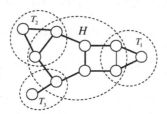

Fig. 1. A tri-comb

Consider the following variations on conditions 2 and 3:

2′. For each i, $\gamma(T_i)$ is a single edge;
3′. Each T_i has, at least, one vertex in H and one vertex not in H.

The sets $\{H, T_1, \ldots, T_{2k+1}\}$ that satisfy conditions 1, 2′, and 3 are the well-known class of *blossoms* and the sets that satisfy conditions 1 and 3′ are the well-known class of *combs*. Hence tri-combs are more general than blossoms and a special case of combs.

For x a variable vector indexed on E, we associate with each tri-comb $C = \{H, T_1, \ldots, T_{2k+1}\}$ the following *tri-comb inequality*:

$$x(\gamma(H)) + \sum_{i=1}^{2k+1} x\left(\gamma(T_i)\right) \leq |H| + \sum_{i=1}^{2k+1} (|T_i| - 1) - (k + 1),$$

which we abbreviate, $a(C)x \leq b(C)$. In general, these inequalities have non-zero variable coefficients in $\{1, 2\}$, with the 2s precisely on the common edges. The tri-comb inequalities can be shown to be feasible for $P^M(G)$ (hence for $P^F(G)$) using standard arguments (e.g., see [5]). For the cases of blossoms and combs, this same inequality is the well-known *blossom inequality* and *comb inequality*, respectively. Observe that for blossoms, the right hand side of the inequality simplifies to $|H|+k$. (We have borrowed the notation for this inequality from [5].)

We are now ready to state our characterization of $P^M(G)$.

Theorem 1. *For a subcubic graph $G = (V, E)$, $P^M(G)$ is determined by*

$$x(\delta(v)) \leq 2 \quad \forall v \in V \tag{1}$$

$$x_e \leq 1 \quad \forall e \in E \tag{2}$$

$$x(T) \leq 2 \quad \forall \text{ triangles } T \tag{3}$$

$$a(C)x \leq b(C) \quad \forall \text{ tri-combs } C \tag{4}$$

$$x_e \geq 0 \quad \forall e \in E \tag{5}$$

Let $S^M(G)$ denote the system in Theorem 1. If we remove inequalities (3) and replace "tri-combs" with "blossoms" in inequality (4), then we obtain the convex hull of incidence vectors of 2-matchings given by Edmonds [10] for general graphs. If, instead, we remove inequalities (4) and replace inequalities (2) with $x_e \leq 2 \ \forall e \in E$, we obtain the convex hull of incidence vectors of C_3-free $\{0,1,2\}$-matchings given by Cornuéjols and Pulleyblank [7] for general graphs.

Fig. 2 contains a subcubic graph that illustrates the need for the tri-comb inequalities in $S^M(G)$. The numbers indicate a fractional solution x that can be seen to be extreme for the system of inequalities: (1), (2), (3), and (5). The tri-comb inequality associated with the entire graph (which is a tri-comb C with indicated handle H) has $b(C) = 6$; however, the fractional solution has $a(C)x = 6.5$. In contrast, one can easily check that no blossom inequality is able to cut off this fractional extreme solution.

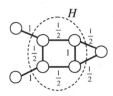

Fig. 2. A fractional extreme point cut off by a tri-comb inequality

Our characterization of $P^F(G)$ follows.

Theorem 2. *For a subcubic graph $G = (V, E)$, $P^F(G)$ is determined by*

$$x(\delta(v)) = 2 \quad \forall v \in V \tag{6}$$

$$x_e \leq 1 \quad \forall e \in E \tag{7}$$

$$x(T) = 2 \quad \forall \text{ triangles } T \tag{8}$$

$$a(C)x \leq b(C) \quad \forall \text{ blossoms } C \tag{9}$$

$$x_e \geq 0 \quad \forall e \in E \tag{10}$$

Let $S^F(G)$ denote the system in Theorem 2. If we remove inequalities (8), then we obtain the convex hull of incidence vectors of 2-factors given by Edmonds [10] for general graphs.

3   The 2-Factor Polytope for Subcubic Graphs

In this section we outline the proof of Theorem 2. The proof uses a triangle shrinking operation and a variation on the Basic Polyhedral Theorem in [6].

Fig. 3 shows the four triangle patterns that can exist in a subcubic graph (i.e., a triangle can have 0, 1, 2, or 3 vertices of degree 2). If two or three vertices of a triangle have degree 2 in the graph, then there is obviously no feasible solution to the triangle-free 2-factor problem. Therefore, we only consider the subcubic graphs with the triangle patterns shown in Fig. 3(a) and Fig. 3(b).

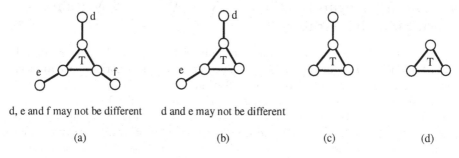

d, e and f may not be different   d and e may not be different

(a)   (b)   (c)   (d)

Fig. 3. Four triangle patterns

Let $G = (V, E)$ be a graph and let $S \subseteq V$, such that $|S| \geq 2$. We let $G \times S$ denote the graph obtained by *shrinking* (or contracting) S. That is, the vertices of $G \times S$ are the vertices of $V \backslash S$, plus a new vertex, say v, obtained by identifying all the vertices in S. The edges of $G \times S$ are the edges in $\gamma(V \backslash S)$ and the edges of $\delta(S)$, each of which now has one vertex v and its original vertex in $V \backslash S$. All edges in $G \times S$ retain their identities from G.

If T is a triangle (which is a set of three edges), we let $V(T)$ denote the vertices of T.

The following lemmas will help us obtain $P^F(G)$.

Lemma 1. *Let G be a subcubic graph and T be a triangle of G that has the pattern of Figure 3(a) or Figure 3(b). A linear system sufficient to define $P^F(G)$ is obtained by taking the union of linear systems sufficient to define $P^F(G \times V(T))$ and $P^F(G \times (V \backslash V(T)))$.*

Lemma 2. $P^F(G \times (V \backslash V(T)))$ *is determined by*

$$x(\delta(v)) = 2 \quad \forall v \in V(T)$$
$$x_e \leq 1 \quad \forall e \in E(G \times (V \backslash V(T)))$$

$$x(T) = 2$$
$$x_e \geq 0 \quad \forall e \in E(G \times (V \backslash V(T)))$$

The proof of Lemma 1 is quite similar to the proof of the Basic Polyhedral Theorem in [6]. That theorem says that one can obtain a linear system sufficient to define the TSP polytope $TSP(G)$ by the union of two linear systems sufficient to define $TSP(G \times S)$ and $TSP(G \times \bar{S})$, where G has a 3-edge cutset with shores S and $\bar{S}$. The proof of Lemma 2 is straightforward.

We can now obtain $P^F(G)$ as follows: First, we iteratively shrink triangles that have three original nodes of G and apply the above two lemmas until every triangle in the graph has at least one shrunk node. Then we apply the polyhedral description of 2-factors [10] to this final graph. Finally, the union of all the linear systems obtained through applying Lemma 1 and Lemma 2 and the linear system sufficient to define the 2-factors of the final graph defines $P^F(G)$.

4 The Algorithm for Finding Max Weight 2-Matchings in Subcubic Graphs

In this section we give an overview of the algorithm, followed by a few details describing how the algorithm is set up. The details of the algorithm are approximately ten pages long and will appear elsewhere.

The algorithm has two main phases: primal and dual. While maintaining primal and dual feasible solutions, the algorithm alternates between these two phases until it produces primal and dual feasible solutions that satisfy complementary slackness, and hence are optimal. The primal phase has two main stages. We call the first stage "triangle alteration." In this stage we identify special triangles in the original graph and alter each by either shrinking the triangle to a vertex, shrinking an edge of the triangle to a vertex, or deleting one of its edges. In some sense, this is an elaborated version of the shrinking operation described in the preceeding section; however, this type of shrinking cannot be done just once at the beginning of the algorithm – it occurs repeatedly throughout. In the second stage of the primal phase we grow an alternating tree looking for ways to satisfy violated complementary slackness conditions. If no such improvement is found, the algorithm moves into its second phase, the dual change. In this phase the dual solution is changed, again in an effort to satisfy violated complementary slackness conditions. The growth and dual change steps are, in a general sense, typical of Edmonds-style matching algorithms; the triangle alteration stage is unique to this problem. The primal growth stage is quite straightforward due to the simple structure of the graphs and our triangle alteration stage; however, the dual change is significantly more complex than is typical of such algorithms.

We next present some details we need to set up the algorithm. For an arbitrary subcubic graph $G = (V, E)$, let $w \in \mathbb{R}^E$ be an arbitrary weight vector, let $\mathcal{T}$ denote the set of all triangles in G, and let $\mathcal{TC}$ denote the set of all tri-combs in G. We let $\mathcal{T}(e)$ denote the triangles of G that contain edge e, let $\mathcal{TC}_1(e)$ denote the tri-combs of G that contain e as a non-common edge, and let $\mathcal{TC}_2(e)$ denote the tri-combs of G that contain e as a common edge.

The *primal LP* is the following:

$$\max \ wx \ \text{s.t.} \ x \in S^M(G).$$

If we associate variable vectors y, z, τ, π with constraints (1), (2), (3), (4), respectively, then we obtain the corresponding *dual LP*:

$$\min \ 2\sum_{v \in V} y_v + \sum_{e \in E} z_e + 2\sum_{T \in \mathcal{T}} \tau_T + \sum_{C \in \mathcal{TC}} b(C)\pi_C$$

s.t.

$$y_u + y_v + z_{uv} + 2\sum_{T \in \mathcal{T}(e)} \tau_T + \sum_{C \in \mathcal{TC}_1(e)} \pi_C + 2\sum_{C \in \mathcal{TC}_2(e)} \pi_C \geq w_e \ \forall e = uv \in E$$

$$y, z, \tau, \pi \geq 0.$$

From linear programming theory, a primal feasible solution x and a dual feasible solution y, z, τ, π are both optimal if and only if they satisfy the following complementary slackness conditions:

$$x_e > 0 \Rightarrow y_u + y_v + z_{uv} + \sum_{C \in \mathcal{TC}_1(e)} \pi_C + 2\Big(\sum_{T \in \mathcal{T}(e)} \tau_T + \sum_{C \in \mathcal{TC}_2(e)} \pi_C\Big) = w_e; \quad (11)$$

$$y_v > 0 \Rightarrow x(\delta(v)) = 2; \quad (12)$$

$$z_e > 0 \Rightarrow x_e = 1; \quad (13)$$

$$\tau_T > 0 \Rightarrow x(T) = 2; \quad (14)$$

$$\pi_C > 0 \Rightarrow a(C)x = b(C). \quad (15)$$

At each stage of the primal-dual algorithm we maintain an integral primal feasible solution x, which is the incidence vector of a triangle-free 2-matching, and a dual feasible solution y, z, τ, π, which satisfies (11), (13), (14) and (15). Condition (12) is not, in general, satisfied. The algorithm modifies the variables x, y, z, τ, π (maintaining primal and dual feasibility as well as conditions (11), (13), (14) and (15)) until condition (12) is satisfied at which point x is the incidence vector of a maximum weight triangle-free 2-matching.

References

1. Berge, C.: Sur le couplate maximum d'un graphe. Comptes Rendus Hebdomadaires des Séances de l'Académie Sciences [Paris] **247** (1958) 258–259
2. Bertram, E., Horak, P.: Decomposing 4-regular graphs into triangle-free 2-factors. SIAM J. Disc. Math. **10** (1997) 309–317
3. Boyd, S.C., Cunningham, W.: Small travelling salesman polytopes. Math. Oper. Res. **16** (1991) 259–271
4. Chvátal, V.,: Edmonds polytopes and weakly hamiltonian graphs. Math. Prog. **5** (1973) 29–40
5. Cook, W.J., Cunningham, W.H., Pulleyblank, W.R., Schrijver, A.: Combinatorial Optimization. John Wiley & Sons, New York (1998)

6. Cornuéjols, G., Naddef, D., and Pulleyblank, W.R.: The traveling salesman problem in graphs with 3-edge cutsets. J.A.C.M. **32** (1985) 383–410
7. Cornuéjols, G., Pulleyblank, W.R.: A matching problem with side constraints. Disc. Math. **29** (1980) 135–159
8. Cunningham, W.H., Wang, Y.: Restricted 2-factor polytopes. Math. Prog. **87** (2000) 87–111
9. Edmonds, J.: Paths, trees, and flowers. Canad. J. Math. **17** (1965) 449–467
10. Edmonds, J.: Maximum matching and a polyhedron with 0,1 vertices. J. of Res. National Bur. of Standards **69** (1965) 125–130
11. Fisher, M.L., Nemhauser, G.L., Wolsey, L.A.: An analysis of approximations for finding a maximum weight Hamiltonian circuit. Oper. Res. **27** (1979) 799–809
12. Gabow, H.N.: An efficient reduction technique for degree-constrained subgraph and bideredted network flow problems. In: Proceedings of the Fifteenth Annual ACM Symposium on Theory of Computing. The Association for Computing Machinery, New York (1983) 448–456
13. Grötschel, M., Padberg, M.W.: On the symmetric travelling salesman problem II: Lifting theorems and facets. Math. Prog. **16** (1979) 282–302
14. Hartvigsen, D.: Extensions of Matching Theory. Ph.D. Thesis, Carnegie-Mellon University (1984); under the supervision of Gérard Cornuéjols.
15. Hartvigsen, D.: Finding maximum square-free 2-matchings in bipartite graphs. J. of Comb. Th. B **96** (2006) 693–705
16. Hell, P., Kirkpatrick, D., Kratochvíl, J., Kříž, I.: On restricted 2-factors. SIAM J. Disc. Math **1** (1988) 472–484
17. Holton, D., Aldred, R.E.L.: Planar graphs, regular graphs, bipartite graphs and Hamiltonicity. Australas. J. Combin. **20** (1999) 111–131
18. Holyer, I.: The NP-completeness of edge coloring. SIAM Journal on Computing **10** (1981) 718–720
19. Johnson, E.: Network Flows, Graphs and Integer Programming. Ph.D. Thesis, University of California, Berkeley (1965)
20. Kaiser, T., Skrekovski, R.: Cycles intersecting cuts of prescribed sizes. Manuscript (2005)
21. Lawler, E.L., Lenstra, J.K., Rinnooy Kan, A.H.G., Shmoys, D.B.: The Traveling Salesman Problem – A Guided Tour of Combinatorial Optimization. Wiley, Chichester (1985)
22. Nam, Y.: Matching Theory: Subgraphs with Degree Constraints and other Properties. Ph.D. Thesis, University of British Columbia (1994); Under the supervision of R. Anstee.
23. Petersen, J.: Die Theorie der regulären graphs. Acta Mathematica **15** (1891) 193–220
24. Petersen, J.: Sur le theoreme de Tait. L'Intermediaire des Mathematiciens **5** (1898) 225–227
25. Schrijver, A.: Combinatorial Optimization, Polyhedra and Efficiency. Springer, Berlin (2003)
26. Tait, P.G.: Remarks on the previous communication. Proceedings of the Royal Society of Edinburgh **10** (1878-80) 729
27. Tutte, W.T.: The factorization of linear graphs. J. London Math. Soc. **22** (1947) 107–111
28. Tutte, W.T.: A short proof of the factor theorem for finite graphs. Canadian J. of Math. **6** (1954) 347–352
29. Vornberger, O.: Easy and hard cycle covers. Manuscript, Universität Paderborn (1980)

The Smoothed Number of Pareto Optimal Solutions in Bicriteria Integer Optimization[*]

Rene Beier[1], Heiko Röglin[2], and Berthold Vöcking[2]

[1] Max-Planck-Institut für Informatik
Saarbrücken, Germany
rbeier@mpi-inf.mpg.de
[2] Department of Computer Science
RWTH Aachen, D-52056 Aachen, Germany
{roeglin,voecking}@cs.rwth-aachen.de

Abstract. A well established heuristic approach for solving various bicriteria optimization problems is to enumerate the set of Pareto optimal solutions, typically using some kind of dynamic programming approach. The heuristics following this principle are often successful in practice. Their running time, however, depends on the number of enumerated solutions, which can be exponential in the worst case.

In this paper, we prove an almost tight bound on the expected number of Pareto optimal solutions for general bicriteria integer optimization problems in the framework of smoothed analysis. Our analysis is based on a semi-random input model in which an adversary can specify an input which is subsequently slightly perturbed at random, e. g., using a Gaussian or uniform distribution.

Our results directly imply tight polynomial bounds on the expected running time of the Nemhauser/Ullmann heuristic for the 0/1 knapsack problem. Furthermore, we can significantly improve the known results on the running time of heuristics for the bounded knapsack problem and for the bicriteria shortest path problem. Finally, our results also enable us to improve and simplify the previously known analysis of the smoothed complexity of integer programming.

1 Introduction

We study integer optimization problems having two criteria, say profit and weight, which are to be optimized simultaneously. A common approach for solving such problems is generating the set of *Pareto optimal* solutions, also known as the *Pareto set*. Pareto optimal solutions are optimal compromises of the two criteria in the sense that any improvement of one criterion implies an impairment to the other. In other words, a solution S^* is Pareto optimal if there exists no other solution S that dominates S^*, i. e., has at least the profit and at most the weight of S^* and at least one inequality is strict. Generating the Pareto set

---

[*] This work was supported by DFG grant VO 889/2 and by the EU within the 6th Framework Programme under contract 001907 (DELIS).

M. Fischetti and D.P. Williamson (Eds.): IPCO 2007, LNCS 4513, pp. 53–67, 2007.

is of great interest in many scenarios and widely used in practice. Unfortunately, this approach fails to yield reasonable results in the worst case because even integer optimization problems with a simple combinatorial structure can have exponentially many Pareto optimal solutions. In practice, however, generating the Pareto set is often feasible since typically the number of Pareto optimal solutions does not attain its worst-case bound.

The discrepancy between practical experience and worst-case results motivates the study of the number of Pareto optimal solutions in a more realistic scenario. One possible approach is to study the average number of Pareto optimal solutions rather than the worst case number. In order to analyze the average, one has to define a probability distribution on the set of instances, with respect to which the average is taken. In most situations, however, it is not clear how to choose a probability distribution that reflects typical inputs. In order to bypass the limitations of worst-case and average-case analysis, Spielman and Teng defined the notion of *smoothed analysis* [15]. They consider a semi-random input model in which an adversary specifies an input which is then randomly perturbed. One can hope that semi-random input models are more realistic than worst-case and average-case input models since the adversary can specify an arbitrary input with a certain structure, and the subsequent perturbation generates an instance which is still close to the adversarial one.

We consider integer optimization problems in a semi-random setting, in which an adversary can specify an arbitrary set $S \subseteq D^n$ of feasible solutions and two objective functions: profit $p \colon S \to \mathbb{R}$ and weight $w \colon S \to \mathbb{R}$, where $D \subset \mathbb{Z}$ denotes a finite set of integers. We assume that the profit is to be maximized and the weight is to be minimized. This assumption is without loss of generality as our results are not affected by changing the optimization direction of any of the objective functions. In our model, the weight function w can be chosen arbitrarily by the adversary, whereas the profit p has to be linear of the form $p(x) = p_1 x_1 + \cdots + p_n x_n$. The adversary can choose an arbitrary vector of profits from $[-1, 1]^n$, but in the second step of the semi-random input model, the profits p_i are randomly perturbed by adding an independent Gaussian random variable with mean 0 and standard deviation σ to each profit p_i. The standard deviation σ can be seen as a parameter measuring how close the analysis is to a worst-case analysis: The smaller σ is chosen, the smaller is the influence of the perturbation and, hence, the closer is the analysis to a worst-case analysis. Our probabilistic analysis is not restricted to Gaussian perturbations but is much more general. In fact, it covers arbitrary probability distributions with a bounded density function and a finite absolute mean value. In particular, if one is interested in obtaining a positive domain for the profits, one can restrict the adversary to profits $p_i \in [0, 1]$ and perturb these profits by adding independent random variables that are distributed uniformly over some interval $[0, c]$.

We present a new method for bounding the expected number of Pareto optimal solutions in the aforementioned scenario which yields an upper bound that depends polynomially on the number of variables n, the integer with the largest absolute value in D, and the reciprocal of the standard deviation σ. This immediately implies polynomial upper bounds on the expected running time of several heuristics for generating the Pareto set of problems like, e. g., the Bounded Knapsack problem. Previous results of this kind were restricted to the case of binary optimization problems. For this special case, our method yields an improved upper bound, which matches the known lower bound. Furthermore, we show that our results on the expected number of Pareto optimal solutions yield a significantly simplified and improved analysis of the smoothed complexity of integer programming.

1.1 Previous Results

Multi-objective optimization is a well studied research area. Various algorithms for generating the Pareto set of various optimization problems like, e. g., the (bounded) knapsack problem [11,8], the bicriteria shortest path problem [4,14] and the bicriteria network flow problem [5,10], have been proposed. The running time of these algorithms depends crucially on the number of Pareto optimal solutions and, hence, none of them runs in polynomial time in the worst case. In practice, however, generating the Pareto set is tractable in many situations. For instance, Müller-Hannemann and Weihe [9] study the number of Pareto optimal solutions in multi-criteria shortest path problems experimentally. They consider examples that arise from computing the set of best train connections (in view of travel time, fare, and number of train changes) and conclude that in this application scenario generating the complete Pareto set is tractable even for large instances. For more examples, we refer the reader to [6].

One way of coping with the bad worst-case behavior is to relax the requirement of finding the complete Pareto set. Papadimitriou and Yannakakis present a general framework for finding *approximate Pareto sets*. A solution S is ε-dominated by another solution S' if $p(S)/p(S') \leq 1 + \varepsilon$ and $w(S')/w(S) \leq 1 + \varepsilon$. We say that $\mathcal{P}_\varepsilon$ is an ε-approximation of a Pareto set $\mathcal{P}$ if for any solution $S \in \mathcal{P}$ there is a solution $S' \in \mathcal{P}_\varepsilon$ that ε-dominates it. Papadimitriou and Yannakakis show that for any Pareto set $\mathcal{P}$, there is an ε-approximation of $\mathcal{P}$ with polynomially many points (w. r. t. the input size and $1/\varepsilon$) [12]. Furthermore they give necessary and sufficient conditions under which there is an FPTAS to generate $\mathcal{P}_\varepsilon$. Vassilvitskii and Yannakakis [16] show how to compute ε-approximate Pareto curves of almost minimal size.

Beier and Vöcking analyze the expected number of Pareto optimal solutions for binary optimization problems [2]. They consider the aforementioned scenario with $\mathcal{D} = \{0,1\}$ and show that the expected number of Pareto optimal solutions is bounded from above by $O(n^4/\sigma)$. This result implies that the Nemhauser/Ullmann algorithm [11] has polynomial expected running time. Furthermore, they also present a lower bound of $\Omega(n^2)$ on the expected number of Pareto optimal solutions for profits that are chosen uniformly from the interval $[0,1]$.

In [3] Beier and Vöcking analyze the smoothed complexity of binary optimization problems. They consider optimization problems with one objective function in which the set of feasible solutions is given as $\mathcal{S} \cap \mathcal{B}_1 \cap \ldots \cap \mathcal{B}_m$, where $\mathcal{S} \subseteq \{0,1\}^n$

denotes a fixed ground set and $\mathcal{B}_i$ denotes a halfspace induced by a linear constraint of the form $w_{i,1}x_1 + \cdots + w_{i,n}x_n \leq t_i$. Similar to the aforementioned model it is assumed that the coefficients $w_{i,j}$ are perturbed by adding independent random variables to them. Based on the probabilistic analysis of certain structural properties, Beier and Vöcking show that a binary optimization problem in this form has polynomial smoothed complexity if and only if there exists a pseudo-polynomial (w. r. t. the $w_{i,j}$) time algorithm for solving the problem. The term *polynomial smoothed complexity* is defined analogously to the way polynomial complexity is defined in average-case complexity theory, adding the requirement that the running time should be polynomially bounded not only in the input size but also in $1/\sigma$. This characterization is extended to the case of integer optimization problems where $\mathcal{D} \subset \mathbb{Z}$ is a finite set of integers by Röglin and Vöcking [13].

1.2   Our Results

In this paper, we present a new approach for bounding the expected number of Pareto optimal solutions for bicriteria integer optimization problems. This approach yields the first bounds for integer optimization problems and improves the known bound for the binary case significantly. We show that the expected number of Pareto optimal solutions is bounded from above by $O(n^2 k^2 \log(k)/\sigma)$ if $\mathcal{D} = \{0, \ldots, k-1\}$. We also present a lower bound of $\Omega(n^2 k^2)$, assuming that the profits are chosen uniformly at random from the interval $[-1, 1]$. For the case in which the adversary is restricted to linear weight functions, we present a lower bound of $\Omega(n^2 k \log k)$. Furthermore, for the binary case $\mathcal{D} = \{0, 1\}$, the upper bound simplifies to $O(n^2/\sigma)$, which improves the previously known bound by a factor of $\Theta(n^2)$ and matches the lower bound in [2] in terms of n. Hence, our method yields tight bounds in terms of n and almost tight bounds in terms of k for the expected number of Pareto optimal solutions and, thereby, even simplifies the proof in [2]. In the following, we list some applications of these results.

Knapsack Problem. The Nemhauser/Ullmann algorithm solves the knapsack problem by enumerating all Pareto optimal solutions [11]. Its running time on an instance with n items is $\Theta(\sum_{i=1}^{n} q_i)$, where q_i denotes the number of Pareto optimal solutions of the knapsack instance that consists only of the first i items. Beier and Vöcking analyze the expected number of Pareto optimal solutions and show that the expected running time of the Nemhauser/Ullmann algorithm is bounded by $O(n^5/\sigma)$ if all profits are perturbed by adding Gaussian or uniformly distributed random variables with standard deviation σ [2]. Based on our improved bounds on the expected number of Pareto optimal solutions, we conclude the following corollary.

Corollary 1. *For semi-random knapsack instances in which the profits are perturbed by adding independent Gaussian or uniformly distributed random variables with standard deviation σ, the expected running time of the Nemhauser/Ullmann algorithm is $O(n^3/\sigma)$.*

For uniformly distributed profits Beier and Vöcking present a lower bound on the expected running time of $\Omega(n^3)$. Hence, we obtain tight bounds on the running time of the Nemhauser/Ullmann algorithm in terms of the number of items n. This lower bound can easily be extended to the case of Gaussian perturbations.

Bounded Knapsack Problem. In the bounded knapsack problem, a number $k \in \mathbb{N}$ and a set of n items with weights and profits are given. It is assumed that k instances of each of the n items are given. In [7] it is described how an instance with n items of the bounded knapsack problem can be transformed into an instance of the (binary) knapsack problem with $\Theta(n \log (k + 1))$ items. Using this transformation, the bounded knapsack problem can be solved by the Nemhauser/Ullmann algorithm with running time $\Theta(\sum_{i=1}^{n \log (k+1)} q_i)$, where q_i denotes the number of Pareto optimal solutions of the binary knapsack instance that consists only of the first i items. Based on our results on the expected number of Pareto optimal solutions, we obtain the following corollary.

Corollary 2. *The expected running time of the Nemhauser/Ullmann algorithm on semi-random bounded knapsack instances in which the profits are perturbed by adding independent Gaussian or uniformly distributed random variables with standard deviation σ is bounded from above by $O(n^3 k^2 (\log^2 (k + 1))/\sigma)$ and bounded from below by $\Omega(n^3 k \log^2 (k + 1))$.*

Hence, our results yield tight bounds in terms of n for the expected running time of the Nemhauser/Ullmann algorithm.

Bicriteria Shortest Path Problem. Different algorithms have been proposed for enumerating the Pareto set in bicriteria shortest path problems [4,14]. In [4] a modified version of the Bellman/Ford algorithm is suggested. Beier shows that the expected running time of this algorithm is $O(nm^5/\sigma)$ for graphs with n nodes and m edges [1]. We obtain the following improved bound.

Corollary 3. *For semi-random bicriteria shortest path problems in which one objective function is linear and its coefficients are perturbed by adding independent Gaussian or uniformly distributed random variables with standard deviation σ, the expected running time of the modified Bellman/Ford algorithm is $O(nm^3/\sigma)$.*

Smoothed Complexity of Integer Programming. We were not able to bound the expected number of Pareto optimal solutions for optimization problems with more than two objective functions. One approach for tackling multicriteria problems is to solve a constrained problem in which all objective functions except for one are made constraints. Our results for the bicriteria case can be used to improve the smoothed analysis of integer optimization problems with multiple constraints. In [13] we show that an integer optimization problem has polynomial smoothed complexity if and only if there exists a pseudo-polynomial time algorithm for solving the problem. To be more precise, we consider integer optimization problems in which an objective function is to be maximized over a feasible region that is defined as the intersection of a fixed ground set $\mathcal{S} \subseteq \mathcal{D}^n$ with halfspaces $\mathcal{B}_1, \ldots, \mathcal{B}_m$ that are induced by m linear constraints of the form $w_{i,1}x_1 + \cdots + w_{i,n}x_n \leq t_i$, where the $w_{i,j}$ are independently perturbed by adding Gaussian or uniformly distributed random variables with standard deviation σ to them.

The term *polynomial smoothed complexity* is defined such that it is robust under different machine models analogously to the way polynomial average-case complexity is defined. One disadvantage of this definition is that polynomial smoothed/average-case complexity does not imply expected polynomial running time. For the binary case it is shown in [3] that problems that admit a pseudo-linear algorithm, i.e., an algorithm whose running time is bounded by $O(\text{poly}(N)W)$, where N denotes the input size and W the largest coefficient $|w_{i,j}|$ in the input, can be solved in expected polynomial time in the smoothed model. Based on our analysis of the expected number of Pareto optimal solutions, we generalize this result to the integer case.

Theorem 4. *Every integer optimization problem that can be solved in time $O(\text{poly}(N)W)$, where N denotes the input size and $W = \max_{i,j} |w_{i,j}|$, allows an algorithm with expected polynomial (in N and $1/\sigma$) running time for perturbed instances, in which an independent Gaussian or uniformly distributed random variables with standard deviation σ is added to each coefficient.*

In the following section, we introduce the probabilistic model we analyze, which is more general than the Gaussian and uniform perturbations described above. After that, in Sections 3 and 4, we present the upper and lower bounds on the expected number of Pareto optimal solutions. Finally, in Section 5, we present the applications of our results to the smoothed analysis of integer programming.

2   Model and Notations

For the sake of a simple presentation, using the framework of smoothed analysis, we described our results in the introduction not in their full generality. Our probabilistic analysis assumes that the adversary can choose, for each p_i, a probability distribution according to which p_i is chosen independently of the other profits. We prove an upper bound that depends linearly on the maximal density of the distributions and on the expected distance to zero. The maximal density of a continuous probability distribution, i.e., the supremum of the density function, is a parameter of the distribution, which we denote by ϕ. Similar to the standard deviation σ for Gaussian random variables, ϕ can be seen as a measure specifying how close the analysis is to a worst-case analysis. The larger ϕ, the more concentrated the probability mass can be. For Gaussian and uniformly distributed random variables, we have $\phi \sim 1/\sigma$.

In the following, we assume that p_i is a random variable with density f_i and that $f_i(x) \leq \phi_i$ for all $x \in \mathbb{R}$. Furthermore, we denote by μ_i the expected absolute value of p_i, i.e., $\mu_i = \mathbf{E}[|p_i|] = \int_{x \in \mathbb{R}} |x| f_i(x) \, dx$. Let $\phi = \max_{i \in [n]} \phi_i$ and $\mu = \max_{i \in [n]} \mu_i$. We denote by $[n]$ the set $\{1, \ldots, n\}$, and we use the notations $d = |\mathcal{D}|$ and $D = \max\{a - b \mid a, b \in \mathcal{D}\}$.

3 Upper Bound on the Expected Number of Pareto Optimal Solutions

While the profit function is assumed to be linear with stochastic coefficients, the weight function $w : \mathcal{S} \to \mathbb{R}$ can be chosen arbitrarily. We model this by assuming an explicit ranking of the solutions in $\mathcal{S}$, which can be chosen by the adversary. This way, we obtain a bicriteria optimization problem that aims at maximizing the rank as well as the profit. Observe that the weight function can map several solutions to the same value whereas the rank of a solution is always unique. This strict ordering, however, can only increase the number of Pareto optimal solutions.

Theorem 5. *Let $\mathcal{S} \subseteq \mathcal{D}^n$ be a set of arbitrarily ranked solutions with a finite domain $\mathcal{D} \subset \mathbb{Z}$. Define $d = |\mathcal{D}|$ and $D = \max\{a - b \mid a, b \in \mathcal{D}\}$. Assume that each profit p_i is a random variable with density function $f_i : \mathbb{R} \to \mathbb{R}_{\geq 0}$. Suppose $\mu_i = \mathbf{E}[|p_i|]$ and $\phi_i = \sup_{x \in \mathbb{R}} f_i(x)$. Let q denote the number of Pareto optimal solutions. Then*

$$\mathbf{E}[q] \leq 2DdH_d \left(\sum_{i=1}^{n} \phi_i \right) \left(\sum_{i=1}^{n} \mu_i \right) + O(dn) ,$$

where H_d is the d-th harmonic number. For $\mathcal{D} = \{0, \ldots, k - 1\}$ and $\mu = \max_{i \in [n]} \mu_i$ and $\phi = \max_{i \in [n]} \phi_i$ the bound simplifies to

$$\mathbf{E}[q] = O(\mu \phi n^2 k^2 \log k) .$$

Note that the number of Pareto optimal solutions is not affected when all profits are scaled by some constant $c \neq 0$. This property is also reflected by the above bound. The random variable cp_i has maximal density ϕ_i/c and the expected absolute value is $c\mu_i$. Hence, the product $\phi\mu$ is invariant under scaling too.

Proof (Theorem 5). We use the following classification of Pareto optimal solutions. We say that a Pareto optimal solution x is of class $c \in \mathcal{D}$ if there exists an index $i \in [n]$ with $x_i \neq c$ such that the succeeding Pareto optimal solution y satisfies $y_i = c$, where succeeding Pareto optimal solution refers to the highest ranked Pareto optimal solution that is lower ranked than x. The lowest ranked Pareto optimal solution, which does not have a succeeding Pareto optimal solution, is not contained in any of the classes. A Pareto optimal solution can be in several classes but it is at least in one class. Let q_c denote the number of Pareto optimal solutions of class c. Since $q \leq 1 + \sum_{c \in \mathcal{D}} q_c$ it holds $\mathbf{E}[q] \leq 1 + \sum_{c \in \mathcal{D}} \mathbf{E}[q_c]$.

Lemma 6 enables us to bound the expected number of class-0 Pareto optimal solutions. In order to bound $\mathbf{E}[q_c]$ for values $c \neq 0$ we analyze a modified sequence of solutions. Starting from the original sequence $\mathcal{S} = x^1, x^2, \ldots, x^l$ ($x^j \in \mathcal{D}^n$), we obtain a modified sequence $\mathcal{S}^c$ by subtracting $(c, \ldots, c)$ from each solution vector x^j. This way, the profit of each solution is reduced by $c \sum p_i$.

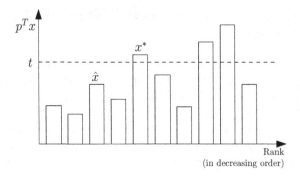

Fig. 1. If $\hat{x}$ is an ordinary class-0 Pareto optimal solution, then there must be an index i with $x_i^* = 0$ and $\hat{x}_i \neq 0$

Observe that this operation does not affect the set of Pareto optimal solutions. A solution x is class-c Pareto optimal in $\mathcal{S}$ if and only if the corresponding solution $x - (c, \dots, c)$ is class-0 Pareto optimal in $\mathcal{S}^c$. Hence, the number of class-c Pareto optimal solutions in $\mathcal{S}$ corresponds to the number of class-0 Pareto optimal solutions in $\mathcal{S}^c$. We apply Lemma 6 for the solution set $\mathcal{S}^c$ with a corresponding domain $\mathcal{D}^c = \{z - c : z \in \mathcal{D}\}$. Since the difference between the largest and the smallest element of the domain does not change, applying Lemma 6 yields that $\mathbf{E}[q]$ is bounded from above by

$$1 + \sum_{c \in \mathcal{D}} \mathbf{E}\left[q_0(\mathcal{S}^c)\right] \leq 1 + \sum_{c \in \mathcal{D}} \left(D \left(\sum_{v \in \mathcal{D}^c \setminus \{0\}} |v|^{-1} \right) \left(\sum_{i=1}^{n} \phi_i \right) \left(\sum_{i=1}^{n} \mu_i \right) + n \right) ,$$

and the theorem follows. □

Lemma 6. *Let $\mathcal{S} \subseteq \mathcal{D}^n$ be a set of arbitrarily ranked solutions with a finite domain $\mathcal{D} \subset \mathbb{Z}$ with $0 \in \mathcal{D}$. Let D denote the difference between the largest and the smallest element in $\mathcal{D}$. Let q_0 denote the number of class-0 Pareto optimal solutions. Then*

$$\mathbf{E}[q_0] \leq D \left(\sum_{v \in \mathcal{D} \setminus \{0\}} |v|^{-1} \right) \left(\sum_{i=1}^{n} \phi_i \right) \left(\sum_{i=1}^{n} \mu_i \right) + n .$$

Proof. The key idea is to prove an upper bound on the probability that there exists a class-0 Pareto optimal solution whose profit falls into a small interval $(t - \varepsilon, t)$, for arbitrary t and ε. We will classify class-0 Pareto optimal solutions to be *ordinary* or *extraordinary*. Considering only ordinary solutions allows us to prove a bound that depends not only on the length ε of the interval but also on $|t|$, the distance to zero. This captures the intuition that it becomes increasingly

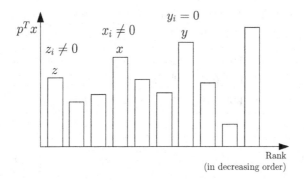

Fig. 2. In this case x is an extraordinary class-0 Pareto optimal solution

unlikely to observe solutions whose profits are much larger than the expected profit of the most profitable solution. The final bound is obtained by observing that there can be at most n extraordinary class-0 Pareto optimal solutions.

We want to bound the probability that there exists an ordinary class-0 Pareto optimal solution whose profit lies in the interval $(t - \varepsilon, t)$. Define x^* to be the highest ranked solution from $\mathcal{S}$ satisfying $p^\mathrm{T} x \geq t$. If x^* exists then it is Pareto optimal. Let $\hat{x}$ denote the Pareto optimal solution that precedes x^*, i.e., $\hat{x}$ has the largest profit among all solutions that are higher ranked than x^* (see Fig. 1). We aim at bounding the probability that $\hat{x}$ is an ordinary class-0 Pareto optimal solution and falls into the interval $(t - \varepsilon, t)$.

We classify solutions to be ordinary or extraordinary as follows. Let x be a class-0 Pareto optimal solution and let y be the succeeding Pareto optimal solution, which must exist as the lowest ranked Pareto optimal solution is not class-0 Pareto optimal. We say that x is extraordinary if for all indices $i \in [n]$ with $x_i \neq 0$ and $y_i = 0$, $z_i \neq 0$ holds for all Pareto optimal solutions z that preceed x. In other words, for those indices i that make x class-0 Pareto optimal, y is the highest ranked Pareto optimal solution that is independent of p_i (see Fig. 2). For every index $i \in [n]$ there can be at most one extraordinary class-0 Pareto optimal solution. In the following we will restrict ourselves to solutions $\hat{x}$ that are ordinary. Define

$$\Lambda(t) = \begin{cases} t - p^\mathrm{T} \hat{x} & \text{if } x^* \text{ and } \hat{x} \text{ exist and } \hat{x} \text{ is ordinary class-0 Pareto optimal} \\ \bot & \text{otherwise.} \end{cases}$$

Let $\mathcal{P}^0$ denote the set of ordinary class-0 Pareto optimal solutions. Whenever $\Lambda(t) < \varepsilon$, then there exists a solution $x \in \mathcal{P}^0$ with $p^\mathrm{T} x \in (t - \varepsilon, t)$, namely $\hat{x}$. The reverse is not true because it might be the case that $\hat{x} \notin \mathcal{P}^0$ but that there exists another solution $x \in \mathcal{P}^0$ with $p^\mathrm{T} x \in (t - \varepsilon, t)$. If, however, ε is smaller than the minimum distance between two Pareto optimal solutions, then the existence of a solution $x \in \mathcal{P}^0$ with $p^\mathrm{T} x \in (t - \varepsilon, t)$ implies $\hat{x} = x$ and hence $\Lambda(t) < \varepsilon$. Let

$\mathcal{A}(t,\varepsilon)$ denote the event that there is at most one Pareto optimal solution with a profit in the interval $(t-\varepsilon,t)$. Then

$$\mathbf{Pr}\left[\Lambda(t)<\varepsilon\right]\geq\mathbf{Pr}\left[(\Lambda(t)<\varepsilon)\wedge\mathcal{A}(t,\varepsilon)\right]$$
$$=\mathbf{Pr}\left[(\exists x\in\mathcal{P}^0:p^{\mathrm{T}}x\in(t-\varepsilon,t))\wedge\mathcal{A}(t,\varepsilon)\right]$$
$$\geq\mathbf{Pr}\left[\exists x\in\mathcal{P}^0:p^{\mathrm{T}}x\in(t-\varepsilon,t)\right]-\mathbf{Pr}\left[\neg\mathcal{A}(t,\varepsilon)\right],$$

and therefore

$$\lim_{\varepsilon\to0}\frac{\mathbf{Pr}\left[\Lambda(t)<\varepsilon\right]}{\varepsilon}\geq\lim_{\varepsilon\to0}\frac{\mathbf{Pr}\left[\exists x\in\mathcal{P}^0:p^{\mathrm{T}}x\in(t-\varepsilon,t)\right]}{\varepsilon}-\lim_{\varepsilon\to0}\frac{\mathbf{Pr}\left[\neg\mathcal{A}(t,\varepsilon)\right]}{\varepsilon}.$$

In the full version we show that for every $t\neq0$ the probability of that two solutions lie in the interval $(t-\varepsilon,t)$ decreases like ε^2 for $\varepsilon\to0$. Hence, for every $t\neq0$, $\lim_{\varepsilon\to0}\frac{\mathbf{Pr}[\neg\mathcal{A}(t,\varepsilon)]}{\varepsilon}=0$. Since the expected number of ordinary class-0 Pareto optimal solutions can be written as

$$\int_{-\infty}^{\infty}\lim_{\varepsilon\to0}\frac{\mathbf{Pr}\left[\exists x\in\mathcal{P}^0:p^{\mathrm{T}}x\in(t-\varepsilon,t)\right]}{\varepsilon}\,dt\leq\int_{-\infty}^{\infty}\lim_{\varepsilon\to0}\frac{\mathbf{Pr}\left[\Lambda(t)<\varepsilon\right]}{\varepsilon}\,dt\,,$$

it remains to analyze the term $\mathbf{Pr}\left[\Lambda(t)<\varepsilon\right]$. In order to analyze this probability we define a set of auxiliary random variables such that $\Lambda(t)$ is guaranteed to always take a value also taken by one of the auxiliary random variables. Then we analyze the auxiliary random variables and use a union bound to conclude the desired bound for $\Lambda(t)$.

Define $\mathcal{D}'=\mathcal{D}\backslash\{0\}$ and $\mathcal{S}^{x_i=v}=\{x\in\mathcal{S}\mid x_i=v\}$ for all $i\in[n]$ and $v\in\mathcal{D}$. Let $x^{*(i)}$ denote the highest ranked solution from $\mathcal{S}^{x_i=0}$ with profit at least t. For each $i\in[n]$ and $v\in\mathcal{D}'$ we define the set $\mathcal{L}^{(i,v)}$ as follows. If $x^{*(i)}$ does not exist or $x^{*(i)}$ is the highest ranked solution in $\mathcal{S}^{x_i=0}$ then we define $\mathcal{L}^{(i,v)}=\emptyset$. Otherwise $\mathcal{L}^{(i,v)}$ consists of all solutions from $\mathcal{S}^{x_i=v}$ that have a higher rank than $x^{*(i)}$. Let $\hat{x}^{(i,v)}$ denote the lowest ranked Pareto optimal solution from the set $\mathcal{L}^{(i,v)}$, i.e., $\hat{x}^{(i,v)}$ has the largest profit among all solutions in $\mathcal{L}^{(i,v)}$. Finally we define for each $i\in[n]$ and $v\in\mathcal{D}'$ the auxiliary random variable

$$\Lambda_i^v(t)=\begin{cases}t-p^{\mathrm{T}}\hat{x}^{(i,v)}&\text{if }\hat{x}^{(i,v)}\text{ exists,}\\\perp&\text{otherwise.}\end{cases}$$

If $\Lambda_i^v(t)\in(0,\varepsilon)$ (which excludes $\Lambda_i^v(t)=\perp$) then the following three events must co-occur:

1. $\mathcal{E}_1$: There exists an $x\in\mathcal{S}^{x_i=0}$ with $p^{\mathrm{T}}x\geq t$.
2. $\mathcal{E}_2$: There exists an $x\in\mathcal{S}^{x_i=0}$ with $p^{\mathrm{T}}x<t$.
3. $\mathcal{E}_3$: $\hat{x}^{(i,v)}$ exists and its profit falls into the interval $(t-\varepsilon,t)$.

The events $\mathcal{E}_1$ and $\mathcal{E}_2$ only depend on the profits p_j, $j\neq i$. The existence and identity of $\hat{x}^{(i,v)}$ is completely determined by those profits as well. Hence, if we fix all profits except for p_i, then $\hat{x}^{(i,v)}$ is fixed and its profit is

$c + vp_i$ for some constant c that depends on the profits already fixed. Observe that the random variable $c + vp_i$ has density at most $\phi_i/|v|$. Hence we obtain$\mathbf{Pr}\left[p^T\hat{x}^{(i,v)} \in (t-\varepsilon, t) \middle| \hat{x}^{(i,v)} \text{ exists}\right] \leq \varepsilon\frac{\phi_i}{|v|}$. Define

$$P^+ = \sum_{j:p_j>0} p_j \quad \text{and} \quad P^- = \sum_{j:p_j<0} p_j \ .$$

Moreover let d^+ and d^- denote the largest and the smallest element in $\mathcal{D}$. For $t \geq 0$, the event $\mathcal{E}_1$ implies $t \leq d^+P^+ + d^-P^-$, and hence $\mathbf{Pr}\left[\mathcal{E}_1\right] \leq \mathbf{Pr}\left[d^+P^+ + d^-P^- \geq t\right]$. For $t \leq 0$, the event $\mathcal{E}_2$ implies $t > d^+P^- + d^-P^+$ and hence $\mathbf{Pr}\left[\mathcal{E}_2\right] \leq \mathbf{Pr}\left[d^+P^- + d^-P^+ \leq t\right]$. By combining these results we get

$$\mathbf{Pr}\left[\Lambda_i^v(t) \in (0,\varepsilon)\right] \leq \begin{cases} \mathbf{Pr}\left[d^+P^+ + d^-P^- \geq t\right]\varepsilon\frac{\phi_i}{|v|}, \text{ for } t \geq 0, \text{ and} \\ \mathbf{Pr}\left[d^+P^- + d^-P^+ \leq t\right]\varepsilon\frac{\phi_i}{|v|}, \text{ for } t \leq 0. \end{cases}$$

Next we argue that $\Lambda(t) < \varepsilon$ implies $\Lambda_i^v(t) \in (0,\varepsilon)$ for at least one pair $(i,v) \in [n] \times \mathcal{D}'$. So assume that $\Lambda(t) < \varepsilon$. By definition, x^* and $\hat{x}$ exist and $\hat{x}$ is an ordinary class-0 Pareto optimal solution. Since $\hat{x}$ is class-0 Pareto optimal and x^* is the succeeding Pareto optimal solution, there exists an index $i \in [n]$ such that

(a) $x_i^* = 0$ and $\hat{x}_i = v \neq 0$ for some $v \in \mathcal{D}'$, and
(b) x^* is not the highest ranked solution in $\mathcal{S}^{x_i=0}$.

The second condition is a consequence of the assumption, that $\hat{x}$ is not extraordinary, i.e., there exists a Pareto optimal solution z with $z_i = 0$ that has higher rank than $\hat{x}$. Recall that $x^{*(i)}$ is defined to be the highest ranked solution in $\mathcal{S}^{x_i=0}$ with $p^Tx \geq t$. As $x^* \in \mathcal{S}^{x_i=0}$, $x^* = x^{*(i)}$. Moreover, $\mathcal{L}^{(i,v)}$ consists of all solutions from $\mathcal{S}^{x_i=v}$ that have a higher rank than x^*. Thus, $\hat{x} \in \mathcal{L}^{(i,v)}$. By construction, $\hat{x}$ has the largest profit among the solutions in $\mathcal{L}^{(i,v)}$ and, therefore $\hat{x}^{(i,v)} = \hat{x}$ and $\Lambda_i^v(t) = \Lambda(t)$. Applying a union bound yields, for all $t \geq 0$,

$$\mathbf{Pr}\left[\Lambda(t) < \varepsilon\right] \leq \sum_{i=1}^{n}\sum_{v\in\mathcal{D}'} \mathbf{Pr}\left[\Lambda_i^v(t) < \varepsilon\right]$$

$$\leq \sum_{i=1}^{n}\sum_{v\in\mathcal{D}'} \mathbf{Pr}\left[d^+P^+ + d^-P^- \geq t\right]\varepsilon\frac{\phi_i}{|v|}$$

$$\leq \mathbf{Pr}\left[d^+P^+ + d^-P^- \geq t\right]\varepsilon\sum_{i=1}^{n}\sum_{v\in\mathcal{D}'}\frac{\phi_i}{|v|} \ .$$

For $t \leq 0$ we get analogously

$$\mathbf{Pr}\left[\Lambda(t) < \varepsilon\right] \leq \mathbf{Pr}\left[d^+P^- + d^-P^+ \leq t\right]\varepsilon\sum_{i=1}^{n}\sum_{v\in\mathcal{D}'}\frac{\phi_i}{|v|} \ .$$

Now we can bound the expected number of class-0 Pareto optimal solutions, taking into account that at most n of them can be extraordinary.

$$\mathbf{E}\left[q_0\right] \leq n + \int_{-\infty}^{\infty} \lim_{\varepsilon \to 0} \frac{\mathbf{Pr}\left[\Lambda(t) \leq \varepsilon\right]}{\varepsilon} \, dt$$

$$\leq n + \int_{0}^{\infty} \lim_{\varepsilon \to 0} \frac{\mathbf{Pr}\left[d^{+} P^{+} + d^{-} P^{-} \geq t\right] \varepsilon \sum_{i=1}^{n} \sum_{v} \frac{\phi_i}{|v|}}{\varepsilon} \, dt$$

$$+ \int_{-\infty}^{0} \lim_{\varepsilon \to 0} \frac{\mathbf{Pr}\left[d^{+} P^{-} + d^{-} P^{+} \leq t\right] \varepsilon \sum_{i=1}^{n} \sum_{v} \frac{\phi_i}{|v|}}{\varepsilon} \, dt$$

$$\leq n + \left(\sum_{v} \frac{1}{|v|}\right) \left(\sum_{i=1}^{n} \phi_i\right) \left(\int_{0}^{\infty} \mathbf{Pr}\left[d^{+} P^{+} + d^{-} P^{-} \geq t\right] dt \right.$$

$$\left. + \int_{0}^{\infty} \mathbf{Pr}\left[-d^{+} P^{-} - d^{-} P^{+} \geq t\right] dt\right)$$

As $0 \in \mathcal{D}$, it holds $d^{+} \geq 0$ and $d^{-} \leq 0$. Hence we have $d^{+} P^{+} + d^{-} P^{-} \geq 0$, $-d^{+} P^{-} - d^{-} P^{+} \geq 0$, and

$$\int_{0}^{\infty} \mathbf{Pr}\left[d^{+} P^{+} + d^{-} P^{-} \geq t\right] dt + \int_{0}^{\infty} \mathbf{Pr}\left[-d^{+} P^{-} - d^{-} P^{+} \geq t\right] dt$$

$$= \mathbf{E}\left[d^{+} P^{+} + d^{-} P^{-}\right] + \mathbf{E}\left[-d^{+} P^{-} - d^{-} P^{+}\right]$$

$$= (d^{+} - d^{-}) \mathbf{E}\left[P^{+} - P^{-}\right] = (d^{+} - d^{-}) \mathbf{E}\left[\sum_{i=1}^{n} |p_i|\right] = D \sum_{i=1}^{n} \mu_i \ . \qquad \square$$

4   Lower Bounds on the Expected Number of Pareto Optimal Solutions

In this section we present a lower bound of $\Omega(n^2 k \log(1 + k))$ on the number of Pareto optimal solutions for $\mathcal{D} = \{0, \ldots, k\}$, generalizing a bound for the binary domain presented in [2]. In Theorem 8 we prove the stronger bound $\Omega(n^2 k^2)$ under slightly stronger assumptions. The weaker bound provides a vector of weights $w_1, \ldots, w_n$, such that the bound holds for a linear weight function $w^{\mathrm{T}} x$. For the stronger bound we can only prove that there is some weight function $w : \mathcal{S} \to \mathbb{R}$ for which the bound holds but this function might not be linear. In combinatorial optimization, however, many problems have linear objective functions. The proofs of the theorems in this section will be contained in the full version of this paper.

Theorem 7. *Let* $\mathcal{D} = \{0, \ldots, k\}$. *Suppose profits are drawn independently at random according to a continuous probability distribution with non-increasing density function* $f : \mathbb{R}_{\geq 0} \to \mathbb{R}_{\geq 0}$. *Let* q *denote the number of Pareto optimal solutions over* $\mathcal{S} = \mathcal{D}^n$. *Then there is a vector of weights* $w_1, \ldots, w_n \in \mathbb{R}_{>0}$ *for which*

$$\mathbf{E}\left[q\right] \geq \frac{H_k}{4} k(n^2 - n) + kn + 1 \ ,$$

where H_k is the k-th harmonic number. If the profits are drawn according to the uniform distribution over some interval $[0, c]$ with $c > 0$ then the above inequality holds with equality.

Similarly, a lower bound of $\Omega(n^2 k \log k)$ can be obtained for the case that f is the density of a Gaussian random variable with mean 0. Since all weights w_i are larger than 0, a solution with a negative profit cannot be contained in a Pareto optimal solution. Hence, we can ignore those items. Restricted to the interval $[0, \infty)$ the density of a Gaussian random variable with mean 0 is non-increasing and, hence, we can apply Theorem 7.

Now we consider general weight functions and show a lower bound of $\Omega(n^2 k^2)$ on the expected number of Pareto optimal solutions for $\mathcal{D} = \{0, \dots, k\}$ and $\mathcal{S} = \mathcal{D}^n$. We assume that k is a function of n with $(5(c+1)+1) \log n \leq k \leq n^c$ for some constant c. We use the probabilistic method to show that, for each sufficiently large $n \in \mathbb{N}$, a ranking exists for which the expected number of Pareto optimal solutions is lower bounded by $n^2 k^2 / \kappa$ for some constant κ depending only on c, that is, we create a ranking at random (but independently of the profits) and show that the expected number of Pareto optimal solutions (where the expectation is taken over both the random ranking and the random profits) satisfies the desired lower bound. This implies that, for each sufficiently large $n \in \mathbb{N}$, there must exist a deterministic ranking on $\{0, \dots, k\}^n$ for which the expected number of Pareto optimal solutions (where the expectation is now taken only over the random profits) is at least $n^2 k^2 / \kappa$.

Theorem 8. *Let $(5(c+1)+1) \log n \leq k \leq n^c$ for some $c \geq 2$ and assume that n is a multiple of $c + 2$. There exists a constant κ depending only on c and a ranking on $\{0, \dots, k\}^n$ such that the expected number of Pareto optimal solutions is lower bounded by $n^2 k^2 / \kappa$ if each profit p_i is chosen independently, uniformly at random from the interval $[-1, 1]$.*

5   Smoothed Complexity of Integer Programming

In [13], we analyze the smoothed complexity of integer programming. We consider integer programs in which an objective function is to be maximized over a feasible region that is defined as the intersection of a fixed ground set $\mathcal{S} \subseteq \mathcal{D}^n$ with a halfspace $\mathcal{B}$ that is induced by a linear constraint $w_1 x_1 + \cdots + w_n x_n \leq t$, where the w_i are independent random variables which can be represented by densities that are bounded by ϕ. We show that an integer optimization problem in this form has polynomial smoothed complexity if and only if there exists a pseudo-polynomial algorithm (w.r.t. the w_i) for solving it.

The main technical contribution in [13] is the analysis of the random variables *loser gap* and *feasibility gap*. The feasibility gap Γ is defined as the slack of the optimal solution from the threshold t. To be more precise, let x^* denote the optimal solution, that is, x^* denotes the solution from $\mathcal{S} \cap \mathcal{B}$ that maximizes the objective function. Then the feasibility gap can be defined as $\Gamma = t - w^{\mathsf{T}} x^*$. A solution $x \in \mathcal{S}$ is called a loser if it has a higher objective value than x^* but is infeasible due to the linear constraint, that is, $w^T x > t$. We denote the set of all losers by $\mathcal{L}$. Furthermore, we define the minimal loser $\overline{x} \in \mathcal{L}$ to be the solution from $\mathcal{L}$ with minimal weight, that is, $\overline{x} = \text{argmin}\{w^T x \mid x \in \mathcal{L}\}$. The loser gap Λ denotes the slack of the minimal loser from the threshold t, i.e., $\Lambda = w^T \overline{x} - t$.

If both the loser and the feasibility gap are not too small, then rounding all weights w_i with sufficient accuracy does not change the optimal solution. Rounding the weights can only affect the optimal solution if either x^* becomes infeasible or a loser x becomes feasible. The former event can only occur if the feasibility gap is small; the latter event can only occur if the loser gap is small. In a rather technical and lengthy analysis we show the following lemma on the probability that the loser or the feasibility gap is small.

Lemma 9. *(Separating Lemma [13]) Let $\mathcal{S} \subseteq \mathcal{D}^n$ with $0^n \notin \mathcal{S}$ be chosen arbitrarily, let $\mu = \max_{i \in [n]} \mathbf{E}\left[|w_i|\right]$, $d = |\mathcal{D}|$, and $d_{\max} = \max\{|a| \mid a \in \mathcal{D}\}$. Then, for all $\varepsilon \in [0, (32\mu n^5 d^7 d_{\max}\phi^2)^{-1}]$,*

$$\mathbf{Pr}\left[\Gamma \le \varepsilon\right] \le 2(\varepsilon \cdot 32\mu n^5 d^7 d_{\max}\phi^2)^{1/3} \text{ and } \mathbf{Pr}\left[\Lambda \le \varepsilon\right] \le 2(\varepsilon \cdot 32\mu n^5 d^7 d_{\max}\phi^2)^{1/3}.$$

In the full version of this paper we present a much simpler proof for the following improved version of the previous lemma.

Theorem 10. *Let $\mathcal{S} \subseteq \mathcal{D}^n$ with $0^n \notin \mathcal{S}$ be chosen arbitrarily, and let $D = \max\{a - b \mid a, b \in \mathcal{D}\} \le 2d_{\max}$. There exists a constant κ such that, for all $\varepsilon \ge 0$,*

$$\mathbf{Pr}\left[\Gamma \le \varepsilon\right] \le \varepsilon \kappa \phi^2 \mu n^3 D d \log^2 d \text{ and } \mathbf{Pr}\left[\Lambda \le \varepsilon\right] \le \varepsilon \kappa \phi^2 \mu n^3 D d \log^2 d .$$

In [13] we show that Lemma 9 can also be used to analyze integer optimization problems with more than one linear constraint. We consider integer optimization problems in which an objective function is to be maximized over a feasible region that is defined as the intersection of a fixed ground set $\mathcal{S} \subseteq \mathcal{D}^n$ with halfspaces $\mathcal{B}_1, \ldots, \mathcal{B}_m$ that are induced by m linear constraints of the form $w_{i,1}x_1 + \cdots + w_{i,n}x_n \le t_i$, where the $w_{i,j}$ are independent random variables which can be represented by densities that are bounded by ϕ.

The feasibility gap Γ for multiple constraints is defined to be the minimal slack of the optimal solution x^* from one of the thresholds, i.e., $\Gamma = \min_{i \in [m]}(t_i - (w_{i,1}x_1 + \cdots + w_{i,n}x_n))$. The loser gap Λ for multiple constraints is defined as $\Lambda = \min_{x \in \mathcal{L}} \max_{i \in [m]}(w_{i,1}x_1 + \cdots + w_{i,n}x_n - t_i)$. In [13] we show how Lemma 9 gives rise to bounds on the sizes of loser and feasibility gap for multiple constraints. Based on this observation we show that an integer optimization problem with multiple constraints has polynomial smoothed complexity if and only if there exists a pseudo-polynomial algorithm (w.r.t. the $w_{i,j}$) for solving it. By applying the same arguments, our bounds in Theorem 10 yield the following corollary.

Corollary 11. *Let $\mathcal{S} \subseteq \mathcal{D}^n$ with $0^n \notin \mathcal{S}$ be chosen arbitrarily, let $D = \max\{a - b \mid a, b \in \mathcal{D}\} \le 2d_{\max}$, and let the set of feasible solutions be given as $\mathcal{S} \cap \mathcal{B}_1 \cap \ldots \cap \mathcal{B}_m$. There exists a constant κ such that, for all $\varepsilon \ge 0$,*

$$\mathbf{Pr}\left[\Gamma \le \varepsilon\right] \le \varepsilon \kappa \phi^2 \mu m n^3 D d \log^2 d \text{ and } \mathbf{Pr}\left[\Lambda \le \varepsilon\right] \le \varepsilon \kappa \phi^2 \mu m n^3 D d \log^2 d .$$

The main improvement upon our previous analysis is that the bounds in Corollary 11 depend only linearly on ε instead of $\varepsilon^{1/3}$. Due to this improvement we can prove Theorem 4 in the same way as its binary version in [3], which is not possible with the bounds derived in [13].

References

1. René Beier. *Probabilistic Analysis of Discrete Optimization Problems*. PhD thesis, Universität des Saarlandes, 2004.
2. René Beier and Berthold Vöcking. Random knapsack in expected polynomial time. *Journal of Computer and System Sciences*, 69(3):306–329, 2004.
3. René Beier and Berthold Vöcking. Typical properties of winners and losers in discrete optimization. *SIAM Journal on Computing*, 35(4):855–881, 2006.
4. H.W. Corley and I.D Moon. Shortest paths in networks with vector weights. *Journal of Optimization Theory and Application*, 46(1):79–86, 1985.
5. Matthias Ehrgott. Integer solutions of multicriteria network flow problems. *Investigacao Operacional*, 19:61–73, 1999.
6. Matthias Ehrgott and Xavier Gandibleux. *Multiple Criteria Optimization*, volume 491 of *Lecture Notes in Economics and Mathematical Systems*, chapter Multiobjective Combinatorial Optimization. Springer-Verlag, 2000.
7. H. Kellerer, U. Pferschy, and D. Pisinger. *Knapsack Problems*. Springer, Berlin, Germany, 2004.
8. Kathrin Klamroth and Margaret M. Wiecek. Dynamic programming approaches to the multiple criteria knapsack problem. *Naval Research Logistics*, 47(1):57–76, 2000.
9. Matthias Müller-Hannemann and Karsten Weihe. Pareto shortest paths is often feasible in practice. In *Proceedings of the 5th International Workshop on Algorithm Engineering (WAE)*, pages 185–198, 2001.
10. Adli Mustafa and Mark Goh. Finding integer efficient solutions for bicriteria and tricriteria network flow problems using dinas. *Computers & OR*, 25(2):139–157, 1998.
11. George L. Nemhauser and Zev Ullmann. Discrete dynamic programming and capital allocation. *Management Science*, 15:494–505, 1969.
12. Christos H. Papadimitriou and Mihalis Yannakakis. On the approximability of trade-offs and optimal access of web sources. In *Proceedings of the 41st Annual Symposium on Foundations of Computer Science (FOCS)*, pages 86–92. IEEE Computer Society, 2000.
13. Heiko Röglin and Berthold Vöcking. Smoothed analysis of integer programming. In *Proceedings of the 11th International Conference on Integer Programming and Combinatorial Optimization (IPCO)*, volume 3509 of *Lecture Notes in Computer Science*, pages 276–290. Springer, 2005.
14. Anders J. V. Skriver and Kim Allan Andersen. A label correcting approach for solving bicriterion shortest-path problems. *Computers & OR*, 27(6):507–524, 2000.
15. Daniel A. Spielman and Shang-Hua Teng. Smoothed analysis of algorithms: Why the simplex algorithm usually takes polynomial time. *Journal of the ACM*, 51(3):385–463, 2004.
16. Sergei Vassilvitskii and Mihalis Yannakakis. Efficiently computing succinct trade-off curves. *Theoretical Computer Science*, 348(2-3):334–356, 2005.

Finding a Polytope from Its Graph in Polynomial Time

Eric J. Friedman

School of Operations Research and Information Engineering, Cornell University
ejf27@cornell.edu
http://www.people.cornell.edu/pages/ejf27/

Abstract. We show that one can compute a (simple) polytope from its graph in Polynomial time. This computation of a polytope from its graph was shown to be solvable by Blind and Mani and more recently Kalai provided a simple proof that leads to an exponential time algorithm. Our proof relies on a Primal-Dual characterization by Joswig, Kaibel and Korner. We describe an exponential Linear Programming which can be used to construct the solution and show that it can be solved in polynomial time.

1 Introduction

In [1] Blind and Mani showed, using tools from homology theory, that one can construct the entire face lattice of a (simple[1]) polytope from its graph. Then in [7], Kalai presented an elementary proof of this result. Whereas Blind and Mani's result was essentially nonconstructive, Kalai's result was constructive but required exponential time (in the size of the graph).

More recently, Joswig, Kaibel and Korner [4] extended Kalai's analysis to provide polynomial certificates for this problem, based on a pair of combinatorial optimization problems that form a primal dual pair. However, they do not provide polynomial algorithms for either of these problems and thus left open the question of whether this problem can be solved in polynomial time.

In this paper, we present a polynomial time algorithm for computing the face lattice of a polytope from its graph, resolving this question. We present a linear program for computing the 2-faces of the polytope from its graph which can be solved in polynomial time. As discussed in [5,6] this resolves the issue, as one can compute the full face lattice from the set of 2 faces.

Our discussion in the remainder of the paper will be self contained, but terse. For more details see the related papers [4,5] and the book [9].

2 2-Systems and Pseudo-polytopes

Let $G = (V, E)$ be the graph of a simple (full dimensional) polytope, P, in $\Re^d$, where V is the set of vertices of the polytope and E are its edges.

---

[1] Note that if the polytope is not simple then it is not uniquely defined by its graph. Thus, we will only consider simple polytopes.

M. Fischetti and D.P. Williamson (Eds.): IPCO 2007, LNCS 4513, pp. 68–73, 2007.

A 2-frame, centered at v, is a set of three distinct nodes, v, v', v'' such that (v, v') and (v, v'') are both elements of E. A 2-system is a set of cycles in G such that every 2-frame is contained in a unique cycle.

Let $\mathcal{O}$ be an acyclic orientation on G. Define $\mathcal{H}(\mathcal{O})$ to be the number of 2-frames that are sinks under $\mathcal{O}$, where the 2-frame (v, v', v'') is a sink if both edges (v, v') and (v, v'') are oriented towards the center of the frame, v.

Our analysis will be based on the following (minor) extension of the main result from [4]. Our modification is that we require a specified vertex not be a source.

Theorem 1 (Joswig, Korner and Kaibel). *Let P be a simple d-polytope. For every 2-system S of G, vertex v, and every acyclic orientation $\mathcal{O}$ of G, such that no 2-frame centered at v is a 2-sink, the inequalities*

$$|S| \leq |V_2(P)| \leq \mathcal{H}(\mathcal{O})$$

hold, where the first inequality holds with equality if and only if $S = V_2(P)$ (the set of 2 faces of P), and the second holds with equality if and only if $\mathcal{O}$ induces precisely one sink on every 2-face of P.

Proof: Our proof is a slight modification of that in [4], since we require that a chosen vertex not be a source. First note that for an acyclic orientation that every cycle must contain a 2-sink. Thus we must have that $|S| \leq \mathcal{H}(\mathcal{O})$. In addition, since $V_2(P)$ is a 2-system this implies that $|V_2(P)| \leq \mathcal{H}(\mathcal{O})$ which in turn implies that $|S| \leq |V_2(P)|$. The second inequality holds with equality when $\mathcal{O}$ is an abstract objective function with v as a source. Such an AOF exists since there exists a linear objective function on the polytope where v is the worst vertex. That $S = V_2(P)$ when the first holds with equality can be shown using the same proof as in [4]. □

Thus, if we can find a 2-system S that maximizes $|S|$ in polynomial time, then we have found $V_2(p)$ and from that one can compute the full face lattice of P in polynomial time. See [5,6] for details.

We use the above theorem to define a "pseudo-polytopal multi-graph" to be a multi-graph G such that there exists a vertex v and "pseudo 2-face set", $V_2(G)$ such that Theorem 1 holds. Clearly the graph of a polytope is pseudo-polytopal; however, as we now show, other multi-graphs (which do not arise from simple polytopes) may also be pseudo-polytopal.

Given a graph G of a polytope P define the contraction of G by a 2-face f to be a new multi-graph $C_f(G)$, where all the nodes in f are contracted to a single node, denoted v. Note that this is a multi-graph as there may be multiple edges connecting v to an adjacent node. We consider each of these to be distinct and may even have a 2-face on only 2 nodes.

Theorem 2. *Let G be the graph of a simple polytope P and F be a 2-face of P. Then $G' = C_f(G)$ is a pseudo-polytopal multi-graph.*

Proof: The proof is identical to the proof of Theorem 1 where we choose $V_2(G')$ to be the $V_2(P) \setminus f$ and $\mathcal{O}$ to be the contraction of an AOF for G where all vertices on the face f are worse than all other vertices. To construct such an AOF simply take the linear objective function with f as a level set and perturb it slightly.   □

In the following section, we will present a binary integer program with an exponential number of variables for computing this 2-system. Somewhat surprisingly, this can be solved in polynomial time.

3   Solving Via Linear Programming

Let T be the set of all 2-frames in G and $t \in T$ be the 2-frame (t_0, t_1, t_2) centered at t_0. Let W be the set of all loops in G. Then to compute $V_2(P)$ we need to solve:

$$max \sum_{w \in W} x_w \quad (IP-S)$$

$$s.t.$$

$$\forall t \in T : \quad \sum_{w \ni t} x_w = 1$$

$$x_w \in \{0, 1\}$$

where we write $w \ni t$ as a shorthand for the 2-frames t contained in w. First we consider the following relaxation of this integer program.

$$max \sum_{w \in W} x_w \quad (\text{LP-S})$$

$$s.t.$$

$$\forall t \in T : \quad \sum_{w \ni t} x_w \le 0$$

$$x_w \ge 0$$

Next, we consider the dual of this LP:

$$min \sum_{t \in T} v_t \quad (\text{LP-SD})$$

$$s.t.$$

$$\forall w \in W : \quad \sum_{t \in w} v_t \ge 1$$

$$v_t \ge 0$$

Let IP-SD be the related binary integer program for LP-SD, i.e., replace $0 \le v_t$ with $v_t \in \{0, 1\}$. Now, consider an acyclic orientation, $\mathcal{O}$ of G and let $v_t = 1$ represent the case when the 2-frame t is a 2-sink. Then the integer program for minimizing $\mathcal{H}(\mathcal{O})$ over all acyclic orientations can be written by adding the constraint that v must arise from an acyclic orientation on G, to IP-SD.

$$min \sum_{t \in T} v_t \quad \text{(IP-}\mathcal{H}\text{)}$$

$$s.t.$$

$$\forall w \in W : \quad \sum_{t \in w} v_t \geq 1$$

$$v_t \geq 0$$

v_t arises from an acyclic orientation of G

This sequence of optimization problems allows us to present our first result:

Theorem 3. *Let P be a simple d-polytope with graph G. Then the following optimization problems for G all have the same optimal value: IP-S, LP-S, LP-SD, IP-SD and IP-$\mathcal{H}$.*

Proof: Let $Opt(problem)$ be the optimal objective value for the optimization problem, "problem". Then it is easy to see that $Opt(IP - S) \leq Opt(LP - S)$ and $Opt(LP - SD) \leq Opt(IP - SD) \leq Opt(IP - \mathcal{H})$ as these are sequences of relaxations. By strong duality, we have $Opt(LP - S) = Opt(LP - SD)$. Now, Theorem 1 completes the proof since it implies that $Opt(IP - S) = Opt(IP - \mathcal{H})$.   □

4   Solution and Integrality of the Linear Program

To complete our analysis we show that LP-SD can be solved in polynomial time and that its solution is actually a solution to IP-S, yielding $V_2(P)$.

Note that even though LP-SD has an exponential number of constraints it can be solved in polynomial time by the ellipsoid method if there exists a polynomial separation algorithm [3]. That is, an algorithm which given a vector v can check whether v is feasible and if not, find a constraint violated by v. In our case such a constraint is a cycle $w \in W$ such that $\sum_{t \in w} w_t < 1$. This can be solved easily in polynomial time via a graphical algorithm.

For example, one can search node by node for a loop starting at that node that violates the constraint. This can be done by finding a shortest path from that node to a copy of itself on a modified version of G where the specified node is doubled and the graph is directed to force any such path to be a cycle.

To complete the analysis one must guarantee that the the solution of the LP is binary. Note that the optimal solution of IP-S is unique, since there is only one true set of 2-faces for a polytope. So it suffices to show that the extreme point solution of LP-S is unique.

Theorem 4. *LP-S has a unique optimal solution.*

Proof: Suppose that LP-S has the binary optimal solution x^* and second extreme point x'. Then there must exist some $w \in W$ such that $x^*_w = 1$ and $x'_w = 0$, otherwise $(1 + \epsilon)x' - \epsilon x^*$ would also be an optimal solution, for small enough $\epsilon > 0$, implying that x'_w is not an extreme point.

Let $f \in F$ denote the face implied by x_w^* and contract the graph G by f, denoting this node by f and the contracted graph by $G' = C_f(G)$.

Now consider IP-S on this graph where we drop the constraints for 2-frames centered at f but require all the remaining ones. Since G' is psuedo-polytopal our previous argument holds for the string of optimization problems induced by G'. In particular, the solution of IP-S must have objective value equal to $|V_2(P)| - 1$; however the projection of x' is feasible for LP-S but has a greater objective value ($|V_2(P)|$), providing a contradiction and proving the theorem. □

Thus, we can find an integral solution of LP-S and hence a solution of IP-S, in polynomial time, and can find the full face lattice in polynomial time.

5   Discussion

Our analysis shows that one can compute the face lattice of a polytope from its graph in polynomial time and suggests that similar techniques might be useful for finding abstract objective functions in polynomial time, an interesting open problem.

One interesting question is whether one can tell whether a graph is polytopal, i.e., arising from a polytope. One way to show that a graph is not polytopal would be to show that the linear program (LP-S) does not have an integral solution or the related existence of a duality gap. However, the existence of psuedo-polytopal graphs shows that this would not be sufficient.

Lastly, we note that our analysis appears to have connections to recent work on unique sink orientations [8,2], as our intermediate integer program (IP-SD) is essentially solving for a unique sink orientation. Thus, minimizing over unique sink orientations is equivalent to minimizing over abstract objective functions. We conjecture that using this equivalence, one could provide an optimization based characterization of unique sink orientations for general polytopes analogous to our definition of pseudo-polytopal graphs.

Acknowledgements

I'd like to thank Adrian Lewis, David Shmoys and Mike Todd for helpful conversations. This research has been supported in part by the NSF under grant ITR-0325453.

References

1. R. Blind and P. Mani-Levitska. On puzzles and polytope isomorphisms. *Aequationes Math.*, 34:287297, 1987.
2. B. Gärtner, W. D. Morris, Jr., and L. Rüst. Unique sink orientations of grids. In *Proc. 11th Conference on Integer Programming and Combinatorial Optimization (IPCO)*, volume 3509 of *Lecture Notes in Computer Science*, pages 210–224, 2005.

3. M. Grötschel, L. Lovász, and A. Schrijver. *Geometric Algorithms and Combinatorial Optimization.* Springer-Verlag, Berlin, 1988.
4. M. Joswig, F. Korner, and V. Kaibel. On the k-systems of a simple polytope. *Isr. J. Math.*, 129:109–118, 2002.
5. V. Kaibel. Reconstructing a simple polytope from its graph. In M. Junger, G. Reinelt, and G. Rinaldi, editors, *Combinatorial Optimization – Eureka, You Shrink!*, pages 105–118. Springer, New York, 2002.
6. V. Kaibel and M. Pfetsch. Computing the face lattice of a polytope from its vertex-facet incidences. *Comput. Geom.*, 23:281–290, 2002.
7. G. Kalai. A simple way to tell a simple polytope from its graph. *J. Comb. Theory, Ser. A*, 49:381–383, 1988.
8. Tibor Szabó and E. Welzl. Unique sink orientations of cubes. In *Proc. 42nd Ann. IEEE Symp. on Foundations of Computer Science (FOCS)*, pages 547–555, New York, 2001.
9. G. M. Ziegler. *Lectures on Polytopes.* Springer Verlag, New York, 1998.

Orbitopal Fixing[*]

Volker Kaibel[1,**], Matthias Peinhardt[1], and Marc E. Pfetsch[2]

[1] Otto-von-Guericke Universität Magdeburg, Fakultät für Mathematik,
Universitätsplatz 2, 39106 Magdeburg, Germany
{kaibel,peinhard}@ovgu.de
[2] Zuse Institute Berlin, Takustr. 7, 14195 Berlin, Germany
pfetsch@zib.de

Abstract. The topic of this paper are integer programming models in which a subset of 0/1-variables encode a partitioning of a set of objects into disjoint subsets. Such models can be surprisingly hard to solve by branch-and-cut algorithms if the order of the subsets of the partition is irrelevant. This kind of symmetry unnecessarily blows up the branch-and-cut tree.

We present a general tool, called orbitopal fixing, for enhancing the capabilities of branch-and-cut algorithms in solving such symmetric integer programming models. We devise a linear time algorithm that, applied at each node of the branch-and-cut tree, removes redundant parts of the tree produced by the above mentioned symmetry. The method relies on certain polyhedra, called orbitopes, which have been investigated in [11]. It does, however, not add inequalities to the model, and thus, it does not increase the difficulty of solving the linear programming relaxations. We demonstrate the computational power of orbitopal fixing at the example of a graph partitioning problem motivated from frequency planning in mobile telecommunication networks.

1 Introduction

Being welcome in most other contexts, symmetry causes severe trouble in the solution of many integer programming (IP) models. This paper describes a method to enhance the capabilities of branch-and-cut algorithms with respect to handling symmetric models of a certain kind that frequently occurs in practice.

We illustrate this kind of symmetry by the example of a graph partitioning problem (another notorious example is the vertex coloring problem). Here, one is given a graph $G = (V, E)$ with nonnegative edge weights $w \in \mathbb{Q}_{\geq 0}^E$ and an integer $q \geq 2$. The task is to partition V into q disjoint subsets such that the sum of all weights of edges connecting nodes in the same subset is minimized.

A straight-forward IP model arises by introducing 0/1-variables x_{ij} for all $i \in [p] := \{1, \ldots, p\}$ and $j \in [q]$ that indicate whether node i is contained in

---

[*] Supported by the DFG Research Center MATHEON *Mathematics for key technologies* in Berlin.

[**] During the research of this work the first author was a visiting professor at Technische Universität Berlin.

M. Fischetti and D.P. Williamson (Eds.): IPCO 2007, LNCS 4513, pp. 74–88, 2007.

subset j (where we assume $V = [p]$). In order to model the objective function, we furthermore need $0/1$-variables y_{ik} for all edges $\{i, k\} \in E$ indicating whether nodes i and k are contained in the same subset. This yields the following model (see, e.g., [5]):

$$\min \sum_{\{i,k\} \in E} w_{ik} \, y_{ik}$$

$$\text{s.t.} \quad \sum_{j=1}^{q} x_{ij} = 1 \qquad \text{for all } i \in [p] \tag{1}$$

$$x_{ij} + x_{kj} - y_{ik} \leq 1 \qquad \text{for all } \{i, k\} \in E, \, j \in [q]$$

$$x_{ij} \in \{0, 1\} \qquad \text{for all } i \in [p], \, j \in [q]$$

$$y_{ik} \in \{0, 1\} \qquad \text{for all } \{i, k\} \in E.$$

The x-variables describe a $0/1$-matrix of size $p \times q$ with exactly one 1-entry per row. They encode the assignment of the nodes to the subsets of the partition. The methods that we discuss in this paper do only rely on this structure and thus can be applied to many other models as well. We use the example of the graph partitioning problem as a prototype application and report on computational experiments in Sect. 5. Graph partitioning problems are discussed in [3, 4, 5], for instance as a relaxation of frequency assignment problems in mobile telecommunication networks. The maximization version is relevant as well [6, 12]. Also capacity bounds on the subsets of the partition (which can easily be incorporated into the model) are of interest, in particular the graph equipartitioning problem [7, 8, 18, 19]. For the closely related clique partitioning problem, see [9, 10].

As it is given above, the model is unnecessarily difficult for state-of-the-art IP solvers. Even solving small instances requires enormous efforts (see Sect. 5). One reason is that every feasible solution (x, y) to this model can be turned into $q!$ different ones by permuting the columns of x (viewed as a $0/1$-matrix) in an arbitrary way, thereby not changing the structure of the solution (in particular: its objective function value). Phrased differently, the symmetric group of all permutations of the set $[q]$ operates on the solutions by permuting the columns of the x-variables in such a way that the objective function remains constant along each orbit. Therefore, when solving the model by a branch-and-cut algorithm, basically the same work will be done in the tree at many places. Thus, there should be potential for reducing the running times significantly by exploiting the symmetry. A more subtle second point is that interior points of the convex hulls of the individual orbits are responsible for quite weak linear programming (LP) bounds. We will, however, not address this second point in this paper.

In order to remove symmetry, the above model for the graph partitioning problem is often replaced by models containing only edge variables, see, e.g. [7]. However, for this to work the underlying graph has to be complete, which might introduce many unnecessary variables. Moreover, formulation (1) is sometimes favorable, e.g., if node-weighted capacity constraints should be incorporated.

One way to deal with symmetry is to restrict the feasible region in each of the orbits to a single representative, e.g., to the lexicographically maximal (with respect to the row-by-row ordering of the x-components) element in the orbit. In fact, this can be done by adding inequalities to the model that enforce the columns of x to be sorted in a lexicographically decreasing way. This can be achieved by $O(pq)$ many *column inequalities*. In [11] even a complete (and irredundant) linear description of the convex hull of all $0/1$-matrices of size $p \times q$ with exactly one 1-entry per row and lexicographically decreasing columns is derived; these polytopes are called *orbitope*. The description basically consists of an exponentially large superclass of the column inequalities, called *shifted column inequalities*, for which there is a linear time separation algorithm available. We recall some of these results in Sect. 2.

Incorporating the inequalities from the orbitope description into the IP model removes symmetry. At each node of the branch-and-cut tree this ensures that the corresponding IP is infeasible as soon as there is no representative in the subtree rooted at that node. In fact, already the column inequalities are sufficient for this purpose.

In this paper, we investigate a way to utilize these inequalities (or the orbitope that they describe) without adding any of the inequalities to the models explicitly. The reason for doing this is the unpleasant effect that adding (shifted) column inequalities to the models results in more difficult LP relaxations. One way of avoiding the addition of these inequalities to the LPs is to derive logical implications instead: If we are working in a branch-and-cut node at which the x-variables corresponding to index subsets I_0 and I_1 are fixed to zero and one, respectively, then there might be a (shifted) column inequality yielding implications for all representatives in the subtree rooted at the current node. For instance, it might be that for some $(i, j) \notin I_0 \cup I_1$ we have $x_{ij} = 0$ for all feasible solutions in the subtree. In this case, x_{ij} can be fixed to zero for the whole subtree rooted at the current node, enlarging I_0. We call the iterated process of searching for such additional fixings *sequential fixing* with (shifted) column inequalities.

Let us mention at this point that deviating from parts of the literature, we do not distinguish between "fixing" and "setting" of variables in this paper.

Sequential fixing with (shifted) column inequalities is a special case of constraint propagation, which is well known from constraint logic programming. Modern IP solvers like SCIP [1] use such strategies also in branch-and-cut algorithms. With orbitopes, however, we can aim at something better: Consider a branch-and-cut node identified by fixing the variables corresponding to sets I_0 and I_1 to zero and one, respectively. Denote by $W(I_0, I_1)$ the set of all vertices x of the orbitope with $x_{ij} = 0$ for all $(i, j) \in I_0$ and $x_{ij} = 1$ for all $(i, j) \in I_1$. Define the sets $I_0^\star$ and $I_1^\star$ of indices of *all* variables, for which no x in $W(I_0, I_1)$ satisfies $x_{ij} = 1$ for some $(i, j) \in I_0^\star$ or $x_{ij} = 0$ for some $(i, j) \in I_1^\star$. Fixing of the corresponding variables is called *simultaneous fixing* at the branch-and-cut node. Simultaneous fixing is always at least as strong as sequential fixing.

Investigations of sequential and simultaneous fixing for orbitopes are the central topic of the paper. The main contributions and results are the following:

○ We present a linear time algorithm for *orbitopal fixing*, i.e., for solving the problem to compute simultaneous fixings for orbitopes (Theorem 4).

○ We show that, for general 0/1-polytopes, sequential fixing, even with complete and irredundant linear descriptions, is weaker than simultaneous fixing (Theorem 2), We clarify the relationships between different versions of sequential fixing with (shifted) column inequalities, where (despite the situation for general 0/1-polytopes) the strongest one is as strong as orbitopal fixing (Theorem 3).

○ We report on computer experiments (Sect. 5) with the graph partitioning problem described above, showing that orbitopal fixing leads to significant performance improvements for branch-and-cut algorithms.

Margot [14, 15, 17] considers a related method for symmetry handling. His approach works for more general types of symmetries than ours. Similarly to our approach, the basic idea is to assure that only (partial) solutions which are lexicographical maximal in their orbit are explored in the branch-and-cut tree. This is guaranteed by an appropriate fixing rule. The fixing and pruning decisions are done by means of a Schreier-Sims table for representing the group action. While Margot's approach is much more generally applicable than orbitopal fixing, the latter seems to be more powerful in the special situation of partitioning type symmetries. One reason is that Margot's method requires to choose the branching variables according to an ordering that is chosen globally for the entire branch-and-cut tree.

Another approach has recently been proposed by Linderoth et al. [13] (in this volume). They exploit the symmetry arising in each node of a branch-and-bound tree when all fixed variables are removed from the model. Thus one may find additional local symmetries. Nevertheless, for partitioning type symmetries one still may miss some part of the (fixed) global symmetry we are dealing with.

We will elaborate on the relations between orbitopal fixing, isomorphism pruning, and orbital branching in more detail in a journal version of the paper.

2 Orbitopes

Throughout the paper, let p and q be integers with $p \geq q \geq 2$. The *orbitope* $O_{p,q}^=$ is the convex hull of all 0/1-matrices $x \in \{0,1\}^{[p] \times [q]}$ with exactly one 1-entry per row, whose columns are in decreasing lexicographical order (i.e., they satisfy $\sum_{i=1}^{p} 2^{p-i} x_{ij} > \sum_{i=1}^{p} 2^{p-i} x_{i,j+1}$ for all $j \in [q-1]$). Let the symmetric group of size q act on $\{0,1\}^{[p] \times [q]}$ via permuting the columns. Then the vertices of $O_{p,q}^=$ are exactly the lexicographically maximal matrices (with respect to the row-by-row ordering of the components) in those orbits whose elements are matrices with exactly one 1-entry per row. As these vertices have $x_{ij} = 0$ for all (i,j) with $i < j$, we drop these components and consider $O_{p,q}^=$ as a subset of the space $\mathbb{R}^{\mathcal{I}_{p,q}}$ with $\mathcal{I}_{p,q} := \{(i,j) \in \{0,1\}^{[p] \times [q]} : i \geq j\}$. Thus, we consider matrices, in which the i-th row has $q(i) := \min\{i, q\}$ components.

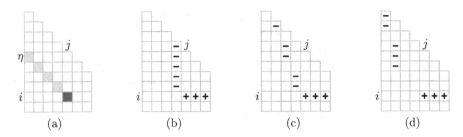

Fig. 1. (a) Example for coordinates $(9,5) = \langle 5,5 \rangle$. (b), (c), (d) Three shifted column inequalities, the left one of which is a column inequality.

In [11], in the context of more general orbitopes, $O_{p,q}^{=}$ is referred to as the *partitioning orbitope with respect to the symmetric group*. As we will confine ourselves with this one type of orbitopes in this paper, we will simply call it *orbitope*.

The main result in [11] is a complete linear description of $O_{p,q}^{=}$. In order to describe the result, it will be convenient to address the elements in $\mathcal{I}_{p,q}$ via a different "system of coordinates": For $j \in [q]$ and $1 \leq \eta \leq p - j + 1$, define $\langle \eta, j \rangle := (j + \eta - 1, j)$. Thus (as before) i and j denote the row and the column, respectively, while η is the index of the diagonal (counted from above) containing the respective element; see Figure 1 (a) for an example.

A set $S = \{\langle 1, c_1 \rangle, \langle 2, c_2 \rangle, \ldots, \langle \eta, c_\eta \rangle\} \subset \mathcal{I}_{p,q}$ with $c_1 \leq c_2 \leq \cdots \leq c_\eta$ and $\eta \geq 1$ is called a *shifted column*. For $(i,j) = \langle \eta, j \rangle \in \mathcal{I}_{p,q}$, a shifted column S as above with $c_\eta < j$, and $B = \{(i,j), (i,j+1), \ldots, (i,q(i))\}$, we call $x(B) - x(S) \leq 0$ a *shifted column inequality*. The set B is called its *bar*. In case of $c_1 = \cdots = c_\eta = j - 1$ the shifted column inequality is called a *column inequality*. See Figure 1 for examples.

Finally, a bit more notation is needed. For each $i \in [p]$, we define $\mathrm{row}_i := \{(i,j) : j \in [q(i)]\}$. For $A \subset \mathcal{I}_{p,q}$ and $x \in \mathbb{R}^{\mathcal{I}_{p,q}}$, we denote by $x(A)$ the sum $\sum_{(i,j) \in A} x_{ij}$.

Theorem 1 (see [11]). *The orbitope $O_{p,q}^{=}$ is completely described by the non-negativity constraints $x_{ij} \geq 0$, the row-sum equations $x(\mathrm{row}_i) = 1$, and the shifted column inequalities.*

In fact, in [11] it is also shown that, up to a few exceptions, the inequalities in this description define facets of $O_{p,q}^{=}$. Furthermore, a linear time separation algorithm for the exponentially large class of shifted column inequalities is given.

3   The Geometry of Fixing Variables

In this section, we deal with general 0/1-integer programs and, in particular, their associated polytopes. We will define some basic terminology used later in the special treatment of orbitopes, and we are going to shed some light on the geometric situation of fixing variables.

We denote by $[d]$ the set of indices of variables, and by $C^d = \{x \in \mathbb{R}^d :$ $0 \le x_i \le 1$ for all $i \in [d]\}$ the corresponding $0/1$-cube. For two disjoint subsets $I_0, I_1 \subseteq [d]$ (with $I_0 \cap I_1 = \varnothing$) we call

$$\{x \in C^d : x_i = 0 \text{ for all } i \in I_0, \ x_i = 1 \text{ for all } i \in I_1\}$$

the *face of* C^d *defined by* (I_0, I_1). All nonempty faces of C^d are of this type.

For a polytope $P \subseteq C^d$ and for a face F of C^d defined by (I_0, I_1), we denote by $\mathrm{Fix}_F(P)$ the smallest face of C^d that contains $P \cap F \cap \{0,1\}^d$ (i.e., $\mathrm{Fix}_F(P)$ is the intersection of all faces of C^d that contain $P \cap F \cap \{0,1\}^d$). If $\mathrm{Fix}_F(P)$ is the nonempty cube face defined by $(I_0^\star, I_1^\star)$, then $I_0^\star$ and $I_1^\star$ consist of all $i \in [d]$ for which $x_i = 0$ and $x_i = 1$, respectively, holds for all $x \in P \cap F \cap \{0,1\}^d$. In particular, we have $I_0 \subseteq I_0^\star$ and $I_1 \subseteq I_1^\star$, or $\mathrm{Fix}_F(P) = \varnothing$. Thus, if I_0 and I_1 are the indices of the variables fixed to zero and one, respectively, in the current branch-and-cut node (with respect to an IP with feasible points $P \cap \{0,1\}^d$), the node can either be pruned, or the sets $I_0^\star$ and $I_1^\star$ yield the maximal sets of variables that can be fixed to zero and one, respectively, for the whole subtree rooted at this node. Unless $\mathrm{Fix}_F(P) = \varnothing$, we call $(I_0^\star, I_1^\star)$ the *fixing of* P *at* (I_0, I_1). Similarly, we call $\mathrm{Fix}_F(P)$ the *fixing* of P *at* F.

Remark 1. If $P, P' \subseteq C^d$ are two polytopes with $P \subseteq P'$ and F and F' are two faces of C^d with $F \subseteq F'$, then $\mathrm{Fix}_F(P) \subseteq \mathrm{Fix}_{F'}(P')$ holds.

In general, it is not clear how to compute fixings efficiently. Indeed, computing the fixing of P at $(\varnothing, \varnothing)$ includes deciding whether $P \cap \{0,1\}^d = \varnothing$, which, of course, is NP-hard in general. Instead, one can try to derive as large as possible subsets of $I_0^\star$ and $I_1^\star$ by looking at relaxations of P. In case of an IP that is based on an intersection with an orbitope, one might use the orbitope as such a relaxation. We will deal with the fixing problem for orbitopes in Sect. 4.

If P is given via an inequality description, one possibility is to use the knapsack relaxations obtained from single inequalities out of the description. For each of these relaxations, the fixing can easily be computed. If the inequality system describing P is exponentially large, and the inequalities are only accessible via a separation routine, it might still be possible to decide efficiently whether any of the exponentially many knapsack relaxations allows to fix some variable (see Sect. 4.2).

Suppose, $P = \{x \in C^d : Ax \le b\}$ and $P_r = \{x \in C^d : a_r^T x \le b_r\}$ is the knapsack relaxation of P for the rth-row $a_r^T x \le b_r$ of $Ax \le b$, where $r = 1, \ldots, m$. Let F be some face of C^d. The face G of C^d obtained by setting $G := F$ and then iteratively replacing G by $\mathrm{Fix}_G(P_r)$ as long as there is some $r \in [m]$ with $\mathrm{Fix}_G(P_r) \subsetneq G$, is denoted by $\mathrm{Fix}_F(Ax \le b)$. Note that the outcome of this procedure is independent of the choices made for r, due to Remark 1. We call the pair $(\tilde{I}_0, \tilde{I}_1)$ defining the cube face $\mathrm{Fix}_F(Ax \le b)$ (unless this face is empty) the *sequential fixing of* $Ax \le b$ *at* (I_0, I_1). In the context of sequential fixing we often refer to (the computation of) $\mathrm{Fix}_F(P)$ as *simultaneous fixing*.

Due to Remark 1 it is clear that $\mathrm{Fix}_F(P) \subseteq \mathrm{Fix}_F(Ax \le b)$ holds.

Theorem 2. *In general, even for a system of facet-defining inequalities describing a full-dimensional 0/1-polytope, sequential fixing is weaker than simultaneous fixing.*

Proof. The following example shows this. Let $P \subset C^4$ be the four-dimensional polytope defined by the trivial inequalities $x_i \geq 0$ for $i \in \{1, 2, 3\}$, $x_i \leq 1$ for $i \in \{1, 2, 4\}$, the inequality $-x_1 + x_2 + x_3 - x_4 \leq 0$ and $x_1 - x_2 + x_3 - x_4 \leq 0$. Let F be the cube face defined by $(\{4\}, \varnothing)$. Then, sequential fixing does not fix any further variable, although simultaneous fixing yields $I_0^\star = \{3, 4\}$ (and $I_1^\star = \varnothing$). Note that P has only 0/1-vertices, and all inequalities are facet defining ($x_4 \geq 0$ and $x_3 \leq 1$ are implied). □

4   Fixing Variables for Orbitopes

For this section, suppose that $I_0, I_1 \subseteq \mathcal{I}_{p,q}$ are subsets of indices of orbitope variables with the following properties:

(P1) $|I_0 \cap \text{row}_i| \leq q(i) - 1$ for all $i \in [p]$
(P2) For all $(i, j) \in I_1$, we have $(i, \ell) \in I_0$ for all $\ell \in [q(i)] \setminus \{j\}$.

In particular, P1 and P2 imply that $I_0 \cap I_1 = \varnothing$. Let F be the face of the 0/1-cube $C^{\mathcal{I}_{p,q}}$ defined by (I_0, I_1). Note that if P1 is not fulfilled, then $O_{p,q}^= \cap F = \varnothing$. The following statement follows immediately from Property P2.

Remark 2. If a vertex x of $O_{p,q}^=$ satisfies $x_{ij} = 0$ for all $(i, j) \in I_0$, then $x \in F$.

We assume that the face $\text{Fix}_F(O_{p,q}^=)$ is defined by $(I_0^\star, I_1^\star)$, if $\text{Fix}_F(O_{p,q}^=)$ is not empty. *Orbitopal fixing* is the problem to compute the simultaneous fixing $(I_0^\star, I_1^\star)$ from (I_0, I_1), or determine that $\text{Fix}_F(O_{p,q}^=) = \varnothing$.

Remark 3. If $\text{Fix}_F(O_{p,q}^=) \neq \varnothing$, it is enough to determine $I_0^\star$, as we have $(i, j) \in I_1^\star$ if and only if $(i, \ell) \in I_0^\star$ holds for for all $\ell \in [q(i)] \setminus \{j\}$.

4.1   Intersection of Orbitopes with Cube Faces

We start by deriving some structural results on orbitopes that are crucial in our context. Since $O_{p,q}^= \subset C^{\mathcal{I}_{p,q}}$ is a 0/1-polytope (i.e., it is integral), we have $\text{conv}(O_{p,q}^= \cap F \cap \{0, 1\}^{\mathcal{I}_{p,q}}) = O_{p,q}^= \cap F$. Thus, $\text{Fix}_F(O_{p,q}^=)$ is the smallest cube face that contains the face $O_{p,q}^= \cap F$ of the orbitope $O_{p,q}^=$.

Let us, for $i \in [p]$, define values $\alpha_i := \alpha_i(I_0) \in [q(i)]$ recursively by setting $\alpha_1 := 1$ and, for all $i \in [p]$ with $i \geq 2$,

$$\alpha_i := \begin{cases} \alpha_{i-1} & \text{if } \alpha_{i-1} = q(i) \text{ or } (i, \alpha_{i-1} + 1) \in I_0 \\ \alpha_{i-1} + 1 & \text{otherwise.} \end{cases}$$

The set of all indices of rows, in which the α-value increases, is denoted by

$$\Gamma(I_0) := \{i \in [p] : i \geq 2, \ \alpha_i = \alpha_{i-1} + 1\} \cup \{1\}$$

(where, for technical reasons 1 is included).

The following observation follows readily from the definitions.

Remark 4. For each $i \in [p]$ with $i \geq 2$ and $\alpha_i(I_0) < q(i)$, the set $S_i(I_0) := \{(k, \alpha_k(I_0) + 1) : k \in [i] \setminus \Gamma(I_0)\}$ is a shifted column with $S_i(I_0) \subseteq I_0$.

Lemma 1. *For each $i \in [p]$, no vertex of $O_{p,q}^= \cap F$ has its 1-entry in row i in a column $j \in [q(i)]$ with $j > \alpha_i(I_0)$.*

Proof. Let $i \in [p]$. We may assume $\alpha_i(I_0) < q(i)$, because otherwise the statement trivially is true. Thus, $B := \{(i,j) \in \mathrm{row}_i : j > \alpha_i(I_0)\} \neq \varnothing$.

Let us first consider the case $i \in \Gamma(I_0)$. As we have $\alpha_i(I_0) < q(i) \leq i$ and $\alpha_1(I_0) = 1$, there must be some $k < i$ such that $k \notin \Gamma(I_0)$. Let k be maximal with this property. Thus we have $k' \in \Gamma(I_0)$ for all $1 < k < k' \leq i$. According to Remark 4, $x(B) - x(S_k(I_0)) \leq 0$ is a shifted column inequality with $x(S_k(I_0)) = 0$, showing $x(B) = 0$ as claimed in the lemma.

Thus, let us suppose $i \in [p] \setminus \Gamma(I_0)$. If $\alpha_i(I_0) \geq q(i) - 1$, the claim holds trivially. Otherwise, $B' := B \setminus \{(i, \alpha_i(I_0) + 1)\} \neq \varnothing$. Similarly to the first case, now the shifted column inequality $x(B') - x(S_{i-1}(I_0)) \leq 0$ proves the claim.   □

For each $i \in [p]$ we define $\mu_i(I_0) := \min\{j \in [q(i)] : (i,j) \notin I_0\}$. Because of Property P1, the sets over which we take minima here are non-empty.

Lemma 2. *If we have $\mu_i(I_0) \leq \alpha_i(I_0)$ for all $i \in [p]$, then the point $x^\star = x^\star(I_0) \in \{0,1\}^{\mathcal{I}_{p,q}}$ with $x^\star_{i,\alpha_i(I_0)} = 1$ for all $i \in \Gamma(I_0)$ and $x^\star_{i,\mu_i(I_0)} = 1$ for all $i \in [p] \setminus \Gamma(I_0)$ and all other components being zero, is contained in $O_{p,q}^= \cap F$.*

Proof. Due to $\alpha_i(I_0) \leq \alpha_{i-1}(I_0) + 1$ for all $i \in [p]$ with $i \geq 2$, the point $x^\star$ is contained in $O_{p,q}^=$. It follows from the definitions that $x^\star$ does not have a 1-entry at a position in I_0. Thus, by Remark 2, we have $x^\star \in F$.   □

We now characterize the case $O_{p,q}^= \cap F = \varnothing$ (leading to pruning the corresponding node in the branch-and-cut tree) and describe the set $I_0^\star$.

Proposition 1.

1. We have $O_{p,q}^= \cap F = \varnothing$ if and only if there exists $i \in [p]$ with $\mu_i(I_0) > \alpha_i(I_0)$.
2. If $\mu_i(I_0) \leq \alpha_i(I_0)$ holds for all $i \in [p]$, then the following is true.
   (a) For all $i \in [p] \setminus \Gamma(I_0)$, we have

   $$I_0^\star \cap \mathrm{row}_i = \{(i,j) \in \mathrm{row}_i : (i,j) \in I_0 \text{ or } j > \alpha_i(I_0)\}.$$

   (b) For all $i \in [p]$ with $\mu_i(I_0) = \alpha_i(I_0)$, we have

   $$I_0^\star \cap \mathrm{row}_i = \mathrm{row}_i \setminus \{(i, \alpha_i(I_0))\}.$$

   (c) For all $s \in \Gamma(I_0)$ with $\mu_s(I_0) < \alpha_s(I_0)$ the following holds: If there is some $i \geq s$ with $\mu_i(I_0) > \alpha_i(I_0 \cup \{(s, \alpha_s(I_0))\})$, then we have

   $$I_0^\star \cap \mathrm{row}_s = \mathrm{row}_s \setminus \{(s, \alpha_s(I_0))\}.$$

   Otherwise, we have

   $$I_0^\star \cap \mathrm{row}_s = \{(s,j) \in \mathrm{row}_s : (s,j) \in I_0 \text{ or } j > \alpha_s(I_0)\}.$$

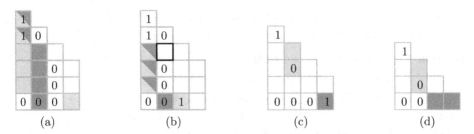

Fig. 2. (a): Example for Prop. 1 (1). Light-gray entries indicate the entries $(i, \mu_i(I_0))$ and dark-gray entries indicate entries $(i, \alpha_i(I_0))$. (b): Example of fixing an entry to 1 for Prop. 1 (2c). As before light-gray entries indicate entries $(i, \mu_i(I_0))$. Dark-gray entries indicate entries $(i, \alpha_i(I_0 \cup \{(s, \alpha_s(I_0))\}))$ with $s = 3$. (c) and (d): Gray entries show the SCIs used in the proofs of Parts 1(a) and 1(b) of Thm. 3, respectively.

Proof. Part 1 follows from Lemmas 1 and 2.

In order to prove Part 2, let us assume that $\mu_i(I_0) \leq \alpha_i(I_0)$ holds for all $i \in [p]$. For Part 2a, let $i \in [p] \setminus \Gamma(I_0)$ and $(i, j) \in \text{row}_i$. Due to $I_0 \subset I_0^\star$, we only have to consider the case $(i, j) \notin I_0$. If $j > \alpha_i(I_0)$, then, by Lemma 1, we find $(i, j) \in I_0^\star$. Otherwise, the point that is obtained from $x^\star(I_0)$ (see Lemma 2) by moving the 1-entry in position $(i, \mu_i(I_0))$ to position (i, j) is contained in $O_{p,q}^= \cap F$, proving $(i, j) \notin I_0^\star$.

In the situation of Part 2b, the claim follows from Lemma 1 and $O_{p,q}^= \cap F \neq \varnothing$ (due to Part 1).

For Part 2c, let $s \in \Gamma(I_0)$ with $\mu_s(I_0) < \alpha_s(I_0)$ and define $I_0' := I_0 \cup \{(s, \alpha_s(I_0))\}$. It follows that we have $\mu_i(I_0') = \mu_i(I_0)$ for all $i \in [p]$.

Let us first consider the case that there is some $i \geq s$ with $\mu_i(I_0) > \alpha_i(I_0')$. Part 1 (applied to I_0' instead of I_0) implies that $O_{p,q}^= \cap F$ does not contain a vertex x with $x_{s,\alpha_s(I_0)} = 0$. Therefore, we have $(s, \alpha_s(I_0)) \in I_1^\star$, and thus $I_0^\star \cap \text{row}_s = \text{row}_s \setminus \{(s, \alpha_s(I_0))\}$ holds (where for "$\subseteq$" we exploit $O_{p,q}^= \cap F \neq \varnothing$ by Part 1, this time applied to I_0).

The other case of Part 2c follows from $s \notin \Gamma(I_0')$ and $\alpha_s(I_0') = \alpha_s(I_0) - 1$. Thus, Part 2a applied to I_0' and s instead of I_0 and i, respectively, yields the claim (because of $(s, \alpha_s(I_0)) \notin I_0^\star$ due to $s \in \Gamma(I_0)$ and $O_{p,a}^= \cap F \neq \varnothing$). □

4.2 Sequential Fixing for Orbitopes

Let us, for some fixed $p \geq q \geq 2$, denote by $\mathcal{S}_{\text{SCI}}$ the system of the nonnegativity inequalities, the row-sum equations (each one written as two inequalities, in order to be formally correct) and all shifted column inequalities. Thus, according to Theorem 1, $O_{p,q}^=$ is the set of all $x \in \mathbb{R}^{\mathcal{I}_{p,q}}$ that satisfy $\mathcal{S}_{\text{SCI}}$. Let $\mathcal{S}_{\text{CI}}$ be the subsystem of $\mathcal{S}_{\text{SCI}}$ containing only the column inequalities (and all nonnegativity inequalities and row-sum equations).

At first sight, it is not clear whether sequential fixing with the exponentially large system $\mathcal{S}_{\text{SCI}}$ can be done efficiently. A closer look at the problem reveals, however, that one can utilize the linear time separation algorithm for shifted column inequalities (mentioned in Sect. 2) in order to devise an algorithm for this sequential fixing, whose running time is bounded by $O(\varrho pq)$, where ϱ is the number of variables that are fixed by the procedure.

In fact, one can achieve more: One can compute sequential fixings with respect to the affine hull of the orbitope. In order to explain this, consider a polytope $P = \{x \in C^d : Ax \leq b\}$, and let $S \subseteq \mathbb{R}^d$ be some affine subspace containing P. As before, we denote the knapsack relaxations of P obtained from $Ax \leq b$ by P_1, ..., P_m. Let us define $\text{Fix}_F^S(P_r)$ as the smallest cube-face that contains $P_r \cap S \cap \{0,1\}^d \cap F$. Similarly to the definition of $\text{Fix}_F(Ax \leq b)$, denote by $\text{Fix}_F^S(Ax \leq b)$ the face of C^d that is obtained by setting $G := F$ and then iteratively replacing G by $\text{Fix}_G^S(P_r)$ as long as there is some $r \in [m]$ with $\text{Fix}_G^S(P_r) \subsetneq G$. We call $\text{Fix}_F^S(Ax \leq b)$ the *sequential fixing* of $Ax \leq b$ at F *relative to* S. Obviously, we have $\text{Fix}_F(P) \subseteq \text{Fix}_F^S(Ax \leq b) \subseteq \text{Fix}_F(Ax \leq b)$. In contrast to sequential fixing, sequential fixing relative to affine subspaces *in general* is NP-hard (as it can be used to decide whether a linear equation has a 0/1-solution).

Theorem 3. *1. There are cube-faces F^1, F^2, F^3 with the following properties:*

    (a) $\text{Fix}_{F^1}(\mathcal{S}_{SCI}) \subsetneq \text{Fix}_{F^1}(\mathcal{S}_{CI})$

    (b) $\text{Fix}_{F^2}^{\text{aff}(O_{p,q}^=)}(\mathcal{S}_{CI}) \subsetneq \text{Fix}_{F^2}(\mathcal{S}_{SCI})$

    (c) $\text{Fix}_{F^3}^{\text{aff}(O_{p,q}^=)}(\mathcal{S}_{SCI}) \subsetneq \text{Fix}_{F^3}^{\text{aff}(O_{p,q}^=)}(\mathcal{S}_{CI})$

2. For all cube-faces F, we have $\text{Fix}_F^{\text{aff}(O_{p,q}^=)}(\mathcal{S}_{SCI}) = \text{Fix}_F(O_{p,q}^=)$.

Proof. For Part 1(a), we chose $p = 5$, $q = 4$, and define the cube-face F_1 via $I_0^1 = \{(3,2),(5,1),(5,2),(5,3)\}$ and $I_1^1 = \{(1,1),(5,4)\}$. The shifted column inequality with shifted column $\{(2,2),(3,2)\}$ and bar $\{(5,4)\}$ allows to fix x_{22} to one (see Fig. 2 (c)), while no column inequality (and no nonnegativity constraint and no row-sum equation) allows to fix any variable.

For Part 1(b), let $p = 4$, $q = 4$, and define F^2 via $I_0^2 = \{(3,2),(4,1),(4,2)\}$ and $I_1^2 = \{(1,1)\}$. Exploiting that $x_{43} + x_{44} = 1$ for all $x \in \text{aff}(O_{p,q}^=) \cap F^2$, we can use the column inequality with column $\{(2,2),(3,2)\}$ and bar $\{(4,3),(4,4)\}$ to fix x_{22} to one (see Fig. 2 (d)), while no fixing is possible with $\mathcal{S}_{SCI}$ only.

For Part 1(c), we can use $F^3 = F^1$. The proof of Part 2 is omitted here. □

The different versions of sequential fixing for partitioning orbitopes are dominated by each other in the following sequence: $\mathcal{S}_{CI} \rightarrow \{\mathcal{S}_{SCI}, \text{affine } \mathcal{S}_{CI}\} \rightarrow$ affine $\mathcal{S}_{SCI}$, which finally is as strong as orbitopal fixing. For each of the arrows there exists an instance for which dominance is strict. The examples in the proof of Theorem 3 also show that there is no general relation between $\mathcal{S}_{SCI}$ and affine $\mathcal{S}_{CI}$.

In particular, we could compute orbitopal fixings by the polynomial time algorithm for sequential fixing relative to $\text{aff}(O_{p,q}^=)$. It turns out, however, that this is not the preferable choice. In fact, we will describe below a linear time algorithm for solving the orbitopal fixing problem directly.

Algorithm 1. Orbitopal Fixing

1: Set $I_0^\star \leftarrow I_0$, $I_1^\star \leftarrow I_1$, $\mu_1 \leftarrow 1$, $\alpha_1 \leftarrow 1$, and $\Gamma = \varnothing$.
2: **for** $i = 2, \ldots, p$ **do**
3:      compute $\mu_i \leftarrow \min\{j : (i, j) \notin I_0\}$.
4:      **if** $\alpha_{i-1} = q(i)$ or $(i, \alpha_{i-1} + 1) \in I_0$ **then**
5:          $\alpha_i \leftarrow \alpha_{i-1}$
6:      **else**
7:          $\alpha_i \leftarrow \alpha_{i-1} + 1$, $\Gamma \leftarrow \Gamma \cup \{i\}$
8:      **if** $\mu_i > \alpha_i$ **then**
9:          return "Orbitopal fixing is empty"
10: Set $I_0^\star \leftarrow I_0^\star \cup \{(i, j) : j > \alpha_i\}$.
11: **if** $|I_0^\star \cap \mathrm{row}_i| = q(i) - 1$ **then**
12:     set $I_1^\star \leftarrow I_1^\star \cup (\mathrm{row}_i \setminus I_0^\star)$.
13: **for all** $s \in \Gamma$ with $(s, \alpha_s) \notin I_1^\star$ **do**
14: Set $\beta_s \leftarrow \alpha_s - 1$.
15:     **for** $i = s + 1, \ldots, p$ **do**
16:         **if** $\beta_{i-1} = q(i)$ or $(i, \beta_{i-1} + 1) \in I_0$ **then**
17:             $\beta_i \leftarrow \beta_{i-1}$
18:         **else**
19:             $\beta_i \leftarrow \beta_{i-1} + 1$
20:         **if** $\mu_i > \beta_i$ **then**
21:             $I_1^\star \leftarrow I_1^\star \cup \{(s, \alpha_s)\}$ and $I_0^\star \leftarrow \mathrm{row}_s \setminus \{(s, \alpha_s)\}$.
22:             Proceed with the next s in Step 13.

4.3   An Algorithm for Orbitopal Fixing

Algorithm 1 describes a method to compute the simultaneous fixing $(I_0^\star, I_1^\star)$ from (I_0, I_1) (which are assumed to satisfy Properties P1 and P2). Note that we use β_i for $\alpha_i(I_0 \cup \{(s, \alpha_s(I_0))\})$.

Theorem 4. *A slight modification of Algorithm 1 solves the orbitopal fixing problem in time* $\mathrm{O}(pq)$.

Proof. The correctness of the algorithm follows from the structural results given in Proposition 1.

In order to prove the statement on the running time, let us assume that the data structures for the sets $I_0, I_1, I_0^\star$, and $I_1^\star$ allow both membership testing and addition of single elements in constant time (e.g., the sets can be stored as bit vectors).

As none of the Steps 3 to 12 needs more time than $\mathrm{O}(q)$, we only have to take care of the second part of the algorithm starting in Step 13. (In fact, used verbatim as described above, the algorithm might need time $\Omega(p^2)$.)

For $s, s' \in \Gamma$ with $s < s'$ denote the corresponding β-values by β_i $(i \geq s)$ and by β_i' $(i \geq s')$, respectively. We have $\beta_i \leq \beta_i'$ for all $i \geq s'$, and furthermore, if equality holds for one of these i, we can deduce $\beta_k = \beta_k'$ for all $k \geq i$. Thus, as soon as a pair (i, β_i) is used a second time in Step 20, we can break the for-loop in Step 15 and reuse the information that we have obtained earlier.

This can, for instance, be organized by introducing, for each $(i, j) \in \mathcal{I}_{p,q}$, a flag $f(i, j) \in \{\text{red}, \text{green}, \text{white}\}$ (initialized by white), where $f(i, j) = \text{red} / \text{green}$

means that we have already detected that $\beta_i = j$ eventually leads to a positive/negative test in Step 20. The modifications that have to be applied to the second part of the algorithm are the following: The selection of the elements in Γ in Step 13 must be done in increasing order. Before performing the test in Step 20, we have to check whether $f(i, \beta_i)$ is green. If this is true, then we can proceed with the next s in Step 13, after setting all flags $f(k, \beta_k)$ to green for $s \leq k < i$. Similarly, we set all flags $f(k, \beta_k)$ to red for $s \leq k \leq i$, before switching to the next s in Step 22. And finally, we set all flags $f(k, \beta_k)$ to green for $s \leq k \leq p$ at the end of the body of the s-loop starting in Step 13.

As the running time of this part of the algorithm is proportional to the number of flags changed from white to red or green, the total running time indeed is bounded by $O(pq)$ (since a flag is never reset).   □

5   Computational Experiments

We performed computational experiments for the graph partitioning problem mentioned in the introduction. The code is based on the SCIP 0.90 framework by Achterberg [1], and we use CPLEX 10.01 as the basic LP solver. The computations were performed on a 3.2 GHz Pentium 4 machine with 2 GB of main memory and 2 MB cache running Linux. All computation times are CPU seconds and are subject to a *time limit of four hours*. Since in this paper we are not interested in the performance of heuristics, we initialized all computations with the *optimal primal solution*. We compare different variants of the code by counting *winning* instances. An instance is a winner for variant A compared to variant B, if A finished within the time limit and B did not finish or needed a larger CPU time; if A did not finish, then the instance is a winner for A in case that B did also not finish, leaving, however, a larger gap than A. If the difference between the times or gaps are below 1 sec. and 0.1, respectively, the instance is not counted.

In all variants, we fix the variables x_{ij} with $j > i$ to zero. Furthermore, we heuristically separate general clique inequalities $\sum_{i,j \in C} y_{ij} \geq b$, where $b = \frac{1}{2}t(t-1)(q-r) + \frac{1}{2}t(t+1)r$ and $C \subseteq V$ is a clique of size $tq + r > q$ with integers $t \geq 1$, $0 \leq r < q$ (see [3]). The separation heuristic for a fractional point $y^\star$ follows ideas of Eisenblätter [5]. We generate the graph $G' = (V, E')$ with $\{i, k\} \in E'$ if and only if $\{i, k\} \in E$ and $y_{ik}^\star < b(b+1)/2$, where $y^\star$ is the y-part of an LP-solution. We search for maximum cliques in G' with the branch-and-bound method implemented in SCIP (with a branch-and-bound node limit of 10 000) and check whether the corresponding inequality is violated.

Our default branching rule combines *first index* and *reliability branching*. We branch on the first fractional x-variable in the row-wise variable order used for defining orbitopes, but we skip columns in which a 1 has appeared before. If no such fractional variable could be found, we perform reliability branching as described by Achterberg, Koch, and Martin [2].

We generated random instances with n vertices and m edges of the following types. For $n = 30$ we used $m = 200$ (*sparse*), 300 (*medium*), and 400 (*dense*).

Table 1. Results of the branch-and-cut algorithm. All entries are rounded averages over three instances. CPU times are given in seconds.

			basic		Iso Pruning		OF		
n	m	q	nsub	cpu	nsub	cpu	nsub	cpu	#OF
30	200	3	1 082	6	821	4	697	5	6
30	200	6	358	1	122	0	57	0	25
30	200	9	1	0	1	0	1	0	0
30	200	12	1	0	1	0	1	0	0
30	300	3	3 470	87	2 729	64	2 796	69	7
30	300	6	89 919	445	63 739	168	8 934	45	353
30	300	9	8 278	19	5 463	5	131	0	73
30	300	12	1	0	1	0	1	0	0
30	400	3	11 317	755	17 433	800	9 864	660	8
30	400	6	458 996	14 400	1 072 649	11 220	159 298	3 142	1 207
30	400	9	2 470 503	14 400	1 048 256	2 549	70 844	450	7 305
30	400	12	3 668 716	12 895	37 642	53	2 098	12	1 269
50	560	3	309 435	10 631	290 603	14 400	288 558	10 471	10
50	560	6	1 787 989	14 400	3 647 369	14 400	1 066 249	9 116	4 127
50	560	9	92	0	2 978	5	10	0	10
50	560	12	1	0	1	0	1	0	0

Additionally, for $n = 50$ we choose $m = 560$ in search for the limits of our approach. For each type we generated three instances by picking edges uniformly at random (without recourse) until the specified number of edges is reached. The edge weights are drawn independently uniformly at random from the integers $\{1, \ldots, 1000\}$. For each instance we computed results for $q = 3, 6, 9$, and 12.

In a first experiment we tested the speedup that can be obtained by performing orbitopal fixing. For this we compare the variant (*basic*) without symmetry breaking (except for the zero-fixing of the upper right x-variables) and the version in which we use orbitopal fixing (*OF*); see Table 1 for the results. Columns *nsub* give the number of nodes in the branch-and-bound tree. The results show that orbitopal fixing is clearly superior (OF winners: 26, basic winners: 3), see also Figure 3.

Table 1 shows that the sparse instances are extremely easy, the instances with $m = 300$ are quite easy, while the dense instances are hard. One effect is that

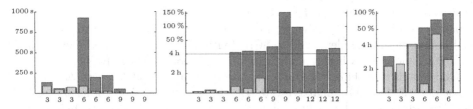

Fig. 3. Computation times/gaps for the basic version (dark gray) and the version with orbitopal fixing (light gray). From left to right: instances with $n = 30$, $m = 300$, instances for $n = 30$, $m = 400$, instances for $n = 50$, $m = 560$. The number of partitions q is indicated on the x-axis. Values above 4 hours indicate the gap in percent.

often for small m and large q the optimal solution is 0 and hence no work has to be done. For $m = 300$ and 400, the hardest instances arise when $q = 6$. It seems that for $q = 3$ the small number of variables helps, while for $q = 12$ the small objective function values help. Of course, symmetry breaking methods become more important when q gets larger.

In a second experiment we investigated the symmetry breaking capabilities built into CPLEX. We suspect that it breaks symmetry within the tree, but no detailed information was available. We ran CPLEX 10.01 on the IP formulation stated in Sect. 1. In one variant, we fixed variables x_{ij} with $j > i$ to zero, but turned symmetry breaking off. In a second variant, we turned symmetry breaking on and did not fix variables to zero (otherwise CPLEX seems not to recognize the symmetry). These two variants performed about equally good (turned-on winners: 13, turned-off winners: 12). The variant with no symmetry breaking and no fixing of variables performed extremely badly. The results obtained by the OF-variant above are clearly superior to the best CPLEX results (CPLEX could not solve 10 instances within the time limit, while OF could not solve 2). Probably this is at least partially due to the separation of clique inequalities and the special branching rule in our code.

In another experiment, we turned off orbitopal fixing and separated shifted column inequalities in every node of the tree. The results are that the OF-version is slightly better than this variant (OF winners: 13, SCI winners: 10), but the results are quite close (OF average time: 1563.3, SCI average time: 1596.7). Although by Part 2 of Theorem 3, orbitopal fixing is not stronger than fixing with SCIs (with the same branching decisions), the LPs get harder and the process slows down a bit.

Finally, we compared orbitopal fixing to the isomorphism pruning approach of Margot. We implemented the *ranked branching rule* (see [16]) adapted to the special symmetry we exploit, which simplifies Margot's algorithm significantly. It can be seen from Table 1 that isomorphism pruning is inferior to both orbitopal fixing (OF winners: 25, isomorphism pruning winners: 3) and shifted column inequalities (26:2), but is still a big improvement over the basic variant (23:7).

6 Concluding Remarks

The main contribution of this paper is a linear time algorithm for the orbitopal fixing problem, which provides an efficient way to deal with partitioning type symmetries in integer programming models. The result can easily be extended to "packing orbitopes" (where, instead of $x(\text{row}_i) = 1$, we require $x(\text{row}_i) \leq 1$). Our proof of correctness of the procedure uses the linear description of $O_{p,q}^=$ given in [11]. However, we only need the validity of the shifted column inequalities in our arguments. In fact, one can devise a similar procedure for the case where the partitioning constraints $x(\text{row}_i) = 1$ are replaced by covering constraints $x(\text{row}_i) \geq 1$, though, for the corresponding "covering orbitopes" no complete linear descriptions are known at this time. A more detailed treatment of this will be contained in a journal version of the paper, which will also include comparisons to the isomorphism pruning method [14, 15, 17] and to orbital branching [13].

References

[1] T. ACHTERBERG, *SCIP – A framework to integrate constraint and mixed integer programming*, Report 04-19, Zuse Institute Berlin, 2004. http://www.zib.de/Publications/abstracts/ZR-04-19/.

[2] T. ACHTERBERG, T. KOCH, AND A. MARTIN, *Branching rules revisited*, Oper. Res. Lett., 33 (2005), pp. 42–54.

[3] S. CHOPRA AND M. RAO, *The partition problem*, Math. Program., 59 (1993), pp. 87–115.

[4] ——, *Facets of the k-partition polytope*, Discrete Appl. Math., 61 (1995), pp. 27–48.

[5] A. EISENBLÄTTER, *Frequency Assignment in GSM Networks: Models, Heuristics, and Lower Bounds*, PhD thesis, TU Berlin, 2001.

[6] J. FALKNER, F. RENDL, AND H. WOLKOWICZ, *A computational study of graph partitioning*, Math. Program., 66 (1994), pp. 211–239.

[7] C. FERREIRA, A. MARTIN, C. DE SOUZA, R. WEISMANTEL, AND L. WOLSEY, *Formulations and valid inequalities of the node capacitated graph partitioning problem*, Math. Program., 74 (1996), pp. 247–266.

[8] ——, *The node capacitated graph partitioning problem: A computational study*, Math. Program., 81 (1998), pp. 229–256.

[9] M. GRÖTSCHEL AND Y. WAKABAYASHI, *A cutting plane algorithm for a clustering problem*, Math. Prog., 45 (1989), pp. 59–96.

[10] ——, *Facets of the clique partitioning polytope*, Math. Prog., 47 (1990), pp. 367–387.

[11] V. KAIBEL AND M. E. PFETSCH, *Packing and partitioning orbitopes*, Math. Program., (2007). In press.

[12] G. KOCHENBERGER, F. GLOVER, B. ALIDAEE, AND H. WANG, *Clustering of microarray data via clique partitioning*, J. Comb. Optim., 10 (2005), pp. 77–92.

[13] J. LINDEROTH, J. OSTROWSKI, F. ROSSI, AND S. SMRIGLIO, *Orbital branching*, in Proceedings of IPCO XII, M. Fischetti and D. Williamson, eds., vol. 4513 of LNCS, Springer-Verlag, 2007, pp. 106–120.

[14] F. MARGOT, *Pruning by isomorphism in branch-and-cut*, Math. Program., 94 (2002), pp. 71–90.

[15] ——, *Exploiting orbits in symmetric ILP*, Math. Program., 98 (2003), pp. 3–21.

[16] ——, *Small covering designs by branch-and-cut*, Math. Program., 94 (2003), pp. 207–220.

[17] ——, *Symmetric ILP: Coloring and small integers*, Discrete Opt., 4 (2007), pp. 40–62.

[18] A. MEHROTRA AND M. A. TRICK, *Cliques and clustering: A combinatorial approach*, Oper. Res. Lett., 22 (1998), pp. 1–12.

[19] M. M. SØRENSEN, *Polyhedral computations for the simple graph partitioning problem*, working paper L-2005-02, Århus School of Business, 2005.

New Variants of Lift-and-Project Cut Generation from the LP Tableau: Open Source Implementation and Testing

Egon Balas[1,*] and Pierre Bonami[2]

[1] Tepper School of Business, Carnegie Mellon University, Pittsburgh PA
eb17@andrew.cmu.edu.
[2] T.J. Watson Research Center, IBM, Yorktown Heights, NY
pbonami@us.ibm.com

Abstract. We discuss an open source implementation and preliminary computational testing of three variants of the Balas-Perregaard procedure for generating lift-and-project cuts from the original simplex tableau, two of which are new. Variant 1 is the original procedure of [6] with minor modifications. Variant 2 uses a new procedure for choosing the pivot element: After identifying the set of row candidates for an improving pivot, the pivot element (and column) is chosen by optimizing over the entries of all candidate rows. Finally, Variant 3 replaces the source row with its disjunctive modularization, and after each pivot it again modularizes the resulting source row. We report on computational results with the above three variants and their combinations on 65 MIPLIB.3 instances.

Keywords: integer programming, branch and cut algorithms.

1 Introduction

The revolution of the last 15 years in the state of the art of integer programming was brought about, besides faster computers and more efficient linear programming codes, also by improved cutting plane techniques. Lift-and-project (L&P) cuts were the first to be generated in rounds and to be embedded into a branch-and-cut framework. They were also the first locally valid cuts lifted into globally valid ones. Soon after the success of L&P cuts [3,4], it was shown [5] that mixed integer Gomory (MIG) cuts used in the same manner could also enhance the performance of MIP solvers. Thus, during the nineties a number of different cut families (cover and flow cover inequalities, MIG cuts, simple disjunctive cuts, MIR cuts etc.) became part of the toolkit of commercial MIP solvers and have led to a radical improvement of their performance. The L&P cuts themselves, however, were found to be computationally too expensive to be incorporated into commercial codes, as each such cut came at the price of solving a Cut Generating Linear Program (CGLP) in a higher dimensional space. It was not until a few

---

* Research supported by the National Science Foundation through grant #DMI-0352885 and by the Office of Naval Research through contract N00014-03-1-0133.

M. Fischetti and D.P. Williamson (Eds.): IPCO 2007, LNCS 4513, pp. 89–103, 2007.

years later, when a way was found [6] to generate L&P cuts by pivoting in the original simplex tableau, without constructing the higher dimensional CGLP, that these cuts became sufficiently cost-effective to be incorporated into a state-of-the-art MIP solver, where they soon proved their value [10] and became the default cut generator.

Although the algorithm for generating L&P cuts from the original simplex tableau is now in practical use and has contributed to solving countless hard integer programs, its implementation was until now commercial property not publicly available, which made it harder for researchers to experiment with different versions of it. In this paper we discuss an implementation of this algorithm in the COIN-OR framework, publicly available [9] since September 2006, and compare three different variants of it. Variant 1 is a slightly modified version of the original algorithm [6] for generating L&P cuts by pivoting in the original LP tableau which incorporates the various improvements proposed in [10,11], whereas the other two variants contain substantial changes in the algorithm, which give rise to different pivot sequences and therefore different cuts. Variant 2 uses a new rule for choosing the entering variable in the pivoting procedure. Instead of first choosing a most promising pivot row and then identifying the best column in that row, this version of the algorithm first identifies all candidate rows for an improving pivot, then chooses the pivot element as the best one among the entries of all the candidate rows. Variant 3 uses recursive disjunctive modularization of the source row. In other words, rather than first generating an unstrengthened "deepest" L&P cut through a sequence of pivots in the original LP tableau and then strengthening the end product by modular arithmetic, this version replaces the source row with its disjunctive modularization, and after each pivot it again applies the disjunctive modularization to the resulting transformed source row. Each of the three Variants give rise to sequences of pivots different from each other. In the case of both Variants 2 and 3, each pivot is guaranteed to produce an improvement in cut strength at least equal to that produced by the corresponding pivot of Variant 1, but this additional improvement comes at some computational cost.

After describing each of the three Variants, we compare them on a battery of MIPLIB test problems and assess the results by trying to identify the merits and demerits of each Variant.

Consider a problem of the form $\min\{cx : x \in P, x_j \in \mathbb{Z}, j = 1, \ldots, p\}$ (MIP) and its linear programming relaxation $\min\{cx : x \in P\}$ (LP), where P is the polyhedron defined by the system

$$
\begin{aligned}
Ax &\geq b \\
-x_j &\geq -1 \quad j = 1, \ldots, p \\
x &\geq 0
\end{aligned}
\tag{1}
$$

Here A is $m \times n$, $1 \leq p \leq n$, and (1) will also be denoted as $\tilde{A}x \geq \tilde{b}$. Note that the vector $s \in \mathbb{R}^{m+p+n}$ of surplus variables has n components of the form $s_{m+p+j} = x_j$, which represent just a set of different names for the structural variables x_j.

Let x^* be an optimal solution to (LP) and let

$$x_k = \bar{a}_{k0} - \sum_{j \in J} \bar{a}_{kj} s_j \qquad (2)$$

be the row of the optimal simplex tableau corresponding to basic variable x_k, with $0 < \bar{a}_{k0} < 1$ and J the index set of nonbasic variables. The *intersection cut* [1] from the convex set $\{x \in \mathbb{R}^n : 0 \leq x_k \leq 1\}$, also known as the *simple disjunctive cut* from the condition $x_k \leq 0 \vee x_k \geq 1$ applied to (2), is $\pi s \geq \pi_0$, where $\pi_0 = \bar{a}_{k0}(1 - \bar{a}_{k0})$ and $\pi_j := \max\{\bar{a}_{kj}(1 - \bar{a}_{k0}), -\bar{a}_{kj}\bar{a}_{k0}\}$, $j \in J$.

This cut can be strengthened [1] using the integrality of some variables in J, by replacing π with $\bar{\pi}$, where $\bar{\pi}_j = \pi_j$ for $j \in J \setminus \{1, \ldots, p\}$, and $\bar{\pi}_j := \min\{f_{kj}(1 - \bar{a}_{k0}), (1 - f_{kj})\bar{a}_{k0}\}, j \in J \cap \{1, \ldots, p\}$, where $f_{kj} = \bar{a}_{kj} - \lfloor \bar{a}_{kj} \rfloor$. This *strengthened intersection cut* or *strengthened simple disjunctive cut* is the same as the *mixed integer Gomory (MIG) cut*.

On the other hand, given the same optimal solution x^* to (LP), a *deepest lift-and-project (L&P) cut* $\alpha x \geq \beta$ is obtained by solving a Cut Generating Linear Program [3] in a higher dimensional space:

$$
\begin{aligned}
\min \; & \alpha x^* - \beta \\
\text{s.t.} \quad & \\
\alpha \quad & - u\tilde{A} \quad\quad + u_0 e_k \quad\quad = 0 \\
\alpha \quad & \quad\quad - v\tilde{A} \quad\quad - v_0 e_k = 0 \\
& - \beta + u\tilde{b} \quad\quad\quad\quad = 0 \\
& - \beta \quad\quad + v\tilde{b} \quad\quad + v_0 \quad = 0 \\
& ue + ve + u_0 \quad + v_0 \quad = 1 \\
& u, v, u_0, v_0 \geq 0
\end{aligned}
\qquad (\text{CGLP})_k
$$

where $e = (1, \ldots, 1)$ and e_k is the k-th unit vector.

While an optimal solution to $(\text{CGLP})_k$ yields a "deepest" cut $\alpha x \geq \beta$, i.e. one that cuts off x^* by a maximum amount, any solution to the constraint set of $(\text{CGLP})_k$ yields a member of the family of L&P cuts. If $(\alpha, \beta, u, v, u_0, v_0)$ is a basic solution to $(\text{CGLP})_k$, the coefficients of the corresponding L&P cut are $\beta = u\tilde{b} = v\tilde{b} + v_0$, $\alpha_k = \max\{u\tilde{A}_k - u_k - u_0, v\tilde{A}_k - v_k + v_0\}$, and $\alpha_j = \max\{u\tilde{A}_j - u_j, v\tilde{A}_j - v_j\}, j \neq k$ where $\tilde{A}_j$ is the j-th column of $\tilde{A}$.

Again, this cut can be strengthened using the integrality of some of the structural variables by replacing α with $\bar{\alpha}$, where $\bar{\alpha}_j = \alpha_j$ for $j = k$ and $j \notin \{1, \ldots, p\}$, and $\bar{\alpha}_j = \min\{u\tilde{A}_j - u_j + u_0 \lceil m_j \rceil, v\tilde{A}_j - v_j - v_0 \lfloor m_j \rfloor\}$, $j \in \{1, \ldots, p\} \setminus \{k\}$, with $m_j = (v\tilde{A}_j - v_j - u\tilde{A}_j + u_j)/(u_0 + v_0)$.

In [6] it was shown that the intersection cut obtained from a given component x_k of a basic feasible solution of (LP) is equivalent to the L&P cut obtained from a basic solution to $(\text{CGLP})_k$, where the bases in question are related to each other in a well defined manner. The same relationship holds between the strengthened version of the intersection cut, i.e. the mixed integer Gomory cut, on the one hand, and the strengthened L&P cut on the other. Furthermore, a strengthened L&P cut is equivalent to a MIG cut from some LP tableau that in general is neither optimal nor feasible, and the search for a deepest L&P cut can be viewed as the search for the appropriate simplex tableau from which to derive the corresponding MIG cut. The next section discusses this connection.

2 The Correspondence Between L&P Cuts and MIG Cuts

Let $\alpha x \geq \beta$ be a L&P cut corresponding to a basic feasible solution $(\alpha, \beta, u, v, u_0, v_0)$ of $(CGLP)_k$, and let $\bar{a}x \geq \beta$ be its strengthened version. Further, let $u_0 > 0$, $v_0 > 0$ (these are known to be the only solutions yielding cuts that differ from the rows of $\tilde{A}x \geq \tilde{b}$), and let M_1 and M_2 be the index sets of the basic components of u and v respectively. Then $M_1 \cap M_2 = 0$, $|M_1 \cup M_2| = n$, and the square submatrix $\hat{A}$ of $\tilde{A}$ whose rows are indexed by $M_1 \cup M_2$ is nonsingular (see [6]). Now define $J := M_1 \cup M_2$. Then letting $\hat{b}$ denote the subvector of $\tilde{b}$ corresponding to $\hat{A}$ and writing s_J for the surplus variables indexed by J, we have

$$\hat{A}x - s_J = \hat{b} \quad \text{or} \quad x = \hat{A}^{-1}\hat{b} - \hat{A}^{-1}s_J \tag{3}$$

and the row of (3) corresponding to x_k (a basic variable, since $k \notin J$) can be written as

$$x_k = \bar{a}_{k0} - \sum_{j \in J} \bar{a}_{kj}s_j, \tag{4}$$

where $\bar{a}_{k0} = e_k\hat{A}^{-1}\hat{b}$ and $\bar{a}_{kj} = -\hat{A}_{kj}^{-1}$. Notice that (4) is the same as (2). Furthermore, it can be shown that $0 < \bar{a}_{k0} < 1$, and we have (from [6])

Theorem 1. *The MIG cut $\pi s \geq \pi_0$ from (4) is equivalent to the strengthened L&P cut $\alpha x \geq \beta$.*

Conversely, suppose (4) is the row associated with x_k in a basic solution to (LP), not necessarily optimal or even feasible, such that $0 < \bar{a}_{k0} < 1$. Then we have

Theorem 2. *Let (M_1, M_2) be any partition of J such that $j \in M_1$ if $\bar{a}_{kj} < 0$ and $j \in M_2$ if $\bar{a}_{kj} > 0$. Then the solution to $(CGLP)_k$ corresponding to the basis with components $(\alpha, \beta, u_0, v_0, \{u_i : i \in M_1\}, \{v_i : i \in M_2\})$ defines a L&P cut $\alpha x \geq \beta$ whose strengthened version $\bar{\alpha}x \geq \beta$ is equivalent to the MIG cut $\pi s \geq \pi_0$ derived from (4).*

Note that the partition (M_1, M_2) of J, and therefore the basis of $(CGLP)_k$ defined by it, is not unique, since the variables $j \in J$ such that $\bar{a}_{kj} = 0$ can be assigned either to M_1 or to M_2. This means that the correspondence between bases described above maps each basis B of (LP) into a set of bases $\varphi(B)$ of $(CGLP)_k$, where typically $|\varphi(B)| > 1$. However, all bases in $\varphi(B)$ correspond to the same solution of $(CGLP)_k$, i.e. they are degenerate, and the correspondence between basic solutions (as opposed to bases) of (LP) and $(CGLP)_k$ is one to one (see [6] for details).

3 The Lift-and-Project Procedure in the Original LP Tableau

The lift-and-project procedure in the (LP) tableau uses the above correspondence to mimic the optimization of $(\text{CGLP})_k$ by the simplex algorithm. Consider the row corresponding to x_k of the form (2) which we call the source row. At each iteration of the procedure, we perform a pivot in a row $i \neq k$, which brings about a linear combination of the source row with row i

$$x_k + \gamma x_i = \overline{a}_{k0} + \gamma \overline{a}_{i0} - \sum_{j \in J}(\overline{a}_{kj} + \gamma \overline{a}_{ij})s_j \qquad (5)$$

such that the intersection cut obtained from this new row is more violated by x^* than the one obtained from the source row. This combination (the choice of the row i and of γ), is guided by the correspondence with $(\text{CGLP})_k$. Each row i of the (LP) simplex tableau corresponds to a pair of columns of $(\text{CGLP})_k$ with associated nonbasic variables u_i, v_i.

The first main step in the procedure is to compute the reduced costs r_{u_i} and r_{v_i} in $(\text{CGLP})_k$ for all $i \notin J \cup \{k\}$. As shown in [6], these reduced costs can be expressed in terms of the entries $\overline{a}_{ij}$ of the (LP) tableau and the solution x^*. We use these expressions in our computations. If there is no negative reduced cost, the current basis is optimal for $(\text{CGLP})_k$ and the optimal strengthened lift-and-project cut is obtained as the MIG cut from the source row of (LP) (using the correspondence of Theorem 2). On the other hand, if at least one negative reduced cost exists, then the cut can be improved by performing a pivot in $(\text{CGLP})_k$ where the corresponding variable u_i or v_i enters the basis. In the (LP) tableau, this negative reduced cost (r_{u_i} or r_{v_i}) corresponds to a basic variable x_i which has to leave the basis.

Choosing the variable x_i to enter the basis is the second main step of the procedure. In [6], two evaluation functions $f^+(\gamma)$ (resp. $f^-(\gamma)$) were defined, which represent the objective function value of $(\text{CGLP})_k$, i.e. the violation of the cut resulting from the combination of row k and row i for positive, respectively negative values of γ. These two functions are minimized to select the variable to enter the basis which leads to the largest improvement in cut violation among all variables that can replace the exiting variable.

Once the exiting and entering variables have been selected, the pivot in (LP) is performed and the procedure is iterated from the new basis until $(\text{CGLP})_k$ is optimized. The pseudo-code of Figure 1 describes this procedure.

As shown in [4], the lift-and-project cuts are more efficiently generated in a subspace where all the non-basic structural variables of (LP) are fixed to their values in the optimal solution. Performing the separation in the subspace while working in the (LP) tableau is done simply by removing all the structural nonbasic variables from it before starting the pivoting procedure. At the end of the procedure a lifting step is performed to obtain a valid cut for the original problem by recomputing the source row in the full space and generating the corresponding MIG cut.

Let x^* be the optimal solution to (LP).
Let $k \in \{1, \ldots, p\}$ with x_k^* fractional.
Let I and J be the index sets of basic and non-basic variables in an optimal basis of (LP).
Let $\overline{A}$ be the optimal tableau.
Let *num_pivots*:= 0.
while *num_pivots* < *pivot_limit* **do**
>   Compute the reduced costs r_{u_i}, r_{v_i} for each $i \notin J \cup \{k\}$
>   **if** There exists i such that $r_{u_i} < 0 \vee r_{v_i} < 0$
>>   **then**
>>>   Let $\hat{i} := \arg \min_{i \notin J \cup \{k\}} \{r_{u_i}, r_{v_i}\}$,
>>>   Let $J' = \{j \in J : |\overline{a}_{\hat{i}j}| \geq \epsilon,\}$ be the set of admissible pivots.
>>>   Let $J^+ = J' \cap \{j \in J : -\overline{a}_{kj}/\overline{a}_{\hat{i}j} < 0\}$.
>>>   Let $\hat{j} := \arg \min \{\arg \min_{j \in J^+} f^+(\gamma_j), \arg \min_{j \in J' \backslash J^+} f^-(\gamma_j)\}$.
>>>   Perform a pivot in (LP) by pivoting out $\hat{i}$ and pivoting in $\hat{j}$.
>>>   Let $I := I \cup \{\hat{j}\} \backslash \{\hat{i}\}$.
>>>   Let $\overline{A}$ be the updated tableau in the new basis.
>>>   Let *num_pivots* += 1.
>>   **else** /* *cut is optimal.* */
>>>   Generate the MIG cut from row k of the current tableau.
>>>   **exit**
>   **fi**
od

Fig. 1. Lift-and-Project Procedure

4   Computation of the Reduced Cost and of the Evaluation Functions

A key point for efficiently implementing the lift-and-project procedure is the computation of the reduced costs and the evaluation functions.

As shown in [11], for a given partition (M_1, M_2) (as defined in section 2) the expressions for the reduced costs depend only linearly on the coefficients of the tableau, and therefore the reduced costs of all non-basic variables in $(CGLP)_k$ can be computed by doing only one multiplication with the basis inverse. The expressions for the reduced costs are

$$r_{u_i} = -\sigma + \overline{a}_{i0}(1 - x_k^*) - \tau_i \quad \text{and} \quad r_{v_i} = -\sigma - \overline{a}_{i0}(1 - x_k^*) + s_i^* + \tau_i$$

where $s^* = \tilde{A}x^* - b$, $\sigma = (\sum_{j \in M_2} \overline{a}_{kj}s_j^* - \overline{a}_{k0}(1 - x_k^*))/(1 + \sum_{j \in J} |\overline{a}_{kj}|)$ is the current objective value of $(CGLP)_k$, and $\tau_i = \sum_{j \in M_1} \sigma \overline{a}_{ij} + \sum_{j \in M_2} (s_j^* - \sigma)\overline{a}_{ij}$.

A critical element in computing the reduced costs is the choice of the partition (M_1, M_2). If for all $j \in J$, $\bar{a}_{kj}$ is non-zero, this partition is uniquely defined; but if this is not the case, several partitions can be chosen. The rule given in [6] is to take $M_1 = \{j \in J : \bar{a}_{kj} < 0 \ \wedge \ (\bar{a}_{kj} = 0 \ \wedge \ \bar{a}_{ij} > 0)\}$ (and $M_2 = J \setminus M_1$) for computing r_{u_i}, and $M_1 = \{j \in J : \bar{a}_{kj} < 0 \ \wedge \ (\bar{a}_{kj} = 0 \wedge \bar{a}_{ij} < 0)\}$ for computing r_{v_i}. This rule has the advantage that if a negative reduced cost is found, then the corresponding pivot leads to a strictly better cut. On the other hand, to determine this partition, one has to compute the coefficients $\bar{a}_{ij}$ for all j such that $\bar{a}_{kj} = 0$ and all i. Therefore we use another rule. Namely, following [11], we randomly assign all the zero elements of the source row to either M_1 or M_2. This rule has the disadvantage that although the reduced cost for the perturbed row is negative, it may happen that all the pivots with the corresponding variable u_i or v_i entering the basis are degenerate in $(CGLP)_k$. Nevertheless, in our experiments, this rule had a clear computational advantage.

The second main step of the procedure is the computation of the evaluation functions f^+ and f^-, given by

$$f^+(\gamma) = \frac{\sum\limits_{j \in J} \max\{\bar{a}_{kj}, -\gamma \bar{a}_{ij}\} s_j^* - \bar{a}_{k0} + (\bar{a}_{k0} + \gamma \bar{a}_{i0}) x_k^*}{1 + \gamma + \sum\limits_{j \in J} |\bar{a}_{kj} + \gamma \bar{a}_{ij}|}$$

and

$$f^-(\gamma) = \frac{\sum\limits_{j \in J} \max\{0, \bar{a}_{kj} + \gamma \bar{a}_{ij}\} s_j^* - (\bar{a}_{k0} + \gamma \bar{a}_{i0})(1 - x_k^*)}{1 - \gamma + \sum\limits_{j \in J} |\bar{a}_{kj} + \gamma \bar{a}_{ij}|}$$

As shown in [11], these functions are unimodal piecewise continuously differentiable and their minimum can be found efficiently, once rows k and i of the tableau are specified, by computing the values of f^+ (resp. f^-) by increasing (resp. decreasing) the value of $\gamma_l = -\frac{\bar{a}_{kl}}{\bar{a}_{il}}$ for valid pivots of the correct sign.

5 Most Violated Cut Selection Rule

Here we present a variant of the lift-and-project procedure which uses a new rule for choosing the leaving and entering variables in the pivot sequence. The lift-and-project procedure in the (LP) tableau usually requires a remarkably small number of pivots to obtain the optimal L&P cut, nevertheless it may be computationally interesting to reduce this number further by studying alternate rules for this choice. The rule discussed here performs, at each iteration, the pivot to the adjacent basis in (LP) for which the objective of $(CGLP)_k$ is decreased by the largest amount or, equivalently, the one for which the intersection cut obtained from the row k of (LP) is the most violated by x^*.

Let us denote by $f_i^+(\gamma)$ (resp. $f_i^-(\gamma)$) the function $f^+(\gamma)$ (resp. $f^-(\gamma)$) defined for source row k and a row i of the tableau. Recall that these functions give the violation of the intersection cut derived from the row obtained by adding γ times row i to row k, depending on the sign of γ. Thus, the violation of the cut in the adjacent basis of (LP) where variable i leaves the basis and variable j enters the basis is given by $f_i^+(\gamma_j)$ if $\gamma_j = -\bar{a}_{kj}/\bar{a}_{ij} > 0$ and $f_i^-(\gamma_j)$ if $\gamma_j = -\bar{a}_{kl}/\bar{a}_{ij} < 0$, and the most violated intersection cut which can be derived from an adjacent basis has violation

$$\hat{\sigma} = \min_{i \in I \setminus \{k\}} \min\{\min_{j \in J^+} f_i^+(\gamma_j), \min_{j \in J^-} f_i^-(\gamma_j)\}$$

where I is the basic index set and J^+, J^- are the index sets for $\gamma_j > 0$ and $\gamma_j < 0$, respectively.

Here the variables $\hat{i}$ and $\hat{j}$ for which this minimum is attained are selected as the leaving and entering variables respectively. By computing the reduced costs r_{u_i} and r_{v_i}, we first identify all the candidate rows for an improving pivot. Then for each such row i we minimize the functions f_i^+ and f_i^-.

This clearly amounts to more computation at each iteration than the selection rule used in Variant 1, where only one minimization of the evaluation function is performed at each pivot. But on the other hand, the cut violation is increased at each iteration by an amount at least as large, and therefore one may expect to obtain in less iterations a cut with a given violation. In particular, in the presence of zero elements in the source row, it presents the advantage that fewer degenerate pivots in $(\text{CGLP})_k$ are performed.

6   Disjunctive Modularization

L&P cuts are obtained from disjunctions of the type

$$(u\tilde{A}x - u_0 x_k \geq u\tilde{b}) \vee (v\tilde{A}x + v_0 x_k \geq v\tilde{b} + v_0)$$

where solving the $(\text{CGLP})_k$ optimizes the multipliers u, u_0, v and v_0. Once the optimal values for these multipliers are obtained, the cut can be further strengthened, as mentioned in section 1, by using modular arithmetic on the coefficients of the integer-constrained components of x. This latter operation can be interpreted (see [4]) as subtracting from x_k on each side of the disjunction a product of the form mx, where m is an integer vector, and then optimizing the components of m over all integer values. In other words, the strengthened deepest intersection cut is the result of a sequence of two optimization procedures, first in the mutipliers u, v, u_0 and v_0, then in the components of m. But this raises the quest for a procedure that would simultaneously optimize both the continuous multipliers and the integer vector m. While this is an intricate task, equivalent to finding an optimal split cut, which has been treated elsewhere [7], the disjunctive modularization procedure described below is meant to approximate this goal.

Consider again the equation of the source row (2) for an intersection cut or a MIG cut. By applying disjunctive modularization to this equation we mean deriving from it the modularized equation

$$y_k = \varphi_{k0} - \sum_{j \in J} \varphi_{kj} s_j \tag{6}$$

where y_k is a new, integer-constrained variable of unrestricted sign, $\varphi_{k0} = \bar{a}_{k0}$,

$$\varphi_{kj} := \begin{cases} \bar{a}_{kj} - \lfloor \bar{a}_{kj} \rfloor, \, j \in J_1^+ := \{j \in J_1 : \bar{a}_{kj} - \lfloor \bar{a}_{kj} \rfloor \leq \bar{a}_{k0}\} \\ \bar{a}_{kj} - \lceil \bar{a}_{kj} \rceil, \, j \in J_1^- := \{j \in J_1 : \bar{a}_{kj} - \lfloor \bar{a}_{kj} \rfloor > \bar{a}_{k0}\} \\ \bar{a}_{kj} \qquad\qquad j \in J_2 := J \setminus J_1 \end{cases}$$

and $J_1 := J \cap \{1, \ldots, p\}$.

Clearly, every set of s_j, $j \in J$, that satisfies (2) with x_k integer, also satisfies (6) with y_k integer; hence the equation (6) is valid. Also, it is easy to see that the intersection cut derived from (6) is the strengthened intersection cut, or MIG cut derived from (2). However, at this point we do not intend to generate a cut. Instead, we append (6) to the optimal (LP) tableau and declare it the source row in place of (2) for the entire pivoting sequence. Further, after each pivot in row $\hat{\imath}$ and column $\hat{\jmath}$ the transformed row of y_k, say $y_k = \varphi'_{k0} - \sum_{j \in J'} \varphi'_{kj} s_j$ where $J' := (J \setminus \{\hat{\jmath}\}) \cup \{\hat{\imath}\}$, is treated again with disjunctive modularization. Namely, this time the row of y_k is replaced with $y_k = \bar{\varphi}_{k0} - \sum_{j \in J'} \bar{\varphi}_{kj} s_j$ where $\bar{\varphi}_{k0} = \varphi'_{k0}$, and

$$\bar{\varphi}_{kj} := \begin{cases} \varphi'_{kj} - \lfloor \varphi'_{kj} \rfloor, \, j \in (J'_1)^+ \\ \varphi'_{kj} - \lceil \varphi'_{kj} \rceil, \, j \in (J'_1)^- \\ \varphi'_{kj} \qquad\qquad j \in J'_2 \end{cases} \tag{7}$$

with $(J'_1)^+$, $(J'_1)^-$ and J'_2 defined analogously to J_1^+, J_1^- and J_2.

The expressions used for calculating the reduced costs r_{u_i}, r_{v_i} and the evaluation functions $f^+(\gamma), f^-(\gamma)$ used for selecting the pivot element at each iteration remain valid, except for the fact that the entries $\bar{a}_{kj}$ of the current row (2) of x_k are replaced (since this is no longer the source row) with the entries $\bar{\varphi}_{kj}$ of the current row of y_k (see [2] for details).

It is clear that the modularized source row, if used for cut generation, would yield a cut that dominates the one from the unmodularized source row. It can also be shown that every iteration of the cut generating algorithm that uses disjunctive modularization improves the cut obtainable from the source row.

7   Computational Results

The algorithm for generating L&P cuts from the (LP) tableau was implemented, in all three of its Variants discussed above, as a cut generator called CglLandP [9] in the COIN-OR framework. This generator is open-source and is available since September 2006 as part of the Cut Generation Library [8]. All the computations have been carried out using the publicly available version of the cut generator and were performed on a computer equipped with a 2 GHz AMD Optetron CPU and 3 GB of RAM.

Before presenting our results, it will be useful to recall a comparison between the computational efforts required by the original procedure that generates L&P cuts by solving the higher dimensional (CGLP), and the new one that pivots in the (LP) tableau. Based on running XPRESS on about 100 test problems with

Table 1. Comparing 10 rounds of different cuts at the root node

| | MIG Cuts | | | Lift-and-Project Cuts | | | | | | | | |
| | | | | Variant 1 | | | Variant 2 | | | Variant 3 | | |
	time (sec)	% gap closed	average cut violation	time (sec)	% gap closed	average cut violation	time (sec)	% gap closed	average cut violation	time (sec)	% gap closed	average cut violation
10teams	1.7	100.00	2.9728e-04	10.3	100.00	5.4248e-03	12.0	100.00	2.9581e-03	10.1	100.00	6.0645e-03
air03	0.3	100.00	1.0055e-04	0.7	100.00	2.5349e-02	0.3	100.00	1.0055e-04	0.8	100.00	2.5638e-02
air04	15.1	13.13	2.6198e-05	176.4	18.17	4.1924e-04	284.8	22.32	1.9377e-04	182.4	19.83	3.9787e-04
air05	11.8	6.89	4.0196e-05	98.9	12.99	1.3294e-03	163.1	14.12	4.6990e-04	114.4	12.65	1.3018e-03
arki001	2.6	52.07	4.2588e-03	6.9	52.89	2.2820e-02	6.4	52.07	3.0722e-02	5.8	43.80	2.6925e-02
bell3a	0.0	72.11	1.0944e-02	0.0	72.04	1.3493e-02	0.0	70.74	1.3322e-02	0.0	71.07	1.19344e-02
bell5	0.0	90.71	2.1735e-02	0.1	92.32	2.1099e-02	0.1	92.62	2.2064e-02	0.1	92.39	2.1817e-02
blend2	0.1	34.14	1.8580e-03	0.3	37.21	2.1410e-02	0.3	34.34	1.8501e-02	0.2	36.14	3.0460e-02
cap6000	0.2	62.50	3.9211e-05	1.7	62.50	7.1012e-05	2.2	62.50	6.3832e-05	2.5	62.50	8.1029e-05
dano3mip	75.2	0.03	2.9685e-03	498.4	0.03	1.2500e-02	223.7	0.03	1.0568e-02	147.6	0.03	1.5132e-02
danoint	0.7	1.74	7.4911e-04	7.5	1.59	1.2143e-02	10.5	1.88	8.8077e-03	9.2	1.38	9.2614e-03
dcmulti	0.4	69.54	2.5191e-02	2.5	78.17	4.5511e-02	2.3	83.10	4.5834e-02	1.5	76.60	4.6794e-02
dsbmip	0.3	no gap	4.3132e-02	0.6	no gap	5.8988e-02	0.5	no gap	6.6541e-02	0.5	no gap	6.1998e-02
egout	0.0	99.83	3.9095e-02	0.0	100.00	7.6798e-02	0.0	100.00	7.6902e-02	0.0	100.00	7.6798e-02
enigma	0.0	no gap	6.6426e-03	0.0	no gap	1.1151e-02	0.0	no gap	8.8013e-03	0.0	no_gap	1.7631e-02
fast0507	80.9	3.45	6.8759e-06	325.2	3.67	7.3836e-04	297.6	4.05	1.6800e-04	357.3	3.40	1.2375e-03
fiber	0.8	79.79	8.8217e-04	1.6	87.07	4.3319e-03	1.9	92.65	5.8489e-03	2.8	88.35	4.2286e-03
fixnet6	0.3	85.77	7.4356e-03	1.3	87.90	3.6747e-02	2.3	89.15	3.7913e-02	1.3	89.09	3.6157e-02
flugpl	0.0	14.05	1.2964e-02	0.0	15.94	1.7391e-02	0.0	16.09	1.5143e-02	0.0	15.94	1.7391e-02
gen	0.1	81.97	3.2112e-03	0.2	81.42	1.5461e-02	0.3	81.97	1.5146e-02	0.2	80.33	1.4654e-02
gesa2	1.0	75.12	5.0931e-03	1.8	76.33	1.3842e-02	2.5	90.64	1.6601e-02	1.3	77.62	1.3775e-02
gesa2_o	1.1	63.28	5.0887e-03	1.5	63.74	1.2591e-02	3.6	63.91	1.4217e-02	2.1	64.40	1.2421e-02
gesa3	1.0	56.16	2.4234e-03	2.0	80.37	7.4798e-03	2.6	84.18	9.6241e-03	2.2	83.16	7.5700e-03
gesa3_o	0.9	58.13	3.1590e-03	2.5	80.62	9.3592e-03	5.2	83.74	8.9960e-03	2.1	77.89	8.4678e-03
gt2	0.0	100.00	5.4967e-03	0.1	100.00	1.0840e-02	0.0	100.00	2.0432e-02	0.0	100.00	2.1752e-02
harp2	0.9	37.29	3.1194e-04	2.0	40.62	3.8258e-03	4.2	45.51	4.8609e-03	3.3	40.02	4.4110e-03
khb05250	0.2	94.34	3.2644e-02	0.3	96.68	5.3187e-02	0.4	97.18	6.1590e-02	0.3	96.44	5.6201e-02
l152lav	1.7	20.78	1.4208e-04	8.4	39.87	3.3926e-03	9.5	40.30	2.0602e-03	13.8	33.10	3.7684e-03
lseu	0.0	85.81	2.8617e-03	0.0	88.83	1.3431e-02	0.1	89.19	1.0132e-02	0.1	85.27	1.5530e-02
markshare1	0.0	0.00	2.4035e-03	0.0	0.00	1.5459e-02	0.0	0.00	7.8762e-03	0.0	0.00	1.0355e-02
markshare2	0.0	0.00	1.7025e-03	0.0	0.00	4.5266e-03	0.0	0.00	5.3009e-03	0.0	0.00	1.0701e-02
mas74	0.1	7.62	3.6506e-04	0.5	8.89	5.3231e-03	0.5	8.75	3.0362e-03	0.4	8.29	7.5558e-03
mas76	0.0	7.40	2.5738e-04	0.4	9.09	2.1974e-03	0.4	8.63	1.6655e-03	0.4	8.84	5.8109e-03
misc03	0.1	20.00	3.7024e-03	0.2	19.44	2.6853e-02	0.7	23.75	1.9687e-02	0.1	17.24	3.0294e-02
misc06	0.1	78.26	1.3575e-03	0.3	90.22	5.6188e-03	0.2	95.65	5.8362e-03	0.3	90.22	6.7744e-03
misc07	0.0	0.72	3.7471e-03	0.1	0.72	2.8068e-02	0.3	2.51	2.9601e-02	0.1	0.72	3.0492e-02
mitre	0.3	100.00	1.5473e-03	0.4	100.00	3.8563e-03	0.6	100.00	5.6427e-03	0.4	100.00	3.8195e-03
mkc	3.3	30.66	2.7229e-03	4.7	49.98	1.7965e-02	4.4	46.18	1.8324e-02	4.7	43.84	1.8470e-02
mod008	0.0	30.44	3.1792e-04	0.1	33.73	5.8576e-03	0.1	41.62	5.4240e-03	0.1	39.35	2.4730e-02
mod010	0.1	100.00	2.2217e-04	0.7	100.00	7.6252e-03	0.1	100.00	2.5233e-04	1.4	94.79	3.9760e-03
mod011	6.4	38.50	3.1465e-02	19.3	39.61	5.7859e-02	60.0	41.42	6.1338e-02	17.6	39.37	5.7518e-02
modglob	0.3	61.05	2.5368e-02	0.5	62.31	4.5024e-02	1.0	58.17	4.4318e-02	0.6	63.89	4.3180e-02
noswot	0.0	no gap	1.3127e-02	0.1	no gap	3.4442e-02	0.1	no_gap	4.1982e-02	0.2	no gap	4.2526e-02
nw04	7.2	100.00	5.7806e-06	6.0	100.00	8.4070e-03	10.2	100.00	5.7806e-06	34.0	100.00	6.1504e-03
p0033	0.0	76.98	9.4281e-03	0.0	75.57	2.3048e-02	0.0	78.38	2.2338e-02	0.0	75.75	1.9114e-02
p0201	0.2	54.97	2.1807e-03	1.1	83.28	1.1511e-02	1.6	79.78	1.1691e-02	1.2	84.92	9.7795e-03
p0282	0.1	24.10	1.0334e-02	0.2	55.66	6.8690e-02	0.3	48.78	5.6026e-02	0.3	59.46	7.2102e-02
p0548	0.2	95.57	7.7912e-03	0.3	97.60	1.7712e-02	0.5	94.83	1.7799e-02	0.3	97.16	1.7990e-02
p2756	0.6	97.90	2.2700e-02	0.8	97.00	4.4864e-02	1.0	97.42	3.6267e-02	1.0	97.16	4.4811e-02
pk1	0.0	0.00	2.9938e-03	0.0	0.00	9.5531e-03	0.1	0.00	1.4450e-02	0.0	0.00	3.7375e-02
pp08a	0.3	90.39	3.5520e-02	0.6	89.49	4.5770e-02	0.5	92.49	5.1440e-02	0.6	92.59	4.6828e-02
pp08aCUTS	0.5	65.48	2.2789e-02	1.0	71.99	3.7008e-02	1.4	75.06	3.4990e-02	1.1	71.49	3.5609e-02
qiu	2.0	8.37	5.6655e-03	23.4	29.18	1.1153e-02	45.5	30.60	1.0984e-02	23.4	29.18	1.1153e-02
qnet1	1.5	36.18	3.5439e-04	4.5	39.39	6.6478e-03	4.6	46.06	5.3104e-03	7.1	42.29	5.8200e-03
qnet1_o	1.2	56.80	9.8367e-04	4.1	67.05	1.4260e-02	3.4	69.15	1.4546e-02	5.0	68.93	1.4840e-02
rentacar	0.3	30.56	2.7062e-02	0.4	37.20	5.3647e-02	5.5	43.26	5.0001e-02	0.4	37.20	5.3647e-02
rgn	0.0	12.30	5.9142e-03	0.0	16.37	4.3256e-02	0.0	19.76	3.9477e-02	0.1	33.62	-0.00558971
rout	0.2	5.19	2.1877e-03	3.4	24.54	1.2801e-02	4.3	35.35	1.0830e-02	3.7	29.17	1.4507e-02
set1ch	0.5	68.44	5.5305e-02	0.9	75.66	7.3559e-02	1.1	75.43	7.6604e-02	1.6	73.00	7.3731e-02
seymour	4.9	14.27	1.0106e-02	21.7	14.60	1.1268e-02	30.7	19.92	1.5199e-02	20.5	15.75	1.2845e-02
stein27	0.0	0.00	3.2688e-02	0.0	0.00	6.0127e-02	0.0	0.00	6.1522e-02	0.0	0.00	6.0228e-02
stein45	0.0	0.00	1.9886e-02	0.3	0.00	5.5584e-02	0.3	0.00	5.3331e-02	0.3	0.00	5.5584e-02
swath	4.1	26.87	1.1443e-04	5.4	27.60	1.0963e-02	6.5	27.20	6.4202e-03	4.6	28.36	1.0804e-02
vpm1	0.0	52.70	8.7372e-03	0.0	75.59	2.0991e-02	0.1	76.82	2.1853e-02	0.0	75.59	2.1079e-02
vpm2	0.0	53.51	8.6445e-03	0.1	61.66	1.7934e-02	0.2	64.13	1.6697e-02	0.1	59.29	2.0777e-02
Average	3.566	48.45	9.515e-3	19.27	53.62	2.232e-2	18.81	55.05	2.164e-2	15.28	53.69	2.297e-2

each of the two methods, Perregaard [10] reported that the new method required
5% of the number of pivots and 1.3% of the time required by the original one
for generating a L&P cut.

Table 1 presents a comparison for 10 rounds of cuts generated at the root node, where a round means a cut for every integer-constrained fractional variable. In this experiment, the problems are preprocessed with COIN `CglPreproces` procedure and then 10 rounds of cuts are generated. The test set consists of 65 problems from the MIPLIB.3 library. The four methods compared are mixed integer Gomory (MIG) cuts, and the three variants of lift-and-project cuts presented in this paper: Variant 1 (Balas and Perregaard's algorithm cf. Sections 3 and 4), Variant 2 (the algorithm using the most violated cut selection rule, cf. Section 5) and Variant 3 (the algorithm using disjunctive modularization cf. Section 6). For each of the methods, we report the running time, the percentage of the initial integrality gap closed, and the average violation for each cut generated in 10 rounds (where the violation is the change in the objective of $(CGLP)_k$ after each cut added).

As can be seen from the table, generating lift-and-project cuts with the three different variants proposed here is not much more expensive than generating MIG cuts. For our test set, it took on the average 3.566 seconds per instance to perform 10 rounds of MIG cuts, while it took 19, 19 and 15 seconds respectively per instance for the three variants of lift-and-project cuts. Considering that cut generation takes less than 5% of the total time needed to solve a mixed integer program (see [10]), this difference is not significant. This extra computational cost made it possible to close a significantly larger fraction of the integrality gap, namely 54%, 55% and 54% with Variants 1, 2 and 3, respectively, versus 48% for the MIG cuts. Of the 65 instances, there are only two (bell3a and p2756) on which the MIG cuts close a slightly larger fraction of the gap than the three flavors of lift-and-project cuts. Even more striking is the difference in the strength of the individual cuts, as measured by the amount of their violation by the current LP solution: it is on the average 2.4 times as large for the lift-and-project cuts as it is for the MIG cuts.

To more thoroughly assess the effectiveness of lift-and-project cuts, it is of course necessary to solve the instances to completion by running a branch-and-cut code and using these cuts to strengthen the LP relaxation. To this end, we present two comparative experiments of complete resolution for the MIPLIB.3 problems. The first experiment, presented in Table 2, consists in performing 10 rounds of cut generation at the root node and then solving the problem by branch-and-bound without further cut generation. In the second experiment, summarized in Table 3, 10 rounds of cuts are generated at the root node and 1 round is performed every 10 nodes of the branch-and-cut tree. Again, the four cut generation methods tested are MIG cuts and the three variants of lift-and-project cuts. For all three variants, the limit on the number of pivots is set to 10.

The branch-and-cut runs are performed by using Cbc (COIN-OR Branch and Cut) with some specific settings: a two hours time limit for solving each problem is imposed; all the default cut generation procedures of Cbc are deactivated; the variable selection strategy used is strong branching with the default parameters of Cbc (i.e. performing strong branching on the 5 most fractional variables); the node selection strategy is best bound.

Table 2. Comparing complete resolutions with cut-and-branch with 10 rounds of cuts. S means solved within the time limit, T means aborted because of the time limit.

| | MIG Cuts | | | Lift-and-Project Cuts | | | | | | | | |
| | | | | Variant 1 | | | Variant 2 | | | Variant 3 | | |
	status	time (sec)	# nodes	status	time (sec)	# nodes	status	time (sec)	# nodes	status	time (sec)	# nodes
	Group A:Instances solved with MIG cuts in less than 10 seconds											
air03	S	0.52	1	S	1.14	1	S	0.82	1	S	1.18	1
dcmulti	S	4.27	57	S	6.46	51	S	6.49	41	S	6.61	51
egout	S	0.04	5	S	0.03	1	S	0.04	3	S	0.02	1
enigma	S	2.91	1168	S	2.62	931	S	3.00	1257	S	0.25	40
fixnet6	S	4.78	49	S	10.26	43	S	27.16	147	S	11.21	67
flugpl	S	0.38	243	S	0.44	309	S	0.49	347	S	0.44	309
gen	S	1.28	35	S	1.92	31	S	0.87	17	S	1.46	35
gt2	S	0.01	1	S	1.16	229	S	2.15	259	S	4.60	1041
khb05250	S	1.28	31	S	1.61	31	S	1.68	17	S	1.64	31
lseu	S	1.18	425	S	1.40	537	S	2.56	611	S	2.11	709
misc03	S	2.92	157	S	5.85	111	S	5.97	73	S	3.59	81
misc06	S	1.08	26	S	0.85	10	S	0.51	7	S	1.06	13
mitre	S	0.59	1	S	1.18	1	S	1.50	1	S	1.34	1
mod008	S	6.65	919	S	7.22	729	S	6.29	691	S	3.06	313
mod010	S	0.59	1	S	1.08	1	S	0.71	1	S	1.32	1
p0033	S	0.24	157	S	0.17	53	S	0.18	71	S	0.05	9
p0201	S	4.07	153	S	10.53	215	S	7.75	65	S	2.63	17
p0282	S	0.50	47	S	1.11	43	S	1.01	55	S	0.92	51
p0548	S	5.24	409	S	11.59	883	S	9.29	657	S	2.60	187
p2756	S	8.39	168	S	20.41	306	S	10.68	158	S	13.88	200
rentacar	S	4.42	13	S	4.81	13	S	5.16	13	S	4.80	13
rgn	S	2.07	527	S	4.93	533	S	2.71	367	S	3.47	363
stein27	S	3.34	873	S	3.33	877	S	3.58	891	S	3.52	893
vpm1	S	4.81	415	S	0.21	5	S	0.27	5	S	0.23	5
	Group B: Instances solved with MIG cuts in a time between 10 seconds and 10 minutes											
10teams	S	301.57	1091	S	561.25	1600	S	129.68	329	S	321.45	1030
bell3a	S	19.69	12871	S	27.31	19765	S	28.30	19205	S	21.42	12927
bell5	S	46.12	22015	S	72.72	29655	S	20.82	9215	S	43.22	17755
blend2	S	27.73	2117	S	25.43	1723	S	7.66	271	S	40.36	2725
cap6000	S	311.07	1557	S	502.07	1923	S	465.71	1853	S	445.67	1825
dsbmip	S	15.26	168	S	16.00	159	S	12.15	145	S	43.14	528
fiber	S	428.58	8339	S	115.94	2607	S	32.48	257	S	109.52	923
gesa3	S	34.59	483	S	20.82	129	S	12.98	87	S	29.17	219
gesa3_o	S	36.45	591	S	48.02	353	S	57.45	319	S	32.43	225
l152lav	S	189.21	657	S	157.63	465	S	214.21	293	S	280.03	439
misc07	S	148.21	3745	S	229.06	4913	S	182.43	4161	S	235.73	4593
nw04	S	10.47	1	S	14.05	1	S	33.34	1	S	71.95	1
qnet1	S	170.77	567	S	121.46	263	S	94.20	287	S	214.00	489
qnet1_o	S	35.52	171	S	80.97	261	S	39.46	131	S	68.18	189
stein45	S	125.43	9819	S	119.04	11767	S	113.12	10093	S	118.13	11381
vpm2	S	401.08	40191	S	165.28	13717	S	267.30	19531	S	309.16	23561
	Group C: Instances solved with MIG cuts in more than 10 minutes or unsolved											
air04	S	4244.27	903	S	2945.54	689	S	6559.62	841	S	2057.31	467
air05	S	1872.60	1199	S	3048.48	1055	S	6301.18	1783	S	6061.42	1795
gesa2	S	1742.22	34263	S	3525.33	92709	S	3574.77	74509	S	3843.77	83425
gesa2_o	T	7200.61	90683	T	7199.81	97291	T	7201.03	79284	T	7200.91	88376
mas76	S	3643.00	765927	S	2729.41	730081	S	1733.07	783863	S	2104.06	731935
mod011	T	7199.75	19457	T	7200.16	17990	T	7199.88	16488	T	7200.20	13504
modglob	S	2140.48	257313	S	714.17	38231	S	1122.85	68141	S	563.46	29151
pk1	S	656.52	318694	S	665.89	321034	S	651.76	328540	S	681.34	357178
pp08a	S	1164.49	55853	S	745.10	30769	S	261.32	12081	S	537.56	26211
pp08aCUTS	S	962.24	45755	S	646.92	20095	S	782.85	29135	S	869.54	23443
qiu	S	3077.94	8505	S	4278.59	4665	T	7200.33	3647	S	3864.11	4665
Average	—	711.1	3.351e+04	—	707.4	2.843e+04	—	870.6	2.883e+04	—	734.1	2.83e+04
Geo. Mean	—	23.226	599.19	—	27.901	508.96	—	25.4	420.4	—	25.897	428.82

In Table 2, for each problem and each method, we indicate the status, the computing time and the number of nodes to solve the problem. Averages and geometric means are reported in the last two lines of the table. Among the 65 problems of the MIPLIB.3, 14 were not solved in the two hours time limit with any of the methods tested (namely arki01, dano3mip, danoint, fast0507, harp2, mas74, markshare1, markshare2, mkc, noswot, rout, set1ch, seymour and swath). We do not include statistics for these 14 problems.

As Table 2 shows, the average size of the branch-and-bound trees generated by each of Variants 1, 2 and 3 is about 15% smaller than the one obtained with the MIG cuts. The average time needed to solve an instance remains roughly the same for Variants 1 and 3 as for the MIG cuts, and increases by a fifth for Variant 2.

The experiment reported in Table 2 was in cut-and-branch mode, in that cuts were only generated at the root node. Our next experiment explores the use of cuts in the branch-and-cut mode: it generates 10 rounds of cuts at the root node, and one round of cuts at every 10-th node of the branch-and-bound tree. A summary of its results are reported in Table 3 (we only report averages by groups of instances the same as the ones constituted in Table 2). The complete results are available at [9]

It is highly edifying to examine the effect of cut generation in the branch-and-bound tree. One would expect these extra cuts to reduce the size of the search tree by making the linear programming relaxation tighter, evidently at some computational cost not so much from generating the cuts as from the increased time needed to solve the linear programs with more constraints. So our expectation was for a decrease in the size of the tree, but an increase of the computing time per node. Surprisingly, a very different picture emerges from comparing Tables 2 and 3. The average number of search tree nodes is indeed smaller for Table 3, but only by 1.4% in the case of MIG cuts and by 7%, 9% and 18% respectively for the three variants of lift-and-project cuts. On the other hand, the total computing time is reduced by 32% in the case of the MIG cuts, and by 31%, 26% and 35% respectively for Variants 1, 2 and 3 of the lift-and-project cuts. In other words, adding cuts at some nodes of the branch-and-bound tree has reduced, rather than increased, the computing time per node. Another aspect of this finding is the fact that in many instances an increase in the number of search tree nodes is accompanied by a decrease in the total number of pivots performed during the procedure (excluding those used for cut generation, see [9]).

In trying to explain this strange phenomenon, we looked in detail at several runs and found that the most likely explanation lies in the fact that the cuts added at some nodes tend to substantially enhance the power of reduced cost fixing. In other words, they help fix more 0-1 variables whose reduced cost exceeds the difference between the value of the incumbent solution and the current upper bound (difference which is reduced as the result of adding the cuts), and thereby they facilitate the solution of the subproblems rather than making it harder. This explanation is partially corroborated by the detailed data that we were able to retrieve for a few instances, in the sense that in all cases the number of variables fixed by reduced cost throughout the run is significantly larger (by anywhere from 15% to 50% to even 400% in one case) for the runs of Table 3

Table 3. Summary of the results comparing complete resolutions with branch-and-cut, generating 10 rounds of cuts at the root node and then one round every 10 nodes

| | | MIG Cuts | | Lift-and-Project Cuts | | | | | |
| | | | | Variant 1 | | Variant 2 | | Variant 3 | |
	# instances	time (sec)	# nodes	time (sec)	# nodes	time (sec)	# nodes	time (sec)	# nodes
Group A	24	2.126	214	2.789	195.9	3.065	177.2	3.042	168.9
Group B	16	77.15	3649	75.56	3276	72.84	3150	123.3	3827
Group C	11	2135	1.475e+05	2154	1.183e+05	2898	1.163e+05	2020	1.013e+05
Average	–	485.8	3.305e+04	489.5	2.663e+04	649.3	2.615e+04	475.9	2.312e+04
Geo. Mean	–	17.307	510.8	19.734	432.86	19.324	383.63	20.243	409.7

than for those of Table 2, but this does not solve the mystery, which requires further study.

As it is well known that cutting planes tend to play a more significant role in solving hard instances than easy ones (easy instances are often solved faster without cutting planes), we turned our attention to the behavior of our procedures on the hardest of the instances that we solved. There were 11 instances whose solution required over 10 minutes, but that were nevertheless solved within our time limit of 2 hours, and their data are collected in Table 4 for the case of branch-and-cut with 10 rounds of cuts at the root node and another round of cuts after every 10-th node (the same runs described in Table 3). Out of these 11 instances, the procedure using MIG cuts was fastest in 3 cases, whereas Variants 1, 2 and 3 of the lift-and project based procedure were fastest in 1, 3 and 4 cases, respectively. Similarly, in terms of the number of branch and bound tree nodes generated, the MIG cuts did best in 2 instances, whereas Variants 1, 2 and 3 were best in 2, 2 and 4 instances, respectively. Table 4 shows in boldface the best

Table 4. Comparing branch-and-cut with 10 rounds of cuts at the root node and one round at every 10 nodes, on the 11 instances requiring more than 10 minutes. The best performers for each instance (in terms of time and nodes) are boldfaced.

| | MIG Cuts | | | Lift-and-Project Cuts | | | | | | | | |
| | | | | Variant 1 | | | Variant 2 | | | Variant 3 | | |
	status	time (sec)	# nodes	status	time (sec)	# nodes	status	time (sec)	# nodes	status	time (sec)	# nodes
air04	S	2310.95	1027	S	1766.50	685	S	**1220.82**	533	S	1289.52	**481**
air05	S	**890.55**	**865**	S	1364.21	1085	S	2688.23	1965	S	1500.95	1115
gesa2	S	1089.67	23899	S	**791.22**	**17767**	S	1834.55	39627	S	1337.58	31899
gesa2_o	S	**3266.18**	**50109**	S	3630.03	58059	T	>7208.71	>76979	S	5962.31	100257
mas76	S	3965.26	609723	S	6293.23	734951	T	>7201.54	>740897	S	**2414.62**	**557405**
mod011	S	**4907.91**	23463	S	4950.25	**22453**	S	6118.13	25097	S	6034.56	23825
modglob	S	2517.83	392369	S	752.74	49065	S	653.60	43677	S	**388.67**	**22811**
pk1	S	764.46	354890	S	805.46	321690	S	777.89	297800	S	752.74	291104
pp08a	S	1575.11	120203	S	1038.75	67785	S	**323.80**	**18977**	S	677.60	43437
pp08aCUTS	S	858.07	40175	S	479.43	**20751**	S	472.34	23113	S	833.71	37959
qiu	S	1344.05	5429	S	1819.18	6649	S	3375.35	10315	S	1030.98	**3561**
Average	—	2135	1.475e+05	—	2154	1.183e+05	—	2898	1.163e+05	—	2020	1.013e+05
Geo. Mean	—	1758.9	33285	—	1543.5	24451	—	1760	25885	—	**1393**	**22853**

performers for each instance. The geometric means of the computing time and of the number of search tree nodes for Variant 3 of the L&P procedure are less than the corresponding means for the MIG cut-based procedure by 48% and 31 %, respectively.

Generating lift-and-project cuts from the LP simplex tableau rather than the higher dimensional Cut Generating Linear Program is a new approach (the correspondence making this possible was discovered around 2002 [6] and its first implementation [10], corresponding more or less to our Variant 1, was done in 2003). Therefore the parameters used in our runs reported in this paper are first choices, to be improved upon by further research and experimentation. It is therefore legitimate to also look at the performance of the best of the three Variants in comparison with the classical MIG cuts on this set of hard problems. The result of that comparison is that the "best of three" is the fastest on 8 of the 11 instances, and generates the fewest search tree nodes on 9 of the 11 instances.

References

1. E. Balas, "Disjunctive Programming." *Annals of Discrete Mathematics, 5,* 1979, 3-51.
2. E. Balas, "Generating deepest mixed integer cuts by disjunctive modularization." Tepper School of Business, Carnegie Mellon University, May 2002.
3. E. Balas, S. Ceria, and G. Cornuéjols, "A lift-and-project cutting plane algorithm for mixed 0-1 programs." *Mathematical Programming, 58* 1993, 295-324.
4. E. Balas, S. Ceria, G. Cornuéjols, "Mixed 0-1 programming by lift-and-project in a branch-and-cut framework." *Man. Science, 112,* 1996, 1229-1246.
5. E. Balas, S. Ceria, G. Cornuéjols, and Natraj, "Gomory cuts revisited." *OR Letters, 19,* 1996, 1-10.
6. E. Balas and M. Perregaard, "A precise correspondence between lift-and-project cuts, simple disjunctive cuts, and mixed integer Gomory cuts for 0-1 programming." *Math. Program. B, 94* 2003, 221-245.
7. E. Balas and A. Saxena, "Optimizing over the split closure." MSRR# 674, Tepper School of Business, Carnegie Mellon University, 2005. To appear in *Math Programming A.*
8. COIN-OR website: `http://www.coin-or.org/`
9. CglLandP: `https://projects.coin-or.org/Cgl/wiki/CglLandP`.
10. M. Perregaard, "A practical implementation of lift-and-project cuts. International Symposium on Mathematical Programming, Copenhagen (2003).
11. M. Perregaard, "Generating Disjunctive Cuts for Mixed Integer Programs." Ph.D. Thesis, Carnegie Mellon University, 2003.

Orbital Branching

James Ostrowski[1], Jeff Linderoth[1], Fabrizio Rossi[2], and Stefano Smriglio[2]

[1] Department of Industrial and Systems Engineering, Lehigh University,
200 W. Packer Ave., Bethlehem, PA 18015, USA
{jao204,jtl3}@lehigh.edu
[2] Dipartimento di Informatica, Università di L'Aquila, Via Vetoio I-67010 Coppito
(AQ), Italy
{rossi,smriglio}@di.univaq.it

Abstract. We introduce *orbital branching*, an effective branching method for integer programs containing a great deal of symmetry. The method is based on computing groups of variables that are equivalent with respect to the symmetry remaining in the problem after branching, including symmetry which is not present at the root node. These groups of equivalent variables, called orbits, are used to create a valid partitioning of the feasible region which significantly reduces the effects of symmetry while still allowing a flexible branching rule. We also show how to exploit the symmetries present in the problem to fix variables throughout the branch-and-bound tree. Orbital branching can easily be incorporated into standard IP software. Through an empirical study on a test suite of symmetric integer programs, the question as to the most effective orbit on which to base the branching decision is investigated. The resulting method is shown to be quite competitive with a similar method known as *isomorphism pruning* and significantly better than a state-of-the-art commercial solver on symmetric integer programs.

1 Introduction

In this work, we focus on packing and covering integer programs (IP)s of the form

$$\max_{x \in \{0,1\}^n} \{e^T x \mid Ax \le e\} \text{ and} \tag{PIP}$$

$$\min_{x \in \{0,1\}^n} \{e^T x \mid Ax \ge e\} , \tag{CIP}$$

where $A \in \{0,1\}^{m \times n}$, and e is a vector of ones of conformal size. Our particular focus is on cases when (CIP) or (PIP) is highly-symmetric, a concept we formalize as follows. Let Π^n be the set of all permutations of $I^n = \{1, \ldots, n\}$. Given a permutation $\pi \in \Pi^n$ and a permutation $\sigma \in \Pi^m$, let $A(\pi, \sigma)$ be the matrix obtained by permuting the columns of A by π and the rows of A by σ, i.e. $A(\pi, \sigma) = P_\sigma A P_\pi$, where P_σ and P_π are permutation matrices. The *symmetry group* $\mathcal{G}$ of the matrix A is the set of permutations

$$\mathcal{G}(A) \stackrel{\text{def}}{=} \{\pi \in \Pi^n \mid \exists \sigma \in \Pi^m \text{ such that } A(\pi, \sigma) = A\} .$$

M. Fischetti and D.P. Williamson (Eds.): IPCO 2007, LNCS 4513, pp. 104–118, 2007.
© Springer-Verlag Berlin Heidelberg 2007

So, for any $\pi \in \mathcal{G}(A)$, if $\hat{x}$ is feasible for (CIP) or (PIP) (or the LP relaxations of (CIP) or (PIP)), then if the permutation π is applied to the coordinates of $\hat{x}$, the resulting solution, which we denote as $\pi(\hat{x})$, is also feasible. Moreover, the solutions $\hat{x}$ and $\pi(\hat{x})$ have equal objective value.

This equivalence of solutions induced by symmetry is a major factor that might confound the branch-and-bound process. For example, suppose $\hat{x}$ is a (non-integral) solution to an LP relaxation of PIP or CIP, with $0 < \hat{x}_j < 1$, and the decision is made to branch down on variable x_j by fixing $x_j = 0$. If $\exists \pi \in \mathcal{G}(A)$ such that $[\pi(\hat{x})]_j = 0$, then $\pi(\hat{x})$ is a feasible solution for this child node, and $e^T \hat{x} = e^T(\pi(\hat{x}))$, so the relaxation value for the child node will not change. If the cardinality of $\mathcal{G}(A)$ is large, then there are many permutations through which the parent solution of the relaxation can be preserved in this manner, resulting in many branches that do not change the bound on the parent node. Symmetry has long been recognized as a curse for solving integer programs, and auxiliary (often extended) formulations are often sought that reduce the amount of symmetry in an IP formulation [1,2,3]. In addition, there is a body of research on valid inequalities that can help exclude symmetric feasible solutions [4,5,6]. Kaibel and Pfetsch [7] formalize many of these arguments by defining and studying the properties of a polyhedron known as an orbitope, the convex hull of lexicographically maximal solutions with respect to a symmetry group. Kaibel et al. [8] then use the properties of orbitopes to remove symmetry in partitioning problems.

A different idea, *isomorphism pruning*, introduced by Margot [9,10] in the context of IP and dating back to Bazaraa and Kirca [11], examines the symmetry group of the problem in order to prune isomorphic subproblems of the enumeration tree. The branching method introduced in this work, *orbital branching*, also uses the symmetry group of the problem. However, instead of examining this group to ensure that an isomorphic node will never be evaluated, the group is used to guide the branching decision. At the cost of potentially evaluating isomorphic subproblems, orbital branching allows for considerably more flexibility in the choice of branching entity than isomorphism pruning. Furthermore, orbital branching can be easily incorporated within a standard MIP solver and even exploit problem symmetry that may only be locally present at a nodal subproblem.

The remainder of the paper is divided into five sections. In Sect. 2 we give some mathematical preliminaries. Orbital branching is introduced and formalized in Sect. 3, and a mechanism to fix additional variables based on symmetry considerations called *orbital fixing* is described there. A more complete comparison to isomorphism pruning is also presented in Sect. 3. Implementation details are provided in Sect. 4, and computational results are presented in Sect. 5. Conclusions about the impact of orbital branching and future research directions are given in Sect. 6.

2 Preliminaries

Orbital branching is based on elementary concepts from algebra that we recall in this section to make the presentation self-contained. Some definitions are made in terms of an arbitrary permutation group Γ, but for concreteness, the reader may consider the group Γ to be the symmetry group of the matrix $\mathcal{G}(A)$.

For a set $S \subseteq I^n$, the *orbit* of S under the action of Γ is the set of all subsets of I^n to which S can be sent by permutations in Γ, i.e.,

$$\text{orb}(S, \Gamma) \stackrel{\text{def}}{=} \{S' \subseteq I^n \mid \exists \pi \in \Gamma \text{ such that } S' = \pi(S)\} .$$

In the orbital branching we are concerned with the orbits of sets of cardinality one, corresponding to decision variables x_j in PIP or CIP. By definition, if $j \in \text{orb}(\{k\}, \Gamma)$, then $k \in \text{orb}(\{j\}, \Gamma)$, i.e. the variable x_j and x_k share the same orbit. Therefore, the union of the orbits

$$\mathcal{O}(\Gamma) \stackrel{\text{def}}{=} \bigcup_{j=1}^{n} \text{orb}(\{j\}, \Gamma)$$

forms a partition of $I^n = \{1, 2, \ldots, n\}$, which we refer to as the orbital partition of Γ, or simply the *orbits* of Γ. The orbits encode which variables are "equivalent" with respect to the symmetry Γ.

The stabilizer of a set $S \subseteq I^n$ in Γ is the set of permutations in Γ that send S to itself.

$$\text{stab}(S, \Gamma) = \{\pi \in \Gamma \mid \pi(S) = S\} .$$

The stabilizer of S is a subgroup of Γ.

We characterize a node $a = (F_1^a, F_0^a)$ of the branch-and-bound enumeration tree by the indices of variables fixed to one F_1^a and fixed to zero F_0^a at node a. The set of free variables at node a is denoted by $N^a = I^n \setminus F_0^a \setminus F_1^a$. At node a, the set of feasible solutions to (CIP) or (PIP) is denoted by $\mathcal{F}(a)$, and the value of an optimal solution for the subtree rooted at node a is denoted as $z^*(a)$.

3   Orbital Branching

In this section we introduce orbital branching, an intuitive way to exploit the orbits of the symmetry group $\mathcal{G}(A)$ when making branching decisions. The classical 0-1 branching variable dichotomy does not take advantage of the problem information encoded in the symmetry group. To take advantage of this information in orbital branching, instead of branching on individual variables, orbits of variables are used to create the branching dichotomy. Informally, suppose that at the current subproblem there is an orbit of cardinality k in the orbital partitioning. In orbital branching, the current subproblem is divided into $k + 1$ subproblems: the first k subproblems are obtained by fixing to one in turn each variable in the orbit while the $(k + 1)^{\text{st}}$ subproblem is obtained by fixing all variables in the orbit to zero. For any pair of variables x_i and x_j in the same orbit, the subproblem created when x_i is fixed to one is essentially equivalent to the subproblem created when x_j is fixed to one. Therefore, we can keep in the subproblem list only *one* representative subproblem, pruning the $(k-1)$ equivalent subproblems. This is formalized below.

Let $A(F_1^a, F_0^a)$ be the matrix obtained by removing from the constraint matrix A all columns in $F_0^a \cup F_1^a$ and either all rows intersecting columns in F_1^a (CIP case) or all columns nonorthogonal to columns in F_1^a (PIP case). Note that if $x \in \mathcal{F}(a)$ and x is feasible with respect to the matrix A, then x is feasible with respect to the matrix $A(F_1^a, F_0^a)$.

Let $O = \{i_1, i_2, \ldots, i_{|O|}\} \subseteq N^a$ be an orbit of the symmetry group $\mathcal{G}(A(F_1^a, F_0^a))$. Given a subproblem a, the disjunction

$$x_{i_1} = 1 \vee x_{i_2} = 1 \vee \ldots x_{i_O} = 1 \vee \sum_{i \in O} x_i = 0 \qquad (1)$$

induces a feasible division of the search space. In what follows, we show that for any two variables $x_j, x_k \in O$, the two children $a(j)$ and $a(k)$ of a, obtained by fixing respectively x_j and x_k to 1 have the same optimal solution value. As a consequence, disjunction (1) can be replaced by the binary disjunction

$$x_h = 1 \vee \sum_{i \in O} x_i = 0 \ , \qquad (2)$$

where h is a variable in O. Formally, we have Theorem 1.

Theorem 1. *Let O be an orbit in the orbital partitioning $\mathcal{O}(\mathcal{G}(A(F_1^a, F_0^a)))$, and let j, k be two variable indices in O. If $a(j) = (F_1^a \cup \{j\}, F_0^a)$ and $a(k) = (F_1^a \cup \{k\}, F_0^a)$ are the child nodes created when branching on variables x_j and x_k, then $z^*(a(j)) = z^*(a(k))$.*

Proof. Let x^* be an optimal solution of $a(j)$ with value $z^*(a(j))$. Obviously x^* is also feasible for a. Since j and k are in the same orbit O, there exists a permutation $\pi \in \mathcal{G}(A(F_1^a, F_0^a))$ such that $\pi(j) = k$. By definition, $\pi(x^*)$ is a feasible solution of a with value $z^*(a(j))$ such that $x_k = 1$. Therefore, $\pi(x^*)$ is feasible for $a(k)$, and $z^*(a(k)) = z^*(a(j))$. $\qquad \square$

The basic *orbital branching* method is formalized in Algorithm 1.

Algorithm 1. Orbital Branching

Input: Subproblem $a = (F_1^a, F_0^a)$, non-integral solution $\hat{x}$.
Output: Two child subproblems b and c.

Step 1.   Compute orbital partition $\mathcal{O}(\mathcal{G}(A(F_1^a, F_0^a))) = \{O_1, O_2, \ldots, O_p\}$.
Step 2.   Select orbit O_{j^*}, $j^* \in \{1, 2, \ldots, p\}$.
Step 3.   Choose arbitrary $k \in O_{j^*}$. Return subproblems $b = (F_1^a \cup \{k\}, F_0^a)$ and $c = (F_1^a, F_0^a \cup O_{j^*})$.

The consequence of Theorem 1 is that the search space is limited, but orbital branching has also the relevant effect of reducing the likelihood of encountering symmetric solutions. Namely, no solutions in the left and right child nodes of the current node will be symmetric with respect to the local symmetry. This is formalized in Theorem 2.

Theorem 2. *Let b and c be any two subproblems in the enumeration tree. Let a be the first common ancestor of b and c. For any $x \in \mathcal{F}(b)$ and $\pi \in \mathcal{G}(A(F_0^a, F_1^a))$, $\pi(x)$ does not belong $\mathcal{F}(c)$.*

Proof. Suppose not, i.e., that there $\exists x \in \mathcal{F}(b)$ and a permutation $\pi \in \mathcal{G}(A(F_0^a, F_1^a))$ such that $\pi(x) \in \mathcal{F}(c)$. Let $O_i \in \mathcal{O}(\mathcal{G}(A(F_1^a, F_0^a)))$ be the orbit chosen to branch on at subproblem a. W.l.o.g. we can assume $x_k = 1$ for some $k \in O_i$. We have that $x_k = [\pi(x)]_{\pi(k)} = 1$, but $\pi(k) \in O_i$. Therefore, by the orbital branching dichotomy, $\pi(k) \in F_0^c$, so $\pi(x) \notin \mathcal{F}(c)$.   □

Note that by using the matrix $A(F_1^a, F_0^a)$, orbital branching attempts to use symmetry found at all nodes in the enumeration tree, not just the symmetry found at the root node. This makes it possible to prune nodes whose corresponding solutions are not symmetric in the original IP.

3.1   Orbital Fixing

In orbital branching, all variables fixed to zero and one are removed from the constraint matrix at every node in the enumeration tree. As Theorem 2 demonstrates, using orbital branching in this way ensures that any two nodes are not equivalent with respect to the symmetry found at their first common ancestor. It is possible however, for two child subproblems to be equivalent with respect to a symmetry group found elsewhere in the tree. In order to combat this type of symmetry we perform *orbital fixing*, which works as follows.

Consider the symmetry group $\mathcal{G}(A(F_1^a, \emptyset))$ at node a. If there exists an orbit O in the orbital partition $\mathcal{O}(\mathcal{G}(A(F_1^a, \emptyset)))$ that contains variables such that $O \cap F_0^a \neq \emptyset$ and $O \cap N^a \neq \emptyset$, then all variables in O can be fixed to zero. In the following theorem, we show that such variable setting (orbital fixing) excludes feasible solutions only if there exists a feasible solution of the same objective value to the left of the current node in the branch and bound tree. (We assume that the enumeration tree is oriented so that the branch with an additional variable fixed at one is the left branch).

To aid in our development, we introduce the concept of a *focus node*. For $x \in \mathcal{F}(a)$, we call node $b(a, x)$ a focus node of a with respect to x if $\exists y \in \mathcal{F}(b)$ such that $e^T x = e^T y$ and b is found to the left of a in the tree.

Theorem 3. *Let $\{O_1, O_2, \ldots O_q\}$ be an orbital partitioning of $\mathcal{G}(A(F_1^a, \emptyset))$ at node a, and let the set*

$$S \stackrel{\text{def}}{=} \{j \in N^a \mid \exists k \in F_0^a \text{ and } j, k \in O_\ell \text{ for some } \ell \in \{1, 2, \ldots q\}\}$$

be the set of free variables that share an orbit with a variable fixed to zero at a. If $x \in \mathcal{F}(a)$ with $x_i = 1$ for some $i \in S$, then there exists a focus node for a with respect to x.

Proof. Suppose that a is the first node in any enumeration tree where S is non-empty. Then, there exist $j \in F_0^a$ and $i \in S$ such that $i \in \mathrm{orb}(\{j\}, \mathcal{G}(A(F_1^a, \emptyset)))$, i.e., there exists a $\pi \in \mathcal{G}(A(F_1^a, \emptyset))$ with $\pi(i) = j$. W.l.o.g., suppose that j is any of the first such variables fixed to zero on the path from the root node to a and let c be the subproblem in which such a fixing occurs. Let $\rho(c)$ be the parent node of c. By our choice of j as the first fixed variable, for all $i \in F_0^a$, we have $x_{\pi(i)} = 0$. Then, there exists $x \in \mathcal{F}(a)$ with $x_i = 1$ such that $\pi(x)$ is not feasible in a (since it does not satisfy the bounds) but it is feasible in $\rho(c)$ and has the same objective value of x. Since j was fixed by orbital branching then the left child of $\rho(c)$ has $x_h = 1$ for some $h \in \mathrm{orb}(\{j\}, \mathcal{G}(A(F_1^{\rho(c)}, F_0^{\rho(c)})))$. Let $\pi' \in \mathcal{G}(A(F_1^{\rho(c)}, F_0^{\rho(c)}))$ have $\pi'(j) = h$. Then $\pi'(\pi(x))$ is feasible in the left node with the same objective value of x. The left child node of $\rho(c)$ is then the focus node of a with respect to x.

If a is not a first node in the enumeration tree one can apply the same argument to the first ancestor b of a such that $S \neq \emptyset$. The focus node of $c = (b, x)$ is then a focus node of (a, x).

$\square$

An immediate consequence of Theorem 3 is that for all $i \in F_0^a$ and for all $j \in \mathrm{orb}(\{i\}, \mathcal{G}(A(F_1^a, \emptyset)))$ one can set $x_j = 0$. We update orbital branching to include orbital fixing in Algorithm 2.

Algorithm 2. Orbital Branching with Orbital Fixing

Input: Subproblem $a = (F_1^a, F_0^a)$ (with free variables $N^a = I^n \setminus F_1^a \setminus F_0^a$), fractional solution $\hat{x}$.

Output: Two child nodes b and c.

Step 1. Compute orbital partition $\mathcal{O}(\mathcal{G}(A(F_1^a, \emptyset))) = \{\hat{O}_1, \hat{O}_2, \ldots, \hat{O}_q\}$. Let $S \overset{\text{def}}{=} \{j \in N^a \mid \exists k \in F_0^a$ and $(j \cap k) \in \hat{O}_\ell$ for some $\ell \in \{1, 2, \ldots q\}\}$.

Step 2. Compute orbital partition $\mathcal{O}(\mathcal{G}(A(F_1^a, F_0^a))) = \{O_1, O_2, \ldots, O_p\}$.

Step 3. Select orbit O_{j^*}, $j^* \in \{1, 2, \ldots, p\}$.

Step 4. Choose arbitrary $k \in O_{j^*}$. Return child subproblems $b = (F_1^a \cup \{k\}, F_0^a \cup S)$ and $c = (F_1^a, F_0^a \cup O_{j^*} \cup S)$.

In orbital fixing, the set S of additional variables set to zero is a function of F_0^a. Variables may appear in F_0^a due to a branching decision or due to traditional methods for variable fixing in integer programming, e.g. reduced cost fixing or implication-based fixing. Orbital fixing, then, gives a way to *enhance* traditional variable-fixing methods by including the symmetry present at a node of the branch and bound tree.

3.2   Comparison to Isomorphism Pruning

The fundamental idea behind isomorphism pruning is that for each node $a = (F_1^a, F_0^a)$, the orbits $\mathrm{orb}(F_1^a, \mathcal{G}(A))$ of the "equivalent" sets of variables to F_1^a are computed. If there is a node $b = (F_1^b, F_0^b)$ elsewhere in the enumeration tree such that $F_1^b \in \mathrm{orb}(F_1^a, \mathcal{G}(A))$, then the node a need not be evaluated—the node a is pruned by isomorphism. A very distinct and powerful advantage of this method is that *no* nodes whose sets of fixed variables are isomorphic will be evaluated. One disadvantage of this method is that computing $\mathrm{orb}(F_1^a, \mathcal{G}(A))$ can require computational effort on the order of $O(n|F_1^a|!)$. A more significant disadvantage of isomorphism pruning is that $\mathrm{orb}(F_1^a, \mathcal{G}(A))$ may contain many equivalent subsets to F_1^a, and the entire enumeration tree must be compared against this list to ensure that a is not isomorphic to any other node b. In a series of papers, Margot offers a way around this second disadvantage [9,10]. The key idea introduced is to declare one *unique representative* among the members of $\mathrm{orb}(F_1^a, \mathcal{G}(A))$, and if F_1^a is not the unique representative, then the node a may safely be pruned. The advantage of this extension is that it is trivial to check whether or not node a may be pruned once the orbits $\mathrm{orb}(F_1^a, \mathcal{G}(A))$ are computed. The disadvantage of the method is ensuring that the unique representative occurs *somewhere* in the branch and bound tree requires a relatively inflexible branching rule. Namely, *all* child nodes at a fixed depth must be created by branching on the *same* variable.

Orbital branching does not suffer from this inflexibility. By not focusing on pruning *all* isomorphic nodes, but rather eliminating the symmetry through branching, orbital branching offers a great deal more flexibility in the choice of branching entity. Another advantage of orbital branching is that by using the symmetry group $\mathcal{G}(A(F_1^a, F_0^a))$, symmetry *introduced* as a result of the branching process is also exploited.

Both methods allow for the use of traditional integer programming methodologies such as cutting planes and fixing variables based on considerations such as reduced costs and implications derived from preprocessing. In isomorphism pruning, for a variable fixing to be valid, it must be that *all* non-isomorphic optimal solutions are in agreement with the fixing. Orbital branching does not suffer from this limitation. A powerful idea in both methods is to combine the variable fixing with symmetry considerations in order to fix many additional variables. This idea is called *orbit setting* in [10] and *orbital fixing* in this work (see Sect. 3.1).

4   Implementation

The orbital branching method has been implemented using the user application functions of MINTO v3.1 [12]. The branching dichotomy of Algorithm 1 or 2 is implemented in the `appl_divide()` method, and reduced cost fixing is implemented in `appl_bounds()`. The entire implementation, including code for all the branching rules subsequently introduced in Sect. 4.2 consists of slightly over 1000 lines of code. All advanced IP features of MINTO were used, including *clique inequalities*, which can be useful for instances of (PIP).

4.1   Computing $\mathcal{G}(\cdot)$

Computation of the symmetry groups required for orbital branching and orbital fixing is done by computing the automorphism group of a related graph. Recall that the automorphism group $\mathrm{Aut}(G(V,E))$ of a graph $G = (V,E)$, is the set of permutations of V that leave the incidence matrix of G unchanged, i.e.

$$\mathrm{Aut}(G(V,E)) = \{\pi \in \Pi^{|V|} \mid (i,j) \in E \Leftrightarrow (\pi(i), \pi(j)) \in E\} \ .$$

The matrix A whose symmetry group is to be computed is transformed into a bipartite graph $G(A) = (N, M, E)$ where vertex set $N = \{1, 2, \ldots, n\}$ represents the variables, and vertex set $M = \{1, 2, \ldots, m\}$ represents the constraints. The edge $(i,j) \in E$ if and only if $a_{ij} = 1$. Under this construction, feasible solutions to (PIP) are subsets of the vertices $S \subseteq N$ such that each vertex $i \in M$ is adjacent to *at most* one vertex $j \in S$. In this case, we say that S *packs* M. Feasible solutions to (CIP) correspond to subsets of vertices $S \subseteq N$ such that each vertex $i \in M$ is adjacent to *at least* one vertex $j \in S$, or S *covers* M. Since applying members of the automorphism group preserves the incidence structure of a graph, if S packs (covers) M, and $\pi \in \mathrm{stab}(M, \mathrm{Aut}(G(A)))$, then there exists a $\sigma \in \Pi^m$ such that $\sigma(M) = M$ and $\pi(S)$ packs (covers) $\sigma(M)$. This implies that if $\pi \in \mathrm{stab}(M, \mathrm{Aut}(G(A)))$, then the restriction of π to N must be an element of $\mathcal{G}(A)$, i.e. using the graph $G(A)$, one can find elements of symmetry group $\mathcal{G}(A)$. In particular, we compute the orbital partition of the stabilizer of the constraint vertices M in the automorphism group of $G(A)$, i.e.

$$\mathcal{O}(\mathrm{stab}(M, \mathrm{Aut}(G(A)))) = \{O_1, O_2, \ldots, O_p\} \ .$$

The orbits $O_1, O_2, \ldots, O_p$ in the orbital partition are such that if $i \in M$ and $j \in N$, then i and j are not in the same orbit. We can then refer to these orbits as *variable* orbits and *constraint* orbits. In orbital branching, we are concerned only with the variable orbits.

There are several software packages that can compute the automorphism groups required to perform orbital branching. The program nauty [13], by McKay, has been shown to be quite effective [14], and we use nauty in our orbital branching implementation.

The complexity of computing the automorphism group of a graph is not known to be polynomial time. However, nauty was able to compute the symmetry groups of our problems very quickly, generally faster than solving an LP at a given node. One explanation for this phenomenon is that the running time of nauty's backtracking algorithm is correlated to the size of the symmetry group being computed. For example, computing the automorphism group of the clique on 2000 nodes takes 85 seconds, while graphs of comparable size with little or no symmetry require fractions of a second. The orbital branching procedure quickly reduces the symmetry group of the child subproblems, so explicitly recomputing the group by calling nauty is computational very feasible. In the table of results presented in the Appendix, we state explicitly the time required in computing automorphism groups by nauty.

4.2   Branching Rules

The orbital branching rule introduced in Sect. 3 leaves significant freedom in choosing the orbit on which to base the partitioning. In this section, we discuss mechanisms for deciding on which orbit to branch. As input to the branching decision, we are given a fractional solution $\hat{x}$ and orbits $O_1, O_2, \ldots O_p$ (consisting of all currently free variables) of the orbital partitioning $\mathcal{O}(\mathcal{G}(A(F_0^a, F_1^a)))$ for the subproblem at node a. Output of the branching decision is an index j^* of an orbit on which to base the orbital branching. We tested six different branching rules.

Rule 1: Branch Largest: The first rule chooses to branch on the largest orbit O_{j^*}:

$$j^* \in \arg \max_{j \in \{1, \ldots p\}} |O_j| \ .$$

Rule 2: Branch Largest LP Solution: The second rule branches on the orbit O_{j^*} whose variables have the largest total solution value in the fractional solution $\hat{x}$:

$$j^* \in \arg \max_{j \in \{1, \ldots p\}} \hat{x}(O_j) \ .$$

Rule 3: Strong Branching: The third rule is a strong branching rule. For each orbit j, two tentative child nodes are created and their bounds z_j^+ and z_j^- are computed by solving the resulting linear programs. The orbit j^* for which the product of the change in linear program bounds is largest is used for branching:

$$j^* \in \arg \max_{j \in \{1, \ldots p\}} (|e^T \hat{x} - z_j^+|)(|e^T \hat{x} - z_j^-|) \ .$$

Note that if one of the potential child nodes in the strong branching procedure would be pruned, either by bound or by infeasibility, then the bounds on the variables may be fixed to their values on the alternate child node. We refer to this as *strong branching fixing*, and in the computational results in the Appendix, we report the number of variables fixed in this manner. As discussed at the end of Sect. 3.1, variables fixed by strong branching fixing may result in additional variables being fixed by orbital fixing.

Rule 4: Break Symmetry Left: This rule is similar to *strong branching*, but instead of fixing a variable and computing the change in objective value bounds, we fix a variable and compute the change in the size of the symmetry group. Specifically, for each orbit j, we compute the size of the symmetry group in the resulting left branch if orbit j (including variable index i_j) was chosen for branching, and we branch on the orbit that reduces the symmetry by as much as possible:

$$j^* \in \arg \min_{j \in \{1, \ldots p\}} (|\mathcal{G}(A(F_1^a \cup \{i_j\}, F_0^a))|) \ .$$

Rule 5: Keep Symmetry Left: This branching rule is the same as **Rule 4**, except that we branch on the orbit for which the size of the child's symmetry group would remain the largest:

$$j^* \in \arg \max_{j \in \{1, \ldots p\}} (|\mathcal{G}(A(F_1^a \cup \{i_j\}, F_0^a))|) \ .$$

Rule 6: Branch Max Product Left: This rule attempts to combine the fact that we would like to branch on a large orbit at the current level and also keep a large orbit at the second level on which to base the branching dichotomy. For each orbit $O_1, O_2, \ldots, O_p$, the orbits $P_1^j, P_2^j, \ldots, P_q^j$ of the symmetry group $\mathcal{G}(A(F_1^a \cup \{i_j\}, F_0^a))$ of the left child node are computed for some variable index $i_j \in O_j$. We then choose to branch on the orbit j^* for which the product of the orbit size and the largest orbit of the child subproblem is largest:

$$j^* \in \arg \max_{j \in \{1, \ldots p\}} \left(|O_j| (\max_{k \in \{1, \ldots q\}} |P_k^j|) \right) \ .$$

5 Computational Experiments

In this section, we give empirical evidence of the effectiveness of orbital branching, we investigate the impact of choosing the orbit on which branching is based, and we demonstrate the positive effect of orbital fixing. The computations are based on the instances whose characteristics are given in Table 1. The instances beginning with cod are used to compute maximum cardinality binary error correcting codes [15], the instances whose names begin with cov are covering designs [16], the instance f5 is the "football pool problem" on five matches [17], and the instances sts are the well-known Steiner-triple systems [18]. The cov formulations have been strengthened with a number of Schöenheim inequalities, as derived by Margot [19]. All instances, save for f5, are available from Margot's web site: http://wpweb2.tepper.cmu.edu/fmargot/lpsym.html.

The computations were run on machines with AMD Opteron processors clocked at 1.8GHz and having 2GB of RAM. The COIN-OR software Clp was used to solve the linear programs at nodes of the branch and bound tree. All code was compiled with the GNU family of compilers using the flags -O3 -m32. For each instance, the (known) optimal solution value was set to aid pruning and reduce the "random" impact of finding a feasible solution in the search. Nodes were searched in a best-first fashion. When the size of the maximum orbit in the orbital partitioning is less than or equal to two, nearly all of the symmetry in the problem has

Table 1. Symmetric Integer Programs

Name	Variables
cod83	256
cod93	512
cod105	1024
cov1053	252
cov1054	2252
cov1075	120
cov1076	120
cov954	126
f5	243
sts27	27
sts45	45

been eliminated by the branching procedure, and there is little use to perform orbital branching. In this case, we use MINTO's default branching strategy. The CPU time was limited in all cases to four hours.

In order to succinctly present the results, we use performance profiles of Dolan and Moré [20]. A performance profile is a relative measure of the effectiveness of one solution method in relation to a group of solution methods on a fixed set of problem instances. A performance profile for a solution method m is essentially a plot of the probability that the performance of m (measured in this case with CPU time) on a given instance in the test suite is within a factor of β of the *best* method for that instance.

Figure 1 shows the results of an experiment designed to compare the performance of the six different orbital branching rules introduced in Sect. 4.2. In this experiment, both reduced cost fixing and orbital fixing were used. A complete table showing the number of nodes, CPU time, CPU time computing automorphism groups, the number of variables fixed by reduced cost fixing, orbital fixing, and strong branching fixing, and the deepest tree level at which orbital branching was performed is shown in the Appendix.

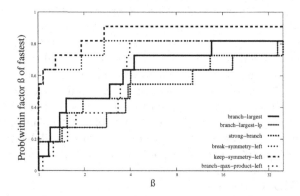

Fig. 1. Performance Profile of Branching Rules

A somewhat surprising result from the results depicted in Fig. 1 is that the most effective branching method was Rule 5, the method that keeps the symmetry group size large on the left branch. (This method gives the "highest" line in Fig. 1). The second most effective branching rule appears to be the rule that tries to *reduce* the group size by as much as possible. While these methods may not prove to be the most robust on a richer suite of difficult instances, one conclusion that we feel safe in making from this experiment is that considering *the impact* on the symmetry of the child node of the current branching decision is important. Another important observation is that for specific instances, the choice of orbit on which to branch can have a huge impact on performance. For example, for the instance cov1054, branching rules 4 and 5 both reduce the number of child nodes to 11, while other mechanisms that do not consider the impact of the branching decision on the symmetry of the child nodes cannot solve the problem in four hours of computing time.

The second experiment was aimed at measuring the impact of performing orbital fixing, as introduced in Sect. 3.1. Using branching rule 5, each instance in Table 1 was run both with and without orbital fixing. Figure 2 shows a performance profile comparing the results in the two cases. The results shows that orbital fixing has a significant positive impact.

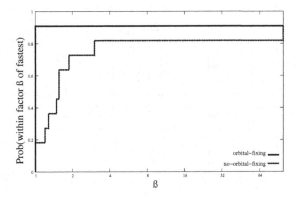

Fig. 2. Performance Profile of Impact of Orbital Fixing

The final comparison we make here is between orbital branching (with keep-symmetry-left branching), the isomorphism pruning algorithm of Margot, and the commercial solver CPLEX version 10.1, which has features for symmetry detection and handling. Table 2 summarizes the results of the comparison. The results for isomorphism pruning are taken directly from the paper of Margot using the most sophisticated of his branching rules "BC4" [10]. The paper [10] does not report results on sts27 or f5. The CPLEX results were obtained on an Intel Pentium 4 CPU clocked at 2.40GHz. Since the results were obtained on three different computer architectures and each used a different LP solver for the child subproblems, the CPU times should be interpreted appropriately.

The results show that the number of subproblems evaluated by orbital branching is smaller than isomorphism pruning in three cases, and in nearly all cases, the number of nodes is comparable. For the instance cov1076, which is not solved by orbital branching, a large majority of the CPU time is spent computing symmetry groups at each node. In a variant of orbital branching that

Table 2. Comparison of Orbital Branching, Isomorphism Pruning, and CPLEX v10.1

	Orbital Branching		Isomorphism Pruning		CPLEX v10.1	
Instance	Time	Nodes	Time	Nodes	Time	Nodes
cod83	2	25	19	33	391	32077
cod93	176	539	651	103	fail	488136
cod105	306	11	2000	15	1245	1584
cov1053	50	745	35	111	937	99145
cov1054	2	11	130	108	fail	239266
cov1075	292	377	118	169	141	10278
cov1076	fail	13707	3634	5121	fail	1179890
cov954	22	401	24	126	9	1514
f5	66	935	–	–	1150	54018
sts27	1	71	–	–	0	1647
sts45	3302	24317	31	513	24	51078

Table 3. Performance of Orbital Branching Rules on Symmetric IPs

Instance	Branching Rule	Time	Nodes	Nauty Time	# Fixed by RCF	# Fixed by OF	# Fixed by SBF	Deepest Orbital Level
cod105	Break Symmetry	305.68	11	22.88	0	1020	0	4
cod105	Keep Symmetry	306.47	11	22.92	0	1020	0	4
cod105	Branch Largest LP Solution	283.54	7	11.87	0	0	0	2
cod105	Branch Largest	283.96	9	18.01	0	0	0	3
cod105	Max Product Orbit Size	302.97	9	17.41	0	920	0	3
cod105	Strong Branch	407.14	7	11.85	0	1024	1532	2
cod83	Break Symmetry	2.35	25	1.09	44	910	0	7
cod83	Keep Symmetry	2.38	25	1.10	44	910	0	7
cod83	Branch Largest LP Solution	8.81	93	2.76	209	534	0	6
cod83	Branch Largest	10.03	113	3.41	183	806	0	14
cod83	Max Product Orbit Size	9.39	115	4.59	109	634	0	11
cod83	Strong Branch	9.44	23	0.97	27	878	394	6
cod93	Break Symmetry	175.47	529	75.15	3382	3616	0	17
cod93	Keep Symmetry	175.58	529	75.31	3382	3616	0	17
cod93	Branch Largest LP Solution	3268.89	12089	1326.26	181790	3756	0	29
cod93	Branch Largest	2385.80	8989	920.90	142351	4986	0	49
cod93	Max Product Orbit Size	587.06	2213	215.68	28035	1160	0	29
cod93	Strong Branch	2333.22	161	19.76	380	2406	13746	14
cov1053	Break Symmetry	50.28	745	27.51	0	836	0	33
cov1053	Keep Symmetry	50.31	745	27.54	0	836	0	33
cov1053	Branch Largest LP Solution	1841.41	23593	990.12	0	5170	0	71
cov1053	Branch Largest	148.37	2051	70.73	0	1504	0	36
cov1053	Max Product Orbit Size	192.18	2659	91.72	0	1646	0	68
cov1053	Strong Branch	1998.55	1455	53.96	0	5484	34208	54
cov1054	Break Symmetry	1.77	11	0.85	0	186	0	4
cov1054	Keep Symmetry	1.76	11	0.85	0	186	0	4
cov1054	Branch Largest LP Solution	14400	54448	7600.80	0	814	0	35
cov1054	Branch Largest	14400	54403	7533.80	0	1452	0	49
cov1054	Max Product Orbit Size	14400	52782	7532.77	0	1410	0	38
cov1054	Strong Branch	14400	621	87.76	0	204	4928	32
cov1075	Break Symmetry	14400	9387	13752.11	37121	0	0	2
cov1075	Keep Symmetry	291.85	377	268.45	379	926	0	15
cov1075	Branch Largest LP Solution	906.48	739	861.57	1632	716	0	23
cov1075	Branch Largest	268.49	267	248.45	793	1008	0	13
cov1075	Max Product Orbit Size	395.11	431	366.24	1060	1066	0	21
cov1075	Strong Branch	223.53	67	60.71	106	128	1838	10
cov1076	Break Symmetry	14400	8381	13853.35	2	0	0	3
cov1076	Keep Symmetry	14400	13707	13818.47	11271	1564	0	26
cov1076	Branch Largest LP Solution	14400	6481	13992.74	10	116	0	14
cov1076	Branch Largest	14400	6622	13988.71	0	176	0	13
cov1076	Max Product Orbit Size	14400	6893	13967.86	71	580	0	14
cov1076	Strong Branch	14400	1581	3255.74	5	164	58	23
cov954	Break Symmetry	21.72	401	14.81	570	1308	0	14
cov954	Keep Symmetry	21.70	401	14.83	570	1308	0	14
cov954	Branch Largest LP Solution	11.30	175	7.03	498	48	0	5
cov954	Branch Largest	15.69	265	10.51	671	212	0	12
cov954	Max Product Orbit Size	14.20	229	9.25	602	212	0	11
cov954	Strong Branch	17.55	45	1.74	50	100	1084	8
f5	Break Symmetry	65.86	935	23.25	2930	2938	0	17
f5	Keep Symmetry	65.84	935	23.26	2930	2938	0	17
f5	Branch Largest LP Solution	91.32	1431	28.95	7395	272	0	8
f5	Branch Largest	100.66	1685	30.75	7078	434	0	11
f5	Max Product Orbit Size	102.54	1691	30.96	7230	430	0	13
f5	Strong Branch	671.51	123	2.59	187	760	8586	15
sts27	Break Symmetry	0.84	71	0.71	0	8	0	10
sts27	Keep Symmetry	0.83	71	0.71	0	8	0	10
sts27	Branch Largest LP Solution	2.33	115	2.12	3	86	0	14
sts27	Branch Largest	0.97	73	0.83	1	28	0	13
sts27	Max Product Orbit Size	2.88	399	2.42	1	888	0	11
sts27	Strong Branch	1.63	75	1.15	2	76	0	14
sts45	Break Symmetry	3302.70	24317	3230.12	12	0	0	4
sts45	Keep Symmetry	3301.81	24317	3229.88	12	0	0	4
sts45	Branch Largest LP Solution	4727.29	36583	4618.66	25	0	0	2
sts45	Branch Largest	4389.80	33675	4289.45	36	0	0	2
sts45	Max Product Orbit Size	4390.39	33675	4289.79	36	0	0	2
sts45	Strong Branch	1214.04	7517	884.79	2	144	45128	21

uses a symmetry group that is smaller but much more efficient to compute (and which space prohibits us from describing in detail here), cov1076 can be solved in 679 seconds and 14465 nodes. Since in any optimal solution to the Steiner triple systems, more than 2/3 of the variables will be set to 1, orbital branching would be much more efficient if all variables were complemented, or equivalently if the orbital branching dichotomy (2) was replaced by its complement. Margot [10] also makes a similar observation, and his results are based on using the complemented instances, which may account for the large gap in performance of the two methods on sts45. We are currently instrumenting our code to deal with instances for which the number of ones in an optimal solution is larger than $1/2$. Orbital branching proves to be faster than CPLEX in six cases, while in all cases the number of evaluated nodes is remarkably smaller.

6   Conclusion

In this work, we presented a simple way to capture and exploit the symmetry of an integer program when branching. We showed through a suite of experiments that the new method, orbital branching, outperforms state-of-the-art solvers when a high degree of symmetry is present. In terms of reducing the size of the search tree, orbital branching seems to be of comparable quality to the isomorphism pruning method of Margot [10]. Further, we feel that the simplicity and flexibility of orbital branching make it an attractive candidate for further study. Continuing research includes techniques for further reducing the number of isomorphic nodes that are evaluated and on developing branching mechanisms that combine the child bound improvement and change in symmetry in a meaningful way.

Acknowledgments

The authors would like to thank Kurt Anstreicher and François Margot for inspiring and insightful comments on this work. In particular, the name *orbital branching* was suggested by Kurt. Author Linderoth would like to acknowledge support from the US National Science Foundation (NSF) under grant DMI-0522796, by the US Department of Energy under grant DE-FG02-05ER25694, and by IBM, through the faculty partnership program. Author Ostrowski is supported by the NSF through the IGERT Grant DGE-9972780.

References

1. Barnhart, C., Johnson, E.L., Nemhauser, G.L., Savelsbergh, M.W.P., Vance, P.H.: Branch and Price: Column generation for solving huge integer programs. Operations Research **46** (1998) 316–329
2. Holm, S., Sørensen, M.: The optimal graph partitioning problem: Solution method based on reducing symmetric nature and combinatorial cuts. OR Spectrum **15** (1993) 1–8
3. Méndez-Díaz, I., Zabala, P.: A branch-and-cut algorithm for graph coloring. Discrete Applied Mathematics **154**(5) (2006) 826–847
4. Macambira, E.M., Maculan, N., de Souza, C.C.: Reducing symmetry of the SONET ring assignment problem using hierarchical inequalities. Technical Report ES-636/04, Programa de Engenharia de Sistemas e Computação, Universidade Federal do Rio de Janeiro (2004)
5. Rothberg, E.: Using cuts to remove symmetry. Presented at the 17th International Symposium on Mathematical Programming
6. Sherali, H.D., Smith, J.C.: Improving zero-one model representations via symmetry considerations. Management Science **47**(10) (2001) 1396–1407

7. Kaibel, V., Pfetsch, M.: Packing and partitioning orbitopes. Mathemathical Programming (2007) To appear.
8. Kaibel, V., Peinhardt, M., Pfetsch, M.: Orbitopal fixing. In: IPCO 2007: The Twelfth Conference on Integer Programming and Combinatorial Optimization, Springer (2007) To appear.
9. Margot, F.: Pruning by isomorphism in branch-and-cut. Mathematical Programming **94** (2002) 71–90
10. Margot, F.: Exploiting orbits in symmetric ILP. Mathematical Programming, Series B **98** (2003) 3–21
11. Bazaraa, M.S., Kirca, O.: A branch-and-bound heuristic for solving the quadratic assignment problem. Naval Research Logistics Quarterly **30** (1983) 287–304
12. Nemhauser, G.L., Savelsbergh, M.W.P., Sigismondi, G.C.: MINTO, a Mixed INTeger Optimizer. Operations Research Letters **15** (1994) 47–58
13. McKay, B.D.: Nauty User's Guide (Version 1.5). Australian National University, Canberra. (2002)
14. Foggia, P., Sansone, C., Vento, M.: A preformance comparison of five algorithms for graph isomorphism. Proc. 3rd IAPR-TC15 Workshop Graph-Based Representations in Pattern Recognition (2001) 188–199
15. Litsyn, S.: An updated table of the best binary codes known. In Pless, V.S., Huffman, W.C., eds.: Handbook of Coding Theory. Volume 1. Elsevier, Amsterdam (1998) 463–498
16. Mills, W.H., Mullin, R.C.: Coverings and packings. In: Contemporary Design Theory: A Collection of Surveys. Wiley (1992) 371–399
17. Hamalainen, H., Honkala, I., Litsyn, S., Östergård, P.: Football pools—A game for mathematicians. American Mathematical Monthly **102** (1995) 579–588
18. Fulkerson, D.R., Nemhauser, G.L., Trotter, L.E.: Two computationally difficult set covering problems that arise in computing the 1-width of incidence matrices of Steiner triples. Mathematical Programming Study **2** (1973) 72–81
19. Margot, F.: Small covering designs by branch-and-cut. Mathematical Programming **94** (2003) 207–220
20. Dolan, E., Moré, J.: Benchmarking optimization software with performance profiles. Mathematical Programming **91** (2002) 201–213

Distinct Triangle Areas in a Planar Point Set

Adrian Dumitrescu[1,*] and Csaba D. Tóth[2]

[1] Deptartment of Computer Science, University of Wisconsin-Milwaukee, WI
53201-0784, USA
ad@cs.uwm.edu
[2] Department of Mathematics, MIT, Cambridge, MA 02139, USA
toth@math.mit.edu

Abstract. Erdős, Purdy, and Straus conjectured that the number of distinct (nonzero) areas of the triangles determined by n noncollinear points in the plane is at least $\lfloor \frac{n-1}{2} \rfloor$, which is attained for $\lceil n/2 \rceil$ and respectively $\lfloor n/2 \rfloor$ equally spaced points lying on two parallel lines. We show that this number is at least $\frac{17}{38}n - O(1) \approx 0.4473n$. The best previous bound, $(\sqrt{2} - 1)n - O(1) \approx 0.4142n$, which dates back to 1982, follows from the combination of a result of Burton and Purdy [5] and Ungar's theorem [23] on the number of distinct directions determined by n noncollinear points in the plane.

1   Introduction

Let S be a finite set of points in the plane. Consider the (nondegenerate) triangles determined by triples of points of S. There are at most $\binom{n}{3}$ triangles, some of which may have the same area. Denote by $g(S)$ the number of *distinct (nonzero) areas* of the triangles determined by S. For every $n \in \mathbb{N}$, let $g(n)$ be the minimum of $g(S)$ over all sets S of n noncollinear points in the plane. The problem of finding $g(n)$ has a long history; the attention it has received is perhaps due to its simplicity and elegance, as well as to its connections to another fundamental problem in combinatorial geometry—that of finding the minimum number of directions spanned by n points in the plane. The problem of *distinct areas* is also similar in nature to a notoriously hard problem of *distinct distances*. It is listed for instance in the problem collection by Croft, Falconer, and Guy [6], and more recently by Braß, Moser, and Pach [3]; see also [12].

The first estimates on $g(n)$ were given in 1976 by Erdős and Purdy [10], who proved that

$$c_1 n^{3/4} \leq g(n) \leq c_2 n,$$

for some absolute constants $c_1, c_2 > 0$. The upper bound follows easily if we consider the points $(i, j) \in \mathbb{N}^2$ for $1 \leq i, j \leq \sqrt{n}$ and observe that every triangle area is a multiple of $\frac{1}{2}$ and bounded by $n/2$. A simple construction that consists of two sets of $\lceil n/2 \rceil$ and respectively $\lfloor n/2 \rfloor$ equally spaced points lying on two

---

* Supported in part by NSF CAREER grant CCF-0444188.

M. Fischetti and D.P. Williamson (Eds.): IPCO 2007, LNCS 4513, pp. 119–129, 2007.
© Springer-Verlag Berlin Heidelberg 2007

parallel lines was found by Burton and Purdy [5], and also by Straus [21]: It gives $\lfloor \frac{n-1}{2} \rfloor$ triangles of distinct areas.

In 1979, Burton and Purdy [5] obtained a linear lower bound, which follows from a linear bound on the number of directions determined by n noncollinear points in the plane. More precisely, denoting by $f(n)$ the minimum number of directions determined by n noncollinear points in the plane, they showed that

$$\left\lfloor \frac{n}{2} \right\rfloor \leq f(n) \leq 2 \left\lfloor \frac{n}{2} \right\rfloor.$$

Using this result, an averaging argument of Burton and Purdy gave

$$0.32n \leq g(n) \leq \left\lfloor \frac{n-1}{2} \right\rfloor.$$

In 1982, Ungar proved a sharp bound

$$f(n) = 2 \left\lfloor \frac{n}{2} \right\rfloor \tag{1}$$

on the minimum number of directions determined by n noncollinear points, using a purely combinatorial approach of *allowable sequences* devised by Goodman and Pollack [14,15]. A combination of Burton and Purdy's argument [5] with Ungar's theorem [23] immediately gives

$$(\sqrt{2} - 1)n - O(1) \leq g(n) \leq \left\lfloor \frac{n-1}{2} \right\rfloor.$$

In this paper, we refine Burton and Purdy's averaging argument by applying yet one more time (and perhaps not for the last time) Ungar's technique on allowable sequences, and further improve the lower bound on distinct triangle areas.

Theorem 1. *The number of triangles of distinct areas determined by n noncollinear points in the plane is at least*

$$g(n) \geq \frac{17}{38}n - O(1) \approx 0.4473n.$$

In fact, we prove Theorem 1 in a stronger form: There are at least $17n/38 - O(1)$ triangles of distinct areas having *a common side*, in other words there are at least this many points of our set at distinct distances from the line determined by a pair of points in the set. One can draw here a parallel with the problem of distinct distances raised by Erdős in 1946: What is the minimum number of distinct distances $t(n)$ determined by n points in the plane? Erdős conjectured that $t(n) = \Omega(n/\sqrt{\log n})$, and moreover, that there is a point in the set which determines this many distinct distances to other points. In a sequence of recent breakthrough developments since 1997, all new lower bounds on $t(n)$ due to Székely [22], Solymosi and C. Tóth [20], and including the current best one due to Katz and Tardos [16], in fact give lower bounds on the maximum number of inter-point distances measured *from a single point*. For triangles areas in the plane, we have a similar phenomenon: By the argument of Burton and Purdy [5], every set S of n noncollinear points in the plane contains two distinct points $p, q \in S$ such that the points of S determine $\Omega(n)$ distinct distances to the line pq, therefore at least this many triangles with distinct areas. As mentioned above, our bound holds also in this stronger sense. A similar example is that of tetrahedra of distinct volumes determined by a set of n points in $\mathbb{R}^3$ (not all in the same plane): we have recently shown [8] that n points determine $\Omega(n)$ tetrahedra of distinct volumes, which share a common side. One exception to this phenomenon is the problem of distinct distances among vertices of a convex polygon, as the results of [1,2,7] show (see also [3]).

2   Proof of Theorem 1

Burton and Purdy's idea. We first review Burton and Purdy's argument [5]. Let S be a set of n noncollinear points in the plane, and let L denote the set of connecting lines (i.e., lines incident to at least 2 points of S). We may assume w.l.o.g. that there is no horizontal line in L. For a line $\ell \in L$, let $\ell_1, \ell_2, \ldots, \ell_r \in L$ be all connecting lines parallel to ℓ (including ℓ) such that ℓ_i lies to the left of ℓ_{i+1} for $1 \leq i < r$. Let $k_i \geq 2$ denote the number of points along $\ell_i \in L$ for $i = 1, \ldots, r$. Let s be the number of *singleton* points of S not covered by any of $\ell_1, \ldots, \ell_r$. We clearly have $\sum_{i=1}^{r} k_i + s = n$. Taking any two points $p, q \in S$ on ℓ_1 or on ℓ_r, the triangles Δpqz_i have different areas for at least $r + \lceil s/2 \rceil - 1$ indices i, where z_i are either singleton points or points on different connecting lines lying all on the same side of pq. Therefore the number m of distinct areas satisfies

$$m \geq r + \lceil s/2 \rceil - 1.$$

The next step is selecting a suitable direction of connecting lines, more precisely, one with a small number of pairs of points, i.e., with a small value of $\sum_{i=1}^{r} \binom{k_i}{2}$. By Ungar's theorem, there is a direction corresponding to the lines $\ell_1, \ldots, \ell_r$, such that

$$\sum_{i=1}^{r} \binom{k_i}{2} \leq \binom{n}{2} \Big/ (n-1) = \frac{n}{2}.$$

After observing that $\sum_{i=1}^{r} \binom{k_i}{2}$ is minimal if the points on these r connecting lines are distributed as evenly as possible, Burton and Purdy derive a quadratic equation whose solution gives (using Ungar's theorem instead of their weaker bound of $\lfloor n/2 \rfloor$ on the number of directions) a lower bound of $m \geq (\sqrt{2} - 1)n - O(1) \approx 0.4142n$ on the number of distinct triangle areas. Detailed calculations show that a configuration attaining the Burton-Purdy bound should have $2 + \sqrt{2}$ points on each connecting line parallel to the certain direction (determined by at most $n/2$ pairs of points), a value which is certainly infeasible.

A tiny improvement. We first formulate a system of linear inequalities (the linear program (LP1) below). Unlike Burton and Purdy's quadratic equation, our linear program imposes an integrality condition on the number of points on each connecting line parallel to a specified direction; which leads to a tiny improvement ($5/12$ versus $\sqrt{2}-1$). More important, our linear system paves the way for a more substantial improvement obtained by two linear programs with additional constraints (to be described later).

Assume that the connecting lines $\ell_1, \ell_2 \ldots, \ell_r \in L$ are vertical and contain at most $n/2$ point pairs (by Ungar's theorem). Every vertical line of L (passing through at least two points) is called a *regular line*. A regular line passing through *exactly* k points ($k \geq 2$) is called a k-line. We call a vertical line passing through exactly one point of S a *singleton line*.

Partition the n points of S as follows. Let s be a real number $0 \leq s < 1$ such that there are sn singleton points to the left of the leftmost regular line ℓ_1. Similarly, let tn be the number of singleton points to the right of ℓ_r, and let $a_1 n$ be the number of remaining singleton points. (See Figure 1.) For $k = 2, 3, \ldots, 8$, let $a_k n$ be the number of points on k-lines. Finally denote by a_9 the total number of points on regular lines with at least 9 points each. We have accounted for all points of S, hence we have

$$s + t + \sum_{k=1}^{9} a_k = 1.$$

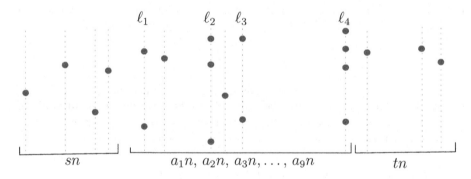

Fig. 1. The orthogonal projection of a point set S in a direction determined by S

Let $xn = \sum_{i=1}^{r} \binom{k_i}{2}$ be the total number of point pairs on vertical lines. Let en denote the number of distinct horizontal distances measured from the leftmost regular line ℓ_1 to its right: Consequently, there are en triangles with distinct areas having a common side along the leftmost regular line. Similarly, let fn denote the number of distinct horizontal distances measured from the rightmost regular line ℓ_r to its left. We can deduce lower bounds on e and f: Since $en \geq tn + a_1 n + a_2 n/2 + a_3 n/3 + \ldots + a_8 n/8 - 1$, we have $e \geq t + a_1 + a_2/2 + a_3/3 + \ldots + a_8/8 - 1/n$, and similarly, $f \geq s + a_1 + a_2/2 + a_3/3 + \ldots + a_8/8 - 1/n$. We can also give a lower bound for x in terms of the previous parameters. We have

$$x \geq \frac{1}{2}a_2 + \frac{2}{2}a_3 + \frac{3}{2}a_4 + \ldots + \frac{8}{2}a_9,$$

since if there are $a_k n$ points on k-lines, then the number of k-lines is $a_k n/k$, and each k-line contains $\binom{k}{2}$ vertical point pairs. Hence, there are $a_k n\binom{k}{2}/k = a_k n(k-1)/2$ pairs of points on k-lines, $k = 2, 3, \ldots, 8$. Similarly there are at least $\frac{8}{2}a_9 n$ pairs of points on lines incident to at least 9 points. Putting all of these equations and inequalities together, we formulate the following linear program.

minimize r $\hspace{3cm}$ (LP1)

subject to $x \leq 0.5$;

$$
\begin{cases}
s + t + a_1 + a_2 + a_3 + a_4 + a_5 + a_6 + a_7 + a_8 + a_9 = 1; \\
\frac{1}{2}a_2 + a_3 + \frac{3}{2}a_4 + 2a_5 + \frac{5}{2}a_6 + 3a_7 + \frac{7}{2}a_8 + 4a_9 \leq x; \\
t + a_1 + \frac{1}{2}a_2 + \frac{1}{3}a_3 + \frac{1}{4}a_4 + \frac{1}{5}a_5 + \frac{1}{6}a_6 + \frac{1}{7}a_7 + \frac{1}{8}a_8 - \frac{1}{n} \leq e; \\
s + a_1 + \frac{1}{2}a_2 + \frac{1}{3}a_3 + \frac{1}{4}a_4 + \frac{1}{5}a_5 + \frac{1}{6}a_6 + \frac{1}{7}a_7 + \frac{1}{8}a_8 - \frac{1}{n} \leq f; \\
e \leq r; \\
f \leq r;
\end{cases}
$$

$$s, t, a_1, a_2, a_3, a_4, a_5, a_6, a_7, a_8, a_9, e, f, r, x \geq 0;$$

The linear system (LP1) does not describe completely a point configuration (e.g., we do not make any distinction among k-lines for $k \geq 9$), but all these inequalities must hold if the variables correspond to a point set S. Let (LP1') be the linear program obtained from (LP1) by removing the two terms $\frac{1}{n}$, and let r be its solution. Since the constraints are linear, the term $\frac{1}{n}$ can only contribute a constant additive blow-up in the LP solution. That is, if r is the solution of (LP1'), the solution of (LP1) is $r - O(1/n)$. We can deduce that there are at least $rn - O(1)$ distinct triangle areas with a common side on either ℓ_1 or ℓ_r.

A solution to (LP1') is $r = 5/12 \approx 0.4166$, attained for $s = t = 1/4$, $a_3 = 1/2$, $a_1 = a_2 = a_4 = a_5 = a_6 = a_7 = a_8 = a_9 = 0$, $e = f = 5/12$, and $x = 1/2$. That is, there are $n/6$ 3-lines in the middle, and $n/4$ singleton lines on each side, and $5n/12 - O(1)$ distinct areas measured from left or right. Another optimal solution that looks similar consists of $n/12$ 4-lines in the middle, and $n/3$ singleton lines on each side, for which the number of distinct areas is also $5n/12 - O(1)$.

Allowable sequences. We now give a very brief account on Ungar's technique (following [23]) and allowable sequences [12], as they are relevant to our proof. Allowable sequences occur in the context of transforming the permutation $1, 2, \ldots, n$ into the reverse permutation $n, n - 1, \ldots, 1$ by going through a sequence of permutations. The operation between two consecutive permutations, called *move*, consists of inverting pairwise disjoint increasing strings. In a geometric context, each symbol corresponds to a point in the plane; each permutation is the left-to-right order in an orthogonal projections of the points on a directed line. The directed line is rotated around the origin, and a move occurs when the normal of this line coincides with a direction of a connecting line (a line in L). An example of a sequence arising in this way is 1(23)4(56), 13(246)5, (136)425, 63(14)25, 6(34)(125), 64(35)21, 6(45)321, and 654321. We have put parentheses around the increasing string (called *blocks*) reversed at the next move. So each permutation with the blocks enclosed in parentheses describes also the next move.

Ungar's theorem states that for even n, going from $1, 2, \ldots, n$ to $n, n-1, \ldots, 1$ but *not* in one move, requires at least n moves (in other words, if every block reversed has fewer than n elements, at least n moves are needed). The general idea in the proof is that building up a long increasing block involves many moves required by dismantling other (possibly long) decreasing blocks formed at earlier moves, and vice versa. More precisely, the moves have the following properties.

(I) In one move, a decreasing string can get shorter by at most one element at each end.

(II) in one move, an increasing string can get longer by at most one element at each end.

For instance, the reason for (I) is that a move reverses increasing strings, and so only the first and the last elements of a decreasing string can be part of a block in a move. We refer the reader to [23] for more details. Properties (I) and (II) further imply that if a block B of size at least 3 is reversed in one move, then all but the two extreme elements of B must be singletons in the next move. Analogously, if a block B of size at least 3 is reversed in a move, then at least one of its elements is a singleton in the previous move.

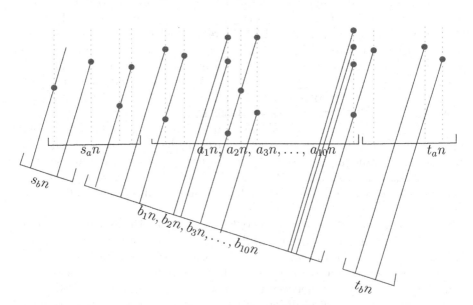

Fig. 2. The orthogonal projection of a point set S in two consecutive directions, a and b, determined by S

New bound. The idea for our new bound is the following. Recall that two optimal solutions of (LP1') we have seen have a similar structure: (A) $n/6$ 3-lines in the middle, and $n/4$ singleton lines on each side, or (B) $n/12$ 4-lines in the middle, and $n/3$ singleton lines on each side. Assume that there are two consecutive moves, π_1 and π_2, in an allowable sequence such that both look like (A) or (B). Notice that our observations regarding the blocks of size at least 3 imply that there cannot be two consecutive such moves, since the first move would force many singletons in the middle segment of π_2 (at least one for each block of π_1). This suggests that one of two consecutive directions of L must give a configuration where the solution of (LP1') is above $5/12$. We follow with the precise technical details in the proof of Theorem 1.

By Ungar's theorem, the average number of pairs determining the same direction is at most $n/2$, so there are two consecutive moves (corresponding to two consecutive directions of lines in L) parallel to at most n pairs of points. We introduce a similar notation as above for a single direction, but we distinguish the notation by indices a and b, respectively (e.g., $s_a n$ and $s_b n$ are the number of points which give singletons at the left side of the first and the second permutation, respectively). This time we count up to 9-lines (rather than 8-lines) and group together the k-lines for $k \geq 10$. We denote by $a_{10} n$ and $b_{10} n$ the total number of points on lines with at least 10 points each. By symmetry, we need to consider only two cases (instead of the four combinations of $s_a \lesseqgtr s_b$ and $t_a \lesseqgtr t_b$).

Case (i): $s_b \leq s_a$ and $t_b \leq t_a$.
Case (ii): $s_a \leq s_b$ and $t_b \geq t_a$.

We are lead to minimizing the following two linear programs (LP2i) and (LP2ii), where (LP2i) corresponds to Case (i) and (LP2ii) corresponds to Case (ii).

Case (i): $s_b \leq s_a$ and $t_b \leq t_a$. We formulate the linear program (LP2i) as follows. We repeat the constraints of (LP1) for both moves, and impose the constraint $x_a + x_b \leq 1$ since the total number of pairs for the two consecutive directions is at most n. We introduce two linear constraints to express $r = \max(r_a, r_b)$. Constraints (α) and (β) are crucial: Constraint (α) indicates that if in the first move, a block B of size at least 3 is reversed, then all but the two extreme elements of B must be singletons in the next move; constraint (β) specifies that each block B of size at least 3 which is reversed in the second move must contain an element which is a singleton in the first move (with the possible exception of two blocks that lie on the boundary of the singletons s_a and t_a).

Here is an example regarding constraint (β). Let π_1 and π_2 denote the two consecutive moves (each represented by pairwise disjoint blocks). The prefixes (resp., suffixes) of length s_b (resp., t_b) coincide, and are made of singletons. So each block of size at least 3 in the second move in between these common prefix and suffix strings (to be reversed in the second move) must pick up at least a singleton in a_1 from π_1 or must be made entirely up of singletons in the $(s_a - s_b)$ and $(t_a - t_b)$ segments of π_1 (except for at most two blocks crossing segment borders). For instance, if a move transforms permutation $\pi_1 = \ldots (47)(359) \ldots$ to $\pi_1' = \ldots 74953 \ldots$, then no triple (or other longer block) may be formed in the next move. But if there was a singleton in between, like in $\pi_1 = \ldots (47)6(359)\ldots$, then a triple may be formed in the next move: For instance, $\pi_2 = \ldots 7(469)53\ldots$.

minimize r (LP2i)

subject to $s_b \leq s_a$;

$t_b \leq t_a$;

$$(LP1)_a \begin{cases} s_a + t_a + a_1 + a_2 + a_3 + a_4 + a_5 + a_6 + a_7 + a_8 + a_9 + a_{10} = 1; \\ \frac{1}{2}a_2 + a_3 + \frac{3}{2}a_4 + 2a_5 + \frac{5}{2}a_6 + 3a_7 + \frac{7}{2}a_8 + 4a_9 + \frac{9}{2}a_{10} \leq x_a; \\ t_a + a_1 + \frac{a_2}{2} + \frac{a_3}{3} + \frac{a_4}{4} + \frac{a_5}{5} + \frac{a_6}{6} + \frac{a_7}{7} + \frac{a_8}{8} + \frac{a_9}{9} - \frac{1}{n} \leq e_a; \\ s_a + a_1 + \frac{a_2}{2} + \frac{a_3}{3} + \frac{a_4}{4} + \frac{a_5}{5} + \frac{a_6}{6} + \frac{a_7}{7} + \frac{a_8}{8} + \frac{a_9}{9} - \frac{1}{n} \leq f_a; \\ e_a \leq r_a; \\ f_a \leq r_a; \end{cases}$$

$$(LP1)_b \begin{cases} s_b + t_b + b_1 + b_2 + b_3 + b_4 + b_5 + b_6 + b_7 + b_8 + b_9 + b_{10} = 1; \\ \frac{1}{2}b_2 + b_3 + \frac{3}{2}b_4 + 2b_5 + \frac{5}{2}b_6 + 3b_7 + \frac{7}{2}b_8 + 4b_9 + \frac{9}{2}b_{10} \leq x_b; \\ t_b + b_1 + \frac{b_2}{2} + \frac{b_3}{3} + \frac{b_4}{4} + \frac{b_5}{5} + \frac{b_6}{6} + \frac{b_7}{7} + \frac{b_8}{8} + \frac{b_9}{9} - \frac{1}{n} \leq e_b; \\ s_b + b_1 + \frac{b_2}{2} + \frac{b_3}{3} + \frac{b_4}{4} + \frac{b_5}{5} + \frac{b_6}{6} + \frac{b_7}{7} + \frac{b_8}{8} + \frac{b_9}{9} - \frac{1}{n} \leq f_b; \\ e_b \leq r_b; \\ f_b \leq r_b; \end{cases}$$

$x_a + x_b \leq 1$;

$r_a \leq r$;

$r_b \leq r$;

$(\alpha)\quad \frac{1}{3}a_3 + \frac{2}{4}a_4 + \frac{3}{5}a_5 + \frac{4}{6}a_6 + \frac{5}{7}a_7 + \frac{6}{8}a_8 + \frac{7}{9}a_9 + \frac{8}{10}a_{10} \leq b_1$;

$(\beta)\quad b_3 + b_4 + b_5 + b_6 + b_7 + b_8 + b_9 + b_{10} - \frac{2}{n} \leq 3a_1 + s_a - s_b + t_a - t_b$;

$s_a, t_a, a_1, a_2, a_3, a_4, a_5, a_6, a_7, a_8, a_9, a_{10}, e_a, f_a, r_a, x_a \geq 0$;

$s_b, t_b, b_1, b_2, b_3, b_4, b_5, b_6, b_7, b_8, b_9, b_{10}, e_b, f_b, r_b, x_b \geq 0$;

$r \geq 0$;

When we ignore the terms $O(\frac{1}{n})$, we get a new system (LP2i') with the following solution: $r = 17/38 \approx 0.4473$, attained for $s_a = t_a = 15/38$, $a_1 = a_2 = a_3 = 0$, $a_4 = 4/19$, $a_5 = a_6 = a_7 = a_8 = a_9 = a_{10} = 0$ for the first permutation, and $s_b = t_b = 3/38$, $b_1 = b_2 = 2/19$, $b_3 = 12/19$, $b_4 = b_5 = b_6 = b_7 = b_8 = b_9 = b_{10} = 0$ for the second permutation; also $x_a = 6/19$, $x_b = 13/19$, $e_a = f_a = r_a = e_b = f_b = r_b = 17/38$.

Case (ii): $s_b \leq s_a$ and $t_b \geq t_a$. The linear program (LP2ii) is very similar to (LP2i). Besides the first two constraints, which are specific to this case, only constraints (γ) and (δ) are different: Constraint (γ) specifies that each block B of size at least 3 which is reversed in the second move must contain at least one singleton in the first move; constraint (δ) specifies the same thing when going back from the second permutation to the first one (by time reversibility).

$$\text{minimize } r \qquad\qquad\qquad\text{(LP2ii)}$$

subject to $s_b \le s_a;$

$\qquad\qquad t_a \le t_b;$

$(\text{LP1})_a \begin{cases} s_a + t_a + a_1 + a_2 + a_3 + a_4 + a_5 + a_6 + a_7 + a_8 + a_9 + a_{10} = 1; \\ \frac{1}{2}a_2 + a_3 + \frac{3}{2}a_4 + 2a_5 + \frac{5}{2}a_6 + 3a_7 + \frac{7}{2}a_8 + 4a_9 + \frac{9}{2}a_{10} \le x_a; \\ t_a + a_1 + \frac{a_2}{2} + \frac{a_3}{3} + \frac{a_4}{4} + \frac{a_5}{5} + \frac{a_6}{6} + \frac{a_7}{7} + \frac{a_8}{8} + \frac{a_9}{9} - \frac{1}{n} \le e_a; \\ s_a + a_1 + \frac{a_2}{2} + \frac{a_3}{3} + \frac{a_4}{4} + \frac{a_5}{5} + \frac{a_6}{6} + \frac{a_7}{7} + \frac{a_8}{8} + \frac{a_9}{9} - \frac{1}{n} \le f_a; \\ e_a \le r_a; \\ f_a \le r_a; \end{cases}$

$(\text{LP1})_b \begin{cases} s_b + t_b + b_1 + b_2 + b_3 + b_4 + b_5 + b_6 + b_7 + b_8 + b_9 + b_{10} = 1; \\ \frac{1}{2}b_2 + b_3 + \frac{3}{2}b_4 + 2b_5 + \frac{5}{2}b_6 + 3b_7 + \frac{7}{2}b_8 + 4b_9 + \frac{9}{2}b_{10} \le x_b; \\ t_b + b_1 + \frac{b_2}{2} + \frac{b_3}{3} + \frac{b_4}{4} + \frac{b_5}{5} + \frac{b_6}{6} + \frac{b_7}{7} + \frac{b_8}{8} + \frac{b_9}{9} - \frac{1}{n} \le e_b; \\ s_b + b_1 + \frac{b_2}{2} + \frac{b_3}{3} + \frac{b_4}{4} + \frac{b_5}{5} + \frac{b_6}{6} + \frac{b_7}{7} + \frac{b_8}{8} + \frac{b_9}{9} - \frac{1}{n} \le f_b; \\ e_b \le r_b; \\ f_b \le r_b; \end{cases}$

$\qquad\qquad x_a + x_b \le 1;$

$\qquad\qquad r_a \le r;$

$\qquad\qquad r_b \le r;$

$(\gamma) \quad a_3 + a_4 + a_5 + a_6 + a_7 + a_8 + a_9 + a_{10} - \frac{1}{n} \le 3b_1 + t_b - t_a;$

$(\delta) \quad b_3 + b_4 + b_5 + b_6 + b_7 + b_8 + b_9 + b_{10} - \frac{1}{n} \le 3a_1 + s_a - s_b;$

$\qquad\qquad s_a, t_a, a_1, a_2, a_3, a_4, a_5, a_6, a_7, a_8, a_9, a_{10}, e_a, f_a, r_a, x_a \ge 0;$

$\qquad\qquad s_b, t_b, b_1, b_2, b_3, b_4, b_5, b_6, b_7, b_8, b_9, b_{10}, e_b, f_b, r_b, x_b \ge 0;$

$\qquad\qquad r \ge 0;$

When we ignore the terms $O(\frac{1}{n})$, we get a new system (LP2ii') with the following solution: $r = 25/54 \approx 0.4629$, attained for $s_a = t_a = 23/54$, $a_1 = 1/27$, $a_2 = a_3 = a_4 = a_5 = a_6 = a_7 = a_8 = a_9 = 0$, $a_{10} = 1/9$, for the first permutation, and $s_b = t_b = 23/54$, $b_1 = 1/27$, $b_2 = b_3 = b_4 = b_5 = b_6 = b_7 = b_8 = b_9 = 0$, $b_{10} = 1/9$, for the second permutation; also $x_a = 1/2$, $x_b = 1/2$, $e_a = f_a = r_a = e_b = f_b = r_b = 25/54$.

Since the solution of (LP2i') is smaller than that of (LP2ii'), i.e., $17/38 < 25/54$, we conclude that there are always $\frac{17}{38}n - O(1) \approx 0.4473n$ triangles of distinct areas.

One may ask if the same result can be obtained using fewer variables in the LPs, or whether a better result can be obtained by increasing the number of variables in the LPs. The answer to both questions is negative.

3   Remarks

In 1982, Erdős, Purdy, and Straus [13] considered the generalization of the problem of distinct triangle areas to higher dimensions and posed the following:

Problem (Erdős, Purdy, and Straus). Let S be a set of n points in $\mathbb{R}^d$ not all in one hyperplane. What is the minimal number $g_d(n)$ of distinct volumes of nondegenerate simplices with vertices in S?

By taking d sets of about n/d equally spaced points on parallel lines through the vertices of a $(d-1)$-simplex, one gets $g_d(n) \leq \lfloor \frac{n-1}{d} \rfloor$. Erdős, Purdy, and Straus conjectured that equality holds at least for sufficiently large n (see also [6]). The first development in this old problem for higher dimensions is only very recent: for $d = 3$ we have shown that the tetrahedra determined by n points in $\mathbb{R}^3$, not all in a plane, have at least $\Omega(n)$ distinct volumes, which thereby confirms the conjecture in 3-space apart from the multiplicative constant [8].

We conclude with two problems on distinct triangle areas. The former is directly related to the original problem of distinct areas studied here, and appears to have been first raised by Erdős and Pach in the 1980s [17], while the latter appears to be new.

Given a planar point set S, consider the set L of connecting lines. A connecting line is called an *ordinary line* if it passes through exactly two points of S. By the well known Sylvester-Gallai theorem [18,3], any finite set of noncollinear points in the plane determines an ordinary line. Consider now the set Θ of directions of lines in L. A direction $\theta \in \Theta$ is called an *ordinary direction* if all connecting lines of direction θ are ordinary lines.

Problem 1. Let S be a set of n noncollinear points in the plane. Is it true that apart from a finite set of values of n, Θ always contains an ordinary direction?

It should be clear that such a direction would be enough to prove the Erdős-Purdy-Strauss conjecture that S determines at least $\lfloor (n-1)/2 \rfloor$ distinct (nonzero) triangle areas — apart from a finite set of exceptions for n. Observe that $n = 7$ is such an exception, since the configuration of 7 points given by the three vertices of a triangle, the midpoints of its three sides, and the triangle center admits no ordinary direction.

Problem 2. Let S be a set of n noncollinear points in the plane. Is it true that each point $p \in S$ is the vertex of $\Omega(n)$ triangles of distinct areas determined by S? In other words, is there a constant $c > 0$ such that for every $p \in S$, the point set S determines at least cn triangles of distinct areas, all incident to p?

References

1. E. Altman, On a problem of Erdős, *American Mathematical Monthly*, **70** (1963), 148–157.
2. E. Altman, Some theorems on convex polygons, *Canadian Mathematical Bulletin*, **15** (1972), 329–340.

3. P. Braß, W. Moser, and J. Pach, *Research Problems in Discrete Geometry*, Springer, New York, 2005.

4. P. Braß, G. Rote, and K. J. Swanepoel, Triangles of extremal area or perimeter in a finite planar point set, *Discrete & Computational Geometry* **26** (2001), 51–58.

5. G. R. Burton and G. Purdy, The directions determined by n points in the plane, *Journal of London Mathematical Society* **20** (1979), 109–114.

6. H. T. Croft, K. J. Falconer, and R. K. Guy, *Unsolved Problems in Geometry*, Springer, New York, 1991.

7. A. Dumitrescu, On distinct distances from a vertex of a convex polygon, *Discrete & Computational Geometry* **36** (2006), 503–509.

8. A. Dumitrescu and Cs. D. Tóth, On the number of tetrahedra with minimum, uniform, and distinct volumes in three-space, in *Proceedings of the 18th ACM-SIAM Symposium on Discrete Algorithms*, ACM Press, 2007, 1114–1123.

9. P. Erdős, On sets of distances of n points, *American Mathematical Monthly* **53** (1946), 248–250.

10. P. Erdős and G. Purdy, Some extremal problems in geometry IV, *Congressus Numerantium* **17** (Proceedings of the 7th South-Eastern Conference on Combinatorics, Graph Theory, and Computing), 1976, 307–322.

11. P. Erdős and G. Purdy, Some extremal problems in geometry V, *Proceedings of the 8th South-Eastern Conference on Combinatorics, Graph Theory, and Computing*, 1977, 569–578.

12. P. Erdős and G. Purdy, Extremal problems in combinatorial geometry. in *Handbook of Combinatorics*, Vol. 1, 809–874, Elsevier, Amsterdam, 1995.

13. P. Erdős, G. Purdy, and E. G. Straus, On a problem in combinatorial geometry, *Discrete Mathematics* **40** (1982), 45–52.

14. J. E. Goodman and R. Pollack, On the combinatorial classification of nondegenerate configurations in the plane, *Journal of Combinatorial Theory Ser. A* **29** (1980), 220–235.

15. J. E. Goodman and R. Pollack, A combinatorial perspective on some problems in geometry, *Congressus Numerantium* **32** (1981), 383–394.

16. N. H. Katz and G. Tardos, A new entropy inequality for the Erdős distance problem, in *Towards a Theory of Geometric Graphs (J. Pach, ed.)*, vol. 342 of Contemporary Mathematics, AMS, Providence, RI, 2004, 119–126.

17. J. Pach, personal communication, January 2007.

18. J. Pach and P. K. Agarwal, *Combinatorial Geometry*, John Wiley, New York, 1995.

19. J. Pach and G. Tardos, Isosceles triangles determined by a planar point set, *Graphs and Combinatorics*, **18** (2002), 769–779.

20. J. Solymosi and Cs. D. Tóth, Distinct distances in the plane, *Discrete & Computational Geometry*, **25** (2001), 629–634.

21. E. G. Straus, Some extremal problems in combinatorial geometry, in *Proceedings of the Conference on Combinatorial Theory*, vol. 686 of Lecture Notes in Mathematics, Springer, 1978, pp. 308–312.

22. L. Székely, Crossing numbers and hard Erdős problems in discrete geometry, *Combinatorics, Probability and Computing* **6** (1997), 353–358.

23. P. Ungar, $2N$ noncollinear points determine at least $2N$ directions, *Journal of Combinatorial Theory Ser. A* **33** (1982), 343–347.

Scheduling with Precedence Constraints of Low Fractional Dimension

Christoph Ambühl[1], Monaldo Mastrolilli[2], Nikolaus Mutsanas[2],
and Ola Svensson[2]

[1] University of Liverpool - Great Britain
christoph@csc.liv.ac.uk
[2] IDSIA- Switzerland
{monaldo,nikolaus,ola}@idsia.ch

Abstract. We consider the single machine scheduling problem to minimize the average weighted completion time under precedence constrains.
Improving on the various 2-approximation algorithms is considered one
of the ten most prominent open problems in scheduling theory. Recently,
research has focused on special cases of the problem, mostly by restricting the set of precedence constraints to special classes such as convex
bipartite, two-dimensional, and interval orders.

In this paper we extend our previous results by presenting a framework
for obtaining $(2 - 2/d)$-approximation algorithms provided that the set
of precedence constraints has fractional dimension d. Our generalized
approach yields the best known approximation ratios for all previously
considered classes of precedence constraints, and it provides the first
results for bounded degree and interval dimension 2 orders.

As a negative result we show that the addressed problem remains
NP-hard even when restricted to the special case of interval orders.

1 Introduction

The problem we consider in this paper is a classical problem in scheduling theory,
known as $1|prec|\sum_j w_j C_j$ in standard scheduling notation (see e.g. Graham et
al. [12]). It is defined as the problem of scheduling a set $N = \{1, \ldots, n\}$ of n jobs
on a single machine, which can process at most one job at a time. Each job j has
a processing time p_j and a weight w_j, where p_j and w_j are nonnegative integers.
Jobs also have precedence constraints between them that are specified in the form
of a *partially ordered set* (*poset*) $\mathbf{P} = (N, P)$, consisting of the set of jobs N and
a partial order i.e. a reflexive, antisymmetric, and transitive binary relation P on
N, where $(i, j) \in P$ $(i \neq j)$ implies that job i must be completed before job j
can be started. The goal is to find a non-preemptive schedule which minimizes
$\sum_{j=1}^{n} w_j C_j$, where C_j is the time at which job j completes in the given schedule.

The described problem was shown to be strongly NP-hard already in 1978
by Lawler [17] and Lenstra & Rinnooy Kan [18]. While currently no inapproximability result is known (other than that the problem does not admit a fully

M. Fischetti and D.P. Williamson (Eds.): IPCO 2007, LNCS 4513, pp. 130–144, 2007.
© Springer-Verlag Berlin Heidelberg 2007

polynomial time approximation scheme), there are several 2-approximation algorithms [26,29,13,6,5,20,2]. Closing this approximability gap is a longstanding open problem in scheduling theory (see e.g. [30]).

Due to the difficulty to obtain better than 2-approximation algorithms, much attention has recently been given to special cases which manifests itself in recent approximation and exact algorithms [16,33,7,2,3].

On the negative side, Woeginger [33] proved that many quite severe restrictions on the weights and processing times do not influence approximability. For example, the special case in which all jobs either have $p_j = 1$ and $w_j = 0$, or $p_j = 0$ and $w_j = 1$, is as hard to approximate as the general case. This suggests that in order to identify classes of instances which allow a better than 2-approximation one has to focus on the precedence constraints rather than the weights and processing times.

Indeed, Lawler [17] gave an exact algorithm for series-parallel orders already in 1978. For interval orders and convex bipartite precedence constraints, Woeginger [33] gave approximation algorithms with approximation ratio arbitrarily close to the golden ratio $\frac{1}{2}(1 + \sqrt{5}) \approx 1.61803$.

Recently, Ambühl & Mastrolilli [2] settled an open problem first raised by Chudak & Hochbaum [6] and whose answer was subsequently conjectured by Correa & Schulz [7]. The results in [2,7] imply that $1|prec|\sum w_j C_j$ is a special case of the weighted vertex cover problem. More precisely, they proved that every instance S of $1|prec|\sum w_j C_j$ can be translated in polynomial time into a weighted graph G_P, such that finding the optimum of S can be reduced to finding an optimum vertex cover in G_P. This result even holds for approximate solutions: Finding an α-approximate solution for S can be reduced to finding an α-approximate vertex cover in G_P.

Based on these results, three of the authors [3] discovered an interesting connection between $1|prec|\sum w_j C_j$ and the dimension theory of posets [32], by observing that the graph G_P is well known in dimension theory as the graph of incomparable pairs of a poset $\mathbf{P}$. Applying results from dimension theory allowed to describe a framework for obtaining simple and efficient approximation algorithms for $1|prec|\sum w_j C_j$ with precedence constraints of low dimension, such as convex bipartite and semi-orders. In both cases, the new 4/3-approximation algorithms outperform the previously known results. The approach even yields a polynomial algorithm for 2-dimensional precedence constraints, based on the fact that the minimum weighted vertex cover on G_P can be solved in polynomial time since G_P is bipartite for a 2-dimensional poset $\mathbf{P}$ [32,7]. This considerably extends Lawler's result [17] for series-parallel orders. Unfortunately, the framework in [3] fails in the case of interval orders (in this case the dimension can be of the order of $\log \log n$ [32]).

The work in this paper originated from the study of $1|prec|\sum w_j C_j$ under interval orders (abbreviated $1|interval\text{-}order|\sum_j w_j C_j$). Interval orders appear in many natural contexts [10]. We provide both positive and negative results.

In the first part of the paper, we further generalize our previous framework [3] such that it can be applied to precedence constraints of low *fractional* dimension [4] (Section 3). The extended framework yields $(2-2/d)$-approximation algorithms whenever precedence constraints have fractional dimension bounded by a constant d and satisfy a mild condition (see Section 3). Since the fractional dimension of interval orders is bounded by 4 (see Section 4.1), this gives a 1.5-approximation algorithm and improves the previous result in [33]. The extended framework can also be applied to interval dimension two posets (Section 4.2), bounded degree posets (Section 4.3), and posets obtained by the lexicographic sums (Section 4.4).

In the second part of the paper, we show that $1|interval\text{-}order|\sum_j w_j C_j$ remains NP-hard (Section 5). This result is rather unexpected as many problems can be solved in polynomial time when restricted to interval orders (see e.g. [25]). The reduction heavily relies on the connection between $1|prec|\sum w_j C_j$ and weighted vertex cover described in [2].

In summary, our results indicate a strong relationship between the approximability of $1|prec|\sum_j w_j C_j$ and the fractional dimension d of the precedence constraints. In particular, it is polynomial for $d = 2$, but NP-hard already for $d \geq 3$. The latter stems from the facts that problem $1|prec|\sum_j w_j C_j$ is strongly NP-hard even for posets with in-degree 2 [17], and the fractional dimension of these posets is bounded by 3 [8]. This leaves the complexity for $2 < d < 3$ as an open question.

2   Definitions and Preliminaries

2.1   Posets and Fractional Dimension

Let $\mathbf{P} = (N, P)$ be a poset. For $x, y \in N$, we write $x \leq y$ when $(x, y) \in P$, and $x < y$ when $(x, y) \in P$ and $x \neq y$. When neither $(x, y) \in P$ nor $(y, x) \in P$, we say that x and y are incomparable, denoted by $x||y$. We call $\text{inc}(\mathbf{P}) = \{(x, y) \in N \times N : x||y \text{ in } P\}$ the set of incomparable pairs of $\mathbf{P}$. A poset $\mathbf{P}$ is a linear order (or a total order) if for any $x, y \in N$ either $(x, y) \in P$ or $(y, x) \in P$, i.e. $\text{inc}(\mathbf{P}) = \emptyset$. A partial order P' on N is an extension of a partial order P on the same set N, if $P \subseteq P'$. An extension that is a linear order is called a linear extension. Mirroring the definition of the fractional chromatic number of a graph, Brightwell & Scheinerman [4] introduce the notion of fractional dimension of a poset. Let $\mathcal{F} = \{L_1, L_2, \ldots, L_t\}$ be a nonempty multiset of linear extensions of $\mathbf{P}$. The authors in [4] call $\mathcal{F}$ a k-fold realizer of $\mathbf{P}$ if for each incomparable pair (x, y), there are at least k linear extensions in $\mathcal{F}$ which reverse the pair (x, y), i.e., $|\{i = 1, \ldots, t : y < x \text{ in } L_i\}| \geq k$. We call a k-fold realizer of size t a k:t-realizer. The fractional dimension of $\mathbf{P}$ is then the least rational number $\text{fdim}(\mathbf{P}) \geq 1$ for which there exists a k:t-realizer of $\mathbf{P}$ so that $k/t \geq 1/\text{fdim}(\mathbf{P})$. Using this terminology, the dimension of $\mathbf{P}$, denoted by $\dim(\mathbf{P})$, is the least t for which there exists a 1-fold realizer of P. It is immediate that $\text{fdim}(\mathbf{P}) \leq \dim(\mathbf{P})$ for any poset $\mathbf{P}$. Furthermore [4], $\text{fdim}(\mathbf{P}) = 1$, or $\text{fdim}(\mathbf{P}) \geq 2$.

2.2   Scheduling, Vertex Cover, and Dimension Theory

In [7,2,3] a relationship between $1|prec|\sum_j w_j C_j$, weighted vertex cover, and the dimension theory of posets is shown. This relationship will turn out to be useful for both improving the approximation ratio for several classes of precedence constraints and establishing the NP-hardness of $1|interval\text{-}order|\sum_j w_j C_j$.

Let $\mathbf{P} = (N, P)$ be any poset, that is not a linear order. Felsner and Trotter [9] associate with $\mathbf{P}$ a hypergraph $\mathcal{H}_\mathbf{P}$, called the *hypergraph of incomparable pairs*, defined as follows. The vertices of $\mathcal{H}_\mathbf{P}$ are the incomparable pairs in $\mathbf{P}$. The edge set consists of those sets U of incomparable pairs such that no linear extension of $\mathbf{P}$ reverses all incomparable pairs in U. Let $G_\mathbf{P}$ denote the ordinary graph, called the *graph of incomparable pairs*, determined by all edges of size 2 in $\mathcal{H}_\mathbf{P}$. In [9,32] it is shown that the dimension of $\mathbf{P}$ is equal to the chromatic number of $\mathcal{H}_\mathbf{P}$, i.e., $dim(\mathbf{P}) = \chi(\mathcal{H}_\mathbf{P}) \geq \chi(G_\mathbf{P})$. In [4], it was noted that the same relationship holds for the fractional versions, i.e., $\text{fdim}(\mathbf{P}) = \chi_f(\mathcal{H}_\mathbf{P}) \geq \chi_f(G_\mathbf{P})$. We refer the reader to [28] for an introduction to fractional graph coloring.

Given an instance S of $1|prec|\sum_j w_j C_j$, we associate with S a weighted vertex cover instance VC_S on $G_\mathbf{P}$, where $G_\mathbf{P}$ is the graph of incomparable pairs of the poset $\mathbf{P}$ representing the precedence constraints and each vertex $(i, j) \in \text{inc}(\mathbf{P})$ has weight $p_i \cdot w_j$. We denote the *value* of a solution s by $val(s)$.

Theorem 1 ([2,3,7]). *Let S be an instance of $1|prec|\sum_j w_j C_j$ where precedence constraints are given by the poset $\mathbf{P} = (N, P)$. Then the following transformations can be performed in polynomial time.*

1. *Any feasible solution s' of S can be turned into a feasible solution c' of VC_S, such that*

$$\text{val}(c') \leq \text{val}(s') - \sum_{(i,j)\in P} p_i \cdot w_j.$$

2. *Any feasible solution c' to VC_S can be turned into a feasible solution s' of S, such that*

$$\text{val}(s') \leq \text{val}(c') + \sum_{(i,j)\in P} p_i \cdot w_j.$$

In particular, if c^* and s^* are optimal solutions to VC_S and S, respectively, we have $\text{val}(c^*) = \text{val}(s^*) - \sum_{(i,j)\in P} p_i \cdot w_j$.

We remark that the term $\sum_{(i,j)\in P} p_i \cdot w_j$ is a *fixed cost* and it is present in all feasible schedules of S. This follows from the facts that a job's processing time is always included in its completion time, and any feasible schedule of S must schedule job i before job j if $i < j$ in P.

3   Scheduling and Fractional Dimension

In this section, we present an algorithmic framework that can be used to obtain better than 2-approximation algorithms provided that the set of precedence constraints has low fractional dimension. Applications that follow this pattern are given in Section 4.

We say that a poset $\mathbf{P}$ admits an *efficiently samplable k:t-realizer* if there exists a randomized algorithm that, in polynomial time, returns any linear extension from a k-fold realizer $\mathcal{F} = \{L_1, L_2, \ldots, L_t\}$ with probability $1/t$.

Let S be an instance of $1|prec|\sum_j w_j C_j$ where precedence constraints are given by a poset $\mathbf{P} = (N, P)$. Assuming that $\mathbf{P}$ admits an efficiently samplable k:t-realizer $\mathcal{F} = \{L_1, \ldots, L_t\}$, we proceed as follows.

Let $V_\mathbf{P}$ and $E_\mathbf{P}$ be the vertex set and edge set, respectively, of the graph of incomparable pairs $G_\mathbf{P}$. Consider the following integer program formulation of the weighted vertex cover VC_S:

$$\min \quad \sum_{i \in V_\mathbf{P}} w_i x_i$$

$$\text{s.t.} \quad x_i + x_j \geq 1 \qquad \{i, j\} \in E_\mathbf{P}$$

$$x_i \in \{0, 1\} \qquad i \in V_\mathbf{P}$$

where w_i denotes the weight of vertex $v_i \in V_\mathbf{p}$, as specified in the definition of VC_S (see Section 2.2). Let [VC-LP] denote the linear relaxation of the integer program above.

Nemhauser & Trotter [23,24] proved that any basic feasible solution to [VC-LP] is *half-integral*, that is $x_i \in \{0, \frac{1}{2}, 1\}$ for all $i \in V$. Let V_i be the set of nodes whose corresponding variables took value $i \in \{0, \frac{1}{2}, 1\}$ in the optimal solution of [VC-LP].

Observe that for any linear extension L, the set of all incomparable pairs that are reversed in L is an independent set in the graph of incomparable pairs $G_\mathbf{P}$. Now, pick uniformly at random a linear extension L of $\mathcal{F}$ in polynomial time. Note that $V_0 \cup (V_{1/2} \setminus L)$ defines an independent set of $G_\mathbf{P}$. Generalizing a result by Hochbaum in [14], we prove that the complement of $V_0 \cup (V_{1/2} \setminus L)$ is a vertex cover whose expected value is within $(2 - 2\frac{k}{t})$ times the weight of an optimum cover. By Theorem 1, we can transform (in polynomial time) the solution of VC_S into a feasible solution of S of expected value at most $(2 - 2\frac{k}{t})$ times the value of an optimum schedule. We summarize the above arguments in the following theorem.

Theorem 2. *The problem* $1|prec|\sum_j w_j C_j$, *whenever precedence constraints admit an efficiently samplable* k:t-realizer, *has a randomized* $(2 - 2\frac{k}{t})$-*approximation algorithm.*

For a proof of this theorem, see Appendix A.1. Following a similar argumentation, Hochbaum's approach [14] for approximating the vertex cover problem can be extended to fractional coloring, yielding the same approximation result.

A natural question is for which posets one can have an efficiently samplable k:t-realizer. In the general case, Jain & Hedge [15] recently proved that it is hard to approximate the dimension of a poset with n elements within a factor $n^{0.5-\epsilon}$, and the same hardness of approximation holds for the fractional dimension. However, for several special cases, including interval orders (Section 4.1) and bounded degree posets (Section 4.3), efficiently samplable k:t-realizers exist.

4   Precedence Constraints with Low Fractional Dimension

4.1   Interval Orders

A poset $\mathbf{P} = (N, P)$ is an *interval order* if there is a function F, which assigns to each $x \in N$ a closed interval $F(x) = [a_x, b_x]$ of the real line $\mathbb{R}$, so that $x < y$ in P if and only if $b_x < a_y$ in $\mathbb{R}$. Interval orders can be recognized in $O(n^2)$ time [21,25]. The dimension of interval orders can be of the order of $\log \log n$ [32], whereas the fractional dimension is known to be less than 4 [4], and this bound is asymptotically tight [8]. In the following we show how to obtain a 1.5-approximation algorithm for $1|interval\text{-}order| \sum_j w_j C_j$. By Theorem 2, it is sufficient to prove that interval orders admit an efficiently sampalble $k{:}t$-realizer with $t/k = 4$.

Given a poset $\mathbf{P} = (N, P)$, disjoint subsets A and B of the ground set N, and a linear extension L of P, we say that B *is over* A *in* L if, for every incomparable pair of elements (a, b) with $a \in A$ and $b \in B$, one has $b > a$ in L. The following property of interval orders is fundamental.

Theorem 3 (Rabinovitch [27,10]). *A poset* $\mathbf{P} = (N, P)$ *is an interval order if and only if for every pair* (A, B) *of disjoint subsets of* N *there is a linear extension* L *of* P *with* B *over* A.

By using this property we can easily obtain a k-fold realizer $\mathcal{F} = \{L_1, \ldots, L_t\}$ with $k = 2^{n-2}$ and $t = 2^n$, where $n = |N|$. Indeed, consider every subset A of N and let L_A be a linear extension of P in which $B = N \setminus A$ is over A. Now let $\mathcal{F}$ be the multiset of all the L_A's. Note that $|\mathcal{F}| = 2^n$. Moreover, for any incomparable pair (x, y) there are at least $k = 2^{n-2}$ linear extensions in $\mathcal{F}$ for which $x \in B$ and $y \in A$. Finally, observe that we can efficiently pick uniformly at random one linear extension from $\mathcal{F}$: for every job $j \in N$ put j either in A or in B with the same probability $1/2$.

By the previous observations and Theorem 2, we have a randomized polynomial time 1.5-approximation for $1|interval\text{-}order| \sum_j w_j C_j$. The described algorithm can easily be derandomized by using the classical method of conditional probabilities.

Theorem 4. *Problem* $1|interval\text{-}order| \sum_j w_j C_j$ *has a deterministic polynomial time 1.5-approximation algorithm.*

4.2   Interval Dimension Two

The *interval dimension* of a poset $\mathbf{P} = (N, P)$, denoted by $\dim_I(\mathbf{P})$, is defined [32] as the least t for which there exist t extensions $Q_1, Q_2, \ldots, Q_t$, so that:

- $P = Q_1 \cap Q_2 \cap \cdots \cap Q_t$ and
- (N, Q_i) is an interval order for $i = 1, 2, \ldots, t$.

Generally $\dim_I(\mathbf{P}) \le \dim(\mathbf{P})$. Obviously, if $\mathbf{P}$ is an interval order, $\dim_I(\mathbf{P}) = 1$.

The class of posets of interval dimension 2 forms a proper superclass of the class of interval orders. Posets of interval dimension two can be recognized in

$O(n^2)$ time due to Ma & Spinrad [19]. Given a poset $\mathbf{P}$ with $\dim_I(\mathbf{P}) = 2$, their algorithm also yields an interval realizer $\{Q_1, Q_2\}$. As described in Section 4.1, we obtain k-fold realizers $\mathcal{F}_1 = \{L_1, L_2, \ldots, L_t\}$ and $\mathcal{F}_2 = \{L'_1, L'_2, \ldots, L'_t\}$ of Q_1 and Q_2, respectively, with $k = 2^{n-2}$ and $t = 2^n$. It is immediate that $\mathcal{F} = \mathcal{F}_1 \cup \mathcal{F}_2$ is a k-fold realizer of $\mathbf{P}$ of size $2t = 2^{n+1}$. Furthermore, we can efficiently pick uniformly at random one linear extension from $\mathcal{F}$: pick uniformly at random a linear extension from either $\mathcal{F}_1$ or $\mathcal{F}_2$ with the same probability $1/2$. Again by using conditional probabilities we have the following.

Theorem 5. *Problem* $1|prec| \sum_j w_j C_j$, *whenever precedence constraints have interval dimension at most* 2, *has a polynomial time* 1.75-*approximation algorithm.*

4.3   Posets of Bounded Degree

In the following we will see how to obtain, using Theorem 2, an approximation algorithm for $1|prec| \sum w_j C_j$ when the precedence constraints form a poset of bounded degree. Before we proceed, we need to introduce some definitions.

Let $\mathbf{P} = (N, P)$ be a poset. For any job $j \in N$, define the *degree of* j, denoted $\deg(j)$, as the number of jobs comparable (but not equal) to j in $\mathbf{P}$. Let $\Delta(\mathbf{P}) = \max\{\deg(j) : j \in N\}$. Given a job j, let $D(j)$ denote the set of all jobs which are less than j, and $U(j)$ those which are greater than j in P. Define $\deg_D(j) = |D(j)|$ and $\Delta_D(\mathbf{P}) = \max\{\deg_D(j) : j \in N\}$. The quantities $\deg_U(j)$ and $\Delta_U(\mathbf{P})$ are defined dually.

We observe that the NP-completeness proof for $1|prec| \sum w_j C_j$ given by Lawler [17] was actually provided for posets $\mathbf{P}$ with $\Delta_D(\mathbf{P}) = 2$. By using fractional dimension we show that these posets (with bounded $\min\{\Delta_D, \Delta_U\}$) allow for better than 2-approximation.

Theorem 6. *Problem* $1|prec| \sum w_j C_j$ *has a polynomial time* $(2 - 2/f)$-*approximation algorithm, where* $f = 1 + \min\{\Delta_D, \Delta_U, 1\}$.

Proof. Let $\mathbf{P} = (N, P)$ be the poset representing the precedence constraints with bounded $\min\{\Delta_D, \Delta_U\}$. Assume, without loss of generality, that $\mathbf{P}$ is *not* decomposable with respect to lexicographic sums (see Section 4.4). Otherwise, a decomposition with respect to lexicographic sums can be done in $O(n^2)$ time (see e.g. [22]), and each component can be considered separately. We call an incomparable pair $(x, y) \in \text{inc}(\mathbf{P})$ a *critical pair* if for all $z, w \in N \setminus \{x, y\}$

1. $z < x$ in P implies $z < y$ in P, and
2. $y < w$ in P implies $x < w$ in P.

Critical pairs play an important role in dimension theory: if for each critical pair (x, y), there are at least k linear extensions in $\mathcal{F}$ which reverse the pair (x, y) then $\mathcal{F}$ is a k-fold realizer of P and vice versa [4].

For any permutation M of N, consider the set $C(M)$ of critical pairs (x, y) that satisfy the following two conditions:

1. $x > (D(y) \cup \{y\})$ in M if $|D(y)| < \Delta_D$
2. $x > D(y)$ in M if $|D(y)| = \Delta_D$

In [8], Felsner & Trotter present an algorithm that converts in polynomial time a permutation M of N to a linear extension L of P so that L reverses all critical pairs in the set $C(M)$. Now set $t = |N|!$ and consider the set $\mathcal{M} = \{M_1, M_2, \ldots, M_t\}$ of all permutations of the ground set N. Observe that for any critical pair (x, y) there are at least $n!/(\Delta_D + 1)$ different permutations $M_i \in \mathcal{M}$, where the critical pair is reversed, i.e., $(y, x) \in C(M_i)$. Applying the algorithm in [8] we obtain a k-fold realizer $\mathcal{F} = \{L_1, \ldots, L_t\}$ of P with $t = n!$ and $k = n!/(\Delta_D + 1)$. Moreover, we can efficiently pick uniformly at random one linear extension from $\mathcal{F}$: generate uniformly at random one permutation of jobs (e.g. by using Knuth's shuffle algorithm) and transform it into a linear extension with the described properties by using the algorithm in [8]. The described algorithm can be derandomized by using the classical method of conditional probabilities. Finally observe that we can repeat a similar analysis by using Δ_U instead of Δ_D.   □

In fact, this result is stronger than the same statement with $d = \Delta(\mathbf{P})$. To see this, consider the *graph poset* $\mathbf{P}(G) = (N, P)$ defined as follows: given an undirected graph $G(V, E)$, let $N = V \cup E$ and for every $v \in V$ and $e = \{v_1, v_2\} \in E$, put $(v, e) \in P$ if and only if $v \in \{v_1, v_2\}$. If $\Delta(G)$ is unbounded, this also holds for $\Delta(\mathbf{P})$. However, since every edge is adjacent to only two vertices, Δ_D is bounded by 2, thus the value $1 + \min\{\Delta_U, \Delta_D\}$ is also bounded. On the other hand, for the complete graph on n nodes, K_n, Spencer [31] showed that $\dim(\mathbf{P}(K_n)) = \Theta(\log \log n)$. Therefore, the poset $\mathbf{P}(K_n)$ is an example where the dimension of the poset is unbounded, while $\min\{\Delta_D, \Delta_U\}$ (and thus also the fractional dimension) is bounded. This means that the fractional dimension approach can yield a substantially better result than the dimension approach used in [3].

4.4 Lexicographic Sums

In this section we show how to use previous results to obtain approximation algorithms for new ordered sets. The construction we use here, *lexicographic sums*, comes from a very simple pictorial idea (see [32] for a more comprehensive discussion). Take a poset $\mathbf{P} = (N, P)$ and replace each of its points $x \in N$ with a partially ordered set $\mathbf{Q_x}$, the *module*, such that the points in the module have the same relation to points outside it. A more formal definition follows. For a poset $\mathbf{P} = (N, P)$ and a family of posets $\mathcal{S} = \{(Y_x, Q_x) \mid x \in N\}$ indexed by the elements in N, the lexicographic sum of $\mathcal{S}$ over (N, P), denoted $\sum_{x \in (N,P)} (Y_x, Q_x)$ is the poset (Z, R) where $Z = \{(x, y) \mid x \in N, y \in Y_x\}$ and $(x_1, y_1) \leq (x_2, y_2)$ in R if and only if one of the following two statements holds:

1. $x_1 < x_2$ in P.
2. $x_1 = x_2$ and $y_1 \leq y_2$ in Q_{x_1}.

We call $\mathcal{P} = P \cup \mathcal{F}$ the *components* of the lexicographic sum. A lexicographic sum is *trivial* if $|N| = 1$ or if $|Y_x| = 1$ for all $x \in N$. A poset is *decomposable with respect to lexicographic sums* if it is isomorphic to a non-trivial lexicographic sum.

In case the precedence constraints of every component admit an efficiently samplable realizer, we observe that this translates into a randomized approximation
algorithm:

Theorem 7. *Problem* $1|prec|\sum_j w_j C_j$, *whenever precedence constraints form
a lexicographic sum whose components* $i \in \mathcal{P}$ *admit efficiently samplable realizers, has a polynomial time randomized* $(2 - \frac{2t}{k}) - approximation$ *algorithm, where*
$t/k = \max_{i \in \mathcal{P}}(t_i/k_i)$.

Finally, we point out that, if the approximation algorithm for each component
can be derandomized, this yields a derandomized approximation algorithm for
the lexicographic sum.

5   NP-Completeness for Interval Orders

In this section we show that $1|prec|\sum_j w_j C_j$ remains NP-complete even in the
special case of interval order precedence constraints. To prove this we exploit
the vertex cover nature of problem $1|prec|\sum w_j C_j$.

Theorem 8. *Problem* $1|interval\text{-}order|\sum_j w_j C_j$ *is NP-complete.*

Proof. A graph G is said to have bounded degree d if every vertex v in G is
adjacent to at most d other vertices. The problem of deciding if a graph G
with bounded degree 3 has a (unweighted) vertex cover of size at most m is
NP-complete [11]. We provide a reduction from the minimum vertex cover on
graphs with bounded degree 3 to $1|interval\text{-}order|\sum_j w_j C_j$.

Given a connected graph $G = (V, E)$ with bounded degree 3, we construct an
instance S of $1|interval\text{-}order|\sum_j w_j C_j$ so that S has a schedule with value less
than $m + c + 1$ if and only if G has a vertex cover of size at most m, where c is
a fixed value defined later (see Equation (1)). We present the construction of S
in two stages.

Stage 1 (Tree-layout of the graph). Starting from any vertex $s \in V$, consider
the tree $T = (V, E_T)$, with $E_T \subseteq E$, rooted at s on the set of nodes reachable
from s by using, for example, breadth-first search. Furthermore, we number the
vertices of T top-down and left-right. Figure 1 shows the breadth-first search
tree T for K_4.

Define $G' = (V', E')$ to be the graph obtained from T in the following way. For
each vertex v_i in T we add two new vertices u_2^i, u_1^i and edges $\{u_2^i, u_1^i\}, \{u_1^i, v_i\}$.
Furthermore, for each edge $\{v_i, v_j\} \in E \setminus E_T$ with $i < j$ we add vertices e_1^{ij}, e_2^{ij}
and edges $\{v_i, e_1^{ij}\}, \{e_1^{ij}, e_2^{ij}\}, \{e_2^{ij}, u_2^j\}$.

The following claim relates the optimum unweighted vertex covers of G and G'.

Claim 1. Let $C_* \subseteq V$ and $C'_* \subseteq V'$ be optimum vertex cover solutions to G and
G', respectively, then $|C_*| = |C'_*| - |V| - |E \setminus E_T|$. (For a proof, see Appendix A.2).

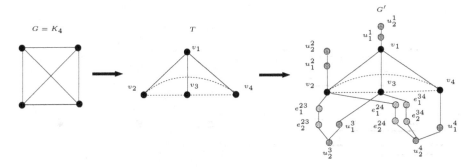

Fig. 1. The breadth first search tree $T = (V, E_T)$ for the graph $G = K_4$, and the graph G'. The solid edges belong to E_T.

Stage 2 (Construction of scheduling instance). Given the vertex cover graph $G = (V, E)$ and its corresponding tree $T = (V, E_T)$, we construct the scheduling instance S with processing times, weights, and precedence constraints to form an interval order I as defined below (see Figure 2 for an example), where k is a value to be determined later.

Job	Interval Repr.	Proc. Time	Weight
s_0	$[-1,0]$	1	0
s_1	$[0, 1]$	$1/k$	1
$s_j, j = 2, \ldots, \|V\|$	$[i, j]$, where $\{v_i, v_j\} \in E_T, i < j$	$1/k^j$	k^i
$m_i, i = 1, \ldots, \|V\|$	$[i - \frac{1}{2}, \|V\| + i]$	$1/k^{(\|V\|+i)}$	k^i
$e_i, i = 1, \ldots, \|V\|$	$[\|V\| + i, \|V\| + i + 1]$	0	$k^{(\|V\|+i)}$
b_{ij}, where $\{v_i, v_j\} \in E \setminus E_T, i < j$	$[i, j - \frac{1}{2}]$	$1/k^j$	k^i

Remark 1. Let i and j be two jobs in S with interval representations $[a, b]$ and $[c, d]$ respectively, where $a \leq d$. By the construction of the scheduling instance S we have $p_i \leq 1/k^{\lceil b \rceil}$ and $w_j \leq k^{\lceil c \rceil}$. It follows that $p_i \cdot w_j = 1$ or $p_i \cdot w_j \leq 1/k$ if i and j are incomparable, since $p_i \cdot w_j \geq k$ implies that $b < c$, i.e., i's interval representation is completely to the left of j's interval representation. Furthermore, if $p_i \cdot w_j = 1$ then $\lceil b \rceil = \lceil c \rceil$.

Let $D = \quad \{(s_0, s_1)\}$
$\qquad \cup \{(s_i, s_j) : v_i$ is the parent of v_j in $T\}$
$\qquad \cup \{(s_i, m_i), (m_i, e_i) : i = 1, 2, \ldots, \|V\|\}$
$\qquad \cup \{(s_i, b_{ij}), (b_{ij}, m_j) : \{v_i, v_j\} \in E \setminus E_T, i < j\}$

By the interval representation of the jobs and the remark above, we have the following:

Claim 2. A pair of incomparable jobs (i,j) has $p_i \cdot w_j = 1$ if $(i, j) \in D$; otherwise if $(i, j) \notin D$ then $p_i \cdot w_j \leq 1/k$.

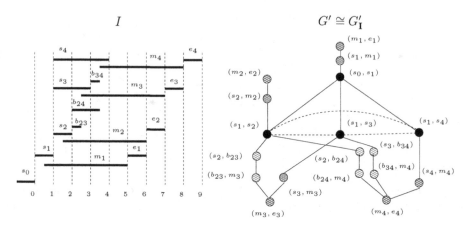

Fig. 2. The interval order I obtained from K_4; $G'_\mathbf{I}$ is the subgraph induced on the graph of incomparable pairs $G_\mathbf{I}$ by the vertex subset D (the vertices with weight 1)

Claim 3. Let $G'_\mathbf{I} = (D, E_I)$ be the subgraph induced on the graph of incomparable pairs $G_\mathbf{I}$ by the vertex subset D. Then G' and $G'_\mathbf{I}$ are isomorphic. (For a proof, see Appendix A.3).

By Claim 2, each incomparable pair of jobs $(i, j) \notin D$ satisfies $p(i) \cdot w(j) \leq 1/k$. Let n be the number of jobs in the scheduling instance S and select k to be $n^2 + 1$. Let C, C_I, and C'_I be optimal vertex cover solutions to G, $G_\mathbf{I}$ and $G'_\mathbf{I}$ (defined as in Claim 3), respectively. Then, by the selection of k and Claim 2, we have $|C'_I| \leq |C_I| \leq |C'_I| + \sum\limits_{(i,j)\in\mathrm{inc}(I)\setminus D} p_i w_j < |C'_I| + 1$. Furthermore, Claims 3 and 1 give us that $|C| + |V| + |E \setminus E_T| \leq |C_I| < |C| + |V| + |E \setminus E_T| + 1$. This, together with Theorem 1, implies that $|C| \leq m$ if and only if there is a schedule of S with value less than $m + c + 1$, where

$$c = |V| + |E \setminus E_T| + \sum_{(i,j)\in I} p_i \cdot w_j. \qquad (1)$$

$\square$

Acknowledgements

We are grateful to Andreas Schulz for many helpful discussions we had with him during his visit at IDSIA. This research is supported by Swiss National Science Foundation project 200021-104017/1, "Power Aware Computing", and by the Swiss National Science Foundation project 200020-109854, "Approximation Algorithms for Machine scheduling Through Theory and Experiments II". The first author is supported by Nuffield Foundation Grant NAL32608.

References

1. P. Alimonti and V. Kann. Some APX-completeness results for cubic graphs. *Theor. Comput. Sci.*, 237(1-2):123–134, 2000.
2. C. Ambühl and M. Mastrolilli. Single machine precedence constrained scheduling is a vertex cover problem. In *Proceedings of the 14th Annual European Symposium on Algorithms (ESA)*, volume 4168 of *Lecture Notes in Computer Science*, pages 28–39. Springer, 2006.
3. C. Ambühl, M. Mastrolilli, and O. Svensson. Approximating precedence-constrained single machine scheduling by coloring. In *Proceedings of APPROX + RANDOM*, volume 4110 of *Lecture Notes in Computer Science*, pages 15–26. Springer, 2006.
4. G. R. Brightwell and E. R. Scheinerman. Fractional dimension of partial orders. *Order*, 9:139–158, 1992.
5. C. Chekuri and R. Motwani. Precedence constrained scheduling to minimize sum of weighted completion times on a single machine. *Discrete Applied Mathematics*, 98(1-2):29–38, 1999.
6. F. A. Chudak and D. S. Hochbaum. A half-integral linear programming relaxation for scheduling precedence-constrained jobs on a single machine. *Operations Research Letters*, 25:199–204, 1999.
7. J. R. Correa and A. S. Schulz. Single machine scheduling with precedence constraints. *Mathematics of Operations Research*, 30(4):1005–1021, 2005. Extended abstract in Proceedings of the 10th Conference on Integer Programming and Combinatorial Optimization (IPCO 2004), pages 283–297.
8. S. Felsner and W. T. Trotter. On the fractional dimension of partially ordered sets. *DMATH: Discrete Mathematics*, 136:101–117, 1994.
9. S. Felsner and W. T. Trotter. Dimension, graph and hypergraph coloring. *Order*, 17(2):167–177, 2000.
10. P. C. Fishburn. *Interval Orders and Interval Graphs*. John Wiley and Sons, 1985.
11. M. R. Garey, D. S. Johnson, and L. J. Stockmeyer. Some simplified NP-complete graph problems. *Theor. Comput. Sci.*, 1(3):237–267, 1976.
12. R. Graham, E. Lawler, J. K. Lenstra, and A. H. G. Rinnooy Kan. Optimization and approximation in deterministic sequencing and scheduling: A survey. In *Annals of Discrete Mathematics*, volume 5, pages 287–326. North–Holland, 1979.
13. L. A. Hall, A. S. Schulz, D. B. Shmoys, and J. Wein. Scheduling to minimize average completion time: off-line and on-line algorithms. *Mathematics of Operations Research*, 22:513–544, 1997.
14. D. S. Hochbaum. Efficient bounds for the stable set, vertex cover and set packing problems. *Discrete Applied Mathematics*, 6:243–254, 1983.
15. K. Jain, R. Hedge. Some inapproximability results for the (fractional) poset dimension *Personal communication*, 2006
16. S. G. Kolliopoulos and G. Steiner. Partially-ordered knapsack and applications to scheduling. In *Proceedings of the 10th Annual European Symposium on Algorithms (ESA)*, pages 612–624, 2002.
17. E. L. Lawler. Sequencing jobs to minimize total weighted completion time subject to precedence constraints. *Annals of Discrete Mathematics*, 2:75–90, 1978.
18. J. K. Lenstra and A. H. G. Rinnooy Kan. The complexity of scheduling under precedence constraints. *Operations Research*, 26:22–35, 1978.
19. T.-H. Ma and J. P. Spinrad. On the 2-chain subgraph cover and related problems. *J. Algorithms*, 17(2):251–268, 1994.

20. F. Margot, M. Queyranne, and Y. Wang. Decompositions, network flows and a precedence constrained single machine scheduling problem. *Operations Research*, 51(6):981–992, 2003.
21. R. H. Möhring. Computationally tractable classes of ordered sets. In I. Rival, editor, *Algorithms and Order*, pages 105–193. Kluwer Academic, 1989.
22. R. H. Möhring. Computationally tractable classes of ordered sets. *Algorithms and Order*, pages 105–194, 1989.
23. G. L. Nemhauser and L. E. Trotter. Properties of vertex packing and independence system polyhedra. *Mathematical Programming*, 6:48–61, 1973.
24. G. L. Nemhauser and L. E. Trotter. Vertex packings: Structural properties and algorithms. *Mathematical Programming*, 8:232–248, 1975.
25. C. H. Papadimitriou and M. Yannakakis. Scheduling interval-ordered tasks. *SIAM Journal of Computing*, 8:405–409, 1979.
26. N. N. Pisaruk. A fully combinatorial 2-approximation algorithm for precedence-constrained scheduling a single machine to minimize average weighted completion time. *Discrete Applied Mathematics*, 131(3):655–663, 2003.
27. I. Rabinovitch. The dimension of semiorders. *J. Comb. Theory, Ser. A*, 25:50–61, 1978.
28. E. R. Scheinerman and D. H. Ullman. *Fractional Graph Theory*. John Wiley and Sons Inc., 1997.
29. A. S. Schulz. Scheduling to minimize total weighted completion time: Performance guarantees of LP-based heuristics and lower bounds. In *Proceedings of the 5th Conference on Integer Programming and Combinatorial Optimization (IPCO)*, pages 301–315, 1996.
30. P. Schuurman and G. J. Woeginger. Polynomial time approximation algorithms for machine scheduling: ten open problems. *Journal of Scheduling*, 2(5):203–213, 1999.
31. J. Spencer. On minimum scrambling sets of simple orders. *Acta Mathematica*, 22:349–353, 1971.
32. W. T. Trotter. *Combinatorics and Partially Ordered Sets: Dimension Theory*. Johns Hopkins Series in the Mathematical Sciences. The Johns Hopkins University Press, 1992.
33. G. J. Woeginger. On the approximability of average completion time scheduling under precedence constraints. *Discrete Applied Mathematics*, 131(1):237–252, 2003.

A   Omitted Proofs

A.1   Proof of Theorem 2

Proof. Let S be an instance of $1|prec|\sum_j w_j C_j$ where precedence constraints are given by a poset $\mathbf{P} = (N, P)$ that admits an efficiently samplable k:t-realizer $\mathcal{F} = \{L_1, L_2, \ldots, L_t\}$. Furthermore, we assume that $\mathrm{fdim}(\mathbf{P}) \geq 2$. The case when $\mathrm{fdim}(\mathbf{P}) = 1$, i.e., $\mathbf{P}$ is a linear order, is trivial.

Let $V_{\mathbf{P}}$ and $E_{\mathbf{P}}$ be the vertex set and edge set, respectively, of the graph of incomparable pairs $G_{\mathbf{P}}$. Consider the weighted vertex cover VC_S on $G_{\mathbf{P}}$ where each vertex (incomparable pair) $(i, j) \in V_{\mathbf{P}}$ has weight $w_{(i,j)} = p_i \cdot w_j$, as specified in the definition of VC_S (see Section 2.2). Solve the [VC-LP] formulation of VC_S (see Section 3) and let V_i be the set of vertices with value i ($i = 0, \frac{1}{2}, 1$) in the optimum solution. Denote by $G_{\mathbf{P}}[V_{1/2}]$ the subgraph of $G_{\mathbf{P}}$ induced by the vertex set $V_{1/2}$.

We consider the linear extensions of $\mathcal{F}$ as outcomes in a uniform sample space. For an incomparable pair (x, y), the *probability that y is over x in $\mathcal{F}$* is given by

$$Prob_{\mathcal{F}}[y > x] = \frac{1}{t}|\{i = 1, \ldots, t : y > x \in L_i\}| \geq \frac{k}{t} \qquad (2)$$

The last inequality holds because every incomparable pair is reversed in at least k linear extensions of $\mathcal{F}$.

Let us pick one linear extension L uniformly at random from $\mathcal{F} = \{L_1, \ldots, L_t\}$. Then, by linearity of expectation, the expected value of the independent set $I_{1/2}$, obtained by taking the incomparable pairs in $V_{1/2}$ that are reversed in L, is

$$E[w(I_{1/2})] = \sum_{(i,j)\in V_{1/2}} Prob_{\mathcal{F}}[j > i] \cdot w_{(i,j)} \geq \frac{k}{t} \cdot w(V_{1/2}) \qquad (3)$$

A vertex cover solution C for the graph $G_{\mathbf{P}}[V_{1/2}]$ can be obtained by picking the nodes that are not in $I_{1/2}$, namely $C = V_{1/2} \setminus I_{1/2}$. The expected value of this solution is

$$E[w(C)] = w(V_{1/2}) - E[w(I_{1/2})] \leq \left(1 - \frac{k}{t}\right) w(V_{1/2})$$

As observed in [14], $V_1 \cup C$ gives a valid vertex cover for graph $G_{\mathbf{P}}$. Moreover, the expected value of the cover is bounded as follows

$$E[w(V_1 \cup C)] \leq w(V_1) + \left(1 - \frac{k}{t}\right) w(V_{1/2}) \qquad (4)$$

$$\leq 2\left(1 - \frac{k}{t}\right)\left(w(V_1) + \frac{1}{2}w(V_{1/2})\right) \qquad (5)$$

$$\leq \left(2 - \frac{2k}{t}\right) OPT \qquad (6)$$

where the last inequality holds since $w(V_1) + \frac{1}{2}w(V_{1/2})$ is the optimal value of [VC-LP]. Note that $t/k \geq \text{fdim}(\mathbf{P}) \geq 2$ was used for the second inequality. Theorem 1 implies that any α-approximation algorithm for VC_S also gives an α-approximation algorithm for S. Thus we obtain a randomized $(2 - 2\frac{k}{t})$-approximation algorithm for S. $\square$

A.2   Proof of Claim 1

This proof is similar to the proof in [1] for proving APX-completeness of vertex cover on cubic graphs.

Proof of Claim. It is easy to see that from every vertex cover $C \subseteq V$ of G we can construct a vertex cover $C' \subseteq V'$ of G' of size exactly $|C| + |V| + |E \setminus E_T|$. In C' we include u_1^i for all $i \in \{i : v_i \in V \setminus C\}$; u_2^i for all $i \in \{i : v_i \in C\}$; e_1^{ij} for each $(v_i, v_j) \in E \setminus E_T$ with $v_i \in V \setminus C$; e_2^{ij} for each $(v_i, v_j) \in E \setminus E_T$ with $v_i \in C$; and every vertex in C.

Given a vertex cover $C' \subseteq V'$ of G' we transform it into a vertex cover $C \subseteq V$ of G in the following manner. Suppose there exists $v_i, v_j \in V$ with $i < j$ such that $\{v_i, v_j\} \in E$ and $v_i \notin C', v_j \notin C'$. Since C' is a feasible vertex cover of G' we have that $\{v_i, v_j\} \in E \setminus E_T$ and either $\{e_1^{ij}, e_2^{ij}, u_1^j\} \subseteq C'$ or $\{e_1^{ij}, u_2^j, u_1^j\} \subseteq C'$. Thus we can obtain a vertex cover $C'' \subseteq V'$ of G' with $|C''| \le |C'|$ by letting $C'' = (C' \setminus \{u_1^j, e_2^{ij}\}) \cup \{v_j, u_2^j\}$. Repeating this procedure will result in a vertex cover $C''' \subseteq V'$ of G' with $|C'''| \le |C'|$ such that $C = C''' \cap V$ is a feasible vertex cover of G. Furthermore it is easy to see that $|C| \le |C'''| - |V| - |E \setminus E_T|$.   □

A.3   Proof of Claim 3

Proof of Claim. We relate the two graphs G_I' and G' by the bijection $f : D \to V'$, defined as follows.

$$f((a,b)) = \begin{cases} v_j, & \text{if } (a,b) = (s_i, s_j), \\ u_1^i, & \text{if } (a,b) = (s_i, m_i), \\ u_2^i, & \text{if } (a,b) = (m_i, e_i), \\ e_1^{ij}, & \text{if } (a,b) = (s_i, b_{ij}), \\ e_2^{ij}, & \text{if } (a,b) = (b_{ij}, m_j). \end{cases}$$

Suppose $\{(a,b), (c,d)\} \in E_I$. Since I is an interval order (does not contain any $\mathbf{2 + 2}$ structures as induced posets [21,32]) and by the definition of D we have that $b = c$. Now consider the possible cases of $\{(a,b), (b,d)\}$.

$(a = s_i, b = s_j, d = s_k, i < j < k)$ By construction of I, v_j is the parent of v_k, i.e., $(f((s_i, s_j)), f((s_j, s_k)) = (v_j, v_k) \in E_T \subseteq E'$.

$(a = s_i, b = s_j, d = b_{jk}, i < j < k)$ Then $f((s_i, s_j)) = v_j$ and $f((s_j, b_{jk})) = e_1^{ij}$ and by definition of G' we have $(v_j, e_1^{jk}) \in E'$.

The remaining cases $(a = s_i, b = s_j, d = m_j, i < j)$, $(a = s_i, b = b_{ij}, d = m_j, i < j)$, $(a = s_i, b = m_i, d = e_i)$, and $(a = b_{ij}, b = m_j, d = e_j, i < j)$ are similar to the two above and it is straightforward to check the implication $\{(a,b), (b,d)\} \in E_I \Rightarrow \{f((a,b)), f((b,c))\} \in E'$.

On the other hand, suppose $(a,b) \in E'$ and again consider the different possible cases.

$(a = v_i, b = v_j, i < j)$ Then v_i is the parent of v_j in T and $f^{-1}(v_i) = (s_k, s_i)$ and $f^{-1}(v_j) = (s_i, s_j)$ for some $k < i < j$. Since s_k's interval representation is completely to the left of s_j's interval representation in I the incomparable pairs (s_k, s_i) and (s_i, s_j) cannot be reversed in the same linear extension, i.e., $\{(s_k, s_i), (s_i, s_j)\} \in E_I$.

$(a = v_i, b = e_1^{ij}, i < j)$ Then $f^{-1}(v_i) = (s_k, s_i)$ and $f^{-1}(e_1^{ij}) = (s_i, b_{ij})$ for some $k < i < j$. Since s_k's interval representation is completely to the left of b_{ij}'s interval representation in I the incomparable pairs (s_k, s_i) and (s_i, b_{ij}) cannot be reversed in the same linear extension, i.e., $\{(s_k, s_i), (s_i, b_{ij})\} \in E_I$.

The remaining cases $(a = e_1^{ij}, b = e_2^{ij}, i < j)$, $(a = e_2^{ij}, b = u_2^j, i < j)$, $(a = u_1^j, b = u_2^j, i < j)$, and $(a = v_j, b = u_1^j, i < j)$ are similar to the two above and omitted.
We have thus proved that $\{(a,b), (b,d)\} \in E_I \Leftrightarrow \{f((a,b)), f((b,c))\} \in E'$, i.e., the function f defines an isomorphism between G_I' and G'.   □

Approximation Algorithms for 2-Stage Stochastic Scheduling Problems

David B. Shmoys[1,*] and Mauro Sozio[2,**]

[1] School of ORIE and Dept. of Computer Science, Cornell University, Ithaca, NY 14853
shmoys@cs.cornell.edu
[2] Dept. of Computer Science, University of Rome "La Sapienza", Italy
sozio@di.uniroma1.it

Abstract. There has been a series of results deriving approximation algorithms for 2-stage discrete stochastic optimization problems, in which the probabilistic component of the input is given by means of "black box", from which the algorithm "learns" the distribution by drawing (a polynomial number of) independent samples. The performance guarantees proved for such problems, of course, is generally worse than for their deterministic analogue. We focus on a 2-stage stochastic generalization of the problem of finding the maximum-weight subset of jobs that can be scheduled on one machine where each job is constrained to be processed within a specified time window. Surprisingly, we show that for this generalization, the same performance guarantee that is obtained for the deterministic case can be obtained for its stochastic extension.

Our algorithm builds on an approach of Charikar, Chekuri, and Pál: one first designs an approximation algorithm for the so-called polynomial scenario model (in which the probability distribution is restricted to have the property that there are only a polynomial number of possible realizations of the input that occur with positive probability); then one shows that by sampling from the distribution via the "black box" to obtain an approximate distribution that falls in this class and approximately solves this approximation to the problem, one nonetheless obtains a near-optimal solution to the original problem. Of course, to follow this broad outline, one must design an approximation algorithm for the stochastic optimization problem in the polynomial scenario model, and we do this by extending a result of Bar-Noy, Bar-Yehuda, Freund, Naor, and Schieber.

Furthermore, the results of Bar-Noy et al. extend to a wide variety of resource-constrained selection problems including, for example, the unrelated parallel-machine generalization $R|r_j| \sum w_j U_j$ and point-to-point admission control routing in networks (but with a different performance guarantee). Our techniques can also be extended to yield analogous results for the 2-stage stochastic generalizations for this class of problems.

1 Introduction

Consider the following 2-stage stochastic optimization problem: there are n users, each of whom might request a particular communication channel, which can serve at most

---

* Research supported partially by NSF grants CCR-0635121 & DMI-0500263.
** This work was done while this author was a visiting student at Cornell University. The work was partially supported by NSF grant CCR-0430682 and by EC project DELIS.

M. Fischetti and D.P. Williamson (Eds.): IPCO 2007, LNCS 4513, pp. 145–157, 2007.

one user at a time, for a specified length of time within a specified time interval; for a given planning period, it is not known which of the n users will actually make their request – all that is known is a probability distribution over the subsets of users indicating which subset might be active; each user has an associated profit for actually being scheduled on the channel; alternatively, the manager of the channel can redirect the user to other providers, thereby obtaining a specified (but significantly smaller) profit; the aim is to decide which users to defer so as to maximize the expected profit over the two stages (where the expectation is with respect to the probability distribution over subsets of active users). Thus, this is a stochastic generalization of the (maximization version) of the single machine scheduling problem that is denoted in the notation of [4] as $1|r_j|\sum w_j U_j$ and we shall refer to this generalization as the *2-stage stochastic* $1|r_j|\sum w_j U_j$. For the deterministic version of this problem, Bar-Noy, Bar-Yehuda, Freund, Naor, & Schieber give a ρ-approximation algorithm for any constant $\rho > 2$; rather surprisingly, we show that the exact same result holds for the stochastic generalization. (A *ρ-approximation algorithm* for an optimization problem is a (randomized) polynomial-time algorithm that finds a feasible solution with (expected) cost within a factor of ρ of optimal.)

Recently, there has been a series of results for 2-stage discrete stochastic optimization problems with recourse, starting with the work of Dye, Stougie, and Tomasgard[3] that addressed a knapsack-like single-node network provisioning problem. That paper made the simplifying assumption of the *polynomial scenario model* in which there are (only) a polynomial number of scenarios that can be realized in the second stage, and thereby derived the first worst-case performance guarantees for polynomial-time algorithms for models of this type. Kong & Schaefer [8] gave an 2-approximation algorithm for a 2-stage variant of the the maximum-weight matching problem, again in a polynomial scenario model. Later, Immorlica, Karger, Minkoff, and Mirrokni [7], and also Ravi and Sinha [9] addressed analogous questions based on deterministic problems such as the vertex cover problem, the set covering problem, the uncapacitated facility location problem, and network flow problems. The former paper also considered the situation when the probability distribution conformed to an *independent activation model* which, in our setting for example, would mean that there is a probability associated with each user and the active set is drawn by assuming that these are independent Bernoulli random events. However, for these latter results they introduced the *proportionality assumption* in which the corresponding costs for an element in the two stages had constant ratio λ for all elements. Gupta, Pál, Ravi, and Sinha [5] proposed a much more general mechanism for specifying the probability distribution, in which one has access to a *black box* from which to generate independent samples according to the distribution, and thereby make use of a polynomial number of samples in the process of computing the first-stage decisions. They gave constant approximation algorithms for a number of 2-stage stochastic optimization problems in this model, most notably the minimum-cost rooted Steiner tree problem and the uncapacitated facility location problem, but they also require the proportionality assumption.

Shmoys & Swamy [10] gave an LP-rounding technique, and showed that one could derive a polynomial-time approximation scheme for the exponentially-large linear programming relaxations in order to derive the first approximation algorithms in the black box model without the proportionality assumption, in particular for a variety of set covering-related problems, the uncapacitated facility location problem, and multi-commodity flow problems. Swamy & Shmoys [11] extend this to constant-stage models, and also show that the so-called sample average approximation yields a polynomial approximation scheme for the LP relaxations. Charikar, Chekuri, and Pál [2] gave a general technique based on the sample average approximation that, for a broad class of 2-stage stochastic minimization problem with recourse, in effect reduced the problem of obtaining a good approximation algorithm for the black box model, to the problem of obtaining the analogous result in the polynomial scenario setting.

We build on these results, by first constructing an approximation algorithm for our maximization problem in the polynomial scenario model, and then derive a maximization variant of the result of [2] (but still specialized to our class of problems) to obtain approximation algorithms in the black box probability model.

We focus on the central model in the class proposed by Bar-Noy, Bar-Yehuda, Freund, Naor, and Schieber [1], who gave primal-dual algorithms for a rich class of deterministic resource allocation and scheduling problems. In their terminology, there is a set of activities, $\{\mathcal{A}_1, \ldots, \mathcal{A}_n\}$; let $\mathcal{N} = \{1, \ldots, n\}$ index this set. For each activity $\mathcal{A}_j$, $j \in \mathcal{N}$, there is a set of possible instances A_j that specify the various ways in which the activity might be handled (so, in the description above, assuming integer data for the input times, for each user we have one instance for each possible integer starting time that would have it complete by the deadline). This approach appears to convert the original input to a new input in which there are a pseudopolynomial number of instances for each activity. However, Bar-Noy et al. also show how to convert their pseudopolynomial-time algorithm into a polynomial-time one, while losing only a $1 + \epsilon$ factor in the performance guarantee.

Our algorithm is a rather natural extension of the approach of Bar-Noy et al. We first run their algorithm on each of the polynomially many scenarios, where the profit of selecting an instance is its contribution to the overall expected second stage profit. For each scenario (which is, after all just an ordinary deterministic input), this generates a feasible dual solution. The deterministic dual variables are of two types: those that are dual to the constraint that says that each activity is scheduled in at most one way (that is, at most one instance of each activity is selected); and those that correspond to the constraint that at each time at most one instance (over all activities) is active. The usual interpretation of dual variables leads us to view the former as providing the marginal expected profit attainable by having this activity on hand in a particular scenario. Thus, we decide to defer an activity $\mathcal{A}_j$, if the total of the corresponding dual variables, summed over all scenarios, is less than the profit collected by actually deferring that activity. This gives the stage I actions. The stage II actions for each scenario are computed by adapting the algorithm of Bar-Noy et al.; we first compute a dual solution that includes even the deferred activities, but then does not select any instance of a deferred activity in constructing the primal solution.

The analysis of our algorithm is also surprisingly simple, and is based on a primal-dual approach using an integer programming formulation of the 2-stage problem. We show that the dual solutions constructed in each scenario can be pieced together to yield a feasible solution for the dual to the linear programming relaxation, and can then show that the expected profit of the primal solution constructed is at least half the value of the feasible dual solution found. This yields that the resulting algorithm is a 2-approximation algorithm. Like the algorithm of Bar-Noy et al., this is a pseudopoly-nomial-time algorithm, but an approach identical to the one they employed yields a polynomial-time algorithm, while losing a factor of $1 + \epsilon$ in the performance guarantee. Although we focus on this single-machine scheduling model, our approach can be generalized to yield analogously strong results for 2-stage stochastic generalization of the class of problems for which the framework of Bar-Noy et al. applies. This will be discussed in detail in the full version of this paper.

There are other potential 2-stage stochastic extensions of the problem of computing a maximum-weight subset of jobs that can be feasible scheduled. One other natural approach is to use the first stage to make initial decisions about which users to service (but to commit to serve them if they are active), and then to allow the possibility of serving additional users in the second stage, once the probabilistic choice of scenario has been made (with correspondingly lesser profit). We show that the maximum independent set problem can be reduced to an extremely restricted special case of this model in an approximation-preserving way, and hence we cannot hope to obtain a good approximation algorithm for this setting (unless $\mathcal{P} = \mathcal{NP}$). There are few (if any) such strong inapproximability results known for stochastic optimization problems for which their deterministic analogue is relatively easily approximable.

2   IP and LP Formulations: 2-Stage Stochastic Models

We start by giving a natural integer (linear) programming formulation (and its dual) for the 2-stage stochastic version of $1|r_j| \sum_j w_j U_j$, in its pseudopolynomial-sized variant.

Let $\mathcal{S}$ be a collection of explicitly given scenarios $\{S_1, \ldots, S_m\}$ that occur with positive probability; in each scenario S, for each activity $\mathcal{A}_j$, there is an associated set of available instances $A_j(S) \subseteq A_j$. For each instance I, there is an associated starting time $s(I)$, and an associated ending time $e(I)$. For each scenario $S \in \mathcal{S}$, there is an associated probability $q(S)$, where $q(S) \geq 0$ and $\sum_{S \in \mathcal{S}} q(S) = 1$. In stage I, we must decide which activities to defer, and thereby obtain a (small) profit of p_j^{I}, or else retain for stage II, in which for each scenario S we can obtain a profit $p_j^{\mathrm{II}}(I, S)$ for assigning this activity using instance $I \in A_j(S)$. We give an integer programming formulation of this problem. For each activity $\mathcal{A}_j$, we have a 0-1 variable x_j that indicates whether activity $\mathcal{A}_j$ is deferred in the first phase or not (where $x_j = 1$ means that it is deferred). For each instance I of activity $A_j(S)$, we have a variable $y_j(I, S)$ whose value is 1 if and only if instance I of this activity is scheduled. Let $\mathcal{T}$ be the set of all start-times and end-times of all instances belonging to all activities and let $\mathcal{T}_I = \{t \in \mathcal{T} | s(I) \leq t < e(I)\}$ for each instance I. Moreover, let $f(I) \in \mathcal{T}$ be maximal such that $f(I) < e(I)$.

We can formulate the 2-stage problem of maximizing the total expected profit as follows:

$$\max \sum_{j \in \mathcal{N}} p_j^I x_j + \sum_{j \in \mathcal{N}} \sum_{S \in \mathcal{S}} \sum_{I \in A_j(S)} q(S) p_j^{II}(I, S) y_j(I, S) \tag{SIP}$$

$$\text{s.t. } x_j + \sum_{I \in A_j(S)} y_j(I, S) \leq 1 \qquad\qquad \forall j \in \mathcal{N}, S \in \mathcal{S}, \tag{1}$$

$$\sum_{j \in \mathcal{N}} \sum_{I \in A_j(S): t \in T_I} y_j(I, S) \leq 1 \qquad\qquad \forall S \in \mathcal{S}, t \in \mathcal{T}, \tag{2}$$

$$x_j, y_j(I, S) \in \{0, 1\}, \qquad\qquad \forall j \in \mathcal{N}, S \in \mathcal{S}, I \in A_j(S). \tag{3}$$

Let (SLP) be the LP obtained by replacing (3) by non-negativity constraints for these variables. If we let $u_j(S)$ be the dual variables corresponding to the constraints (1), and let $v_t(S)$ denote the dual variables corresponding to the constraints (2), then we can write the LP dual of (SLP) as:

$$\min \sum_{j \in \mathcal{N}} \sum_{S \in \mathcal{S}} u_j(S) + \sum_{S \in \mathcal{S}} \sum_{t \in \mathcal{T}} v_t(S) \tag{SD}$$

$$\text{s.t. } \sum_{S \in \mathcal{S}} u_j(S) \geq p_j^I, \qquad\qquad \forall j \in \mathcal{N}, \tag{4}$$

$$u_j(S) + \sum_{t \in T_I} v_t(S) \geq q(S) p_j^{II}(I, S), \qquad \forall j \in \mathcal{N}, S \in \mathcal{S}, I \in A_j(S), \tag{5}$$

$$u_j(S), v_t(S) \geq 0. \tag{6}$$

It is important to note that our algorithm will not need to solve any of these linear programs! We will simply apply an algorithm for the deterministic variant (for which a performance guarantee relative the optimal value of the deterministic LP is known) to an input based on each scenario $S \in \mathcal{S}$, and then use the linear programs to analyze the performance of the resulting algorithm.

3   An Algorithm for the Polynomial Scenario Model

We shall show how to adapt the primal-dual algorithmic framework of Bar-Noy, Bar-Yehua, Freund, Naor, & Schieber [1] to yield an approximation algorithm with the identical performance guarantee for the 2-stage stochastic variant of $1|r_j| \sum w_j U_j$, in the polynomial scenario model. For this model, it is straightforward to derive a constant approximation algorithm. The simplest approach is to randomize, and with probability 1/2 to defer all jobs, and otherwise, to run the 2-approximation algorithm of Bar-Noy et al. on the active jobs in the second stage; this is a randomized 4-approximation algorithm. In the polynomial scenario model, one can improve upon this by comparing the benefit of deferring all users with the expected profit obtained by the Bar-Noy algorithm based on not deferring anyone, and then selecting the better of the two. This is easily shown to be a 3-approximation algorithm (and can be extended to the black box model while losing only a factor of $1 + \epsilon$). Thus, the surprising aspect of our result is that it is in fact possible to obtain an algorithm for the 2-stage generalization without degrading the performance guarantee at all.

The framework of Bar-Noy et al. works in two phases: a pushing phase in which a dual solution is constructed along with a stack of instances that might be selected to be scheduled; and a popping phase in which elements of the stack are popped off, and accepted for scheduling provided that they do not conflict with activities already scheduled by this procedure.

The algorithm for the 2-stage problem proceeds as follows. For each scenario $S \in \mathcal{S}$, the deterministic profit $p_j(I)$ is $q(S)p_j^{\mathrm{II}}(I, S)$ for each $j \in \mathcal{N}$, and each $I \in A_j(S)$. We execute the pushing procedure of the algorithm proposed in Bar-Noy et al. for each scenario $S \in \mathcal{S}$. Algorithm 1 shows the pseudocode for this procedure. We let $u_j(S)$ denote the dual variable corresponding to the deterministic analogue of (1) computed by this procedure. Then, for each activity A_j, $j \in \mathcal{N}$, we check if

$$p_j^{\mathrm{I}} \geq \sum_{S \in \mathcal{S}} u_j(S), \qquad (7)$$

and defer each activity A_j that satisfies this condition. This completes the first stage action. We shall also denote this solution by setting $\bar{x}_j = 1$ for each deferred activity A_j, and setting $\bar{x}_j = 0$ otherwise.

In what follows, we shall say that an instance I is *uncovered* if constraint (5) for instance I is not satisfied and we say that I is *tight* if this constraint is satisfied with equality.

For the second stage, for a given scenario $S \in \mathcal{S}$, we recompute the execution of the pushing procedure. Then we compute a feasible schedule by executing the popping procedure of the algorithm of Bar-Noy et al., but we delete each activity that was deferred in the first phase. We denote this solution by setting $\bar{y}_j(I, S) = 1$ for each scheduled instance I, and setting $\bar{y}_j(I, S) = 0$ otherwise. Algorithm 2 shows the pseudocode for the second phase for a given scenario.

The main intuition behind the deferring rule is the following. Suppose at the end of the pushing phase the total value of variables u of an activity A_j is "small". There are two possible reasons for this. The total profit of all instances of A_j is smaller than p_j^{I}. In this case, it is clear that deferring the activity is the best we can do. If the total profit P of instances of A_j is greater than p_j^{I}, then since u is "small", there are many other instances of other activities which are in conflict with instances of A_j. Hence, P can be "replaced" by the profit of these instances, and we can gain other profit by deferring A_j. More generally, the value of the sum reflects the total expected marginal value of the activity A_j; if this is less than the (sure) profit gained by deferring it, then certainly deferring it is a good thing to do.

We shall prove that the performance guarantee of the two-phase algorithm is 2. The main idea behind this proof is the following. Each instance increases the total value of the dual variables by some amount 2δ. For instances that belong to a non-deferred activity, we are able to charge δ to a scheduled instance. For instances that belong to a deferred activity, we charge this amount to the profit gained by deferring that activity.

Given a scenario S we say that $I \in A_j(S)$ and $\hat{I} \in A_l(S)$ are *incompatible* if $j = l$ or their time intervals overlap. For each instance $I \in A_j(S)$, we refer to the variables which occur in the constraint (5) for I, as "the variables of I".

Algorithm 1. Pushing procedure for the first phase in scenario S

1: $Stack(S)=\emptyset$;
2: $u_j(S) \leftarrow 0 \quad \forall j \in N$;
3: $v_t(S) \leftarrow 0 \quad \forall t \in T$;
4: **while** no uncovered instance is left **do**
5:     select an uncovered instance $I \in A_j(S)$, $j \in \mathcal{N}$ with minimum end-time;
6:     push(I,$Stack(S)$);
7:     let $\delta(I,S) = (q(S)p_j^{\mathrm{II}}(I,S) - u_j(S) - \sum_{t \in T_I} v_t(S))/2$;
8:     $u_j(S) \leftarrow u_j(S) + \delta(I,S)$;
9:     $v_{f(I)}(S) \leftarrow v_{f(I)}(S) + \delta(I,S)$;
10: **end while**

Algorithm 2. The algorithm for the second phase in scenario S

1: /* pushing procedure */
2: $Stack(S)=\emptyset$;
3: $u_j(S) \leftarrow 0 \quad \forall j \in N$;
4: $v_t(S) \leftarrow 0 \quad \forall t \in T$;
5: **while** no uncovered instance is left **do**
6:     select an uncovered instance $I \in A_j(S)$, $j \in \mathcal{N}$ with minimum end-time;
7:     push(I,$Stack(S)$);
8:     let $\delta(I,S) = (q(S)p_j^{\mathrm{II}}(I,S) - u_j(S) - \sum_{t \in T_I} v_t(S))/2$;
9:     $u_j(S) \leftarrow u_j(S) + \delta(I,S)$;
10:     $v_{f(I)}(S) \leftarrow v_{f(I)}(S) + \delta(I,S)$;
11: **end while**
12: /* scheduling procedure */
13: **while** $Stack(S)$ is not empty **do**
14:     I=pop($Stack(S)$);
15:     Let $j \in \mathcal{N} : I \in A_j(S)$;
16:     **if** A_j is not deferred and I is not in conflict with other scheduled instances **then**
17:         schedule I and set $\bar{y}_j(I,S) = 1$;
18:     **end if**
19: **end while**

Theorem 1. *For the 2-stage stochastic maximization version of $1|r_j| \sum w_j U_j$, there is a $(2 + \epsilon)$-approximation algorithm in the polynomial scenario model.*

Proof. We shall consider only the version of the problem in which we have a pseudopolynomial representation of the input: that is, for each activity, we have an explicitly given set of allowed starting times. However, for each scenario, this is exactly the algorithm of Bar-Noy et al. (on a carefully constructed input), who show that it can be converted to run in polynomial time for $1|r_j| \sum w_j U_j$, while losing a factor of $1 + \epsilon$ in the performance guarantee. This will thereby yield the theorem in the form stated above.

Let $\bar{u}_j(S)$ and $\bar{v}_t(S)$ be the value of the dual variables u and v at the end of the algorithm. First consider the constraints (5); the algorithm ensures that these are satisfied by the dual solution computed. This is a consequence of the fact that as long as there exists an uncovered instance, the algorithm pushes an instance in the stack and increases its dual variables making a constraint (5) tight. Hence, at the end of the algorithm, there does not exist an uncovered instance, and each constraint (5) is satisfied. On the other hand, constraint (4) can be violated by any deferred activity. In order to satisfy this constraint, we increase the value of dual variables in the following way. Let

$$\delta_j = p_j^I - \sum_{S \in \mathcal{S}} \bar{u}_j(S) \quad j = 1, \ldots, n$$

and let $\bar{S} \in \mathcal{S}$, be an arbitrarily chosen scenario. For each activity $\mathcal{A}_j$, we increase the value of $\bar{u}_j(\bar{S})$ by δ_j. Clearly, this maintains that the other constraints are satisfied, and ensures that constraint (4) is satisfied now as well.

We now prove the performance guarantee of the algorithm is 2. The essence of the proof is as follows. In each scenario S, for each instance I of a non-deferred activity, we charge $\delta(I, S)$ to some scheduled instance. For each instance I of a deferred activity $\mathcal{A}_j$, we charge δ_j and $\delta(I, S)$ to the profit p_j^I. Hence, at the end of the algorithm, all amounts δ are "charged" to some profit. Moreover, the sum of all these δ, multiplied by 2, gives a bound on the total value of the dual variables. The theorem then follows from weak duality.

Consider a scenario S. Let $\hat{I} \in A_j(S)$ be an instance scheduled in S such that $\mathcal{A}_j$ is not deferred, $j \in \mathcal{N}$. Let $B_{\hat{I}}(S)$ be a set which contains $\hat{I}$ and as well as instances that are:

– incompatible with $\hat{I}$ and
– pushed onto $Stack(S)$ before $\hat{I}$.

Consider each instance I in $B_{\hat{I}}(S)$. When I is placed on the stack, there are two dual variables that are increased by $\delta(I, S)$. For each such I, one of these two variables are variables of $\hat{I}$. If $I \in A_j(S)$, then the variable $u_j(S)$ occurs in constraint (5) for $\hat{I}$. Otherwise, since $e(\hat{I}) \geq e(I)$, then the variable $v_{f(I)}(S)$ occurs in this constraint. Let $\hat{u}$ and $\hat{v}$ be the value of dual variables u and v at the time $\hat{I}$ is pushed in the stack. We have that:

$$\sum_{I \in B_{\hat{I}}(S)} \delta(I, S) \leq \hat{u}_j(S) + \sum_{t \in \mathcal{T}_{\hat{I}}} \hat{v}_t(S) \leq q_S p_j^{II}(\hat{I}, S) \tag{8}$$

where last inequality follows from the fact that $\hat{I}$ is uncovered before being pushed on the stack and after that, its variables are increased in order to make constraint (5) tight.

Note that each instance I of a non-deferred activity belongs to the set $B_{\hat{I}}(S)$ for some instance $\hat{I}$. This follows from the fact that either I is scheduled or there is another instance $\hat{I}$ pushed after I in the stack, which has been scheduled instead of I. This implies that for each scenario $S \in \mathcal{S}$

$$\sum_{\substack{j \in \mathcal{N}: \\ \bar{x}_j = 0}} \sum_{I \in A_j(S)} \delta(I, S) = \sum_{\substack{j \in \mathcal{N}: \\ \bar{x}_j = 0}} \sum_{\substack{\hat{I} \in A_j(S): \\ y_j(\hat{I}, S) = 1}} \sum_{I \in B_{\hat{I}}(S)} \delta(I, S)$$

$$\leq \sum_{\substack{j \in \mathcal{N}: \\ \bar{x}_j = 0}} \sum_{\hat{I} \in A_j(S)} q_S p_j^{II}(\hat{I}, S) \bar{y}_j(\hat{I}, S) \tag{9}$$

For each deferred activity $\mathcal{A}_j$, we have that:

$$\delta_j + \sum_{S \in \mathcal{S}} \sum_{I \in A_j(S)} \delta(I, S) = \sum_{S \in \mathcal{S}} \bar{u}_j(S) = p_j^{\mathrm{I}} \tag{10}$$

By combining Equation (9) and Equation (10), we obtain

$$\sum_{j \in \mathcal{N}} \left(\delta_j + \sum_{\substack{S \in \mathcal{S} \\ I \in A_j(S)}} \delta(I, S) \right) = \sum_{S \in \mathcal{S}} \sum_{\substack{j \in \mathcal{N}: \\ \bar{x}_j = 0}} \sum_{I \in A_j(S)} \delta(I, S) + \sum_{\substack{j \in \mathcal{N}: \\ \bar{x}_j = 1}} \left(\delta_j + \sum_{\substack{S \in \mathcal{S} \\ I \in A_j(S)}} \delta(I, S) \right)$$

$$\le \sum_{S \in \mathcal{S}} \sum_{\substack{j \in \mathcal{N}: \\ \bar{x}_j = 0}} \sum_{I \in A_j(S)} q_S p_j^{\mathrm{II}}(I, S) \bar{y}_j(I, S) + \sum_{\substack{j \in \mathcal{N}: \\ \bar{x}_j = 1}} p_j^{\mathrm{I}}$$

$$\le \sum_{j \in \mathcal{N}} p_j^{\mathrm{I}} \bar{x}_j + \sum_{j \in \mathcal{N}} \sum_{\substack{S \in \mathcal{S} \\ I \in A_j(S)}} q(S) p_j^{\mathrm{II}}(I, S) \bar{y}_j(I, S) \tag{11}$$

Since the initial value of each dual variable is zero, and each instance $I \in A_j(S)$ increases the total value of the dual variables by at most $2\delta(I, S)$, we can sum over all such δ to bound the total value of the dual variables:

$$\sum_{j \in \mathcal{N}} \sum_{S \in \mathcal{S}} \bar{u}_j(S) + \sum_{S \in \mathcal{S}} \sum_{t \in T} \bar{v}_t(S) \le 2 \left(\sum_{j \in \mathcal{N}} \left(\delta_j + \sum_{S \in \mathcal{S}} \sum_{I \in A_j(S)} \delta(I, S) \right) \right) \tag{12}$$

Equations (11) and (12), together with the weak duality theorem, immediately imply the claimed result.

4   An Algorithm for the Black Box Model

We show next that we can adapt the algorithm derived in the previous section for the polynomial scenario setting to the black box model, where the probability distribution is specified only by allowing access to an oracle from which independent samples according the distribution can be drawn. We show that applying the previous algorithm to an approximate version of the distribution based on sampling can be shown to still yield the same performance guarantee. Our analysis uses the structure of the analysis used for the previous algorithm, and builds on the general result for minimization 2-stage stochastic problems derived by Charikar, Chekuri, and Pál [2].

We shall make use of the following version of the Chernoff bound.

Lemma 1. *Let $X_1, \ldots X_N$ be independent random variables with $X_i \in [0, 1]$ and let $X = \sum_{i=1}^{N} X_i$. Then, for any $\epsilon \ge 0$, we have $Pr\left[|X - E[X]| > \epsilon N \right] \le 2 \exp(-\epsilon^2 N)$.*

We assume that there is an inflation factor $\lambda \ge 1$ such that $p_j^{\mathrm{II}}(I, S) \le \lambda p_j^{\mathrm{I}}$, $\forall j \in \mathcal{N}$, $\forall S \in \mathcal{S}$, $\forall I \in A_j(S)$.

The algorithm first takes a polynomial-sized sample from the set of scenarios and then proceeds just as the Algorithm 1 in Section 3 while using a slightly different deferring rule.

More precisely, it takes $N = \Theta(\frac{\lambda^2}{\epsilon^2} \log \frac{n}{\gamma})$ independent random samples $S_1, \ldots, S_N$ from the black box, where n is the number of activities, ϵ will be the allowed additional relative error, and γ is the confidence parameter (that is, we shall obtain that the desired approximation is found with probability at least $1 - \gamma$). Then the algorithm executes the pushing procedure (see Algorithm 1) for each scenario that occurs in the polynomial sample. Observe that the data used by this algorithm for scenario S is described to be $q(S)p_j^{II}(I, S)$. At first glance, this might be worrying, but of course the value $q(S)$ is just a uniform scalar multiple for all profits, and so it makes sense to define $\tilde{u}$ and $\tilde{v}$ as the dual variables computed after executing this algorithm with inputs $p_j^{II}(I, S)$. Observe that the values $\bar{u}$ and $\bar{v}$ for a scenario S from our exact distribution are equal to $q(S)\tilde{u}$ and $q(S)\tilde{v}$, respectively. Given $\epsilon > 0$, we shall defers an activity $\mathcal{A}_j$, $j \in \mathcal{N}$, if and only if:

$$(1 + \epsilon)p_j^I \geq \frac{1}{N} \sum_{i=1}^{N} \tilde{u}_j(S_i) \tag{13}$$

This is the deferring rule for the black box model.

This concludes the description of the first stage action. For the second stage, for a given scenario $S \in \mathcal{S}$, we execute Algorithm 2 for scenario S. (Again, note that the linearity effect of $q(S)$ implies that we can run the algorithm with inputs $p_j^{II}(I, S)$ instead.)

Let us analyze the performance guarantee of this algorithm. The proof proceeds by showing that, under the assumption that there is an inflation factor λ, equation (13) is a good approximation for equation (7). This approach is inspired by the proof in [2] for "low scenarios".

Theorem 2. *For any $\epsilon > 0$ and $\gamma > 0$, with probability at least $1 - \gamma$, the proposed deferring rule is a $(2 + \epsilon)$-approximation algorithm for the 2-stage stochastic variant of the problem $1|r_j| \sum w_j U_j$ in the black box model.*

Proof. Suppose we run Algorithm 1 in each of the exponentially-many scenarios and let $\bar{u}$ and $\bar{v}$ be the value of dual variables computed in this way. Consider activity $\mathcal{A}_j$. Let

$$r = \sum_{S \in \mathcal{S}} \bar{u}_j(S) = \sum_{S \in \mathcal{S}} q(S)\tilde{u}_j(S) \qquad \hat{r} = \frac{1}{N} \sum_{i=1}^{N} \tilde{u}_j(S_i).$$

We will prove that, with "high" probability, $\hat{r}$ is "close" to r. We can view $\hat{r}$ as the arithmetic mean of N independent copies $Q_1, \ldots, Q_N$ of the random variable Q defined as

$$Q = \tilde{u}_j(S).$$

Note that $E[Q] = r$. Let Y_i be the variable Q_i/M where $M = \lambda p_j^I$ and let $Y = \sum_i Y_i$. Note that for each activity $\mathcal{A}_j$ and for each scenario $S \in \mathcal{S}$, there exists some $I \in A_j(S)$ such that $\tilde{u}_j(S) \leq p_j^{II}$. This implies that $Y_i \in [0, 1]$. Moreover, $Y = \sum_i Q_i/M = \frac{N}{M}\hat{r}$ and $E[Y] = \sum_i E[Q_i]/M = \frac{N}{M}r$. By applying the Chernoff bound, we obtain the following:

$$Pr\left[|Y - E[Y]| > \frac{\epsilon}{\lambda}N\right] \leq 2\exp\left(-\frac{\epsilon^2}{\lambda^2}N\right) \Leftrightarrow Pr\left[|r - \hat{r}| > \epsilon p_j^I\right] \leq \frac{\gamma}{n}, \tag{14}$$

where the last inequality follows from the choice of the value of N. By taking the union bound over all activities, we obtain that r is "close" to $\hat{r}$ for all activities, with probability at least $1 - \gamma$.

We use the same argument as we used in the polynomial scenario model to show that constraint (5) is satisfied. Consider constraint (4) for some scenario; it may be violated by any activity. We show that it is satisfied, with high probability, by a non-deferred activity. For a deferred activity, we shall increasing the value of its dual variables, as we did in the polynomial scenario model so that the corresponding constraint is also satisfied with high probability. (It is important to note that this increase in the dual variables is not performed by the algorithm; it is only used for the analysis.)

For each deferred activity $\mathcal{A}_j$, let

$$\delta_j = p_j^{\mathrm{I}} - \sum_{S \in \mathcal{S}} \bar{u}_j(S) \quad j = 1, \ldots, \mathcal{N}$$

and let $\overline{S} \in \mathcal{S}$ be an arbitrarily selected scenario. We increase the value of $\bar{u}_j(\overline{S})$ by δ_j for each deferred activity $\mathcal{A}_j$. From the fact that r is a good approximation of $\hat{r}$, it follows that, for each activity $\mathcal{A}_j$, if

$$\frac{1}{N} \sum_{i=1}^{N} \tilde{u}_j(S_i) \le (1 + \epsilon)p_j^{\mathrm{I}},$$

then with probability at least $1 - \gamma$,

$$\sum_{S \in \mathcal{S}} \bar{u}_j(S) \le (1 + 2\epsilon)p_j^{\mathrm{I}}. \tag{15}$$

This implies that with high probability, for each deferred activity $\mathcal{A}_j$

$$\delta_j + \sum_{S \in \mathcal{S}} \sum_{I \in A_j(S)} \delta(I, S) = \sum_{S \in \mathcal{S}} \bar{u}_j(S) \le (1 + 2\epsilon)p_j^{\mathrm{I}} \tag{16}$$

In a similar way, if for an activity $\mathcal{A}_j$

$$\frac{1}{N} \sum_{i=1}^{N} \bar{u}_j(S_i) > (1 + \epsilon)p_j^{\mathrm{I}}$$

then with probability at least $1 - \gamma$, it follows that

$$\sum_{S \in \mathcal{S}} \bar{u}_j(S) > p_j^{\mathrm{I}}.$$

Hence, the new solution is dual feasible with high probability. Note that Equation (16) is an approximation to Equation (10). This implies that by replacing this new equation in the previous proof we obtain

$$\sum_{j \in \mathcal{N}} \sum_{S \in \mathcal{S}} \bar{u}_j(S) + \sum_{S \in \mathcal{S}} \sum_{t \in \mathcal{T}} \bar{v}_t(S) \le 2(1 + 2\epsilon) \sum_{j \in \mathcal{N}} p_j^{\mathrm{I}} \bar{x}_j +$$

$$+ 2(1 + 2\epsilon) \sum_{j \in \mathcal{N}} \sum_{S \in \mathcal{S}} \sum_{I \in A_j(S)} q(S)p_j^{\mathrm{II}}(I, S)\bar{y}_j(I, S), \tag{17}$$

which completes the proof.

5   An $\mathcal{NP}$-Hardness of Approximation Result

We show that, in contrast to the results of the previous sections, another natural 2-stage stochastic generalization of the problem $1|r_j|\sum w_j U_j$ (even in a very simple case) can not be approximated. Suppose that in the first phase, we select a set of activities that we are committed to serve. In the second phase, for a given scenario, we must schedule exactly one instance of each activity selected in the first phase, and we may augment this solution by scheduling other instances of additional activities. We wish to maximize is the total expected profit (where it is now natural to assume that the profit obtained for an instance in the second phase is less than the corresponding profit in the first). We will refer to this problem as the *augmentation 2-stage stochastic* $1|r_j|\sum w_j U_j$.

An integer programming formulation for this problem is obtained by changing (SIP) in the following way: a 0-1 variable x_j indicates (with value 1) that activity $\mathcal{A}_j$ is selected in the first phase; constraint (1) is replaced by the following two constraints:

$$\sum_{I \in A_j(S)} y_j(I, S) \geq x_j \qquad \forall S \in \mathcal{S}, j \in \mathcal{N} : A_j(S) \neq \emptyset \qquad (18)$$

$$\sum_{I \in A_j(S)} y_j(I, S) \leq 1 \qquad \forall j \in \mathcal{N}, S \in \mathcal{S} \qquad (19)$$

Unfortunately, it is straightforward to show that selecting a feasible set of activities in the first phase can be used to model the maximum independent set problem. This is formalized in the following lemma.

Lemma 2. *If there is a ρ-approximation algorithm for the augmentation 2-stage stochastic $1|r_j|\sum w_j U_j$, then there is a ρ-approximation algorithm for maximum independent set problem.*

Proof Sketch. We give an approximation-preserving reduction from the maximum independent set problem. Given a graph G, we build the following input for the augmentation 2-stage stochastic $1|r_j|\sum w_j U_j$. For each vertex v_j, there is an activity $\mathcal{A}_j$, $j = 1, \ldots, n$, each activity is always released at time 0, has deadline time 1, and takes one time unit to complete; each activity has first-stage profit 1, and second-stage profit 0. For each edge $e_i = (v_j, v_k)$, there is a scenario S_i in which only the activities $\mathcal{A}_j$ and $\mathcal{A}_k$ are active. Each scenario S_i occurs with positive probability, and hence our first stage selection must contain at most one of the endpoints of e_i. Thus, there is a one-to-one correspondence between independent sets in G and feasible first-stage decisions. Furthermore, the objective function value of any first-stage decision is exactly the number of activities selected (since the second stage does not contribute any expected profit). Hence, we see that the two optimization problems are identical.   ■

From Lemma (2) and the result in [6] we obtain the following theorem.

Theorem 3. *For any $\epsilon > 0$, there does not exist a polynomial-time algorithm that approximates the augmentation 2-stage stochastic $1|r_j|\sum w_j U_j$ within a factor $n^{1/2-\epsilon}$, unless $\mathcal{P} = \mathcal{NP}$.*

References

1. A. Bar-Noy, R. Bar-Yehuda, A. Freund, J. Naor, and B. Schieber. A unified approach to approximating resource allocation and scheduling. *Journal of the ACM*, 48:1069–1090, 2001.
2. M. Charikar, C. Chekuri, and M. Pál. Sampling bounds for stochastic optimization. In *Proceedings of APPROX-RANDOM 2005*, pages 257–269, 2005.
3. S. Dye, L. Stougie, and A. Tomasgard. The stochastic single resource service-provision problem. *Naval Research Logistics*, 50:869–887, 2003.
4. R. L. Graham, E. L. Lawler, J. K. Lenstra, and A. H. G. Rinnooy Kan. Optimization and approximation in deterministic sequencing and scheduling: A survey. *Ann. Discrete Math.*, 5:287–326, 1979.
5. A. Gupta, M. Pál, R. Ravi, and A. Sinha. Boosted sampling: approximation algorithms for stochastic optimization. In *Proceedings of the 36th Annual ACM Symposium on Theory of Computing*, pages 265–274, 2004.
6. J. Håstad. Clique is hard to approximate within $n^{1-\epsilon}$. *Acta Mathematica*, 182:105–142, 1999.
7. N. Immorlica, D. Karger, M. Minkoff, and V. S. Mirrokni. On the costs and benefits of procrastination: approximation algorithms for stochastic combinatorial optimization problems. In *Proceedings of the 16th ACM-SIAM Symposium on Discrete Algorithms*, pages 691–700, 2004.
8. N. Kong and A. J. Schaefer. A factor 1/2 approximation algorithm for two-stage stochastic matching problems. *European Journal of Operational Research*, 172:740–746, 2006.
9. R. Ravi and A. Sinha. Hedging uncertainty: Approximation algorithms for stochastic optimization problems. In D. Bienstock and G. Nemhauser, editors, *Integer Programming and Combinatorial Optimization: 10th International IPCO Conference*, number 3064 in Lecture Notes in Computer Science, pages 101–115. Springer-Verlag, 2004.
10. D. B. Shmoys and C. Swamy. Stochastic optimization is (almost) as easy as deterministic optimization. In *Proceedings of the 45th Annual Symposium on Foundations of Computer Science*, pages 228–237, 2004.
11. C. Swamy and D. B. Shmoys. The sampling-based approximation algorithms for multi-stage stochastic optimization. In *Proceedings of the 46th Annual Symposium on Foundations of Computer Science*, pages 357–366, 2005.

On Integer Programming and the Branch-Width of the Constraint Matrix

William H. Cunningham and Jim Geelen

Department of Combinatorics and Optimization
University of Waterloo
Waterloo, Canada N2L 3G1
{whcunnin,jfgeelen}@uwaterloo.ca
http://www.math.uwaterloo.ca/C_andO_Dept/index.shtml

Abstract. Consider an integer program $\max(c^t x : Ax = b, x \geq 0, x \in \mathbf{Z}^n)$ where $A \in \mathbf{Z}^{m \times n}$, $b \in \mathbf{Z}^m$, and $c \in \mathbf{Z}^n$. We show that the integer program can be solved in pseudo-polynomial time when A is non-negative and the column-matroid of A has constant branch-width.

1 Introduction

For positive integers m and n, let $A \in \mathbf{Z}^{m \times n}$, $b \in \mathbf{Z}^m$, and $c \in \mathbf{Z}^n$. Consider the following integer programming problems:

(IPF) Find $x \in \mathbf{Z}^n$ satisfying $(Ax = b, x \geq 0)$.
 (IP) Find $x \in \mathbf{Z}^n$ maximizing $c^t x$ subject to $(Ax = b, x \geq 0)$.

Let $M(A)$ denote the column-matroid of A. We are interested in properties of $M(A)$ which lead to polynomial-time solvability for (IPF) and (IP). Note that, even when A (or, equivalently, $M(A)$) has rank one, the problems (IPF) and (IP) are NP-hard. Papadimitriou [9] considered these problems for instances where A has constant rank.

Theorem 1 (Papadimitriou). *There is a pseudopolynomial-time algorithm for solving (IP) on instances where the rank of A is constant.*

Robertson and Seymour [10] introduced the parameter "branch-width" for graphs and also, implicitly, for matroids. We postpone the definition until Section 2. Our main theorem is the following; a more precise result is given in Theorem 6.

Theorem 2. *There is a pseudopolynomial-time algorithm for solving (IP) on instances where A is non-negative and the branch-width of $M(A)$ is constant.*

The branch-width of a matroid M is at most $r(M)+1$. Theorem 2 does not imply Papadimitriou's theorem, since we require that A is non-negative. In Section 6 we show that the non-negativity can be dropped when we have bounds on the variables. However, the following result shows that we cannot just relax the non-negativity.

M. Fischetti and D.P. Williamson (Eds.): IPCO 2007, LNCS 4513, pp. 158–166, 2007.
© Springer-Verlag Berlin Heidelberg 2007

Theorem 3. *(IPF) is NP-hard even for instances where $M(A)$ has branch-width ≤ 3 and the entries of A are in $\{0, \pm 1\}$.*

We also prove the following negative result.

Theorem 4. *(IPF) is NP-hard even for instances where the entries of A and b are in $\{0, \pm 1\}$ and $M(A)$ is the cycle matroid of a graph.*

We find Theorem 4 somewhat surprising considering the fact that graphic matroids are regular. Note that, if A is a $(0, \pm 1)$-matrix and $M([I, A])$ is regular, then A is a totally unimodular matix and, hence, we can solve (IP) efficiently. It seems artificial to append the identity to the constraint matrix here, but for inequality systems it is more natural.

Recall that $M(A)$ is regular if and only if it has no $U_{2,4}$-minor (see Tutte [13] or Oxley [8], Section 6.6). Moreover, Seymour [12] found a structural characterization of the class of regular matroids. We suspect that the class of **R**-representable matroids with no $U_{2,l}$- or $U_{2,l}^*$-minor is also "highly structured" for all $l \geq 0$ (by which we mean that there is likely to be a reasonable analogue to the graph minors structure theorem; see [11]). Should such results ever be proved, one could imagine using the structure to solve the following problem.

Problem 1. Given a non-negative integer $l \geq 0$, is there a polynomial-time algorithm for solving $\max(c^t x : Ax \leq b, x \geq 0, x \in \mathbf{Z}^n)$ on instances where A is a $(0, \pm 1)$-matrix and $M([I, A])$ has no $U_{2,l}$- or $U_{2,l}^*$-minor?

2   Branch-Width

For a matroid M and $X \subseteq E(M)$, we let $\lambda_M(X) = r_M(X) + r_M(E(M) - X) - r(M) + 1$; we call λ_M the *connectivity function* of M. Note that the connectivity function is *symmetric* (that is, $\lambda_M(X) = \lambda_M(E(M) - X)$ for all $X \subseteq E(M)$) and *submodular* (that is, $\lambda_M(X) + \lambda_M(Y) \geq \lambda_M(X \cap Y) + \lambda_M(X \cup Y)$ for all $X, Y \subseteq E(M)$).

Let $A \in \mathbf{R}^{m \times n}$ and let $E = \{1, \ldots, n\}$. For $X \subseteq E$, we let

$$S(A, X) := \operatorname{span}(A|X) \cap \operatorname{span}(A|(E - X)),$$

where $\operatorname{span}(A)$ denotes the subspace of $\mathbf{R}^m$ spanned by the columns of A and $A|X$ denotes the restriction of A to the columns indexed by X. By the modularity of subspaces,

$$\dim S(A, X) = \lambda_{M(A)}(X) - 1.$$

A tree is *cubic* if its internal vertices all have degree 3. A *branch-decomposition* of M is a cubic tree T whose leaves are labelled by elements of $E(M)$ such that each element in $E(M)$ labels some leaf of T and each leaf of T receives at most one label from $E(M)$. The *width* of an edge e of T is defined to be $\lambda_M(X)$ where $X \subseteq E(M)$ is the set of labels of one of the components of $T - \{e\}$. (Since λ_M is symmetric, it does not matter which component we choose.) The *width* of T

is the maximum among the widths of its edges. The *branch-width* of M is the minimum among the widths of all branch-decompositions of M.

Branch-width can be defined more generally for any real-valued symmetric set-function. For graphs, the branch-width is defined using the function $\lambda_G(X)$; here, for each $X \subseteq E(G)$, $\lambda_G(X)$ denotes the number of vertices incident with both an edge in X and an edge in $E(G) - X$. The branch-width of a graph is within a constant factor of its tree-width. Tree-width is widely studied in theoretical computer science, since many NP-hard problems on graphs can be efficiently solved on graphs of constant tree-width (or, equivalently, branch-width). The most striking results in this direction were obtained by Courcelle [1]. These results have been extended to matroids representable over a finite field by Hliněný [4]. They do not extend to all matroids or even to matroids represented over the reals.

Finding Near-Optimal Branch-Decompositions

For any integer constant k, Oum and Seymour [7] can test, in polynomial time, whether or not a matroid M has branch-width k (assuming that the matroid is given by its rank-oracle). Moreover their algorithm *finds* an optimal branch-decomposition in the case that the branch-width is at most k. The algorithm is not practical; the complexity is $O(n^{8k+13})$. Fortunately, there is a more practical algorithm for finding a near-optimal branch-decomposition. For an integer constant k, Oum and Seymour [6] provide an $O(n^{3.5})$ algorithm that, for a matroid M with branch-width at most k, finds a branch-decomposition of width at most $3k - 1$. The branch decomposition is obtained by solving $O(n)$ matroid intersection problems. When M is represented by a matrix $A \in \mathbf{Z}^{m \times n}$, each of these matroid intersection problems can be solved in $O(m^2 n \log m)$ time; see [2]. Hence we can find a near-optimal branch-decomposition for $M(A)$ in $O(m^2 n^2 \log m)$ time.

3   Linear Algebra and Branch-Width

In this section we discuss how to use branch decompositions to perform certain matrix operations more efficiently. This is of relatively minor significance, but it does improve the efficiency of our algorithms.

Let $A \in \mathbf{Z}^{m \times n}$ and let $E = \{1, \ldots, n\}$. Recall that, for $X \subseteq E$, $S(A, X) = \mathrm{span}(A|X) \cap \mathrm{span}(A|(E - X))$ and that $\dim S(A, X) = \lambda_{M(A)}(X) - 1$. Now let T be a branch-decomposition of $M(A)$ of width k, let e be an edge of T, and let X be the label-set of one of the two components of $T - e$. We let $S_e(A) := S(A, X)$. The aim of this section is to find bases for each of the subspaces $(S_e(A) : e \in E(T))$ in $O(km^2 n)$ time.

Converting to Standard Form

Let $B \subseteq E$ be a basis of $M(A)$. Now let $A_B = A|B$ and $A' = (A_B)^{-1} A$. Therefore $M(A) = M(A')$ and $S_e(A) = \{A_B v : v \in S_e(A')\}$. Note that we can find B

and A' in $O(m^2 n)$ time. Given a basis for $S_e(A')$, we can determine a basis for $S_e(A)$ in $O(km^2)$ time. Since T has $O(n)$ edges, if we are given bases for each of $(S_e(A') : e \in E(T))$ we can find bases for each of $(S_e(A) : e \in E(T))$ in $O(km^2 n)$ time.

Matrices in Standard Form

Henceforth we suppose that A is already in standard form; that is $A|B = I$ for some basis B of $M(A)$. We will now show the stronger result that we can find a basis for each of the subspaces $(S_e(A) : e \in E(T))$ in $O(k^2 mn)$ time (note that $k \leq m + 1$).

We label the columns of A by the elements of B so that the identity $A|B$ is labelled symmetrically. For $X \subseteq B$ and $Y \subseteq E$, we let $A[X, Y]$ denote the submatrix of A with rows indexed by X and columns indexed by Y.

Claim. *For any partition (X, Y) of E,*

$$\lambda_{M(A)}(X) = rank\, A[X \cap B, X - B] + rank\, A[Y \cap B, Y - B] + 1.$$

Moreover $S(A, X)$ is the column-span of the matrix

$$
\begin{array}{cc}
 & \begin{array}{cc} X - B & \qquad Y - B \end{array} \\
\begin{array}{c} X \cap B \\ Y \cap B \end{array} &
\left(\begin{array}{cc}
A[X \cap B, X - B] & 0 \\
0 & A[Y \cap B, Y - B]
\end{array} \right).
\end{array}
$$

Proof. The formula for $\lambda_{M(A)}(X)$ is straightforward and well known. It follows that $S(A, X)$ has the same dimension as the column-space of the given matrix. Finally, it is straightforward to check that each column of the given matrix is spanned by both $A|X$ and $A|(E - X)$.

Let (X, Y) be a partition of E. Note that $B \cap X$ can be extended to a maximal independent subset B_X of X and $B \cap Y$ can be extended to a maximal independent subset B_Y of Y. Now $S(A, X) = S(A|(B_X \cup B_y), B_X)$. Then, by the claim above, given B_X and B_Y we can trivially find a basis for $S(A, X)$.

Finding Bases

A set $X \subseteq E$ is called *T-branched* if there exists an edge e of T such that X is the label-set for one of the components of $T - e$. For each T-branched set X we want to find a maximal independent subset $B(X)$ of X containing $X \cap B$. The number of T-branched sets is $O(n)$, and we will consider them in order of non-decreasing size. If $|X| = 1$, then we can find $B(X)$ in $O(m)$ time. Suppose then that $|X| \geq 2$. Then there is a partition (X_1, X_2) of X into two smaller T-branched sets. We have already found $B(X_1)$ and $B(X_2)$. Note that X is spanned by $B(X_1) \cup B(X_2)$. Moreover, for any T-branched set Y, we have
$$r_{M(A)}(Y) - |Y \cap B| \leq r_{M(A)}(Y) + r_{M(A)}(E - Y) - r(M(A)) = \lambda_{M(A)}(Y) - 1.$$
Therefore $|(B(X_1) \cup B(X_2)) - (B \cap X)| \leq 2(k - 1)$. Recall that $A|B = I$. Then in $O(k^2 m)$ time ($O(k)$ pivots on an $m \times k$-matrix) we can extend $B \cap X$ to a basis $B(X) \subseteq B(X_1) \cup B(X_2)$. Thus we can find all of the required bases in $O(k^2 mn)$ time.

4   The Main Result

In this section we prove Theorem 2. We begin by considering the feasibility version.

IPF(k).
INSTANCE: Positive integers m and n, a non-negative matrix $A \in \mathbf{Z}^{m \times n}$, a non-negative vector $b \in \mathbf{Z}^m$, and a branch-decomposition T of $M(A)$ of width k.
PROBLEM: Does there exist $x \in \mathbf{Z}^n$ satisfying $(Ax = b, x \geq 0)$?

Theorem 5. *IPF(k) can be solved in $O((d+1)^{2k}mn + m^2 n)$ time, where $d = \max(b_1, \ldots, b_m)$.*

Note that for many combinatorial problems (like the set partition problem), we have $d = 1$. For such problems the algorithm requires only $O(m^2 n)$ time (considering k as a constant). Recall that $S(A, X)$ denotes the subspace $\mathrm{span}(A|X) \cap \mathrm{span}(A|(E - X))$, where E is the set of column-indices of A.

The following lemma is the key.

Lemma 1. *Let $A \in \{0, \ldots, d\}^{m \times n}$ and let $X \subseteq \{1, \ldots, n\}$ such that $\lambda_{M(A)}(X) = k$. Then there are at most $(d+1)^{k-1}$ vectors in $S(A, X) \cap \{0, \ldots, d\}^m$.*

Proof. Since $\lambda_{M(A)}(X) \leq k$, $S(A, X)$ has dimension $k-1$; let $a_1, \ldots, a_{k-1} \in \mathbf{R}^m$ span $S(A, X)$. There is a $(k-1)$-element set $Z \subseteq \{1, \ldots, n\}$ such that the matrix $(a_1|Z, \ldots, a_{k-1}|Z)$ is non-singular. Now any vector $x \in \mathbf{R}$ that is spanned by $(a_1, \ldots, a_{k-1})$ is uniquely determined by $x|Z$. So there are at most $(d+1)^{k-1}$ vectors in $\{0, \ldots, d\}^m$ that are spanned by $(a_1, \ldots, a_{k-1})$.

Proof (Proof of Theorem 5.). Let $A' = [A, b]$, $E = \{1, \ldots, n\}$, and $E' = \{1, \ldots, n+1\}$. Now, let T be a branch-decomposition of $M(A)$ of width k and let T' be a branch-decomposition of $M(A')$ obtained from T by subdividing an edge and adding a new leaf-vertex, labelled by $n + 1$, adjacent to the degree 2 node. Note that T' has width $\leq k + 1$. Recall that a set $X \subseteq E$ is T-*branched* if there is an edge e of T such that X is the label-set of one of the components of $T - e$. By the results in the previous section, in $O(m^2 n)$ time we can find bases for each subspace $S(A', X)$ where $X \subseteq E$ is T'-branched.

For $X \subseteq E$, we let $\mathcal{B}(X)$ denote the set of all vectors $b' \in \mathbf{Z}^m$ such that

(1) $0 \leq b' \leq b$,
(2) there exists $z \in \mathbf{Z}^X$ with $z \geq 0$ such that $(A|X)z = b'$, and
(3) $b' \in \mathrm{span}(A'|(E' - X))$.

Note that, if $b' \in \mathcal{B}(X)$, then, by (2) and (3), $b' \in S(A', X)$. If $\lambda_{M(A')}(X) \leq k + 1$, then, by Lemma 1, $|\mathcal{B}(X)| \leq (d+1)^k$. Moreover, we have a solution to the problem (IPF) if and only $b \in \mathcal{B}(E)$.

We will compute $\mathcal{B}(X)$ for all T'-branched sets $X \subseteq E$ using dynamic programming. The number of T'-branched subsets of E is $O(n)$, and we will consider them in order of non-decreasing size. If $|X| = 1$, then we can easily find $\mathcal{B}(X)$ in $O(dm)$ time. Suppose then that $|X| \geq 2$. Then there is a partition (X_1, X_2) of X into two smaller T'-branched sets. We have already found $\mathcal{B}(X_1)$ and $\mathcal{B}(X_2)$. Note that $b' \in \mathcal{B}(X)$ if and only if

(a) there exist $b'_1 \in \mathcal{B}(X_1)$ and $b'_2 \in \mathcal{B}(X_2)$ such that $b' = b'_1 + b'_2$,
(b) $b' \leq b$, and
(c) $b' \in S(A', X)$.

The number of choices for b' generated by (a) is $O((d + 1)^{2k})$. For each such b' we need to check that $b' \leq b$ and $b' \in S(A', X)$. Since we have a basis for $S(A', X)$ and since $S(A', X)$ has dimension $\leq k$, we can check whether or not $b' \in S(A', X)$ in $O(m)$ time (considering k as a constant). Therefore we can find $\mathcal{B}(E)$ in $O((d + 1)^{2k}mn + m^2n)$ time.

We now return to the optimization version.

IP(k).
INSTANCE: Positive integers m and n, a non-negative matrix $A \in \mathbf{Z}^{m \times n}$, a non-negative vector $b \in \mathbf{Z}^m$, a vector $c \in \mathbf{Z}^n$, and a branch-decomposition T of $M(A)$ of width k.
PROBLEM: Find $x \in \mathbf{Z}^n$ maximizing $c^t x$ subject to $(Ax = b, x \geq 0)$.

Theorem 6. *IP(k) can be solved in* $O((d + 1)^{2k}mn + m^2n)$ *time, where* $d = \max(b_1, \ldots, b_m)$.

Proof. The proof is essentially the same as the proof of Theorem 5, except that for each $b' \in \mathcal{B}(X)$ we keep a vector $x \in \mathbf{Z}^X$ maximizing $\sum(c_i x_i : i \in X)$ subject to $((A|X_e)x = b', x \geq 0)$. The details are easy and left to the reader.

Theorem 6 implies Theorem 2.

5   Hardness Results

In this section we prove Theorems 3 and 4. We begin with Theorem 3. The reduction is from the following problem, which is known to be NP-hard; see Lueker [5].

Single Constraint Integer Programming Feasibility (SCIPF).
INSTANCE: A non-negative vector $a \in \mathbf{Z}^n$ and an integer b.
PROBLEM: Does there exist $x \in \mathbf{Z}^n$ satisfying $(a^t x = b, x \geq 0)$?

Proof (Proof of Theorem 3.). Consider an instance (a, b) of (SCIPF). Choose an integer k as small as possible subject to $2^{k+1} > \max(a_1, \ldots, a_n)$. For each $i \in \{1, \ldots, n\}$, let $(\alpha_{i,k}, \alpha_{i,k-1}, \ldots, \alpha_{i,0})$ be the binary expansion of a_i. Now consider the following system of equations and inequalities:

(1) $\displaystyle\sum_{i=1}^{n} \sum_{j=0}^{k} \alpha_{ij} y_{ij} = b.$

(2) $y_{ij} - x_i - \sum_{l=0}^{i-1} y_{i,l} = 0$, for $i \in \{1, \ldots, n\}$ and $j \in \{0, \ldots, k\}$.
(3) $x_i \geq 0$ for each $i \in \{1, \ldots, n\}$.

If $(y_{ij} : \in \{1,\ldots,n\}, j \in \{0,\ldots,k\})$ and $(x_1,\ldots,x_n)$ satisfy (2), then $y_{ij} = 2^j x_i$, and (1) simplifies to $\sum(a_i x_i : i \in \{1,\ldots,n\}) = b$. Therefore there is an integer solution to (1), (2), and (3) if and only if there is an integer solution to $(a^t x = b, \ x \geq 0)$.

The constraint matrix B for system (2) is block diagonal, where each block is a copy of the matrix:

$$
C = \begin{array}{c} \\ 1 \\ 2 \\ \vdots \\ k+1 \end{array}
\begin{array}{cccccc}
1 & 2 & 3 & \cdots & k+1 & k+2 \\
\left(\begin{array}{cccccc}
1 & -1 & -1 & \cdots & -1 & -1 \\
0 & 1 & -1 & & -1 & -1 \\
& & \ddots & \ddots & & \\
0 & 0 & 0 & \cdots & 1 & -1
\end{array}\right).
\end{array}
$$

It is straightforward to verify that $M(C)$ is a circuit and, hence, $M(C)$ has branch-width 2. Now $M(B)$ is the direct sum of copies of $M(C)$ and, hence, $M(B)$ has branch-width 2. Appending a single row to B can increase the branch-width by at most one.

Now we turn to Theorem 4. Our proof is by a reduction from 3D Matching which is known to be NP-complete; see Garey and Johnson [3], pp. 46.

3D Matching.
INSTANCE: Three disjoint sets X, Y, and Z with $|X| = |Y| = |Z|$ and a collection $\mathcal{F}$ of triples $\{x, y, z\}$ where $x \in X$, $y \in Y$, and $z \in Z$.
PROBLEM: Does there exist a partition of $X \cup Y \cup Z$ into triples, each of which is contained in $\mathcal{F}$?

Proof (Proof of Theorem 4.). Consider an instance $(X, Y, Z, \mathcal{F})$ of 3D Matching. For each triple $t \in \mathcal{F}$ we define elements u_t and v_t. Now construct a graph $G = (V, E)$ with

$$V = X \cup Y \cup Z \cup \{u_t : t \in \mathcal{F}\} \cup \{v_t : t \in \mathcal{F}\}, \text{ and}$$

$$E = \bigcup_{t=\{x,y,z\}\in\mathcal{F}} \{(u_t, x), (u_t, y), (u_t, v_t), (v_t, z)\}.$$

Note that G is bipartite with bipartition $(X \cup Y \cup \{v_t : t \in \mathcal{F}\}, Z \cup \{u_t : t \in \mathcal{F}\})$.

Now we define $b \in \mathbf{Z}^V$ such that $b_{u_t} = 2$ for each $t \in \mathcal{F}$ and $b_w = 1$ for all other vertices w of G. Finally, we define a matrix $A = (a_{ve}) \in \mathbf{Z}^{V \times E}$ such that $a_{ve} = 0$ whenever v is not incident with e, $a_{ve} = 2$ whenever $v = u_t$ and $e = (u_t, v_t)$ for some $t \in \mathcal{F}$, and $a_{ve} = 1$ otherwise; see Figure 1.

It is straightforward to verify that $(X, Y, Z, \mathcal{F})$ is a YES-instance of the 3D Matching problem if and only if there exists $x \in \mathbf{Z}^E$ satisfying $(Ax = b, \ x \geq 0)$. Now A and b are not $(0, \pm 1)$-valued, but if, for each $t \in \mathcal{F}$, we subtract the v_t-row from the u_t-row, then the entries in the resulting system $A'x = b'$ are in $\{0, \pm 1\}$.

It remains to verify that $M(A)$ is graphic. It is straightforward to verify that A is equivalent, up to row and column scaling, to a $\{0, 1\}$-matrix A''. Since G

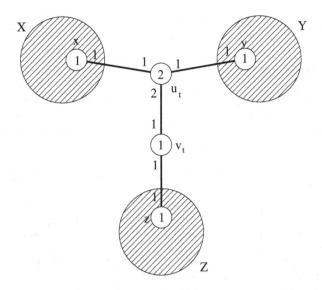

Fig. 1. The reduction

is bipartite, we can scale some of the rows of A'' by -1 to obtain a matrix B with a 1 and a -1 in each column. Now $M(B) = M(A)$ is the cycle-matroid of G and, hence, $M(A)$ is graphic.

6   Bounded Variables

In this section we consider integer programs with bounds on the variables.

Integer Programming with Variable Bounds (BIP)
INSTANCE: Positive integers m and n, a matrix $A \in \mathbf{Z}^{m \times n}$, a vector $b \in \mathbf{Z}^m$, and vectors $c, d \in \mathbf{Z}^n$.
PROBLEM: Find $x \in \mathbf{Z}^n$ maximizing $c^t x$ subject to $(Ax = b, 0 \leq x \leq d)$.

We can rewrite the problem as: *Find $y \in \mathbf{Z}^{2n}$ maximizing $\hat{c}^t y$ subject to* $(\hat{A}y = \hat{b}, y \geq 0)$, where

$$\hat{A} = \begin{bmatrix} A & 0 \\ I & I \end{bmatrix}, \hat{b} = \begin{bmatrix} b \\ d \end{bmatrix}, \text{ and } \hat{c} = \begin{bmatrix} c \\ 0 \end{bmatrix}.$$

Note that, for $i \in \{1, \ldots, n\}$, the elements i and $i+n$ are in series in $M(\hat{A})$, and, hence, $M(\hat{A})$ is obtained from $M(A)$ by a sequence of series-coextensions. Then it is easy to see that, if the branch-width of $M(A)$ is k, then the branch-width of $M(\hat{A})$ is at most $\max(k, 2)$.

Now note that the all-ones vector is in the row-space of $\hat{A}$. Therefore, by taking appropriate combinations of the equations $\hat{A}y = \hat{b}$, we can make an equivalent system $\tilde{A}y = \tilde{b}$ where $\tilde{A}$ is non-negative. Therefore, we obtain the following corollary to Theorem 2.

Corollary 1. *There is a pseudopolynomial-time algorithm for solving (BIP) on instances where the branch-width of $M(A)$ is constant.*

Acknowledgements

We thank Bert Gerards and Geoff Whittle for helpful discussions regarding the formulation of Problem 1 and the proof of Theorem 4. This research was partially sponsored by grants from the Natural Science and Engineering Research Council of Canada.

References

1. B. Courcelle, "Graph rewriting: An algebraic and logical approach", in: *Handbook of Theoretical Computer Science, vol. B*, J. van Leeuwnen, ed., North Holland (1990), Chapter 5.
2. W.H. Cunningham, *Improved bounds for matroid partition and intersection algorithms*, SIAM J. Comput. **15** (1986), 948-957.
3. M.R. Garey and D.S. Johnson, *Computers and Intractability. A guide to the theory of NP-completeness*, A series of books in the mathematical sciences, W.H. Freeman and Co., San Francisco, California, 1979.
4. P. Hliněný, Branch-width, parse trees and monadic second-order logic for matroids, manuscript, 2002.
5. G.S. Lueker, *Two NP-complete problems in non-negative integer programming*, Report No. 178, Department of Computer Science, Princeton University, Princeton, N.J., (1975).
6. S. Oum and P. D. Seymour, *Approximating clique-width and branch-width*, J. Combin. Theory, Ser. B **96** (2006), 514-528.
7. S. Oum and P. D. Seymour, *Testing branch-width*, to appear in J. Combin. Theory, Ser. B.
8. J. G. Oxley, *Matroid Theory*, Oxford University Press, New York, 1992.
9. C.H. Papadimitriou, *On the complexity of integer programming*, J. Assoc. Comput. Mach. **28** (1981), 765-768.
10. N. Robertson and P. D. Seymour, *Graph Minors. X. Obstructions to tree-decomposition*, J. Combin. Theory, Ser. B **52** (1991), 153–190.
11. N. Robertson and P. D. Seymour, *Graph Minors. XVI. Excluding a non-planar graph*, J. Combin. Theory, Ser. B **89** (2003), 43-76.
12. P. D. Seymour, *Decomposition of regular matroids*, J. Combin. Theory, Ser. B **28** (1980), 305–359.
13. W. T. Tutte, *A homotopy theorem for matroids, I, II*, Trans. Amer. Math. Soc. **88** (1958), 144–174.

Matching Problems in Polymatroids
Without Double Circuits*

Márton Makai, Gyula Pap, and Jácint Szabó

MTA-ELTE Egerváry Research Group on Combinatorial Optimization
{marci,gyuszko,jacint}@cs.elte.hu
http://www.cs.elte.hu/egres

Abstract. According to the present state of the theory of the matroid
matching problem, the existence of a good characterization to the size of
a maximum matching depends on the behavior of certain substructures,
called double circuits. In this paper we prove that if a polymatroid has
no double circuits at all, then a partition-type min-max formula characterizes the size of a maximum matching. We provide applications of this
result to parity constrained orientations and to a rigidity problem.

A polynomial time algorithm is constructed by generalizing the principle of shrinking blossoms used in Edmonds' matching algorithm [2].

Keywords: matroids and submodular functions.

1 Introduction

Polymatroid matching is a combinatorial optimization problem which is concerned with parity and submodularity. Early well-solved special cases are the
matching problem of graphs and the matroid intersection problem, which have
in fact motivated Lawler to introduce the matroid and polymatroid matching
problems. Jensen, Korte [6], and Lovász [9] have shown that, in general, the
matroid matching problem is of exponential complexity under the independence
oracle framework. The major breakthrough came when Lovász gave a good characterization to the size of a maximum matching and also a polynomial algorithm
for linearly represented matroids [12,9]. Lovász [10], and Dress and Lovász [1]
observed that the solvability of the linear case is due to the fact that these matroids can be embedded into a matroid satisfying the so-called *double circuit
property*, or *DCP* for short. It was also shown that full linear, full algebraic, full
graphic, and full transversal matroids are DCP matroids [1]. The disadvantage of
this approach is that, due to the embedding into a bigger matroid, the min-max
formula is rather difficult to interpret in a combinatorial way, and often does not
even imply a good characterization. However, the diversity and the importance
of solvable special cases of the matroid matching problem is a motivation to
explore those techniques implying a combinatorial characterization.

* Research is supported by OTKA grants K60802, T037547 and TS049788, by
European MCRTN Adonet, Contract Grant No. 504438.

In this paper we investigate the class of those polymatroids having no nontrivial compatible double circuits, called ntcdc-free for short, defined later. We prove that in these polymatroids a partition-type combinatorial formula characterizes the maximum size of a matching. We remark that in the min-max formula for DCP matroids, for example representable matroids, we have to take a partition and a projection into consideration. Contrarily, in ntcdc-free polymatroids, it suffices to consider partitions in the min-max formula. As an application, we show that two earlier results are special cases of this approach. The first application is that the parity constrained orientation problem of Király and Szabó [7] can be formulated as a matching problem in a ntcdc-free polymatroid, which implies the partition-type formula given in [7]. Second, we deduce a result of Fekete [3] on the problem of adding a clique of minimum size to a graph to obtain a graph that is generically rigid in the plane.

1.1   The Partition Formula

To formulate our main result, some definitions are in order. We denote by $\mathbb{R}_+$ and $\mathbb{N}$ the set of non-negative reals and non-negative integers, respectively. Let S be a finite ground set. A set-function $f : 2^S \to \mathbb{Z}$ is called *submodular* if

$$f(X) + f(Y) \geq f(X \cap Y) + f(X \cup Y) \tag{1}$$

holds whenever $X, Y \subseteq S$. b is called *supermodular* if $-b$ is submodular. The set-function f is said to be *non-decreasing* if $f(X) \leq f(Y)$ for every $\emptyset \neq X \subseteq Y \subseteq S$, and we say that f is *non-increasing* if $-f$ is non-decreasing. A non-decreasing submodular set-function $f : 2^S \to \mathbb{N}$ with $f(\emptyset) = 0$ is called a *polymatroid function*. A polymatroid function $f : 2^S \to \mathbb{Z}_+$ induces a *polymatroid* $P(f)$ and a *base polyhedron* $B(f)$ defined by

$$P(f) := \{x \in \mathbb{R}^S : x \geq 0, x(Z) \leq f(Z) \text{ for all } Z \subseteq S\}, \tag{2}$$

$$B(f) := \{x \in \mathbb{R}^S : x(S) = f(S), \text{ and } x \geq 0, x(Z) \leq f(Z) \text{ for all } Z \subseteq S\}, \tag{3}$$

where $x(Z) := \sum_{i \in Z} x_i$ for some $Z \subseteq S$. A vector $m \in \mathbb{Z}^S$ is called *even* if m_i is even for every $i \in S$. The even vectors $m \in P(f)$ are called the *matchings* of f. The *size* of a matching is $m(S)/2$. The *polymatroid matching problem* is to find a *maximum matching*, i.e. a matching of maximum size

$$\nu(f) = \max\{m(S)/2 : m \text{ is a matching of } f\}.$$

We will investigate the polymatroid matching problem in ntcdc-free polymatroids, defined below. Our main result goes as follows.

Theorem 1. *Let $f : 2^S \to \mathbb{N}$ be a ntcdc-free polymatroid function. Then*

$$\nu(f) = \min \sum_{j=1}^{t} \left\lfloor \frac{f(U_j)}{2} \right\rfloor,$$

where the minimum is taken over all partitions $U_1, U_2, \ldots, U_t$ of S.

We propose two different proofs. In the first proof we exploit a theorem of Lovász, and a couple of polymatroid operations. The second proof relies on a (semi-strongly) polynomial time algorithm, which is based on a generalization of the contraction of blossoms in Edmonds' matching algorithm [2].

1.2   Circuits and Compatible Double Circuits in Polymatroids

Consider a polymatroid function $f : 2^S \to \mathbb{N}$, and a vector $x \in \mathbb{N}^S$. For a set $Z \subseteq S$, we call $\mathrm{def}_{f,x}(Z) := x(Z) - f(Z)$ the *deficiency of set* Z with respect to f, x. A set is called k-*deficient* with respect to f, x if $\mathrm{def}_{f,x}(Z) = k$. The *deficiency of a vector* x is defined by $\mathrm{def}_f(x) := \max_{Z \subseteq S} \mathrm{def}_{f,x}(Z)$, which is non-negative. Notice that $\mathrm{def}_{f,x}(\cdot)$ is a supermodular set-function, hence the family of sets Z such that $\mathrm{def}_{f,x}(Z) = \mathrm{def}_f(x)$ is closed under taking unions and intersections.

Consider a 1-deficient vector x. x is called a *circuit* if $supp(x)$ is equal to the unique inclusionwise minimal 1-deficient set.

Consider a 2-deficient vector $x \in \mathbb{N}^S$, and let $W := supp(x)$. x is called a *compatible double circuit* (or *cdc*, for short), if W is the unique inclusionwise minimal 2-deficient set, and there is a partition $\pi = \{W_1, \cdots, W_k\}$ of W such that $k \geq 2$ and $\{W - W_i : i = 1, \cdots, k\}$ is equal to the family of all inclusionwise minimal 1-deficient sets. We remark that if x is a cdc, then π is uniquely determined – let it be called the *principal partition* of x. If $k = 2$, then x is called a *trivial cdc*. If $k \geq 3$, then x is called a *non-trivial compatible double circuit*, or *ntcdc*, for short.

A polymatroid is called *ntcdc-free* if there is no ntcdc.

2   First Proof of the Partition Formula

For some well-known notions and results on the theory of matroids, polymatroids and matroid matching, see [14]. We need some more preparation.

2.1   Preliminaries

There is a close relation between polymatroid functions and matroids. First, if $M = (T, r)$ is a matroid and $\varphi : T \to S$ is a function then $f : 2^S \to \mathbb{N}$, $X \mapsto r(\varphi^{-1}(X))$ is a polymatroid function, the *homomorphic image of* M *under* φ. Second, for any polymatroid function f it is possible to define a matroid M, the homomorphic image of which is f, in such a way that M is "most independent" in some sense. The ground set T of M is the disjoint union of sets T_i for $i \in S$ of size $|T_i| \geq f(\{i\})$. If $X \subseteq T$ then we define the vector $\chi^X \in \mathbb{N}^S$ with $\chi_i^X = |X \cap T_i|$ for $i \in S$. With this notation, a set $X \subseteq T$ is defined to be independent in M if $\chi^X \in P(f)$. It is routine to prove that M is indeed a matroid with rank function $r(X) = \min_{Y \subseteq X}(|Y| + f(\varphi(X - Y)))$, where $\varphi : T \to S$ maps t to i if $t \in T_i$. This M is called a *prematroid* of f. Note that a prematroid M is uniquely determined by f and by the sizes $|T_i|$, $i \in S$. If M is a matroid with rank function r then the prematroids of r are the parallel extensions of M. If we consider a prematroid M then we tacitly assume that $M = (T, r)$ and that the function $\varphi : T \to S$ is given with $t \mapsto i$ if $t \in T_i$.

If f is a polymatroid function and $x \in \mathbb{Z}^S$ then we define the *rank* of x as $r_f(x) = \min_{U \subseteq S}(x(S - U) + f(U))$. If $x \in \mathbb{N}^S$ then $r_f(x) = x(S)$ if and only if $x \in P(f)$. Besides, if $M = (T, r)$ is a prematroid of f and $X \subseteq T$ then $r_f(\chi^X) = r(X)$. The *span* of $x \in \mathbb{N}^S$ is defined by $\mathrm{sp}_f(x) = \{i \in S : r_f(x + \chi_i) = r_f(x)\}$. If M is a prematroid of f and $X \subseteq T$ then $\mathrm{sp}_f(\chi^X) = \{i \in S : T_i \subseteq \mathrm{sp}_M(X)\}$.

2.2   Circuits and Double Circuits in Matroids

Let $M = (T, r)$ be a matroid. A set $C \subseteq T$ is said to be a *circuit* if $r(C - x) = r(C) = |C| - 1$ for every $x \in C$. A set $D \subseteq T$ is a *double circuit* if $r(D - x) = r(D) = |D| - 2$ for every $x \in D$. If D is a double circuit then the dual of $M|D$ is a matroid of rank 2 without loops, that is a line, showing that there exists a *principal partition* $D = D_1 \dot\cup D_2 \dot\cup \ldots \dot\cup D_d$, $d \geq 2$, such that the circuits of D are exactly the sets of the form $D - D_i$, $1 \leq i \leq d$. We say that D is *non-trivial* if $d \geq 3$, and *trivial* otherwise. A trivial double circuit is simply the direct sum of two circuits.

Analogously, we define circuits and double circuits of the polymatroid function $f : 2^S \to \mathbb{N}$. For a vector $x \in \mathbb{R}_+^S$ let $\mathrm{supp}(x) = \{i \in S : x_i > 0\}$. A vector $c \in \mathbb{N}^S$ is a *circuit* of f if $r_f(c - \chi_i) = r_f(c) = c(S) - 1$ for every $i \in \mathrm{supp}(c)$. A vector $w \in \mathbb{N}^S$ is a *double circuit* of f if $r_f(w - \chi_i) = r_f(w) = w(S) - 2$ for every $i \in \mathrm{supp}(w)$. It is also easy to see the exact relation between matroidal and polymatroidal double circuits, which is given as follows.

Lemma 1. *Let M be a prematroid of f, $D \subseteq T$ and $\chi^D = w$. Then D is a double circuit of M if and only if w is a double circuit of f.*

Recall that we have already defined cdc's and ntcdc's. Next we add another definition, which is easily seen to be equivalent with those above. For $x \in \mathbb{R}^S$ and $U \subseteq S$ we introduce the notation $x|_U$ for the vector by $(x|_U)_i := x_i$ for $i \in U$ and $(x|_U)_i := 0$ for $i \in S - U$. Let M be a prematroid of f and w be a double circuit of f such that there is a set $D \subseteq T$ with $\chi^D = w$. By Lemma 1, D is a double circuit of M, thus it has a principal partition $D = D_1 \dot\cup D_2 \dot\cup \ldots \dot\cup D_{d'}$. We define the principal partition of w as follows. Due to the structure of prematroids it is easy to check that $\mathrm{supp}(w)$ has a partition $W_0 \dot\cup W_1 \dot\cup \ldots \dot\cup W_d$ with the property that each set D_j is either a singleton belonging to some T_i with $w_i \geq 2$ and $i \in W_0$, or is equal to $D \cap \bigcup_{i \in W_h} T_i$ for some $1 \leq h \leq d$. Note that a partition $W_0 \dot\cup W_1 \dot\cup \ldots \dot\cup W_d$ of $\mathrm{supp}(w)$ is the principal partition of w if and only if $w - \chi_i$ is a circuit of f and $w_i \geq 2$ whenever $i \in W_0$, moreover, $w|_{W - W_i}$ is a circuit of f for each $1 \leq i \leq d$. A double circuit w is said to be *compatible* if $W_0 = \emptyset$, and it is *trivial* if D is trivial. We remark that these definitions are easily see equivalent with the above ones.

We shortly mention what is the double circuit property, or DCP, for short. If $M = (T, r)$ is a prematroid of the polymatroid function f and $Z \subseteq T$ then $\varphi(M/Z)$ is called a *contraction* of f. A polymatroid function f is said to have the

DCP if whenever w is a non-trivial compatible double circuit in a contraction f' of f with principal partition $W_1 \dot\cup \ldots \dot\cup W_d$ then $f'(\bigcap_{1 \le i \le d} \mathrm{sp}(w|_{W-W_i})) > 0$, [1]. A polymatroid function without non-trivial compatible double circuits has not necessarily the DCP, as its contractions may have many non-trivial compatible double circuits.

Note that every polymatroid function has double circuits, say $(f(\{i\}) + 2)\chi_i$ for some $i \in S$. However, these are not compatible, as $W_0 = \{i\}$.

Lemma 2. *If $w \in \mathbb{N}^S$ is a double circuit of the polymatroid function $f : 2^S \to \mathbb{N}$ with principal partition $W = W_0 \dot\cup W_1 \dot\cup \ldots \dot\cup W_d$ then $f(W) = w(W) - 2$ and $f(W - W_i) = w(W - W_i) - 1$ for $1 \le i \le d$.*

Proof. We prove that if $x \in \mathbb{N}^S$ is a vector with the property that $r_f(x) = r_f(x - \chi_i)$ for all $i \in \mathrm{supp}(x)$ then $f(\mathrm{supp}(x)) = r_f(x)$. By definition, $r_f(x) = x(S-Y) + f(Y)$ for some $Y \subseteq S$. Note that $r_f(x - \chi_i) \le (x - \chi_i)(S-Y) + f(Y) = r_f(x) - 1$ for all $i \in \mathrm{supp}(x) - Y$. Thus $\mathrm{supp}(x) \subseteq Y$. Finally, $f(Y) = r_f(x) \le f(\mathrm{supp}(x)) \le f(Y)$, since f is non-decreasing. If x is a circuit or a double circuit then $r_f(x) = r_f(x - \chi_i)$ for all $i \in \mathrm{supp}(x)$, we are done.

2.3 Polymatroid Operations

Next we investigate how two polymatroid operations (translation, deletion) effect double circuits. If $f : 2^S \to \mathbb{N}$ is a function and $n \in \mathbb{Z}^S$ then define $f + n : 2^S \to \mathbb{N}$ by $X \mapsto f(X) + n(X)$. If f is a polymatroid function and $n \in \mathbb{N}^S$ then $f + n$ is clearly a polymatroid function, too.

Lemma 3. *If $n \in \mathbb{Z}^S$ and f and $f + n$ are polymatroid functions then a vector w is a double circuit of f with $W = \mathrm{supp}(w)$ if and only if $w + n|_W$ is a double circuit of $f + n$. In this case their principal partition coincide.*

Proof. Clearly, $r_{f+n}(x + n) - (x + n)(S) = r_f(x) - x(S)$ for all $x \in \mathbb{Z}^S$. Thus by symmetry, it is enough to prove that if w is a double circuit of f with support W then $w_i + n_i > 0$ for every $i \in W$. Otherwise by Lemma 2 we would have $w(W - i) - n_i \ge w(W) = f(W) + 2 \ge f(W - i) - n_i + 2$, which is impossible.

Let $u \in \mathbb{N}^S$ be a bound vector and define $f \backslash u = \varphi(r_{M|Z})$ where M is a prematroid of f and $Z \subseteq T$ with $\chi^Z = u$. The matroid union theorem asserts that $(f \backslash u)(X) = \min_{Y \subseteq X}(u(Y) + f(X - Y))$. If M is a matroid with rank function r then $r \backslash u$ is the rank function of $M|\mathrm{supp}(u)$.

Lemma 4. *Let $u \in \mathbb{N}^S$. If $w \in \mathbb{N}^S$ is a double circuit of $f' := f \backslash u$ then w is either a double circuit of f with the same principal partition, or trivial, or non-compatible.*

Proof. Let $M = (T, r)$ be a prematroid of f and $Z \subseteq T$ with $\chi^Z = u$. If $w \le \chi^Z$ then w is a double circuit of f with the same principal partition by Lemma 1. Observe that $w_i \le f'(\{i\}) + 2$ and $f'(\{i\}) \le u_i$ for every $i \in S$. Thus if $w \not\le \chi^Z$ then there exists an $i \in S$ such that $w_i - f'(\{i\}) \in \{1, 2\}$. If $w_i = f'(\{i\}) + 2$

then $r_{f'}(w_i \chi_i) = w_i - 2$, thus $W_0 = \text{supp}(w) = \{i\}$, implying that w is non-compatible. If $w_i = f'(\{i\}) + 1$ then $w_i \chi_i$ is a circuit of f' thus if $W_0 \neq \emptyset$ then w is non-compatible and if $W_0 = \emptyset$ then w is trivial.

Finally we cite Lovász's deep and important theorem on 2-polymatroids, which can be translated to arbitrary polymatroids as follows. This theorem will be a key to our first proof below.

Theorem 2 (Lovász [10]). *If $f : 2^S \to \mathbb{N}$ is a polymatroid function then at least one of the following cases holds.*

1. $f(S) = 2\nu(f) + 1$.
2. *There exists a partition $S = S_1 \dot\cup S_2$, $S_i \neq \emptyset$, s.t. $\nu(f) = \nu(f|_{2^{S_1}}) + \nu(f|_{2^{S_2}})$.*
3. *There exists an $i \in S$, $f(i) \geq 2$ such that for each maximum matching m we have $i \in \text{sp}_f(m)$.*
4. *There exists a certain substructure, called ν-double flower in f, which we do not define here, but which always contains a non-trivial compatible double circuit.*

Proof (First proof of Theorem 1). It is easy to see that $\nu(f) \leq \sum_{j=1}^{t} \left\lfloor \frac{f(U_j)}{2} \right\rfloor$ holds for every partition $U_1, U_2, \ldots, U_t$ of S. For the other direction we argue by induction on the pair $(|S|, |K(f)|)$, where $K(f) = \{s \in S : s \in \text{sp}_f(m)$ for each maximum matching m of $f\}$. If $S = \emptyset$ then the statement is trivial. If $K(f) = \emptyset$ then either *1.* or *2.* holds in Theorem 2. If *1.* holds then the trivial partition will do, while if *2.* holds then we can use our induction hypothesis applied to $f|_{2^{S_1}}$ and $f|_{2^{S_2}}$.

Next, let $K(f) \neq \emptyset$. We prove that if m is a maximum matching of $f + 2\chi_s$ then $m(s) \geq 2$. Indeed, assume that $m(s) = 0$. As m is a maximum matching, there exists a set $s \in U \subseteq S$ with $m(U) \geq (f + 2\chi_s)(U) - 1$. Thus $m(U - s) = m(U) \geq (f + 2\chi_s)(U) - 1 \geq f(U - s) + 1$, which is a contradiction. It is also clear that $m + 2\chi_s$ is a matching of $f + 2\chi_s$ for each matching m of f. Therefore, m is a maximum matching of f if and only if $m + 2\chi_s$ is a maximum matching of $f + 2\chi_s$.

Let $s \in K(f)$. Clearly, $\nu(f) \leq \nu(f + \chi_s) \leq \nu(f + 2\chi_s) = \nu(f) + 1$ and we claim that in fact, $\nu(f + \chi_s) = \nu(f)$ holds. Indeed, if $\nu(f + \chi_s) = \nu(f) + 1$ and m is a maximum matching of $f + \chi_s$ then m is also a maximum matching of $f + 2\chi_s$, thus $m(s) \geq 2$. Then $m - 2\chi_s$ is a maximum matching of f and, as $s \in \text{sp}_f(m - 2\chi_s)$, there exists a set $s \in U \subseteq S$ with $(m - 2\chi_s)(U) = f(U)$. This implies $m(U) = f(U) + 2$, contradicting to that m is a matching of $f + \chi_s$.

So if m is a maximum matching of f then m is a maximum matching of $f + \chi_s$, too, and clearly, $\text{sp}_f(m) = \text{sp}_{f + \chi_s}(m) - s$. Thus we have $K(f + \chi_s) \subseteq K(f) - s$. By Lemma 3, $f + \chi_s$ has no non-trivial compatible double circuits, so we can apply induction to $f + \chi_s$. This gives a partition $U_1, U_2, \ldots, U_t$ of S such that $\nu(f + \chi_s) = \sum_{j=1}^{t} \left\lfloor \frac{1}{2}(f + \chi_s)(U_j) \right\rfloor$. But then, $\nu(f) = \nu(f + \chi_s) = \sum_{j=1}^{t} \left\lfloor \frac{1}{2}(f + \chi_s)(U_j) \right\rfloor \geq \sum_{j=1}^{t} \left\lfloor \frac{f(U_j)}{2} \right\rfloor$.

3   Second, Constructive Proof of the Partition Formula

The second proof is based on projections of blossoms, which is the generalization of the principle in Edmonds' matching algorithm [2]. For this, of course, we need some more definitions and direct observations concerning projections.

3.1   Projections

Consider a polymatroid function f on groundset S, as above. For a subset $B \subseteq S$ we define the *projection* $f^B : 2^{S-B} \to \mathbb{N}$ by $f^B(X) := \min\{ f(X), f(X \cup B) - f(B) + 1 \}$ for $X \subseteq S - B$. It is easy to see that f^B is a polymatroid function, and its induced polymatroid is equal to

$$P(f^B) = \{y \in \mathbb{R}^{S-B} : \text{ there is } [z, y] \in P(f) \text{ s.t. } z(B) = f(B) - 1\}. \quad (4)$$

For $x \in \mathbb{R}^S$, $Z \subseteq S$ we introduce the notation $x\|_Z \in \mathbb{R}^Z$ for the vector such that $(x\|_Z)_i = x_i$ for all $i \in Z$.

Consider a family $\mathcal{H} = \{H_1, \cdots, H_m\}$ of disjoint subsets of S. Assume that there is a vector $x \in P(f)$ such that for all $i = 1, \cdots, m$, we have $x(H_i) = f(H_i) - 1$, and there is an element $h_i \in H_i$ such that $x + \chi_{h_i} \in P(f)$. By (4) we get that $x\|_{S-H_i} \in P(f^{H_i})$, thus $f^{H_i}(H_j) = f(H_j)$ for all $i \neq j$. This implies that we obtain the same polymatroid function on groundset $S - \cup\mathcal{H}$ no matter which order the sets H_i are projected. Let $f^{\mathcal{H}}$ denote the unique polymatroid function obtained by projecting all the members of $\mathcal{H}$. Then

$$P(f^{\mathcal{H}}) = \{y \in \mathbb{R}^{S-\cup\mathcal{H}} : \text{ there is } [z, y] \in P(f) \text{ s.t. } z(H_i) = f(H_i) - 1\}, \quad (5)$$

and we get that for any $X \subseteq S - \bigcup\mathcal{H}$,

$$f^{\mathcal{H}}(X) = \min\{f(X \cup \bigcup\mathcal{H}') - x(\bigcup\mathcal{H}') : \mathcal{H}' \subseteq \mathcal{H}\}. \quad (6)$$

We remark without proof that $f^{\mathcal{H}}$ may be evaluated in strongly polynomial time.

3.2   Blossoms

The notion of blossoms comes from an algorithmic point of view, which is the analogue of Edmonds' blossoms in the matching algorithm. An ear-decomposition of a matching is constructed by finding a circuit induced in the matching, and iterating this procedure after the projection. More precisely, the definition is the following. If $y \in P(f)$, $y + \chi_u \in P(f)$, $y + 2\chi_u \notin P(f)$, $u \in C \subseteq S$, and C is the unique inclusionwise minimal 1-deficient set for $y + 2\chi_u$, then we say that "u *induces a circuit on C in y*".

Consider a matching x with respect to a polymatroid function $f : 2^S \to \mathbb{N}$. Consider a laminar family $\mathcal{F} = \{B_1, \cdots, B_k\}$ of subsets of S, that is, any two members of $\mathcal{F}$ are either disjoint or one contains the other. For indices $i = 1, \cdots, k$, let $\mathcal{F}_i$ denote the family of inclusionwise maximal proper subsets of B_i in $\mathcal{F}$, and let $G_i := B_i - \cup\mathcal{F}_i$. Consider a set $U = \{u_1, \cdots, u_k\} \subseteq S$ such that $u_i \in G_i$. Hence $\mathcal{F}, U$ is called an *x-ear-decomposition* if

(a) $x(B_i) = f(B_i) - 1$, and
(b) u_i induces a circuit on G_i in $x||_{S-\cup\mathcal{F}_i}$ with respect to $f^{\mathcal{F}_i}$.

Notice that the above definition implies that $x + \chi_{u_i} \in P(f)$ holds whenever B_i is an inclusionwise minimal member of $\mathcal{F}$. This implies that the projection of $\mathcal{F}$, or $\mathcal{F}_i$ satisfies the assumption in the previous section, and thus the projection may be performed in arbitrary order. Notice, if we drop an inclusionwise maximal member $B_i \in \mathcal{F}$ together with u_i, we retain another ear-decomposition. A set B appearing in the family $\mathcal{F}$ of some ear-decomposition is called an x-*blossom*. An ear-decomposition of a blossom B is an ear-decomposition $\mathcal{F}, U$ such that B is the unique inclusionwise maximal member of $\mathcal{F}$.

The following Lemma 5 will be our crucial inductive tool to deal with ear-decompositions by extending a matching with respect to $f^{\mathcal{F}}$ to a matching with respect to f.

Lemma 5. *Suppose we are given a matching x, an x-blossom B together with an x-ear-decomposition, and a vector $y \in P(f^B)$. There is a polynomial time algorithm to find either*

(A) a ntcdc, or
(B) an even vector $z \in (2\mathbb{N})^B$ such that $z(B) = f(B) - 1$ and $[z, y] \in P(f)$.

Proof. Let us use notation from above. The algorithm is recursive on the number k of ears. Firstly, notice that $\mathrm{def}_f([x||_B, y]) \leq 1$. If $\mathrm{def}_f([x||_B, y]) = 0$, then (B) holds for $z = x||_B$, and we are done. Henceforth we suppose that $\mathrm{def}_f([x||_B, y]) = 1$, and let D denote the inclusionwise minimal 1-deficient set for $[x||_B, y]$. Say $B = B_k$ and $G = G_k$.

We claim that either $[x||_G, y] \in P(f^{\mathcal{F}_k})$, or $D \subseteq (S - B) \cup G$. Suppose $[x||_G, y] \notin P(f^{\mathcal{F}_k})$. By (4), there is a set Q such that $\mathrm{def}_{f,[x||_B,y]}(Q) \geq 1$, and for all $B_i \in \mathcal{F}_k$ we have $Q \cap B_i = \emptyset$ or $Q \supseteq B_i$. Clearly, $\mathrm{def}_{f,[x||_B,y]}(B) = -1$. Since $y \in P(f^B)$, we get that $\mathrm{def}_{f,[x||_B,y]}(B \cup Q) \leq 0$. Thus, by supermodularity of deficiency, $0 \leq \mathrm{def}_{f,[x||_B,y]}(B \cap Q) = \mathrm{def}_{f,x}(B \cap Q)$. Recall that for every inclusionwise minimal set $B_i \in \mathcal{F}$ we have $x + \chi_{u_i} \in P(f)$ for $u_i \in B_i$. Thus, $u_i \notin B \cap Q$, which implies that $D \subseteq Q \subseteq (S - B) \cup G$.

Now suppose that $[x||_G, y] \in P(f^{\mathcal{F}_k})$. Thus, by (4), there is a (not necessarily even) vector $z' \in \mathbb{N}^{\cup \mathcal{F}_k}$ such that $[z', x||_G, y] \in P(f)$, and $z'(B_i) = b(B_i) - 1$ for all $B_i \in \mathcal{F}_k$. Then we apply the algorithm recursively for $B_i \in \mathcal{F}_k$ and $[z', x||_G, y]$, that is, we replace $z'||_{B_i}$ step-by-step by an even vector retaining the above properties – or we find a ntcdc.

Finally suppose that $D \subseteq (S - B) \cup G$. Notice that $y \in P(f^B)$ implies $D \cap B \neq \emptyset$. Also, $x \in P(f)$ implies $D - B \neq \emptyset$. Moreover, $y \in P(f^B)$ implies $\mathrm{def}_{f,[x||_B,y]}(B \cup D) \leq 0$. Recall that $\mathrm{def}_{f,[x||_B,y]}(D) = 1$ and $\mathrm{def}_{f,[x||_B,y]}(B) = -1$. By supermodularity of deficiency, $\mathrm{def}_{f,[x||_B,y]}(B \cap D) \geq 0$. Thus, by (b) we get that $u_k \notin D$. Consider an arbitrary element $d \in D \cap B$. By (b), $[x||_G + 2\chi_{u_k} - \chi_d, 0] \in P(f^{\mathcal{F}_k})$. By applying the algorithm recursively for $[x||_G + 2\chi_{u_k} - \chi_d, 0]$ one can find either a ntcdc, or an even vector $q \in (2\mathbb{N})^{\cup \mathcal{F}_k}$ such that $[q, x||_G + 2\chi_{u_k} - \chi_d, 0] \in P(f)$. Next, we will find out whether there is an element e such that $z = [q, x||_G + 2\chi_{u_k} - 2\chi_e]$ satisfies (B). Clearly, all these vectors are even. It is easy to see that $\operatorname{def}_f([q, x||_G + 2\chi_{u_k}, y])$ is 1 or 2. If $\operatorname{def}_f([q, x||_G + 2\chi_{u_k}, y]) = 1$, then for some element e we get that $[q, x||_G + 2\chi_{u_k} - 2\chi_e, y]$, and we are done. If $\operatorname{def}_f([q, x||_G + 2\chi_{u_k}, y]) = 2$, then let W denote the unique minimal 2-deficient set. If there is an element $e \in W$ such that all the 1-deficient sets contain e, then $[q, x||_G + 2\chi_{u_k} - 2\chi_e, y] \in P(f)$, and we are done. Otherwise, if for every element e there is a 1-deficient set $e \notin W_e$, then $[q, x||_G + 2\chi_{u_k}, y]|_W \in \mathbb{N}^S$ is a cdc. Notice that B and D are circuits in $[q, x||_G + 2\chi_{u_k}, y]$, thus $W - B \in \pi$ and $W - D \in \pi$. Since $d \in B \cap D \neq \emptyset$, this implies $|\pi| \geq 3$.

3.3   A Semi-strongly Polynomial Time Algorithm

We construct a semi-strongly polynomial time algorithm which either returns a ntcdc, or returns a maximum matching x and a partition certifying its maximality. The algorithm maintains a matching, and iteratively augments its size by one, until it either finds a certifying partition, or a ntcdc. We may initiate x as a basis of $P(f)$, rounded down to the closest even vector. This initialization may be performed in semi-strongly polynomial time, where "semi-" comes only from the fact that we have to take lower integer part to detect parity. The remaining part of the algorithm may be performed in strongly polynomial time.

The idea behind the algorithm is the following. If our matching x is a basis in the polymatroid, then we are done. Thus we find an element $u \in S$ such that $x + \chi_u \in P(f)$. If $x + 2\chi_u \in P(f)$, then that gives a larger matching, and we are done. Otherwise, we may assume that $x + \chi_u \in P(f)$ and $x + 2\chi_u \notin P(f)$, i.e. u induces a circuit in x, which can be used building blossoms and projections. If we find a larger matching in the projection, then we use Lemma 5 to expand blossoms and retain a larger matching over the original groundset. This idea is developed in detail below.

Consider a matching x. Define $\mathcal{C} := \emptyset$. In a general step of the algorithm, $\mathcal{C} = \{B_1, \cdots, B_k\}$ is a family of disjoint x-blossoms. This implies that $x||_{S - \bigcup \mathcal{C}} \in P(f^{\mathcal{C}})$. We distinguish three cases on how close $x||_{S - \bigcup \mathcal{C}}$ is to a basis of $P(f^{\mathcal{C}})$.

Case 1. Suppose that $x(S - \bigcup \mathcal{C}) = f^{\mathcal{C}}(S - \bigcup \mathcal{C})$. Then, by claim (6), there is a set $\mathcal{C}' \subseteq \mathcal{C}$ such that $f(S - \bigcup \mathcal{C} + \bigcup \mathcal{C}') = x(S - \bigcup \mathcal{C}' + \bigcup \mathcal{C}')$. Then $\mathcal{C}' = \emptyset$, since for all blossoms $B_i \in \mathcal{C}$ there is an element $t \in B_i$ such that $x + \chi_t \in P(f)$. We conclude that x is a maximum matching, certified by the partition $\mathcal{C} \cup \{S - \bigcup \mathcal{C}\}$.

Case 2. Suppose that $x||_{S - \bigcup \mathcal{C}} + \chi_u \in P(f^{\mathcal{C}})$, but $x||_{S - \bigcup \mathcal{C}} + 2\chi_u \notin P(f^{\mathcal{C}})$. Then there is a set $u \in Z \subseteq S - \bigcup \mathcal{C}$ such that u induces a circuit on Z in $x||_{S - \bigcup \mathcal{C}}$ with respect to $f^{\mathcal{C}}$. By claim (6) there is a set $\mathcal{C}' \subseteq \mathcal{C}$ such that $f(Z \cup \bigcup \mathcal{C}') = x(Z \cup \bigcup \mathcal{C}') + 1$. Thus, $\mathcal{C} - \mathcal{C}' + \{Z \cup \bigcup \mathcal{C}'\}$ is a blossom family.

Case 3. Suppose that $x||_{S - \bigcup \mathcal{C}} + 2\chi_u \in P(f^{\mathcal{C}})$. In this case, by applying Lemma 5 for members of $\mathcal{C}$, we construct either a matching larger than x, or a ntcdc. This is done as follows. By assertion (5), there is a (not necessarily even) vector $z \in \mathbb{N}^{\bigcup \mathcal{C}}$ such that $x' := [z, x||_{S - \bigcup \mathcal{C}} + 2\chi_u] \in P(f)$, and $z(B_i) = f(B_i) - 1$ for $i = 1, \cdots, k$. Thus, for an arbitrary index $i \in \{1, \cdots, k\}$ we get that $x'||_{S - B_i} \in P(f^{B_i})$. By applying Lemma 5 for B_i, we either construct a ntcdc, or we may replace entries of x' in B_i with even numbers, and retain the above properties. By repeating this procedure for $i = 1, \cdots, k$ we retain a matching x' that is larger than x.

4 Applications

4.1 A Parity Constrained Orientation Theorem

Frank, Jordán and Szigeti [4] proved that the existence of a k-rooted-connected orientation with prescribed parity of in-degrees can be characterized by a partition type condition. Recently, Király and Szabó [7] proved that the connectivity requirement in this parity constrained orientation problem can be given by a more general non-negative intersecting supermodular function. It is well-known that all these problems can be formalized as polymatroid parity problems. In this section we show that it is possible to formalize the problem of Király and Szabó in such a way that the arising polymatroid function has no non-trivial double circuits. So Theorem 1 can be applied to yield the result in [7].

$H = (V, \mathcal{E})$ is called a *hypergraph* if V is a finite set and $\emptyset \notin \mathcal{E}$ is a collection of multisets of V, the set of *hyperedges* of H. If in every hyperedge $h \in \mathcal{E}$ we designate a vertex $v \in h$ as the *head vertex* then we get a *directed hypergraph* $D = (V, \mathcal{A})$, called an *orientation* of H. For a set $X \subseteq V$, let $\delta_D(X)$ denote the set of directed hyperedges *entering* X, that is the set of hyperedges with head in X and at least one vertex in $V - X$.

Let $p : 2^V \to \mathbb{N}$ be a function with $p(\emptyset) = p(V) = 0$. An orientation D of a hypergraph $H = (V, \mathcal{E})$ *covers* p if $|\delta_D(X)| \geq p(X)$ for every $X \subseteq V$. In a *connectivity orientation problem* the question is the existence of an orientation covering p. When we are talking about *parity constrained orientations*, we are looking for connectivity orientations such that the out-degree at each vertex is of prescribed parity. Now define $b : 2^V \to \mathbb{Z}$ by

$$b(X) = \sum_{h \in \mathcal{E}} h(X) - |\mathcal{E}[X]| - p(X) \quad \text{for } X \subseteq V, \tag{7}$$

where $\mathcal{E}[X]$ denotes the set of hyperedges $h \in \mathcal{E}$ with $h \cap (V - X) = \emptyset$, and h equivalently stands for the hyperedge and its multiplicity function. It is clear that if $x : V \to \mathbb{N}$ is the out-degree vector of an orientation covering p then $x \in B(b)$. The contrary is also easy to prove, see e.g. in [14]:

Lemma 6. *Let $H = (V, \mathcal{E})$ be a hypergraph, $p : 2^V \to \mathbb{N}$ be a function with $p(\emptyset) = p(V) = 0$, and $x : V \to \mathbb{N}$. Then H has an orientation covering p such that the out-degree of each vertex $v \in V$ is $x(v)$ if and only if $x \in B(b)$.*

The function $b : 2^V \to \mathbb{Z}$ is said to be *intersecting submodular* if (1) holds whenever $X \cap Y \neq \emptyset$. Similarly, $p : 2^V \to \mathbb{Z}$ is *intersecting supermodular* if $-p$ is intersecting submodular. If $b : 2^V \to \mathbb{N}$ is a non-negative, non-decreasing intersecting submodular function then we can define a polymatroid function $\widehat{b} : 2^V \to \mathbb{N}$ by $\widehat{b}(X) = \min \left\{ \sum_{i=1}^t b(X_i) \ : \ X_1 \dot{\cup} X_2 \dot{\cup} \ldots \dot{\cup} X_t = X \right\}$ for $X \subseteq V$, which is called the *Dilworth truncation* of b. It is also well-known that, if p : $2^V \to \mathbb{N}$ is intersecting supermodular with $p(V) = 0$, then p is non-increasing.

Thus if $p : 2^V \to \mathbb{N}$ is an intersecting supermodular function with $p(\emptyset) = p(V) = 0$ then $b : 2^V \to \mathbb{Z}$, as defined in (7), is a non-decreasing intersecting submodular function, but it is not necessarily non-negative. The following theorem can be proved using basic properties of polymatroid functions.

Theorem 3. *Let $H = (V, \mathcal{E})$ be a hypergraph and $p : 2^V \to \mathbb{N}$ be an intersecting supermodular function with $p(\emptyset) = p(V) = 0$. Define b as in (7). Then H has an orientation covering p if and only if $b(V) \le \sum_{j=1}^t b(U_j)$ holds for every partition $U_1, U_2, \ldots, U_t$ of V.*

Let $H = (V, \mathcal{E})$ be a hypergraph and $T \subseteq V$. Our goal is to find an orientation of H covering p, where the set of odd out-degree vertices is as close as possible to T.

Theorem 4 (Király and Szabó [7]). *Let $H = (V, \mathcal{E})$ be a hypergraph, $T \subseteq V$, $p : 2^V \to \mathbb{N}$ be an intersecting supermodular function with $p(\emptyset) = p(V) = 0$, and assume that H has an orientation covering p. Define b as in (7). For an orientation D of H let $Y_D \subseteq V$ denote the set of odd out-degree vertices in D. Then*

$$\min\left\{ |T \triangle Y_D| \ : \ D \text{ is an orientation of } H \text{ covering } p \right\} =$$

$$\max\left\{ b(V) - \sum_{j=1}^t b(U_j) + |\{j : b(U_j) \not\equiv |T \cap U_j| \bmod 2\}| \right\}, \quad (8)$$

where the maximum is taken on partitions $U_1, U_2, \ldots, U_t$ of V.

An interesting corollary is the following non-defect form, which is again a generalization of Theorem 3.

Theorem 5. *Let $H = (V, \mathcal{E})$ be a hypergraph, $T \subseteq V$, and let $p : 2^V \to \mathbb{N}$ be an intersecting supermodular function with $p(\emptyset) = p(V) = 0$. Then, H has an orientation covering p with odd out-degrees exactly in the vertices of T, if and only if*

$$b(V) \le \sum_{j=1}^t b(U_j) - |\{j : b(U_j) \not\equiv |T \cap U_j| \bmod 2\}| \quad (9)$$

holds for every partition $U_1, U_2, \ldots, U_t$ of V.

Proof. For every $v \in T$ add a loop $2\chi_v$ to $\mathcal{E}$, resulting in the hypergraph $H' = (V, \mathcal{E}')$. Define b' as in (7), w.r.t. H'. As there is a straightforward bijection between the orientations of H and H', we have $\min\{|T \triangle Y_D| : D \text{ is an orientation of } H \text{ covering } p\} = \min\{|Y_{D'}| : D' \text{ is an orientation of } H' \text{ covering } p\}$, and $b(V) - \sum_{j=1}^t b(U_j) + |\{j : b(U_j) \not\equiv |T \cap U_j| \bmod 2\}| = b'(V) - \sum_{j=1}^t b'(U_j) + |\{j : b'(U_j) \text{ is odd }\}|$. Thus we can assume that $T = \emptyset$.

By Lemma 6, the integer vectors of $B(b)$ are exactly the out-degree vectors of the orientations of H covering p. Thus the $\ge$ direction is easy to check. Now we prove the other direction. As H has an orientation covering p, if $\emptyset \subseteq U \subseteq V$ then $b(U) + b(V - U) \ge b(V)$ by Theorem 3, implying that $b(U) \ge b(V) - b(V - U) \ge 0$.

Thus, b is non-decreasing, and we can define the polymatroid function $f = \widehat{b}$. We claim that it is enough to prove that $\nu(f) = \min \sum_{i=1}^{s} \lfloor \frac{1}{2} f(V_i) \rfloor$, where the minimum is taken over all partitions $V_1, V_2, \ldots, V_s$ of V. Indeed, using the definition of the Dilworth-truncation and that $b(V) = f(V)$ by Theorem 3, we get

$$\min\{|Y_D| : D \text{ is an ori. of } H \text{ covering } p\} = f(V) - 2\nu(f) =$$

$$= b(V) - \min \left\{ \sum_{i=1}^{s} f(V_i) - |\{i : f(V_i) \text{ is odd}\}| \; : \; V_1, \ldots, V_s \text{ partitions } V \right\} \leq$$

$$\leq b(V) - \min \left\{ \sum_{j=1}^{t} b(U_j) - |\{j : b(U_j) \text{ is odd}\}| \; : \; U_1, \ldots, U_t \text{ partitions } V \right\}.$$

Thus by Theorem 1 it is enough to prove that $\widehat{b}$ has no non-trivial compatible double circuits. The next lemma does the job.

Lemma 7. *Let $H = (V, \mathcal{E})$ be a hypergraph and let $p : 2^V \to \mathbb{N}$ an intersecting supermodular function with $p(\emptyset) = 0$. Suppose moreover that $b : 2^V \to \mathbb{Z}$ defined by (7) is non-negative and non-decreasing. Then the polymatroid function $f := \widehat{b}$ has no non-trivial compatible double circuits.*

Proof. Assume that $w : V \to \mathbb{N}$ is a non-trivial compatible double circuit of f with principal partition $W = W_1 \dot\cup W_2 \dot\cup \ldots \dot\cup W_d$. Clearly, $b(W) \geq w(W) - 2$. Let $1 \leq i < j \leq d$ and $Z = W - W_i$. As $w|_Z$ is a circuit, Lemma 2 yields that $w(Z) - 1 = f(Z) = \min \sum \{b(X_i) : X_1, \ldots, X_k \text{ partitions } Z\}$. However, if a non-trivial partition with $k \geq 2$ gave equality here, then we would have $f(Z) = \sum b(X_i) \geq \sum f(X_i) \geq \sum w(X_i) = w(Z) > f(Z)$, because $w|_{X_i} \in P(f)$. Thus $w(W - W_i) - 1 = b(W - W_i)$, and similarly, $x(W - W_j) - 1 = b(W - W_j)$. By applying intersecting submodularity to $W - W_i$ and $W - W_j$, and using that $w|_{W - W_i - W_j} \in P(f)$, we get $0 \geq b(W) - b(W - W_i) - b(W - W_j) + b(W - W_i - W_j) \geq (w(W) - 2) - (w(W - W_i) - 1) - (w(W - W_j) - 1) + w(W - W_i - W_j) = 0$, so equality holds throughout. As a corollary, each hyperedge $e \in \mathcal{E}[W]$ is spanned by one of the W_i's, and

$$\binom{d-1}{2}(b(W) + 2) = \binom{d-1}{2} w(W) =$$

$$= \sum_{1 \leq i < j \leq d} w(W - W_i - W_j) = \sum_{1 \leq i < j \leq d} b(W - W_i - W_j). \quad (10)$$

On the other hand,

$$\binom{d-1}{2} \sum_{h \in \mathcal{E}} h(W) = \sum_{1 \leq i < j \leq d} \sum_{h \in \mathcal{E}} h(W - W_i - W_j),$$

since $\sum_{h \in \mathcal{E}} h$ is modular, and

$$\binom{d-1}{2} p(W) \leq \sum_{1 \leq i < j \leq d} p(W - W_i - W_j),$$

since p is non-negative and non-increasing. Finally,

$$\binom{d-1}{2}|\mathcal{E}[W]| = \binom{d-1}{2}\sum_{i=1}^{d}|\mathcal{E}[W_i]| = \sum_{1 \le i < j \le d}|\mathcal{E}[W - W_i - W_j]|.$$

By the definition of b, the last 3 equalities together contradict (10).

Let us give an example showing that polymatroids without non-trivial compatible double circuits are not closed under contractions. Let $V = \{v_1, v_2, v_3, v_4\}$, $\mathcal{E} = \{v_1v_i, v_iv_i : i \in \{2, 3, 4\}\}$, $p(\{v_1\}) = 1$ and $p(U) = 0$ for the other sets. Then, by Lemma 7, $\hat{b}$ has no non-trivial compatible double circuits, while the polymatroid obtained from $\hat{b}$ by contracting an element in the prematroid from the preimage of v_1 has the non-trivial compatible double circuit $(1, 2, 2, 2)$.

4.2 A Planar Rigidity Problem

If $G = (V, E)$ is a graph and $p : V \to \mathbb{R}^2$ is an embedding into the Euclidean plane then (G, p) is said to be a *framework*. We think of the edges of G as rigid bars with flexible joins at the vertices. An *infinitesimal motion* means an assignment of velocities $x(v) \in \mathbb{R}^2$ to each vertex $v \in V$ such that the bar lengths are preserved, that is $(p(u)-p(v)) \perp (x(u)-x(v))$. The framework (G, p) is called *rigid* if all infinitesimal motions of (G, p) correspond to isometries of $\mathbb{R}^2$. The question of pinning down a minimum vertex set resulting a rigid framework was solved by Lovász in his seminal paper [10] about matroid parity. We say that $G = (V, E)$ is *generic rigid* if all frameworks (G, p) with *algebraically independent* coordinates p are rigid. The problem of finding a vertex set $Z \subseteq V$ of minimum size such that $G + K_Z$ is generic rigid is left open by [10], and it was solved recently by Fekete [3]. For more on the 2-dimensional rigidity see Laman [8] and Lovász and Yemini [11].

The setup of [3] puts the problem into a bit more general setting. Let $G = (V, E)$ be a graph, and for $l \in \{2, 3\}$ let $M_{2,l}$ be the matroid on ground set E such that $F \subseteq E$ is independent in $M_{2,l}$ if and only if $|F[X]| \le 2|X| - l$ for all $X \subseteq V$, $|X| \ge 2$. It can be proved that M is really a matroid. For clarity, $M_{2,2}$ is two times the cycle matroid of G, and so G has two edge-disjoint spanning trees if and only if $r_{2,2}(E) = 2|V| - 2$. As $M_{2,3}$ is the rigidity matroid of G, the graph G is generic rigid if and only if $r_{2,3}(E) = 2|V| - 3$. For $Z \subseteq V$ let $K_Z = (Z, E_Z)$ be the graph with vertex set Z having $4 - l$ parallel edges between any two vertices of Z. Our goal is to find a set $Z \subseteq V$ of minimum size such that $E + E_Z$ has rank $2|V| - l$. For $l = 2$, this is equivalent to shrinking a minimum vertex set Z such that G/Z has two edge-disjoint spanning trees.

We assume that E is independent in $M_{2,l}$, since if E is replaced by one of its bases then the solution set does not change. Fekete [3] proved the following lemma. For $X \subseteq V$ let $e(X)$ denote the number of edges having at least one end vertex in X.

Lemma 8 ([3]). *Let $l \in \{2, 3\}$. Assume that E is independent in $M_{2,l}$ and that $r_{2,l}(E) < 2|V| - l$. Let $Z \subseteq V$. Then $r(E + E_Z) = 2|V| - l$ if and only if $e(Y) \ge 2|Y|$ for every $Y \subseteq V - Z$.*

Therefore, the goal is to find a set $Z \subseteq V$ of minimum size such that $e(Y) \geq 2|Y|$ for every $Y \subseteq V - Z$. Let $f : 2^V \to \mathbb{N}$ be the polymatroid function with $f(X) = \min_{Y \subseteq X} 2|Y| + e(X - Y)$, i.e. f is obtained from the polymatroid function $X \mapsto e(X)$ by deleting with the vector $(2, 2, \ldots, 2)$. Hence for $l = 2$ the value $|V| - \nu(f)$ means the minimum size of a set Z whose contraction results in a graph with two edge-disjoint spanning trees, and for $l = 3$ it is the minimum size of a set Z such that $G + K_Z$ is generic rigid. In [10] the computation of $\nu(f)$ is reduced to the matching problem of graphs, yielding a partition type characterization. This characterization follows from the previous results of this paper, too. First, by Lemma 7 with the choice $p = 0$, the polymatroid function $X \mapsto e(X)$ has no non-trivial compatible double circuits. As f is obtained from $X \mapsto e(X)$ by deletion, Claim 4 yields that nor f has. Thus, $\nu(f) = \min \sum_{j=1}^{t} \lfloor \frac{1}{2} f(U_j) \rfloor$, where the minimum is taken over all partitions $U_1, U_2, \ldots, U_t$ of V. By the definition of f, we get the following.

Theorem 6 (Fekete, [3]). *Let $l \in \{2, 3\}$. Assume that E is independent in $M_{2,l}$ and that $r_{2,l}(E) < 2|V| - l$. Then the minimum size of a set $Z \subseteq V$ such that $r(E + E_Z) = 2|V| - l$ is $|V| - \nu(f)$, where*

$$\nu(f) = \min \left| V - \bigcup_{j=1}^{t} U_j \right| + \sum_{j=1}^{t} \left\lfloor \frac{e(U_j)}{2} \right\rfloor,$$

where the minimum is taken over all subpartitions $U_1, U_2, \ldots, U_t$ of V.

Acknowledgments. The authors are grateful for the support of András Frank, and discussions with Zsolt Fekete.

References

1. Dress, A. and Lovász, L., On some combinatorial properties of algebraic matroids, *Combinatorica*, (1987), **7/1**, 39–48
2. J. Edmonds, Paths, trees, and flowers, *Canadian Journal of Mathematics*, **17** (1965) 449–467
3. Fekete, Z., Source location with rigidity and tree packing requirements, www.cs.elte.hu/egres, TR-2005-04
4. Frank, A. and Jordán, T. and Szigeti, Z., An orientation theorem with parity conditions, *Discrete Appl. Math.*, (2001), **115/1-3**, 37–47
5. Frank, A. and Király, T. and Király, Z., On the orientation of graphs and hypergraphs, *Discrete Appl. Math.*, (2003), **131/2**, 385–400
6. Jensen, P.M. and Korte, B., Complexity of matroid property algorithms, *SIAM Journal on Computing*, (1982), **11/1**, 184–190
7. Király, T. and Szabó, J., A note on parity constrained orientations, www.cs.elte.hu/egres, TR-2003-11
8. Laman, G., On graphs and rigidity of plane skeletal structures, *J. Engrg. Math.*, (1970), **4**, 331–340
9. Lovász, L., The matroid matching problem, Algebraic methods in graph theory, Vol. I, II (Szeged, 1978), Colloq. Math. Soc. János Bolyai, **25**, 495–517

10. Lovász, L., Matroid matching and some applications, *J. Combin. Theory Ser. B*, (1980), **28/2**, 208–236
11. Lovász, L. and Yemini, Y., On generic rigidity in the plane, *SIAM J. Algebraic Discrete Methods*, (1982) **3/1**, 91–98
12. Lovász, L., Selecting independent lines from a family of lines in a space, *Acta Sci. Math. (Szeged)*, (1980), **42/1-2**, 121–131
13. Nebeský, L., A new characterization of the maximum genus of a graph, *Czechoslovak Math. J.*, (1981), **31(106)**, 604–613
14. Schrijver, A., Combinatorial optimization. Polyhedra and efficiency, Springer-Verlag, Berlin, 2003

Maximizing a Submodular Set Function Subject to a Matroid Constraint (Extended Abstract)

Gruia Calinescu[1,**], Chandra Chekuri[2], Martin Pál[3], and Jan Vondrák[4]

[1] Computer Science Dept., Illinois Institute of Technology, Chicago, IL
calinescu@iit.edu.
[2] Dept. of Computer Science, University of Illinois, Urbana, IL 61801
chekuri@cs.uiuc.edu.
[3] Google Inc., 1440 Broadway, New York, NY 10018
mpal@google.com.
[4] Dept. of Mathematics, Princeton University, Princeton, NJ 08544
jvondrak@math.princeton.edu.

Abstract. Let $f : 2^N \to \mathcal{R}^+$ be a non-decreasing submodular set function, and let $(N, \mathcal{I})$ be a matroid. We consider the problem $\max_{S \in \mathcal{I}} f(S)$. It is known that the greedy algorithm yields a 1/2-approximation [9] for this problem. It is also known, via a reduction from the max-k-cover problem, that there is no $(1 - 1/e + \epsilon)$-approximation for any constant $\epsilon > 0$, unless $P = NP$ [6]. In this paper, we improve the 1/2-approximation to a $(1 - 1/e)$-approximation, when f is a sum of weighted rank functions of matroids. This class of functions captures a number of interesting problems including set coverage type problems. Our main tools are the pipage rounding technique of Ageev and Sviridenko [1] and a probabilistic lemma on monotone submodular functions that might be of independent interest.

We show that the generalized assignment problem (GAP) is a special case of our problem; although the reduction requires $|N|$ to be exponential in the original problem size, we are able to interpret the recent $(1 - 1/e)$-approximation for GAP by Fleischer *et al.* [10] in our framework. This enables us to obtain a $(1 - 1/e)$-approximation for variants of GAP with more complex constraints.

1 Introduction

This paper is motivated by the following optimization problem. We are given a ground set N of n elements and a non-decreasing submodular set function $f : 2^N \to \mathcal{R}^+$. The function f is *submodular* iff $f(A) + f(B) \geq f(A \cup B) + f(A \cap B)$ for all $A, B \subseteq N$. We restrict attention to non-decreasing (or monotone) submodular set functions, that is $f(A) \geq f(B)$ for all $B \subseteq A$ and $f(\emptyset) = 0$. An independence family $\mathcal{I} \subseteq 2^N$ is a family of subsets that is downward closed, that

---

** Research partially supported by NSF grant CCF-0515088.

is, $A \in \mathcal{I}$ and $B \subseteq A$ implies that $B \in \mathcal{I}$. A set A is *independent* iff $A \in \mathcal{I}$. A family $\mathcal{I}$ is a p-independence family for an integer $p \geq 1$ if for all $A \in \mathcal{I}$ and $e \in N$ there exists a set $B \subseteq A$ such that $|B| \leq p$ and $A \setminus B + e$ is independent. For computational purposes we will assume that f and $\mathcal{I}$ are specified as oracles although in many specific settings of interest, an explicit description is often available. The problem (or rather class of problems) of interest in this paper is the following: $\max_{S \in \mathcal{I}} f(S)$. We will be mostly interested in the special case when $\mathcal{I}$ consists of the independent sets of a matroid on N. The problem of maximizing a submodular set function subject to independence constraints has been studied extensively. A number of interesting and useful combinatorial optimization problems, including NP-hard problems, are special cases. Some notable examples are maximum independent set in a matroid, weighted matroid intersection, and maximum coverage. Below we describe some candidates for f and $\mathcal{I}$ that arise frequently in applications.

Modular functions: A function $f : 2^N \rightarrow \mathcal{R}^+$ is modular iff $f(A) + f(B) = f(A \cup B) + f(A \cap B)$. If f is modular then there is a weight function $w : N \rightarrow \mathcal{R}^+$ such that $f(A) = w(A) = \sum_{e \in A} w(e)$.

Set Systems and Coverage: Given a universe U and n subsets $S_1, S_2, \ldots, S_n$ of U we obtain several natural submodular functions on the set $N = \{1, 2, \ldots, n\}$. First, the coverage function f given by $f(A) = |\cup_{i \in A} S_i|$ is submodular. This naturally extends to the weighted coverage function; given a non-negative weight function $w : U \rightarrow \mathcal{R}^+$, $f(A) = w(\cup_{i \in A} S_i)$. We obtain a multi-cover version as follows. For $x \in U$ let $k(x)$ be an integer. For each $x \in U$ and S_i let $c(S_i, x) = 1$ if $x \in S_i$ and 0 if $x \notin S_i$. Given $A \subseteq N$, let $c'(A, x)$, the coverage of x under A, be defined as $c'(A, x) = \min\{k(x), \sum_{i \in A} c(S_i, x)\}$. The function f with $f(A) = \sum_{x \in U} c'(A, x)$ is submodular. A related function defined by $f(A) = \sum_{x \in U} \max_{i \in A} w(S_i, x)$ is also submodular where $w(S_i, x)$ is a non-negative weight for S_i covering x.

Weighted rank functions of matroids and their sums: The rank function of a matroid $\mathcal{M} = (N, \mathcal{I})$, $r_{\mathcal{M}}(A) = \max\{|S| : S \subseteq A, S \in \mathcal{I}\}$, is submodular. Given $w : N \rightarrow \mathcal{R}^+$, the weighted rank function defined by $r_{\mathcal{M},w}(A) = \max\{w(S) : S \subseteq A, S \in \mathcal{I}\}$ is a submodular function. A sum of weighted rank functions is also submodular. Functions arising in this way form a rich class of submodular functions. In particular, all the functions on set systems and coverage mentioned above are captured by this class. However, the class does not include all monotone submodular functions; one notable exception is multi-cover by multisets.

Matroid Constraint: An independence family of particular interest is one induced by a matroid $\mathcal{M} = (N, \mathcal{I})$. A very simple matroid constraint that is of much importance in applications [5,14,2,3,10] is the partition matroid; N is partitioned into ℓ sets $N_1, N_2, \ldots, N_\ell$ with associated integers $k_1, k_2, \ldots, k_\ell$, and a set $A \subseteq N$ is independent iff $|A \cap N_i| \leq k_i$. In fact even the case of $\ell = 1$ (the uniform matroid) is of interest. Laminar matroids generalize partition matroids. We have a laminar family of sets on N and each set S in the family has an integer value k_S. A set $A \subseteq N$ is independent iff $|A \cap S| \leq k_S$ for each S in the family.

Intersection of Matroids: A natural generalization of the single matroid case is obtained when we consider intersections of different matroids $\mathcal{M}_1, \mathcal{M}_2, \ldots, \mathcal{M}_p$ on the same ground set N. That is, $\mathcal{I} = \cap_i \mathcal{I}_i$ where $\mathcal{I}_i$ is the independence family of $\mathcal{M}_i$. A simple example is the family of hypergraph matchings in a p-partite graph ($p = 2$ is simply the family of matchings in a bipartite graph).

Matchings: Given a general graph $G = (V, N)$ the set of matchings forms a 2-independent family. Given a hypergraph $G = (V, N)$ such that each edge $e \in N$ is of cardinality at most p, the set of matchings in G induce a p-independent family. Note that matchings in general graphs are not captured as intersections of matroids.

The Greedy Algorithm: A simple greedy algorithm is quite natural for this problem. The algorithm incrementally builds a solution (without backtracking) starting with the empty set. In each iteration it adds an element that most improves the current solution (according to f) while maintaining independence of the solution. The greedy algorithm yields a $1/p$-approximation for maximizing a modular function subject to a p-independence constraint [12,13]. For submodular functions, the greedy algorithm yields a ratio of $1/(p+1)$ [9]. [1] These ratios for greedy are tight for all p even when the p-independent system is obtained as an intersection of p matroids. For large but fixed p, the p-dimensional matching problem is NP-hard to approximate to within an $\Omega(\log p/p)$ factor [11].

For the problem of maximizing a submodular function subject to a matroid constraint (special case of $p = 1$), the greedy algorithm achieves a ratio of $1/2$. When the matroid is the simple uniform matroid ($S \subseteq N$ is independent iff $|S| \le k$) the greedy algorithm yields a $(1 - 1/e)$-approximation [14]. This special case already captures the maximum coverage problem for which it is shown in [6] that, unless $P = NP$, no $1 - 1/e + \epsilon$ approximation is possible for any constant $\epsilon > 0$. This paper is motivated by the following question. Is there a $(1-1/e)$-approximation algorithm for maximizing a submodular function subject to (any given) matroid constraint? We resolve this question for a subclass of monotone submodular functions, which can be expressed as a sum of weighted rank functions of matroids. The following is our main result.

Theorem 1. *Given a ground set N, let $f(S) = \sum_{i=1}^{m} g_i(S)$ where $g_1, \ldots, g_m :$ $2^N \to \mathcal{R}^+$ are weighted rank functions, g_i defined by a matroid $\mathcal{M}_i = (N, \mathcal{X}_i)$ and weight function $w_i : N \to \mathcal{R}^+$. Given another matroid $\mathcal{M} = (N, \mathcal{I})$ and membership oracles for $\mathcal{M}_1, \mathcal{M}_2, \ldots, \mathcal{M}_m$ and $\mathcal{M}$, there is a polynomial time $(1 - 1/e)$-approximation for the problem $\max_{S \in \mathcal{I}} f(S)$.*

As immediate corollaries we obtain a $(1 - 1/e)$-approximation for a number of coverage problems under a matroid constraint. It is known that there exist submodular monotone functions that cannot be expressed as a sum of weighted rank functions of matroids (see [16], 44.6e). For such functions, our framework

---

[1] We give a somewhat new proof of this result in the full version of the paper. If only an α-approximate oracle ($\alpha \le 1$) is available for the function evaluation, the ratio obtained is $\alpha/(p+\alpha)$. Several old and recent applications of greedy can be explained using this observation.

does not seem to apply at this moment. We leave it as an open question whether a $(1 - 1/e)$-approximation is possible for all monotone submodular functions.

Our main tools are the the pipage rounding technique of Ageev and Sviridenko [1], and the following useful lemma.

Lemma 1. *Let $f : 2^N \to \mathcal{R}^+$ be a monotone submodular function and let $f^* : [0,1]^N \to \mathcal{R}^+$ be defined as $f^*(y) = \min_S(f(S) + \sum_i y_i(f(S + i) - f(S)))$. For $y \in [0,1]^N$, let $\hat{y}$ denote a random vector in $\{0,1\}^N$ obtained by independently setting $\hat{y}_i = 1$ with probability y_i and 0 otherwise. Then, $\mathbf{E}[f(\hat{y})] \geq (1-1/e)f^*(y)$.*

We give a non-trivial application of Theorem 1 to variants of the generalized assignment problem (GAP). In GAP we are given n bins and m items. Each item i specifies a size s_{ji} and a value (or profit) v_{ji} for each bin j. Each bin has capacity 1 and the goal is to assign a subset of items to bins such that the bin capacities are not violated and the profit of the assignment is maximized. Recently Fleischer *et al.* [10] gave a $(1 - 1/e)$-approximation for this problem, improving upon a $1/2$-approximation [4]. We rederive the same ratio casting the problem as a special case of submodular function maximization. Moreover our techniques allow us to obtain a $(1-1/e)$-approximation for GAP even under any given laminar matroid constraint on the bins. A simple and easy to understand example is GAP with the added constraint that at most k of the n bins be used.

Theorem 2. *Let A be an instance of GAP with n bins and m items and let B be the set of bins. Let $\mathcal{M} = (B, \mathcal{I})$ be a laminar matroid on B. There is a polynomial time $(1 - 1/e)$-approximation to find a maximum profit assignment to bins such that the subset $S \subseteq B$ of bins that are used in the assignment satisfy the constraint $S \in \mathcal{I}$.*

We note that the approximation ratio for GAP has been improved to $1-1/e+\delta_1$ for a small $\delta_1 > 0$ in [8] using the same LP as in [10]. However, the algorithm in [10] extends to even more general assignment problems in which the sets of items allowed in a bin are further constrained; for such allocation problems it is shown in [10] that it is NP-hard to obtain an approximation ratio of $1 - 1/e + \epsilon$ for any constant $\epsilon > 0$. Our framework also extends to this wider class of assignment problems and hence $1 - 1/e$ is the best approximation factor one can achieve with this approach.

1.1   Preliminaries

Given a submodular function $f : N \to \mathcal{R}^+$ and $A \subset N$, the function f_A defined by $f_A(S) = f(S \cup A) - f(A)$ is also submodular. Further, if f is monotone, f_A is also monotone. For $i \in N$, we abbreviate $S \cup \{i\}$ by $S + i$. By $f_A(i)$, we denote the "marginal value" $f(A+i) - f(A)$. Submodularity is equivalent to $f_A(i)$ being non-increasing as a function of A for every fixed i.

Given a matroid $\mathcal{M} = (N, \mathcal{I})$, we denote by $r_\mathcal{M}$ the rank function of $\mathcal{M}$ where $r_\mathcal{M}(A) = \max\{|S| : S \subseteq A, S \in \mathcal{I}\}$. The rank function is monotone and submodular. We denote by $P(\mathcal{M})$ the polytope associated with $\mathcal{M}$; this is the set of all real vectors $y \in [0,1]^N$ that satisfy the constraints: $y(S) \leq r_{\mathcal{M}}(S) \; \forall S \subseteq N$, where $y(S) = \sum_{i \in S} y_i$. Edmonds showed that the vertices of $P(\mathcal{M})$ are precisely the characteristic vectors of the independent sets of $\mathcal{M}$. Further, given a membership oracle for $\mathcal{M}$ (that is given $S \subseteq N$, the oracle answers if $S \in \mathcal{I}$ or not), one can optimize linear functions over $P(\mathcal{M})$.

A base of $\mathcal{M}$ is a set $S \in \mathcal{I}$ such that $r_{\mathcal{M}}(S) = r_{\mathcal{M}}(N)$. The base polytope $B(\mathcal{M})$ of $\mathcal{M}$ is given by $\{y \in P(\mathcal{M}) \mid y(N) = r_{\mathcal{M}}(N)\}$. The extreme points of $B(\mathcal{M})$ are the characteristic vectors of the bases of $\mathcal{M}$. Given the problem $\max_{S \in \mathcal{I}} f(S)$, where $\mathcal{M} = (N, \mathcal{I})$ is a matroid, there always exists an optimum solution S^* where S^* is a base of $\mathcal{M}$. Note that this is false if f is not monotone. Thus, for monotone f, it is equivalent to consider the problem $\max_{S \in \mathcal{B}} f(S)$ where $\mathcal{B}$ is the set of bases of $\mathcal{M}$. See [16] for more details on matroids and polyhedral aspects.

2   Pipage Rounding Framework

Ageev and Sviridenko [1] developed an elegant technique for rounding solutions of linear and non-linear programs that they called "pipage rounding". Subsequently, Srinivasan [17] and Gandhi *et al.* [15] interpreted some applications of pipage rounding as a deterministic variant of dependent randomized rounding. In a typical scenario, randomly rounding a fractional solution of a linear program does not preserve the feasibility of constraints, in particular equality constraints. Nevertheless, the techniques of [1,17,15] show that randomized rounding can be applied in a certain controlled way to guide a solution that respects certain class of constraints. In particular these techniques were used to round fractional solutions to the generalized assignment problem. In this paper we show that the rounding framework applies quite naturally to our problem. Further, our analysis also reveals the important role of submodularity in this context.

We now describe the pipage rounding framework as adapted to our problem. We follow [1] in spirit although our notation and description is somewhat different and tailored to our application: given a monotone submodular function $f : 2^N \to \mathcal{R}^+$ and a matroid $\mathcal{M} = (N, \mathcal{I})$, we wish to solve $\max_{S \in \mathcal{I}} f(S)$. Let $y_i \in \{0,1\}$ be a variable that indicates whether i is picked in a solution to the problem. Then $\max_{S \in \mathcal{I}} f(S)$ can be written as the following problem: $\max\{f(y) : y \in P(\mathcal{M}), y \in \{0,1\}^N\}$. As we observed in Section 1.1, this is equivalent to $\max\{f(y) : y \in B(\mathcal{M}), y \in \{0,1\}^N\}$ where $B(\mathcal{M})$ is the base polytope of $\mathcal{M}$.

The framework relies on the ability to solve a relaxation of the problem in polynomial time. To obtain a relaxation we let $y \in [0,1]^N$. This also requires us to find an *extension* of f to a function $\tilde{f} : [0,1]^N \to \mathcal{R}^+$ such that the problem $\max\{\tilde{f}(y) : y \in P(\mathcal{M})\}$ can be solved in polynomial time. We require two properties of the extension: (i) $\tilde{f}(y) = f(y)$ for all $y \in \{0,1\}^N$, and (ii) monotonicity, that is $\tilde{f}(y) \geq \tilde{f}(z)$, for all $y \geq z$; $y, z \in [0,1]^N$. Note that the optimum value of the relaxation is at least the integral optimum solution denoted by OPT. Given an optimum fractional solution y^* to the relaxation, our goal is to round y^* to an integer solution z such that $f(z) \geq \alpha \tilde{f}(y^*) \geq \alpha$OPT. Clearly the quality of the relaxation depends on the extension function $\tilde{f}$. The rounding framework relies on a potential function $F : [0, 1]^N \to \mathcal{R}^+$, derived from f, that guides the rounding and at the same time allows one to derive bounds on the quality of the approximation. The reason to consider $\tilde{f}$ and F separately will become clear later. Assuming the existence of $\tilde{f}$ and F, we describe the pipage rounding algorithm for our problem.

Given $y \in [0, 1]^n$ we say that i is fractional in y if $0 < y_i < 1$. For $y \in P(\mathcal{M})$, a set $A \subseteq N$ is *tight* if $y(A) = r_{\mathcal{M}}(A)$. The following useful proposition follows easily from the submodularity of the rank function $r_{\mathcal{M}}$.

Proposition 1. *If A and B are two tight sets with respect to y then $A \cap B$ and $A \cup B$ are also tight with respect to y.*

The monotonicity of $\tilde{f}$ also implies the following.

Proposition 2. *There exists an optimum solution y^* to $\max\{\tilde{f}(y) : y \in P(\mathcal{M})\}$ such that $y^*(N) = \sum_{i \in N} y_i^* = r_{\mathcal{M}}(N)$.*

Alternatively we can solve the problem $\max\{\tilde{f}(y) : y \in B(\mathcal{M})\}$ which would automatically ensure that $y^*(N) = r_{\mathcal{M}}(N)$. We are interested in tight sets that contain a fractional variable. Observe that a tight set with a fractional variable has at least two fractional variables. Given a tight set A with fractional variables i, j, we let $y_{ij}(\epsilon)$ be the vector obtained by adding ϵ to y_i and subtracting ϵ from y_j and leaving the other values unchanged. Let $\epsilon_{ij}^+(y) = \max\{\epsilon \geq 0 \mid y_{ij}(\epsilon) \in P(\mathcal{M})\}$. Similarly we let $\epsilon_{ij}^-(y) = \min\{\epsilon \leq 0 \mid y_{ij}(\epsilon) \in P(\mathcal{M})\}$. We let $y_{ij}^+ = y_{ij}(\epsilon_{ij}^+)$ and $y_{ij}^- = y_{ij}(\epsilon_{ij}^-)$. For a given y and $i, j \in N$, we define a real-valued function $F_{ij}^y : [\epsilon_{ij}^-(y), \epsilon_{ij}^+(y)] \to \mathcal{R}^+$ where $F_{ij}^y(\delta) = F(y_{ij}(\delta))$.

Algorithm **PipageRound**(y)**:**
    While (y is not integral) do
        Let A be a minimal tight set containing fractional $i, j \in A$
        If $(F(y_{ij}^+) \geq F(y_{ij}^-))$ $y \leftarrow y_{ij}^+$
        Else $y \leftarrow y_{ij}^-$
    EndWhile
    Output y, $f(y)$.

Lemma 2. *The pipage rounding algorithm outputs an integral feasible y in $O(n^2)$ iterations. Given an oracle access to F and a membership oracle for $\mathcal{M}$, the algorithm can be implemented in polynomial time.*

Proof (sketch). Using Proposition 2, we assume that N is tight with respect to y. Since y_{ij}^+ and y_{ij}^- both belong to $P(\mathcal{M})$, the algorithm maintains the invariant that $y \in P(\mathcal{M})$ and that N is tight. Thus there is always a tight set with two fractional variables as long as y is not integral. We observe that the algorithm does not alter a variable y_i once $y_i \in \{0, 1\}$. To simplify the algorithm's analysis we can alter it slightly so that the set A that is picked in each iteration is not only minimal but also of minimum cardinality among such minimal sets. Let $y(h)$ be the vector y at the beginning of iteration h. We claim that $y(h + n - 1)$

has at least one more integral variable than $y(h)$. This will give us the desired bound of $O(n^2)$ on the total number of iterations.

To prove the claim, let A_h be the tight set picked by the algorithm, and $i_h, j_h \in A_h$ the two fractional variables modified in iteration h. If one of them becomes integral in $y(h+1)$, we are done. Otherwise we claim that $|A_{h+1}| < |A_h|$, hence after $n-1$ iterations we are guaranteed to have one more integral variable. To see that $|A_{h+1}| < |A_h|$, assume wlog that $y(h+1) = y(h)^+_{i_h j_h}$; since i_h, j_h are still fractional, there is a new tight set B with respect to $y(h+1)$, which prevented us from going further. B contains exactly one of i_h, j_h, otherwise $y(B)$ does not change in iteration h. From Proposition 1, it follows that $B \cap A_h$ is also tight, it contains a fractional variable, and $|B \cap A_h| < |A_h|$. In the next iteration, we can use $A_{h+1} = B \cap A_h$. To implement an iteration, we need to compute y^+_{ij}, y^-_{ij} and the new tight set in polynomial time. These can be done by appealing to known methods [16]. We defer the details to a full version of the paper.

To obtain a guarantee on the quality of the solution, F needs to satisfy some properties, as suggested in [1].

– F is an extension of f and $F(y) \geq \alpha \tilde{f}(y)$ for all $y \in [0, 1]^N$.
– F^y_{ij} is *convex* for all y and i, j.

Given the above two conditions, it is shown in [1] that the pipage rounding algorithm yields the following: given an optimum fractional solution y^*, the rounding yields an integral solution z such that $F(z) \geq F(y^*)$. This follows from the convexity requirement on F^y_{ij}; either $F(y^+_{ij}) \geq F(y)$ or $F(y^-_{ij}) \geq F(y)$ and the choice of the algorithm ensures that in each iteration the value of F does not decrease. Therefore we can conclude that $f(z) = F(z) \geq F(y^*) \geq \alpha \tilde{f}(y^*)$. Since $\tilde{f}(y^*) \geq$ OPT, we have $f(z) \geq \alpha$OPT.

3   Extensions of Submodular Functions

In this section, we address the issue of extending a monotone submodular function $f : 2^N \to \mathcal{R}^+$ to continuous functions $\tilde{f}, F : [0, 1]^N \to \mathcal{R}^+$, as required by the framework.

F as the expected value of f: We consider a simple and natural candidate for F that is implicitly generated from f. Define $F(y) = \mathbf{E}[f(\hat{y})]$ where $\hat{y}$ is a random integer vector obtained from y by independently rounding each i to 1 with probability y_i and to 0 with probability $1 - y_i$. In shorthand, we write $F = \mathbf{E}f$. We can evaluate $F = \mathbf{E}f$ to any desired accuracy by taking several independent samples. We defer details that show that a polynomial number of samples suffice to obtain a $(1 - 1/poly(n))$-approximation to $F(y)$. Alternatively we could use a randomized version of the pipage rounding that does not require us to evalute F explicitly.

In [1], F was given as an explicit function for some simple functions and the convexity of F^y_{ij} was explicitly shown. A nice feature of $F = \mathbf{E}f$ is that the convexity requirement is satisfied for all submodular f.

Lemma 3. *For any submodular f, if $F = \mathbf{E}f$, then F_{ij}^y is convex for all $y \in [0,1]^N$ and $i, j \in N$.*

Proof. Let $F = \mathbf{E}f$. For $S \subseteq N \setminus \{i,j\}$ and $y \in [0,1]^N$, let $p_y(S) = \prod_{l \in S} y_l \prod_{l \in N \setminus \{i,j\} \setminus S}(1 - y_l)$ be the probability that S is precisely the set obtained by randomized rounding on $N \setminus \{i,j\}$. Then

$$F(y) = \sum_{S \subseteq N \setminus \{i,j\}} p_y(S) \left((1 - y_i)(1 - y_j)f(S) + (1 - y_i)y_j f(S+j) \right.$$
$$\left. + y_i(1 - y_j)f(S + i) + y_i y_j f(S + i + j) \right).$$

We have $F_{ij}^y(\delta) = F(y_{ij}(\delta))$. Let $x = y_{ij}(\delta)$, i.e. $x_i = y_i + \delta$, $x_j = y_j - \delta$ and $x_l = y_l$ for all $l \in N \setminus \{i,j\}$. Hence it follows that $p_x(S) = p_y(S)$ for $S \subseteq N \setminus \{i,j\}$. It can be seen that $F(y_{ij}(\delta)) = F(x) = c_2\delta^2 + c_1\delta + c_0$ where c_2, c_1, c_0 do not depend on δ (they depend only on y and f). Thus to show that $F_{ij}^y(\delta)$ is convex in δ, it is sufficient to prove that $c_2 \geq 0$. It is easy to check that

$$c_2 = \sum_{S \subseteq N \setminus \{i,j\}} p_y(S)(-f(S) + f(S + j) + f(S + i) - f(S + i + j)).$$

By submodularity, $f(S+i)+f(S+j) \geq f((S+i)\cap(S+j))+f((S+i)\cup(S+j)) = f(S) + f(S + i + j)$ which proves that $c_2 \geq 0$.

Next, we need an extension $\tilde{f}$ such that $\max\{\tilde{f}(y) : y \in P(\mathcal{M})\}$ can be solved in polynomial time. The approximation guarantee is the largest α such that $F(y) \geq \alpha \tilde{f}(y)$.

Extension f^+: Our first candidate for $\tilde{f}$ is an extension similar to the objective function of the "Configuration LP" [10,7,8].

$$- \ f^+(y) = \max \left\{ \sum_{S \subseteq N} \alpha_S f(S) : \sum_S \alpha_S \leq 1, \alpha_S \geq 0 \ \& \ \forall j; \sum_{S: j \in S} \alpha_S \leq y_j \right\}.$$

Extension f^*: Another candidate is a function appearing in [14] and subsequently [9,18,19], where it is used indirectly in the analysis of the greedy algorithm for submodular function maximization:

$$- \ f^*(y) = \min \left\{ f(S) + \sum_{j \in N} f_S(j)y_j : S \subseteq N \right\}.$$

Unfortunately, as the theorem below shows, it is NP-hard to evaluate $f^+(y)$ and $f^*(y)$ and also to optimize them over matroid polytopes.

Theorem 3. *It is NP-hard to compute $f^+(y)$ or $f^*(y)$ for a given $y \in [0,1]^n$ and a given monotone submodular function f. Also, there is $\delta > 0$ such that for a given matroid $\mathcal{M}$ it is NP-hard to find any point $z \in P(\mathcal{M})$ such that $f^+(z) \geq (1-\delta)\max\{f^+(y) : y \in P(\mathcal{M})\}$. Similarly, it is NP-hard to find any point $z \in P(\mathcal{M})$ such that $f^*(z) \geq (1-\delta)\max\{f^*(y) : y \in P(\mathcal{M})\}$. These results hold even for coverage-type submodular functions and partition matroids.*

We defer the proof to a full version of the paper; the authors are unaware of prior work that might have addressed this question. Still, both $f^+(y)$ and $f^*(y)$ will be useful in our analysis. We remark that for any class of submodular functions where either $f^+(y)$ or $f^*(y)$ is computable in polynomial time, we obtain a $(1 - 1/e)$-approximation for our problem.

It is known and easy to see that for $y \in \{0, 1\}^N$, both f^+ and f^* functions coincide with f and thus they are indeed extensions of f. For any $y \in [0, 1]^N$, we first show the following.

Lemma 4. *For any monotone submodular f, $F(y) \le f^+(y) \le f^*(y)$.*

Proof. To see the first inequality, let $\alpha_S = \prod_{i \in S} y_i \prod_{i \notin S}(1 - y_i)$ be the probability that we obtain $\hat{y} = \chi_S$ by independent rounding of y. Since $\sum_{S:j \in S} \alpha_S = \Pr[\hat{y}_j = 1] = y_j$, this is a feasible solution for $f^+(y)$ and therefore $f^+(y) \ge \sum_S \alpha_S f(S) = \mathbf{E}[f(\hat{y})] = F(y)$.

For the second inequality, consider any feasible vector α_S and any set $T \subseteq N$:

$$\sum_S \alpha_S f(S) \le \sum_S \alpha_S \left(f(T) + \sum_{j \in S} f_T(j) \right) \le f(T) + \sum_{j \in N} y_j f_T(j)$$

using submodularity and the properties of α_S. By taking the maximum on the left and the minimum on the right, we obtain $f^+(y) \le f^*(y)$.

It is tempting to conjecture that $f^+(y)$ and $f^*(y)$ are in fact equal, due to some duality relationship. However, this is not the case: both inequalities in Lemma 4 can be sharp and both gaps can be close to $1 - 1/e$. For the first inequality, consider the submodular function $f(S) = \min\{|S|, 1\}$ and $y_j = 1/n$ for all j; then $F(y) = 1 - (1 - 1/n)^n$ and $f^+(y) = 1$. For the second inequality, choose a large but fixed k, $f(S) = 1 - (1 - |S|/n)^k$ and $y_j = 1/k$ for all j. The reader can verify that $f^+(y) = 1 - (1 - 1/k)^k$, while $f^*(y) \ge 1 - k/n \to 1$ as $n \to \infty$. We prove that $1 - 1/e$ is the worst possible gap for both inequalities. Moreover, even the gap between $F(y)$ and $f^*(y)$ is bounded by $1 - 1/e$.

Lemma 5. *For any monotone submodular f, $F(y) \ge \left(1 - \frac{1}{e}\right) f^*(y)$.*

Proof. For each element $j \in N$, set up an independent Poisson clock $\mathcal{C}_j$ of rate y_j, i.e. a device which sends signals at random times, in any infinitesimal time interval of size dt independently with probability $y_j dt$. We define a random process which starts with an empty set $S(0) = \emptyset$ at time $t = 0$. At any time when the clock $\mathcal{C}_j$ sends a signal, we include element j in S, which increases its value by $f_S(j)$. (If j is already in S, nothing happens; the marginal value $f_S(j)$ is zero in this case.) Denote by $S(t)$ the random set we have at time t. By the definition of a Poisson clock, $S(1)$ contains element j independently with probability $1 - e^{-y_j} \le y_j$. Since such a set can be obtained as a subset of the random set defined by $\hat{y}$, we have $\mathbf{E}[f(S(1))] \le F(y)$ by monotonicity. We show that $\mathbf{E}[f(S(1))] \ge (1 - 1/e)f^*(y)$ which will prove the claim.

Let $t \in [0, 1]$. Condition on $S(t) = S$ and consider how $f(S(t))$ changes in an infinitesimal interval $[t, t + dt]$. The probability that we include element j is $y_j dt$.

Since dt is very small, the events for different elements j are effectively disjoint. Thus the expected increase of $f(S(t))$ is (up to $O(dt^2)$ terms)

$$\mathbf{E}[f(S(t+dt)) - f(S(t)) \mid S(t) = S] = \sum_{j \in N} f_S(j) y_j dt \geq (f^*(y) - f(S))dt$$

using the definition of $f^*(y)$. We divide by dt and take the expectation over S:

$$\frac{1}{dt}\mathbf{E}[f(S(t+dt)) - f(S(t))] \geq f^*(y) - \mathbf{E}[f(S(t))].$$

We define $\phi(t) = \mathbf{E}[f(S(t))]$, i.e. $\frac{d\phi}{dt} \geq f^*(y) - \phi(t)$. We solve this differential inequality by considering $\psi(t) = e^t\phi(t)$ and $\frac{d\psi}{dt} = e^t(\frac{d\phi}{dt} + \phi(t)) \geq e^t f^*(y)$. Since $\psi(0) = \phi(0) = 0$, this implies

$$\psi(x) = \int_0^x \frac{d\psi}{dt} dt \geq \int_0^x e^t f^*(y)dt = (e^x - 1)f^*(y)$$

for any $x \geq 0$. We conclude that $\mathbf{E}[f(S(t))] = \phi(t) = e^{-t}\psi(t) \geq (1 - e^{-t})f^*(y)$ and $F(y) \geq \mathbf{E}[f(S(1))] \geq (1 - 1/e)f^*(y)$.

We remark that we did not actually use submodularity in the proof of Lemma 5! Formally, it can be stated for all monotone functions f. However, $f^*(y)$ is not a proper extension of f when f is not submodular (e.g., $f^*(y)$ is identically zero if $f(S) = 0$ for $|S| \leq 1$). So the statement of Lemma 5 is not very meaningful in this generality.

To summarize what we have proved so far, we have two relaxations of our problem:

- $\max\{f^+(y) : y \in P(\mathcal{M})\}$
- $\max\{f^*(y) : y \in P(\mathcal{M})\}$

Our framework together with Lemma 4 and Lemma 5 implies that both of these relaxations have integrality gap at most $1 - 1/e$. Theorem 3 shows NP-hardness of solving the relaxations. We show how to use the framework efficiently in a restricted case of interest which is described in the following section.

4   Sums of Weighted Rank Functions

We achieve a $(1 - 1/e)$-approximation, under a matroid constraint $\mathcal{M}$, for any submodular function f that can be expressed as a sum of "weighted rank functions" of matroids. This is the most general subclass of submodular functions for which we are able to use the framework outlined in Section 2 in an efficient way. Here we describe this in detail.

Weighted rank functions of matroids: Given a matroid $(N, \mathcal{X})$ and a weight function $w : N \to \mathcal{R}^+$, we define a *weighted rank function* $g : 2^N \to \mathcal{R}^+$,

$$g(S) = \max\{\sum_{j \in I} w_j : I \subseteq S \ \& \ I \in \mathcal{X}\}.$$

It is well known that such a function is monotone and submodular. A simple special case is when $\mathcal{X} = \{I \mid |I| = 1\}$. Then $g(S)$ returns simply the maximum-weight element of S; this will be useful in our application to GAP.

Sums of weighted rank functions: We consider functions $f : 2^N \to \mathcal{R}^+$ of the form $f(S) = \sum_{i=1}^m g_i(S)$ where each g_i is a weighted rank function for matroid $(N, \mathcal{X}_i)$ with weights w_{ij}. Again, $f(S)$ is monotone and submodular.

The functions that can be generated in this way form a fairly rich subclass of monotone submodular functions. In particular, they generalize submodular functions arising from coverage systems. Coverage-type submodular functions can be obtained by considering a simple uniform matroid $(N, \mathcal{X})$ with $\mathcal{X} = \{I \subseteq N \mid |I| \leq 1\}$. For a collection of sets $\{A_j\}_{j \in N}$ on a ground set $[m]$, we can define m collections of weights on N, where $w_{ij} = 1$ if A_j contains element i, and 0 otherwise. Then the weighted rank function $g_i(S) = \max\{w_{ij} : j \in S\}$ is simply an indicator of whether $\bigcup_{j \in S} A_j$ covers element i. The sum of the rank functions $g_i(S)$ gives exactly the size of this union $f(S) = \sum_{i=1}^m g_i(S) = \left| \bigcup_{j \in S} A_j \right|$. Generalization to the weighted case is straightforward.

LP formulation for sums of weighted rank functions: For a submodular function given as $f(S) = \sum_{i=1}^m g_i(S)$ where $g_i(S) = \max\{w_i(I) : I \subseteq S, I \in \mathcal{X}_i\}$, consider an extension $g_i^+(y)$ for each g_i, as defined in Section 3:

$$g_i^+(y) = \max\{\sum_{S \subseteq N} \alpha_S g_i(S) : \sum_S \alpha_S \leq 1, \alpha_S \geq 0 \ \& \ \forall j; \sum_{S:j \in S} \alpha_S \leq y_j\}.$$

Here, we can assume without loss of generality that α_S is nonzero only for $S \in \mathcal{X}_i$ (otherwise replace each S by a subset $I \subseteq S, I \in \mathcal{X}_i$, such that $g_i(S) = w_i(I)$). Therefore, g_i^+ can be written as

$$g_i^+(y) = \max\{\sum_{I \in \mathcal{X}_i} \alpha_I \sum_{j \in I} w_{ij} : \sum_{I \in \mathcal{X}_i} \alpha_I \leq 1, \alpha_I \geq 0 \ \& \ \forall j; \sum_{I \in \mathcal{X}_i : j \in I} \alpha_I \leq y_j\}.$$

We can set $x_{ij} = \sum_{I \in \mathcal{X}_i : j \in I} \alpha_I$ and observe that a vector $x_i = (x_{ij})_{j \in N}$ can be obtained in this way if and only if it is a convex linear combination of independent sets; i.e., if it is in the matroid polytope $P(\mathcal{X}_i)$. The objective function becomes $\sum_{j \in N} w_{ij} \sum_{I \in \mathcal{X}_i : j \in I} \alpha_I = \sum_{j \in N} w_{ij} x_{ij}$ and so we can write equivalently

$$g_i^+(y) = \max\{\sum_{j \in N} w_{ij} x_{ij} : x_i \in P(\mathcal{X}_i) \ \& \ \forall j; x_{ij} \leq y_j\}.$$

We sum up these functions to obtain an extension $\tilde{f}(y) = \sum_{i=1}^m g_i^+(y)$. This leads to the following LP formulation for the problem $\max\{\tilde{f}(y) : y \in P(\mathcal{M})\}$:

We can solve the LP using the ellipsoid method, since a separation oracle can be efficiently implemented for each matroid polytope, and therefore also for this LP. To obtain a $(1-1/e)$-approximation (Theorem 1) via the above LP using the pipage rounding framework from Section 2, it is sufficient to prove the following lemma.

$$\max \sum_{i=1}^{m} \sum_{j \in N} w_{ij} x_{ij};$$
$$\forall i, j; x_{ij} \leq y_j,$$
$$\forall i; x_i \in P(\mathcal{X}_i),$$
$$y \in P(\mathcal{M}).$$

Lemma 6. *For any sum of weighted rank functions* f, $F(y) \geq (1 - 1/e)\tilde{f}(y)$.

Proof. By Lemma 5, $F(y) \geq (1 - 1/e)f^*(y)$ and hence it suffices to prove that $f^*(y) \geq \tilde{f}(y)$. By Lemma 4, $g_i^+(y) \leq g_i^*(y)$ where $g_i^*(y) = \min_{S_i}(g_i(S_i) + \sum_j y_j g_{i,S_i}(j))$. (Here, $g_{i,S_i}(j) = g_i(S_i + j) - g_i(S_i)$.) Consequently,

$$\tilde{f}(y) = \sum_{i=1}^{m} g_i^+(y) \leq \sum_{i=1}^{m} \min_{S_i}(g_i(S_i) + \sum_{j \in N} y_j g_{i,S_i}(j))$$

$$\leq \min_{S} \sum_{i=1}^{m}(g_i(S) + \sum_{j \in N} y_j g_{i,S}(j)) = \min_{S}(f(S) + \sum_{j \in N} y_j f_S(j)) = f^*(y).$$

5   The Generalized Assignment Problem

Here we consider an application of our techniques to the Generalized Assignment Problem ("GAP"). An instance of GAP consists of n bins and m items. Each item i has two non-negative numbers for each bin j; a value v_{ji} and a size s_{ji}. We seek an assignment of items to bins such that the total size of items in each bin is at most 1, and the total value of all items is maximized.

In [10], a $(1-1/e)$-approximation algorithm for GAP has been presented. The algorithm uses LP_1.

In LP_1, $\mathcal{F}_j$ denotes the collection of all feasible assignments for bin j, i.e. sets satisfying $\sum_{i \in S} s_{ji} \leq 1$. The variable $y_{j,S}$ represents bin j receiving a set of items S. Although this is an LP of exponential size, it is shown in [10] that it can be solved to an arbitrary precision in polynomial time. Then the fractional solution can be rounded to an integral one to obtain a $(1 - 1/e)$ approximation.

$$LP_1: \quad \max \sum_{j, S \in \mathcal{F}_j} y_{j,S} v_j(S);$$
$$\forall j; \sum_{S \in \mathcal{F}_j} y_{j,S} \leq 1,$$
$$\forall i; \sum_{j, S \in \mathcal{F}_j : i \in S} y_{j,S} \leq 1,$$
$$\forall j, S; \ y_{j,S} \geq 0.$$

We show in this section that this $(1 - 1/e)$-approximation algorithm can be interpreted as a special case of submodular maximization subject to a matroid constraint[2], and this framework also allows some generalizations of GAP[3]. For this purpose, we reformulate the problem as follows.

We define $N = \{(j, S) \mid 1 \le j \le n, \ S \in \mathcal{F}_j\}$ and a submodular function $f : 2^N \to \mathcal{R}^+$,

$$f(\mathcal{S}) = \sum_{i=1}^{m} \max\{v_{ji} : \exists(j, S) \in \mathcal{S}, i \in S\}.$$

We maximize this function subject to a matroid constraint $\mathcal{M}$, where $\mathcal{S} \in \mathcal{M}$ iff $\mathcal{S}$ contains at most one pair (j, S) for each j. Such a set $\mathcal{S}$ corresponds to an assignment of set S to bin j for each $(j, S) \in \mathcal{S}$. This is equivalent to GAP: although the bins can be assigned overlapping sets in this formulation, we only count the value of the most valuable assignment for each item. We can write $f(\mathcal{S}) = \sum_{i=1}^{m} g_i(\mathcal{S})$ where $g_i(\mathcal{S}) = \max\{v_{ji} : \exists(j, S) \in \mathcal{S}, i \in S\}$ is a weighted rank function of a matroid $\mathcal{X}_i$ on N. In the matroid $\mathcal{X}_i$ an element $(j, S) \in N$ has weight v_{ji} if $i \in S$ and 0 otherwise. A set is independent in $\mathcal{X}_i$ iff its cardinality is at most 1. Therefore the problem falls under the umbrella of our framework.

We now write explicitly the LP arising from interpreting GAP as a submodular function problem. We have variables $y_{j,S}$ for each j and $S \in \mathcal{F}_j$. In addition, for each matroid $\mathcal{X}_i$, we define copies of these variables $x_{i,j,S}$. The resulting linear program is given as LP_2.

LP_2 has exponentially many variables and exponentially many constraints. However, observe that a feasible solution $y_{j,S}$ for LP_1 is also feasible for LP_2, when we set $x_{i,j,S} = y_{j,S}$ for $i \in S$ and 0 otherwise. This is because the constraint $\sum_{j,S:i \in S} y_{j,S} \le 1$ in LP_1 implies $x_i \in P(\mathcal{X}_i)$, and the constraint $\sum_S y_{j,S} \le 1$ implies $y \in P(\mathcal{M})$.

$$LP_2 : \quad \max \sum_{j,S \in \mathcal{F}_j, i \in S} v_{ji} x_{i,j,S};$$
$$\forall i, j, S; x_{i,j,S} \le y_{j,S},$$
$$\forall i; \ x_i \in P(\mathcal{X}_i),$$
$$y \in P(\mathcal{M}).$$

Therefore, we can solve LP_1 using the techniques of [10] and then convert the result into a feasible solution of LP_2. Finally, we can apply the pipage rounding technique to obtain a $(1 - 1/e)$-approximation.

This is simply a reformulation of the algorithm from [10]. However, the flexibility of our framework allows a more complicated matroid constraint $\mathcal{M}$ than each bin choosing at most one set. We briefly discuss this below.

Laminar matroid constraints on the bins: Let B be the set of bins in a GAP instance. Consider a *laminar* matroid $\mathcal{M}$ on B. We consider the problem of assigning items to a subset of bins $B' \subseteq B$ such that B' is independent in $\mathcal{M}$. An example is when $\mathcal{M}$ is the simple uniform matroid; that is B' is independent iff $|B'| \le k$. This gives rise to a variant of GAP in which at most k of the n bins

---

[2] This formulation of GAP is also described in [10] as a personal communication from an author of this paper.

[3] In [10] more general allocation problems are considered that allow constraints on the sets of items packable within a bin. Our approach also works for such problems but in this extended abstract we limit our discussion to GAP.

can be used. One can modify LP_1 by adding a new constraint: $\sum_{j,S\in\mathcal{F}_j} y_{j,S} \le k$, to obtain a relaxation LP_3 for this new problem.

Using the same ideas as those in [10], one can solve LP_3 to an arbitrary precision in polynomial time. The simple rounding scheme of [10] for LP_1 does not apply to LP_3. However, as before, we can see that a solution to LP_3 is feasible for LP_2 where the matroid $\mathcal{M}$ now also enforces the additional constraint that at most k elements from N are chosen. Thus pipage rounding can be used to obtain a $(1-1/e)$-approximation. A similar reasoning allows us to obtain a $(1-1/e)$-approximation for any laminar matroid constraint on the bins

$$LP_3 : \quad \max \sum_{j,S\in\mathcal{F}_j} y_{j,S}v_j(S);$$

$$\forall j; \sum_{S\in\mathcal{F}_j} y_{j,S} \le 1,$$

$$\forall i; \sum_{j,S\in\mathcal{F}_j : i\in S} y_{j,S} \le 1,$$

$$\sum_{j,S\in\mathcal{F}_j} y_{j,S} \le k,$$

$$\forall j, S; \ y_{j,S} \ge 0.$$

B. We defer the details to a full version of the paper.

6   Conclusions

We obtained a $(1 - 1/e)$-approximation for an interesting and useful class of submodular functions. We note that the methods in the paper apply to some interesting submodular functions that are not in the class. An example is the maximum multiset multicover problem which generalizes the multicover problem defined in Section 1. The difference between multicover and multiset multicover is that a set can cover an element multiple times (at most the requirement of the element). We can obtain a $(1-1/e)$ approximation for this problem even though this function cannot be expressed as a weighted sum of matroid rank functions. We defer the details. It would be of much interest to prove or disprove the existence of a $(1 - 1/e)$-approximation for all monotone submodular functions. Note that our hardness results (Theorem 3) hold even when f can be expressed as a sum of weighted rank functions of matroids, yet we can obtain a $(1 - 1/e)$-approximation in this case.

The unconstrained problem $\max_{S\subseteq N} f(S)$ is NP-hard and hard to approximate if f is a *non-monotone* submodular set function; the Max-Cut problem is a special case. However, the pipage rounding framework is still applicable to non-monotone functions (as already shown in [1]). For non-monotone functions, the problem we need to consider is $\max_{S\in\mathcal{B}} f(S)$ where $\mathcal{B}$ is the set of bases of $\mathcal{M}$. It is easy to see that Lemma 2 and Lemma 3 still apply. Thus, the approximation ratio that can be guaranteed depends on the extension $\tilde{f}$.

Pipage rounding [1] and dependent randomized rounding [17,15] are based on rounding fractional solutions to the assignment problem into integer solutions while maintaining the quality of a solution that is a function of the variables on the edges of the underlying bipartite graph. A number of applications are given in [1,17,15]. This paper shows that submodularity and uncrossing properties of solutions to matroids and other related structures are the basic ingredients in the applicability of the pipage rounding technique. We hope this insight will lead to more applications in the future.

References

1. A. Ageev and M. Sviridenko. Pipage rounding: a new method of constructing algorithms with proven performance guarantee. *J. of Combinatorial Optimization*, 8:307–328, 2004.
2. C. Chekuri and A. Kumar. Maximum coverage problem with group budget constraints and applications. *Proc. of APPROX*, Springer LNCS, 72–83, 2004.
3. C. Chekuri and M. Pál. A recursive greedy algorithm for walks in directed graphs. *Proc. of IEEE FOCS*, 2005.
4. C. Chekuri and S. Khanna. A PTAS for the multiple knapsack problem. *SIAM J. on Computing*, 35(3):713–728, 2004.
5. G. Cornuejols, M. Fisher and G. Nemhauser. Location of bank accounts to optimize float: an analytic study of exact and approximate algorithms. *Management Science*, 23: 789–810, 1977.
6. U. Feige. A threshold of $\ln n$ for approximating set cover. *JACM*, 45(4):634–652, 1998
7. U. Feige. On maximizing welfare when utility functions are subadditive. *Proc. of ACM STOC*, 41–50, 2006.
8. U. Feige and J. Vondrák. Approximation algorithms for allocation problems: Improving the Factor of $1 - 1/e$. *Proc. of IEEE FOCS*, 667–676, 2006.
9. M. L. Fisher, G. L. Nemhauser and L. A. Wolsey. An analysis of approximations for maximizing submodular set functions - II. *Math. Prog. Study*, 8:73–87, 1978.
10. L. Fleischer, M.X. Goemans, V.S. Mirrokni and M. Sviridenko. Tight approximation algorithms for maximum general assignment problems. *Proc. of ACM-SIAM SODA*, 611–620, 2006.
11. E. Hazan, S. Safra and O. Schwartz. On the complexity of approximating k-set packing. *Proc. of APPROX*, 2003.
12. T. A. Jenkyns. The efficiency of the "greedy" algorithm. *Proc. of 7th South Eastern Conference on Combinatorics, Graph Theory and Computing*, 341–350, 1976.
13. B. Korte and D. Hausmann. An analysis of the greedy heuristic for independence systems. *Annals of Discrete Math.*, 2:65–74, 1978.
14. G. L. Nemhauser, L. A. Wolsey and M. L. Fisher. An analysis of approximations for maximizing submodular set functions - I. *Math. Prog.*, 14:265–294, 1978.
15. R. Gandhi, S. Khuller, S. Parthasarathy and A. Srinivasan. Dependent rounding and its applications to approximation algorithms. *JACM*, 53(3):324–360, 2006.
16. A. Schrijver. Combinatorial optimization - polyhedra and efficiency. Springer, 2003.
17. A. Srinivasan. Distributions on level-sets with applications to approximation algorithms. *Proc. of IEEE FOCS*, 588–597, 2001.
18. L. Wolsey. An analysis of the greedy algorithm for the submodular set covering problem. *Combinatorica*, 2:385–393, 1982.
19. L. Wolsey. Maximizing real-valued submodular functions: Primal and dual heuristics for location Problems. *Math. of Operations Research*, 7:410–425, 1982.

On a Generalization of the Master Cyclic Group Polyhedron

Sanjeeb Dash[1], Ricardo Fukasawa[2,*], and Oktay Günlük[1]

[1] Mathematical Sciences Department
IBM T. J. Watson Research Center, Yorktown Heights, NY, 10598
`sanjeebd@us.ibm.com, oktay@watson.ibm.com`
[2] H. Milton Stewart School of Industrial and Systems Engineering
Georgia Institute of Technology, Atlanta, GA, 30332
`rfukasaw@isye.gatech.edu`

Abstract. We study the Master Equality Polyhedron (MEP) which generalizes the Master Cyclic Group Polyhedron and the Master Knapsack Polyhedron.

We present an explicit characterization of the nontrivial facet-defining inequalities for MEP. This result generalizes similar results for the Master Cyclic Group Polyhedron by Gomory [9] and for the Master Knapsack Polyhedron by Araoz [1]. Furthermore, this characterization also gives a polynomial time algorithm for separating an arbitrary point from the MEP.

We describe how facet defining inequalities for the Master Cyclic Group Polyhedron can be lifted to obtain facet defining inequalities for the MEP, and also present facet defining inequalities for the MEP that cannot be obtained in such a way. Finally, we study the mixed-integer extension of the MEP and present an interpolation theorem that produces valid inequalities for general Mixed Integer Programming Problems using facets of the MEP.

Keywords: integer programming, polyhedral combinatorics.

1 Introduction

We study the Master Equality Polyhedron (MEP), which we define as:

$$K(n, r) = conv \left\{ (x, y) \in \mathbb{Z}_+^n \times \mathbb{Z}_+^n : \sum_{i=1}^{n} ix_i - \sum_{i=1}^{n} iy_i = r \right\} \qquad (1)$$

where $n, r \in \mathbb{Z}$ and $n > 0$. Without loss of generality we assume that $r \geq 0$. To the best of our knowledge, $K(n, r)$ was first defined by Uchoa [14] in a slightly different form and described as an important object for study.

---

* Work developed while at IBM Research.

M. Fischetti and D.P. Williamson (Eds.): IPCO 2007, LNCS 4513, pp. 197–209, 2007.

As lower dimensional faces, MEP contains two well known polyhedra from the literature: The Master Cyclic Group Polyhedron (MCGP), which is defined as

$$P(n,r) = conv\left\{(x,y) \in \mathbb{Z}_+^{n-1} \times \mathbb{Z}_+ : \sum_{i=1}^{n-1} ix_i - ny_n = r\right\}, \tag{2}$$

where $r, n \in \mathbb{Z}$, and $0 \leq r < n$; and the Master Knapsack Polyhedron (MKP), which is defined as

$$K(r) = conv\left\{x \in \mathbb{Z}_+^r : \sum_{i=1}^{r} ix_i = r\right\}, \tag{3}$$

where $r \in \mathbb{Z}$ and $r > 0$.

Facets of $P(n,r)$ are a useful source of cutting planes for general MIPs. The Gomory mixed-integer cut (also known as the mixed-integer rounding (MIR) inequality) can be derived from a facet of $P(n,r)$ [10]. Other facets and studies related to the Master Cyclic Group Polyhedron can be found in [2,4,5,6,7,8,11,12,13]. In particular, several relationships between facet-defining inequalities of the MCGP and facet-defining inequalities of the MKP were established in [2]. We note that the Master Cyclic Group Polyhedron is usually presented as

$$P'(n,r) = conv\left\{x \in \mathbb{Z}_+^{n-1} : \sum_{i=1}^{n-1} ix_i \equiv r \mod n\right\}$$

which is the projection of $P(n,r)$ in the space of x variables. We use (2) as it makes the comparison to $K(n,r)$ easier and clearer.

Gomory [9] and Araoz [1] give an explicit characterization of the polar of the nontrivial facets of $P(n,r)$ and $K(r)$. In this paper, we give a similar description of the nontrivial facets of $K(n,r)$, yielding as a consequence a polynomial time algorithm to separate over it. We also analyze some structural properties of the MEP and relate it to the MCGP.

In addition, we describe how to obtain valid inequalities for general MIPs using facet defining inequalities for the MEP.

Finally, we remark that another motivation to study the MEP is that it also arises as a natural structure in a reformulation of the Fixed-Charge Network Flow problem, which has recently been used in [15] to derive strong cuts for the Capacitated Minimum Spanning Tree Problem and can also be used in other problems such as the Capacitated Vehicle Routing Problem.

2   Polyhedral Analysis of $K(n,r)$

From this point until the end of section 2.1 we assume $0 < r \leq n$. In subsections 2.2 and 2.3, we consider the cases $r = 0$ and $r > n$. We start with some basic polyhedral properties of $K(n,r)$.

Lemma 1. $dim(K(n,r)) = 2n - 1$.

Lemma 2. *The nonnegativity constraints of $K(n,r)$ are facet-defining if $n \geq 2$.*

Let e_i denote the unit vector with a one in the component corresponding to variable x_i and f_i denote the unit vector with a one in the component corresponding to variable y_i.

Clearly, $K(n,r)$ is an unbounded polyhedron. We next characterize all the extreme rays (unbounded one-dimensional faces) of $K(n,r)$. We represent an extreme ray $\{u + \lambda v : u, v \in \mathbb{R}_+^{2n}, \lambda \geq 0\}$ of $K(n,r)$ simply by the vector v. Let $r_{ij} = je_i + if_j$ for any $i, j \in \{1, \ldots, n\}$.

Lemma 3. *The set of extreme rays of $K(n,r)$ is given by $R = \{r_{ij} : 1 \leq i, j \leq n\}$.*

As $K(n,r)$ is not a full-dimensional polyhedron, any valid inequality $\pi x + \rho y \geq \pi_o$ for $K(n,r)$ has an equivalent representation with $\rho_n = 0$. If a valid inequality does not satisfy this condition, one can add an appropriate multiple of the equation $\sum_{i=1}^{n} ix_i - \sum_{i=1}^{n} iy_i = r$ to it. Therefore, without loss of generality, we may assume that all valid inequalities for $K(n,r)$ satisfy $\rho_n = 0$.

We classify the facets of $K(n,r)$ as *trivial* and *non-trivial* facets.

Definition 1. *The following facet-defining inequalities of $K(n,r)$ are called trivial:*

$$x_i \geq 0, \forall i = 1, \ldots, n$$

$$y_i \geq 0, \forall i = 1, \ldots, n-1$$

All other facet-defining inequalities of $K(n,r)$ are called nontrivial.

Notice that we left inequality $y_n \geq 0$ out of the trivial set. That happens just because of technical details to simplify the statement of our theorems and lemmas. In fact there is nothing particularly special about the $y_n \geq 0$ inequality other than it is the only nonnegativity constraint that does not comply directly with the $\rho_n = 0$ assumption.

Let $N = \{1, \ldots, n\}$. We next state our main result:

Theorem 1. *The inequality $\pi x + \rho y \geq \pi_o$ defines a nontrivial facet of $K(n,r)$ if and only if it can be represented as an extreme point of $T \subseteq \mathbb{R}^{2n+1}$ where T is defined by the following linear equations and inequalities:*

$$
\begin{array}{llll}
\pi_i + \rho_j \geq \pi_{i-j}, & \forall i, j \in N, & i > j, & \text{(F1)} \\
\pi_i + \pi_j \geq \pi_{i+j}, & \forall i, j \in N, & i + j \leq n, & \text{(F2)} \\
\rho_k + \pi_i + \pi_j \geq \pi_{i+j-k}, & \forall i, j, k \in N, & 1 \leq i + j - k \leq n, & \text{(F3)} \\
\pi_i + \pi_{r-i} = \pi_o, & \forall i \in N, & i < r, & \text{(EP1)} \\
\pi_r = \pi_o, & & & \text{(EP2)} \\
\pi_i + \rho_{i-r} = \pi_o, & \forall i \in N & i > r, & \text{(EP3)} \\
\rho_n = 0, & & & \text{(N1)} \\
\pi_o = 1. & & & \text{(N2)}
\end{array}
$$

This theorem implies that for $0 < r \leq n$, the separation problem over $K(n,r)$ can be solved in polynomial time. Although the restriction that $0 < r \leq n$ might seem undesirable, later in Sect. 2.3 we show that the separation can be done for every value of r.

Note that the definition of T in Theorem 1 is similar to that of a polar of $K(n,r)$. However, T is not a polar, as it does not contain extreme points of the polar that correspond to the trivial facet-defining inequalities. In addition, some of the extreme rays of the polar are not present in T. It is possible to interpret T as an important subset of the polar that contains all extreme points of the polar besides the ones that lead to the trivial inequalities.

2.1   Facet Characterization

In this section we develop the required analysis to prove Theorem 1. We start by noting some necessary conditions for validity, which arise by looking at points and rays of $K(n,r)$:

Observation 2. *Let $\pi x + \rho y \geq \pi_o$ be a valid inequality for $K(n,r)$, then the following holds:*

$$j\pi_i + i\rho_j \geq 0, \forall i,j \in N \tag{R1}$$
$$\pi_i + \pi_{r-i} \geq \pi_o, \forall 1 \leq i < r \tag{P1}$$
$$\pi_r \geq \pi_o \tag{P2}$$
$$\pi_i + \rho_{i-r} \geq \pi_o, \forall r < i \leq n \tag{P3}$$

Note that (R1) is obtained by considering the extreme rays of $K(n,r)$ and (P1)-(P3) are obtained by considering the following feasible points of $K(n,r)$:

$$\{e_i + e_{r-i}, \forall 1 \leq i < r\} \cup e_r \cup \{e_i + f_{i-r}, \forall r < i \leq n\}$$

We call these points the *Elementary points* of $K(n,r)$. Note that there are $n - \lfloor \frac{r-1}{2} \rfloor$ Elementary points.

We next present some conditions satisfied by all nontrivial facet defining inequalities.

Lemma 4. *Let $\pi x + \rho y \geq \pi_o$ be a nontrivial facet-defining inequality of $K(n,r)$, then it satisfies* (F1)-(F3) *as well as* (EP1)-(EP3).

Proof. (F1): Pick a point (x^*, y^*) tight at $\pi x + \rho y \geq \pi_o$ such that $x^*_{i-j} > 0$. Note that $(x^*, y^*) + (e_i + f_j - e_{i-j})$ is a point of $K(n,r)$. Thus, (F1) holds.

The proofs of (F2) and (F3) are analogous.

(EP1): Pick points (x', y') and (x'', y'') tight at (π, ρ, π_o) such that $x'_i > 0$ and $x''_{r-i} > 0$. Then $(x''', y''') = (x', y') + (x'', y'') - e_i - e_{r-i} \in K(n,r)$, thus $(\pi, \rho)^T (x''', y''') = (\pi, \rho)^T (x', y') + (\pi, \rho)^T (x'', y'') - \pi_i - \pi_{r-i} = 2\pi_o - \pi_i - \pi_{r-i} \geq \pi_o \Rightarrow \pi_i + \pi_{r-i} \leq \pi_o$. So (P1) $\Rightarrow$ (EP1).

Proofs of (EP2) and (EP3) are analogous, using (P2) and (P3) instead of (P1). ∎

It is worth mentioning that conditions (EP1)-(EP3) imply that all nontrivial facets intersect at a nonempty lower dimensional face of $K(n, r)$. Note that all Elementary Points of $K(n, r)$ are in this lower-dimensional face, which has therefore dimension at least $n - \lfloor \frac{r-1}{2} \rfloor - 1$.

In the following Lemma we show that a subset of the conditions presented in Theorem 1 suffices to ensure the validity.

Lemma 5. *Let (π, ρ, π_o) satisfy (EP2), (F1), (F2) and (F3). Then $\pi x + \rho y \geq \pi_o$ defines a valid inequality for $K(n, r)$.*

Proof. We will prove this by contradiction. Assume that $\pi x + \rho y \geq \pi_o$ satisfies (EP2), (F1), (F2) and (F3) but $\pi x + \rho y \geq \pi_o$ does not define a valid inequality for $K(n, r)$, $r > 0$. Let (x^*, y^*) be an integer point in $K(n, r)$ that has minimum L_1 norm amongst all points violated by $\pi x + \rho y \geq \pi_o$. Note that since $r > 0$, then $x^* \neq 0$.

If $||(x^*, y^*)||_1 = 0$ then $(x^*, y^*) = 0 \notin K(n, r)$. If $||(x^*, y^*)||_1 = 1$ then clearly $x^* = e_r$ and $y^* = 0$ but as $\pi_r = \pi_o$, (x^*, y^*) does not violate the inequality. Therefore $||(x^*, y^*)||_1 \geq 2$. We next consider three cases.

Case 1: Assume that $y^* = 0$. In this case, $\sum_{i=1}^{n} i x_i^* = r$. By successively applying (F2), we obtain

$$\pi_o > \sum_{i=1}^{n} \pi_i x_i^* \geq \sum_{i=1}^{n} \pi_{i x_i^*} \geq \pi_{\sum_{i=1}^{n} i x_i^*} = \pi_r$$

which contradicts (EP2). Therefore $y^* \neq 0$.

Case 2: Assume that $x_i^* > 0$ and $y_j^* > 0$ for some $i > j$. Let $(x', y') = (x^*, y^*) + (e_{i-j} - e_i - f_j)$. Note that $(x', y') \in K(n, r)$, and $||(x', y')||_1 = ||(x^*, y^*)||_1 - 1$. Moreover, since $\pi x + \rho y \geq \pi_o$ satisfies (F1), $\pi x' + \rho y' = \pi x^* + \rho y^* + \pi_{i-j} - \pi_i - \rho_j \leq \pi x^* + \rho y^* < \pi_o$, which contradicts the choice of (x^*, y^*). Therefore $i \leq j$ whenever $x_i^* > 0$ and $y_j^* > 0$.

Case 3: Assume that for any $i, j \in N$, if $x_i^* > 0$ and $y_j^* > 0$, then $i \leq j$. Suppose there exists $i, j \in N$ such that $x_i^* > 0$, $x_j^* > 0$ or $x_i^* \geq 2$ (in which case, we let $j = i$). If $i + j \leq n$, let $(x', y') = (x^*, y^*) + (e_{i+j} - e_i - e_j)$. If $i + j > n$, since $y^* \neq 0$ there exists k such that $y_k^* > 0$ and $k \geq i$, thus $i + j - k \leq n$. So let $(x', y') = (x^*, y^*) + (e_{i+j-k} - e_i - e_j - f_k)$.

Note that in either case $(x', y') \in K(n, r)$ and $||(x', y')||_1 < ||(x^*, y^*)||_1$. Moreover, since (π, ρ, π_o) satisfy (F2) and (F3), in either case $\pi x' + \rho y' \leq \pi x^* + \rho y^* < \pi_o$, which contradicts the choice of (x^*, y^*). ∎

One condition that so far has not been mentioned is (N2), which is a normalization condition like (N1). The following Lemma states that we are not eliminating any nontrivial facets by making such an assumption.

Lemma 6. *Let $\pi x + \rho y \geq \pi_o$ be a nontrivial facet-defining inequality of $K(n, r)$, that satisfies $\rho_n = 0$. Then $\pi_o > 0$.*

Combining Lemmas 4-6 with some more technical observations it is possible to prove Theorem 1. As a corollary of the theorem, we also make the following observation:

Observation 3. *Let* (π, ρ, π_o) *be an extreme point of* T, *then for all* $k \in N$:

$$\lceil k/r \rceil \geq \pi_k \geq 0$$
$$\lceil n/r \rceil \geq \rho_k \geq -\lceil k/r \rceil$$

As a final remark, it is interesting to note that conditions (R1) do not appear in the description of T even though they are necessary for any valid inequality. This happens because conditions (R1) are implied by (F1), (F2) and (F3). We formally state this fact in the next observation:

Observation 4. *Let* $(\pi, \rho, \pi_o) \in T$. *Then:*

$$j\pi_i + i\rho_j \geq 0, \quad \forall 1 \leq i, j \leq n$$

2.2   Facets of $K(n, 0)$

Observe that $LK(n, 0)$, the linear relaxation of $K(n, 0)$, is a cone and is pointed (as it is contained in the nonnegative orthant) and has a single extreme point $(x, y) = (0, 0)$. Therefore $LK(n, 0)$ equals its integer hull, i.e., $LK(n, 0) = K(n, 0)$. In Lemma 3, we characterized the extreme rays of $K(n, r)$ and thereby showed that the characteristic cone of $K(n, r)$ is generated by the vectors $\{r_{ij}\}$. But the characteristic cone of $K(n, r)$ for some $r > 0$ is just $K(n, 0)$. Therefore, $LK(n, 0)$ is generated by the vectors $\{r_{ij}\}$, and the next result follows.

Theorem 5. *The inequality* $\pi x + \rho y \geq \pi_o$ *is facet defining for* $K(n, 0)$ *if and only if* (π, ρ, π_o) *is a minimal face of*

$$T_o = \begin{cases} j\pi_i + i\rho_j \geq 0, \forall i, j \in N, \\ \pi_o = 0. \end{cases}$$

In his work on the MCGP, Gomory also studied the convex hull of non-zero integral solutions in $P(n, 0)$ and gave a dual characterization of its facets. We now consider a similar modification of $K(n, 0)$ and study the set:

$$\bar{K}(n, 0) = conv\left\{(x, y) \in \mathbb{Z}_+^n \times \mathbb{Z}_+^n : \sum_{i=1}^n ix_i - \sum_{i=1}^n iy_i = 0, (x, y) \neq 0\right\}$$

By an analysis similar to the case where $r > 0$, it is possible to prove the following theorem:

Theorem 6. *The inequality $\pi x + \rho y \geq \pi_o$ defines a nontrivial facet of $\bar{K}(n,0)$ if and only if it can be represented as an extreme point of $\bar{T}_o$, where $\bar{T}_o$ is defined by the following linear equations and inequalities:*

$$\pi_i + \rho_j \geq \pi_{i-j}, \qquad \forall i,j \in N, \quad i > j, \tag{F1}$$

$$\pi_i + \rho_j \geq \rho_{j-i}, \qquad \forall i,j \in N, \quad i < j, \tag{F1'}$$

$$\pi_i + \rho_i = \pi_o, \qquad \forall i \in N, \tag{EP1-R0}$$

$$\pi_o = 1, \tag{N1-R0}$$

$$\rho_n = 0. \tag{N2-R0}$$

2.3   Separating over $K(n,r)$

We finish this section by presenting the following theorem stating that separation over $K(n,r)$ can be done in polynomial time when r is bounded by a polynomial function of n and pseudo-polynomial time otherwise. This theorem is an immediate consequence of Theorems 1 and 5.

Theorem 7. *Given $(x^*, y^*) \in \mathbb{R}^n \times \mathbb{R}^n$, the problem of separating (x^*, y^*) from $K(n,r)$ can be solved in time polynomial in $\max\{n, r\}$.*

Proof. If $0 < r \leq n$, the separation problem can be solved in time polynomial in n by first checking if (x^*, y^*) violates any nonnegativity constraint or the constraint $\sum_{i=1}^n i x_i - \sum_{i=1}^n i y_i = r$ and if not, solve:

$$\min\{(\pi, \rho, \pi_o)^T (x^*, y^*, 0) : (\pi, \rho, \pi_o) \in T\}$$

If there exists $(\pi, \rho, \pi_o) \in T$ such that $(\pi, \rho)^T (x^*, y^*) < 1$, then $\pi x + \rho y \geq \pi_o$ defines a hyperplane that separates (x^*, y^*) from $K(n,r)$. Otherwise, (x^*, y^*) is in the same affine subspace as $K(n,r)$ and satisfies all nontrivial and trivial facets of $K(n,r)$, thus $(x^*, y^*) \in K(n,r)$.

If $r > n$, then define $(x', y') \in \mathbb{R}^r \times \mathbb{R}^r$ such that $x_i' = x_i^*; y_i' = y_i^*, \forall 1 \leq i \leq n$ and $x_i' = y_i' = 0, \forall n < i \leq r$. and note that $(x', y') \in K(r,r) \iff (x^*, y^*) \in K(n,r)$, so the separation can be done in time polynomial in r.

In the case where $r = 0$, we can solve $\min\{(\pi, \rho)^T (x^*, y^*) : (\pi, \rho) \in T_o\}$ and we'll know $(x^*, y^*) \in K(n,0)$ if and only if the optimum is 0. Otherwise, the problem is unbounded, in which case the ray which proves unboundedness gives us a valid inequality separating (x^*, y^*) from $K(n,0)$.   ∎

3   Lifting Facets of $P(n,r)$

Lifting is a general principle for constructing valid (facet defining) inequalities for higher dimensional sets using valid (facet defining) inequalities for lower dimensional sets. Starting with the early work of Gomory [9], this approach was generalized by Wolsey [16], Balas and Zemel [3] and Gu et. al [17], among others.

In this section we discuss how facets of $P(n,r)$ can be lifted to obtain facets of $K(n,r)$. $P(n,r)$ can also be considered as an $n-1$ dimensional face of $K(n,r)$

obtained by setting n variables to their lower bounds. Throughout this section we assume that $n > r > 0$.

We start with a result of Gomory [9] that gives a complete characterization of the nontrivial facets (i.e., excluding the non-negativity inequalities) of $P(n, r)$.

Theorem 8 (Gomory [9]). *Inequality $\bar{\pi} x \geq 1$ defines a non-trivial facet of $P(n, r)$ if and only if $\bar{\pi} \in \mathbb{R}^{n-1}$ is an extreme point of*

$$
Q = \begin{cases}
\pi_i + \pi_j \geq \pi_{(i+j) \bmod n} & \forall i, j \in \{1, \ldots, n-1\}, \\
\pi_i + \pi_j = \pi_r & \forall i, j \text{ such that } r = (i+j) \bmod n, \\
\pi_j \geq 0 & \forall j \in \{1, \ldots, n-1\}, \\
\pi_r = 1.
\end{cases}
$$

Given a non-trivial facet defining inequality for $P(n, r)$

$$
\sum_{i=1}^{n-1} \bar{\pi}_i x_i \geq 1 \tag{4}
$$

it is possible to *lift* this inequality to obtain a facet-defining inequality

$$
\sum_{i=1}^{n-1} \bar{\pi}_i x_i + \pi'_n x_n + \sum_{i=1}^{n-1} \rho'_i y_i \geq 1 \tag{5}
$$

for $K(n, r)$. We call inequality (5) a *lifted* inequality and note that in general for a given starting inequality there might be an exponential number of lifted inequalities, see [16].

3.1   The Restricted Coefficient Polyhedron $T^{\bar{\pi}}$

First note that a non-trivial facet of $P(n, r)$ can only yield a non-trivial facet of $K(n, r)$. This, in turn, implies that $(\bar{\pi}, \pi'_n, \rho', 0)$ has to be an extreme point of the coefficient polyhedron T. Therefore, the lifting procedure can also be seen as a way of extending an extreme point of Q to obtain an extreme point of T.

Let $p = (\bar{\pi}, \pi'_n, \rho', 0)$ be an an extreme point of T. Then, p also has to be an extreme point of the lower dimensional polyhedron

$$
T^{\bar{\pi}} = T \cap \left\{ \pi_i = \bar{\pi}_i, \ \forall i \in \{1, \ldots, n-1\} \right\}
$$

obtained by fixing some of the coordinates.

Let $L = \{n - r + 1, \ldots, n - 1\}$.

Lemma 7. *If inequality (4) defines a non-trivial facet of $P(n, r)$, then $T^{\bar{\pi}} \neq \emptyset$ and it has the form*

$$T^{\bar{\pi}} = \begin{cases} \tau & \geq & \pi_n \geq 0 & \\ \rho_k & \geq l_k & \forall\, k \in L \\ \rho_k + \pi_n & \geq t_k & \forall\, k \in L \\ \rho_k - \pi_n & \geq f_k & \forall\, k \in L \\ \pi_n + \rho_{n-r} & = 1 & \\ \rho_n & = 0 & \\ \rho_k & = \bar{\pi}_{n-k} & \forall k \in \{1,\dots,n-r-1\} \\ \pi_i & = \bar{\pi}_i & \forall i \in \{1,\dots,n-1\} \end{cases}$$

where numbers l_k, t_k, f_k and τ can be computed easily using $\bar{\pi}$.

We next make a simple observation that will help us show that $T^{\bar{\pi}}$ has a small (polynomial) number of extreme points.

Lemma 8. *If $p = (\bar{\pi}, \pi'_n, \rho', 0)$ is an extreme point of $T^{\bar{\pi}}$, then*

$$\rho'_k = \max\left\{ l_k, t_k - \pi'_n, f_k + \pi'_n \right\}$$

for all $k \in L$.

We next characterize the set possible values π'_n can take at an extreme point of $T^{\bar{\pi}}$.

Lemma 9. *Let $p = (\bar{\pi}, \pi'_n, \rho', 0)$ be an extreme point of $T^{\bar{\pi}}$, if $\pi'_n \notin \{0, \tau\}$, then*

$$\pi'_n \in \Lambda = \left(\bigcup_{k \in L_1} \left\{ t_k - l_k, l_k - f_k \right\} \right) \bigcup \left(\bigcup_{k \in L_2} \left\{ (t_k - f_k)/2 \right\} \right)$$

where $L_1 = \{k \in L : t_k + f_k < 2l_k\}$ and $L_2 = L \setminus L_1$.

Combining the previous Lemmas, we have the following result:

Theorem 9. *Given a non-trivial facet defining inequality (4) for $P(n,r)$, there are at most $2r$ lifted inequalities that define facets of $K(n,r)$.*

Proof. The set L in the proof of Lemma 9 has $r - 1$ members and therefore together with 0 and τ, there are at most $2r$ possible values for π'_n in a facet defining lifted inequality (5). As the value of π'_n uniquely determines the remaining coefficients in the lifted inequality, by Lemma 8, the claim follows. ∎

Note that, in general determining all possible lifted inequalities is a hard task. However, the above results show that obtaining all possible facet-defining inequalities lifted from facets of $P(n,r)$ is straightforward and can be performed in polynomial time. We conclude this section with a result on sequential lifting.

Lemma 10. *If variable x_n is lifted before all y_k for $k \in \{n-r,\dots,n-1\}$, then independent of the rest of the lifting sequence the lifted inequality is*

$$\sum_{i=1}^{n-1} \bar{\pi}_i x_i + \sum_{i=1}^{n-1} \bar{\pi}_{n-i} y_i \geq 1.$$

4 Mixed Integer Rounding Inequalities

In this section we study MIR inequalities in the context of $K(n,r)$. Our analysis also provides an example that shows that lifting facets of $P(n,r)$ cannot give all facets of $K(n,r)$. Throughout, we will use the notation $\hat{x} := x - \lfloor x \rfloor$. Recall that, for a general single row system of the form: $\{w \in \mathbb{Z}_+^p : \sum_{i=1}^p a_i w_i = b\}$ where $\hat{b} > 0$, the MIR inequality is:

$$\sum_{i=1}^p \left(\lfloor a_i \rfloor + \min\left(\hat{a}_i/\hat{b}, 1\right) \right) w_i \geq \lceil b \rceil.$$

We define the $\frac{1}{t}$-MIR (for $t \in \mathbb{Z}_+$) to be the MIR inequality obtained from the following equivalent representation of $K(n,r)$:

$$K(n,r) = \left\{ (x,y) \in \mathbb{Z}_+^n \times \mathbb{Z}_+^n : \sum_{i=1}^n (i/t)x_i - \sum_{i=1}^n (i/t)y_i = r/t \right\}.$$

Lemma 11. *Given $t \in \mathbb{Z}$ such that $2 \leq t \leq n$, the $\frac{1}{t}$-MIR inequality*

$$\sum_{i=1}^n \left(\left\lfloor \frac{i}{t} \right\rfloor + \min\left(\frac{i \mod t}{r \mod t}, 1 \right) \right) x_i +$$

$$\sum_{i=1}^n \left(-\left\lceil \frac{i}{t} \right\rceil + \min\left(\frac{(t-i) \mod t}{r \mod t}, 1 \right) \right) y_i \geq \left\lceil \frac{r}{t} \right\rceil$$

is facet defining for $K(n,r)$ provided that $r/t \notin \mathbb{Z}$.

It is easy to check that if $t > n$, then the $\frac{1}{t}$-MIR is not facet defining for $K(n,r)$. Moreover, note that if $r/t \in \mathbb{Z}$, then the condition that $\hat{b} > 0$ is not satisfied, thus the $\frac{1}{t}$-MIR inequalities are not facet defining unless they satisfy the conditions of Lemma 11.

By using the $\frac{1}{t}$-MIR as an example, one can then show the following corollary:

Corollary 1. *Not all facet-defining inequalities of $K(n,r)$ can be obtained from lifting facet-defining inequalities of $P(n,r)$, for $0 < r \leq n-2$, $n \geq 9$*

For $r = n-1$, it is harder to say, since in this case all points in T automatically satisfy all equations in Q. So every facet-defining inequality of $K(n,r)$ can be obtained by lifting a valid inequality for $P(n,r)$ corresponding to a point in Q. However, this point is not necessarily an extreme point of Q, and thus the corresponding valid inequality is not necessarily a facet of $P(n,r)$.

5 Mixed-Integer Extension

Consider the mixed-integer extension of $K(n,r)$:

$$K'(n,r) = \left\{ (v_+, v_-, x, y) \in \mathbb{R}^2 \times \mathbb{Z}^{2n} : v_+ - v_- + \sum_{i=1}^n i x_i - \sum_{i=1}^n i y_i = r \right\}$$

where $n, r \in \mathbb{Z}$ and $n > r > 0$. As with the mixed-integer extension of the master cyclic group of Gomory studied by Gomory and Johnson [10], the facets of $K'(n, r)$ can easily be derived from the facets of $K(n, r)$ when r is an integer.

Proposition 1. *All non-trivial facet defining inequalities for $K'(n, r)$ have the form*

$$\pi_1 v_+ + \rho_1 v_- + \sum_{i=1}^{n} \pi_i x_i + \sum_{i=1}^{n} \rho_i y_i \geq \pi_0. \tag{6}$$

Furthermore, inequality (6) is facet defining if and only if $\pi x + \rho y \geq \pi_o$ defines a non-trivial facet of $K(n, r)$.

5.1   General Mixed-Integer Sets

Gomory and Johnson used facets of $P(n, r)$ to derive valid inequalities for knapsack problems. In particular, they derived *subadditive functions* from facet coefficients via *interpolation*. We show here how to derive valid inequalities for knapsack problems from facets of $K(n, r)$.

Definition 2. *Given a facet defining inequality $\pi x + \rho y \geq \pi_o$ for $K(n, r)$, let $f^z : \mathbb{Z} \cap [-n, n] \to \mathbb{R}$ be defined as:*

$$f^z(s) = \begin{cases} \pi_s & \text{if } s > 0 \\ 0 & \text{if } s = 0 \\ \rho_{-s} & \text{if } s < 0 \end{cases}$$

We say $f : [-n, n] \to \mathbb{R}$ where

$$f(v) = (1 - \hat{v}) f^z(\lfloor v \rfloor) + \hat{v} f^z(\lceil v \rceil)$$

is a facet-interpolated function *derived from (π, ρ, π_0).*

Proposition 2. *Let f be a facet-interpolated function derived from a facet of $K(n, r)$. Consider the set*

$$Q = \left\{ (s, w) \in R_+^q \times Z_+^p : \sum_{i=1}^{q} c_i s_i + \sum_{i=1}^{p} a_i w_i = b \right\},$$

where the coefficients of the knapsack constraint defining Q are rational numbers. Let t be such that $t a_i, t b \in [-n, n]$ and $t b > 0$. Then

$$f(1) \sum_{i=1}^{q} (t c_i)^+ s_i + f(-1) \sum_{i=1}^{q} (-t c_i)^+ s_i + \sum_{i=1}^{p} f(t a_i) w_i \geq f(t b)$$

where $(\alpha)^+ = \max(\alpha, 0)$, is a valid inequality for Q.

6   Conclusion

We studied a generalization of the Master Cyclic Group Polyhedron and presented an explicit characterization of the polar of its nontrivial facet-defining inequalities. We also showed that one can obtain valid inequalities for a general MIP that cannot be obtained from facets of the Master Cyclic Group Polyhedron.

In addition, for mixed-integer knapsack sets with rational data and nonnegative variables without upper bounds, our results yield a pseudo-polynomial time algorithm to separate and therefore optimize over their convex hull. This can be done by scaling their data and aggregating variables to fit into the Master Equality Polyhedron framework.

Our characterization of the MEP can also be used to find violated Homogeneous Extended Capacity Cuts efficiently. These cuts were proposed in [15] for solving Capacitated Minimum Spanning Tree problems and Capacitated Vehicle Routing problems.

An interesting topic for further study is the derivation of "interesting" classes of facets for the MEP, i.e., facets which cannot be derived trivially from facets of the MCGP or as rank one mixed-integer rounding inequalities.

References

1. J. Araoz. *Polyhedral Neopolarities*. Phd thesis, University of Waterloo, Department of Computer Sciences, 1974.
2. J. Araoz, L. Evans, R. E. Gomory, and E. Johnson. Cyclic group and knapsack facets. *Mathematical Programming Ser. B*, 96(2):377–408, 2003.
3. E. Balas and E. Zemel. Facets of the knapsack polytope from minimal covers. *SIAM Journal of Applied Mathematics*, 34:119–148, 1978.
4. S. Dash and O. Günlük. On the strength of gomory mixed-integer cuts as group cuts. Technical Report RC23967, IBM Research Division, Yorktown Heights, NY 10598, 2006.
5. S. Dash and O. Günlük. Valid inequalities based on simple mixed-integer sets. *Mathematical Programming*, 105:29–53, 2006.
6. S. Dash and O. Günlük. Valid inequalities based on the interpolation procedure. *Mathematical Programming*, 106:111–136, 2006.
7. M. Fischetti and M. Monaci. How tight is the corner relaxation? *Discrete Optimization*, 2007. To appear.
8. M. Fischetti and C. Saturni. Mixed-integer cuts from cyclic groups. *Mathematical Programming A*, 109(1):27–53, 2007.
9. R. Gomory. Some polyhedra related to combinatorial problems. *Journal of Linear Algebra and its Applications*, 2:451–558, 1969.
10. R. Gomory and E. Johnson. Some continuous functions related to corner polyhedra I. *Mathematical Programming*, 3:23–85, 1972.
11. R. Gomory and E. Johnson. Some continuous functions related to corner polyhedra II. *Mathematical Programming*, 3:359–389, 1972.
12. R. Gomory and E. Johnson. T-space and cutting planes. *Mathematical Programming*, 96:341–375, 2003.
13. R. Gomory, E. Johnson, and L. Evans. Cyclic group and knapsack facets. *Mathematical Programming*, 96:321–339, 2003.
14. E. Uchoa. Robust branch-and-cut-and-price for the CMST problem and extended capacity cuts. Presentation in the MIP 2005 Workshop, Minneapolis (2005). Available at http://www.ima.umn.edu/matter/W7.25-29.05/activities/Uchoa-Eduardo/cmst-ecc-IMA.pdf

15. E. Uchoa, R. Fukasawa, J. Lysgaard, A. Pessoa, M. Poggi de Aragão, and D. Andrade. Robust branch-cut-and-price for the capacitated minimum spanning tree problem over a large extended formulation. *Mathematical Programming*, To appear.
16. L. Wolsey. Facets and strong valid inequalities for integer programs. *Oper. Res.*, 24:367–372, 1976.
17. G. Nemhauser Z. Gu and M. Savelsbergh. Sequence independent lifting in mixed integer programming. *J. Comb. Optim.*, 4:109–129, 2000.

A Framework to Derive Multidimensional Superadditive Lifting Functions and Its Applications*

Bo Zeng and Jean-Philippe P. Richard

School of Industrial Engineering, Purdue University, 315 N. Grant Street, West Lafayette, IN 47907-2023.

Abstract. In this paper, we present a systematic method to derive strong superadditive approximations of multidimensional lifting functions using single-dimensional superadditive functions. This constructive approach is based on the observation that, in many cases, the lifting function of a multidimensional problem can be expressed or approximated through the single-dimensional lifting function of some of its components. We then apply our approach to two variants of classical models and show that it yields an efficient procedure to derive strong valid inequalities.

1 Introduction

Lifting is the process of deriving valid inequalities for a complex mixed integer program (MIP) from valid inequalities of a simple restriction. Lifting, in its common acception, was introduced by Padberg [15] and generalized by Wolsey [23]. It was used to study the polyhedral structure of many mixed integer programs such as $0-1$ knapsack sets (see Balas [4], Hammer et al. [11] and Wolsey [22]) and single node flow sets (see Gu et al. [9] and Atamtürk [1]). More importantly, cutting planes obtained through lifting have been proven to be very effective at reducing solution times for $0-1$ MIPs; see Crowder et al. [7], Gu et al. [8] and Van Roy and Wolsey [20]. As a consequence, lifted cuts generated from simple substructures of MIPs have been implemented in various commercial software, including CPLEX and X-Press.

Given a valid inequality (*seed inequality*) that is strong for the restriction of a set of interest, lifting is typically implemented sequentially, i.e. fixed variables are reintroduced into the inequality one at a time (or one group at a time). Furthermore, to determine the lifting coefficient of a variable, it is necessary to obtain an optimal solution of the lifting problem which is itself an MIP. Because the lifting problems to be solved are different for each lifted variable, lifting can rapidly become prohibitive. Nevertheless, when the lifting function of a seed inequality is well-structured, lifting can be performed efficiently. In particular, Wolsey [24], Gu et al. [10] and Atamtürk [3] showed that if the lifting function of the seed inequality is superadditive, then all the lifting coefficients can be obtained from the first lifting problem.

---

* This research is supported by NSF Grant DMI-03-48611.

M. Fischetti and D.P. Williamson (Eds.): IPCO 2007, LNCS 4513, pp. 210–224, 2007.
© Springer-Verlag Berlin Heidelberg 2007

There are various inequalities whose lifting functions are naturally superadditive. Examples include some flow covers inequalities (see Gu et al. [10]), and mixed integer cover inequalities for knapsack problems with a single continuous variable (see Marchand and Wolsey [13]). However, most often, lifting functions are not superadditive. In these cases, a superadditive lower approximation of the exact lifting function can be used to generate strong cuts [3, 10]. This idea was successfully used by Gu et al. [10] for 0–1 knapsack problems, by Gu et al. [9] and Louveaux and Wolsey [12] for single node flow models, by Shebalov and Klabjan [18] for mixed-integer programs with variable upper bounds and by Atamtürk [2] for general mixed integer knapsack sets. We note however that the lifting functions used in all of these cases are single-dimensional.

In one dimension, constructing a high-quality superadditive approximation of a lifting function is typically difficult and verifying that it is superadditive is often cumbersome. Although Atamtürk [3] proved that multidimensional superadditive lifting functions yield sequence independent lifting for general MIPs, a practical implementation of the idea seems to be difficult at first because in addition to the difficulties mentioned for single-dimensional problems, the derivation of exact multidimensional lifting functions is difficult and the proof that approximations are of good quality is hard. To the best of our knowledge, all but one of the superadditive lifting functions that were investigated to date are single-dimensional. The only exception is our study of the 0−1 knapsack problem with disjoint cardinality constraints [26] where we derived provably strong superadditive approximations of the multidimensional lifting function of cover inequalities. In [26], we observed that the high-dimensional exact lifting function of a minimal cover inequality could be represented using a composition of lower-dimensional exact lifting functions. We used this observation to build multidimensional superadditive lifting functions from the known superadditive approximations of the lower-dimensional lifting functions.

In this paper, we generalize these results to typical 0–1 MIP sets and propose a framework to construct high-dimensional superadditive lifting functions using known lower-dimensional superadditive lifting functions. We also show how this approach can be applied to variants of the knapsack and single node flow models with additional constraints. In particular, we obtain with our approach various families of strong inequalities for MIPs that are difficult to study using traditional tools.

The paper is organized as follows. In Section 2, after briefly reviewing sequence independent and superadditive lifting, we describe a way to represent / approximate high-dimensional exact lifting functions of valid inequalities for 0−1 MIP sets using the exact lifting functions of simpler 0−1 MIP sets. Then, we propose a framework to construct high-dimensional superadditive approximations of lifting functions using this representation. In Section 3, we apply our framework to the precedence-constrained knapsack model (PCKP). In particular, we build strong multidimensional superadditive lifting functions and derive strong lifted inequalities. Similarly, in Section 4, we obtain a family of facet-defining inequalities for the single node flow model with disjoint cardinality constraints (SNFCC). In Section 5, we give a conclusion and discuss future directions of research.

2   Constructing Multidimensional Superadditive Lifting Functions

In this section, we first review basic results about lifting. Then, we give a method to represent or approximate high-dimensional exact lifting functions using lower-dimensional ones. Finally, we describe an approach to build high-dimensional superadditive lifting functions that is based on the previous representation.

2.1   Lifting and Superadditive Lifting Functions

In this section, we briefly review lifting concepts and techniques. We focus on $0-1$ MIP models in which the continuous variables have variable upper bounds. The description for pure $0-1$ integer program is simpler and can be obtained similarly.

Let $N = \{1, \ldots, n\}$. Consider $S = \{(x, y) \in \{0, 1\}^n \times \mathbb{R}^n_+ : Ax + By \leq d, y_j \leq u_j x_j, \forall j \in N\}$. We define PS to be the convex hull of S and define $PS(N_0, N_1) = conv\{(x, y) \in S : x_j = 0 \ \forall j \in N_0, x_j = 1 \ \forall j \in N_1, y_j = 0 \ \forall j \in N_0, y_j = u_j \ \forall j \in N_1\}$. We use a similar notation for pure $0-1$ sets.

Assume that

$$\sum_{j \in \hat{N}} \alpha_j x_j + \sum_{j \in \hat{N}} \beta_j y_j \leq \alpha_0 \tag{1}$$

is a strong valid inequality for $PS(N_0, N_1)$ with $\hat{N} = N \backslash (N_0 \cup N_1)$. We wish to reintroduce (*lift*) the fixed variables (x_j, y_j) for $j \in N_0 \cup N_1$ into the seed inequality (1). Without loss of generality, we denote $N_0 \cup N_1 = \{1, \ldots, \hat{n}\}$ and assume that (x_1, y_1) is the first pair of variables to be lifted. Define $l_0 = l_1 = 0$ if (x_1, y_1) is lifted from $(0, 0)$ and define $l_0 = 1$ and $l_1 = u_1$ if (x_1, y_1) is lifted from $(1, u_1)$. The inequality obtained through lifting is

$$\sum_{j \in \hat{N}} \alpha_j x_j + \sum_{j \in \hat{N}} \beta_j y_j + \alpha_1(x_1 - l_0) + \beta_1(y_1 - l_1) \leq \alpha_0 \tag{2}$$

where α_1 and β_1 are chosen in such a way that

$$\alpha_1(x_1 - l_0) + \beta_1(y_1 - l_1) \leq f(A_1(x_1 - l_0) + B_1(y_1 - l_1)) \tag{3}$$

for $(x_1, y_1) \in \{(s, t) \in \{0, 1\} \times [0, u_l] : t \leq s u_l\}$ and where

$$f(z) = \min \alpha_0 - \sum_{j \in \hat{N}} \alpha_j x_j + \sum_{j \in \hat{N}} \beta_j y_j$$

$$\text{s.t.} \sum_{j \in \hat{N}} (A_j x_j + B_j y_j) \leq d - z, y_j \leq u_j x_j, \forall j \in \hat{N}. \tag{4}$$

By sequentially applying the above lifting operation with respect to the variables $(x_1, y_1), \ldots, (x_{\hat{n}}, y_{\hat{n}})$, the seed inequality (1) is progressively converted into a strong valid inequality for PS. It is proven in Nemhauser and Wolsey [14] that if $PS(N_0, N_1)$ is full-dimensional, (1) is facet-defining for $PS(N_0, N_1)$ and (3) is satisfied at equality by two new affinely independent solutions, then (2) is facet-defining for $PS(N_0 \backslash \{1\}, N_1)$ if lifted from $(0, 0)$ or for $PS(N_0, N_1 \backslash \{1\})$ if lifted from $(1, u_1)$.

Usually $f(z)$ in (4) is referred to as the *exact lifting function* (or *lifting function*) of the seed inequality (1). As we mentioned in Section 1, generating strong cuts through sequential lifting is typically computationally intensive. Wolsey [24], Gu et al. [10], and Atamtürk [3] showed that if the lifting function f is superadditive, i.e. $f(z_1) + f(z_2) \leq f(z_1 + z_2)$ for $z_1, z_2, z_1 + z_2$ in the domain of the lifting function, then lifting coefficients are independent of the lifting sequence and can be directly obtained from f. Since most lifting functions are not superadditive, superadditive lower approximations are often used to generate strong cuts [10, 3]. We use the criteria of non-dominance and maximality proposed by Gu et al. [10] to measure the strength of superadditive approximations.

2.2   Representation of High-Dimensional Lifting Function

In this section, we give a representation of high-dimensional lifting functions of given seed inequalities using low-dimensional lifting functions. Although the method does not always describe the high-dimensional lifting functions exactly, it has two advantages. First, it significantly reduces the difficulties associated with describing high-dimensional exact lifting functions. Second, it can be used to derive a superadditive approximation of the initial high-dimensional function using superadditive approximation of the lower-dimensional lifting functions.

An intuitive explanation of our scheme is as follows. When a new constraint is introduced into the initial constraint matrix, it forces some variables to become 0 or 1. If these variables are known, we can use the lifting function associated with the initial set of constraints to represent or approximate the exact lifting function of the new system.

Consider $PS = conv\{x \in \{0, 1\}^n : Ax \leq b\}$ with $A = \{A_1, \ldots, A_n\} \in \mathbb{R}^{m \times n}$ and $b \in \mathbb{R}^m$. Let $\sum_{j \in \hat{N}} \pi_j x_j \leq \pi_0$ be a valid inequality for $PS(N \backslash \hat{N}, \emptyset)$ and denote its lifting function by f. Assume now that the constraint $\sum_{j \in N} p_j x_j \leq b_{m+1}$ is added to A and denote the augmented constraint matrix by A'. Then, define f' to be the lifting function based on A'. Clearly, $f : \mathbb{R}^m \to \mathbb{R}$ and $f' : \mathbb{R}^{m+1} \to \mathbb{R}$. Note that, the objective functions of the lifting problems defining f and f' are identical. It is also clear that $f(z) = f'\binom{z}{-\infty}$ for $z \in \mathbb{R}^m$.

Proposition 1. *Let $j^* \in \hat{N}$ and assume that $\hat{x}$ is an optimal solution to $f'\binom{z}{p}$.*

(i) *If $\hat{x}_{j^*} = 0$, then*

$$f'\binom{z}{p} \geq \max\{f(z - A_{j^*}) + \pi_{j^*}, f(z)\}; \tag{5}$$

(ii) *If $\hat{x}_{j^*} = 1$, then*

$$f'\begin{pmatrix} z \\ p \end{pmatrix} \geq \max\{f(z + A_{j^*}) - \pi_{j^*}, f(z)\}. \tag{6}$$

□

The conclusion of Proposition 1 is very general since it is independent of the structure of the $0-1$ set and of the seed inequality. It is helpful because in many lifting functions of $0-1$ sets with multiple constraints, it is easy to determine the variables that are forced to 0 or 1 by the addition of a constraint. In such cases, we can derive a representation of f' for all $\begin{pmatrix} z \\ p \end{pmatrix} \in \mathbb{R}^{m+1}$. Furthermore, we observe that the expressions of Proposition 1 can be recursively used to approximate complicated situations. For example, if $\{x_1, x_2\}$ are forced to 0 and 1 respectively at $\begin{pmatrix} z \\ p \end{pmatrix}$ in an optimal solution to f', then we can write $f'\begin{pmatrix} z \\ p \end{pmatrix} \geq \max\{f(z - A_1 + A_2) + \pi_1 - \pi_2, f(z - A_1) + \pi_1, f(z + A_2) - \pi_2, f(z)\}$. We also observe that in various types of multidimensional lifting functions, the inequalities in (5) and (6) can be proven to be satisfied at equality, i.e. we can use the low-dimensional lifting functions through (5) and (6) to represent the high-dimensional lifting functions exactly; see Zeng and Richard [26] for a proof in the case of cover inequalities for knapsack problems with disjoint cardinality constraints.

The situation for general mixed integer program is more difficult than that presented in Proposition 1. Next, we generalize these results to describe the effect on the lifting function of adding constraints to mixed integer sets of the flow type, which form an important class of MIPs.

Consider $PS = conv\{(x, y) \in \mathbb{R}^n_+ \times \{0, 1\}^n : Ax + By \leq d, y_j \leq u_j x_j, \forall j\}$ with $A = \{A_1, \ldots, A_n\}, B = \{B_1, \ldots, B_n\} \in \mathbb{R}^{m \times n}$. Let $\sum_{j \in \hat{N}} \alpha_j x_j + \sum_{j \in \hat{N}} \beta_j y_j \leq \alpha_0$ be a valid inequality for $PS(N_0, N_1)$ and denote its lifting function by g. Assume that the constraint $\sum_{j \in N} p_j x_j \leq d_{m+1}$ is added to $[AB]$ and denote the augmented constraint matrix by $[A'B]$. Then, define g' to be the lifting function based on $[A'B]$.

Proposition 2. *Let $j^* \in N\backslash(N_0 \cup N_1)$ and assume that $(\hat{x}, \hat{y})$ is an optimal solution to $g'\begin{pmatrix} z \\ p \end{pmatrix}$.*

(i) *If $\hat{x}_{j^*} = 0$, then*

$$g'\begin{pmatrix} z \\ p \end{pmatrix} \geq \max\{g(z - A_{j^*} - B_{j^*} u_{j^*}) + \alpha_{j^*} + \beta_{j^*} u_{j^*}, g(z)\}; \tag{7}$$

(ii) *If $\hat{x}_{j^*} = 1$, then*

$$g'\begin{pmatrix} z \\ p \end{pmatrix} \geq \max\{g(z + A_{j^*} + B_{j^*} \hat{y}_{j^*}) - \alpha_{j^*} - \beta_{j^*} \hat{y}_{j^*}, g(z)\}. \tag{8}$$

where $\hat{y}_{j^*}$ is the value of the ith element of $\hat{y}$.

□

Note that Proposition 2 can also be applied recursively to approximate more complicated situations.

2.3 A Framework to Build Multidimensional Superadditive Functions

In this subsection, we propose a constructive framework to build high-dimensional superadditive lifting functions from lower-dimensional ones. This framework is composed of a general scheme that can be enhanced by several simple rules. These rules can be used alone or in combination to create new superadditive lifting functions.

First observe that f' and g' in Proposition 1 and Proposition 2 contain forms such as $f(z - A_{j*}) + \pi_{j*}$ and $g(z - A_{j*} - B_{j*}u_{j*}) + \alpha_{j*} + \beta_{j*}u_{j*}$. In Theorem 1, we show how to find superadditive approximations for a generalization of these forms and show how to combine them to obtain a multidimensional superadditive approximation of the lifting function. We first generalize the concept of *superadditivity*. Let $\varphi : \mathbb{R}^{m_1} \longmapsto \mathbb{R}^{m_2}$. We say that φ is non-decreasing if $\varphi(\boldsymbol{x}) \geq \varphi(\boldsymbol{y})$ when $\boldsymbol{x} \geq \boldsymbol{y}$ for $\boldsymbol{x}, \boldsymbol{y} \in \mathbb{R}^{m_1}$. We say that φ is superadditive if $\varphi(\boldsymbol{x}) + \varphi(\boldsymbol{y}) \leq \varphi(\boldsymbol{x} + \boldsymbol{y})$ for all $\boldsymbol{x}, \boldsymbol{y} \in \mathbb{R}^{m_1}$.

Theorem 1. *Let* $\pi_1 : \mathbb{R}^{m_1} \longmapsto \mathbb{R}$, $\pi_2 : \mathbb{R}^{m_1} \times \mathbb{R}^{m_2} \longmapsto \mathbb{R}$, *and* $\pi_3 : \mathbb{R}^{m_1} \times \mathbb{R}^{m_2} \longmapsto \mathbb{R}^{m_1}$ *be superadditive functions over their domains, and assume that* $\pi_1(\boldsymbol{0}) = \pi_2(\boldsymbol{0}, \boldsymbol{0}) = 0$ *and* $\pi_3(\boldsymbol{0}, \boldsymbol{0}) = \boldsymbol{0}$. *Assume that* π_1 *is non-decreasing. The function* $\kappa : \mathbb{R}^{m_1} \times \mathbb{R}^{m_2} \longmapsto \mathbb{R}$ *defined as*

$$\kappa(\boldsymbol{x}, \boldsymbol{y}) = \pi_1(\boldsymbol{x} + \pi_3(\boldsymbol{x}, \boldsymbol{y})) + \pi_2(\boldsymbol{x}, \boldsymbol{y}) \tag{9}$$

is superadditive over $\mathbb{R}^{m_1} \times \mathbb{R}^{m_2}$ *with* $\kappa(\boldsymbol{0}, \boldsymbol{0}) = 0$. *Furthermore, if* π_2 *and* π_3 *are non-decreasing, then* κ *is non-decreasing.* □

Next, we present several simple rules to compose superadditive functions. Rule 1 presents a way to extend an existing superadditive function to a larger domain. Rule 2 and Rule 3 are adapted from Nemhauser and Wolsey [14].

Rule 1. *Let* $\pi_1(\boldsymbol{x}) : \mathbb{D}_1 \subseteq \mathbb{R}^{m_1} \to \mathbb{R}$ *be a superadditive function. Let* $\boldsymbol{y} \in \mathbb{R}^{m_1} \backslash \mathbb{D}_1$. *Then, the function*

$$\kappa(\boldsymbol{x}) = \begin{cases} \pi_1(\boldsymbol{x}), & \text{if } \boldsymbol{x} \in \mathbb{D}_1 \\ \sup\{\pi_1(\boldsymbol{x}_1) + \pi_1(\boldsymbol{x}_2) : \boldsymbol{x} = \boldsymbol{x}_1 + \boldsymbol{x}_2, \boldsymbol{x}_1, \boldsymbol{x}_2 \in \mathbb{D}_1, \}, & \text{if } \boldsymbol{x} = \boldsymbol{y} \end{cases}$$

is superadditive over $\mathbb{D}_1 \cup \{\boldsymbol{y}\}$. □

Assume now that $\gamma_i : \mathbb{R}^m \longmapsto \mathbb{R}$ are superadditive functions for $i = 1, 2$.

Rule 2. *The function* κ *defined as* $\kappa(\boldsymbol{x}) = \min\{\gamma_1(\boldsymbol{x}), \gamma_2(\boldsymbol{x})\}$ *is superadditive over* $\mathbb{R}^m$. □

Rule 3. *The function* κ *defined as* $\kappa(\boldsymbol{x}) = \gamma_1(\boldsymbol{x}) + \gamma_2(\boldsymbol{x})$ *is superadditive over* $\mathbb{R}^m$. □

In Section 3 and in Section 4, we show how to apply Theorem 1 and Rules 1−3 to build strong multidimensional superadditive lifting functions for specific MIPs.

3 Superadditive Lifting Functions for $0-1$ PCKP

Let $N = \{1,\ldots,n\}$ and $0 \le a_j \le b$ for $j \in N$. The precedence-constrained knapsack model is defined as

$$X_{PCKP} = \{x \in \{0,1\}^n : \sum_{j\in N} a_j x_j \le b, x_{t(i)} \le x_{h(i)}, i = 1,\ldots,r\} \qquad (10)$$

where $t(i) \ne h(i)$ for $i = 1,\ldots,r$. We denote the convex hull of X_{PCKP} as P_{PCKP} and the convex hull of the classical knapsack set as P_{KP}.

Precedence-constrained knapsack problems arise frequently in planning, scheduling and project management. The polyhedral structure of PCKP has been studied by various authors; see Boyd [6], Park and Park [17], van de Leensel et al. [19] and Boland et al. [5]. In particular, van de Leensel et al. [19] proved that lifting *minimal induced cover inequality* is a NP-hard problem in general. In this section, we focus on deriving strong inequalities from minimal cover inequalities of the knapsack constraint using superadditive lifting functions. Note that precedence constraints are one of the many additional features that we can incorporate into the lifting function using the framework we proposed in Section 2. Another variant of knapsack problem, the $0-1$ knapsack problem with disjoint cardinality constraints was studied in Zeng and Richard [25, 26]. These papers are the roots of the results of Section 2.

For the traditional $0-1$ knapsack polytope P_{KP}, we say that a set $C \subseteq N$ is a cover if $\sum_{j\in C} a_j > b$. Furthermore, we say that a cover C is minimal if, for all $j \in C$, $C\backslash\{j\}$ is not a cover. Given a minimal cover C, the cover inequality

$$\sum_{j\in C} x_j \le |C| - 1 \qquad (11)$$

is facet-defining for $P_{KP}(N\backslash C, \emptyset)$.

We now use cover inequality (11) to derive strong valid inequality for P_{PCKP}. To simplify the exposition, we assume in this paper that the precedence constraints are organized into s disjoint paths such that $x_{j_{i,1}} \ge \cdots \ge x_{j_{i,|N_i|}}$ where $N_i = \{j_{i,1},\ldots,j_{i,|N_i|}\}$ is the ith path. It is not restrictive to assume that $N = N_1 \cup \cdots \cup N_s$ since path can have length one. Furthermore, because the precedence constraint structure of any PCKP problem can be relaxed into a set of disjoint paths, our results are applicable to the general case.

Define $i[j]$ to be the index of the path x_j belongs to, i.e. $j \in N_{i[j]}$. Also denote $C \cap N_i = C_i$ for $i = 1,\ldots,s$ and define $F(j)$ for $j \in N\backslash C$ to be the set all ancestors of x_j in its path. We next present necessary and sufficient conditions for (11) to be strong for P_{PCKP}.

Proposition 3. *Let C be a minimal cover. The cover inequality (11) is facet-defining for $P_{PCKP}(N\backslash C, \emptyset)$ if and only if $C_i = \{j_{i,1}\}$ or $C_i = \emptyset$ for $i = 1,\ldots,s$.* □

Next, we describe how to lift minimal cover inequality. First observe that the lifting of variables in a given path is fixed because of the precedence constraints.

In fact, when lifting x_j from 0 to 1, all the ancestors of j are forced to 1 because of the precedence constraints. Based on this observation and using Theorem 1 and Rule 1, we can easily derive a multidimensional superadditive lifting function using the single-dimensional superadditive approximation θ of the lifting function for a minimal cover inequality proposed in Gu et al. [10]. We denote this multidimensional function by $\phi\binom{z}{v}$ with $(z, v) \in [0, b] \times \mathbb{D}$ where $\mathbb{D} = \{0, 1\}^s$.

Proposition 4. *The function*

$$\phi\binom{z}{v} = \begin{cases} \theta(z) & \text{if } v = 0 \\ \theta(z) & \text{if } v = e_i \text{ and } C_i = \emptyset \\ \max\{\theta(z + a_{j_{i,1}}) - 1, \theta(z)\} & \text{if } v = e_i \text{ and } C_i = \{j_{i,1}\} \\ \displaystyle\sup_{v=\sum_{k \in V} e_k, z = \sum_{k \in V} z_k,\ z_k \geq 0\ \forall k} \left\{ \sum_{k \in V} \phi\binom{z_k}{e_k} \right\} & \text{if } v \notin \{0, e_1, \ldots, e_s\} \end{cases}$$

for $(z, v) \in [0, b] \times \mathbb{D}$ is a valid superadditive approximation of the lifting function of (11) that is non-dominated and maximal. □

In the next theorem, we present the lifted cover inequality for P_{PCKP} that can be obtained using the multidimensional superadditive lifting function ϕ.

Theorem 2. *Assume that C is a minimal cover from the knapsack constraint of P_{PCKP} that satisfies the condition of Proposition 3, then*

$$\sum_{j \in C} x_j + \sum_{j \in N \setminus C} \left[\phi\binom{p_j^+ + a_j}{e_{i[j]}} - \phi\binom{p_j^+}{e_{i[j]}} \right] x_j \leq |C| - 1. \tag{12}$$

with $p_j^+ = \sum_{k \in F(j) \setminus C_{i[j]}} a_k$ is a valid inequality for P_{PCKP}. □

Note that the inequality (12) can be derived very efficiently since the function θ (and therefore the function ϕ) is known in closed form. Next, we show in Example 1 that (12) is strong by comparing it to the lifted cover inequality obtained using the single-dimensional superadditive lifting function θ from [10].

Example 1. Let

$$S = \{x \in \{0, 1\}^8 : 8x_1 + 7x_2 + 6x_3 + 4x_4 + 6x_5 + 6x_6 + 5x_7 + 8x_8 \leq 22,$$
$$x_6 \leq x_5 \leq x_2,\ x_8 \leq x_7 \leq x_2\}.$$

The cover inequality $x_1 + x_2 + x_3 + x_4 \leq 3$ is facet-defining for $PS(\{5, 6, 7, 8\}, \emptyset)$.

Using the traditional single-dimensional superadditive lifting function of the cover inequality, we obtain the following lifted inequality

$$x_1 + x_2 + x_3 + x_4 + 0.5x_5 + 0.5x_6 + x_8 \leq 3. \tag{13}$$

We now show that we can obtain a stronger inequality using the precedence structure. First, as illustrated in Figure 1, we relax the precedence constraints into two disjoint paths. We then apply the results of Proposition 4 to obtain the multidimensional superadditive approximation $\phi\binom{z}{v}$ for $(z, v) \in [0, b] \times \{0, e_1\} \times \{0, e_2\}$ that is illustrated in Figure 2. The solid line in Figure 2 shows the exact lifting function and the doted line describes the superadditive approximation. Lifting the variables x_5, x_6, x_7 and x_8, we obtain $\alpha_5 = \phi\binom{6}{e_1} = \theta(13) - 1 = 1$, $\alpha_6 = \phi\binom{12}{e_1} - \phi\binom{6}{e_1} = \theta(19) - \theta(13) = 1$, $\alpha_7 = \phi\binom{5}{e_2} = \theta(5) = 0$, $\alpha_8 = \phi\binom{13}{e_2} - \phi\binom{5}{e_2} = \theta(13) - \theta(5) = 2$. Therefore, the lifted cover inequality is

$$x_1 + x_2 + x_3 + x_4 + x_5 + x_6 + 2x_8 \leq 3. \tag{14}$$

which clearly dominates (13).   □

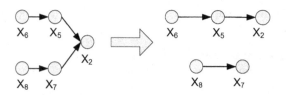

Fig. 1. Relaxing the precedence structure into disjoint paths

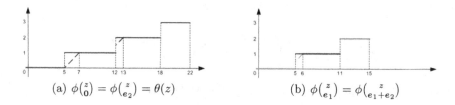

(a) $\phi\binom{z}{0} = \phi\binom{z}{e_2} = \theta(z)$ (b) $\phi\binom{z}{e_1} = \phi\binom{z}{e_1 + e_2}$

Fig. 2. Exact lifting function and superadditive approximation

4   Superadditive Lifting Functions for SNFCC

The single node flow model is a relaxation of numerous logistics, transportation and telecommunication network design problems. Research on single node flow model is very extensive; see Padberg et al. [16], Van Roy and Wolsey [21] and Gu et al. [9, 10] among many others. In this paper, we consider a variant of this model with disjoint cardinality constraints. It is defined as

$$X_{SNFCC} = \{(x, y) \in \{0, 1\}^n \times \mathbb{R}_+^n : \sum_{j \in N} y_j \leq b, y_j \leq a_j x_j, \forall j \in N,$$

$$\sum_{j \in N_i} x_j \leq K_i, i = 1, \ldots, r\}$$

where $a_1 \geq \cdots \geq a_n > 0$, $K_i \geq 1$ for $i = 1, \ldots, r$, $N_i \cap N_j = \emptyset$ if $i \neq j$ and $N = N_0 \cup \cdots \cup N_r$.

We denote the convex hull of SNFCC as P_{SNFCC}. Again, cardinality constraints are one of the side constraints that can easily be incorporated into the lifting function in our scheme and help us illustrate our multidimensional lifting approach. A flow cover is a set $C \subseteq N$ such that $\sum_{j \in C} a_j - b = \lambda > 0$. Let $C^+ = \{j \in C : a_j > \lambda\}$. The corresponding flow cover inequality is

$$\sum_{j \in C} y_j + \sum_{j \in C^+} (a_j - \lambda)(1 - x_j) \leq b. \tag{15}$$

It is proven in Nemhauser and Wolsey [14] that this valid inequality is facet-defining for $P_{SNF}(N \backslash C, \emptyset)$. Gu et al. [10] studied the problem of lifting (15) and proved that the lifting function $\psi(z)$ of (15) is superadditive over $[-\lambda, +\infty)$. Therefore, the lifting of variables (x_j, y_j) for $j \in N \backslash C$ is sequence independent because $a_j > 0$ for $j \in N$. We generalize these results for the cardinality constrained cases. For the sake of brevity, we consider here the most general case where C is not a subset of N_i for $i = 0, \dots, r$. Next, we present a set of sufficient conditions under which (15) is strong for P_{SNFCC}.

Proposition 5. *The flow cover inequality* (15) *is facet-defining for* $P_{SNFCC}(N \backslash C, \emptyset)$ *if* $|C \cap N_i| \leq K_i$ *for* $i = 1, \dots, r$.   □

Define now $C \cap N_i = C_i$, $i_M \in \arg\max\{a_j : j \in C_i, j \in C^+\}$ and $i_m \in \arg\min\{a_j : j \in C_i, j \notin C^+\}$. Also, assume that $C_i = \{j_{i,1}, \dots, j_{i,|C_i|}\}$ with $a_{j_{i,1}} \geq \dots \geq a_{j_{i,|C_i|}}$. To distinguish it from the traditional lifting function ψ of the flow cover, we refer to the lifting function of the flow cover inequality with the consideration of cardinality constraints as $\Psi\binom{z}{v}$ for $v \in \mathbb{D}' = \{0, e_1, \dots, e_r\}$. There are three nontrivial cases that we need to consider to approximate Ψ as a function of ψ using Proposition 2.

Theorem 3. *The lifting function* $\Psi\binom{z}{v}$ *for* $(z, v) \in \mathbb{R}_+ \times \mathbb{D}'$ *is* $\Psi\binom{z}{0} = \psi(z)$ *and*

(i) *if* $C_i \subseteq C^+$, *then*

$$\Psi\binom{z}{e_i} \geq \begin{cases} \psi(z) & \text{if } |C_i| \leq K_i - 1, \\ \max\{\psi(z - a_{i_M}) + \lambda, \psi(z)\} & \text{if } |C_i| = K_i. \end{cases} \tag{16}$$

(i) *if* $C_i \cap C^+ = \emptyset$ *and* $C_i \neq \emptyset$, *then*

$$\Psi\binom{z}{e_i} \geq \begin{cases} \psi(z) & \text{if } |C_i| \leq K_i - 1, \\ \max\{\psi(z - a_{i_m}) + a_{i_m}, \psi(z)\} & \text{if } |C_i| = K_i. \end{cases} \tag{17}$$

(iii) *if* $C_i \not\subseteq C^+$ *and* $C_i \cap C^+ \neq \emptyset$, *then*

$$\Psi\binom{z}{e_i} \geq \begin{cases} \psi(z) & \text{if } |C_i| \leq K_i - 1, \\ \min\{\psi_1(z), \psi_2(z)\} & \text{if } |C_i| = K_i \end{cases} \tag{18}$$

where $\psi_1(z) = \max\{\psi(z - a_{i_M}) + \lambda, \psi(z)\}$ *and* $\psi_2(z) = \max\{\psi(z - a_{i_m}) + a_{i_m}, \psi(z)\}$.   □

In fact, we can further prove that (16)-(18) are satisfied at equality. Then, we use Theorem 1, Rule 1 and Rule 2 to verify that Ψ is naturally superadditive.

Theorem 4. *The function* $\Psi\left(\genfrac{}{}{0pt}{}{z}{v}\right)$ *defined as*

$$\Psi\left(\begin{matrix} z \\ v \end{matrix}\right) = \begin{cases} \Psi\left(\begin{matrix} z \\ v \end{matrix}\right) & \text{if } v \in \mathbb{D}' \\ \displaystyle\sup_{\{z=\sum_{i=1}^{s} z_i,\ z_i \geq 0\ \forall i\}} \{\sum_{i=1}^{s} \Psi\left(\begin{matrix} z_i \\ e_i \end{matrix}\right)\} & \text{if } v \notin \mathbb{D}' \end{cases} \tag{19}$$

is superadditive over $[0,b] \times \mathbb{D}$ *where* $\mathbb{D} = [0,b] \times \{0,\ldots,K_1\} \times \cdots \times \{0,\ldots,K_r\}$. □

It follows from Theorem 4 that we can apply sequence independent lifting to obtain the lifted flow cover inequality

$$\sum_{j \in C} y_j + \sum_{j \in C^+} (a_j - \lambda)(1 - x_j) + \sum_{j \in N\backslash C} (\alpha_j x_j + \beta_j y_j) \leq b. \tag{20}$$

To derive the coefficients α_j and β_j for $j \in N\backslash C$, we define $A_j = \sum_{h=1}^{j} a_h$ for $j = 1,\ldots,n$ and $A_0 = 0$. Also, we let $s^+ = |C^+|$. In Theorem 5, we present the lifting coefficients for the lifted flow cover inequality using $\Psi\left(\genfrac{}{}{0pt}{}{z}{v}\right)$. We do not list here the results for the case where $C_i \not\subseteq C^+$ and $C_i \cap C^+ \neq \emptyset$ because the expressions for lifting coefficients are similar but more complicated.

Theorem 5. *Inequality (20) is facet-defining for* PS_{SNFCC} *if* $(\alpha_j, \beta_j) \in H_j$ *for* $j \in N\backslash C$ *where* H_j *is defined as follows when* $j \in N_i$:

(i) *When (1) $i = 0$, or (2) $|C_i| \leq K_i - 1$, or (3) $|C_i| = K_i$ and $a_{i_M} = a_1$.*
If $a_j \leq a_1 - \lambda$, we $H_j = \{(0,0)\}$. Otherwise, let $l = \arg\max_{0 \leq h \leq s^+}\{a_j \geq A_h - \lambda\}$ and define $H_j = \{(0,0)\} \cup H_j^1 \cup H_j^2$ with

$$H_j^1 = \{(\lambda(k-1) - \frac{\lambda(A_k - \lambda)}{a_k}, \frac{\lambda}{a_k}) : k = 2,\ldots,l\}$$

and

$$H_j^2 = \begin{cases} \emptyset & \text{if } a_j = A_l - \lambda \\ \{(l\lambda - A_l, 1)\} & \text{if } A_l - \lambda < a_j \leq A_l \text{ or } a_j > A_{s^+} \\ \{(l\lambda - a_j\rho, \rho)\} & \text{if } a_j < A_{s^+} \text{ and } A_l < a_j < A_{l+1} - \lambda \end{cases}$$

where $\rho = \frac{\lambda}{a_j + \lambda - A_l}$.

(ii) *When $|C_i| = K_i$ and $C_i \subseteq C^+$.*
If $a_j \leq a_{i_M} - \lambda$, define $H_j = \{(0,0)\}$. Otherwise, let $l = \arg\max_{1 \leq h \leq s^+}\{a_j \geq \min\{A_h - \lambda, A_{h-1} + a_{i_M} - \lambda\}\}$ and define $H_j = \{(0,0)\} \cup H_j^1 \cup H_j^2$ with

$$H_j^1 = \{(\lambda(k-1) - \frac{\lambda(\min\{A_k, A_{k-1} + a_{i_M}\} - \lambda)}{\min\{A_k, A_{k-1} + a_{i_M}\} - \min\{A_{k-1}, A_{k-2} + a_{i_M}\}},$$
$$\frac{\lambda}{\min\{A_k, A_{k-1} + a_{i_M}\} - \min\{A_{k-1}, A_{k-2} + a_{i_M}\}}) : k = 2,\ldots,l\}$$

and H_j^2 is equal to (5) with A_l replaced by $\min\{A_l, A_{l-1} + a_{i_M}\}$.

(iii) *When $|C_i| = K_i$ and $C_i \cap C^+ = \emptyset$.*
$\quad$ *If $0 < a_j \leq a_{i_m} + a_1 - \lambda$, define $H_j = \{0, \frac{a_{i_m}}{\max\{a_j, a_{i_m}\}}\}$. Otherwise, let*
$l = \arg\max_{1 \leq h \leq s^+}\{a_j \geq A_h + a_{i_m} - \lambda\}$ *and define $A_j^+ = A_j + a_{i_m}$ for*
$j = 1, \ldots, s^+$ *and $H_j = \{(0, \frac{a_{i_m}}{a_{i_m} + a_1 - \lambda})\} \cup H_j^1 \cup H_j^2$ with*

$$H_j^1 = \{((k-1)\lambda + a_{i_m} - \frac{\lambda(A_k - \lambda + a_{i_m})}{a_k}, \frac{\lambda}{a_k}) : k = 2, \ldots, l\}$$

and

$$H_j^2 = \begin{cases} \emptyset & \text{if } a_j = A_l^+ - \lambda \\ \{(l\lambda - A_l, 1)\} & \text{if } A_l^+ - \lambda < a_j \leq A_l^+ \text{ or } a_j > A_{s^+}^+ \\ \{(l\lambda + a_{i_m} - a_j\rho^+, \rho^+)\} & \text{if } a_j < A_{s^+}^+ \text{ and } A_l^+ < a_j < A_{l+1}^+ - \lambda \end{cases}$$

where $\rho^+ = \frac{\lambda}{a_j + \lambda - A_l^+}$. $\quad\square$

We note that part (i) of Theorem 5 is identical to Theorem 9 in [10]. However, Theorem 5 in general yields stronger coefficients as illustrated in Example 2.

Example 2. Let

$$S = \{(x, y) \in \{0, 1\}^8 \times \mathbb{R}_+^8 : y_1 + y_2 + y_3 + y_4 + y_5 + y_6 + y_7 + y_8 \leq 24,$$
$$y_1 \leq 9x_1, y_2 \leq 7x_2, y_3 \leq 6x_3, y_4 \leq 3x_4,$$
$$y_5 \leq 2x_5, y_6 \leq 2x_6, y_7 \leq 3x_7, y_8 \leq 12x_8,$$
$$x_6 + x_7 + x_8 \leq 1\}.$$

The flow cover inequality

$$\sum_{j=1}^{6} y_j + (9 - 5)(1 - x_1) + (7 - 5)(1 - x_2) + (6 - 5)(1 - x_3) \leq 24$$

based on $C = \{1, 2, 3, 4, 5, 6\}$ is facet-defining for $PS(\{7, 8\}, \emptyset)$ since $\lambda = 5$.
$\quad$ In Figure 3, we show the multidimensional lifting function Ψ and the single-dimensional function ψ. If we use the single-dimensional lifting function ψ, we obtain the inequalities

$$\sum_{j=1}^{6} y_j + 4(1 - x_1) + 2(1 - x_2) + (1 - x_3) + \begin{Bmatrix} 0 \\ -\frac{20}{7} \\ -6 \end{Bmatrix} x_8 + \begin{Bmatrix} 0 \\ \frac{5}{7} \\ 1 \end{Bmatrix} y_8 \leq 24.$$

Using the results of Theorem 5(iii), we obtain

$$\sum_{j=1}^{6} y_j + 4(1 - x_1) + 2(1 - x_2) + 1(1 - x_3) + \frac{2}{3}y_7 + \begin{Bmatrix} 0 \\ -3 \end{Bmatrix} x_8 + \begin{Bmatrix} \frac{1}{3} \\ \frac{5}{6} \end{Bmatrix} y_8 \leq 24. \quad (21)$$

All the inequalities represented by (21) are facet-defining for the convex hull of S. $\quad\square$

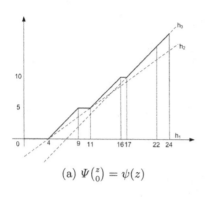
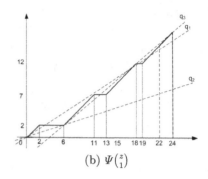

(a) $\Psi\binom{z}{0} = \psi(z)$ (b) $\Psi\binom{z}{1}$

Fig. 3. Superadditive lifting function Ψ

5 Conclusion

In this paper we propose a novel approach to construct multidimensional superadditive lifting functions and apply it to study two variants of classical MIP
models. This approach is based on the observation that it is usually possible
to represent high-dimensional lifting functions using lower-dimensional lifting
functions. The approach we propose is systematic, constructive and the multidimensional superadditive lifting functions obtained yield strong inequalities for
models in which a direct lifting approach would have been difficult. In particular,
we obtained multidimensional superadditive lifting functions for the precedence-
constrained knapsack model and for the single node flow model with disjoint
cardinality constraints. For these models, we presented a set of cutting planes
that are stronger than those obtained from the knapsack or flow constraint only.
To the best of our knowledge, our framework is the first attempt to construct
multidimensional superadditive lifting functions.

We are currently generalizing the procedure to generate more complicated
multidimensional superadditive lifting functions for unstructured MIPs with
multiple constraints. We are also considering several more general MIP models
that have practical significance in transportation and network design. Finally,
we are currently carrying an empirical evaluation of the cutting planes produced
by the multidimensional lifting techniques presented in this paper.

References

[1] A. Atamtürk. Flow pack facets of the single node fixed-charge flow polytope.
    Operations Research Letters, 29:107–114, 2001.
[2] A. Atamtürk. On the facets of the mixed-integer knapsack polyhedron.
    Mathematical Programming, 98:145–175, 2003.
[3] A. Atamtürk. Sequence independent lifting for mixed-integer programming.
    Operations Research, 52:487–490, 2004.
[4] E. Balas. Facets of the knapsack polytope. *Mathematical Programming*, 8:
    146–164, 1975.

[5] N. Boland, C. Fricke, G. Froylandz, and R. Sotirov. Clique-based facets for the precedence constrained knapsack polyhedron. Technical report, The University of Melbourne , Australia, 2005.

[6] E. Boyd. Polyhedral results for the precedence-constrained knapsack problem. *Discrete Applied Mathematics*, 41:185–201, 1993.

[7] H. Crowder, E. Johnson, and M. Padberg. Solving large scale zero-one linear programming problem. *Operations Research*, 31:803–834, 1983.

[8] Z. Gu, G. Nemhauser, and M. Savelsbergh. Lifted cover inequalities for 0-1 integer programs: computation. *INFORMS Journal on Computing*, 10: 427–437, 1998.

[9] Z. Gu, G. Nemhauser, and M. Savelsbergh. Lifted flow cover inequalities for mixed 0-1 integer programs. *Mathematical Programming*, 85:439–468, 1999.

[10] Z. Gu, G. Nemhauser, and M. Savelsbergh. Sequence independent lifting in mixed integer programming. *Journal of Combinatorial Optimization*, 4: 109–129, 2000.

[11] P. Hammer, E. Johnson, and U.Peled. Facets of regular 0-1 polytopes. *Mathematical Programming*, 8:179–206, 1975.

[12] Q. Louveaux and L. Wolsey. Lifting, superadditivity, mixed integer rounding and single node flow sets revisited. *4OR*, 1:173–207, 2003.

[13] H. Marchand and L. Wolsey. The 0-1 knapsack problem with a single continuous variable. *Mathematical Programming*, 85:15–33, 1999.

[14] G. Nemhauser and L. Wolsey. *Integer and Combinatorial Optimization*. Wiley, 1988.

[15] M. Padberg. On the facial structure of set packing polyhedra. *Mathematical Programming*, 5:199–215, 1973.

[16] M. Padberg, T. Van Roy, and L. Wolsey. Valid inequalities for fixed charge problems. *Mathematical Programming*, 33:842–861, 1985.

[17] K. Park and S. Park. Lifting cover inequalities for the precedence-constrained knapsack problem. *Discrete Applied Mathematics*, 72:219–241, 1997.

[18] S. Shebalov and D. Klabjan. Sequence independent lifting for mixed integer programs with variable upper bounds. *Mathematical Programming*, 105: 523–561, 2006.

[19] R.L.M.J. van de Leensel, C.P.M. van Hoesel, and J.J. van de Klundert. Lifting valid inequalities for the precedence constrained knapsack problem. *Mathematical Programming*, 86:161–185, 1999.

[20] T. Van Roy and L. Wolsey. Solving mixed integer programming problems using automatic reformulation. *Operations Research*, 35:45–57, 1987.

[21] T. Van Roy and L. Wolsey. Valid inequalities for mixed 0-1 programs. *Discrete Applied Mathematics*, 14:199–213, 1986.

[22] L. Wolsey. Faces for a linear inequality in 0-1 variables. *Mathematical Programming*, 8:165–178, 1975.

[23] L. Wolsey. Facets and strong valid inequalities for integer programs. *Operations Research*, 24:367–372, 1976.

[24] L. Wolsey. Valid inequalities and superadditivity for 0/1 integer programms. *Mathematics of Operations Research*, 2:66–77, 1977.

[25] B. Zeng and J.-P.P Richard. Sequentially lifted valid inequalities for $0-1$ knapsack problem with disjoint cardinality constraints. Technical report, Purdue University, 2006.

[26] B. Zeng and J.-P.P Richard. Sequence independent lifting for 0–1 knapsack problem with disjoint cardinality constraints. Technical report, Purdue University, 2006.

On the Exact Separation of Mixed Integer Knapsack Cuts

Ricardo Fukasawa[1] and Marcos Goycoolea[2]

[1] H. Milton Stewart School of Industrial and Systems Engineering
Georgia Institute of Technology
rfukasaw@isye.gatech.edu
[2] School of Business
Universidad Adolfo Ibañez
marcos.goycoolea@uai.cl

Abstract. During the last decades, much research has been conducted deriving classes of valid inequalities for single-row mixed integer programming polyhedrons. However, no such class has had as much practical success as the MIR inequality when used in cutting plane algorithms for general mixed integer programming problems. In this work we analyze this empirical observation by developing an algorithm which takes as input a point and a single-row mixed integer polyhedron, and either proves the point is in the convex hull of said polyhedron, or finds a separating hyperplane. The main feature of this algorithm is a specialized subroutine for solving the Mixed Integer Knapsack Problem which exploits cost and lexicographic dominance. Separating over the entire closure of single-row systems allows us to establish natural benchmarks by which to evaluate specific classes of knapsack cuts. Using these benchmarks on Miplib 3.0 instances we analyze the performance of MIR inequalities. Computations are performed in exact arithmetic.

Keywords: cutting plane algorithms, integer programming.

1 Introduction

Consider positive integers n, m and let $d \in \mathbb{Q}^m$, $D \in \mathbb{Q}^{m \times n}$, $l \in \{\mathbb{Q} \cup \{-\infty\}\}^n$ and $u \in \{\mathbb{Q} \cup \{+\infty\}\}^n$. Let $I \subseteq N := \{1, \ldots, n\}$ and consider the mixed integer set:

$$P = \{x \in \mathbb{R}^n : Dx \leq d, \ l \leq x \leq u, \ x_i \in \mathbb{Z}, \ \forall i \in I\}.$$

We say that a mixed integer knapsack set of the form,

$$K = \{x \in \mathbb{R}^n : ax \leq b, \ l \leq x \leq u, \ x_i \in \mathbb{Z}, \ \forall i \in I\}$$

with $b \in \mathbb{Q}$, $a \in \mathbb{Q}^n$ is *implied* by P if (a, b) is a non-negative linear combination of rows obtained from (D, d). Observe that if K is implied by P, then $P \subseteq K$. Hence, any inequality which is valid for K is also valid for P. We henceforth call such inequalities *knapsack cuts* derived from K.

M. Fischetti and D.P. Williamson (Eds.): IPCO 2007, LNCS 4513, pp. 225–239, 2007.
© Springer-Verlag Berlin Heidelberg 2007

Deriving strong knapsack cuts is of great practical importance to Mixed Integer Programming (MIP). In fact, most cutting planes known for general mixed integer programming are knapsack cuts. For example, Gomory Mixed Integer cuts [19,28] are knapsack cuts derived from the tableaus of linear programming relaxations, and Lifted Cover Inequalities [12,23] are knapsack cuts derived from the original rows of P. Other classes of knapsack cuts include mixed-integer-rounding (MIR) cuts and their variations [11,26,28], split cuts [10], lift-and-project cuts [4], and group cuts [15,20] – to name but a few.

In this paper we discuss an empirical methodology for evaluating sub-classes of knapsack cuts. Formally, consider P as defined above, $c \in \mathbb{Q}^n$, and $\mathcal{C}$ a set of valid inequalities for P. Define,

$$z^*(\mathcal{C}) = \min\{cx : Dx \le d, \, l \le x \le u, \, \pi x \le \pi_o \, \forall (\pi, \pi_o) \in \mathcal{C}\}.$$

Observe that the value $z^*(\mathcal{C})$ defines a benchmark by which to evaluate classes of cuts that are subsets of $\mathcal{C}$. For example, consider a family of implied knapsack sets $\mathcal{K}$ and let $\mathcal{C}^{\mathcal{K}}$ represent the set of all knapsack cuts which can be derived from some set $K \in \mathcal{K}$. Likewise, let $\mathcal{M}^{\mathcal{K}}$ represent the set of all MIR inequalities which can be derived from some set $K \in \mathcal{K}$. Given that $\mathcal{M}^{\mathcal{K}} \subseteq \mathcal{C}^{\mathcal{K}}$ it is easy to see that $z^*(\mathcal{C}^{\mathcal{K}}) \ge z^*(\mathcal{M}^{\mathcal{K}})$ and that the proximity of these two values gives an indication of the strength of MIR inequalities derived from that particular family $\mathcal{K}$.

In our computational experiments we will consider two specific families of implied knapsack sets: The set $\mathcal{F}$ of all formulation rows of P ; and, given a basic solution of the simplex algorithm, the set $\mathcal{T}$ of all tableau rows.

Boyd [8] and Yan and Boyd [30] compute $z^*(\mathcal{C}^{\mathcal{F}})$ for a subset of pure and mixed 0-1 instances in MIPLIB 3.0 [7]. Fischetti and Lodi [18] extend this result by computing $z^*(\mathcal{C}^{\mathcal{A}})$, where $\mathcal{A}$ is the set of all implied knapsack polyhedra, for a similar test set of pure 0-1 problems.

In this paper we compute the values $z^*(\mathcal{C}^{\mathcal{F}})$ and $z^*(\mathcal{C}^{\mathcal{T}})$ for a larger subset of MIPLIB 3.0 instances, including general mixed integer problems. We compare these values to estimates of $z^*(\mathcal{M}^{\mathcal{F}})$ and $z^*(\mathcal{M}^{\mathcal{T}})$ (i.e., the bounds obtained by using MIR inequalities) and attempt to address the well acknowledged observation that it is difficult to identify classes of knapsack inequalities which systematically outperform the MIR inequality in broad test sets. Recently, Dash and Günlük [15] also try to analyze this issue in terms of cuts from the cyclic group problem.

The organization of this paper is as follows. In the next section, we discuss how to solve the problem of separating over a single mixed integer knapsack set. This methodology described requires the use of a subroutine for solving the mixed integer knapsack problem. An algorithm for solving this problem is discussed in Sect. 3. Computational results are presented in Sect. 4, while final remarks and a discussion ensues in Sect. 5.

2   Identifying Violated Knapsack Cuts

Consider $x^* \in \mathbb{R}^n$ and a mixed integer knapsack set K. In this section we address the following questions: Is $x^* \in conv(K)$? If not, can we find an inequality $\pi x \le \pi_o$ which is valid for K, and such that $\pi x^* > \pi_o$?

We assume that K has no free variables, since it is easy to substitute a free variables by two non-negative variables. Let $\{x^1, x^2, \ldots, x^q\}$ and $\{r^1, r^2, \ldots, r^t\}$ represent the extreme points and extreme rays of $conv(K)$. The following proposition, which follows from the work of Applegate et. al [1], allows us to address this question.

Proposition 1. *Consider the following linear programming (LP) problem with variables* $u, v, \pi \in \mathbb{R}^n$, *and* $\pi_o \in \mathbb{R}$:

$$LP_1 : \min \sum_{i=1}^{n}(u_i + v_i)$$

$$s.t.$$

$$
\begin{array}{llll}
\pi x^k - \pi_o & \leq 0 & \forall k = 1 \ldots q & (C1) \\
\pi r^k & \leq 0 & \forall k = 1 \ldots t & (C2) \\
\pi x^* - \pi_o & = 1 & & (C3) \\
\pi + u - v & = 0 & & (C4) \\
u \geq 0, v \geq 0. & & &
\end{array}
$$

If this problem is infeasible, then $x^* \in conv(K)$, *and thus there exists no knapsack cut violated by* x^*. *Otherwise, this problem admits an optimal solution* (u, v, π, π_o) *such that inequality* $\pi x \leq \pi_o$ *is a valid knapsack cut maximizing:*

$$\frac{\pi x^* - \pi_o}{||\pi||_1}$$

That is, the hyperplane defined by (π, π_o) *maximizes the* L_1 *distance to* x^*.

Because LP_1 has an exponential number of constraints, we use a dynamic cut generation algorithm to solve the problem. We begin with constraints $(C3)-(C4)$ and a subset of constraints $(C1) - (C2)$. The cut generation algorithm requires solving the problem $\max\{\pi x : x \in K\}$ at each iteration. If this problem is unbounded at any given iteration, then there exits an extreme ray r^j of $conv(K)$ such that $\pi r^j > 0$. That is, we have identified a violated constraint. If this problem is not unbounded, then there exists an optimal solution corresponding to an extreme point x^k of $conv(K)$. If $\pi x^k > \pi_o$ then we have found a violated constraint. Otherwise, it means that all constraints of the problem are satisfied. Solving the oracle problem is discussed in Sect. 3.

Notice that in general, it is not possible to assure that the solution of $\max\{\pi x : x \in K\}$ given by the oracle will correspond to an extreme point or ray of $conv(K)$. However, constraints $(C1) - (C2)$ can be re-defined in terms of all points/rays of K without affecting the correctness of Proposition 1. Even though this would result in an infinite number of constraints, under very mild assumptions [17], the dynamic cut generation algorithm will still converge in a finite number of iterations.

In order to speed up the solution of LP_1 we make use of certain characterizations of violated knapsack cuts.

Let $K = \{x \in \mathbb{R}^n : ax \leq b, \ l \leq x \leq u, \ x_i \in \mathbb{Z}, \ \forall i \in I\}$. We may assume without loss of generality [21] that the bound constraints are tight. Say that a knapsack cut for K is *trivial* if it is implied by the linear programming relaxation of K. A proof of the following result concerning non-trivial knapsack cuts can be found in Atamtürk [3].

Proposition 2. *Every non-trivial facet-defining knapsack cut $\pi x \leq \pi_o$ of $conv(K)$ satisfies the following properties:*

(i) *If $a_i > 0$, $\pi_i \geq 0$*
(ii) *If $a_i < 0$, $\pi_i \leq 0$*
(iii) *$\pi_i = 0$ for all $i \notin I$ such that $a_i > 0$ and $u_i = +\infty$.*
(iv) *$\pi_i = 0$ for all $i \notin I$ such that $a_i < 0$ and $l_i = -\infty$.*
(v) *There exists a constant $\alpha > 0$ such that $\pi_i = \alpha a_i$ for all $i \notin I$ such that $a_i > 0$ and $l_i = -\infty$, and for all $i \notin I$ such that $a_i < 0$ and $u_i = +\infty$.*

The following result concerning violated and non-trivial knapsack cuts is a simple generalization of a technique employed in Boyd [8].

Proposition 3. *Consider $x^* \notin conv(K)$. Let $H^+ = \{i \in N : a_i > 0, x_i^* = l_i\}$ and $H^- = \{i \in N : a_i < 0, x_i^* = u_i\}$. If there does not exist a trivial inequality separating x^* from $conv(K)$, then there exists a knapsack cut $\pi x \leq \pi_o$ such that $\pi_i = 0, \forall i \in H^+ \cup H^-$.*

We make use of Propositions 2 – 3 in the following way: We restrict the signs of coefficients according to Proposition 2 items (i) and (ii). Coefficients π_i with $i = 1, \ldots, n$ which can be assumed to be zero are eliminated from LP_1. Further, a single variable is used for all coefficients π_i with $i = 1, \ldots, n$ for which we know that $\pi_i = \alpha a_i$. Note that this last reduction is equivalent to aggregating the unbounded continuous variables into a single variable.

Two other techniques are used to speed up the separation process. The first one uses the fact that MIR inequalities are knapsack cuts. With that in mind, we first apply an MIR separation heuristic to try to find violated knapsack cuts and only use the above separation procedure if the MIR heuristic fails.

The other technique relies on the following simple observation. Let $U = \{i \in N : x_i^* = u_i\}$ and $L = \{i \in N : x_i^* = l_i\}$. If we define,

$$K^* = K \cap \{x : x_i = u_i \; \forall i \in U\} \cap \{x : x_i = l_i \; \forall i \in L\},$$

we know that $x^* \in conv(K)$ iff $x^* \in conv(K^*)$. Thus, answering the question: "Is $x^* \in conv(K)$?" can be done in a space of usually much smaller dimension by testing instead if $x^* \in conv(K^*)$.

If our test shows that $x^* \in conv(K^*)$, we are done with the separation since we know that in this case $x^* \in conv(K)$. However, if $x^* \notin conv(K^*)$ we still need to get a cut separating x^* from $conv(K)$ and thus we have to run our separation algorithm in the original space. Notice, however, that if $x^* \notin conv(K^*)$, our separation algorithm will return a cut separating x^* from $conv(K^*)$, so one could potentially lift this cut to obtain a cut separating x^* from $conv(K)$. We have not implemented this feature yet, but we expect that it will significantly speed up our algorithm.

To summarize, we outline the complete algorithm below:

Algorithm 1. Outline of knapsack separation process

Input: x^* and K
Output: $x^* \in conv(K)$ or a cut separating x^* from $conv(K)$

begin
    Run the MIR separation heuristic
    if *cut found* **then**
        | **return** the MIR cut separating x^* from $conv(K)$
    else
        Apply Propositions 1 and 2 to simplify LP_1
        Solve LP_1 in a reduced space to separate x^* from $conv(K^*)$
        if $x^* \in conv(K^*)$ **then**
            | **return** $x^* \in conv(K)$
        else
            Solve LP_1 in the original variable space to separate x^* from $conv(K)$

end

---

3   Solving the Mixed Integer Knapsack Problem

In this section we are concerned with the problem of solving the Mixed Integer Knapsack Problem (MIKP),

$$\max\{cx : x \in K\} \tag{1}$$

We will assume that the problem is feasible, and are interested in either (a) proving that the problem is unbounded by finding an extreme ray r^* of $conv(K)$, or (b) computing the optimal value of the problem by finding the optimal solution $x^* \in K$.

Variants of MIKP have long been studied in the research literature. In these it is typically assumed that all coefficients defining the problem are integer, that all variables must take integer values (i.e. no continuous variables are allowed), and that $l_i = 0$ for all $i = 1, \ldots, n$. In addition: In the Knapsack Problem (KP) $u_i = 1$ for all $i = 1, \ldots, n$, in the Bounded Knapsack Problem (BKP) $u_i < \infty$ for all $i = 1, \ldots, n$, and in the Unbounded Knapsack Problem (UKP) $u_i = \infty$ for all $i = 1, \ldots, n$. Most modern algorithms for solving KP, BKP, and UKP are based either on branch and bound (following the work of Horowitz and Sahni [24]) and on dynamic programming (following the work of Bellman [6]). However, the most efficient codes seldom make explicit use of Linear Programming and in addition, they never consider the use of both integer and continuous variables. For excellent surveys describing the rich literature on this topic, the reader is advised to consult Kellerer et al [25] and Martello and Toth [27].

While it is reasonable to expect that many of these algorithms could be adapted for solving our general case with a mix of continuous, integer, bounded and unbounded variables, the fact that they are designed to work with integer co-efficients raises certain concerns with regards to the application discussed in this paper. In fact, part of our motivation is to study the efficacy of cuts derived from tableau rows. However, these rows are rarely are made up of integer coefficients, and whats more, they are typically very ill conditioned. Thus, scaling them so as to obtain integers may result in extremely large numbers. Considering this important shortcoming, and the need to further study these algorithms in order to account for the mixed use of bounded, unbounded, continuous and integer variables, our approach has been to pursue an LP-based branch and bound approach, which seems naturally suited to mixed integer programming problems. This issue, however, is one which merits further research. In what follows we describe our algorithm for solving MIKP.

Detecting Unbounded Solutions

For each $i \in 1, \ldots, n$ define the *efficiency* of variable x_i as $e_i = c_i/a_i$ if $a_i \neq 0$, as $e_i = +\infty$ if $a_i = 0$ and $c_i > 0$, and as $e_i = -\infty$ if $a_i = 0$ and $c_i < 0$. In addition, we say that x_i is a *potentiator* if,

$$(a_i \leq 0, \ c_i > 0, \ u_i = +\infty) \text{ or } (a_i \geq 0, \ c_i < 0, \ l_i = -\infty).$$

We say that x_i is an *incrementor* if,

$$(a_i > 0, \ c_i > 0, \ u_i = +\infty) \text{ or } (a_i < 0, \ c_i < 0, \ l_i = -\infty).$$

We say that x_i is a *decrementor* if,

$$(a_i > 0, \ c_i \geq 0, \ l_i = -\infty) \text{ or } (a_i < 0, \ c_i \leq 0, \ u_i = +\infty).$$

By identifying a potentiator, or instead, by identifying the most efficient incrementor and the least efficient decrementor, it is possible to easily establish if a problem is unbounded, as shown by the following Proposition:

Proposition 4. *MIKP is unbounded if and only if one of the following conditions hold,*

- *MIKP admits a potentiator x_j.*
- *MIKP admits an incrementor x_i and a decrementor x_j such that $e_i > e_j$.*

Note that Proposition 4 implies that it can be determined if MIKP is unbounded in linear time. Note also that once the potentiator, or instead, the incrementor and decrementor have been identified, it is easy to construct an extreme ray of $conv(K)$.

Preprocessing

We consider the following four-step preprocessing algorithm (see [21],[29]) which assumes the problem is not unbounded.

1. Fix to u_i all variables x_i such that $c_i \geq 0$ and $a_i \leq 0$. Fix to l_i to all variables x_i such that $c_i \leq 0$ and $a_i \geq 0$.

2. Make all bounds as tight as possible.
3. Aggregate variables. If two variables x_i, x_j of the same type (integer or continuous) are such that $a_i = a_j$ and $c_i = c_j$ aggregate them into a new variable x_k of the same type such that $a_k = a_i = a_j$, $c_k = c_i = c_j$, $l_k = l_i + l_j$ and $u_k = u_i + u_j$.
4. Sort variables in order of decreasing efficiency. Break ties checking for variable types (integer or continuous).

Branch and Bound

We use a depth-first-search branch and bound algorithm which always branches on the unique fractional variable. We use a simple linear programming algorithm, a variation of Dantzig's algorithm [13] , which runs in linear time by taking advantage of the fact that variables are sorted by decreasing efficiency. We do not use any cutting planes in the algorithm, nor any heuristics to generate feasible solutions. The algorithm uses variable reduced-cost information to improve variable bounds at each node of the tree.

Domination

Consider x^1 and x^2, two feasible solutions of MIKP. We say that x^1 *cost-dominates* x^2 if $cx^1 > cx^2$ and $ax^1 \leq ax^2$. On the other hand, we say that x^1 *lexicographically-dominates* x^2 if $cx^1 = cx^2$ and $ax^1 \leq ax^2$, and if in addition, there exists $i \in 1, \ldots, n$ such that $x_i^1 < x_i^2$ and $x_k^1 = x_k^2$, $\forall k \in 1, \ldots, (i-1)$. We say that a solution is *dominated* if it is cost-dominated or lexicographically-dominated. Observe that there exists a unique non-dominated optimal solution (or none at all).

Traditional branch and bound algorithms work by pruning nodes when (a) they are proven infeasible, or (b) when it can be shown that the optimal solution in those nodes has value worse than a bound previously obtained. In our implementation, we additionally prune nodes when (c) it can be shown that every optimal solution in those nodes is dominated.

Using dominance to improve the branch and bound search can have an important impact on the effectiveness of the search. In fact, lexicographic and cost dominance allow us to disregard feasible solutions that are not the unique lexicographically smallest optimum solution, hence significantly reducing the search space.

In general, the problem of detecting if a solution is dominated can be extremely difficult. In what follows we describe a simple methodology for identifying specific cases of domination.

Consider indices $i, j \in I$, and non-zero integers k_i, k_j. If $a_i k_i + a_j k_j \geq 0$ and $c_i k_i + c_j k_j < 0$ we say that (i, j, k_i, k_j) defines an integer cost-domination tuple. If $k_i \geq 0$, $a_i k_i + a_j k_j \geq 0$ and $c_i k_i + c_j k_j = 0$ we say that (i, j, k_i, k_j) defines an integer lexicographic-domination tuple. Observe that whenever (c_i, a_i) and (c_j, a_j) are linearly independent there exist an infinite amount of cost-domination pairs. Likewise, there exist an infinite amount of lexicographic-domination tuples in the linear dependence case. However, in each case, there always exists a *minimal domination tuple*. That is, a domination tuple (i, j, k_i, k_j) such that all other domination tuples (i, j, k_i', k_j') defined for the same variables, satisfy $|k_i| \leq |k_i'|$ and $|k_j| \leq |k_j'|$. The propositions below show how domination tuples allow for the easy identification of dominated solutions.

Proposition 5. *Consider an integer cost-domination tuple (i, j, k_i, k_j) and let x be a feasible MIKP solution. If any of the following three conditions hold:*

- $k_i > 0$, $k_j > 0$, $x_i \geq l_i + k_i$ and $x_j \geq l_j + k_j$,
- $k_i < 0$, $k_j > 0$, $x_i \leq u_i + k_i$ and $x_j \geq l_j + k_j$,
- $k_i < 0$, $k_j < 0$, $x_i \leq u_i + k_i$ and $x_j \leq u_j + k_j$.

Then x is cost-dominated.

Proposition 6. *Consider an integer lexicographic-domination tuple (i, j, k_i, k_j) and let x be a feasible MIKP solution. If either of the following conditions hold:*

- $k_j > 0$, $x_i \geq l_i + k_i$, and $x_j \geq l_j + k_j$,
- $k_j < 0$, $x_i \geq l_i + k_i$, and $x_j \leq u_j + k_j$,

then x is lexicographically-dominated.

To see that these propositions are true, it is simply a matter of observing that if the conditions hold for a feasible x, then defining x' so that $x_i' = x_i - k_i$, $x_j' = x_j - k_j$ and $x_k' = x_k$ for $k \neq i, j$, we have x' is feasible and dominates x.

The following propositions illustrate how domination tuples can be used to strengthen branch and bound algorithm. This is achieved by preventing nodes with dominated solutions from being created through additional enforced bound changes.

Proposition 7. *Consider two integer type variables x_i and x_j and a domination tuple (i, j, k_i, k_j) such that $k_i > 0$. If in some node of the branch and bound tree we impose $x_i \geq l_i + \alpha_i$, where $\alpha_i \geq k_i$, then:*

- *If $k_j > 0$ we can impose the constraint $x_j \leq l_j + k_j - 1$ in that node.*
- *If $k_j < 0$ we can impose the constraint $x_j \geq u_j + k_j + 1$ in that node.*

The case $k_i < 0$ is analogous.

In order to use the above propositions in the branch and bound algorithm we compute what we call a *domination table* before initiating the solve. This table is defined as a list of all possible (minimal) domination tuples. Observe that we only need store domination tuples (i, j, k_i, k_j) such that $|k_i| \leq (u_i - l_i)$ and $|k_j| \leq (u_j - l_j)$. In order to compute domination tuples we perform a simple enumeration algorithm which uses bounds to identify where to start and stop the enumerations.

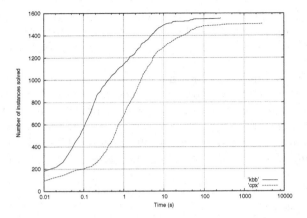

Fig. 1. Histogram comparing KBB algorithm with CPLEX

4   Computational Experiments

In this section, our computational experiments are described. All implementations were compiled using the "C" and "C++" programming languages, using the Linux operating system (v2.4.27) and Intel Xeon dual-processor computers (2GB of RAM, at 2.66GHz). Since generating cuts which are invalid is a real point of concern, we found it appropriate to use the exact arithmetic, both for solving LP_1, and for the MIKP oracle. Thus, we used Applegate et al. [2] exact LP solver for LP_1, and the GNU Multiple Precision (GMP) Arithmetic library [22] to implement the MIKP algorithm.

4.1   The Optimization Oracle

We first compare the performance of our MIKP algorithm ("kbb") with the performance of CPLEX 9.0 ("cpx"), the only alternative for MIKP we know of to date. Note that CPLEX was ran with all its default settings, except for the tolerance, which was set to 10^{-6}. Note also that our MIKP algorithm was ran using double floating arithmetic, with a tolerance of 10^{-6}.

In our first implementation of the separation algorithm we had incorporated a version of kbb which did not use domination branching. We quickly realized that this algorithm was not efficient enough. When running this version of the code, we saved all problems which took our algorithm more than 2.0 seconds to solve. These are the 1,556 problems that we now use to compare cpx with the full version of kbb. It is important to note that by the nature of the way these instances were generated, there might be some of instances that are very similar to each other.

In Fig. 1 we present a histogram summarizing the running times of kbb and cpx. Each point in the curves represents the number of instances which were solved within a given maximum time. For instance, note that the number of instances solved to optimality by kbb within a second is roughly 1150, whereas the number of instances solved to optimality by cpx is roughly 700. Note that the time is represented in logarithmic scale. Further, observe that the hardest instance for kbb takes several hundred seconds – roughly ten times less than the hardest instance for cpx.

It is clear from this histogram that the kbb algorithm outperforms cpx in the instance set. Note that this does not necessarily mean that kbb solves every instance faster than cpx, but rather, that cumulatively, kbb performs better. In fact, on average, kbb takes 81% less time than cpx, and explores 37.5% less branch-and-bound nodes. Moreover, in 49 instances, CPLEX fails to find the optimum solution since it runs out of memory after creating too large a branch and bound tree.

4.2   Knapsack Cuts

We next use an implementation of the algorithms presented in Sect. 2 and Sect. 3 to compare the practical performance of MIR cuts against the performance of separating all possible knapsack cuts. As detailed in Sect. 1, given a family $\mathcal{K}$ of knapsack sets implied by P, such a comparison can be made by comparing the values $z^*(\mathcal{C}^\mathcal{K})$ and $z^*(\mathcal{M}^\mathcal{K})$. In this article we only consider the set $\mathcal{K} = \mathcal{F}$, i.e., the family of knapsack sets induced by the original formulation rows, and the set $\mathcal{K} = \mathcal{T}$, i.e., the family of knapsack sets induced by the simplex tableau rows of the optimal LP solution for the original LP relaxation.

Computing $z^*(\mathcal{M}^\mathcal{K})$ is NP-hard [9], so instead we approximate this value using an MIR separation heuristic. Given a point x^*, for every $K \in \mathcal{K}$ we try to find MIR inequalities that are violated by x^*. We add these inequalities to the LP relaxation of P and repeat the process until no more MIR inequalities are found. The MIR inequalities for each K are derived by a separation heuristic which combines scaling and variable complementation techniques (for details see [21], [26], and [14]). Denote by $z_M^\mathcal{K}$ the objective function value at the end of the procedure. Since this is just a heuristic, after completing a run, there may be violated MIR inequalities which have not been identified. Therefore $z_M^\mathcal{K}$ should be considered an estimate of $z^*(\mathcal{M}^\mathcal{K})$.

Note that though the MIR separation problem is NP-hard, one could use the approaches of Balas and Saxena [5] or Dash, Günlük and Lodi [16] to better approximate $z^*(\mathcal{M}^\mathcal{K})$.

To compute $z^*(\mathcal{C}^\mathcal{K})$, we proceed as follows. Given a fractional solution, we loop through all of the mixed integer knapsack sets $K \in \mathcal{K}$. For each of these we invoke the procedure outlined in Sect. 2 and identify a violated cut if such exists. After completing this loop we add the cuts to the problem and repeat. The procedure ends when for every K we can prove that there is no violated knapsack cut.

Computational tests are performed on all MIPLIB 3.0 instances using the mixed integer knapsack sets $\mathcal{K} = \mathcal{F}$ and $\mathcal{K} = \mathcal{T}$. For each problem instance let z_{UB}^* represent the value of the optimal (or best known) solution and z_{LP}^* the LP relaxation value. For each set $\mathcal{K}$ and each instance we compute the following performance measures:

LP-PERF: Performance of the original LP formulation. That is, the value of the LP relaxation gap:

$$\frac{z^*_{UB} - z^*_{LP}}{|z^*_{UB}|}.$$

KNAP-PERF: Performance of the knapsack cuts. That is, how much of the LP gap was closed by the knapsack cuts:

$$\frac{z^*(\mathcal{C}^\mathcal{K}) - z^*_{LP}}{z^*_{UB} - z^*_{LP}}.$$

MIR-PERF: Performance of MIR separation heuristic. That is, how much of the LP gap closed by the knapsack cuts was closed by the MIR cuts:

$$\frac{z^\mathcal{K}_M - z^*_{LP}}{z^*(\mathcal{C}^\mathcal{K}) - z^*_{LP}}$$

Knapsack Cuts Derived from Formulation Rows

In this section we analyze the performance of knapsack and MIR inequalities on formulation rows of MIPLIB 3.0 instances. Results are summarized in Table 1. Of the 59 instances in the library, we eliminated eight instances which were unfinished at the time of writing the article (arki001, cap6000, dano3mip, harp2, mitre, mod008, pk1 and rout), three for which LP-PERF was equal to 0.0 (dsbmip, enigma, and noswot), and thirty two for which KNAP-PERF and MIR-PERF were both equal to 0.0.

Table 1. Benchmarks for Formulation Closure

Instance	LP-PERF	KNAP-PERF	MIR-PERF
fiber	61.55%	93.82%	97.06 %
gen	0.16%	99.78%	100.00 %
gesa2	1.18%	71.03%	98.48 %
gesa3	0.56%	49.33%	96.90 %
gt2	36.41%	94.52%	97.93 %
l152lav	1.39%	1.36%	0.41 %
lseu	25.48%	76.09%	88.25 %
mod010	0.24%	18.34%	100.00 %
p0033	18.40%	87.42%	87.31 %
p0201	9.72%	33.78%	100.00 %
p0282	31.56%	98.59%	95.42 %
p0548	96.37%	84.34%	62.76 %
p2756	13.93%	86.35%	51.49 %
qnet1	10.95%	89.06%	56.68 %
qnet1_o	24.54%	95.12%	88.65 %
rgn	40.63%	57.49%	100.00 %

First, note that knapsack cuts alone can considerably close the remaining LP gap in some problems (column KNAP-PERF). In fact, in 9 problems out of the 16 problems in which knapsack cuts improved the gap, over 84% of the gap was closed, and in 14 out of 16 problems, over 50 % of the gap was closed. On average, the GAP closed by the knapsack cuts among these 16 instances is around 71%. It is interesting, however, that in thirty two instances knapsack cuts should do nothing to improve the gap. If in addition we consider in our average the thirty two instances for which KNAP-PERF is 0.0%, this drops to 23.66%.

Second, consider the column MIR in which we can get an idea of how well the mixed integer rounding cut closure compares to the knapsack cut closure. Observe that of the 16 problems, in 12 of them, by using the MIR cuts alone, we close over 87% of the GAP closed by the knapsack cuts. This indicates that MIR inequalities are a very important subset of knapsack inequalities; at least for the instances considered. A natural question is the following: How much could we improve the value of MIR-PERF if we used an exact MIR separation algorithm as opposed to a heuristic? In an attempt to answer this question we fine-tuned the settings of the MIR heuristic for the problems p0033 and qnet1. In these, we managed to improve the value of MIR-PERF from 87.31% to 100% and from 56.68% to 77.27% respectively.

Knapsack Cuts Derived from Tableau Rows

In this section we analyze the performance of knapsack and MIR inequalities on tableau rows of MIPLIB 3.0 instances. For this we compute z_{LP}^* and store the tableau rows in the set of knapsack polyhedra $\mathcal{K} = \mathcal{T}$, which we use for all subsequent computations. Results are summarized in Table 2. Of the 59 instances in the library, we eliminated thirty two instances which were unfinished at the time of writing the article, three for which LP-PERF was equal to 0.0 (dsbmip, enigma, and noswot), and two for which KNAP-PERF and MIR-PERF were both equal to 0.0 (stein27 and stein45).

First, it is important to remark that separating knapsack cuts from tableau rows is considerable more difficult than separating knapsack cuts from original formulation rows. This is due to several reasons: Tableau rows are typically much more dense, coefficients tend to be numerically very bad, and rows tend to have a lot of continuous variables. This added difficulty is reflected in the fact that out of 59 instances, in two days of runs we just managed to solve 24 instances to completion, as opposed to the 48 which we solved when considering formulation rows.

Second, it is interesting to note that the value KNAP-PERF is very erratic, uniformly ranging in values from 100% to 0.0%. In contrast to the case of formulation rows, only two instances are such that KNAP-PERF is 0.0%.

The last, and perhaps most startling observation, is that the MIR-PERF is always at 100%, if not very close. If this result were true in general, it would be very surprising. However, because there are still thirty two instances which have not been solved one must be very careful. Because of the way in which we computed these numbers, it could be the case that those instances with MIR-PERF close to 100% are easier for our methodology to solve. It is very reasonable

Table 2. Benchmarks for Tableau Closure

Instance	LP-PERF	KNAP-PERF	MIR-PERF
air03	0.38 %	100.00 %	100.00%
bell3a	1.80 %	60.15 %	100.00%
bell5	3.99 %	14.68 %	98.94%
dcmulti	2.24 %	50.49 %	99.94%
egout	73.67 %	55.33 %	100.00%
fixnet6	69.85 %	11.08 %	100.00%
flugpl	2.86 %	11.74 %	100.00%
gesa2	1.18 %	28.13 %	99.98%
gesa2_o	1.18 %	29.65 %	99.67%
khb05250	10.31 %	75.14 %	100.00%
misc03	43.15 %	7.24 %	100.00%
misc06	0.07 %	26.98 %	100.00%
misc07	49.64 %	0.72 %	100.00%
modglob	1.49 %	18.05 %	100.00%
p0033	18.40 %	74.71 %	100.00%
p0201	9.72 %	34.36 %	100.00%
pp08a	62.61 %	50.97 %	100.00%
qiu	601.15 %	3.47 %	100.00%
rgn	40.63 %	9.78 %	100.00%
set1ch	41.31 %	39.18 %	100.00%
vpm1	22.92 %	49.09 %	96.30%
vpm2	28.08 %	19.39 %	98.85%

to expect that instances with MIR-PERF well below 100% are more difficult to solve as they require more iterations of the knapsack separation algorithm as opposed to iterations of the MIR separation heuristic.

5   Final Remarks

It is important to note that these results are very preliminary. We put great care into ensuring that the generated cuts are valid and that the procedure runs correctly, but this makes the methodology very slow. For example, some of the KNAP-PERF values computed took as much as 5 days to obtain. Some of the unsolved instances have been ran for over a week without a final answer being reported. We are currently developing further techniques by which these computations can be accelerated. Part of the difficulty arises from the fact that exact arithmetic is being employed. In average, we have observed that performing exact arithmetic computations take 100 times longer than floating point arithmetic computations.

One of the main goals of this study has been to assess the overall effectiveness of MIR inequalities relative to knapsack cuts. The motivation being the empirical observation that though much research has been conducted studying inequalities derived from single row systems, no such class of inequalities has been able to systematically improve upon the performance of MIRs. In this regard, the results we present are surprising. We observe that in most test problems, the bound obtained by optimizing over the MIR closure is very similar in value (if not equal) to the bound obtained optimizing over the knapsack closure. Though it is important to note that this observation is limited in the number of test problems considered, it does help explain the lack of success in generating other cuts from tableau and formulation rows, and, suggests that for further bound improvements we might have to consider new row aggregation schemes, or cuts derived from multiple row systems.

References

1. D. Applegate, R. E. Bixby, V. Chvátal, and W. Cook. TSP cuts which do not conform to the template paradigm. In *Computational Combinatorial Optimization, Optimal or Provably Near-Optimal Solutions [based on a Spring School]*, pages 261–304, London, UK, 2001. Springer-Verlag GmbH.
2. D. Applegate, W. Cook, S. Dash, and D. Espinoza. Exact solutions to linear programming problems. *Submitted to Operations Research Letters*, 2006.
3. A. Atamtürk. On the facets of the mixed–integer knapsack polyhedron. *Mathematical Programming*, 98:145–175, 2003.
4. E. Balas and M. Perregaard. A precise correspondence between lift-and-project cuts, simple disjuntive cuts, and mixed integer Gomory cuts for 0-1 programming. *Mathematical Programming*, 94:221–245, 2003.
5. E. Balas and A. Saxena. Optimizing over the split closure. *Mathematical Programming*, To appear.
6. R. E. Bellman. *Dynamic Programming*. Princeton University Press, 1957.
7. R. E. Bixby, S. Ceria, C. M. McZeal, and M. W. P Savelsbergh. An updated mixed integer programming library: MIPLIB 3.0. *Optima*, (58):12–15, June 1998.
8. A. E. Boyd. Fenchel cutting planes for integer programs. *Operations Research*, 42:53–64, 1992.
9. A. Caprara and A. Letchford. On the separation of split cuts and related inequalities. *Mathematical Programming*, 94(2-3):279–294, 2003.
10. W. Cook, R. Kannan, and A. Schrijver. Chvátal closures for mixed integer programming problems. *Mathematical Programming*, 47:155–174, 1990.
11. G. Cornuéjols, Y. Li, and D. Vanderbussche. K-cuts: A variation of gomory mixed integer cuts from the LP tableau. *Informs Journal On Computing*, 15:385–396, 2003.
12. H. Crowder, E.L. Johnson, and M. Padberg. Solving large-scale zero-one linear-programming problems. *Operations Research*, 31:803–834, 1983.
13. G. B. Dantzig. Discrete variable extremum problems. *Operations Research*, 5(2):266–277, 1957.
14. S. Dash, M. Goycoolea, and O. Günlük. Two-step mir inequalities for mixed-integer programs. *Optimization Online*, Jul 2006.
15. S. Dash and O. Günlük. On the strength of gomory mixed-integer cuts as group cuts. *IBM research report RC23967*, 2006.
16. S. Dash, O. Günlük, and A. Lodi. MIR closures of polyhedral sets. Available online at http://www.optimization-online.org/DB_HTML/2007/03/1604.html.

17. D. G. Espinoza. *On Linear Programming, Integer Programming and Cutting Planes*. PhD thesis, School of Industrial and Systems Engineering, Georgia Institute of Technology, March 2006.

18. M. Fischetti and A. Lodi. On the knapsack closure of 0-1 integer linear problems. Presentation at 10th International Workshop on Combinatorial Optimization, Aussois (2006). Available at http://www-id.imag.fr/IWCO2006/slides/Fischetti.pdf.

19. R. E. Gomory. Early integer programming (reprinted). *Operations Research*, 50: 78–81, Jan 2002.

20. R. E. Gomory and E.L. Johnson. Some continuous functions related to corner polyhedra I. *Mathematical Programming*, 3:23–85, 1972.

21. M. Goycoolea. *Cutting Planes for Large Mixed Integer Programming Models*. PhD thesis, Georgia Institute of Technology, 2006.

22. T. Granlund. The GNU multiple precision arithmetic library. Available on-line at http://www.swox.com/gmp/.

23. Z. Gu, G. L. Nemhauser, and M. W. P. Savelsbergh. Lifted cover inequalities for 0-1 integer programs: Computation. *INFORMS Journal on Computing*, 10:427–437, 1998.

24. E. Horowitz and S. Sahni. Computing partitions with applications to the knapsack problem. *Journal of the ACM*, 21:277–292, 1974.

25. H. Kellerer, U. Pferschy, and D. Pisinger. *Knapsack Problems*. Springer, Berlin, Germany, 2004.

26. H. Marchand and L.A. Wolsey. Aggregation and mixed integer rounding to solve MIPs. *Operations Research*, 49:363–371, 2001.

27. S. Martello and P. Toth. *Knapsack Problems: Algorithms and Computer Implementations*. J. Wiley, New York, 1990.

28. G. L. Nemhauser and L. A. Wolsey. A recursive procedure for generating all cuts for 0-1 mixed integer programs. *Mathematical Programming*, 46:379–390, 1990.

29. M.W.P. Savelsbergh. Preprocessing and probing for mixed integer programming problems. *ORSA Journal on Computing*, 6:445–454, 1994.

30. X. Q. Yan and E. A. Boyd. Cutting planes for mixed-integer knapsack polyhedra. *Mathematical Programming*, 81:257–262, 1998.

A Faster Strongly Polynomial Time Algorithm for Submodular Function Minimization

James B. Orlin

Sloan School of Management, MIT
Cambridge, MA 02139
jorlin@mit.edu

Abstract. We consider the problem of minimizing a submodular function f defined on a set V with n elements. We give a combinatorial algorithm that runs in $O(n^5 \, EO + n^6)$ time, where EO is the time to evaluate $f(S)$ for some $S \subseteq V$. This improves the previous best strongly polynomial running time by more than a factor of n.

1 Introduction

Let $V = \{1, 2, \ldots, n\}$. A set function f on V is said to be *submodular* if the following is true:

$$f(X) + f(Y) \geq f(X \cup Y) + f(X \cap Y) \quad \text{for all subsets } X, Y \subseteq V. \tag{1}$$

Here we consider the problem of Submodular Function Minimization (SFM), that is, determining a subset $S \subseteq V$ that minimizes $f(\ )$. Our contribution is to develop a strongly polynomial time algorithm for SFM that improves upon the best previous time bounds by a factor greater than n.

For a given subset $X \subseteq V$, let $f_X(Y) = f(X \cup Y) - f(X)$. It is elementary and well known that for fixed X, the function $f_X(\ )$ is submodular whenever $f(\ )$ is submodular. An equivalent way of defining submodularity is as follows.

For all subsets X, Y of V, and for each element $v \notin (X \cup Y)$, if $X \subseteq Y$ then $f_Y(v) \leq f_X(v)$.

In this way, submodular functions model decreasing marginal returns, and are economic counterparts of concave functions. Nevertheless, Lovasz [11] showed that they behave algorithmically more similarly to convex functions, and provided analysis on why this is true.

Examples of submodular functions include cut capacity functions, matroid rank functions, and entropy functions. For additional examples of submodular functions and for applications of SFM see McCormick [12], Fleischer [4], Fushishige [6], and Schrijver [14].

We assume without loss of generality that $f(\varnothing) = 0$. Otherwise, if $f(\varnothing) \neq 0$, we can subtract $f(\varnothing)$ from $f(S)$ for all $S \subseteq V$.

Grotschel, Lovasz, and Schrijver [7] and [8] gave the first polynomial time and strongly polynomial time algorithms for minimizing a submodular function. Their

M. Fischetti and D.P. Williamson (Eds.): IPCO 2007, LNCS 4513, pp. 240–251, 2007.

algorithms rely on the ellipsoid algorithm. Schrijver [13] and Iwata, Fleischer, and Fujishige [10] independently developed strongly polynomial time combinatorial algorithms for minimizing a submodular function. Both algorithms build on the work of Cunningham [1], who developed a pseudo-polynomial time algorithm for minimizing a submodular function.

Let EO be the maximum amount of time it takes to evaluate $f(S)$ for a subset $S \subseteq V$. EO stands for evaluation of the oracle function, as per McCormick [12]. In general, one expects EO to be at least n since the input size is $\Omega(n)$; however, this running time can sometimes be improved in an amortized sense if one is evaluating EO multiple times consecutively, as is done by many of the SFM algorithms including the one presented here. Let M be an upper bound on $|f(S)|$ for all $S \subseteq V$.

The running times of the algorithms of Schrijver [13] and Iwata, Fleischer, and Fujishige [10] were shown to be $O(n^8 \text{ EO} + n^9)$. Fleischer and Iwata [5] improved the running time of the combinatorial algorithms to $O(n^7 \text{ EO} + n^8)$. Vygen [15] showed that the running time of Schrijver's original algorithm was also $O(n^7 \text{ EO} + n^8)$. Subsequently Iwata [9] developed a scaling based algorithm whose running time is $O(n^4 \text{ EO} \log M + n^5 \log M)$. To date, the best strongly polynomial time combinatorial algorithm for SFM was the strongly polynomial version of Iwata's algorithm, which runs in $O((n^6 \text{ EO} + n^7) \log n)$ time.

We present a new approach for solving submodular minimization. As have previous approaches, our algorithm relies on expressing feasible points in the base polyhedron as a convex combination of extreme points. However, our algorithm works directly with vectors of the base polyhedron rather than relying on an auxiliary network, or on augmenting paths, or on flows.

We present a strongly polynomial time algorithm that runs in $O(n^5 \text{ EO} + n^6)$ steps, thus improving upon Iwata's time bound by a factor of $n \log n$. This also improves upon the best strongly polynomial time implementation of the ellipsoid algorithm for SFM, which runs in $\tilde{O}(n^5 \text{ EO} + n^7)$ as reported by McCormick [12], where $\tilde{O}$ indicates that factors of $\log n$ may have been omitted from the time bound. Most of the proofs in this manuscript are omitted. A complete draft including the proofs is available on the author's website.

2 The Base Polyhedron

For a vector $x \in \mathbb{R}^{|V|}$, let $x(v)$ denote the v-th component. We let $x^-(v) = \min\{0, x(v)\}$. For a subset $S \subseteq V$, , we let $x(S) = \sum_{v \in S} x(v)$.

The *base polyhedron* is

$$B(f) = \{x \mid x \in \mathbb{R}^n, x(V) = f(V), \forall S \subseteq V : x(S) \le f(S)\}.$$

A vector in $B(f)$ is called a *base*. An extreme point of $B(f)$ is called an *extreme base*. Edmonds 3 established the following duality theorem, which Cunningham 1 used to develop a pseudo-polynomial time algorithm for SFM. Subsequently all other efficient algorithms for SFM use the following duality theorem or a closely related result.

Theorem 1 (*Edmonds*). *For a submodular function* $f : 2^V \to \mathbb{R}$.

$$\max\{x^-(V) : x \in B(f)\} = \min\{f(S) : S \subseteq V\}. \tag{2}$$

The function $x^-()$ is not linear, and the optimizer of $\max\{x^-(V) : x \in B(f)\}$ is not, in general, an extreme point of the base polyhedron. The polynomial time algorithms in [9] and [10] proceed by representing vectors in the base polyhedron as a convex combination of extreme bases of the base polyhedron.

An extreme base can be computed by the greedy algorithm of Edmonds and Shapley [3] as follows: Let $L = \{v_1, \ldots, v_n\}$ be any linear ordering (permutation) of the elements of V. In our notation, for each j, v_j is in the j-th position of the permutation. The extreme base y_L *induced* by L is obtained by letting

$$y_L(v_j) = f(\{v_1, \ldots, v_j\}) - f(\{v_1, \ldots, v_{j-1}\}) \text{ for } j = 1 \text{ to } n.$$

If $P(j) = \{v_1, \ldots, v_j\}$, then we can also write $y_L(v_j) = f_{P(j-1)}(v_j)$, which is the marginal contribution for f of adding v_j to $\{v_1, \ldots, v_{j-1}\}$.

3 Distance Functions and Optimality Conditions

A *distance function* is a mapping $d : V \to \{0, 1, \ldots, n\}$. Each distance function d induces a linear order $L(d)$ (denoted as $\prec_d$) of V as follows: $u \; \mathsf{p}_d \; v$ if $d(u) < d(v)$ or if $d(u) = d(v)$ and $u < v$. The extreme base induced by the order $L(d)$ will be denoted as y_d.

In the algorithm presented in Section 5, at each iteration of the algorithm, we will maintain a collection D of $O(n)$ different distance functions of V, a vector x in the base polyhedron, and a vector λ. The vectors x and λ satisfy the following:

$$x = \sum_{d \in D} \lambda_d y_d, \ \sum_{d \in D} \lambda_d = 1, \text{ and } \lambda \geq 0. \tag{3}$$

We also write this as $x = \lambda_D y_D$, where $y_D = \{y_d : d \in D\}$. We let $D_{\min}(v)$ be shorthand for $\min\{d(v) : d \in D\}$. We say that the triple (x, λ, D) is *valid* if the following is true:

1. If $x(v) < 0$, then $d(v) = 0$ for all $d \in D$;
2. $d(v) \leq D_{\min}(v) + 1$ for all $d \in D$ and $v \in V$;

The algorithm will maintain a valid triple (x, λ, D) throughout all iterations. Sometimes, we will just say that the collection D of distance functions is valid.

Definition. We say that the quadruple (D, λ, x, S) satisfies the *optimality conditions* if it satisfies (3) and if it satisfies (4–6).

$$x(v) \leq 0 \ \text{ for } v \in S \tag{4}$$

$$x(v) \geq 0 \ \text{ for } v \in V \setminus S \tag{5}$$

$$v \prec_d w \text{ for all } d \in D, v \in S \text{ and } w \in V \setminus S. \tag{6}$$

We will also say that the triple (D, λ, x) satisfies the optimality conditions if there is a subset $S \subseteq V$ such that (D, λ, x, S) satisfies the optimality conditions. By (6), given any element $d \in V$, one can narrow the choice of S to n possibilities.

Lemma 1 (Sufficiency of Optimality Conditions). If the quadruple (D, λ, x, S) for SFM satisfies the optimality conditions, then S is a minimum cost set, and x is an optimal base in the base polyhedron.

Proof. By assumption, x is in the base polyhedron. Moreover, suppose without loss of generality that the elements are reordered so that $S = \{1, 2, ..., |S|\}$. Then

$$x^-(V) = x^-(S) = x(S) = \sum_{d \in D} \lambda_d y_d(S) = \sum_{d \in D} \lambda_d f(S) = f(S). \qquad (7)$$

Thus $x^-(V) = f(S)$, and by Theorem 1, S is optimal.   ◆

Lemma 2 (Existence of Optimality Conditions). If S is a minimum cost set for SFM, then there is a quadruple (D, λ, x, S) for SFM that satisfies the optimality conditions.

Proof. Let x be an optimal base in the base polyhedron. Moreover, suppose without loss of generality that the elements are reordered so that $S = \{1, 2, ..., |S|\}$. Then

$$x^-(V) \leq x^-(S) \leq x(S) = \sum_{d \in D} \lambda_d y_d(S) = \sum_{d \in D} \lambda_d f(S) = f(S). \qquad (8)$$

Since $x^-(V) = f(S)$, it follows that (D, λ, x, S) satisfies the optimality conditions. (We have not established that D is valid, but our algorithm will produce a valid D as well).   ◆

Definition. We say that the quadruple (D, λ, x, S) satisfies the *partial optimality conditions* if it satisfies (3) and if it satisfies (5) and (6).

Lemma 3 (Partial Optimality Conditions). If the quadruple (D, λ, x, S) for SFM satisfies the partial optimality conditions, then there is a minimum cost set $S^* \subseteq S$.

We first claim that if the partial optimality conditions are satisfied, then $\emptyset$ is an optimal set for f_S among subsets of $V \backslash S$. If the claim is true then for any subset T of V,

$$f(T) \geq f(S \cap T) + f(S \cup T) - f(S) = f(S \cap T) + f_S(T \backslash S) \geq f(S \cap T).$$

So, if the claim is true, then the Lemma is true. We next prove the claim.

For each $d \in D$, let y'_d be the extreme base induced by d for the base polyhedron $B(f_S)$ defined over the set of elements $u \in V \backslash S$, and let $x' = \sum_{d \in D} \lambda_d y'_d$. We will show that for each $d \in D$ and for each $u \in V \backslash S$, $y'_d(u) = y_d(u)$. If this statement is true, it follows that $x'(u) = x(u)$ for $u \in V \backslash S$, and thus $(D, \lambda, x', \emptyset)$ satisfies the optimality conditions for f_S over the set $V \backslash S$ and thus the claim is true.

So, suppose that $d \in D$ and $u \in V \backslash S$. Let $P(d, u) = \{v \in V : v \, \mathsf{p}_d \, u\}$. By (6), $S \subseteq P(d, u)$. Thus

$$y'_d(u) = f_S(u + P(d,u) \setminus S) - f_S(P(d,u) \setminus S)$$
$$= [f(u + P(d,u)) - f(S)] - [f(P(d,u)) - f(S)] = y_d(u).$$

This establishes that the claim is true, and thus the lemma is true.   ♦

We will also say that the triple (D, λ, x) satisfies the partial optimality conditions if there is a subset $S \subseteq V$ such that (D, λ, x, S) satisfies the partial optimality conditions.

A sufficient condition for the partial optimality conditions to hold for valid distance functions D is the presence of a *distance gap at level k*, which is value k with $0 < k < n$ such that

1. there is some v with $D_{\min}(v) = k$, and
2. there is no u with $D_{\min}(u) = k - 1$.

By letting $S = \{u \in V$ with $D_{\min}(u) < k\}$, it is easy to verify that (D, λ, x, S) will satisfy the partial optimality conditions. In such a case, we will eliminate all elements in $V \setminus S$ from the problem. It would be possible to maintain these elements if we wanted to determine an optimal base, but they are not needed if we just want to determine a minimum cost set.

4 Distance Functions and Extreme Vectors

Suppose that d is a distance function. We let $\text{INC}(d, v)$ be the distance function obtained by incrementing the distance label of v by 1 and keeping all other distance labels the same. That is, if $d' = \text{INC}(d, v)$, then

$$d'(u) = \begin{cases} d(v)+1 & \text{if } u = v \\ d(u) & \text{if } u \neq v \end{cases}.$$

Lemma 4. Suppose that $d' = \text{INC}(d, v)$. Then

1. $y_{d'}(v) \leq y_d(v)$,
2. $y_{d'}(u) \geq y_d(u)$ if $u \neq v$.

Proof. For each $u \in V$, let $P(u) = \{w \in V : w \prec_d u\}$. Let $P'(u) = \{w \in V : w \prec_{d'} u\}$. Note that $u \notin P(u)$, and $u \notin P'(u)$. Then for all $u \in V$, $y_{d'}(u) - y_d(u) = f_{P'(v)}(u) - f_{P(v)}(u)$.

Since $P(v) \subseteq P'(v)$, it follows from the submodularity of f that $f_{P'(v)}(v) \leq f_{P(v)}(v)$, and so $y_{d'}(v) - y_d(v) \leq 0$. Similarly, for $u \neq v$, $P'(u) \subseteq P(u)$, and so $f_{P(v)}(u) \leq f_{P'(v)}(u)$.   ♦

For any subset $S \subseteq V$, We let $d(S) = \sum_{v \in S} d(v)$. We will maintain the distance functions in D in non-decreasing order of $d(V)$.

For each $v \in V$, we will maintain a *primary distance function* $p(v) \in D$, which is the first element d of D such that $d(v) = D_{\min}(v)$. By the way that we ordered the elements of D, the primary distance function for v will minimize $d(V)$ among all $d \in D$ with $d(v) = D_{\min}(v)$. In addition, for every $v \in V$, we will maintain a *secondary distance function* $s(v) = \text{INC}(p(v), v)$. Our algorithm modifies x by increasing $\lambda_{s(v)}$ and simultaneously decreasing $\lambda_{p(v)}$ for $v \in V$.

We maintain the order of D, and the functions $p(v)$ and $s(v)$ for all v by running the Procedure Update as follows:

Procedure Update(D, p, s)
begin
   $D := \{d : \lambda_d > 0\}$;
   order the vectors in D in non-decreasing order of $d(V)$;
   for each $v \in V$, let $p(v)$ be the first element of D with $d(v) = D_{\min}(v)$;
   for each $v \in V$, let $s(v) = \text{INC}(p(v), v)$;
end

5   A Strongly Polynomial Algorithm for SFM

In this section, we present the strongly polynomial time algorithm for SFM. But first, we point out that occasionally the size of D grows too large and we want to decrease its size. Accordingly, we run a procedure called Reduce(x, λ, D) to reduce the size of D without affecting the base vector x.

Procedure Reduce(x, λ, D)
INPUT: a collection D of distance functions, a non-negative vector λ such that $\sum_{d \in D} \lambda_d = 1$. Let $x = \sum_{d \in D} \lambda_d y_d$.
OUTPUT: a subset $D' \subseteq D$ and a vector λ' such that

1. $\sum_{d \in D'} \lambda'_d = 1$ and $\lambda' \geq 0$, and $x = \sum_{d \in D'} \lambda'_d y_d$, and

2. the set $\{y_d : d \in D'\}$ is linear independent.

We will call the procedure when $3n \leq |D| < 4n$, and so the running time will be $O(n^3)$ using standard techniques from linear programming. For details on how to carry out Reduce, see Schrijver [13] or McCormick [12].

In the following procedure, let $V^0 = \{v \in V : x(v) = 0\}$. Let $V^+ = \{v \in V : x(v) > 0\}$.

Algorithm SFM
begin
   $d := 0$;
   $D = \{d\}$; $\lambda_d := 1$; $x := y_d$;
   while the optimality conditions are not satisfied
   begin
     choose an element $v^* \in V^+$;
     choose a vector $\gamma \geq 0$ with $\gamma \neq 0$ so that
$$\sum_{v \in V^0 + v^*} \gamma(v)[y_{s(v)}(u) - y_{p(v)}(u)] = 0 \text{ for all } u \in V^0 ;$$

let $x' := \sum_{v \in V^0 + v^*} \gamma(v)[y_{s(v)} - y_{p(v)}]$;

choose α maximum so that $x(u) + \alpha x'(u) \geq 0$ for all $u \in V^+$, and

$$\alpha \sum_{u:p(u)=d} \gamma(u) \leq \lambda_d \text{ for all } d \in D;$$

$x := x + \alpha x'$;

$\lambda_d := \lambda_d + \alpha \sum_{u:s(u)=d} \gamma(u) - \alpha \sum_{u:p(u)=d} \gamma(u)$ for all $d \in D \cup \{s(u): u \in V\}$;

Update(D, p, s);

if $|D| \geq 3n$, **then Reduce**(x, λ, D);

if there is a distance gap at level k, then $V := \{v \in V : D_{\min}(v) \leq k\}$;

end while

end

The algorithm initializes by letting $x = y_d$, where $d(v) = 0$ for all $v \in V$. Subsequently, the algorithm continues until the optimality conditions are satisfied.

At each iteration, the algorithm selects a non-zero vector $\gamma \geq 0$ with the property that one can modify x by increasing $y_{s(v)}$ by $\gamma(v)$ and decreasing $y_{p(v)}$ by $\gamma(v)$ for all v so that the following is true: if $x(v) = 0$ prior to the modification, then $x(v) = 0$ after the modification. It is not obvious that such a vector γ exists. We prove its existence in the next section, and show that it can be determined easily by solving a system of linear equations.

Once we determine the vector γ, we modify λ and x. After the modification, at least one of the following changes takes place: either V^0 increases in size or there is some primary vector $p(v)$ that leaves D because $d_{p(v)} = 0$ after the modification. In fact, α is chosen sufficiently large so that one of these two events occur and so that no element ever leaves V^0, and so that any element leaving V^+ must enter V^0.

We reduce the size of D whenever $|D| \geq 3n$, and we eliminate elements from V whenever a distance gap is found.

In Section 7, we will show that the algorithm terminates in $O(n^6)$ steps with an optimal set. The proof of the time bound relies on a potential function argument.

6 The Auxiliary Matrix and How to Choose γ

In this section, we show how to choose γ by solving a system of at most n equations.

One of the key steps of the algorithm is as follows: choose a vector $\gamma \geq 0$ with $\gamma \neq 0$ so that

$$\sum_{v \in V^0 + v^*} \gamma(v)[y_{s(v)}(u) - y_{p(v)}(u)] = 0 \text{ for all } u \in V^0;$$

We consider two separate cases.

Case 1. $\gamma(v^*) = 0$.

In this case, we need to solve $\sum_{v \in V^0} \gamma(v)[y_{s(v)}(u) - y_{p(v)}(u)] = 0$ for all $u \in V^0$;

Case 2. $\chi(v^*) = 1$. (We can always scale γ so that this is true whenever $\chi(v^*) \neq 0$). In this case, we need to solve $\sum_{v \in V^0} \gamma(v)[y_{s(v)}(u) - y_{p(v)}(u)] = y_{p(v^*)}(u) - y_{s(v^*)}(u)$ for all $u \in V^0$.

Suppose that the rows and columns of the constraint matrix are both indexed by the elements of V^0. Then the constraint matrices for Cases 1 and 2 are identical. The right hand side b in Case 2 may be non-zero; however, by Lemma 4, $b \leq 0$.

We refer to the constraint matrix A^* for Cases 1 and 2 as the *auxiliary matrix*. By Lemma 4, the auxiliary matrix satisfies the following properties:

6.1. A^* is an $| V^0 | \times | V^0 |$ matrix.
6.2.   The diagonal elements of A^* are non-positive.
6.3.   All non-diagonal elements of A^* are non-negative.
6.4.   Each column sum of A^* is non-positive.

In the case that A^* is invertible, it is the negative of what is known in the literature as an M-matrix, and thus the inverse of A^* is non-positive. See, for example, [1] for results on M-matrices.

Theorem 2. Let A^* be an auxiliary matrix. If A^* is singular, then there is a vector $w' \neq 0$, such that $w' \geq 0$, and $A^*w' = 0$. If A^* is non-singular then $(A^*)^{-1} \leq 0$, and thus the solution to $A^*w' = b$ is non-positive whenever b is non-negative.

Proof. The second half of the theorem is well known. The first half can easily be derived from [1], but we include a proof for completeness. Suppose that A^* is singular. Choose $w \neq 0$ so that $Aw = 0$. If $w \geq 0$, there is nothing to prove. Similarly if $w \leq 0$, then we can replace w by $-w$ and there is nothing to prove. So, suppose that there are $k < n$ positive coefficients of w. Without loss of generality assume that $w(v) > 0$ for $v = 1$ to k. (Otherwise, one can simultaneously reorder the rows and columns so that this is true.)

Let us write $A^* = \begin{bmatrix} A_{11} & A_{12} \\ A_{21} & A_{22} \end{bmatrix}$, where A_{11} denotes the first k rows and columns of A^*. Let us rewrite w as $w = \begin{bmatrix} w_1 \\ w_2 \end{bmatrix}$, where w_1 denotes the first k components of w. By assumption, $A_{11}w_1 + A_{12}w_2 = 0$ and $A_{11}w_1 + A_{21}w_2 = 0$. By 6.3, $A_{12} \geq 0$. By assumption, $w_2 \leq 0$. Therefore, $A_{11}w_1 \geq 0$. We will next show that $A_{11}w_1 = 0$.

Let 1 denote a row vector of k ones. Then $1A_{11} \leq 0$ by 6.3 and 6.4. If $1A_{11} \neq 0$, then $1A_{11}w_1 < 0$, contradicting that $A_{11}w_1 \geq 0$. We conclude that $1A_{11} = 0$. It follows that $1A_{11}w_1 = 0$, which combined with $A_{11}w_1 \geq 0$ shows that $A_{11}w_1 = 0$. In addition, by 6.1c and 6.1d, $A_{21} = 0$.

Finally, we extend w to a vector w' of $| V^0 |$ components by letting

$$w' = \begin{bmatrix} w_1 \\ 0 \end{bmatrix}.$$

Then $Aw' = 0$, which is what we wanted to prove.   ◆

By Theorem 2, the solution for γ in cases 1 and 2 can both be found by solving a system of equations on the auxiliary matrix, which takes $O(|V^0|^3) = O(n^3)$ time. Moreover, the running time is faster when the auxiliary matrix only changes by q columns in an iteration. In this case, the time to solve the system of equations at a given iteration is $O(qn^2)$.

We note that occasionally a column of the auxiliary matrix is 0, in which case it is trivial to find a non-zero vector w' with $Aw' = 0$. However, this speedup does not affect the worst case analysis.

7 Proof of Correctness and Time Bound

In this section we establish the correctness of the SFM algorithm and show that it runs in $O(n^5 \text{ EO} + n^6)$ time.

We first establish that the following remain true throughout the execution of the algorithm:

  7.1. At each iteration, there is a set D of valid distance functions, an element $x \in B(f)$ and a vector λ such that (3) is satisfied.
  7.2. If $x(v) = 0$ at some iteration, then $x(v) = 0$ at all subsequent iterations;
  7.3. $D_{\min}(v)$ is non decreasing over all iterations for all $v \in V$.
  7.4. If $x(v) < 0$, then $D_{\min}(v) = 0$;

Theorem 3. Conditions 7.1 to 7.4 are satisfied at each stage of the algorithm SFM.

Proof. Conditions 7.1 to 7.4 are all satisfied immediately subsequent to the initialization. Suppose inductively that they are satisfied at some iteration of the algorithm, and we consider what happens after some procedure is called.

We first consider the procedure **Reduce**. This procedure maintains (3) and eliminates a number of elements of D. It is easy to verify that 7.1-7.4 remain true subsequent to the call of **Reduce**.

Next, we consider eliminating elements when a distance gap is found. This results in eliminating components from y_d for all d and from x, and also changes the base polyhedron. However, it is easy to see that 7.1 to 7.4 remain satisfied with respect to the new base polyhedron.

Finally, we consider changes that occur in Procedure SFM. When we modify λ, note that every increase in $\lambda_{s(v)}$ is matched by a decrease in $\lambda_{p(v)}$. For this reason, if $\sum_{d \in D} \lambda_d = 1$ holds prior to modifying λ, it also holds afterwards. Also, by our choice of α we modify λ in such a way that it is always non-negative, and so (3.1) is still satisfied.

The solution to the system of linear equations yields a vector x' with the property that $x'(v) = 0$ for all $v \in V^0$. So, 7.2 is true after we replace x by $x + \alpha x'$.

We next consider 7.3. The only distance functions added to D are of the form $s(v)$. If $u \neq v$, then $D_{\min}(u)$ is unchanged if $s(v)$ is added to D. As for $D_{\min}(v)$, the vector $d = p(v)$ is chosen so that $D_{\min}(v) = d(v)$. Accordingly, if $d' = s(v)$, then $d'(v) = D_{\min}(v) +1$, and so 7.3 remains satisfied.

7.4 also remains satisfied. If $x(v) < 0$, then we do not create any distance functions with $d(v) \geq 1$. This completes the proof.   ◆

Theorem 4. The SFM algorithm terminates with a set S that minimizes the submodular function and finds an optimum solution x in the base polyhedron. The algorithm runs in $O(n^5 \, EO + n^6)$ time.

Prior to proving the main theorem, we state our potential function, and prove three lemmas.

For $v \in V$, let $h(v) = d(V)$, where $d = p(v)$. Thus $h(v)$ is the sum of the distances in $p(v)$. Let $\hat{h}(v) = \sum_{u \in V} (d(u) - D_{\min}(u))$. Since D is valid, it follows that $0 \le \hat{h}(v) \le n$ for all $v \in V$. Moreover, $h(v) - \hat{h}(v) = \sum_{v \in V} D_{\min}(v)$.

Let $H(v) = \{d \in D : d(V) = h(v)$ and $D_{\min}(v) = d(v)\}$. Note that any distance functions in $H(v)$ could have been chosen as a primary distance function for v if we had broken ties differently in ordering the elements of D.

We define the potential function Φ as follows:

$$\Phi(v) = |H(v)| \text{ and } \Phi = \sum_{v \in V^{in}} \Phi(v).$$

The next two lemmas concern $h(v)$ and $H(v)$.

Lemma 5. For each $v \in V$, the number of times that $h(v)$ changes over all iterations of the algorithm is $O(n^2)$.

Proof. We will actually bound the number of changes in $\hat{h}(v)$. Note that it is possible for $h(v)$ to change while $\hat{h}(v)$ stays constant if $D_{\min}(u)$ increases. But the number of changes in $D_{\min}(\ )$ over all iterations is $O(n^2)$. If the number of changes of $\hat{h}(v)$ is $O(n^2)$, then so is the number of changes of $h(v)$.

Recall that $0 \le \hat{h}(v) \le n$. We first consider changes in $\hat{h}(v)$ in between successive changes in $D_{\min}(v)$, and we refer to this set of iterations as a *phase*. The value $\hat{h}(v)$ cannot decrease during a phase unless $D_{\min}(u)$ increases for some $u \in V$, in which case $\hat{h}(v)$ can decrease by at most 1. All other changes in $h(v)$ during the phase are increases. So the total number of changes in $\hat{h}(v)$ is at most n plus the two times the number of increases in $D_{\min}(u)$ for some u. Suppose that we "charge" the latter changes in $\hat{h}(v)$ to changes in $D_{\min}$. In this case, the number of charged changes in $\hat{h}(v)$ over all iterations is $O(n^2)$, and the number of other changes in $\hat{h}(v)$ is at most n per phase. So the number of changes in $\hat{h}(v)$ is $O(n^2)$ over all phases.   ◆

Lemma 6. The distance function $s(v) \notin H(u)$ for any $u \in V$.

Proof. Let $d = p(v)$, and let $d' = s(v)$. We note that $d' \notin H(v)$ because $d'(V) = h(v) + 1$. So, we consider $u \ne v$. If $D_{\min}(u) = d'(u)$, then $D_{\min}(u) = d(u)$. In this case $h(u) \le d(V) < d'(V)$, and so $d' \notin H(u)$.   ◆

We next prove a lemma concerning the potential function Φ. We note that Φ decreases at some iterations and increases at others. By the *total decrease* in Φ over all iterations, we mean the sum of the decreases in Φ as summed over all iterations at which Φ decreases. We define *total increase* analogously.

Lemma 7. The total increase in Φ over all iterations is $O(n^4)$, and the total decrease in Φ over all iterations is also $O(n^4)$.

Proof. Given that $\Phi = O(n^2)$, it suffices to show that the total increase over all iterations is $O(n^4)$ after which the $O(n^4)$ bound on the total decrease will follow.

We first note that the only vectors that are added to D are vectors $d = s(v)$ for some $v \in V^0$. By Lemma 6, these additions to D do not change the potential function (until $p(v)$ is deleted from D). The potential function changes only when one of the following two steps takes place:

1. changes in $H(v)$ while $h(v)$ remains constant;
2. changes in $H(v)$ when $h(v)$ also changes.

By Lemma 6, each change in $H(v)$ while $h(v)$ remains constant can only result in a decrease in $\Phi(v)$. So, we only need to bound increases in changes in Φ when $h(v)$ changes for some v.

Each change in $h(v)$ can lead to an increase of at most $|D| = O(n)$ in $\Phi(v)$. By Lemma 5, there are $O(n^2)$ changes in $h(v)$ for each v and thus the total increase in Φ over all iterations due changes in $h(\ )$ is $O(n^4)$.   ♦

We are now ready to prove Theorem 4.

Proof of Theorem 4. We first note that if the algorithm terminates, then it must terminate with an optimal solution since satisfying the optimality conditions is the only termination criterion.

The bottlenecks of the algorithm are the following:

1. Adding columns $A(v) = s(v) - p(v)$ to the auxiliary matrix.
2. Solving a system of equations $A^*w = b$ or $A^*w = 0$;
3. Reducing the number of columns in D via Procedure Reduce.

We add a column to $A(v)$ only when $p(v)$ was deleted from D. A deletion of $p(v)$ for some v either leads to a change in $h(v)$ or else it leads to a decrease in $|H(v)|$. The former can happen $O(n^3)$ times by Lemma 5. We now consider the latter case.

Deleting a single element $d = p(v)$ can result in several columns needing to be added to A^*. In particular, it is possible that $d = p(u)$ for a subset $U \subseteq V$. If d is deleted from D, then we need to replace $|U|$ different columns of A^*. But in this case, deleting d from D reduces $|H(u)|$ for all $u \in U$, and thus reduces Φ by $|U|$. We conclude that the number of columns added to A^* is at most the total decrease in Φ over all iterations, which is $O(n^4)$ by Lemma 7.

Thus the running time for adding columns to the auxiliary matrix is $O(n^5 \text{ EO})$ since determining the values for a column takes $O(n \text{ EO})$ steps. For each column added to the auxiliary matrix, it takes $O(n^2)$ time to carry out elementary row operations to get A^* into canonical form for solving the system of equations. This takes $O(n^6)$ time over all iterations. Thus the running time for adding columns to A^* and carrying out elementary row operations is $O(n^5 \text{ EO} + n^6)$.

We call the procedure **Reduce** when $|D| \geq 3n$, and we eliminate at least $2n$ elements of D. The running time is thus $O(n^3)$ for each call of **Reduce**. Each distance function d that is deleted from D must have been added as a vector of the form $s(v)$ at some iteration, and this happens only $O(n^4)$ times. Thus the total time to carry out Reduce is $O(n^6)$.

We conclude that the total running time is $O(n^5 EO + n^6)$ time.   ◆

We have developed a strongly polynomial time algorithm for SFM that dominates previous strongly polynomial time algorithms by a factor greater than n. Moreover, whereas other algorithms rely on the combinatorics of paths and flows, our algorithm relies on an iterative local search plus a combinatorial potential function argument.

Acknowledgments. This research was supported by the Office of Naval Research under Grant N00014-98-1-0317. I also thank Professors Satoru Iwata and Satoru Fujishige for their constructive comments on an earlier draft of this manuscript.

References

1. Berman, A., and Plemmons R. J.: *Nonnegative Matrices in the Mathematical Sciences*, SIAM, 1994.
2. Cunningham, W. H.. On Submodular Function Minimization: *Combinatorica* 3 (1985) 185-192.
3. Edmonds, J.: Submodular Functions, Matroids, and Certain Polyhedra. In Combinatorial Structures and their Applications, R. Guy, H. Hanani, N. Sauer, and J. Schönheim, eds., Gordon and Breach (1970) 69-87.
4. Fleischer, L. K.: Recent Progress in Submodular Function Minimization. *Optima*, 2000, 1-11.
5. Fleischer, L.K. and Iwata, S.: Improved Algorithms for Submodular Function Minimization and Submodular Flow. Proceedings of the 32th Annual ACM Symposium on Theory of Computing (2000) 107–116.
6. Fujishige, S.: *Submodular Functions and Optimization.* Second Edition. North-Holland (2005).
7. The Ellipsoid Algorithm and its Consequences in Combinatorial Optimization. *Combinatorica*, 1 (1981), 499–513.
8. Grötschel, M. Lovász, L.and Schrijver, A.: *Geometric Algorithms and Combinatorial Optimization.* Springer-Verlag. (1988).
9. Iwata, S.. A Faster Scaling Algorithm for Minimizing Submodular Functions. *SIAM J. on Computing* 32 (2002) 833–840.
10. Iwata, S., Fleischer, L., and. Fujishige, S: A Combinatorial, Strongly Polynomial-Time Algorithm for Minimizing Submodular Functions. *J. ACM* 48 (2001) 761–777.
11. Lovász, L.: Submodular Functions and Convexity. In *Mathematical Programming — The State of the Art*, A. Bachem, M. Gr¨otschel, B. Korte eds., Springer, Berlin (1983) 235–257.
12. McCormick, S.T.: Submodular Function Minimization. In *Discrete Optimization*, K. Aardal, G. Nemhauser, and R. Weismantel, eds. Handbooks in Operations Research and Management Science, Volume 12. Elsevier. (2005).
13. Schrijver, A.: A Combinatorial Algorithm Minimizing Submodular Functions in Strongly Polynomial Time. *J. Combin. Theory Ser. B* 80 (2000) 346–355.
14. Schrijver, A.: *Combinatorial Optimization: Polyhedra and Efficiency.* Springer, Berlin (2003).
15. Vygen, J.. A Note on Schrijver's Submodular Function Minimization Algorithm. *Journal of Combinatorial Theory B* 88 (2003) 399–402.

On Convex Minimization over Base Polytopes

Kiyohito Nagano*

University of Tokyo, Tokyo 113-8656, Japan, and
Kyoto University, Kyoto 606-8502, Japan
kiyohito_nagano@mist.i.u-tokyo.ac.jp

Abstract. This note considers convex optimization problems over base polytopes of polymatroids. We show that the decomposition algorithm for the separable convex function minimization problems helps us give simple sufficient conditions for the rationality of optimal solutions and that it leads us to some interesting properties, including the equivalence of the lexicographically optimal base problem, introduced by Fujishige, and the submodular utility allocation market problem, introduced by Jain and Vazirani. In addition, we develop an efficient implementation of the decomposition algorithm via parametric submodular function minimization algorithms. Moreover, we show that, in some remarkable cases, non-separable convex optimization problems over base polytopes can be solved in strongly polynomial time.

Keywords: submodular functions, convex optimization.

1 Introduction

This note considers convex optimization problems over base polytopes of polymatroids, which is associated with monotone submodular functions. Submodular functions appear in the systems of graphs and networks. Besides, they naturally model economies of scale. In fact, convex optimization problems over base polytopes have numerous applications. As a generalization of lexicographically optimal flow introduced by Megiddo [18], Fujishige [7] defined the concept of lexicographically optimal base of a polymatroid and showed that finding that point is equivalent to minimizing a separable convex quadratic function over the base polytope. Although they came from different backgrounds, it is known that the egalitarian solution of Dutta and Ray [4] in a convex game is essentially the same concept as the lexicographically optimal base. Jain and Vazirani [15] introduced the submodular utility allocation (SUA) market, in which a submodular function specifies the maximum utility, and captured an equilibrium for the SUA market as an optimal solution to the maximization of the sum of logarithm functions over a base polytope.

Let us see some indirect applications in which such problems will appear as subproblems. Fujishige [8] showed the minimum norm point in the base polytope can be utilized for submodular function minimization (SFM). For

---

* Supported by Grant-in-Aid for JSPS Fellows.

M. Fischetti and D.P. Williamson (Eds.): IPCO 2007, LNCS 4513, pp. 252–266, 2007.

uncapacitated facility location problems with submodular penalties and other problems, Chudak and Nagano [2] designed approximation algorithms which require solving convex optimization problems over submodular constraints iteratively. With the aid of the lexicographically optimal base, the minimum ratio problem $\min_X\{\widetilde{f}(X)/\sum_{v\in X} w_v : \varnothing \neq X \subseteq V\}$ can be solved immediately, where $\widetilde{f}$ is a submodular set function defined on subsets of a finite set $V = \{1, \ldots, n\}$ and $w = (w_v : v \in V)$ is a positive vector in $\mathbb{R}^n$ (though it can be solved in a more direct manner by the discrete Newton method). In the greedy algorithm for the set covering problem with submodular costs due to Hayrapetyan, Swamy and Tardos [12], several minimum ratio problems have to be solved to find a subset which has the smallest cost-effectiveness. Besides, the primal-dual algorithm for the prize collecting forest problems with submodular penalties given by Sharma, Swamy and Williamson [23] repeatedly solves minimum ratio problems in order to determine the next dual constraint that will go tight at each step. For other applications of minimum ratio problems, see Fujishige's book [9, §7.2(b.3)].

Fujishige [7] presented a decomposition algorithm to find the lexicographically optimal base by $O(n)$ calls of SFM, where n is the number of elements of the ground set. In [11], Groenevelt extended Fujishige's algorithm to solve a general separable convex minimization, though explicit running time was not given and the rationality of values in the algorithm was not considered. On the other hand, Fleischer and Iwata [6] extended their push-relabel algorithm for SFM to solve the parametric minimization problem for a strong map sequence of submodular functions and they noted that, in a way similar to the parametric maximum flow algorithm of Gallo, Grigoriadis and Tarjan [10], their algorithm can be applied to solve the lexicographically optimal base problem efficiently. Taking a different approach, Hochbaum [13] proposed scaling-based algorithms for separable convex minimization over submodular constraints. Her algorithm calls a membership oracle for the base polytope, that is to say, an SFM oracle as a basic operation.

In this note, we mainly consider the minimization of separable and strictly convex functions. By describing the decomposition algorithm in a simplified form, we reveal the running time and give simple sufficient conditions for the rationality of the optimal solution to the minimization of the separable convex function over the base polytope. Furthermore, we illustrate some interesting properties, including a new remark that the lexicographically optimal base problem and the SUA market problem are equivalent. A part of these nice properties can also be derived from the result on the universal bases given by Murota [19]. At the same time, his approach is different from ours. Besides, by refining and generalizing the discussion in [6], we develop an efficient implementation of the decomposition algorithm via the Fleischer-Iwata push/relabel algorithm for SFM. We believe that a parametric minimization version of Orlin's new algorithm [21] could be developed and our framework would also work in that case. Finally, we deal with non-separable convex functions and show that in some remarkable cases the minimization can be carried out in strongly polynomial time.

This note is organized as follows. In Section 2, we define the main problem and see optimality conditions. In Section 3, we will see about related problems, give examples of objective functions and check the rationality of optimal solutions to subproblems. Section 4 describes the decomposition algorithm in a simplified form and discusses the rationality and the equivalence of some problems. In Section 5, we review the basic framework of combinatorial algorithms for SFM and develop an efficient implementation of the decomposition algorithm via parametric SFM algorithms. Lastly we consider the minimization of non-separable convex functions in Section 6.

2   Preliminaries

Let V be a finite nonempty set with $|V| = n$. Suppose that $V = \{1, \ldots, n\}$. A set function $\widetilde{f}$ defined on 2^V is submodular if $\widetilde{f}(X) + \widetilde{f}(Y) \geq \widetilde{f}(X \cup Y) + \widetilde{f}(X \cap Y)$ for each $X, Y \subseteq V$ and monotone if $\widetilde{f}(X) \leq \widetilde{f}(Y)$ for each $X, Y \subseteq V$ with $X \subseteq Y$. It is easy to see that the minimizers of submodular function $\widetilde{f}$ are closed under union and intersection, and thus there exist the (unique) minimal minimizer and the (unique) maximal minimizer. For a vector $x \in \mathbb{R}^V$ and an element $v \in V$, we denote by $x(v)$ the component of x on v.

Let $\widetilde{f} : 2^V \to \mathbb{R}$ be a submodular function with $\widetilde{f}(\varnothing) = 0$. We assume that $\widetilde{f}$ is given by a value-giving oracle. With such a function $\widetilde{f}$, the base polytope $\mathbf{B}(\widetilde{f})$ is defined by

$$\mathbf{B}(\widetilde{f}) = \{\, x \in \mathbb{R}^V : x(X) \leq \widetilde{f}(X)\ (\forall X \subseteq V),\ x(V) = \widetilde{f}(V)\} \subseteq \mathbb{R}^V$$

where $x(X) = \sum_{v \in X} x(v)$. It is known that $\mathbf{B}(\widetilde{f})$ is nonempty and bounded. A vector in $\mathbf{B}(\widetilde{f})$ is called a base and an extreme point of $\mathbf{B}(\widetilde{f})$ is called an extreme base. Consider any total order $\prec$ in V. The greedy algorithm [5] gives an extreme base $b^\prec \in \mathbb{R}^V$ by setting $b^\prec(v) = \widetilde{f}(L_\prec(v) \cup \{v\}) - \widetilde{f}(L_\prec(v))$ for each $v \in V$, where $L_\prec(v) = \{u \in V : u \prec v\}$. Conversely, it is known that each extreme base can be obtained in this way. For submodular functions f_1, f_2, if $Y \supseteq X$ implies $f_1(Y) - f_1(X) \geq f_2(Y) - f_2(X)$, we write $f_1 \to f_2$ or $f_2 \leftarrow f_1$. We call the relation $f_1 \to f_2$ a strong map.

Throughout this note, we suppose the function $f : 2^V \to \mathbb{R}$ is rational, submodular, monotone and satisfies $f(\varnothing) = 0$, $f(\{v\}) > 0$ for each $v \in V$. In other words, f is a rank function of a polymatroid. So we have $\mathbf{B}(f) \subseteq \mathbb{R}_{\geq 0}^V$ and $\mathbf{B}(f) \cap \mathbb{R}_{>0}^V$ is nonempty. Let $g_v : \mathbb{R} \to \mathbb{R} \cup \{+\infty\}$ be a convex function on $\mathrm{dom}\, g_v$ for each $v \in V$. In this note, we mainly consider the separable convex function minimization problem over the base polytope:

$$\min_x \{g(x) : x \in \mathbf{B}(f)\} \quad \text{where} \quad g(x) = \sum_{v \in V} g_v(x(v)). \tag{1}$$

For each $v \in V$, let $e_v \in \mathbb{R}^V$ be the characteristic vector that has value 1 on v and 0 elsewhere. The following theorem states that the local optimality with respect to directions of the form $e_u - e_v$ for $u, v \in V$ implies the global optimality.

Theorem 1 ([11], [9, Theorem 8.1]). *For $x \in \mathbf{B}(f)$, x is an optimal solution of (1) if and only if for each $u, v \in V$ such that $x + \varepsilon(e_u - e_v) \in \mathbf{B}(f)$ for some*

$\varepsilon > 0$, we have $D_u^+(x(u)) \geq D_v^-(x(v))$ where D_u^+ is the right derivative of g_u and D_u^- is the left derivative of g_v.

For any base x and each subset X with $\varnothing \neq X \subset V$, by standard arguments about tight subsets, one can show that $x(X) = f(X)$ if and only if $x + \varepsilon(e_u - e_v) \notin \mathbf{B}(f)$ for any $\varepsilon > 0$ and each pair (u, v) with $u \in X$, $v \in V \setminus X$. In particular, we have:

Corollary 2 ([9, Theorem 8.2]). *Suppose that g_v is differentiable and strictly convex for each $v \in V$. Let $x \in \mathbf{B}(f)$ and $\xi_1 < \cdots < \xi_\ell$ denote the distinct values of $g_v'(x(v))$. Let $H_s = \{v \in V : g_v'(x(v)) \leq \xi_s\}$ for $s = 1, \ldots, \ell$. Then, x is the optimal solution of problem (1) if and only if $x(H_s) = f(H_s)$ for each $s = 1, \ldots, \ell$.*

To simplify the discussion, we mainly assume that g_v is differentiable and strictly convex and define the interval $J \subseteq \mathbb{R}$ by $J := \bigcap_{v \in V} \{g_v'(x(v)) : x(v) \in \mathrm{dom}\, g_v\}$. For any $B \geq 0$ (or > 0) and each nonempty subset $U \subseteq V$, we suppose the problem

$$\min_{(x(v) : v \in U)} \left\{ \sum_{v \in U} g_v(x(v)) : x(U) = B \right\} \qquad (2)$$

has the optimal solution $x_U \in \mathbb{R}^U$ such that there exists $\alpha \in J$ such that $x_U(v) = (g_v')^{-1}(\alpha)$ for each $v \in U$. In Section 4, we will see problem (1) has the rational optimal solution if the optimal solution to (2) is always rational for any rational B and each nonempty subset U.

3   Examples of Problems

We review some related problems. After that, we give examples of function g and check the rationality of optimal solutions to (2) for rational B and $U \subseteq V$.

3.1   Related Problems

Let $w \in \mathbb{R}_{>0}^V$ be a positive vector. For $x \in \mathbb{R}^V$, let $T_w(x) = \left(\frac{x(v_1)}{w(v_1)}, \ldots, \frac{x(v_n)}{w(v_n)} \right)$ be the real n-sequence such that $\frac{x(v_1)}{w(v_1)} \leq \cdots \leq \frac{x(v_n)}{w(v_n)}$ where $\{v_1, \ldots, v_n\} = V$. For two real n-sequences $\rho = (\rho_1, \ldots, \rho_n)$ and $\sigma = (\sigma_1, \ldots, \sigma_n)$, we write $\rho \geq_{\mathrm{LEX}} \sigma$ if $\rho = \sigma$ or $\rho \neq \sigma$ and $\rho_i > \sigma_i$ for the minimum index i such that $\rho_i \neq \sigma_i$.

Lexicographically Optimal Bases. A base x is called a lexicographically optimal (lex-optimal) base with respect to w if $T_w(x) \geq_{\mathrm{LEX}} T_w(y)$ for all y in the base polytope. Fujishige [7] showed that such a base x_{LEX} of $\mathbf{B}(f)$ is unique and coincides with the optimal solution of (1) with $g_v(x(v)) = x(v)^2/w(v)$. Now the monotonicity of f is not crucial because, for any $M \in \mathbb{R}$, $x_{\mathrm{LEX}} + M \cdot w$ is lexicographically optimal in $\mathbf{B}(f + M \cdot w) (= \mathbf{B}(f) + M \cdot w)$ w.r.t. w.

In Section 6, we will use the following characterization of the lex-optimal bases, which is a special case of Theorem 1.

Corollary 3. *For $x \in \mathbf{B}(f)$, x minimizes $\sum_{v \in V} \frac{x(v)^2}{w(v)}$ over the base polytope $\mathbf{B}(f)$ if and only if for each elements $u, v \in V$ such that $x + \varepsilon(e_u - e_v) \in \mathbf{B}(f)$ for some $\varepsilon > 0$, we have $\frac{x(u)}{w(u)} \geq \frac{x(v)}{w(v)}$.*

Minimum Ratio Problems. Consider the minimum ratio problem which asks for a subset $X \in 2^V \setminus \{\varnothing\}$ minimizing $f(X)/w(X)$. Now the monotonicity of f is not essential again. Let $x_{\mathrm{LEX}} \in \mathbf{B}(f)$ be the lex-optimal base w.r.t. w, $\xi_1 = \min_v \frac{x_{\mathrm{LEX}}(v)}{w(v)}$ and $X_1 = \{v : \frac{x_{\mathrm{LEX}}(v)}{w(v)} = \xi_1\}$. For any $X \subseteq V$ with $X \neq \varnothing$, we have $\xi_1 w(X) \leq x_{\mathrm{LEX}}(X) \leq f(X)$ and so $\xi_1 \leq f(X)/w(X)$. On the other hand, by Corollary 2, $x_{\mathrm{LEX}}(X_1) = f(X_1)$ and thus $\xi_1 = f(X_1)/w(X_1)$. Therefore, using the lex-optimal base, the minimum ratio problem can be easily solved. This problem, however, can be solved more directly via the discrete Newton method. See, e.g., [6, §4.1].

Egalitarian Allocations. Imagine that V is a set of players. We assume that set function $\mathrm{val} : 2^V \to \mathbb{R}$ is a convex game, that is, $-\mathrm{val}$ is submodular and it satisfies $\mathrm{val}(\varnothing) = 0$. It is pointed out that the egalitarian allocation in a convex game [3,4], which is often called the Dutta-Ray solution, is essentially the same concept as the lex-optimal base w.r.t. $\mathbf{1} = (1, \ldots, 1)$. To be precise, it is the lex-optimal base of $\mathbf{B}(\widetilde{f})$ w.r.t. $\mathbf{1}$ where $\widetilde{f}(X) = \mathrm{val}(V) - \mathrm{val}(V \setminus X)$ ($X \subseteq V$).

Submodular Utility Allocation Markets. Let $m(v) > 0$ be the money possessed by buyer $v \in V$. The maximization of $\sum_v m(v) \ln x(v)$ over the polytope $P = \{x \in \mathbb{R}^V : x(X) \leq f(X) \ (X \subseteq V), \ x \geq \mathbf{0}\}$ is called the submodular utility allocation (SUA) market problem [15]. As the base polytope $\mathbf{B}(f)$ is the set of all the maximal points of P with respect to the partial order $\leq$ among vectors in $\mathbb{R}^V$, this problem is a special case of problem (1).

The Minimum Norm Point. Let $\widetilde{f} : 2^V \to \mathbb{R}$ be a submodular function with $\widetilde{f}(\varnothing) = 0$ and let x_{MN} be the point that minimize $\|x\|$ over $\mathbf{B}(\widetilde{f})$ where $\|.\|$ is the Euclidean norm, that is, $\|x\| = \sqrt{\sum_v x(v)^2}$ for $x \in \mathbb{R}^V$. By Corollary 2, it is easy to see $X_{<0} := \{v \in V : x_{\mathrm{MN}}(v) < 0\}$ is the unique minimal minimizer and $X_{\leq 0} := \{v \in V : x_{\mathrm{MN}}(v) \leq 0\}$ is the unique maximal minimizer of $\widetilde{f}$.

3.2   Examples of Objective Functions

Let w and m be vectors in $\mathbb{Q}_{>0}^V$, a be a vector in $\mathbb{Q}^V$, q be a vector in $\mathbb{R}_{>0}^V$, $p \in \mathbb{R}$ be a number such that $p \neq 0, -1$ and let $g_0 : \mathbb{R} \to \mathbb{R} \cup \{+\infty\}$ be a differentiable and strictly convex function with $\mathrm{dom}\, g_0 \supseteq \mathbb{R}_{\geq 0}$. Define function $\mathrm{sgn} : \mathbb{R} \to \mathbb{R}$ by $\mathrm{sgn}(\tau) = 0$ if $\tau = 0$ and $\mathrm{sgn}(\tau) = \tau/|\tau|$ otherwise. For example, we consider the convex functions

$$g^{\mathrm{MN}}(x) = \tfrac{1}{2}\|x\|^2, \quad g^{\mathrm{MN}'}(x) = \tfrac{1}{2}\|x + a\|^2$$

and

$$
\begin{aligned}
&g_v^{\mathrm{Lex}}(x(v)) = \tfrac{1}{2w(v)}x(v)^2, && g_v^{\mathrm{Pow}}(x(v)) = \tfrac{\mathrm{sgn}(p)\, x(v)^{p+1}}{(p+1)\, w(v)^p}, \\
&g_v^{\mathrm{SUA}}(x(v)) = -m(v) \ln x(v), && g_v^{\mathrm{exp}}(x(v)) = \exp(x(v) + a(v)), \\
&g_v^{\mathrm{S}}(x(v)) = g_0(x(v)), && g_v^{\mathrm{W}}(x(v)) = x(v) g_0\!\left(\tfrac{w(v)}{x(v)}\right), \\
&g_v^{\mathrm{Bad}}(x(v)) = \tfrac{x(v)}{q(v)}(\ln x(v) - 1),
\end{aligned}
$$

for each $v \in V$. Functions g^{Pow} and g^{S} are defined on $\mathbb{R}_{\geq 0}^V$, and g^{SUA}, g^{W} and g^{Bad} are defined on $\mathbb{R}_{>0}^V$.

Among these functions, let us see that it suffices to examine g^{S}, g^{W} and g^{Bad} as objective functions of (1) and (2). Trivially, $g^{\mathrm{Lex}} = g^{\mathrm{Pow}}$ if $p = 1$, and $g^{\mathrm{MN}} = g^{\mathrm{S}}$ if $g_0(\tau) = \frac{\tau^2}{2}$. By resetting $f := f - a + (\max_v a(v)) \cdot \mathbf{1}$, f is still monotone and the minimization of g^{exp} and $g^{\mathrm{MN'}}$ can be reduced to the problem with $g(x) = g^{\mathrm{S}}(x)$. The function g^{Pow} is a special case of g^{W} where $g_0(\tau) = \frac{\mathrm{sgn}(p)}{p+1}\tau^{-p}$. If $g_0(\tau) = \tau \ln \tau$ and $w = m$, we have $g_v^{\mathrm{W}}(x(v)) = g_v^{\mathrm{SUA}}(x(v)) + m(v) \ln m(v)$ for each v. Thus g^{SUA} is a special case of g^{W} ignoring the constant term. In some sense, we may view g^{SUA} as g^{Pow} with $p = -1$ because $g_v^{\mathrm{SUA}}(x(v)) = \lim_{p \to -1}(g_v^{\mathrm{Pow}}(x(v)) + \frac{1}{(p+1)w(v)^p})$.

Let $B \geq 0$ (or > 0) and $U \in 2^V \setminus \{\varnothing\}$. We denote the optimal solutions of problem (2) with $g = g^{\mathrm{S}}$, g^{W} and g^{Bad} by x_U^{S}, x_U^{W} and $x_U^{\mathrm{Bad}} \in \mathbb{R}^U$, respectively. Easily we have

$$x_U^{\mathrm{S}}(v) = \frac{1}{|U|} \cdot B, \quad \text{for each } v \in U. \tag{3}$$

So, if B is rational, x_U^{S} is a rational vector and the size of vector x_U^{S} is polynomially bounded by n and the size of B. Now we let $g = g^{\mathrm{W}}$. Then $g_v'(x(v)) = g_0(\frac{w(v)}{x(v)}) - \frac{w(v)}{x(v)}g_0'(\frac{w(v)}{x(v)})$. Thus, for vector $(x(v) : v \in U)$ if there exists a number $\alpha \in J$ such that $g_v'(x(v)) = \alpha$ for each $v \in U$, we can write $x(v) = C \cdot w(v)$ for each $v \in U$, where C is some constant. Therefore we have

$$x_U^{\mathrm{W}}(v) = \frac{w(v)}{w(U)} \cdot B, \quad \text{for each } v \in U. \tag{4}$$

Thus the size of x_U^{W} is polynomially bounded by the sizes of w and B.

Finally, let us see a bad example in which the rationality of the optimal solution of (2) does not hold. Now we let $g = g^{\mathrm{Bad}}$. Then $J = \mathbb{R}$ and $(g_v')^{-1}(\alpha) = (\exp(\alpha))^{q(v)}$ for any $\alpha \in \mathbb{R}$. So we can write $x_U^{\mathrm{Bad}}(v) = \beta^{q(v)}$ for each $v \in U$ where $\beta > 0$ is the unique positive solution of the equation $\sum_{v \in U} \beta^{q(v)} = B$. In general, we cannot give β or x_U^{Bad} in a closed form. Moreover x_U^{Bad} is not necessarily rational.

4   The Decomposition Algorithm

We describe the decomposition algorithm [7,11] for separable convex minimization problems over base polytopes in a quite simplified form. The point of this section is that the correctness of the algorithm gives us sufficient conditions for the rationality of optimal solutions and some good properties. In Sections 4 and 5, we assume that g_v is differentiable and strictly convex for each $v \in V$. Though each g_v is not necessarily differentiable, a similar algorithm also works and thus things essentially do not change.

4.1   The Decomposition Algorithm and the Rationality

Let x^* be the optimal solution of (1). We denote the distinct values of $g'_v(x^*(v))$ by $\xi_1^* < \cdots < \xi_\ell^*$ and let $H_s^* = \{v \in V : g'_v(x^*(v)) \leq \xi_s^*\}$ for $s = 1, \ldots, \ell$. For convenience, we let $\xi_0 = -\infty$, $\xi_{\ell+1} = +\infty$, $H_0^* = \varnothing$ and $\mathcal{H}^* := \{H_0^*, \ldots, H_\ell^*\}$. For $\alpha \in J$, define $\overline{x}_\alpha \in \mathbb{R}^V$ as

$$\overline{x}_\alpha(v) = (g'_v)^{-1}(\alpha) \tag{5}$$

for each $v \in V$. If $\alpha < \alpha'$, we have $\overline{x}_\alpha < \overline{x}_{\alpha'}$. For $\alpha \in J$ and $U \subseteq V$, the vector $(\overline{x}_\alpha(v) : v \in U)$ is the unique optimal solution to (2) with $B = \overline{x}_\alpha(U)$. Remark that set function $f_\alpha := f - \overline{x}_\alpha$ is submodular for $\alpha \in J$ and that we have the relation $f_\alpha \to f_{\alpha'}$ if $\alpha < \alpha'$.

Lemma 4. *Let $\alpha \in J$. If $\xi_s^* < \alpha < \xi_{s+1}^*$, H_s^* is the unique minimizer of f_α. If $\alpha = \xi_s^*$, H_{s-1}^* is the unique minimal minimizer and H_s^* is the unique maximal minimizer of f_α.*

Proof. Suppose $\xi_s^* < \alpha < \xi_{s+1}^*$. As g'_v is strictly increasing, $x^*(v) - \overline{x}_\alpha(v) < 0$ if $v \in H_s^*$ and $x^*(v) - \overline{x}_\alpha(v) > 0$ otherwise. Thus using Corollary 2, for each $X \subseteq V$ with $X \neq H_s$, we have

$$f(X) - \overline{x}_\alpha(X) \geq x^*(X) - \overline{x}_\alpha(X) > x^*(H_s^*) - \overline{x}_\alpha(H_s^*) = f(H_s^*) - \overline{x}_\alpha(H_j^*).$$

Suppose $\alpha = \xi_s$ holds. Then $v \in H_{s-1}^*$ iff $x^*(v) - \overline{x}_\alpha(v) < 0$ and $v \in H_s^*$ iff $x^*(v) - \overline{x}_\alpha(v) \leq 0$. So H_{s-1}^* and H_s^* minimize f_α and any minimizer X satisfies $H_{s-1}^* \subseteq X \subseteq H_s^*$. □

This lemma implies that problem (1) can be reduced to the parametric problem:

$$\min_X \{f_\alpha(X) : X \subseteq V\} \text{ for all } \alpha \in J. \tag{6}$$

This fact leads us to the framework of the decomposition algorithm [7,11]. By successively computing $\overline{x}_\alpha$ for some appropriately chosen $\alpha \in J$ and minimizing f_α, we find $H_s^* \in \mathcal{H}^*$ one by one and finally we obtain the chain $H_0^* \subset \cdots \subset H_\ell^*$ and the point x^*.

The algorithm is recursive and now the description is simplified to reveal the running time explicitly. First we know that $H_0^* = \varnothing$ and $H_\ell^* = V$, although we do not know how much ℓ is. Let $S = H_s^*$ and $T = H_t^*$ for some s and t such that $0 \leq s < t \leq \ell$. Now we give the procedure DA(S, T) which returns a vector $x_{T\setminus S}^* \in \mathbb{R}^{T\setminus S}$. We denote by $\alpha_{S,T}$ the number $\alpha \in J$ satisfying $\overline{x}_\alpha(T\setminus S) = f(T) - f(S)$. We let $\alpha = \alpha_{S,T}$. Note that $\alpha = \xi_t$ if $s+1 = t$ and that $\xi_{s+1} < \alpha < \xi_t$ if $s + 1 < t$. Moreover $\overline{x}_{S,T} := (\overline{x}_\alpha(v) : v \in T \setminus S)$ is the optimal solution of problem (2) with $U = T \setminus S$ and $B = f(T) - f(S)$. Define the submodular function $\overline{f}_{S,T} : 2^{T\setminus S} \to \mathbb{R}$ by

$$\overline{f}_{S,T}(W) = f_{\alpha_{S,T}}(W \cup S) - f_{\alpha_{S,T}}(S)$$
$$= f(W \cup S) - f(S) - \overline{x}_{S,T}(W) \tag{7}$$

for each $W \subseteq T \setminus S$. The procedure computes the maximal minimizer R of f_α. Since any minimizer R' of f_α satisfies $S \subseteq R' \subseteq T$, it suffices to compute $\overline{x}_{S,T}$ and minimize $\overline{f}_{S,T}$. If $R = T$, that is, $t = s + 1$, then $x^*(v) = \overline{x}_\alpha(v)$ for each $v \in T \setminus S$ and $\mathsf{DA}(S, T)$ returns $x^*_{T \setminus S} := \overline{x}_{S,T}$. Next, consider the case where $R \subset T$ and so $S \subset R$. By Lemma 4, $R = H^*_r$ for some r with $s + 1 \leq r < t$. Let $x_1 = x^*_{R \setminus S}$ and $x_2 = x^*_{T \setminus R}$ be vectors returned by $\mathsf{DA}(S, R)$ and $\mathsf{DA}(R, T)$, respectively. The procedure $\mathsf{DA}(S, T)$ returns the vector $x^*_{T \setminus S}$ obtained by setting $x^*_{T \setminus S}(v) := x^*_1(v)$ for each $v \in R \setminus S$ and $x^*_{T \setminus S}(v) := x^*_2(v)$ for each $v \in T \setminus R$. By induction, we can see that $x^*_{T \setminus S} = (x^*(v) : v \in T \setminus S)$. Thus $\mathsf{DA}(\varnothing, V)$ returns the optimal solution of problem (1).

This algorithm implies sufficient conditions for the rationality of the optimal solution of problem (1).

Theorem 5. *Suppose problem* (2) *has a rational optimal solution for any rational number $B \geq 0$ and each subset $U \in 2^V \setminus \{\varnothing\}$. Then the optimal solution of problem* (1) *is rational.*

Proof. Let $v \in H^*_s \setminus H^*_{s-1}$. The optimal solution to (2) with $U = H^*_s \setminus H^*_{s-1}$ and $B = f(H^*_s) - f(H^*_{s-1})$ is rational. So $x^*(v)$ is also rational. $\qquad \square$

Note that this theorem can also be shown directly from Corollary 2. Moreover, we can immediately get the following useful observation.

Corollary 6. *Suppose the problems of the form* (2) *with objective function g and $\widetilde{g}$ always have the same optimal solution for any $B \geq 0$ and each U. Then the optimal solutions to the problems of the form* (1) *are also the same.*

Additionally, we assume that the optimal solution x_U to (2) is rational for any $B \in \mathbb{Q}_{\geq 0}$ and each $U \subseteq V$ and that arithmetic operations involving values $x_U(v)$ $(v \in U)$ can be carried out as basic steps. In the algorithm, we compute $\overline{x}_{S,T}$ and minimize $\overline{f}_{S,T}$ for some $S, T \in \mathcal{H}^*$ at most n times respectively. Since f and $\overline{x}_{S,T}$ are rational, $\overline{f}_{S,T}$ is also rational. Therefore we can utilize a submodular function minimization algorithm to minimize $\overline{f}_{S,T}$ in polynomial time. Notice that we need the maximal minimizer of $\overline{f}_{S,T}$. This point will be discussed in the next section.

Theorem 7. *The optimal solution of problem* (1) *can be obtained by solving problem* (2) *and performing submodular function minimization at most n times respectively.*

4.2   Equivalence of Problems

The decomposition algorithm directly leads us to some interesting facts about the convex minimization over $\mathbf{B}(f)$. Recall the functions introduced in Section 3. By Corollary 6, in view of (3) and (4), we have:

Corollary 8. *The minimum norm point x_{MN} of $\mathbf{B}(f)$ is the (unique) optimal solution to problem $\min_x \{\sum_v g_0(x(v)) : x \in \mathbf{B}(f)\}$.*

Corollary 9. *The following are equivalent:*

(9.a) $x \in \mathbb{R}^V$ *minimizes* $\sum_v \frac{1}{2w(v)} x(v)^2$ *over* $\mathbf{B}(f)$;

(9.b) $x \in \mathbb{R}^V$ *minimizes* $\sum_v \frac{\mathrm{sgn}(p)}{(p+1)\, w(v)^p} x(v)^{p+1}$ *over* $\mathbf{B}(f)$ *where* $p \neq 0, -1$;

(9.c) $x \in \mathbb{R}^V$ *maximizes* $\sum_v w(v) \ln x(v)$ *over* $\mathbf{B}(f)$;

(9.d) $x \in \mathbb{R}^V$ *minimizes* $\sum_v x(v) g_0\left(\frac{w(v)}{x(v)}\right)$ *over* $\mathbf{B}(f)$.

The equivalence of (9.a) and (9.d), that is, Corollary 9 itself can also be derived from a general result on the universal bases of Murota [19], which is obtained in a different way from our approach. In view of (9.a) and (9.c), however, it is still a somewhat surprising fact and a new remark that the lexicographically optimal base problem [7] and the SUA market problem [15] are equivalent.

Corollary 8 can be slightly generalized using Theorem 1. Let $g_N : \mathbb{R} \to \mathbb{R} \cup \{\infty\}$ be a convex function which is not necessarily differentiable. For example, $g_N(\tau) = |\tau|$ or $g_N(\tau) = \max\{-\tau, 0\}$. Note that many submodular function minimization algorithms are based on the maximization of $\sum_v \min\{0, x(v)\}$ over base polytopes (see §5.1).

Corollary 10. *Let* $x_{\mathrm{MN}} \in \mathbb{R}^V$ *be the minimum norm point of* $\mathbf{B}(f)$. *Then* x_{MN} *also minimizes* $\sum_v g_N(x(v))$ *over* $\mathbf{B}(f)$. *Besides, if* $g_N(x(v))$ *is strictly convex,* x_{MN} *is the unique minimizer.*

5   An Efficient Implementation

For submodular function minimization (SFM), Fleischer and Iwata [6] developed a push/relabel algorithm using Schrijver's subroutine [22]. They also extended their algorithm to parametric minimization for a strong map sequence of submodular functions. In addition, they noted that their algorithm can be used to solve the lexicographically optimal base problem efficiently (though their discussion includes some minor errors). By refining and generalizing the discussion of [6], we propose an efficient implementation of the decomposition algorithm via the Fleischer-Iwata algorithm. We also believe that our framework could be extended via (possible) parametric variants of Orlin's new algorithm [21] for SFM.

In this section, if $\alpha = \alpha_{S,T}$ for some subsets $S, T \subseteq V$ with $S \subset T$, we assume that the time of function evaluation of f_α is bounded by EO and that arithmetic operations involving values $f_\alpha(X)$ ($X \subseteq V$) can be regarded as basic steps. Let $\prec_\circ$ be a total order in $V = \{1, \ldots, n\}$ such that $1 \prec_\circ \cdots \prec_\circ n$. Let $\widetilde{f} : 2^V \to \mathbb{R}$ be any submodular function with $\widetilde{f}(\varnothing) = 0$ and $\min_X \widetilde{f}(X) =: \widetilde{f}^*$. We also denote by EO the upper bound on the time to evaluate $\widetilde{f}$.

5.1   Submodular Function Minimization

We briefly review the basic framework of combinatorial algorithms for SFM. See McCormick [17] for a nice survey on SFM. For $x \in \mathbb{R}^V$, define the vector $x^- \in \mathbb{R}^V$ by $x^-(v) = \min\{0, x(v)\}$ for each $v \in V$. For any base $x \in \mathbf{B}(\widetilde{f})$

and each $X \subseteq V$, we have $x^-(V) \leq x(X) \leq \widetilde{f}(X)$. Furthermore, the result of Edmonds [5] immediately implies that $\max_x \{x^-(V) : x \in \mathbf{B}(\widetilde{f})\} = \widetilde{f}^*$.

In Schrijver's algorithm [22] and the Fleischer-Iwata algorithm [6], at any step, we keep a point $x \in \mathbf{B}(\widetilde{f})$ as a convex combination $x = \sum_{i \in I} \lambda_i b^{\prec_i}$ of extreme bases where I is a finite set of indices with $|I| = O(n)$ and each $\prec_i$ is a total order in V. With such a point $x \in \mathbf{B}(\widetilde{f})$, we consider a digraph $D = (V, A_I)$ where $A_I = \{(u, v) : u \prec_i v \text{ for some } i \in I\}$. Let $P = \{v \in V : x(v) > 0\}$ and $N = \{v \in V : x(v) < 0\}$. We iteratively update x until D has no path from P to N. Then let R_1 be the vertices that can reach N and R_2 be the vertices that are not reachable from P in D. Clearly, we have $N \subseteq R_h \subseteq V \setminus P$ for $h = 1, 2$. Both of the algorithms [6,22], which can be implemented in $O(n^8 + n^7 \text{ EO})$ time, return R_1 or R_2 and terminate. It is easy to see that x is an optimal solution to $\max_x \{x^-(V) : x \in \mathbf{B}(\widetilde{f})\}$ and $\widetilde{f}(R_1) = \widetilde{f}(R_2) = \widetilde{f}^*$.

Remark that the maximal minimizer of a submodular function is needed in the decomposition algorithm and little attention was paid to this point in the algorithm of [6]. The following example shows that R_1 and R_2 are not necessarily maximal. Suppose that $n = 5$ and $a = (0, -1, 0, 1, 0) \in \mathbb{R}^V$ and the function $\widetilde{f} : 2^V \to \mathbb{R}$ is defined by $\widetilde{f}(X) = \sum_{v \in X} a(v)$. Then the minimal minimizer is $\{2\}$ and the maximal minimizer is $\{1, 2, 3, 5\}$. By initially setting $I = \{1\}$, $\prec_1 = \prec_o$ and $\lambda_1 = 1$, we get the point $x = b^{\prec_o} = (0, -1, 0, 1, 0)$ and the digraph D has no directed path from $P = \{4\}$ to $N = \{2\}$. So we obtain $R_1 = \{1, 2\}$ and $R_2 = \{1, 2, 3\}$, neither of which is maximal. Fortunately, it is known that we can compute the maximal and the minimal minimizer in $O(n^3 \text{EO})$ additional time using the result of [1]. See e.g. Note 10.11 of [20] for details.

Lemma 11. *Given a maximizer $x = \sum_{i \in I} \lambda_i b^{\prec_i}$ of $\max_x \{x^-(V) : x \in \mathbf{B}(\widetilde{f})\}$ with $|I| = O(n)$, the unique maximal minimizer and the unique minimal minimizer of $\widetilde{f}$ can be found in $O(n^3 \text{EO})$ time.*

5.2   The Fleischer-Iwata Algorithm for Parametric SFM

We review the computation of a minimizer of every function in a strong map sequence of submodular functions, $f_1 \to \cdots \to f_k$, via the ordinary and the reverse push/relabel algorithms for SFM of Fleischer and Iwata [6].

Consider a base $x = \sum_{i \in I} \lambda_i b^{\prec_i} \in \mathbf{B}(\widetilde{f})$ and the digraph D for x. We need a concept of a valid labeling on V. A labeling $d : V \to \mathbb{Z}$ is valid for x if $d(v) = 0$ for $v \in N$, $d(u) \leq d(v) + 1$ for each $(u, v) \in A_I$ and $d(v) \leq n$ for $v \in V$. A labeling $d^R : V \to \mathbb{Z}$ is r-valid for x if $d^R(v) = 0$ for $v \in P$, $d^R(v) \leq d^R(u) + 1$ for each $(u, v) \in A_I$ and $d^R(v) \leq n$ for $v \in V$. Obviously, the labeling $d_o := \mathbf{0}$ on V is always valid and r-valid for any x.

Let $f^{(1)}$, $f^{(2)}$ and $f^{(3)}$ be submodular functions defined on 2^V such that $f^{(h)}(\varnothing) = 0$ for $h = 1, 2, 3$ and $f^{(1)} \to f^{(2)} \to f^{(3)}$. For each total order $\prec$ in V, we denote the outputs of the greedy algorithm w.r.t. $f^{(h)}$ by $b_h^{\prec} \in \mathbf{B}(f^{(h)})$ for each h. Notice that $b_1^{\prec} \geq b_2^{\prec} \geq b_3^{\prec}$ for each total order $\prec$. To simplify the notation, we use subindex 1, 2 and 3 for digraphs associated with $\mathbf{B}(f^{(1)})$, $\mathbf{B}(f^{(2)})$, and $\mathbf{B}(f^{(3)})$ respectively. Suppose that we have bases $x_1 \in \mathbf{B}(f^{(1)})$ and

$x_3 \in \mathbf{B}(f^{(3)})$ with $x_1 = \sum_{i \in I_1} \lambda_i b_1^{\prec i}$ and $x_3 = \sum_{j \in I_3} \mu_j b_3^{\prec j}$ where $|I_1|$ and $|I_3|$ are $O(n)$.

First we let d be a valid labeling for x_3. By setting $x_2 := \sum_{j \in I_3} \mu_j b_2^{\prec j}$, a base of $\mathbf{B}(f^{(2)})$ can be obtained in $O(n^2 EO)$ time. From the construction, the digraphs D_2 and D_3 have the same edges, $x_2 \geq x_3$ and $N_2 \subseteq N_3$. Therefore d is still valid for $x_2 \in \mathbf{B}(f^{(2)})$. The ordinary push/relabel (OPR) algorithm in [6] finds a maximizer $\tilde{x}_2 = \sum_{j \in \tilde{I}} \tilde{\mu}_j b_2^{\prec j}$ of $\max_x \{x^-(V) : x \in \mathbf{B}(f^{(2)})\}$ with $|\tilde{I}| = O(n)$ and a valid labeling $\tilde{d}$ for $\tilde{x}_2$ such that $\tilde{d} \geq d$ in $O((\tilde{d}(V) - d(V))(n^6 + n^5 EO))$ time. Thus, using the OPR algorithm, each function in $f_k \leftarrow \cdots \leftarrow f_1$ can be minimized in this order in $O(n^8 + (n^7 + kn^2)EO)$ time in total. Note that the minimization of f_k can be started with the exreme base $b^{\prec \circ} \in \mathbf{B}(f_k)$ and the labeling $d = 0$.

Next we let d^R be an r-valid labeling for x_1. As above, one can show that d^R is also r-valid for $x_2 := \sum_{i \in I_1} \lambda_i b_2^{\prec i} \in \mathbf{B}(f^{(2)})$. Starting with x_2 and d^R, the reverse push/relabel (RPR) algorithm [6] finds a maximizer $\tilde{x}_2$ and an r-valid labeling $\tilde{d}^R$ for $\tilde{x}_2$ with $\tilde{d}^R \geq d^R$ in $O((\tilde{d}^R(V) - d^R(V))(n^6 + n^5 EO))$ time. So, every function in $f_1 \to \cdots \to f_k$ can be minimized in this order in $O(n^8 + (n^7 + kn^2)EO)$ time.

5.3   An Efficient Implementation of the Decomposition Algorithm

If we know vectors $\{\bar{x}_{\alpha_1}, \ldots, \bar{x}_{\alpha_{\ell-1}}\}$ such that $\xi_j^* < \alpha_j < \xi_{j+1}^*$ for each $j = 1, \ldots, \ell - 1$, it follows from the relation $f_{\alpha_1} \to \cdots \to f_{\alpha_{\ell-1}}$ that algorithms for parametric SFM solve problem (1) in the same asymptotic running time as a single SFM. In the decomposition algorithm, however, each vector $\bar{x}_\alpha$ is obtained in an on-line manner. As in [10] and [6], we introduce the procedure Slice, which plays the same role as the procedure DA in Section 4. We can assume that $f_{-\infty}$ has the unique minimizer $\varnothing$ and $f_{+\infty}$ has the unique minimizer V. For $S, T \subseteq V$ with $S \subset T$, define the function $f_{S,T} : 2^{T \setminus S} \to \mathbb{R}$ by $f_{S,T}(W) = f(W \cup S) - f(S)$ for each $W \subseteq T \setminus S$.

Let $S, T \in \mathcal{H}^*$ be subsets with $S \subset T$ such that S is the maximal minimizer of $f^{(1)} := f_{\alpha_1}$ and T is the maximal minimizer of $f^{(3)} := f_{\alpha_3}$ for $\alpha_1, \alpha_3 \in J \cup \{\pm\infty\}$ with $\alpha_1 < \alpha_3$. In addition, we are given subsets $S', T' \in \mathcal{H}^*$ such that $S' \subseteq S$ and $T \subseteq T'$. Now we regard $V' := T' \setminus S'$ as the universe. Define the function $f' : 2^{V'} \to \mathbb{R}$ by $f' := f_{S',T'}$ and define f'_α in the same way for any $\alpha \in J$, that is, $f'_\alpha := (f_\alpha)_{S',T'}$. Suppose that a labeling d_1^R on V' is r-valid for $x_1 = \sum_{i \in I_1} \lambda_i b_1^{\prec i} \in \mathbf{B}(f'_{\alpha_1})$ and a labeling d_3 on V' is valid for $x_3 = \sum_{j \in I_3} \mu_j b_3^{\prec j} \in \mathbf{B}(f'_{\alpha_3})$.

In the procedure Slice$(f', \alpha_1, \alpha_3, x_1, x_3, d_1^R, d_3)$ for finding vector $x^*_{T \setminus S}$ in $\mathbb{R}^{T \setminus S}$, we set $\alpha_2 := \alpha_{S,T}$, $x'_2 := \sum_{i \in I_1} \lambda_i b_2^{\prec i}$, $x''_2 := \sum_{j \in I_3} \mu_j b_2^{\prec j} \in \mathbf{B}(f'_{\alpha_2})$, and try to minimize $f^{(2)} := f_{\alpha_2}$ (or, equivalently, f'_{α_2}) by concurrently running the RPR algorithm, starting with x'_2, d_1^R, and the OPR algorithm, starting with x''_2, d_3. Suppose the RPR algorithm stops first with the base $x_2 = \sum_{i \in I_2} \lambda_i b_2^{\prec i}$ of $\mathbf{B}(f'_{\alpha_2})$ and the r-valid labeling d_2^R on V' for x_2. (The other case is symmetric.) By lemma 11, the maximal minimizer R of $f^{(2)}$ can be found in $O(n^3 EO)$

additional time and it holds that $S \subset R \subseteq T$. If $R = T$, the procedure Slice returns $x^*_{T \setminus S} := \overline{x}_{S,T}$. So we consider the case where $S \subset R \subset T$. If $2|R| < |S| + |T|$, we perform Slice$(f', \alpha_2, \alpha_3, x_2, x_3, d_2^R, d_3)$ and Slice$(f'', \alpha_1, \alpha_2, b^{\prec_\circ}, b^{\prec_\circ}, d_\circ, d_\circ)$ where $f'' := f_{S,R}$, $d_\circ = \mathbf{0} \in \mathbb{R}^{R \setminus S}$ and $\prec_\circ$ is some total order of $R \setminus S$. If $2|R| \geq |S| + |T|$, we continue the OPR algorithm until it stops. Then, replacing x_2 by the resulting base and letting d_2 be the valid labeling for x_2, we perform Slice$(f', \alpha_1, \alpha_2, x_1, x_2, d_1^R, d_2)$ and Slice$(f'', \alpha_2, \alpha_3, b^{\prec_\circ}, b^{\prec_\circ}, d_\circ, d_\circ)$ where $f'' := f_{R,T}$. Remark that in any case the problem is divided into two problems, the larger one of size n' and the quite smaller one, whose size is less than or equal to $n'/2$.

By running Slice$(f, -\infty, +\infty, b^{\prec_\circ}, b^{\prec_\circ}, d_\circ, d_\circ)$, problem (1) can be solved. Now we show this algorithm can be implemented to run in $\mathrm{O}(n^8 + n^7 \mathrm{EO})$ time. The running time analysis is quite similar to that of the lexicographically optimal flow algorithm proposed by Gallo $et\ al.$ [10]. Let RPR(n) (OPR(n)) denote the time to minimize every function $f_{k'} : 2^V \to \mathbb{R}$ in a strong map sequence $f_1 \to \cdots \to f_k$ with $k \leq n$ via the RPR (OPR) algorithm. That is, RPR(n) and OPR(n) are $\mathrm{O}(n^8 + n^7 \mathrm{EO})$. Once we incur $2(\mathrm{RPR}(n) + \mathrm{OPR}(n))$ time (the factor 2 comes from the fact that we concurrently run the two push/relabel algorithms), then, in the larger subproblems of the procedure Slice, subsequent implementations of the two push/relabel algorithms can be regarded as being free. We denote by $T(n')$ the upper bound on the time to run the procedure Slice$(f', \alpha_1, \alpha_2, x_1, x_2, d_1^R, d_2)$. We obtain the equalities $T(1) = \mathrm{O}(1)$ and

$$T(n) = \mathrm{O}(n^8 + n^7 \mathrm{EO}) + \max\{T(n_1) + T(n_2) + \cdots + T(n_h) :$$
$$n_1 + \cdots + n_h < n; \ n_1, \ldots, n_h \geq 1;$$
$$n_{h'} \leq \tfrac{1}{2}(n - n_1 - \cdots - n_{h'-1}) \text{ for } h' = 1, \ldots, h\}.$$

From these formulas, it is not difficult to verify that $T(n) = \mathrm{O}(n^8 + n^7 \mathrm{EO})$.

Theorem 12. *The optimal solution of problem* (1) *can be obtained in the same asymptotic running time as a single implementation of the push/relabel algorithm for SFM by Fleischer and Iwata, that is,* $\mathrm{O}(n^8 + n^7 \mathrm{EO})$ *time.*

This time complexity is better than the bound $\mathrm{O}((n^8 + n^7 \mathrm{EO}) \log n)$ which is obtained by performing Iwata's algorithm [14] iteratively in the decomposition algorithm. Quite recently, however, Orlin [21] developed a faster algorithm for SFM which runs in $\mathrm{O}(n^6 + n^5 \mathrm{EO})$ time. Therefore, using Orlin's algorithm, problem (1) can be solved in $\mathrm{O}(n^7 + n^6 \mathrm{EO})$ time by Theorem 7. It is not certain but we believe that parametric minimization versions of Orlin's algorithm could be developed and our framework would also work in that case.

6   Non-separable Convex Functions

This section deals with non-separable convex objective functions. We will see that, using the characterization of the lex-optimal base, some non-separable convex optimization problems can be solved in strongly polynomial time.

For $x \in \mathbb{R}^V$ and $u, v \in V$, let $x^{u,v} \in \mathbb{R}^V$ be the vector defined by

$$x^{u,v}(v') = \begin{cases} x(v) & \text{if } v' = u, \\ x(u) & \text{if } v' = v, \\ x(v') & \text{if } v' \in V \setminus \{u, v\}. \end{cases}$$

We let $g^P : \mathbb{R}^V \to \mathbb{R} \cup \{+\infty\}$ be a strictly convex function with $\mathrm{dom}\, g^P \supseteq \mathbb{R}_{\geq 0}^V$ which has invariant value under permutation of indices. That is, we have $g^P(x^{u,v}) = g^P(x)$ for any x and each u, v. If $V = \{1, 2\}$, $x(1)^2 + x(2)^2 + x(1)x(2)$ is such an example. Consider the minimization of g^P over the base polytope $\mathbf{B}(f)$. The following observation, which was originally shown by Maruyama [16] about three decades ago, is a generalization of Corollary 8.

Corollary 13. *The minimum norm point $x_{\mathrm{MN}} \in \mathbb{R}^V$ of $\mathbf{B}(f)$ is the (unique) minimizer of $\min_x\{g^P(x) : \mathbf{B}(f)\}$ and so this problem can be solved in strongly polynomial time.*

Proof. Let x be the (unique) optimal solution of $\min\{g^P(x) : \mathbf{B}(f)\}$. Let δ be any positive number and (u, v) be any ordered pair of elements in V such that $x + \delta(e_u - e_v) \in \mathbf{B}(f)$. By Corollary 3 and the uniqueness of the minimum norm point, in order to verify the statement, it suffices to show the inequality $x(u) \geq x(v)$. Assume, to the contrary, $x(u) < x(v)$. Put $\beta = x(v) - x(u) > 0$ and $y = x + \beta(e_u - e_v)$. By the definition of g^P, we have $g^P(x) = g^P(y)$. Let $\varepsilon = \frac{1}{2}\min\{\delta, \beta\}$. Then we have $0 < \varepsilon < \beta$ and $x + \varepsilon(e_u - e_v) \in \mathbf{B}(f)$. The strict convexity of g^P implies $g^P(x) = (1 - \frac{\varepsilon}{\beta})g^P(x) + \frac{\varepsilon}{\beta}g^P(y) > g^P(x + \varepsilon(e_u - e_v))$, a contradiction to the definition of x. □

Let w be a positive vector. Next we consider a class of problems whose optimal solutions are the lex-optimal bases. For distinct elements $u, v \in V$, define a cone $E(u, v) \subset \mathbb{R}^V$ by $E(u, v) = \{x \in \mathbb{R}_{\geq 0}^V : x(u)/w(u) = x(v)/w(v)\}$. Let $g^L(x) : \mathbb{R}^V \to \mathbb{R} \cup \{+\infty\}$ be a differentiable and strictly convex function with $\mathrm{dom}\, g^L \supseteq \mathbb{R}_{\geq 0}^V$. For any $x \in \mathbb{R}_{\geq 0}^V$ and each $v \in V$, let $\mathrm{D}_v(x) = \frac{\partial g^L(x)}{\partial x(v)}$. Now we consider the minimization of g^L over $\mathbf{B}(f)$ in the case where g^L satisfies

    L1. $\mathrm{D}_u(x) = \mathrm{D}_v(x)$, for each $u, v \in V$ and any $x \in E(u, v)$;

    L2. For any $x \in \mathbb{R}_{\geq 0}^V$ and $u, v \in V$, function $c_{u,v}(\lambda) = \mathrm{D}_u(x + \lambda(e_u - e_v))$
        is strictly increasing on the interval $\{\lambda \in \mathbb{R} : x + \lambda(e_u - e_v) \in \mathbb{R}_{\geq 0}^V\}$.

Clearly, the separable function $g^W(x)$ defined in Section 3 satisfies these properties. For example, if $V = \{1, 2, 3\}$ and $w = (2, 2, 3)$, then $x(1)^2 + x(2)^2 + \frac{3}{4}x(3)^2 + \frac{1}{2}x(1)x(2) + \frac{1}{2}x(1)x(3) + \frac{1}{2}x(2)x(3)$ satisfies L1 and L2. Using Corollary 3, we can generalize Corollary 9 as follows.

Corollary 14. *Suppose that g^L satisfies properties L1 and L2 and $x_L \in \mathbb{R}^V$ is the lex-optimal base of $\mathbf{B}(f)$ w.r.t. w. Then x_L is the (unique) optimal solution to problem $\min_x\{g^L(x) : x \in \mathbf{B}(f)\}$.*

Proof. Let x be the optimal solution to $\min\{g^L(x) : \mathbf{B}(f)\}$, and let $\delta > 0$ be any number and (u, v) be any pair of elements such that $x + \delta(e_u - e_v) \in \mathbf{B}(f)$. It suffices to show $\frac{x(u)}{w(u)} \geq \frac{x(v)}{w(v)}$. Assume that $\frac{x(u)}{w(u)} < \frac{x(v)}{w(v)}$. We let $\beta > 0$ be a number such that $\frac{x(u)+\beta}{w(u)} = \frac{x(v)-\beta}{w(v)}$. Put $y = x + \beta(e_u - e_v)$. Property L1 implies that $D_u(y) = D_v(y)$, and property L2 implies that $D_u(x) < D_u(y)$ and $D_v(y) < D_v(x)$. So we have $D_u(x) < D_v(x)$. Thus, $\exists \varepsilon > 0$ such that $x + \varepsilon(e_u - e_v) =: \widetilde{x} \in \mathbf{B}(f)$ and $g^L(x) > g^L(\widetilde{x})$, which contradicts the definition of x. $\qquad\square$

Acknowledgements

I thank Satoru Iwata for a number of useful suggestions and I am grateful to Tom McCormick for helpful comments. I also thank Satoru Fujishige, who informed me of [16].

References

1. R. E. Bixby, W. H. Cunningham and D. M. Topkis: The partial order of a polymatroid extreme point. *Mathematics of Operations Research*, **10** (1985), pp. 367–378.
2. F. A. Chudak and K. Nagano: Efficient solutions to relaxations of combinatorial problems with submodular penalties via the Lovász extension and non-smooth convex optimization. *Proceedings of the 18th Annual ACM-SIAM Symposium on Discrete Algorithms* (2007), pp. 79–88.
3. B. Dutta: The egalitarian solution and reduced game properties in convex games. *International Journal of Game Theory*, **19** (1990), pp. 153–169.
4. B. Dutta and D. Ray: A Concept of egalitarianism under participation constraints. *Econometrica*, **57** (1989), pp. 615–635.
5. J. Edmonds: Submodular functions, matroids, and certain polyhedra. In R. Guy, H. Hanai, N. Sauer, and J. Schönheim, editors, *Combinatorial Structures and Their Applications*, Gordon and Breach, New York, 1970, pp. 69–87.
6. L. Fleischer and S. Iwata: A push-relabel framework for submodular function minimization and applications to parametric optimization. *Discrete Applied Mathematics*, **131** (2003), pp. 311–322.
7. S. Fujishige: Lexicographically optimal base of a polymatroid with respect to a weight vector. *Mathematics of Operations Research*, **5** (1980), pp. 186–196.
8. S. Fujishige: Submodular systems and related topics. *Mathematical Programming Study*, **22** (1984), pp. 113–131.
9. S. Fujishige: *Submodular Functions and Optimization (Second Edition)*. Elsevier, Amsterdam, 2005.
10. G. Gallo, M. D. Grigoriadis and R. E. Tarjan: A fast parametric maximum flow algorithm and applications. *SIAM Journal on Computing*, **18** (1989), pp. 30–55.
11. H. Groenevelt: Two algorithms for maximizing a separable concave function over a polymatroid feasible region. *European Journal of Operational Research*, **54** (1991), pp. 227–236.
12. A. Hayrapetyan, C. Swamy and É. Tardos: Network design for information networks. *Proceedings of the 16th Annual ACM-SIAM Symposium on Discrete Algorithms* (2005), pp. 933–942.

13. D. S. Hochbaum: Lower and upper bounds for the allocation problem and other nonlinear optimization problems. *Mathematics of Operations Research*, **19** (1994), pp. 390–409.

14. S. Iwata: A faster scaling algorithm for minimizing submodular functions. *SIAM Journal on Computing*, **32** (2003), pp. 833–840.

15. K. Jain and V. V. Vazirani: Eisenberg-Gale markets: algorithms and structural properties. *Proceedings of the 39th ACM Symposium on Theory of Computing* (2007), to appear.

16. F. Maruyama: A unified study on problems in information theory via polymatroids. Graduation Thesis, University of Tokyo, Japan, 1978. (In Japanese.)

17. S. T. McCormick: Submodular function minimization. In K. Aardal, G. L. Nemhauser, and R. Weismantel, editors, *Discrete Optimization* (Handbooks in Operations Research and Management Science 12), Elsevier, Amsterdam, 2005, Chapter 7, pp. 321–391.

18. N. Megiddo: Optimal flows in networks with multiple sources and sinks. *Mathematical Programming*, **7** (1974), pp. 97–107.

19. K. Murota: Note on the universal bases of a pair of polymatroids. *Journal of Operations Research Society of Japan*, **31** (1988), pp. 565–573.

20. K. Murota: *Discrete Convex Analysis*. SIAM, Philadelphia, 2003.

21. J. B. Orlin: A faster strongly polynomial time algorithm for submodular function minimization. *Proceedings of the 12th IPCO Conference* (2007). This proceedings.

22. A. Schrijver: A combinatorial algorithm minimizing submodular functions in strongly polynomial time. *Journal of Combinatorial Theory (B)*, **80** (2000), pp. 346–355.

23. Y. Sharma, C. Swamy and D. P. Williamson: Approximation algorithms for prize-collecting forest problems with submodular penalty functions. *Proceedings of the 18th Annual ACM-SIAM Symposium on Discrete Algorithms* (2007), pp. 1275–1284.

Computational Geometric Approach to Submodular Function Minimization for Multiclass Queueing Systems

Toshinari Itoko[1] and Satoru Iwata[2]

[1] Tokyo Research Laboratory, IBM Japan
itoko@jp.ibm.com
[2] Research Institute for Mathematical Sciences, Kyoto University
iwata@kurims.kyoto-u.ac.jp

Abstract. This paper presents an efficient algorithm for minimizing a certain class of submodular functions that arise in analysis of multiclass queueing systems. In particular, the algorithm can be used for testing whether a given multiclass M/M/1 achieves an expected performance by an appropriate control policy. With the aid of the topological sweeping method for line arrangement, our algorithm runs in $O(n^2)$ time, where n is the cardinality of the ground set. This is much faster than direct applications of general submodular function minimization algorithms.

1 Introduction

Let V be a finite set of cardinality n. For a vector $\boldsymbol{x} := [x_i]_{i \in V}$ indexed by V and a subset $X \subseteq V$, we denote $\sum_{i \in X} x_i$ by $x(X)$. Let h be a nonnegative nondecreasing convex function. This paper deals with the problem of finding a subset $X \subseteq V$ that minimizes

$$f(X) := z(X) - y(X)\,h(x(X)) \qquad (X \subseteq V) \tag{1}$$

for given nonnegative vectors $\boldsymbol{x}, \boldsymbol{y}, \boldsymbol{z}$ indexed by V. Such a minimization problem arises in performance analysis of the most fundamental multiclass queueing system — multiclass M/M/1 (see Section 2).

This problem is a special case of submodular function minimization. A set function f is called *submodular* if it satisfies

$$f(X) + f(Y) \geq f(X \cap Y) + f(X \cup Y), \qquad \forall X, Y \subseteq V.$$

It can be shown that the function f in (1) is submodular (see Appendix). Recent results on submodular functions are expounded in Fujishige [10] and in McCormick [16].

A number of strongly polynomial algorithms have been devised for general submodular function minimization. The first one due to Grötschel, Lovász, and Schrijver [12,13] is based on the ellipsoid method, which is not efficient in practice. Combinatorial strongly polynomial algorithms are devised independently

M. Fischetti and D.P. Williamson (Eds.): IPCO 2007, LNCS 4513, pp. 267–279, 2007.

by Schrijver [19] and by Iwata, Fleischer, and Fujishige [14]. However, these combinatorial algorithms are not yet very fast. Even a very recent algorithm of Orlin [17], which is currently the fastest combinatorial strongly polynomial algorithm, runs in $O(n^5\gamma + n^6)$ time, where γ designates the time required for computing the function value of f. Thus, it is still desirable to have a faster algorithm for minimizing a specific class of submodular functions that naturally arise in applications.

Instead of applying an algorithm for general submodular function minimization, we take a completely different approach based on computational geometry. The first step is to interpret our problem in a three-dimensional space as follows. Each subset $X \subseteq V$ corresponds to a point $(x(X), y(X), z(X))$ in the three-dimensional space. The original problem is then equivalent to finding the minimum value of $\hat{f}(x, y, z) := z - y\,h(x)$ among all such points (x, y, z) corresponding to the subsets of V.

The convex hull of these 2^n points forms a special polytope called zonotope. It will be shown that the minimizer of $\hat{f}$ is among the lower extreme points of the zonotope, i.e., extreme points that are visible from below (see Section 3). The number of such lower extreme points are bounded by $O(n^2)$. Furthermore, exploiting the duality relation between a zonotope in a three-dimensional space and a line arrangement in a plane, we are able to enumerate all the lower extreme points in $O(n^2)$ time with the aid of the topological sweeping method of Edelsbrunner and Guibas [6,7]. Thus our algorithm finds a minimizer of f in $O(n^2)$ time and $O(n)$ space, if the function value of h is computable in a constant time. This is substantially more efficient than direct applications of general submodular function minimization algorithms.

In terms of the application to multiclass M/M/1, the above algorithm provides an efficient way of testing if a given performance specification is achievable by some control policy. Designing an appropriate control policy in the achievable case is another issue. Extending our approach, we also devise an algorithm for doing this in $O(n^4)$ time, which is yet faster than general submodular function minimization algorithms (see Section 4).

2   Multiclass Queueing Systems

This section is devoted to a brief exposition on a connection between our minimization problem and the performance analysis of the fundamental multiclass queueing system called multiclass M/M/1. For comparison, we also give a brief description of the same type of problems for nonpreemptive case.

2.1   Preemptive M/M/1

Multiclass M/M/1 is a system which deals with various types of jobs whose arrival interval and service time follow exponential distributions. Each job of different classes wait in different queues and the server chooses the job to serve the next by a *control policy*. A queueing system allowing preemptive control policies is called *preemptive*. In the following, the set of classes is denoted by $V = \{1, 2, \ldots, n\}$.

In a multiclass M/M/1, when the expected arrival rates and the expected service rates of the job classes are given, the performance of the system depends only on the control policy. A region of performance-measuring vectors achieved by all control policies is called *achievable region* (see e.g. [4]). The performance of a multiclass M/M/1 is often measured by the expected staying time vector $T := [T_i]_{i \in V}$, where T_i is the expected staying time in the system for class i jobs. For preemptive multiclass M/M/1, achievable region of T is known as follows.

Theorem 1 ([3]). *Consider a preemptive multiclass M/M/1 whose mean arrival rates are $\lambda := [\lambda_i]_{i \in V}$ and mean service rates are $\mu := [\mu_i]_{i \in V}$. Let ρ_i be the utilization λ_i / μ_i of the server for class i jobs and assume $\sum_{i \in V} \rho_i < 1$ to ensure the existence of equilibrium. The achievable region of the expected staying time vector $T := [T_i]_{i \in V}$ is a polyhedron represented by 2^n inequalities:*

$$\sum_{i \in X} \rho_i T_i \geq \frac{\sum_{i \in X} \dfrac{\rho_i}{\mu_i}}{1 - \sum_{i \in X} \rho_i}, \quad \forall X \subseteq V. \tag{2}$$

Given a target expected staying time vector $\check{T}$, it is important for system designers to *check performance achievability*: whether $\check{T}$ is in the achievable region ($\check{T}$ is achieved by some control policy) or not. This problem was posed by Federgruen and Groenevelt [8]. They provided an efficient algorithm for the special case of identical service time distribution. This assumption is too restrictive in practice, as we usually classify the jobs by their properties including expected service time.

If we define $x_i := \rho_i$, $y_i := \dfrac{\rho_i}{\mu_i}$ and $h(x) := \dfrac{1}{1-x}$, then $y(X) h(x(X))$ coincides with the right-hand side function of (2). Furthermore, if we define $z_i := \rho_i \check{T}_i$, then the problem of checking performance achievability of preemptive multiclass M/M/1 is reduced to our minimization problem. The target expected staying time vector $\check{T}$ is achievable if and only if the minimum value of f is equal to zero.

For preemptive multiclass M/M/1, there is an another representation of achievable region. Bertsimas, Paschalidis, and Tsitsiklis [2] and Kumar and Kumar [15] independently observed that the achievable region is the projection of a polyhedron in a higher dimensional space. This makes it possible to check the achievability by solving linear programming problem, which however involves $O(n^2)$ variables and $O(n^2)$ inequalities.

2.2 Nonpreemptive M/G/1

For nonpreemptive M/M/1, which does not allow preemption, the performance achievability can be tested by a simpler method. The achievable region of the expected waiting time in queue $\boldsymbol{W} := [W_i]_{i \in V}$ is a polyhedron represented by 2^n inequalities:

$$\sum_{i \in X} \rho_i W_i \geq \left(\sum_{i \in V} \frac{\rho_i}{\mu_i} \right) \frac{\displaystyle\sum_{i \in X} \rho_i}{1 - \displaystyle\sum_{i \in X} \rho_i}, \qquad \forall X \subseteq V.$$

This is obtained as a special case of the fact shown in Gelenbe and Mitrani [11] that the achievable region of the nonpreemptive M/G/1, which admits general service time distributions, is characterized by

$$\sum_{i \in X} \rho_i W_i \geq \left(\frac{1}{2} \sum_{i \in V} \lambda_i M_i^2 \right) \frac{\displaystyle\sum_{i \in X} \rho_i}{1 - \displaystyle\sum_{i \in X} \rho_i}, \qquad \forall X \subseteq V,$$

where M_i^2 denotes the second order moment of the service time distribution for class i.

Let $\check{\boldsymbol{W}}$ be a target expected waiting time vector. The problem of checking performance achievability is reduced to minimizing a submodular function b in the form of

$$b(X) := z(X) - h(x(X)) \qquad (X \subseteq V),$$

where $x_i := \rho_i$, $z_i := \rho_i \check{W}_i$, and $h(x) := \dfrac{cx}{1-x}$ with $c = \dfrac{1}{2} \sum_{i \in V} \lambda_i M_i^2$. This is much simpler than our problem. In fact, it can be solved by sorting job classes in the order of z_i/x_i. For $k = 0, 1, \ldots, n$, let Y_k denote the set of k jobs with smallest values of z_i/x_i. Then the minimizer of b is among the candidates Y_k for $k = 0, 1, \ldots, n$. See Federgruen and Groenevelt [9] for validity.

3   Geometric Approach

In this section, we present an algorithm for finding a minimum value of f defined in (1). The problem can be seen in a three-dimensional space as follows. Each subset $X \subseteq V$ corresponds to a point $(x(X), y(X), z(X))$ in the three-dimensional space. Let $\mathbb{R}_+$ denote the set of nonnegative reals. We also write $\boldsymbol{u}(X) := (x(X), y(X), z(X)) \in \mathbb{R}_+^3$ and $\boldsymbol{U} := \{\boldsymbol{u}(X) \mid X \subseteq V\}$. Then our problem is equivalent to finding a point $(x, y, z) \in \boldsymbol{U}$ that attains the minimum value of $\widehat{f}(x, y, z) = z - y\, h(x)$. An example of a contour surface of $\widehat{f}$ is shown in Fig. 1.

The convex hull of $\boldsymbol{U}$ forms a special polytope called *zonotope*, which is defined by the bounded linear combination of vectors (for example, see Fig. 2), and we denote the zonotope by Z, namely

$$Z := \mathrm{conv}(\boldsymbol{U}) = \left\{ \sum_{i \in V} \eta_i \boldsymbol{u}_i \,\middle|\, 0 \leq \eta_i \leq 1 \ (\forall i \in V) \right\},$$

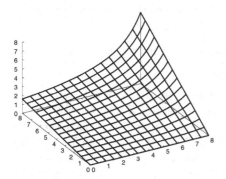

Fig. 1. A contour surface of $\widehat{f}$ in the case of $h(x) = 2/(10 - x)$

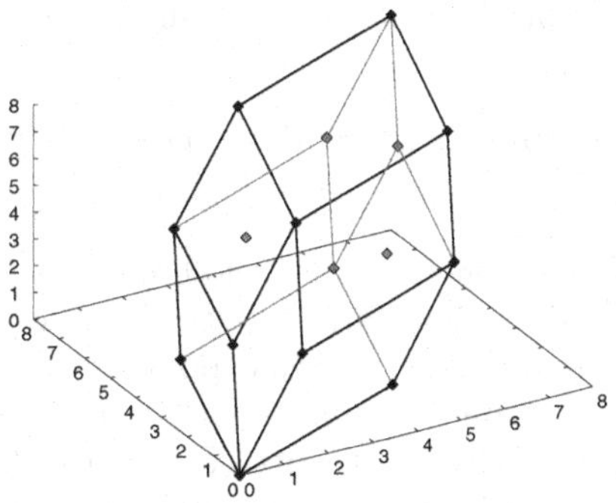

Fig. 2. A zonotope generated by $(4, 1, 1)$, $(2, 1, 3)$, $(1, 2, 3)$, $(1, 4, 1)$

where $\boldsymbol{u}_i := \boldsymbol{u}(\{i\})$ $(\forall i \in V)$.

A point (x, y, z) in Z is called a lower point if $(x, y, z') \notin Z$ for any $z' < z$. If in addition (x, y, z) is an extreme point of Z, it is called a lower extreme point. The number of lower extreme points of Z is known to be $O(n^2)$, which is clarified in Section 3.3.

Our algorithm enumerates all the lower extreme points of Z, and then it identifies a lower extreme point that attains the minimum value of $\widehat{f}$. It will be shown in Section 3.2 that the minimum value among these lower extreme points is in fact the minimum value of $\widehat{f}$ among all the points in $\boldsymbol{U}$. How to enumerate the lower extreme points will be described in Section 3.3. The total running time of this algorithm is $O(n^2)$.

3.1   Lower Extreme Points

Every lower point of Z is described as a maximizer of a certain linear objective function whose coefficient of z is negative. For any $\alpha, \beta \in \mathbb{R}$, we denote by $F(\alpha, \beta)$ the set of maximizers for $(\alpha, \beta, -1)$ direction, namely

$$F(\alpha, \beta) := \mathrm{Argmax}\{\alpha\, x + \beta\, y - z \mid (x, y, z) \in Z\}.$$

For a fixed (α, β), elements in V are classified by the sign of $\alpha\, x_i + \beta\, y_i - z_i$, namely

$$\begin{aligned}
S^+(\alpha, \beta) &:= \{\, i \in V \mid \alpha\, x_i + \beta\, y_i - z_i > 0\,\}, \\
S^\circ(\alpha, \beta) &:= \{\, i \in V \mid \alpha\, x_i + \beta\, y_i - z_i = 0\,\}, \\
S^-(\alpha, \beta) &:= \{\, i \in V \mid \alpha\, x_i + \beta\, y_i - z_i < 0\,\}.
\end{aligned} \tag{3}$$

Then $F(\alpha, \beta)$ is given by

$$F(\alpha, \beta) = \left\{ \boldsymbol{u}(S^+(\alpha, \beta)) + \sum_{i \in S^\circ(\alpha, \beta)} \eta_i\, \boldsymbol{u}_i \,\middle|\, \forall i \in S^\circ(\alpha, \beta),\ 0 \le \eta_i \le 1 \right\}. \tag{4}$$

This implies the following lemma that characterizes the lower extreme points of Z.

Lemma 1. *A vector v is a lower extreme point of Z if and only if $v = u(S^+(\alpha, \beta))$ for some (α, β).*

Proof. Since $u_i \geq 0$, it follows from (4) that $u(S^+(\alpha, \beta))$ is an extreme point of $F(\alpha, \beta)$. Hence $u(S^+(\alpha, \beta))$ is an lower extreme point of Z. Conversely, suppose v is an lower extreme point of Z. There exists a pair (α, β) such that v is the unique maximizer of $\alpha x + \beta y - z$ in Z. Note that $v = u(X)$ for some $X \subseteq V$. Then we have $X = S^+(\alpha, \beta) \cup Y$ for some $Y \subseteq S^\circ(\alpha, \beta)$. Furthermore, since v is the unique maximizer, $u_i = 0$ holds for any $i \in Y$, which implies $u(X) = u(S^+(\alpha, \beta))$.

We denote $\{S^+(\alpha, \beta) \mid \alpha, \beta \in \mathbb{R}\}$ by $\mathcal{L}$. Then Lemma 1 asserts that the set of lower extreme points are given by $\{u(X) \mid X \in \mathcal{L}\}$. The following two lemmas concerning lower points of Z will be used in the proof of the validity of our algorithm in Section 3.2.

Lemma 2. *Any lower point v that is on an edge of Z is a convex combination of two lower extreme points $u(X_1)$ and $u(X_2)$ with $X_1 \subseteq X_2$.*

Proof. There exists a pair (α, β) such that $F(\alpha, \beta)$ is the edge that contains v. Then it follows from (4) that $F(\alpha, \beta)$ is a line segment between $u(X_1)$ and $u(X_2)$, where $X_1 = S^+(\alpha, \beta)$ and $X_2 = S^+(\alpha, \beta) \cup S^\circ(\alpha, \beta)$.

Lemma 3. *Any lower point v of Z is a convex combination of some at most three lower extreme points $u(X_0)$, $u(X_1)$, and $u(X_2)$ with $X_0 \subseteq X_1 \subseteq X_2$.*

Proof. There exists a pair (α, β) such that $F(\alpha, \beta)$ is the minimal face that contains v. Then $u(X_0)$ with $X_0 = S^+(\alpha, \beta)$ is an extreme point of $F(\alpha, \beta)$. Let t be the intersection of the half line from $u(X_0)$ through v and the boundary of $F(\alpha, \beta)$. Note that v is a convex combination of $u(X_0)$ and t. Since t is on an edge of Z, Lemma 2 implies that t is a convex combination of lower extreme points $u(X_1)$ and $u(X_2)$ with $X_1 \subseteq X_2$. Furthermore, since $u(X_1)$ and $u(X_2)$ are extreme points of $F(\alpha, \beta)$, we have $X_0 \subseteq X_1, X_2$. Therefore, v is a convex combination of $u(X_0)$, $u(X_1)$, and $u(X_2)$ with $X_0 \subseteq X_1 \subseteq X_2$.

3.2 Finding the Minimum Value

The following theorem shows that it suffices to examine the lower extreme points of Z on behalf of the points in U. This leads us to an efficient algorithm for finding the minimum value of f, provided that an enumeration algorithm for the lower extreme points is available.

Theorem 2. *The minimum value of f is attained by a member of $\mathcal{L}$, i.e.,*

$$\min\{f(X) \mid X \subseteq V\} = \min\{f(X) \mid X \in \mathcal{L}\}.$$

Proof. Let $\bar{\boldsymbol{v}} = (\bar{x}, \bar{y}, \bar{z})$ be a lower point of Z such that $\bar{x} = x(Y)$ and $\bar{y} = y(Y)$ for $Y \subseteq V$. By Lemma 3, there exist three lower extreme points $\boldsymbol{u}(X_0)$, $\boldsymbol{u}(X_1)$, and $\boldsymbol{u}(X_2)$ of Z with $X_0 \subseteq X_1 \subseteq X_2$ such that

$$\bar{\boldsymbol{v}} = \sigma_0 \boldsymbol{u}(X_0) + \sigma_1 \boldsymbol{u}(X_1) + \sigma_2 \boldsymbol{u}(X_2)$$

for some $\sigma_0, \sigma_1, \sigma_2 \geq 0$ with $\sigma_0 + \sigma_1 + \sigma_2 = 1$. We denote $\boldsymbol{u}(X_j)$ by (x_j, y_j, z_j) for $j = 0, 1, 2$. Then we have

$$\begin{aligned}
\bar{y}\,h(\bar{x}) &= (\sigma_0 y_0 + \sigma_1 y_1 + \sigma_2 y_2)\,h(\sigma_0 x_0 + \sigma_1 x_1 + \sigma_2 x_2) \\
&\leq (\sigma_0 y_0 + \sigma_1 y_1 + \sigma_2 y_2)(\sigma_0 h(x_0) + \sigma_1 h(x_1) + \sigma_2 h(x_2)) \\
&= \sigma_0 y_0 h(x_0) + \sigma_1 y_1 h(x_1) + \sigma_2 y_2 h(x_2) - \sigma_0 \sigma_1 (y_1 - y_0)(h(x_1) - h(x_0)) \\
&\quad - \sigma_1 \sigma_2 (y_2 - y_1)(h(x_2) - h(x_1)) - \sigma_0 \sigma_2 (y_2 - y_0)(h(x_2) - h(x_0)) \\
&\leq \sigma_0 y_0 h(x_0) + \sigma_1 y_1 h(x_1) + \sigma_2 y_2 h(x_2),
\end{aligned}$$

where the first inequality follows from the convexity of h and the second one from the monotonicity. Since $z(Y) \geq \bar{z} = \sigma_0 z_0 + \sigma_1 z_1 + \sigma_2 z_2$, we obtain

$$\begin{aligned}
f(Y) &= z(Y) - y(Y)\,h(x(Y)) \\
&\geq \bar{z} - \bar{y}\,h(\bar{x}) \\
&\geq \sigma_0 (z_0 - y_0 h(x_0)) + \sigma_1 (z_1 - y_1 h(x_1)) + \sigma_2 (z_2 - y_2 h(x_2)) \\
&= \sigma_0 f(X_0) + \sigma_1 f(X_1) + \sigma_2 f(X_2).
\end{aligned}$$

Therefore, if $f(Y)$ attains the minimum value, then X_0, X_1, and X_2 must attain the minimum value as well. Thus the minimum value of f is attained by a member of $\mathcal{L}$.

3.3   Duality Between Zonotope and Hyperplane Arrangement

In this section, we discuss how to enumerate all the lower extreme points of Z. A one-to-one correspondence has been established between zonotopes in the d-dimensional space and hyperplane arrangements in the $d - 1$-dimensional space (see e.g. [5,20]). We exploit this duality principle with $d = 3$.

To visualize the transition of $S^+(\alpha, \beta)$ in $\mathcal{L}$ with respect to α and β, we consider the arrangement of n lines $l_i : \alpha x_i + \beta y_i - z_i = 0$, for $i \in V$ in the (α, β)-plane. Then it follows from Lemma 1 that the lower extreme points of Z corresponds to the cells in this line arrangement. Note that the number of cells in the line arrangement is $O(n^2)$, and so is the number of lower extreme points of Z. Further correspondence between the lower faces of Z and the components of the line arrangement are summarized in Table 1 (see also Fig. 3).

Based on this duality, it suffices to enumerate all the cells of the line arrangement. Since $\boldsymbol{x}$ and $\boldsymbol{y}$ are nonnegative, algorithms for sweeping the arrangement along α or β axis keep a maximal chain on V that corresponds to n cells and enumerate all the cells one by one with updating the chain. Our algorithm maintain not only the chain but also the vectors $\boldsymbol{u}(S^+)$ for all S^+ in the chain to

Table 1. Correspondence between lower faces of Z and components of line arrangement

Lower faces of Z	Components of line arrangement
extreme point	cell
edge	line segment
facet	intersection point

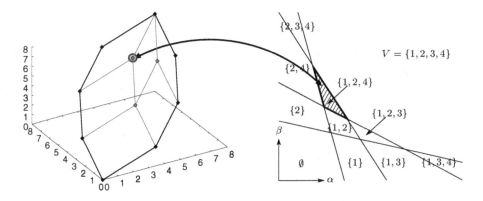

Fig. 3. Correspondence between a lower extreme point of Z and a cell of line arrangement

compute the value of $f(S^+)$ in a constant time on expected. This is achieved by minor modifications of existing algorithms at no additional expense of running time bound and space complexity.

For sweeping line arrangement, the topological sweeping method [6,7] is the most efficient algorithm, which runs in $O(n^2)$ time and $O(n)$ space. Thus we obtain an algorithm to solve the minimization problem in $O(n^2)$ time and $O(n)$ space. An implementation of the topological sweeping method is available from the Web [18].

4   Extension

The algorithm presented in Section 3 enables us to determine whether a preemptive multiclass M/M/1 achieves a performance by some control policy or not. However, even if a given performance specification turns out to be achievable, the algorithm does not yield a concrete control policy. In the real application, an efficient algorithm for finding such an *achieving control policy* is essential. In order to find an achieving control policy in the achievable case, we discuss the following problem.

Let g be a set function defined by

$$g(X) := y(X) h(x(X))   (X \subseteq V) \tag{5}$$

with nonnegative vectors x, y indexed by V and a nonnegative nondecreasing convex function h. Consider a polytope (i.e. bounded polyhedron)

$$B(g) := \{z \mid z(X) \geq g(X), \forall X \subset V \text{ and } z(V) = g(V)\}. \qquad (6)$$

We now deal with the problem of finding a convex combination of some extreme points of $B(g)$ for a given point on $B(g)$. Generally, a polytope in the form of $B(g)$ is called a *base polytope* if g is supermodular i.e.,

$$g(X) + g(Y) \leq g(X \cap Y) + g(X \cup Y), \qquad \forall X, Y \subseteq V.$$

Since the function g in (5) is supermodular (see Appendix), this problem is a membership problem on base polytope for a specific class of supermodular function.

Recall Theorem 1 and define g with $x_i := \rho_i$, $y_i := \dfrac{\rho_i}{\mu_i}$ and $h(x) := \dfrac{1}{1-x}$ as in Section 2.1. For simplicity, we introduce a new performance vector z by $z_i := \rho_i T_i$. Without loss of generality, assume $z(V) = g(V)$ for any performance z. This assumption means that we consider only *work-conserving* system, which never idle if there exists a job in the queue. Then the achievable region of z forms a base polytope in the form of (6), and each of its extreme points is achieved by some absolute priority policy; An *absolute priority policy* is a control policy defined by a total order $\prec$ on V which gives preference to jobs of class i over jobs of class j if $i \prec j$.

We provide an achieving control policy as a random mixture of absolute priority policies. For any achievable performance $\check{z}$, our algorithm finds a convex combination $\check{z} = \sum_{k=1}^{n} a_k \, \pi^{(k)}$ where $\pi^{(k)}$ is an extreme point of $B(g)$. An achieving control policy is derived as a control policy which follows the kth corresponding absolute priority policy with probability a_k.

Note that such a policy is not the only achieving control policy, i.e. there can be other policies that achieve a given vector of performance. For example, in the case of a nonpreemptive multiclass M/G/c with identical servers, Federgruen-Groenevelt [9] provided an algorithm for finding another type of achieving control policy, which is a slightly generalized *dynamic* or *delay dependent* control policy, where a job's priority is proportional to its time spent in the queue with the coefficients being class dependent.

Generally, minimizing a submodular function f defined by $f(X) := z(X) - g(X)$ $(X \subseteq V)$ enables us to determine whether a point z is in $B(g)$ or not. The point z is in $B(g)$ if and only if the minimum value of f, as well as $f(V)$, is equal to zero. Most combinatorial algorithms for general submodular function minimization yield a convex combination for z when z is in $B(g)$. However the algorithm shown in Section 3 does not.

By extending geometric approach, we present an efficient algorithm finding a convex combination for a given point on $B(g)$. With the aid of the topological sweeping method for *plane* arrangement, the algorithm runs in $O(n^4)$ time and $O(n^2)$ space, that is still faster than general submodular function minimization algorithms.

Algorithm for Finding a Convex Combination

As shown in Fig. 4, our algorithm finds extreme points contributing to a convex combination by successively projecting a considering point into lower dimensional faces of $B(g)$. First, the algorithm restricts the considering face F into the minimal face that includes the current point (Step 1). Note that the current point is initialized by a given point on $B(g)$. Second, it selects an appropriate extreme point of F as an observer's position for the next step (Step 2). If the dimension of F is zero i.e. F is a point, it terminates. Third, with the observer, it projects the current point onto a proper face of F (Step 3). The algorithm terminates after at most n iterations (of Steps 1–3) since the dimension of the considering face decreases at least one by one. The selected observers (extreme points of $B(g)$) are the solution.

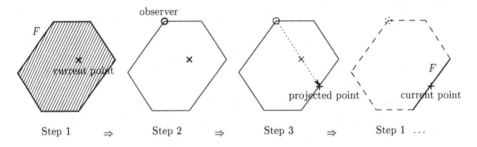

Fig. 4. Frame advance of the algorithm in the case of $n = 3$

The bottle neck of the algorithm is finding a projected point in Step 3. As shown below, it can be reduced to sweeping cells of a plane arrangement in the three-dimensional space. According to [1], this is possible in $O(n^3)$ time and $O(n^2)$ space, which indicates that the entire algorithm runs in $O(n^4)$ time and $O(n^2)$ space.

In Section 3, we proved that, for checking whether a point z is in $B(g)$ or not, it is sufficient to enumerate all the cells of an arrangement defined by n lines in the (α, β)-plane:

$$\alpha x_i + \beta y_i - z_i = 0 \qquad (i \in V). \tag{7}$$

This is because each of the cells corresponds to $X \subseteq V$ for which the validity of an inequality $z(X) \geq g(X)$ should be checked. The set of those subsets was denoted by $\mathcal{L}$. Note that all the definitions of z, u, Z and $\mathcal{L}$ in Section 3 depends on the point z to be checked. If z varies, so do the three-dimensional points $u(X)$ ($X \subseteq V$). As the three-dimensional points moving, their convex hull Z and its lower extreme points changes. We denote by $\mathcal{L}(z)$ the family of sets that correspond to the lower extreme points of Z depending on z.

In Step 3, a point to be checked moves along the half line from an observer vertex π through the current point $\check{z}$. Any point to be checked is defined by $z(t) := \pi + t(\check{z} - \pi)$ with $t \in [1, +\infty)$. Consider z_i in (7) as a variable with t, and replace z_i with $z_i(t) = \pi_i + t(\check{z}_i - \pi_i)$ for all $i \in V$. We obtain n planes in the (α, β, t)-space:

$$\alpha\, x_i + \beta\, y_i + t\,(\pi_i - \check{z}_i) - \pi_i = 0 \qquad (i \in V). \tag{8}$$

All the cells of the arrangement defined by (8) corresponds to the set family

$$\mathcal{L}' := \{S^+(\alpha, \beta, t) \mid \alpha, \beta \in \mathbb{R},\ t \in [1, +\infty)\} \tag{9}$$

where $S^+(\alpha, \beta, t) := \{i \in V \mid \alpha\, x_i + \beta\, y_i - z_i(t) > 0\}$. Since $\mathcal{L}'$ is the union of $\mathcal{L}(z(t))$ for all $t \in [1, +\infty)$, checking validity of the inequalities $z(X) \geq g(X)$ ($X \in \mathcal{L}'$) is sufficient to determine whether $z(t) \in B(g)$ for any $t \in [1, +\infty)$. This readily follows from the proofs in Section 3 for the case of a fixed t. By selecting the maximal t such that $z(X) \geq g(X)$ is valid for all $X \in \mathcal{L}'$ and denoting it by $\bar{t}$, we can find the projected point as $z(\bar{t})$.

5 Conclusion

We have presented an efficient algorithm for minimizing a class of submodular functions that arises in queueing analysis: checking performance achievability of preemptive multiclass M/M/1. With the aid of the topological sweeping method for line arrangement, our algorithm runs in $O(n^2)$ time, which is much faster than previously known methods. We have also presented a fast algorithm for finding a concrete control policy in the achievable case.

Acknowledgments

We thank Takeshi Tokuyama of Tohoku University for helpful information on topological sweeping method. We also thank Takayuki Osogami of Tokyo Research Laboratory, IBM Japan, for helpful information on performance analysis of priority queues.

References

1. E. Anagnostou, V. G. Polimenis, and L. J. Guibas. Topological sweeping in three dimensions. In *Proceedings of International Symposium on Algorithms*, pages 310–317. Springer-Verlag, 1990.
2. D. Bertsimas, I. Ch. Paschalidis, and J. N. Tsitsiklis. Optimization of multiclass queueing networks: Polyhedral and nonlinear characterizations of achievable performance. *The Annals of Applied Probability*, 4:43–75, 1994.
3. E. G. Coffman, Jr. and I. Mitrani. A characterization of waiting time performance realizable by single-server queues. *Operations Research*, 28:810–821, 1980.
4. M. Dacre, K. Glazebrook, and J. Niño-Mora. The achievable region approach to the optimal control of stochastic systems. *Journal of the Royal Statistical Society, B*, 61:747–791, 1999.
5. H. Edelsbrunner. *Algorithms in Combinatorial Geometry*. Springer-Verlag, 1987.

6. H. Edelsbrunner and L. J. Guibas. Topologically sweeping an arrangement. *Journal of Computer and Systems Sciences*, 38:165–194, 1989.
7. H. Edelsbrunner and L. J. Guibas. Topologically sweeping an arrangement — a correction. *Journal of Computer and Systems Sciences*, 42:249–251, 1991.
8. A. Federgruen and H. Groenevelt. Characterization and optimization of achievable performance in general queueing systems. *Operations Research*, 36:733–741, 1988.
9. A. Federgruen and H. Groenevelt. M/G/c queueing systems with multiple customer classes: Characterization and control of achievable performance under nonpreemptive priority rules. *Management Science*, 34:1121–1138, 1988.
10. S. Fujishige. *Submodular Function and Optimization*. North-Holland, 2005.
11. E. Gelenbe and I. Mitrani. *Analysis and Synthesis of Computer Systems*. Academic Press, 1980.
12. M. Grötschel, L. Lovász, and A. Schrijver. The ellipsoid method and its consequences in combinatorial optimization. *Combinatorica*, 1:169–197, 1981.
13. M. Grötschel, L. Lovász, and A. Schrijver. *Geometric Algorithms and Combinatorial Optimization*. Springer-Verlag, 1988.
14. S. Iwata, L. Fleischer, and S. Fujishige. A combinatorial strongly polynomial algorithm for minimizing submodular functions. *Journal of the ACM*, 48:761–777, 2001.
15. S. Kumar and P. R. Kumar. Performance bounds for queueing networks and scheduling policies. *IEEE Transactions on Automatic Control*, 39:1600–1611, 1994.
16. S. T. McCormick. Submodular function minimization. In K. Aardal, G. Nemhauser, and R. Weismantel, editors, *Handbook on Discrete Optimization*. Elsevier, 2005.
17. J. B. Orlin. A faster strongly polynomial time algorithm for submodular function minimization. In *Proceedings of the Twelfth International Conference on Integer Programming and Combinatorial Optimization*. Springer-Verlag, to appear.
18. E. Rafalin, D. Souvaine, and I. Streinu. Topological sweep in degenerate cases. In *Proceedings of the 4th International Workshop on Algorithm Engineering and Experiments (ALENEX)*, pages 155–165, 2002. (Their implementation is avialable from http://www.cs.tufts.edu/research/geometry/sweep/).
19. A. Schrijver. A combinatorial algorithm minimizing submodular functions in strongly polynomial time. *Journal of Combinatorial Theory, B*, 80:346–355, 2000.
20. G. M. Ziegler. *Lectures on Polytopes*. Springer-Verlag, 1995.

Appendix

This Appendix is devoted to showing that the function f is a submodular function. For this purpose, it suffices to show that the function g defined by $g(X) := y(X) h(x(X))$ is supermodular, i.e.,

$$g(X) + g(Y) \leq g(X \cap Y) + g(X \cup Y), \qquad \forall X, Y \subseteq V.$$

Since h is convex, for any $X \subseteq V$ and $i, j \in V$, we have

$$h\big(x(X \cup \{i\})\big) \leq \frac{x_j}{x_i + x_j} h\big(x(X)\big) + \frac{x_i}{x_i + x_j} h\big(x(X \cup \{i, j\})\big),$$

$$h\big(x(X \cup \{j\})\big) \leq \frac{x_i}{x_i + x_j} h\big(x(X)\big) + \frac{x_j}{x_i + x_j} h\big(x(X \cup \{i, j\})\big).$$

By adding these two inequalities, we obtain

$$h\big(x(X \cup \{i\})\big) + h\big(x(X \cup \{j\})\big) \le h\big(x(X)\big) + h\big(x(X \cup \{i, j\})\big),$$

which implies that $h(x(\cdot))$ is a supermodular function. Because of this supermodularity, the nonnegativity of x and y, and the monotonicity of h, we have

$$
\begin{aligned}
&g(X \cup Y) + g(X \cap Y) - g(X) - g(Y) \\
&= h(x(X \cup Y))\, y(X \cup Y) + h(x(X \cap Y))\, y(X \cap Y) - h(X)\, y(X) - h(Y)\, y(Y) \\
&= \big(h(x(X \cup Y)) + h(x(X \cap Y)) - h(x(X)) - h(x(Y))\big)\, y(X \cap Y) \\
&\quad + \big(h(x(X \cup Y)) - h(X)\big)\, y(X \setminus Y) + \big(h(x(X \cup Y)) - h(Y)\big)\, y(Y \setminus X) \\
&\ge 0
\end{aligned}
$$

for any $X, Y \subseteq V$. Thus g is shown to be supermodular. In addition, it is easy to see that $g(X) \ge 0$ for any $X \subseteq V$ and $g(\emptyset) = 0$ hold.

Generating Multiple Solutions for
Mixed Integer Programming Problems

Emilie Danna, Mary Fenelon, Zonghao Gu, and Roland Wunderling

ILOG, Inc. 889 Alder Avenue, Suite 200, Incline Village, NV 89451
{edanna,mfenelon,gu,rwunderling}@ilog.com

Abstract. As mixed integer programming (MIP) problems become easier to solve in pratice, they are used in a growing number of applications where producing a unique optimal solution is often not enough to answer the underlying business problem. Examples include problems where some optimization criteria or some constraints are difficult to model, or where multiple solutions are wanted for quick solution repair in case of data changes. In this paper, we address the problem of effectively generating multiple solutions for the same model, concentrating on optimal and near-optimal solutions. We first define the problem formally, study its complexity, and present three different algorithms to solve it. The main algorithm we introduce, the one-tree algorithm, is a modification of the standard branch-and-bound algorithm. Our second algorithm is based on MIP heuristics. The third algorithm generalizes a previous approach that generates solutions sequentially. We then show with extensive computational experiments that the one-tree algorithm significantly outperforms previously known algorithms in terms of the speed to generate multiple solutions, while providing an acceptable level of diversity in the solutions produced.

1 Introduction

Solving a standard mixed-integer programming (MIP) model $\min_{x \in X} c^T x$ where $X = \{x \in \mathbb{R}^d : Ax \leq b, x_i \in \mathbb{Z}, \ \forall i \in I \subseteq \{1, \dots, d\}\}$ usually means finding a solution x^* that is feasible: $x^* \in X$, and optimal: $c^T x^* \leq c^T x, \ \forall x \in X$. However, there might exist not only one but several different solutions that fit those two criteria. In this paper, we address the problem of generating multiple feasible solutions effectively for the same model, concentrating on optimal and near-optimal solutions.

1.1 Motivation

The three main reasons that motivate generating multiple solutions instead of only one come from the applications of mixed-integer linear programming. MIP is used extensively in industry to make short-term and long-term decisions, such as scheduling operations on various machines in a factory, deciding how much of each product to manufacture, choosing new locations for additional factories, etc.

M. Fischetti and D.P. Williamson (Eds.): IPCO 2007, LNCS 4513, pp. 280–294, 2007.

However, the mathematical model given to a MIP solver is often a simplification of the real business problem. Such a model may leave out details that are difficult to express as linear expressions or that make the model hard to solve. The data used in the mathematical model are also often an estimate of the fluctuating real data. More importantly, some optimization criteria are inherently subjective and difficult to quantify. For example, Schittekat and Sorensen [19] studied the problem of choosing subcontractors in the automotive industry. The obvious choice criterion is the price each subcontractor demands for the same job, but there are other factors to consider, such as the quality of work and service, and these factors are difficult to quantify. Because of these differences between the mathematical model and the real business problem, it is interesting to generate multiple optimal or near-optimal solutions for the mathematical model so that the decision maker can examine them, and, in the end, choose the best solution overall, i.e., the one that also performs best for the criteria that could not be expressed in the MIP model.

The second reason for generating multiple solutions is that MIP is increasingly used beyond the simple framework of formulating a model, solving it, and implementing the solution. If the data have changed between the moment the model was written and the moment the solution is to be implemented (for example in scheduling, if a machine has broken down), then it is valuable to have immediately at hand an alternative solution that does not use this machine, or a variety of solutions that can be used to repair the current solution. Another application mentioned by Schittekat and Sorensen [19] is that being able to show a subcontractor alternative solutions that have approximately the same cost and use the subcontractor's competitors was a very effective tool in negotiating.

Finally, a more technical application is that MIP is increasingly used to solve subproblems inside the branch-and-cut framework itself. For example, several approaches [9,10,2,8] formulate cut separation as a MIP where the objective is to maximize the cut violation, and each solution corresponds to a cut. Although the efficiency of a cut can be predicted approximately by its violation, more complex measures [14,5] could be useful to evaluate efficiency more accurately, but they are too complex to express in the aforementioned MIP models. Moreover, adding several cuts at a time is a well known technique to obtain quick progress in the best bound. MIP models for cut separation would therefore be an interesting application for generating multiple solutions.

1.2   Related Work

Generating multiple solutions to optimization problems has been the subject of few papers: Lee et al. [15] generate multiple solutions for LP; Bacchus solves #SAT in [1]; Schittekat and Sorensen [19] use metaheuristics to generate solutions for a logistics problem. As for generating multiple solutions for MIP, Glover et al. [11] present an interesting approach based on MIP heuristics. However, because of its heuristic nature, this approach lacks the capacity of proving how many different solutions exist for a given problem, and does not guarantee generating all possible solutions. The work closest to our approach is by

Greisdorfer et al. [12]. The authors compute two solutions to a MIP model, either by solving two successive MIP models (sequential approach), or by solving a MIP twice as big as the original model (simultaneous approach). Their paper compares both algorithms in terms of performance and diversity of solutions obtained, showing that the sequential approach outperforms the simultaneous approach. The problem with both algorithms is that, although they can be generalized to p solutions instead of two, they do not scale well when p becomes large, as we will show in Sec. 4.1.

1.3 Outline of the Paper

The remainder of the paper is organized as follows. Sec. 2 formally defines the problems we are going to solve, examines their complexity and presents our measure for solution diversity. Sec. 3 describes our three algorithms for generating multiple solutions. Sec. 4 presents computational results. Sec. 5 concludes with a summary and directions for future work.

2 Definitions

2.1 Problem Definition and Complexity

Given a mixed integer programming model $P = \min_{x \in X} c^T x$ where $X = \{x \in \mathbb{R}^d : Ax \le b, x_i \in \mathbb{Z}, \forall i \in I \subseteq \{1, \ldots, d\}\}$, for which an optimal solution is x^*, we define the following problems:

- MIP(p): Generate p different feasible solutions for P
- #MIP: How many different feasible solutions does P have?
- MIP(p, q): Generate p different feasible solutions $x(1), \ldots, x(n)$ for P within $q\%$ of the optimum, i.e., such that $c^T x(i) \le c^T x^* + q|c^T x^*|/100, \forall i = 1, \ldots, n$
- #MIP(q): How many different feasible solutions within $q\%$ of the optimum does P have?

We consider two solutions to be different if and only if they differ by at least one integer variable. The first reason for not taking into account continuous variables is that the main decision variables are integer, whereas continuous variables usually are less important. Secondly, there might exist an infinite number of solutions that differ only by continuous variables, especially if there is no constraint on the objective value of the solution. Thirdly, the computer representation of real values is not exact, therefore it is difficult to say in pratice that two continuous variables are different without resorting to numerical tolerances.

The four problems are at least as hard as MIP(1), therefore are $\mathcal{NP}$-hard. In addition, #MIP and #MIP(q) belong to $\#\mathcal{P}$, the class of counting problems [20,21]. It follows from the polynomial reduction from SAT to MIP that #MIP and #MIP(q) are at least as hard as #SAT, therefore are $\#\mathcal{P}$-complete.

In the rest of this paper, we will focus on solving the objective-controlled version of the problems: MIP(p, q) and #MIP(q), because, however imprecise the objective modeling can be, it is still very important.

2.2   Solution Diversity

Next to objective value, diversity is the most important characteristic to take into account when comparing sets of solutions produced by different algorithms [12]. Indeed, multiple solutions are mostly useful if they are structurally different from each other [19].

We define the diversity of a set S of solutions as the average pairwise distance: $D(S) = \frac{1}{|S|^2} \sum_{s,s' \in S} d(s,s')$, where the distance between two solutions is the Hamming distance on the set B of binary variables: $d(s,s') = \frac{1}{|B|} \sum_{i \in B} |s_i - s'_i|$. Our measure generalizes to $|S| > 2$ the diversity used in Greisdorfer et al. [12].

$D(S) \leq \frac{1}{2}$ for all sets S of solutions. Indeed, $D(S) = \frac{1}{|S|^2|B|} \sum_{i \in B} \sum_{s,s' \in S} |s_i - s'_i|$. Looking at each variable $i \in B$ individually, it is clear that $\sum_{s,s' \in S} |s_i - s'_i|$ is maximal if $\lfloor |S|/2 \rfloor$ of the solutions have $s_i = 0$ and the remaining solutions have $s_i = 1$. In that case, at most half of the addends $|s_i - s'_i|$ are equal to one, while at least half of them are zero. It follows that $D(S) \leq \frac{1}{|S|^2|B|} \sum_{i \in B} \frac{1}{2} |S|^2 = \frac{1}{2}$.

3   Algorithms

We now describe the three algorithms considered in this paper in detail.

3.1   The One-Tree Algorithm

The standard branch-and-bound algorithm for solving integer programming models aims at progressively reducing the search space as quickly and as much as possible so that it is easier both to find the optimal solution and to prove that it is optimal. However, when the aim is to generate multiple solutions, the perspective needs to be different: if the search space is reduced too much, it will not contain enough solutions. The one-tree algorithm we propose is adapted from the standard branch-and-bound algorithm (outlined in Algorithm 1) for this purpose[1]. It proceeds in two phases. During the first phase (outlined in Algorithm 2), the branch-and-bound tree is constructed and explored to find the optimal solution, and its nodes are kept for the second phase. During the second phase (outlined in Algorithm 3), the tree built in the first phase is reused and explored in a different way to yield multiple solutions. The differences with the standard branch-and-bound algorithm relate to storing integer solutions, fathoming nodes, branching, and dual tightening.

In standard branch-and-bound, an integer solution is stored only if it improves on the incumbent. When generating solutions in the second phase, we store in the set S all integer solutions that are within $q\%$ of the optimum value.

In standard branch-and-bound, a node is fathomed when the sub-model it defines cannot yield any improving integer solution, i.e., when its LP solution is integer-valued or has an objective value worse than the incumbent. In the first

---

[1] Algorithm 1 is of course a very rudimentary outline of branch-and-bound. We left out many techniques, such as cuts and heuristics, and many implementation details to concentrate on the features that differ in the one-tree algorithm.

Algorithm 1. Outline of standard branch-and-bound algorithm

```
Preprocessing
Set of open nodes: N_open ← {rootnode}
Objective value of the incumbent: z* ← +∞
while N_open ≠ ∅ do
    Choose a node n from N_open
    Solve LP at node n. Solution is x(n) with objective z(n).
    if z(n) ≥ z* then
        Fathom the node: N_open ← N_open \ {n}
    else
        if x(n) is integer-valued then
            x(n) becomes new incumbent: x* ← x(n); z* ← z(n)
            Do reduced cost fixing
            Fathom the node: N_open ← N_open \ {n}
        else
            Choose branching variable i such that x_i(n) is fractional
            Build children nodes n_1 = n ∩ {x_i ≤ ⌊x_i(n)⌋} and n_2 = n ∩ {x_i ≥ ⌊x_i(n)⌋ + 1}
            N_open ← N_open ∪ {n_1, n_2} \ {n}
        end if
    end if
end while
```

Algorithm 2. Outline of one-tree algorithm: phase I

```
Preprocessing with only primal reductions
Set of open nodes: N_open ← {rootnode}
Set of stored nodes: N_stored ← ∅
Objective value of the incumbent: z* ← +∞
while N_open ≠ ∅ do
    Choose a node n from N_open
    Solve LP at node n. Solution is x(n) with objective z(n).
    if z(n) ≥ z* then
        Fathom the node and keep it for phase II: N_open ← N_open \ {n}; N_stored ← N_stored ∪ {n}
    else
        if x(n) is integer-valued then
            x(n) becomes new incumbent: x* ← x(n); z* ← z(n)
            Fathom the node and keep it for phase II: N_open ← N_open \ {n}; N_stored ← N_stored ∪ {n}
        else
            Choose branching variable i such that x_i(n) is fractional
            Build children nodes n_1 = n ∩ {x_i ≤ ⌊x_i(n)⌋} and n_2 = n ∩ {x_i ≥ ⌊x_i(n)⌋ + 1}
            N_open ← N_open ∪ {n_1, n_2} \ {n}
        end if
    end if
end while
```

Algorithm 3. Outline of one-tree algorithm: phase II

```
Reuse tree from phase I: N_open ← N_stored
Reuse incumbent from phase I: Set of solutions: S ← {x*}
while N_open ≠ ∅ do
    Choose a node n from N_open
    Solve LP at node n. Solution is x(n) with objective z(n)
    if z(n) > z* + q|z*|/100 then
        Fathom the node: N_open ← N_open \ {n}
    else
        if x(n) is integer-valued then
            x(n) is added to the pool of solutions if it is not a duplicate: if x(n) ∉ S, then S ← S ∪ {x(n)}
        end if
        Choose branching variable i such that it is not fixed by the local bounds of node n: lb_i(n) < ub_i(n)
        Build children nodes n_1 = n ∩ {x_i ≤ ⌊x_i(n)⌋} and n_2 = n ∩ {x_i ≥ ⌊x_i(n)⌋ + 1}
        N_open ← N_open ∪ {n_1, n_2} \ {n}
    end if
end while
```

phase of the one-tree algorithm, nodes are fathomed by the same criterion but instead of being discarded, they are stored for further examination during the second phase. During the second phase, a node is fathomed if it cannot yield any additional integer solution within $q\%$ of the optimum value, i.e., if its LP solution is integer-valued and all integer variables have been fixed by the local bounds of the node, or if the objective value of its LP value is strictly more than $q\%$ worse than the optimum value.

In standard branch-and-bound, only variables that are fractional in the node LP solution are branched on. When generating solutions in the second phase, we also branch on variables that are integral in the LP node solution if they are not fixed by local bounds. Suppose the node LP solution is integral and binary variable $x_i = 0$ at the node. There might exist some solutions with $x_i = 1$; for this reason, we need to create the right child node with x_i fixed to 1. But there might also exist solutions with $x_i = 0$ and different values for other variables; for this reason, we need to create the left child node with $x_i = 0$. This branch, however, contains the same integer solution as the parent. So, in order not to count the same solution twice, either we check that it is not a duplicate of a previous solution before adding it to the set S of solutions; or, we keep the LP value of the branching variable in the parent node and discard the solution if this value satisfies the local bounds of the variable at the node. Either way, it is best to explore first the branch that does not contain the solution of the parent node in order to get more diverse solutions.

In standard branch-and-bound, the search space is pruned because of considerations on the objective function. This pruning takes place during preprocessing [13,6], and during the exploration of the tree through reduced cost fixing [22]. In both phases of the one-tree algorithm, we do not want to eliminate potential solutions because they are suboptimal, so in theory we should turn off dual tightening completely, as outlined in Algorithm 2 and 3. However, this choice has a significant impact on performance, and turning off dual tightening completely is not required in practice to obtain a large enough number of solutions. In practice, we need to choose a reasonable tradeoff between speed and number of solutions we wish to obtain. Not fathoming a node even if its objective value exceeds the incumbent, as we explained above, can be seen as a first level of turning dual tightening off. It does not cost anything in terms of speed; it is just expensive in terms of memory. The second level is to turn off dual reductions during reduced cost fixing. This setting is the level we used in our first set of experiments. We will show in Sec. 4.1 how it impacts performance. The third level, which is needed for exhaustive enumeration (see the experiments in Sec. 4.3), is to also turn off dual presolve reductions.

Let us note that one could enumerate solutions in a single phase, simply by fathoming nodes and branching as in the second phase, and turning off some dual tightening. We chose to separate the algorithm into two phases for the following reasons. The first reason is computational: generating multiple solutions carries a performance penalty. So, if we are solving $\text{MIP}(p, q)$ or $\#\text{MIP}(q)$, we want to avoid spending time generating solutions within $q\%$ of a suboptimal incumbent which will be discarded later when the optimal solution is found, and it turns out those solutions are more than $q\%$ above the optimum. The second reason is that the one-tree algorithm is to be used as an interactive discovery tool, where the tree is built once during phase I and explored many times during successive invocations of the phase II algorithm — possibly with different additional constraints, different q values, or different stopping criteria — until the user finds suitable solutions for his business problem. This is also why we chose to turn off dual tightening completely instead of carrying out dual reductions based on the incumbent value augmented by $q\%$, which is also a possible implementation.

Finally, let us mention the stopping criterion for both phases. As with the standard branch-and-bound algorithm, the algorithm can be stopped in both phases before the set of open nodes is empty for reasons such as time limit, node limit, number of solutions generated, etc. An interesting case is the gap criterion. During standard branch-and-bound, the best bound value (minimum of objective value over open nodes) is always less than the incumbent. During the second phase, it will happen, however, that the best bound value becomes greater than the incumbent as nodes get fathomed. If the gap becomes less than $-q\%$, then no additional solutions within $q\%$ of the optimum can be generated, and the algorithm can be stopped.

3.2 Heuristics

Heuristics are a natural way to generate multiple solutions [11,19]. The algorithm we propose, outlined in Algorithm 4, is similar to the one-tree algorithm. The difference is first that solutions are generated mainly by MIP heuristics such as fix-and-dive heuristics [3], RINS [7], and solution polishing [17,18], instead of relying only on the integral node LP solutions. In addition, heuristics can generate solutions that violate local bounds; therefore it is not necessary to store the fathomed nodes as in the first phase of the one-tree algorithm. Finally, it should be noted that this algorithm is not efficient for exhaustive enumeration (#MIP and #MIP(q)), as heuristics do not explore the search space systematically like the one-tree algorithm and risk generating the same solutions many times over.

Algorithm 4. Outline of the algorithm using heuristics for MIP(p, q)

Phase I:
Solve the model with standard branch-and-bound
Optimal solution is x^* with objective value z^*
Set of solutions: $S \leftarrow \{x^*\}$

Phase II:
Start a new tree: $N_{open} \leftarrow \{rootnode\}$
while $|S| < p$ **do**
  Choose a node n from N_{open}
  Solve LP at node n.
  Run fix-and-dive heuristics, RINS, and solution polishing with an objective cutoff of $z^* + q|z^*|/100$
  If the solutions found are not duplicate of already stored solutions, they are added to S.

  Rules for fathoming nodes, branching, and dual tigthening are the same as for the second phase of the one-tree algorithm.
end while

3.3 The Sequential Algorithm

The last algorithm we present in this paper is a simple generalization of the sequential generation of Greisdorfer et al. [12]. We present it mainly for comparison with previous work and for its variant that maximizes diversity. Instead of generating multiple solutions using a unique branch-and-bound tree as in the one-tree algorithm and in heuristics, we solve here a sequence of integer programming models, each providing one solution. The details are given in Algorithm 5.

Algorithm 5. Sequential algorithm for #MIP(q)

Solve P with standard branch-and-bound. Optimal solution is x^* of cost z^*.
Set of solutions: $S \leftarrow \{x^*\}$
B = set of binary variables
Add constraint on objective value: $X \leftarrow X \cap \{c^T x \leq z^* + q|z^*|/100\}$
while P is feasible **do**
  Change objective function of P to maximizing distance to already discovered solutions:
  $\max \sum_{s \in S} \sum_{i \in B: s_i = 0} x_i + \sum_{i \in B: s_i = 1} (1 - x_i)$
  Add diversity constraint to exclude the previously found solution:
  $X \leftarrow X \cap \{\sum_{i \in B: x_i^* = 0} x_i + \sum_{i \in B: x_i^* = 1} (1 - x_i) \geq 1\}$
  Solve P with standard branch-and-bound. Optimal solution is x^*.
  Store the new solution: $S \leftarrow S \cup \{x^*\}$
end while

It is easy to see that this sequential algorithm will be slow to generate a large number of solutions, as no information is reused from one iteration to the next. However, the advantage of this algorithm is that any objective function can be used once the optimal solution of the original problem has been obtained. Our computational experience is that the most effective way is to use the original objective. In the rest of the paper, we will refer to this algorithm as the plain sequential algorithm. But, as outlined in Algorithm 5, we can also try to maximize the distance to already discovered solutions. This algorithm is a greedy procedure that aims at maximizing the diversity of the set of solutions obtained in the end, as defined in Sec. 2.2.

4   Computational Results

When we evaluate the performance of algorithms that generate multiple solutions, several dimensions, possibly mutually conflicting, need to be considered: the number of solutions generated, the solving time needed to generate these solutions, the objective value and the diversity of the solutions generated. For a comparison of algorithms to be valid, it is best to control as many dimensions as possible and to let only one or two vary at a time. This consideration is an additional reason why all our experiments are about the objective-controlled version of the problems: we solve MIP($10,1$), #MIP(1), and #MIP(0).

All experiments were carried out with CPLEX 10.1 on a 3.4 GHz GNU/Linux machine with 2 GB memory. We experimented with models from MIPLIB3.0 [4] and MIPLIB2003 [16] that can be solved to optimality within half an hour.

4.1   Comparison of Performance

The first set of experiments answers the question of how fast each algorithm generates solutions. We compare the time needed for each algorithm to solve MIP($10,1$), i.e., to generate 10 solutions within 1% of the optimum. For this experiment, we also enforce a time limit of 1 hour, except for the sequential algorithm maximizing diversity. Additionally, we answer the question in a slightly different way by comparing how many solutions within 1% of the optimum each algorithm generates in 1 hour. For this experiment, we also enforce a solution limit of 100000. The results, detailed model by model, are given in Table 2, page 290, and Table 3, page 291. Since the limits of 10 solutions or one hour are rather arbitrary, we also present graphically in Fig. 1 the evolution of the number of solutions in function of the time needed to generate them for model 10teams.

In summary, when generating 10 solutions, the first phase of the one-tree algorithm is on average 2.2 times slower than the first phase of heuristics[2]. This difference is due to the fact that dual tightening is turned off during the tree exploration for the one-tree algorithm. But, during the second phase, the one-tree algorithm is on average 10.9 times faster than heuristics to generate solutions. When comparing the total time, the one-tree algorithm has clearly the best performance: it is on average 2.1 times faster than heuristics, 5.5 times faster than the plain sequential algorithm and 20.2 times faster than the sequential algorithm maximizing diversity. These results are corroborated by the number of solutions that each algorithm can generate in one hour: the one-tree algorithm generates on average 2.5 times more solutions than heuristics, 18.2 times more solutions than the plain sequential algorithm, and 52.1 times more solutions than the sequential algorithm maximizing diversity.

4.2   Comparison of Diversity

Only one of the algorithms we have presented explicitly tries to maximize the diversity of the set of solutions to be obtained. However, the experiments of the previous section showed that the sequential algorithm maximizing diversity is much too slow to be practical. We were, therefore, curious to know whether the better performance of the one-tree algorithm, heuristics, and, to a lesser extent, the plain sequential algorithm, were obtained at the expense of a smaller diversity in the solutions these two algorithms generate. Table 4 at page 292 compares the diversity obtained by each algorithm when solving MIP(10, 1). On average, the diversity of the one-tree algorithm, of heuristics, and of the plain sequential algorithm are respectively 3.5, 5.1, and 3.3 smaller than the diversity obtained by the sequential algorithm when maximizing diversity. Given that the first two algorithms are significantly faster and have much room for improvement in solution diversity, we believe that these results are encouraging and show an interesting trade-off between performance and diversity. Our future work will be directed at improving the diversity of the solutions they produce.

4.3   Exhaustive Enumeration of Optimal Solutions

Our last set of experiments concerns solving #MIP(0), i.e., enumerating all possible optimal solutions. We have restricted ourselves to models that contain

---

[2] When presenting average numbers, we compute the geometric mean of the ratio of the performance of the two algorithms compared.

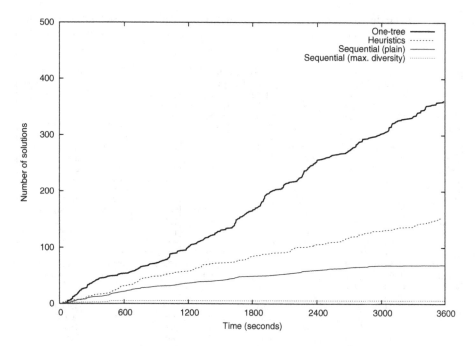

Fig. 1. Number of solutions generated over time for model 10teams

Table 1. Enumeration of all optimal solutions for pure binary models

Model	Number of optimal solutions	One-tree time (in seconds)	Sequential time (in seconds)
10teams	≥ 14764	> 1 day (found 14764 sol.)	> 1 day (found 470 sol.)
air03	1	49.96	1.54
air04	8	37.70	166.41
air05	2	115.32	51.95
cap6000	1	7178.83	6.81
disctom	≥ 547	> 1 day (found 547 sol.)	> 1 day (found 130 sol.)
enigma	3	0.90	1.24
l152lav	1	3.92	1.89
lseu	2	0.27	0.37
mitre	80	234.82	993.45
mod008	6	1.83	1.46
mod010	128	255.81	353.89
nw04	1	499.62	74.74
p0033	9	0.01	0.11
p0201	4	0.79	1.68
p0282	1	0.73	0.31
p0548	≥ 100000	> 551.66 (found 100000 sol.)	> 1 day (found 2940 sol.)
p2756	≥ 100000	> 13519.22 (found 100000 sol.)	> 1 day (found 1401 sol.)
stein27	2106	5.60	19819.2
stein45	70	50.71	1679.87

only binary variables because of the ambiguity of what all solutions mean when continuous variables are involved (see Sec. 2.2), and because the sequential algorithm cannot handle general integer variables (although the one-tree algorithm can). Table 1 shows that we successfully enumerate all optimal solutions for 16

Table 2. Time (in seconds) to enumerate 10 solutions within 1% of the optimum. The ‡ symbol means that the sequential algorithm maximizing diversity had to be limited to two hours for each iteration in order to produce 10 solutions in a reasonable time.

Model	One-tree phase I	One-tree phase II	One-tree total	Heuristics phase I	Heuristics phase II	Heuristics total	Sequential plain	Sequential max diversity
10teams	68.79	54	122.79	31.32	215.6	246.92	260.89	72103.9 ‡
aflow30a	104.85	17.97	122.82	23.57	72.44	96.01	375.26	1221.3
air03	0.58	1.04	1.62	0.55	2.78	3.33	24.41	35.98
air04	41.28	2.74	44.02	17.1	19.84	36.94	248.41	38772.7 ‡
air05	41.31	6.52	47.83	19.42	125.67	145.09	281	7307.85 ‡
arki001	3305.11	0.92	3306.03	27.43	3.62	31.05	2634.18	64954.7 ‡
bell3a	9.17	0.07	9.24	2.76	0.53	3.29	5.64	2.85
bell5	0.14	0.01	0.15	0.13	0.07	0.2	2.17	2.79
blend2	8.17	1.18	9.35	2.56	14.15	16.71	28.47	70.57
cap6000	12.38	0.51	12.89	0.56	0.82	1.38	14.8	49.18
dcmulti	0.69	0.07	0.76	0.54	0.59	1.13	6.42	46.18
disctom	362.47	387.55	750.02	363.48	>3600	>3600	2255.54	54780.6 ‡
dsbmip	0.44	0.38	0.82	0.35	1.35	1.7	3	17.86
egout	0.01	0.01	0.02	0.01	1.69	1.7	0.05	0.04
enigma	0.16	0.79	0.95	0.2	6.74	6.94	2.24	1.47
fiber	1.04	0.24	1.28	0.22	1052.65	1052.87	11.99	18.65
fixnet6	1.44	0.08	1.52	1.78	15.87	17.65	20.11	16.14
flugpl	0.01	>3600	>3600	0.01	>3600	>3600	>3600	>3600
gen	0.02	0.07	0.09	0.02	1.04	1.06	0.49	1.78
gesa2	0.99	0.15	1.14	0.4	0.33	0.73	5.4	105.1
gesa2_o	4.04	0.09	4.13	2.34	0.28	2.62	48.82	3266.6
gesa3	1.13	0.23	1.36	0.68	1.52	2.2	72.03	864.76
gesa3_o	0.89	0.13	1.02	0.8	1.01	1.81	17.97	48543.2 ‡
gt2	0.01	0.04	0.05	0.01	0.07	0.08	0.17	0.44
khb05250	0.14	0.2	0.34	0.11	66.15	66.26	2.44	4.97
l152lav	1.4	0.19	1.59	0.94	4.03	4.97	16.21	53.93
lseu	0.23	0.18	0.41	0.1	2407.72	2407.82	0.97	1.11
mas76	213.48	26.61	240.09	81.94	16.75	98.69	2188.01	1146.94
misc03	0.53	0.02	0.55	0.2	1.16	1.36	9.96	36.36
misc06	0.1	0.04	0.14	0.1	0.6	0.7	1.94	5.23
misc07	86.35	0.73	87.08	8.45	3.75	12.2	891.92	3119.26
mitre	0.64	20.57	21.21	0.59	5.64	6.23	14.78	27.7
mod008	1.17	0.7	1.87	0.17	10.83	11	2.41	28.63
mod010	0.42	0.15	0.57	0.27	0.73	1	4.48	13.69
mod011	125.32	82.58	207.9	56.47	>3600	>3600	1417.06	10956.5 ‡
modglob	0.24	0.04	0.28	0.16	0.19	0.35	3.15	45.94
mzzv11	253.15	45.38	298.53	179.42	42.88	222.3	3249.93	21817.4 ‡
mzzv42z	122.4	3.61	126.01	68.99	43.01	112	921.46	51016.4 ‡
nw04	744.85	23.7	768.55	29.11	106.88	135.99	1337.25	1479.38
p0033	0.01	0.01	0.02	0.01	0.15	0.16	0.12	0.16
p0201	1.02	0.16	1.18	0.24	7.61	7.85	6.1	4.95
p0282	0.74	0.12	0.86	0.16	0.97	1.13	2.41	5
p0548	0.12	0.06	0.18	0.06	0.08	0.14	0.99	3.42
p2756	0.66	0.11	0.77	0.36	0.29	0.65	4.82	10.78
pk1	187.93	7.37	195.3	89.98	>3600	>3600	269.59	206.46
pp08aCUTS	4.26	0.47	4.73	2.62	7.77	10.39	37.52	111.83
pp08a	1.6	0.46	2.06	0.92	8.28	9.2	22.2	62.06
qiu	250.26	0.24	250.5	53.06	73.46	126.52	2450.64	1311.36
qnet1	2.72	0.74	3.46	2.26	1.24	3.5	35.46	68.27
qnet1_o	1.72	0.29	2.01	1.69	0.69	2.38	26.46	65.8
rgn	1.26	0.02	1.28	0.76	0.33	1.09	8.54	1
rout	975.64	4.41	980.05	34.49	16.95	51.44	2940.5	2097.27
set1ch	0.54	0.08	0.62	0.58	0.57	1.15	9.75	51.31
stein27	0.79	0.01	0.8	0.4	0.2	0.6	7.16	2.51
stein45	20.56	3.26	23.82	17.19	9.83	27.02	177.87	65.5
vpm1	0.01	0.05	0.06	0.01	0.06	0.07	0.35	1.91
vpm2	1.66	0.54	2.2	0.9	1.6	2.5	14.8	101.48

Table 3. Number of solutions within 1% of the optimum enumerated in one hour. The (p) symbol means the algorithm has proved that no other solutions exist within 1% of the optimum.

Model	One-tree	Heuristics	Sequential plain	Sequential max. diversity
10teams	363	93	68	5
aflow30a	84	61	68	20
air03	938	425	165	126
air04	8015	268	80	1
air05	6728	207	56	7
arki001	12493	20964	13	1
bell3a	≥ 100000	221	1	1
bell5	≥ 100000	7115	16	134
blend2	10 (p)	10 (p)	10 (p)	10 (p)
cap6000	79112	36648	225	23
dcmulti	≥ 100000	1541	640	145
disctom	37	3	25	1
dsbmip	≥ 100000	98385	1366	378
egout	2 (p)	2 (p)	2 (p)	2 (p)
enigma	3 (p)	2	3 (p)	3 (p)
fiber	136	10	279	160
fixnet6	28426	97	555	185
flugpl	5	4	1	1
gen	≥ 100000	≥ 100000	717	371
gesa2	≥ 100000	13164	505	59
gesa2_o	≥ 100000	16268	308	3
gesa3	28989	1361	118	17
gesa3_o	≥ 100000	2051	430	5
gt2	≥ 100000	≥ 100000	13	13
khb05250	28 (p)	28 (p)	28 (p)	28 (p)
l152lav	15958	8101	233	56
lseu	5 (p)	5 (p)	5 (p)	5 (p)
mas76	49	11	14	1
misc03	24 (p)	24 (p)	24 (p)	24 (p)
misc06	≥ 100000	778	781	382
misc07	72	72	42	11
mitre	10308	10091	114	11
mod008	68 (p)	25	68 (p)	68 (p)
mod010	21263	10612	350	180
mod011	49	8	18	5
modglob	≥ 100000	4151	661	43
mzzv11	562	48703	11	1
mzzv42z	1076	43908	30	1
nw04	86	70	28	16
p0033	15 (p)	15 (p)	15 (p)	15 (p)
p0201	44 (p)	44 (p)	44 (p)	44 (p)
p0282	≥ 100000	≥ 100000	1042	623
p0548	≥ 100000	18670	921	515
p2756	10164	92586	412	227
pk1	1 (p)	1 (p)	1 (p)	1 (p)
pp08aCUTS	64 (p)	57	64 (p)	64 (p)
pp08a	64 (p)	56	64 (p)	64 (p)
qiu	144	108	14	17
qnet1	30067	14834	147	97
qnet1_o	28196	13266	138	94
rgn	720	720	720	720
rout	3393	574	11	10
set1ch	≥ 100000	6096	313	112
stein27	2106 (p)	2106 (p)	904	793
stein45	70 (p)	70	70 (p)	70 (p)
vpm1	≥ 100000	18729	1393	1344
vpm2	33 (p)	33	33 (p)	33 (p)

Table 4. Diversity of solutions obtained when solving MIP(10, 1). The ‡ symbol means that the sequential algorithm maximizing diversity had to be limited to two hours for each iteration in order to produce 10 solutions in a reasonable time.

Model	One-tree	Heuristics	Sequential plain	Sequential max. diversity
10teams	0.021	0.029	0.037	0.040 ‡
aflow30a	0.032	0.030	0.038	0.056
air03	0.001	0.001	0.001	0.003
air04	0.004	0.001	0.003	0.011 ‡
air05	0.003	0.002	0.001	0.011 ‡
arki001	0.038	0.022	0.059	0.434 ‡
bell3a	0.000	0.000	0.000	0.000
bell5	0.042	0.102	0.063	0.331
blend2	0.010	0.010	0.010	0.010
cap6000	0.002	0.001	0.002	0.028
dcmulti	0.039	0.044	0.025	0.253
disctom	0.109	0.137	0.423	0.476 ‡
dsbmip	0.028	0.044	0.173	0.237
egout	0.018	0.018	0.018	0.018
enigma	0.062	0.060	0.062	0.062
fiber	0.003	0.015	0.004	0.019
fixnet6	0.006	0.017	0.027	0.046
flugpl	0.000	0.000	0.000	0.000
gen	0.030	0.039	0.036	0.318
gesa2	0.034	0.012	0.008	0.151
gesa2_o	0.006	0.011	0.007	0.159
gesa3	0.006	0.000	0.009	0.086
gesa3_o	0.002	0.000	0.005	0.132 ‡
gt2	0.027	0.028	0.074	0.074
khb05250	0.138	0.119	0.138	0.200
l152lav	0.010	0.009	0.003	0.020
lseu	0.110	0.110	0.110	0.110
mas76	0.067	0.041	0.066	0.000
misc03	0.071	0.075	0.075	0.078
misc06	0.030	0.023	0.023	0.334
misc07	0.056	0.052	0.057	0.063
mitre	0.002	0.002	0.003	0.043
mod008	0.012	0.015	0.012	0.019
mod010	0.011	0.003	0.005	0.022
mod011	0.049	0.067	0.035	0.097 ‡
modglob	0.063	0.063	0.023	0.421
mzzv11	0.006	0.004	0.009	0.026 ‡
mzzv42z	0.006	0.004	0.013	0.033 ‡
nw04	0.000	0.000	0.000	0.000
p0033	0.158	0.228	0.108	0.242
p0201	0.121	0.077	0.144	0.153
p0282	0.014	0.020	0.006	0.164
p0548	0.007	0.018	0.015	0.080
p2756	0.001	0.017	0.009	0.083
pk1	0.000	0.000	0.000	0.000
pp08aCUTS	0.130	0.053	0.100	0.207
pp08a	0.069	0.103	0.100	0.207
qiu	0.333	0.239	0.346	0.370
qnet1	0.007	0.005	0.004	0.026
qnet1_o	0.005	0.006	0.003	0.026
rgn	0.062	0.055	0.066	0.071
rout	0.084	0.058	0.077	0.095
set1ch	0.034	0.034	0.020	0.145
stein27	0.363	0.353	0.393	0.444
stein45	0.383	0.346	0.415	0.429
vpm1	0.048	0.038	0.046	0.120
vpm2	0.052	0.048	0.058	0.067

models. As expected, the one-tree algorithm is generally faster than the sequential algorithm, but it is slower when there is a very small number of solutions to be enumerated because it requires all dual tightening (in presolve and during the tree) to be turned off. Note that for models with a very large number of solutions, there are probably smarter ways to enumerate all of them, for example, by taking into account symmetries.

5 Conclusion and Future Work

In this paper, we have formally introduced four problems representative of the issues of generating multiple solutions for mixed integer programming problems. We have presented three new algorithms to solve them, and we have shown with extensive computational experiments on the MIPLIB model library that it is within our reach to generate multiple solutions effectively. In particular, the main algorithm we introduced in this paper, the one-tree algorithm, improves significantly over previously known algorithms. Unlike previous approaches such as heuristics, this algorithm is able to compute all solutions for a model and prove that no other solutions exist. It also performs on average significantly faster than previously known algorithms, such as heuristics and sequential enumeration, especially when a large number of solutions is requested.

We have also studied the diversity of the solutions produced, as this characteristic is very important for applications. We have presented a variation of the sequential algorithm that explicitly maximizes diversity. This algorithm is very slow but useful to compare the diversity of the solutions produced by our other algorithms. Our preliminary results are encouraging, as the one-tree algorithm and, to a lesser extent, our MIP heuristics are significantly faster but still provide an acceptable level of diversity. We will work in the future on improving the diversity of the solutions produced by the one-tree algorithm. We will also work on taking into account general integer variables and continuous variables in diversity measures.

Acknowledgments

We would like to thank David Woodruff for awakening our interest in generating multiple and diverse solutions, and Tobias Achterberg, Kathleen Callaway, John Gregory, and Andrea Lodi for their proofreading.

References

1. Tian Sang, Fahiem Bacchus, Paul Beame, Henry Kautz, and Toniann Pitassi. Combining Component Caching and Clause Learning for Effective Model Counting, SAT 2004.
2. Egon Balas, Anuret Saxena. Optimizing over the split closure, Technical Report 2006-E5, Tepper School of Business, CMU, 2005.

3. Robert E. Bixby, Mary Fenelon, Zonghao Gu, Edward Rothberg, and Roland Wunderling. MIP: Theory and practice — closing the gap. In M. J. D. Powell and S. Scholtes (eds.), *System Modelling and Optimization: Methods, Theory, and Applications*, pages 19–49. Kluwer Academic Publishers, 2000.

4. Robert E. Bixby, S. Ceria, C. M. McZeal, M. W. P Savelsbergh. An updated mixed integer programming library: MIPLIB 3.0. *Journal Optima*: 58, 12–15, 1998.

5. William Cook, Ricardo Fukasawa, and Marcos Goycoolea. Choosing the best cuts. Workshop on mixed integer programming, MIP 2006.

6. CPLEX 10.0 Manual, Ilog Inc., 2006.

7. Emilie Danna, Edward Rothberg, Claude Le Pape. Exploring relaxation induced neighborhoods to improve MIP solutions. *Mathematical Programming*: 102(1), 71–90, 2005.

8. Sanjeeb Dash, Oktay Günlük, and Andrea Lodi. Separating from the MIR closure of polyhedra. Workshop on mixed integer programming, MIP 2006.

9. Matteo Fischetti, Andrea Lodi. Optimizing over the first Chvátal closure, in M. Jünger and V. Kaibel (eds.), Integer Programming and Combinatorial Optimization - IPCO 2005, LNCS 3509, Springer-Verlag, 12–22, 2005.

10. Matteo Fischetti, Andrea Lodi. MIP models for MIP separation. Workshop on mixed integer programming, MIP 2006.

11. Fred Glover, Arne Løkketangen, and David L. Woodruff. Scatter search to generate diverse MIP solutions. In *OR computing tools for modeling, optimization and simulation: interfaces in computer science and operations research*, M. Laguna and J.L. González-Velarde (eds.), Kluwer Academic Publishers, 299–317, 2000.

12. Peter Greistorfer, Arne Løkketangen, Stephan Voß, and David L. Woodruff. Experiments concerning sequential versus simultaneous maximization of objective function and distance. Submitted to *Journal of Heuristics*, 2006.

13. Karla Hoffman and Manfred Padberg. Improving Representations of Zero-one Linear Programs for Branch-and-Cut, *ORSA Journal of Computing*:3, 121–134, 1991.

14. Miroslav Karamanov and Gerard Cornuejols. Cutting Planes Selection. Workshop on mixed integer programming, MIP 2006.

15. Sangbum Lee, Chan Phalakornkule, Michael M. Domach, and Ignacio E. Grossmann. Recursive MILP model for finding all the alternate optima in LP models for metabolic networks. *Computers and Chemical Engineering*, 24:711–716, 2000.

16. MIPLIB 2003, http://miplib.zib.de/

17. Edward Rothberg. It's a beautiful day in the neighborhood — Local search in mixed integer programming. Workshop on mixed integer programming, MIP 2005.

18. Edward Rothberg. An evolutionary algorithm for polishing mixed integer programming solutions. To appear in INFORMS Journal on Computing.

19. Patrick Schittekat and Kenneth Sorensen. Coping with unquantifiable criteria by generating structurally different solutions — Applications to a large real-life location-routing problem in the automative industry. ISMP 2006.

20. L. G. Valiant. The complexity of computing the permanent. *Theoretical Computer Science*, 8:189–201, 1979.

21. L. G. Valiant. The complexity of enumeration and reliability problems. *SIAM Journal of Computing*, 9:410–421, 1979.

22. Laurence A. Wolsey, *Integer Programming*, Wiley, New York, 1998.

A Branch and Bound Algorithm for Max-Cut Based on Combining Semidefinite and Polyhedral Relaxations[*]

Franz Rendl[1], Giovanni Rinaldi[2], and Angelika Wiegele[1]

[1] Alpen-Adria-Universität Klagenfurt, Institut für Mathematik,
Universitätsstr. 65-67, 9020 Klagenfurt, Austria
`franz.rendl@uni-klu.ac.at, angelika.wiegele@uni-klu.ac.at`
[2] Istituto di Analisi dei Sistemi ed Informatica "Antonio Ruberti" – CNR,
Viale Manzoni, 30, 00185 Roma, Italy
`rinaldi@iasi.cnr.it`

Abstract. In this paper we present a method for finding exact solutions of the Max-Cut problem $\max x^T L x$ such that $x \in \{-1, 1\}^n$. We use a semidefinite relaxation combined with triangle inequalities, which we solve with the bundle method. This approach is due to Fischer, Gruber, Rendl, and Sotirov [12] and uses Lagrangian duality to get upper bounds with reasonable computational effort. The expensive part of our bounding procedure is solving the basic semidefinite programming relaxation of the Max-Cut problem.

We review other solution approaches and compare the numerical results with our method. We also extend our experiments to unconstrained quadratic 0-1 problems and to instances of the graph bisection problem.

The experiments show, that our method nearly always outperforms all other approaches. Our algorithm, which is publicly accessible through the Internet, can solve virtually any instance with about 100 variables in a routine way.

1 Introduction

The Max-Cut problem is one of the fundamental NP-hard combinatorial optimization problems. It corresponds to unconstrained quadratic optimization in binary variables. We will present an exact method for this problem, which allows us to solve instances of modest size (about 100 binary variables) in a routine manner.

Since the late 1980's a systematic investigation based on polyhedral combinatorics was carried out to get exact solutions of the Max-Cut problem (see, e.g., [2, 3, 9, 11, 23, 1]). This approach is quite successful on sparse instances (e.g., in [9] the solution of toroidal grid graphs of sizes up to 22 500 nodes is reported), but it becomes no more usable for dense instances with more than, say, 50 nodes.

---

[*] Supported in part by the EU project Algorithmic Discrete Optimization (ADONET), MRTN-CT-2003-504438.

M. Fischetti and D.P. Williamson (Eds.): IPCO 2007, LNCS 4513, pp. 295–309, 2007.

A major theoretical break-through occured in the early 1990's, when Goemans and Williamson [16] showed that a semidefinite programming (SDP) relaxation of Max-Cut has an error of no more than about 14%, independent of the density of the underlying problem, provided the edge weights in the problem are all nonnegative. This raised the hope that the use of this relaxation might open the way to deal also with dense instances. Unfortunately, this SDP bound is still too weak, see [28]. Closing an initial gap of more than 10% by Branch and Bound is very likely to produce a huge number of subproblems to be investigated, leading to excessive computation times.

In this paper we take up the approach from Helmberg and Rendl [18] of using this SDP bound tightened by the inclusion of triangle inequalities in a Branch and Bound framework. The major improvement as compared to [18] consists in the way we compute the resulting relaxation. We use the approach of Fischer et al. [12], which combines an interior-point method to compute the basic SDP relaxation with the bundle method to handle the triangle inequalities, and which we tuned for the Branch and Bound setting. A similar approach, but based on a pure polyhedral relaxation, was used quite successfully by Frangioni, Lodi, and Rinaldi [13] to compute the bound based on the triangle inequalities very effectively. We report computational results with this approach on a wide variety of instances and compare with virtually all existing methods. With the exception of very sparse graphs, our approach is a substantial improvement over all existing methods to solve the Max-Cut problem to optimality.

The paper is organized as follows. After a quick introduction to the problem (Sect. 2), we describe the SDP bound enhanced with triangle inequalities in Sect. 3. In Sect. 4 we briefly touch the other features of our Branch and Bound approach. We test our approach on a variety of data sets. Some characteristics of these data along with their origin are given in Sect. 5. In Sect. 6 we compare our approach with existing exact methods. Finally we discuss some extensions of our approach to the graph equipartition problem.

Notation. We use standard notation from graph theory. The vector of all ones (of appropriate dimension) is denoted by e, $\mathcal{A}$ is a linear operator mapping symmetric matrices to vectors in $\mathbb{R}^m$, and $\mathcal{A}^T$ is its adjoint operator. For a vector v of size n we denote by $\mathrm{Diag}(v)$ the matrix D of order n with $D_{ii} = v_i$ and with all the off-diagonal elements equal to zero. For a matrix D of order n, $\mathrm{diag}(D)$ denotes the n-dimensional vector v with $v_i = D_{ii}$. Finally, $\mathrm{tr}\, D$ denotes the trace of the square matrix D, i.e., the sum of its diagonal elements.

2   The Max-Cut Problem

The Max-Cut problem is one of the basic NP-hard problems and has attracted scientific interest from the combinatorial optimization community, and also from people interested in nonlinear optimization. There are two essentially equivalent formulations of the problem.

Max-Cut in a Graph. Given an undirected graph $G = (V, E)$ on $|V| = n$ vertices with edge weights w_e for $e \in E$, every bipartition (S, T) of V (where S or T can be empty) defines a cut $(S : T) = \{ij \in E : i \in S, j \in T\}$. The problem is to find a bipartition (S, T) such that the weight of the corresponding cut

$$w(S, T) := \sum_{e \in (S:T)} w_e$$

is maximized. It will be convenient to use matrix notation and introduce the weighted adjacency matrix $A = (a_{ij})$ with $a_{ij} = a_{ji} = w_e$ for edge $e = [ij] \in E$ and $a_{ij} = 0$ if $[ij] \notin E$. Given A we also introduce the matrix L defined by $L = \text{Diag}(Ae) - A$, often called the Laplacian, associated to A.

If we represent bipartitions (S, T) by vectors $x \in \{-1, 1\}^n$ with $x_i = 1$ exactly if $i \in S$, then it is easy to show that $w(S, T) = \frac{1}{4} x^T L x$. Hence finding a cut in a graph with maximum weight is equivalent to solving the following quadratic optimization problem.

$$\text{(MC)} \quad z_{MC} = \max\{x^T L x : x \in \{-1, 1\}^n\}.$$

Quadratic 0-1 Minimization. Given a matrix Q of order n and a vector c, let $q(y) := y^T Q y + c^T y$. We consider the following problem.

$$\text{(QP)} \quad \min\{q(y) : y \in \{0, 1\}^n\}.$$

It is not difficult to show that solving (QP) is equivalent to solving (MC) (see for instance [3]). We consider both models, as both are dealt with in the literature.

3   Semidefinite Relaxations of (MC)

The following semidefinite relaxation of (MC) uses $x^T L x = \text{tr} L(x x^T)$ and introduces a new matrix variable X taking the role of $x x^T$.

$$z_{SDP} = \max\{\text{tr } LX : \text{diag}(X) = e, \ X \succeq 0\}. \tag{1}$$

Its dual form

$$\min\{e^T u : \text{Diag}(u) - L \succeq 0\} \tag{2}$$

was introduced by Delorme and Poljak [10] as the (equivalent) eigenvalue optimization problem

$$\min\{n\lambda_{\max}(L - \text{Diag}(u)) : u \in \mathbb{R}^n, u^T e = 0\}. \tag{3}$$

The primal version (1) can be found in [28]. In [16] it is shown that this relaxation has an error of no more than 13.82%, i.e.,

$$\frac{z_{SDP}}{z_{MC}} \leq 1.1382,$$

provided there are non-negative weights on the edges ($w_e \geq 0$). This relaxation can be further tightened by including the following triangle inequalities (that define the *semimetric polytope*, the basic polyhedral relaxation of Max-Cut).

$$\begin{pmatrix} -1 & -1 & -1 \\ -1 & 1 & 1 \\ 1 & -1 & 1 \\ 1 & 1 & -1 \end{pmatrix} \begin{pmatrix} x_{ij} \\ x_{ik} \\ x_{jk} \end{pmatrix} \leq \begin{pmatrix} 1 \\ 1 \\ 1 \\ 1 \end{pmatrix} \quad 1 \leq i < j < k \leq n.$$

We abbreviate all $4\binom{n}{3}$ of these constraints as $\mathcal{A}(X) \leq e$. Hence we get

$$z_{SDPMET} = \max\{\text{tr } LX : \text{diag}(X) = e, \ \mathcal{A}(X) \leq e, \ X \succeq 0\}. \tag{4}$$

Helmberg and Rendl [18] apply this semidefinite relaxation (solved by an interior point code) in a Branch and Bound scheme. Later on, Helmberg [17] improved this algorithm by fixing variables. The experiments in [18] clearly indicate that an efficient computation of this relaxation is crucial for further computational improvements.

Instead of solving this relaxation with a limited number of inequality constraints by interior point methods, as done in [18], we use the bundle approach, suggested in [12], which we modify to gain computational efficiency in the Branch and Bound process.

The set $\mathcal{E} := \{X : \text{diag}(X) = e, X \succeq 0\}$ defines the feasible region of (1). Therefore (4) can compactly be written as

$$z_{SDPMET} = \max\{\langle L, X \rangle : X \in \mathcal{E}, \mathcal{A}(X) \leq e\}. \tag{5}$$

We now briefly recall the approach from [12] to approximate z_{SDPMET} (from above). Let us introduce the Lagrangian with respect to $\mathcal{A}(X) \leq e$

$$\mathcal{L}(X, \gamma) := \langle L, X \rangle + \gamma^T(e - \mathcal{A}(X)) \tag{6}$$

and the associated dual function

$$f(\gamma) := \max_{X \in \mathcal{E}} \mathcal{L}(X, \gamma) = e^T \gamma + \max_{X \in \mathcal{E}} \langle L - \mathcal{A}^T(\gamma), X \rangle. \tag{7}$$

We get for any $\hat{\gamma} \geq 0$ that

$$z_{SDPMET} = \max_{X \in \mathcal{E}} \min_{\gamma \geq 0} L(X, \gamma) = \min_{\gamma \geq 0} f(\gamma) \leq f(\hat{\gamma}).$$

The problem now consists in finding a 'good' approximation $\hat{\gamma}$ to the correct minimizer of f.

The function f is well-known to be convex but non-smooth. Evaluating f for some $\gamma \geq 0$ amounts to solving a problem of type (1), which can be done easily for problem sizes of our interest. We use a primal-dual interior-point method to solve it, which also provides an optimality certificate X_γ, u_γ (optimal solutions to (1) and (2)). The primal matrix X_γ will turn out to be useful in our algorithmic setup. We have, in particular that

$$f(\gamma) = \mathcal{L}(X_\gamma, \gamma).$$

Moreover, a subgradient of f at γ is given by $e - \mathcal{A}(X_\gamma)$.

Dualizing all triangle constraints would result in a dual problem of dimension roughly $\frac{2}{3}n^3$. We prefer a more economical approach where inequalities are included only if they are likely to be active at the optimum.

Let I be a subset of the triangle inequalities, hence $\mathcal{A}_I(X) \leq e_I$. We also write γ_I for the variables dual to the inequalities in I. Setting the dual variables not in I to zero, it is clear that for any I and any $\gamma_I \geq 0$, we get an upper bound on z_{SDPMET}. Approximating the value z_{SDPMET} therefore breaks down into the following two independent tasks:

1. Identify a subset I of triangle inequalities.
2. For a given set I of inequalities, determine an approximate minimizer $\gamma_I \geq 0$ of f.

The second step can be carried out with any of the subgradient methods for convex nonsmooth functions. For computational efficiency we use the bundle method with a limit on the number of function evaluations.

Carrying out the first step is less obvious. We are interested in constraints which are active at the optimum, but this information is in general not available. Therefore we use the optimizer X_{γ_I}, corresponding to an approximate minimizer γ_I of f, and add to the current set I of constraints the t triangle inequalities most violated by X_{γ_I}. (Here t is a parameter which is dynamically chosen.) Thus we can identify promising new inequalities to be added to I.

On the other hand, we remove any constraint from I where the dual multiplier is close to zero, as this is an indication that the constraint is unlikely to be binding. We iterate this process of selecting and updating a set of triangle inequalities, and then solving the respective relaxation, as long as the decrease of the upper bound is sufficiently large.

4 Branching Rules and Heuristics

4.1 Branching Strategies

We subdivide the set of feasible solutions by simply separating, or merging two vertices i, j. This results again in an instance of (MC), see [27]. There are several natural choices for such a pair i, j for branching.

Easy First. A first idea is to branch on pairs i, j where the decision seems to be obvious. We choose i and j such that their rows are 'closest' to a $\{-1, 1\}$ vector, i.e., they minimize $\sum_{k=1}^{n}(1 - |x_{ik}|)^2$. We may assume, that for these two very well articulated nodes the value $|x_{ij}|$ is also very large. Setting x_{ij} opposite to its current sign should lead to a sharp drop of the optimal solution in the corresponding subtree. Hoping that the bound also drops as fast, we will, presumably, be able to cut off this subtree quickly. This rule has been used also in [18] and called R2.

Difficult First. Another possibility for branching is to fix the hard decisions first. We branch on the pair i, j which minimizes $|x_{ij}|$. This means, we fix the most difficult decisions and hope that the quality of the bound gets better fast and that the subproblems become easier. Following [18] we call this rule R3.

Depending on the class of problems, either rule R2 or R3 was more efficient than the other. We also experimented with the so-called *strong branching*, as this strategy is quite successful for linear programming based relaxations. Unfortunately, sensitivity information, necessary for selecting the branching pair, is much harder to get in the case of semidefinite relaxations, hence there is no computational trade off. Consequently, we did not pursue this strategy any further.

4.2   Generating Feasible Solutions

Generating feasible solutions is done iteratively in basically three steps:

1. Apply the Goemans-Williamson hyperplane rounding technique [16] to the primal matrix X obtained from solving the SDP during the bundle iterations. This gives a cut vector $\bar{x}$.
2. Cut $\bar{x}$ is locally improved by checking all possible moves of a single vertex to the opposite partition block. This gives a cut $\tilde{x}$.
3. Bring the matrix X towards a good cut by using a convex-combination of X and $\tilde{x}\tilde{x}^T$. With this new matrix go to 1. and repeat as long as one finds better cuts.

It turned out, that with this heuristic for most of the instances the optimal cut was found at the root node of the Branch and Bound tree.

5   Random Data for (MC) and (QP)

In this section some random data for presenting numerical results of our algorithm are specified. All the data sets can be downloaded from **http://www.math. uni-klu.ac.at/or/Software**. These instances are taken from various sources. Here we provide some of the characteristics of the data sets.

5.1   Max-Cut Instances

Instances by the Graph Generator 'rudy'. The first group of instances follows [18] and consists of random graphs (of specified edge density) with various types of random edge weights. All graphs were produced by the graph generator 'rudy' [30]. For a detailed description and a list of the rudy-calls the reader is referred to the dissertation of Wiegele [31]. We generated ten instances of size $n = 100$ and given density d of the following types of graphs:

- $G_{0.5}$: unweighted graphs with density $d = 0.5$.
- $G_{-1/0/1}$: complete graphs with edge weights chosen uniformly from $\{-1, 0, 1\}$.

- $G_{[-10,10]}$: Graphs with integer edge weights chosen from $[-10, 10]$ and $d \in \{0.5, 0.9\}$.
- $G_{[0,10]}$: Graphs with integer edge weights chosen from $[0, 10]$ and $d \in \{0.5, 0.9\}$.

Applications in Statistical Physics: Ising Instances. We also consider a set of test-problems of Frauke Liers [personal communication, 2005] coming from physical applications. The first group consists of two- and three-dimensional grid graphs with Gaussian distributed weights (zero mean and variance one). The second group consists of dense Ising instances which are obtained in the following way: all nodes lie evenly distributed on a cycle. The weights of the edges depend on the Euclidean distance between two nodes and a parameter σ, such that the proportion $c_{ij} \sim \frac{\epsilon_{ij}}{r_{ij}^{\sigma}}$ holds (ϵ_{ij} is chosen according to a Gaussian distribution with zero mean and variance one and r_{ij} is the Euclidean distance between nodes i and j).

5.2 (QP) Instances

Pardalos and Rodgers [25] have proposed a test problem generator for unconstrained quadratic binary programming. Their routine generates a symmetric integer matrix Q to define the objective function for (QP), with the linear term c represented by the main diagonal of Q, and has several parameters to control the characteristics of the problem. These parameters are the number n of variables, the density d, i.e., the probability that a nonzero will occur in the off-diagonal part of Q, the lower and upper bounds of the main diagonal of Q are given by c^-, c^+. The lower and upper bounds for the off-diagonal part of Q are given by q^-, q^+. Furthermore we have $q_{ii} \sim$ discrete uniform in (c^-, c^+) and $q_{ij} = q_{ji} \sim$ discrete uniform in (q^-, q^+).

Several test problems generated this way are provided in the OR-library [4], [5]. We have chosen all the problems of sizes of our interest, which are the data sets bqpgka, due to [14] and bqp100 and bqp250, see [6]. Furthermore, in [7] the sets c and e of bqpgka are extended. We call these instances bqpbe.

The characteristics are as follows:
- bqpgka:

	n	d	c^-	c^+	q^-	q^+
bqpgka, set a	$30, \ldots, 100$	$0.0625, \ldots, 0.5$	-100	100	-100	100
bqpgka, set b	$20, \ldots, 120$	1.0	0	63	-100	0
bqpgka, set c	$40, \ldots, 100$	$0.1, \ldots, 0.8$	-100	100	-50	50
bqpgka, set d	100	$0.1, \ldots, 1.0$	-75	75	-50	50
bqpgka, set e	200	$0.1, \ldots, 0.5$	-100	100	-50	50

- bqpbe
  Size ranging from $n = 100$ to $n = 250$ nodes; density ranging from $d = 0.1$ to $d = 1.0$; $c^- = -100$; $c^+ = 100$; $q^- = -50$ and $q^+ = 50$.
- beasley
  Two sizes of $n = 100$ and $n = 250$ nodes; $d = 0.1$; $c^- = -100$; $c^+ = 100$; $q^- = -100$ and $q^+ = 100$.

6   Numerical Results

The algorithm was implemented in C and made publicly available for experimental runs as "Biq Mac" – a solver for binary quadratic and Max-Cut problems, see [29]. If not stated otherwise, test runs were performed on a Pentium IV, 3.6 GHz and 2 GB RAM, operating system Linux. For a more detailed study of the numerical results the reader is referred to the dissertation [31].

6.1   Summarizing Existing Methods and Their Limits

Before we present our computational results, we summarize existing exact methods for (MC) together with their limits, as reported in the publications underlying these approaches.

LP: Linear programming based Branch and Bound approaches go back to Barahona et al. [3]. Liers et al. [23] enhance the algorithm and focus on solving toroidal grid graphs arising from physical applications, the so-called Ising model.

V: Linear programming combined with volume algorithm has been investigated by Barahona and Ladányi [1]. Also in this work, there is an emphasis on toroidal grid graphs.

EO: An exact approach using eigenvalue optimization based on (3) has been first investigated by Poljak and Rendl [27].

QP: The recent work of Billionnet and Elloumi [7] presents an approach based on convex quadratic optimization. This algorithm convexifies the objective function and uses a mixed-integer quadratic programming solver to obtain an exact solution of the problem.

SDPMET: An approach based on SDP and the triangle inequalities was first investigated by Helmberg and Rendl [18]. They solve (4) by an interior point algorithm.

PP: Pardalos and Rodgers [25], [26] solve the quadratic program by Branch and Bound using a preprocessing phase where they try to fix some of the variables. The test on fixing the variables exploits information of the partial derivatives of the cost function.

SOCP: Kim and Kojima [21] and, later on, Muramatsu and Suzuki [24] use a second-order cone programming (SOCP) relaxation as bounding routine in a Branch and Bound framework to solve Max-Cut problems. However, the basic SDP relaxation performs better than their SOCP relaxation and the algorithm is capable of solving very sparse instances only. Therefore we omit comparing with this algorithm in the subsequent sections.

In Table 1 we give a very naïve overview of the capability of these approaches. We consider different types of instances and use the following symbols. A ✔ means, that the approach can solve instances of this type in a routine way. A ♨ indicates that one can have (at least) one cup of coffee while waiting for the solution and maybe there are instances that cannot be solved at all. The ☙ suggests to have some holidays and come back in a couple of days to see whether the job is finished and the ☺ indicates that the chances for solving the problem

Table 1. Who can do what?

	LP	V	EO	QP	SDP	MET	PP	Biq Mac
quadr 0-1, $n = 100, d = .1$	✔	✔	☺	✔	☕	✔		✔
quadr 0-1, $n = 250, d = .1$	?	?	☺	☺	☺	☺		☕
2-dim. torus, $n = 20 \times 20$	✔	✔	☺	☺	☺	?		🚲
3-dim. torus, $n = 7 \times 7 \times 7$	✔	✔	☺	☺	☺	?		☕
$G_{0.5}, n = 100$	☺	?	☺	🚲	🚲	?		✔
$G_{-1/0/1}, n = 100$	☺	?	🚲	🚲	🚲	?		✔

Table 2. Average Biq Mac results for Max-Cut problems. Run times on a Pentium IV, 3.6 GHz, 2GB RAM.

graph	n	d	solved	min avg max time (h:min)	min avg max nodes
$G_{0.5}$	100	0.5	10	5 50 3:44	65 610 2925
$G_{-1/0/1}$	100	0.99	10	7 56 2:31	79 651 1811
$G_{[-10,10]}$	100	0.5	10	9 38 1:13	97 435 815
$G_{[-10,10]}$	100	0.9	10	5 57 3:12	51 679 2427
$G_{[1,10]}$	100	0.5	10	7 48 2:02	111 576 1465
$G_{[1,10]}$	100	0.9	10	12 40 1:26	155 464 1007

with this method are very low. If we do not know, whether an algorithm can solve certain classes of instances or not, we indicate this with a question mark. Most likely, we could place ☺ instead of a question mark.

6.2 Numerical Results of Max-Cut Instances

Instances by the Graph Generator 'rudy'. Table 2 lists the computation times (minimum, average and maximum) and the number of nodes (minimum, average, maximum) of the resulting Branch and Bound (B&B) tree. The branching rule used for this kind of instances is R2.

The average computation time for all instances is approximately one hour. Nevertheless, instances may also be solved within some minutes, and it could also take more than three hours for some graphs to obtain a solution.

The results show that on these classes of instances we outperform all other solution approaches known so far. The currently strongest results on these graphs are due to Billionnet and Elloumi [7]. They are not able to solve instances $G_{-1/0/1}$ of size $n = 100$ at all. Also, they could solve only two out of ten instances of $G_{0.5}$, $n = 100$.

Applications in Statistical Physics: Ising Instances. As explained in Sect. 5.1, we consider two kinds of Ising instances: toroidal grid graphs and complete graphs.

Table 3. Test runs on torus graphs with Gaussian distribution. Branch and Cut algorithm run on 1.8 GHz machine, Biq Mac done on a Pentium IV, 3.6 GHz. Time in seconds.

Problem number	n	[23] time	Biq Mac time	Problem number	n	[23] time	Biq Mac time
2 dimensional				3 dimensional			
g10_5555	100	0.15	10.12	g5_5555	125	2.68	18.01
g10_6666	100	0.14	15.94	g5_6666	125	3.29	24.52
g10_7777	100	0.18	14.89	g5_7777	125	3.07	26.00
g15_5555	225	0.44	304.03	g6_5555	216	20.56	280.85
g15_6666	225	0.78	359.87	g6_6666	216	37.74	2025.74
g15_7777	225	0.67	346.89	g6_7777	216	27.30	277.95
g20_5555	400	1.70	6690.99	g7_5555	343	95.25	432.71
g20_6666	400	3.50	35205.95	g7_6666	343	131.34	550.12
g20_7777	400	2.61	8092.80	g7_7777	343	460.01	117782.75

Instances of the first kind can be solved efficiently by an LP-based Branch and Cut algorithm (see [23]). The computation times of [23] and our algorithm are reported in Table 3. As can be seen, on these sparse instances the LP-based method clearly outperforms our algorithm. However, we find a solution within a gap of 1% in reasonable time for all these samples.

The run time of the Branch-Cut & Price algorithm [22] developed for the second kind of problems depends strongly on the parameter σ. For σ close to zero, we have a complete graph with Gaussian distributed weights. But for σ chosen suitably large, some of the edges become 'unimportant' and the pricing works very well for these graphs. In Table 4 the computation times of [22] and our algorithm are given. For $\sigma = 3.0$, we have roughly speaking the same computation times on the smallest instances. For the biggest ones, our approach clearly dominates. For $\sigma = 2.5$, the Branch-Cut & Price algorithm already takes more than 20 hours for instances of size $n = 150$, whereas our algorithm needs almost similar computation times as in the $\sigma = 3.0$ case.

For both kinds of instances we used branching rule R3.

6.3   Numerical Results of (QP) Instances

In this section we report the results for the instances derived from (QP). Best known lower and upper bounds for `bqpgka` and `beasley` data are reported at the pseudo-Boolean website [8]. Our results are as follows:

- `bqpgka`.
  - **Set a.** All problems are solved in the root node of the B&B tree within seconds.
  - **Set b.** These instances could all be solved, but were extremely challenging for our algorithm. The reason is, that the objective value in the Max-Cut formulation is of magnitude 10^6, and therefore even a relative

Table 4. Test runs on Ising instances (complete graphs). Branch-Cut & Price on a 1.8 GHz machine, Biq Mac on a 3.6 GHz PC. Times in hours:minutes:seconds.

Problem number	n	[22] time	Biq Mac time	Problem number	n	[22] time	Biq Mac time
$\sigma = 3.0$				$\sigma = 2.5$			
100_5555	100	4:52	1:36	100_5555	100	18:22	1:32
100_6666	100	0:24	0:34	100_6666	100	6:27	1:06
100_7777	100	7:31	0:48	100_7777	100	10:08	0:47
150_5555	150	2:36:46	4:38	150_5555	150	21:28:39	4:25
150_6666	150	4:49:05	3:55	150_6666	150	23:35:11	5:39
150_7777	150	3:48:41	6:06	150_7777	150	31:40:07	9:19
200_5555	200	9:22:03	10:07	200_5555	200	–	10:05
200_6666	200	32:48:03	18:53	200_6666	200	–	17:55
200_7777	200	8:53:26	22:42	200_7777	200	–	21:38
250_5555	250	21:17:07	1:46:29	250_5555	250	–	3:00:28
250_6666	250	7:42:25	15:49	250_6666	250	–	1:17:04
250_7777	250	17:30:13	57:24	250_7777	250	–	1:10:50
300_5555	300	17:20:54	2:20:14	300_5555	300	–	6:43:47
300_6666	300	10:21:40	1:32:22	300_6666	300	–	9:04:38
300_7777	300	18:33:49	3:12:13	300_7777	300	–	13:00:10

gap of 0.1% does not allow to fathom the node. However, by allowing a relative error of at most 0.1%, we can solve all problems in the root node of the B&B tree.

- **Set c.** Similar to set a, also these instances were solved within a few seconds in the root node of the B&B tree.
- **Set d.** Here $n = 100$. The problems of set d could be solved within at most 7 minutes.
- **Set e.** We recall $n = 200$. The instances with densities 0.1, 0.2, 0.3 and 0.4 could all be solved within 2 hours of computation time. The instance with $d = 0.5$ has been solved after 35 hours. According to [8], none of these problems were solved before.

- bqpbe.

We report the results of Billionnet and Elloumi [7] and our results in Table 5. As is shown in this table, [7] could not solve all out of the ten problems from the $n = 120$ variables and density 0.8 instances on, whereas our method still succeeded to solve them all. From the instances $n = 150, d = 0.8$ on, the convex-quadratic approach failed to solve any instance within their time limit of 3 hours. We still managed to obtain solutions to all of these instances (although for one graph it took about 54 hours to prove the optimality of the solution).

- beasley.

Solving the 10 problems of size $n = 100$ can be done in the root node within one minute. Regarding the $n = 250$ instances, only two out of the ten problems have been solved before (see [8]), for the other eight problems we

Table 5. Comparison between [7] and Biq Mac. Computation times of the convex-quadratic algorithm were obtained on a laptop Pentium IV, 1.6 GHz (time limit 3 hours), our results were computed on a Pentium IV of 3.6 GHz.

			[7]			Biq Mac		
			CPU time (sec)			CPU time (sec)		
n	d	solved	min	avg.	max	solved	min	avg.	max
100 1.0		10	27	372	1671	10	86	178	436
120 0.3		10	168	1263	4667	10	29	162	424
120 0.8		6	322	3909	9898	10	239	1320	3642
150 0.3		1		6789		10	1425	2263	2761
150 0.8		0		–		10	1654	1848	2133
200 0.3		0		–		10	7627	37265	193530
200 0.8		0		–		10	5541	47740	148515
250 0.1		0		–		10	12211	13295	16663

could prove optimality for the first time. Six out of these eight were solved within 5 hours, the other two needed 15 and 80 hours, respectively.

Deciding which branching rule is advisable for these instances is not so obvious anymore. Tentatively, for sparse problems R3 is superior, but the denser the instances are, the better is the performance of R2. A general recipe or an intelligent way of deciding at the top levels of the B&B tree which rule to follow would be very useful.

7 Equipartition

Finding a bisection of a graph such that each of the sets S and T have equal cardinality is often called *equipartition*. It is also customary to minimize the weight of edges in the cut. Hence the problem is a minor extension of (MC).

$$z_{EP} = \min\{x^T L x : e^T x = 0, x \in \{-1,1\}^n\} \qquad (8)$$

This leads to the following semidefinite relaxation.

$$z_{EP-SDP} = \min\{\mathrm{tr} LX : \mathrm{tr} JX = 0, \mathrm{diag}(X) = e, X \succeq 0\}, \qquad (9)$$

where $J = ee^T$. Let A be the adjacency matrix of the given graph. We consider the Max-Cut instance with cost matrix $B = -A + J$. The "−" in $B = -A + J$ arises, because we minimize instead of maximizing, and the J comes from the constraint $\mathrm{tr} JX = 0$, that comes with a Lagrange multiplier (set equal to 1 for unweighted instances) into the objective function.

We consider the instances introduced in [19] of size $n = 124$ and $n = 250$ and summarize in Table 6 the best results for these instances known so far (see [20]). With our algorithm we prove optimality of the known lower bounds of all instances of size $n = 124$, and one of the instances of size $n = 250$. To the

Table 6. Best known results of the bisection problem for the Johnson graphs and the new gap obtained by Biq Mac

d	best known bound	$|E_{cut}|$	gap	new gap	d	best known bound	$|E_{cut}|$	gap	new gap
$n = 124$					$n = 250$				
0.02	12.01	13	0	0	0.01	26.06	29	2	0
0.04	61.22	63	1	0	0.02	103.61	114	10	8
0.08	170.93	178	7	0	0.04	327.88	357	29	22
0.16	440.08	449	8	0	0.08	779.55	828	48	35

best of our knowledge, these exact solutions were obtained for the first time. The improved gap for the instances of size $n = 250$ and densities 0.02, 0.04 and 0.08 were obtained after a time limit of 32 hours cpu-time.

8 Summary

In this paper we have presented an algorithm, that uses a Branch and Bound framework to solve the Max-Cut and related problems. At each node of the tree we calculate the bound by using a dynamic version of the bundle method that solves the basic semidefinite relaxation for Max-Cut strengthened by triangle inequalities. We conclude, that

- our approach solves any instance of all the test-bed considered with $n \approx 100$ nodes in a routine way. To the best of our knowledge, no other algorithm can manage these instances in a similar way.
- we solve problems of special structure and sparse problems up to $n = 300$ nodes.
- for the first time optimality could be proved for several problems of the OR-library. All problems that are reported at the Pseudo-Boolean website [8] with dimensions up to $n = 250$ are now solved.
- for the first time optimality of the bisection problem for some of the Johnson graphs has been proved, for those where we could not close the gap we reduced the best known gap significantly.
- for sparse problems it is not advisable to use our approach. Since linear programming based methods are capable of exploiting sparsity, solutions might be obtained much faster when applying these methods to sparse data.

Using our algorithm to solve this problem has been made publicly available [29].

References

[1] F. Barahona and L. Ladányi. Branch and cut based on the volume algorithm: Steiner trees in graphs and max-cut. *RAIRO Oper. Res.*, 40(1): 53–73, 2006.

[2] F. Barahona, M. Grötschel, M. Jünger, and G. Reinelt. An application of combinatorial optimization to statistical physics and circuit layout design. *Operations Research*, 36:493–513, 1988.

[3] F. Barahona, M. Jünger, and G. Reinelt. Experiments in quadratic 0-1 programming. *Math. Programming*, 44(2, (Ser. A)):127–137, 1989.

[4] J. E. Beasley. Or-library: distributing test problems by electronic mail. *J. Oper. Res. Soc.*, 41(11):1069–1072, 1990.

[5] J. E. Beasley. Or-library, 1990. http://people.brunel.ac.uk/~mastjjb/jeb/info.html.

[6] J. E. Beasley. Heuristic algorithms for the unconstrained binary quadratic programming problem. Technical report, The Management School, Imperial College, London SW7 2AZ, England, 1998.

[7] A. Billionnet and S. Elloumi. Using a mixed integer quadratic programming solver for the unconstrained quadratic 0-1 problem. *Math. Programming*, 109(1, Ser. A):55–68, 2007.

[8] E. Boros, P. L. Hammer, and G. Tavares. The pseudo-boolean optimization website, 2005. http://rutcor.rutgers.edu/~pbo/.

[9] C. De Simone, M. Diehl, M. Jünger, P. Mutzel, G. Reinelt, and G. Rinaldi. Exact ground states of Ising spin glasses: New experimental results with a branch-and-cut algorithm. *J. Statist. Phys.*, 80(1-2):487–496, 1995.

[10] C. Delorme and S. Poljak. Laplacian eigenvalues and the maximum cut problem. *Math. Programming*, 62(3, Ser. A):557–574, 1993.

[11] M. Elf, M. Jünger, and G. Rinaldi. Minimizing breaks by maximizing cuts. *Operations Research Letters*, 31:343–349, 2003.

[12] I. Fischer, G. Gruber, F. Rendl, and R. Sotirov. Computational experience with a bundle approach for semidefinite cutting plane relaxations of Max-Cut and equipartition. *Math. Programming*, 105(2-3, Ser. B):451–469, 2006.

[13] A. Frangioni, A. Lodi, and G. Rinaldi. New approaches for optimizing over the semimetric polytope. *Math. Program.*, 104(2-3, Ser. B):375–388, 2005.

[14] F. Glover, G. Kochenberger, and B. Alidaee. Adaptative memory tabu search for binary quadratic programs. *Management Sci.*, 44(3):336–345, 1998.

[15] M. X. Goemans and D. P. Williamson. .878-approximation algorithms for max cut and max 2sat. In *Proceedings of the Twenty-Sixth Annual ACM Symposium on the Theory of Computing*, pages 422–431, Montreal, Quebec, Canada, 1994.

[16] M. X. Goemans and D. P. Williamson. Improved approximation algorithms for maximum cut and satisfiability problems using semidefinite programming. *J. Assoc. Comput. Mach.*, 42(6):1115–1145, 1995. preliminary version see [15].

[17] C. Helmberg. Fixing variables in semidefinite relaxations. *SIAM J. Matrix Anal. Appl.*, 21(3):952–969 (electronic), 2000.

[18] C. Helmberg and F. Rendl. Solving quadratic (0, 1)-problems by semidefinite programs and cutting planes. *Math. Programming*, 82(3, Ser. A):291–315, 1998.

[19] D. S. Johnson, C. R. Aragon, L. A. McGeoch, and C. Schevon. Optimization by simulated annealing: an experimental evaluation. part i, graph partitioning. *Oper. Res.*, 37(6):865–892, 1989.

[20] S. E. Karisch and F. Rendl. Semidefinite programming and graph equipartition. In *Topics in semidefinite and interior-point methods (Toronto, ON, 1996)*, volume 18 of *Fields Inst. Commun.*, pages 77–95. Amer. Math. Soc., Providence, RI, 1998.

[21] S. Kim and M. Kojima. Second order cone programming relaxation of nonconvex quadratic optimization problems. *Optim. Methods Softw.*, 15 (3-4):201–224, 2001.

[22] F. Liers. *Contributions to Determining Exact Ground-States of Ising Spin-Glasses and to their Physics.* PhD thesis, Universität zu Köln, 2004.

[23] F. Liers, M. Jünger, G. Reinelt, and G. Rinaldi. Computing exact ground states of hard ising spin glass problems by branch-and-cut. In A. Hartmann and H. Rieger, editors, *New Optimization Algorithms in Physics*, pages 47–68. Wiley, 2004.

[24] M. Muramatsu and T. Suzuki. A new second-order cone programming relaxation for MAX-CUT problems. *J. Oper. Res. Soc. Japan*, 46(2): 164–177, 2003.

[25] P. M. Pardalos and G. P. Rodgers. Computational aspects of a branch and bound algorithm for quadratic zero-one programming. *Computing*, 45(2): 131–144, 1990.

[26] P. M. Pardalos and G. P. Rodgers. Parallel branch and bound algorithms for quadratic zero-one programs on the hypercube architecture. *Ann. Oper. Res.*, 22(1-4):271–292, 1990.

[27] S. Poljak and F. Rendl. Solving the max-cut problem using eigenvalues. *Discrete Appl. Math.*, 62(1-3):249–278, 1995.

[28] S. Poljak and F. Rendl. Nonpolyhedral relaxations of graph-bisection problems. *SIAM J. Optim.*, 5(3):467–487, 1995.

[29] F. Rendl, G. Rinaldi, and A. Wiegele. Biq Mac – a solver for binary quadratic and max-cut problems, 2006. http://BiqMac.uni-klu.ac.at/.

[30] G. Rinaldi. Rudy, 1998. http://www-user.tu-chemnitz.de/~helmberg/rudy.tar.gz.

[31] A. Wiegele. *Nonlinear optimization techniques applied to combinatorial optimization problems.* PhD thesis, Alpen-Adria-Universität Klagenfurt, 2006.

DINS, a MIP Improvement Heuristic

Shubhashis Ghosh*

Department of Computing Science, University of Alberta, Canada
sghosh@ualberta.ca

Abstract. We introduce DISTANCE INDUCED NEIGHBOURHOOD SEARCH (DINS), a MIP improvement heuristic that tries to find improved MIP feasible solutions from a given MIP feasible solution. DINS is based on a variation of local search that is embedded in an exact MIP solver, namely a branch-and-bound or a branch-and-cut MIP solver. The key idea is to use a distance metric between the linear programming relaxation optimal solution and the current MIP feasible solution to define search neighbourhoods at different nodes of the search tree generated by the exact solver. DINS considers each defined search neighbourhood as a new MIP problem and explores it by an exact MIP solver with a certain node limit. On a set of standard benchmark problems, DINS outperforms the MIP improvement heuristics Local Branching due to Fischetti and Lodi and Relaxation Induced Neighbourhood Search due to Danna, Rothberg, and Pape, as well as the generic commercial MIP solver Cplex.

1 Introduction

Mixed integer programs (MIPs) arise in many contexts; they are often intractable and NP-hard, even for feasibility [14]. Therefore, there is interest in designing effective heuristic methods for MIPs. Recently MIP heuristic development has specialized into finding better feasibility heuristic (that tries to find an initial MIP feasible solution), and improvement heuristic (that tries to find improved MIP feasible solutions from a given MIP feasible solution). In this paper, we present a new improvement heuristic.

Recent improvement heuristics such as LOCAL BRANCHING (LB), introduced by Fischetti et al. [9] and re-engineered by Danna et al. [5], and RELAXATION INDUCED NEIGHBOURHOOD SEARCH (RINS), introduced by Danna et al. [5], work in tandem with a state-of-the-art exact solver such as Cplex MIP solver as follows. The exact solver generates a search tree using either branch-and-bound or branch-and-cut approach; the new heuristics periodically select nodes of the search tree at which to perform a localized search. Our heuristic also follows this approach. The heuristics differ primarily in the definition of the search neighbourhood; in LB the search neighbourhood is defined by restricting the number of 0-1 variables to switch their bounds from the known MIP feasible solution (referred as soft fixing), and in RINS it is defined by fixing some variables at their current values in the known MIP feasible solution (referred as hard fixing).

---

* The research support of NSERC is gratefully acknowledged.

M. Fischetti and D.P. Williamson (Eds.): IPCO 2007, LNCS 4513, pp. 310–323, 2007.
© Springer-Verlag Berlin Heidelberg 2007

Our search neighbourhood is defined in terms of a distance metric between a relaxation solution and the current MIP feasible solution, where the distance metric comes from the intuition that improved solutions are more likely to be close to the relaxation solution at the nodes of the search tree.

On a set of standard benchmark MIP instances, DINS outperforms Cplex, RINS, and LB with respect to the quality of solutions obtained within a time limit.

2   Related Previous Work

In order to show the strength of our heuristic, we compare it against Cplex, the exact solver in which it is embedded, and LB and RINS, the two recent improvement heuristics that are most similar in design.

Much research has been done in other kinds of MIP heuristics. There are several heuristics, introduced by Balas et al. [1], Faaland et al. [7], Hillier [12], and Ibaraki et al. [13], that incorporate some form of neighbourhood search, and most of them do so from the relaxation solution of MIP in order to find a MIP feasible solution.

There are also several pivot based heuristics, introduced by Balas et al. [2,3], Løkketangen et al. [15], Nediak et al. [18], and Løkketangen et al. [16], for MIP that try to obtain a MIP solution by performing pivots on the simplex tableau of the relaxation of MIP. Another heuristic introduced by Balas et al. [4], starting from the relaxation solution of MIP, tries to find a MIP solution by first using some pivoting on the simplex tableau and then doing some form of neighbourhood search. Recently Fischetti et al. [8] introduce another heuristic to find a MIP solution from the relaxation solution of MIP, where they solve a sequence of linear programs in the process of finding a MIP feasible solution.

3   Methods

We assume that the input program P is a generic MIP of the form shown below, where c, x, b, A have dimensions $n, n, m, m \times n$ respectively, $N = \{1, \ldots, n\}$ is the set of variable indices of P which is partitioned into $(\mathcal{B}, \mathcal{G}, \mathcal{C})$ with $\mathcal{B}$, $\mathcal{G}$, and $\mathcal{C}$ denoting the indices of 0-1, general integer, and continuous variables respectively. An integer variable is any variable in $\mathcal{B} \cup \mathcal{G}$.

$P : min \ \{ \ c^T x \mid Ax \geq b, \ x_i \in \{0, 1\} \ \forall i \in \mathcal{B},$
$x_j \geq 0 \ \text{and integer} \ \forall j \in \mathcal{G}, \ x_j \geq 0 \ \forall j \in \mathcal{C}\}$

Since we compare DINS with LB and RINS, we describe LB and RINS in some details.

3.1   Local Branching

LB defines the neighbourhood of a feasible solution x^* by limiting at some integer p the number of 0-1 variables currently at 0 or 1 that can switch their bounds. This is achieved by adding to the instance the *LB inequality* $D(x, x^*) \leq p$, where

$$D(x, x^*) := \sum_{j \in V_0} x_j + \sum_{j \in V_1} (1 - x_j),$$

and where V_0 and V_1 are the index sets of the 0-1 variables that are at 0 and 1 respectively in x^*.

LB has been implemented in two different ways. Originally, Fischetti and Lodi [9] treated it as an external branching framework (i.e., creates branches in the search tree by $D(x, x^*) \leq p$ and $D(x, x^*) \geq p + 1$ as opposed to the standard branching which are done on the variables in the branch-and-bound framework) in addition to an heuristic and obtained the diversification (i.e., switching the search in a different region of the MIP feasible space) by defining the neighbourhoods with a change in the value of the parameter p. Later, Danna et al. [5] implemented LB solely as a heuristic and obtained the diversification by defining the neighbourhoods on the new solutions found during the MIP search tree exploration. Danna et al. showed that their implementation of LB outperformed the original. For this reason, we choose the Danna et al. version of LB to compare against our DINS.

3.2 Relaxation Induced Neighbourhood Search

During the exploration of the MIP search tree, the relaxation solution at successive nodes (that are not pruned by infeasibility or bound) provides a better objective value than the objective value of the current MIP solution. Using this, Danna et al. introduce RINS making the intuition that, in improved MIP solutions, it is more likely for the variables to stay at the same values those agree in the current MIP solution and current node relaxation solution. Thus RINS defines the promising neighbourhood fixing all variables whose values at the current MIP solution are equal to their respective values at the current node relaxation solution.

In the implementation of RINS[1], the procedure for exploring the RINS defined neighbourhood is invoked at a particular node of the MIP search tree. At the termination of the procedure, the MIP search tree is resumed, and if the procedure finds a new MIP solution, the MIP solution at the MIP search tree is updated.

As noted by Danna et al. in [5], consecutive nodes of the MIP search tree provide almost identical relaxation solution. Therefore, the RINS procedure is called only every f nodes for some reasonably large f.

3.3 Distance Induced Neighbourhood Search

In contrast to RINS, which performs only hard fixing of variables, and LB, which performs only soft fixing of variables, our DINS incorporates some hard fixing, some soft fixing, and some rebounding (changing lower and upper bounds

---

[1] ILOG Cplex 9.13 comes with an implementation of RINS and can be invoked by setting the Cplex parameter *IloCplex::MIPEmphasis* to 4 [5].

of the variables), all based on a distance metric. In the next sections we show that DINS outperforms both RINS and LB[2] on an instance test bed that includes all the instances studied in [5,9] as well as some other hard instances from other sources.

Like RINS, DINS also rely on the fact that, during exploring the MIP search tree, the relaxation solution at successive nodes (those are not pruned by infeasibility or bound) provides a better objective value compared to the objective value provided by the current MIP solution.

But unlike RINS, the intuition in DINS is that the improved MIP solutions are more likely to be the close ones to the current relaxation solution. An exact modeling of this intuition would require inclusion of the following quadratic inequality which unfortunately cannot be expressed as a linear constraint.

$$\sum_{j \in N} (x_j - x_{j(node)})^2 \leq \sum_{j \in N} (x_{j(mip)} - x_{j(node)})^2,$$

where x_{mip} and x_{node} denote the current MIP solution and the current relaxation solution, and for a variable x_j, $x_{j(mip)}$ and $x_{j(node)}$ denote the values of x_j in x_{mip} and x_{node} respectively.

DINS relaxes the intuition by considering that the improved MIP solutions are close to x_{node} only with respect to the integer variables and choosing the following inequality based on absolute differences as the measure of close ones.

$$\sum_{j \in \mathcal{B} \cup \mathcal{G}} |x_j - x_{j(node)}| \leq \sum_{j \in \mathcal{B} \cup \mathcal{G}} |x_{j(mip)} - x_{j(node)}|.$$

DINS then partially captures this inequality (the chosen distance metric) by defining a neighbourhood with some rebounding, some hard fixing, and some soft fixing of the integer variables.

We notice that if an integer variable x_j, for which the absolute difference, $|x_{j(mip)} - x_{j(node)}|$, is less than 0.5, takes a different value than $x_{j(mip)}$ in an improved solution, the absolute difference increases. On the contrary, if an integer variable, for which the absolute difference is greater or equal to 0.5, takes a different value than $x_{j(mip)}$ in an improved solution, the absolute difference may not increase.

DINS computes new lower and upper bounds of an integer variable x_j, for which the absolute difference is greater or equal to 0.5, so that at an improved solution the absolute difference does not increase. Considering l_j^{old} and u_j^{old} as the existing lower and upper bounds of x_j, DINS computes the new lower and upper bound l_j^{new} and u_j^{new} respectively as follows:

if $(x_{j(mip)} \geq x_{j(node)})$ then
  $l_j^{new} \leftarrow \max(l_j^{old}, \lceil x_{j(node)} - (x_{j(mip)} - x_{j(node)}) \rceil)$, $u_j^{new} \leftarrow x_{j(mip)}$
elsif $(x_{j(mip)} < x_{j(node)})$ then
  $l_j^{new} \leftarrow x_{j(mip)}$, $u_j^{new} \leftarrow \min(u_j^{old}, \lfloor x_{j(node)} + (x_{j(node)} - x_{j(mip)}) \rfloor)$.

---

[2] In [5], Danna et al. have tried two hybrid strategies of RINS and LB and concluded that their performance were not better than RINS alone.

We refer it as rebounding; the rebounding does not change existing bounds for all the variables that fall in this category (for example, no 0-1 variable in this category change its bounds). If all the integer variables, for which $|x_{j(mip)} - x_{j(node)}| < 0.5$, are fixed to their respective current values, then any solution found from this neighbourhood exploration will obviously be a closer one to x_{node} in terms of the chosen distance metric. But the sum of absolute differences can also decrease if the total decrease d in the sum of absolute differences caused by the integer variables for which $|x_{j(mip)} - x_{j(node)}| \geq 0.5$ is greater than the total increase d' in the sum of absolute differences caused by the integer variables for which $|x_{j(mip)} - x_{j(node)}| < 0.5$.

DINS partially captures this observation by allowing the integer variables x_j, for which $|x_{j(mip)} - x_{j(node)}| < 0.5$, to change their values in x_{mip} so that d' is not larger than a chosen small number p. It does this by performing some soft fixing and some hard fixing of these variables. DINS performs soft fixing through the LB inequality which requires introduction of new variables when general integer variables are considered. As in [9] and [5], DINS constructs LB inequality using only 0-1 variables. Therefore, all the general integer variables x_j with $|x_{j(mip)} - x_{j(node)}| < 0.5$ are fixed (hard fixing) at $x_{j(mip)}$.

Among the 0-1 variables with $|x_{j(mip)} - x_{j(node)}| < 0.5$, DINS performs some hard fixing like RINS, but incorporates some more intuition in this process. Like RINS, DINS chooses the same set of variables, that agree in both the current MIP solution and the current node relaxation solution, as the primary candidates for hard fixing. Then it applies a filtering step to this primary candidate set using two information. First information comes from the intuition that if an integer variable, in the primary candidate set, takes the same value in the root relaxation solution of MIP search tree and current node relaxation solution, is more likely to take the same value in improved MIP feasible solutions. The second information comes from the intuition that if an integer variable, in the primary candidate set, takes the same value in the previously encountered MIP solutions, is more likely to take the same value in improved MIP feasible solutions. This two information actually gather knowledge from both the relaxation solutions and previously encountered MIP solutions. DINS uses an array of flag for the integer variables to keep track which variables have taken different values in the previously encountered MIP solutions. Thus the hard fixing in DINS can be stated more explicitly in the following way: let x_{mip}, x_{node}, and x_{root} denote the current MIP solution, the current node relaxation solution, and the root relaxation solution respectively. Also let Δ is an array where $\Delta[j]$ is set if x_j has taken different values in previously encountered MIP solutions. Therefore, a variable x_j is fixed (hard fixing) at value $x_{j(mip)}$ if $x_{j(mip)} = x_{j(node)} = x_{j(root)}$ and $\Delta[j]$ is clear.

Consider $\mathcal{F}$ and $\mathcal{H}$ denote the set of variables for which rebounding and hard fixing has been performed respectively. Now assume $\mathcal{R}$ be the set of variables where $\mathcal{R} = (\mathcal{B} \cup \mathcal{G}) - \mathcal{F} - \mathcal{H}$. According to our construction $\mathcal{R}$ contains only 0-1 variables.

DINS now performs soft fixing on the variables in $\mathcal{R}$, when $|\mathcal{R}| \neq \phi$, by adding the following LB inequality:

$$\sum_{j \in \mathcal{R} \,\wedge\, x_{j(mip)}=0} x_j + \sum_{j \in \mathcal{R} \,\wedge\, x_{j(mip)}=1} (1 - x_j) \leq p$$

As noted earlier, our intuition is that improved feasible solutions are more likely to be obtained by getting close to the current relaxation solution from the current MIP solution. Therefore, DINS generates the promising neighbourhood taking small value for p which means that a solution, in this defined neighbourhood, can have a sum of absolute differences increased by at most p.

Whenever DINS procedure is invoked at a particular node of MIP search tree, it creates the described neighborhood with the initial chosen value of p and explores it using a branch-and-bound or a branch-and-cut solver with a specified node limit nl. If the exploration reaches the node limit without finding a new solution, DINS reduces p by 5 and explores a new neighbourhood. This continues until $p < 0$, or the neighbourhood exploration finds a new solution or the neighbourhood is explored completely without finding a new solution. Whenever the neighbourhood exploration finds a new solution, p is reset to its initial chosen value and continues in the same fashion. The procedure in Figure 1 describes the operation sequence of DINS at a particular node of the MIP search tree. At the termination of the procedure, the MIP search tree is resumed and, if the procedure finds a new MIP solution, the MIP solution at the MIP search tree is updated.

Like RINS, the DINS procedure is called first when the MIP search tree finds its first MIP solution and, thereafter, at every f nodes of the MIP search tree.

4 Computational Results

4.1 Experimental Setup and Instance Test Bed

We implement LB, RINS, and DINS in the C programming language with the MIP search tree generated by Cplex 9.13 MIP solver. All experiments are run on an 2403 MHz AMD Athlon processor with 128 MByte of memory under Redhat Linux 9.0. An implementation of DINS is available at [11].

We compose a benchmark test bed of MIP instances with the property that the test bed excludes the instances which default Cplex either solves to optimality or fails to find a MIP solution in one CPU-hour. With this criteria we have 64 MIP instances (all have some 0-1 variables), described in [10], from the following sources commonly used as benchmark instances for MIP solvers.

- Twenty six instances used in the local branching paper [9]. These instances have been collected from the instance set maintained by DEIS operations research group [6].
- Twelve more instances from the instance set maintained by DEIS operations research group [6].

- Eleven instances from MIPLIB 2003 [17].
- Five job-shop scheduling instances with earliness and tardiness costs used in [8].
- Eleven network design and multi-commodity routing instances used in [5].

Procedure DINS_at_tree_node
INPUT: a 0-1 mixed integer problem P, the current MIP solution x_{mip},
     the current node relaxation solution x_{node}, the root relaxation solution x_{root},
     parameter p, node limit nl, and the flag array Δ.
OUTPUT: A new MIP solution x^* (x_{mip} in case of failure in finding a new solution).

1. if (x_{mip} is a new MIP solution compared to the MIP solution
        at the termination of last call of this procedure)
            update the array Δ accordingly
2. $x^* \leftarrow x_{mip}$, $p_{current} \leftarrow p$, exploreAndNoSolution $\leftarrow$ false
3. repeat
4.   construct $P+$ from P as follows:
            (i) perform rebounding on the variables x_j for which $|x_j^* - x_{j(node)}| \geq 0.5$,
            (ii) perform hard fixing of the general integer variables x_j for which
                 $|x_j^* - x_{j(node)}| < 0.5$,
            (iii) perform hard fixing of the 0-1 integer variables x_j for which
                 $|x_j^* - x_{j(node)}| < 0.5$ and $x_j^* = x_{j(node)} = x_{j(root)}$ and $\Delta[j]$ is clear,
            (iv) let $\mathcal{R}$ be the set of remaining 0-1 integer variables.
                if ($\mathcal{R} \neq \phi$) perform soft fixing by adding the inequality
                $$\sum_{j \in \mathcal{R} \,\wedge\, x_j^*=0} x_j + \sum_{j \in \mathcal{R} \,\wedge\, x_j^*=1} (1 - x_j) \leq p$$
5.   Apply black-box MIP solver to $P+$ with node limit nl and
            an objective cutoff equal to the objective value provided by x^*
6.   if (a new solution x_{new} is obtained) then
7. $x^* \leftarrow x_{new}$, $p_{current} \leftarrow p$, update the array Δ
8.   elsif (node limit reached without having a new solution) then
9.   if($|\mathcal{R}| = \phi$) $p_{current} = -1$
10   else $p_{current} \leftarrow p_{current} - 5$
11.   else exploreAndNoSolution $\leftarrow$ true
12. until ($p_{current} < 0$ or exploreAndNoSolution)
13. return x^*

Fig. 1. Procedure DINS_at_tree_node

4.2   Comparison Among Methods

We compare DINS against RINS, LB, and Cplex in its default setup (default Cplex). One CPU-hour is set to be the execution time for each method and it seems to be sufficient to distinguish the effectiveness of all the methods.

Default Cplex is used for exploring the neighbourhoods generated in LB, RINS, and DINS. The three methods namely LB, RINS, and DINS have a set of parameters which need to be set. As used in [5], for LB, we set $p = 10$ and $nl = 1000$, and for RINS, we use Cplex 9.13 with the parameter *IloCplex::MIPEmphasis* set to 4 where, according to [5], $f = 100$ and $nl = 1000$. For

DINS, we set $p = 5$ (different from LB to relax our intuition a little as well as to make the neighbourhood small), $f = 100$ and $nl = 1000$.

Following Danna et al. [5], we carry out two set of experiments; in one set of experiments we invoke all four methods with a presumably poor solution at the root node of the MIP search tree, and in the other we invoke all four methods with a presumably good solution at the root node of the MIP search tree. Although there is no exact way to distinguish a good and a bad MIP solution, following Danna et al. [5], we presume that the first MIP solution found by the default Cplex MIP solver represents a poor solution, and the solution obtained by default Cplex in one CPU-hour represents a good solution.

In order to capture the quality of obtained solution by each method, we use the measure *percentage of gap* defined by 100*|(obj. value of obtained solution - obj. value of the best known solution) /obj. value of the best known solution|. Table 1 and Table 2 show the percentage of gap obtained at the end of one CPU-hour by all the four methods considered in this paper, where the bold face identifies the best method for the corresponding instance (multiple bold faces appear if there are multiple methods obtaining the same solution).

Following Danna et al. [5], we group the instances into three different sets so that the effectiveness of different methods in different groups becomes visible. According to [5], the groups are defined as 'small spread', 'medium spread', and 'large spread' instances where the gap between the worst solution found by any of the four methods considered in this paper and the best known solution is less than 10%, between 10% and 100%, and larger than 100% respectively. The percentage of gap shown in Table 1 and Table 2 are used to group the instances.

We use three measures to evaluate the performance of different methods.

Our first measure is *best in number of instances*, which represents the number of instances at which a method finds the best solution among the solutions obtained by all the four methods. If multiple methods find the same best solution for an instance, then the instance contributes one in the measures for all the corresponding methods.

Our second measure is the *average percentage of gap*, which represents the arithmetic mean of the percentage of gaps obtained by a method on a group of instances at a certain point of execution.

Our third measure is the *average percentage of improvement*, which represents the arithmetic mean of percentage of improvements obtained by a method on a group of instances at a certain point of execution. In order to visualize how much improvement has been obtained by different methods starting from a presumably poor and good solution, we define the *percentage of improvement* for an instance as 100*|(obj. value of the initial solution - obj. value of the obtained solution) /obj. value of the initial solution|.

Table 3 represents the comparative results of four different methods for both set of experiments.

As expected, DINS, comparing against all other three methods in both set of experiments, has higher percentage of improvement and lower percentage of gap for each of the categorized group of instances, and obtains best solution in

Table 1. Percentage of Gap = 100 ∗ |(obj. value of obtained solution - obj. value of the best known solution)/obj. value of the best known solution| in one CPU-hour

problem	Percentage of Gap			
	Default Cplex	LB	RINS	DINS
Small spread instances				
a1c1s1	2.347	0.250	**0.000**	0.079
a2c1s1	2.978	1.889	**0.000**	0.024
b1c1s1	5.977	1.786	**0.933**	4.444
b2c1s1	4.240	2.701	**0.559**	1.010
biella1	**0.309**	0.806	0.426	0.739
danoint	**0.000**	**0.000**	**0.000**	**0.000**
mkc	0.180	0.049	0.043	**0.021**
net12	**0.000**	**0.000**	**0.000**	**0.000**
nsrand-ipx	0.625	0.625	0.313	**0.000**
rail507	**0.000**	**0.000**	**0.000**	**0.000**
rail2586c	2.518	2.204	1.994	**1.574**
rail4284c	1.774	1.867	**1.027**	**1.027**
rail4872c	1.742	1.290	1.097	**1.032**
seymour	0.473	0.473	**0.000**	0.236
sp97ar	0.428	0.513	0.335	**0.000**
sp97ic	0.793	0.642	0.551	**0.000**
sp98ar	0.184	**0.106**	0.177	0.228
sp98ic	0.270	0.146	0.204	**0.072**
tr12-30	**0.000**	0.024	**0.000**	**0.000**
arki001	0.003	0.003	0.004	**0.002**
roll3000	0.543	0.303	**0.070**	**0.070**
umts	0.013	0.049	0.022	**0.002**
berlin-5-8-0	**0.000**	**0.000**	**0.000**	**0.000**
bg512142	7.257	5.192	0.161	**0.000**
blp-ic97	0.779	0.653	0.358	**0.000**
blp-ic98	0.961	1.056	0.746	**0.515**
blp-ar98	0.655	0.060	0.461	**0.000**
cms750-4	2.372	**0.791**	1.186	**0.791**
dc11	2.018	8.166	6.994	**1.572**
railway-8-1-0	0.250	**0.000**	0.250	0.250
usabbrv-8-25-70	3.306	2.479	**0.000**	1.653
aflow40b	0.257	1.455	**0.000**	**0.000**
dano3mip	2.602	3.595	4.724	**2.230**
fast0507	**0.000**	0.575	0.575	**0.000**
harp2	0.001	0.001	0.023	**0.000**
t1717	7.948	**1.939**	5.979	7.948
noswot	**0.000**	**0.000**	**0.000**	**0.000**
timtab1	7.469	7.779	**0.000**	**0.000**
ljb2	**0.256**	3.329	1.576	3.329
rococoB10-011000	0.802	2.848	**0.437**	**0.437**
rococoB11-010000	5.039	5.839	**1.768**	2.196
rococoB12-111111	5.204	4.489	3.738	**2.541**

Table 2. Percentage of Gap $= 100 * |$(obj. value of obtained solution - obj. value of the best known solution)/obj. value of the best known solution| in one CPU-hour

Continued from Table 1				
problem	Percentage of Gap			
	Default Cplex	LB	RINS	DINS
Small spread instances				
rococoC10-001000	0.044	0.113	0.044	**0.000**
rococoC11-011100	6.018	9.991	9.244	**5.879**
rococoC12-111100	5.188	5.188	**1.298**	4.016
Medium spread instances				
glass4	13.014	7.534	**2.740**	4.794
swath	18.067	5.679	8.089	**4.622**
dg012142	17.457	25.984	4.963	**3.943**
liu	**2.475**	10.066	3.465	5.281
timtab2	16.373	18.484	3.188	**0.912**
ljb7	7.424	21.834	**4.367**	8.908
ljb9	50.717	70.866	55.074	**50.690**
ljb10	**0.807**	13.929	13.693	8.578
rococoB10-011001	7.660	5.309	**5.220**	10.082
rococoB11-110001	9.994	19.558	**4.267**	6.894
rococoC10-100001	16.041	**7.387**	13.316	10.070
rococoC11-010100	27.431	13.615	10.546	**9.029**
rococoC12-100000	12.928	10.090	5.623	**2.799**
Large spread instances				
markshare1	500.000	**400.00**	**400.00**	500.00
markshare2	1300.000	**1100.000**	2000.000	1800.000
dc1c	695.213	2.353	**0.296**	0.773
trento1	**0.000**	193.118	1.912	0.402
ds	11.226	945.745	11.226	**6.119**
ljb12	**39.273**	323.183	49.599	64.987

higher number of instances. It is to be noted that, starting from a presumably good solution, DINS has become best in more number of instances than the number of instances in which it has been best in the experimentation with bad solution.

Furthermore, for different group of instances in Figure 2– 4, we sketch how different methods improve the solution quality (average percentage of gap) over time starting from presumably poor solutions. We can draw some basic conclusions analyzing these figures. For all three group of instances, DINS performance is worse comparing to that of RINS at the initial level of computation, but DINS performance becomes better as the computation progresses and once it becomes better, it maintains its lead over RINS for the remaining part of the computation. For small and large spread instances, DINS obtains the lead over RINS earlier than in medium spread instances. Similarly in medium and large spread instances, DINS performance is worse comparing to that of default Cplex at the initial level of computation, but DINS outperforms default Cplex as the

Table 3. A comparative performance summary for different methods

Average % of improvement			
Group of Instances (# of instances)	Default Cplex	LB	RINS	DINS
experiments from the presumably poor solutions				
all instances (64)	36.19	35.49	38.01	38.05
small spread (45)	23.41	23.61	23.90	23.92
medium spread (13)	60.43	60.25	62.05	62.29
large spread (6)	80.78	70.90	91.64	91.66
experiments from the presumably good solutions				
all instances (64)	2.35	3.04	3.45	3.96
small spread (45)	0.45	0.78	1.26	1.29
medium spread (13)	2.50	4.91	5.10	6.57
large spread (6)	16.31	15.96	16.27	18.47
Average % of gap				
Group of Instances (# of instances)	Default Cplex	LB	RINS	DINS
experiments from the presumably poor solutions				
all instances (64)	44.22	51.19	41.33	39.73
small spread (45)	1.86	1.81	1.05	0.97
medium spread (13)	15.41	17.72	10.35	9.74
large spread (6)	424.28	494.07	410.51	395.38
experiments from the presumably good solutions				
all instances (64)	32.43	31.67	31.21	29.14
small spread (45)	1.41	1.07	0.56	0.54
medium spread (13)	13.57	10.63	10.46	8.59
large spread (6)	305.92	306.77	306.06	288.17
Best in # of instances				
Group of Instances (# of instances)	Default Cplex	LB	RINS	DINS
experiments from the presumably poor solutions				
all instances (64)	13	12	25	39
small spread (45)	9	9	19	32
medium spread (13)	2	1	4	6
large spread (6)	2	2	2	1
experiments from the presumably good solutions				
all instances (64)	16	23	29	48
small spread (45)	13	17	26	35
medium spread (13)	1	5	2	8
large spread (6)	2	1	1	5

computation progresses. LB is always worse than RINS and DINS where, at the end of time limit, LB has an edge over default Cplex only in small spread instances.

In an attempt to see how good intuition DINS has made, we provide some statistical measures from our experimental results. It has been seen that, the

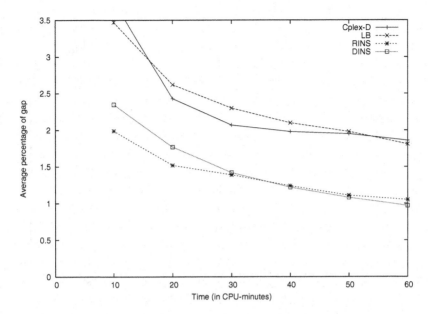

Fig. 2. progress of different methods in reducing percentage of gap on the 45 small spread instances

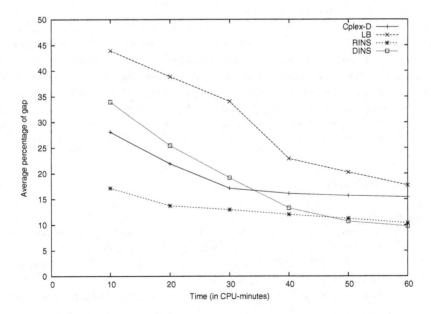

Fig. 3. progress of different methods in reducing percentage of gap on the 13 medium spread instances

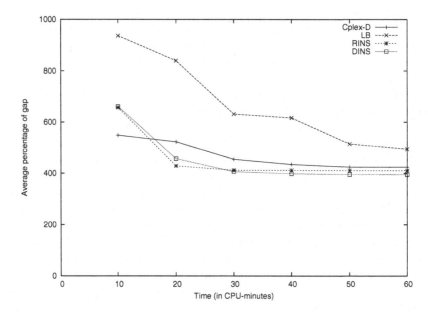

Fig. 4. progress of different methods in reducing percentage of gap on the 6 large spread instances

number of times neighbourhood exploration finds a new solution in all the instances, the chosen distance metric was satisfied in 80.89% occurrences, and the quadratic distance metric was satisfied in 80.5% occurrences. These experimental results support our intuition that improved solutions are more likely to be close to the node relaxation solutions, and also support our choice of distance metric. Moreover, relaxing the chosen distance metric a little bit gives DINS the extra power of finding those improved solutions that do not satisfy the chosen distance metric at the node at which the solution has been obtained, but probably would satisfy the chosen distance metric at some deeper nodes of the MIP search tree.

5   Conclusions

We have introduced DINS, a heuristic to find improved MIP feasible solutions from a known MIP feasible solution, based on a distance metric between the current MIP solution and the current node relaxation solution.

A comparison of DINS against existing neighbourhood search based heuristics shows that it outperforms both RINS and LB in obtaining good MIP solutions within a certain time limit and in the power of improving both poor and good MIP solutions.

Unlike RINS, DINS uses the change of relaxation solution between the root and the node and the change in the encountered MIP solutions in guiding the hard fixing of 0-1 variables; this has an effect in finding the good MIP solutions as the computation progresses. This has been experimentally visualized by having a comparatively worse performance on the benchmark instances by running a modified DINS where the hard fixing of 0-1 variables are carried out according to the hard fixing of RINS.

Acknowledgements. We thank Emilie Danna for the useful email discussions during the implementation and analysis of the methods.

References

1. E. Balas, S. Ceria, M. Dawande, F. Margot, and G. Pataki. Octane: a new heuristic for pure 0-1 programs. *Operations Research*, 49(2):207–225, 2001.
2. E. Balas and C.H. Martin. Pivot and complement – a heuristic for 0-1 programming. *Management Science*, 26(1):86–96, 1980.
3. E. Balas and C.H. Martin. Pivot and shift – a heuristic for mixed integer programming. Technical report, GSIA, Carnegie Mellon University, 1986.
4. E. Balas, S. Schmieta, and C. Wallace. Pivot and shift – a mixed integer programming heuristic. *Discrete Optimization*, 1:3–12, 2004.
5. E. Danna, E. Rothberg, and C.L. Pape. Exploring relaxation induced neighborhhods to improve mip solutions. *Mathematical Programming*, 102:71–90, 2005.
6. DEIS. Library of instances. www.or.deis.unibo.it/research_pages/ORinstances/.
7. B.H. Faaland and F.S. Hillier. Interior path methods for heuristic integer programming procedures. *Operations Research*, 27(6):1069–1087, 1979.
8. M. Fischetti, F. Glover, and A. Lodi. The feasibilty pump. to be appeared on Mathematical Programming.
9. M. Fischetti and A. Lodi. Local branching. *Mathematical Programming B*, 98: 23–49, 2003.
10. S. Ghosh. Description of all the used benchmark instances in this paper. www.cs.ualberta.ca/~shubhash/dins/benchmarks.ps.
11. S. Ghosh. Implementation of DINS. www.cs.ualberta.ca/~shubhash/codes.html.
12. F.S. Hillier. Efficient heuristic procedures for integer linear programming with an interior. *Operations Research*, 17(4):600–637, 1969.
13. T. Ibaraki, T. Ohashi, and H. Mine. A heuristic algorithm for mixed-integer programming problems. *Math. Program. Study.*, 2:115–136, 1974.
14. R. M. Karp. Reducibility among combinatorial problems. In R. E. Miller and J. W. Thatcher, editors, *Complexity of Computer Computations*, pages 85–103. Plenum Press, New York, 1972.
15. A. Løkketangen, K. Jörnsten, and S. Storøy. Tabu search within a pivot and complement framework. *International Transactions in Operational Research*, 1(3): 305–317, 1994.
16. A. Løkketangen and D.L. Woodruff. Integrating pivot based search with branch and bound for binary mip's. *Control and Cybernetics, Special issue on Tabu Search*, 29(3):741–760, 2001.
17. A. Martin, T. Achterberg, and T. Koch. Miplib 2003. http://miplib.zib.de.
18. M. Nediak and J. Eckstein. Pivot, cut, and dive: A heuristic for mixed 0-1 integer programming. *RUTCOR Research Report*, RRR 53-2001, 2001.

Mixed-Integer Vertex Covers on Bipartite Graphs

Michele Conforti[1], Bert Gerards[2,3], and Giacomo Zambelli[1]

[1] Dipartimento di Matematica Pura e Applicata, Universitá di Padova, Via Trieste 63, 35121 Padova, Italy
conforti@math.unipd.it, giacomo@math.unipd.it
[2] Centrum voor Wiskunde en Informatica, Kruislaan 413, 1098 SJ Amsterdam, The Netherlands
Bert.Gerards@cwi.nl
[3] Technische Universiteit Eindhoven, Den Dolech 2, Eindhoven, The Netherlands

Abstract. Let A be the edge-node incidence matrix of a bipartite graph $G = (U, V; E)$, I be a subset of the nodes of G, and b be a vector such that $2b$ is integral. We consider the following mixed-integer set:

$$X(G, b, I) = \{x \ : \ Ax \geq b, \ x \geq 0, \ x_i \text{ integer for all } i \in I\}.$$

We characterize $\operatorname{conv}(X(G, b, I))$ in its original space. That is, we describe a matrix (C, d) such that $\operatorname{conv}(X(G, b, I)) = \{x : Cx \geq d\}$. This is accomplished by computing the projection onto the space of the x-variables of an extended formulation, given in [1], for $\operatorname{conv}(X(G, b, I))$. We then give a polynomial-time algorithm for the separation problem for $\operatorname{conv}(X(G, b, I))$, thus showing that the problem of optimizing a linear function over the set $X(G, b, I)$ is solvable in polynomial time.

1 Introduction

Given a bipartite graph $G = (U, V; E)$, a vector $b = (b_e)_{e \in E}$, with the property that b is *half-integral*, i.e. $2b_e \in \mathbb{Z}$, $e \in E$, and a set $I \subseteq (U \cup V)$, we consider the problem of characterizing the convex hull of all nonnegative $x \in \mathbb{R}^{U \cup V}$ such that

$$x_i + x_j \geq b_{ij} \text{ for every } ij \in E,$$
$$x_i \in \mathbb{Z} \quad \text{for every } i \in I.$$

That is, given the edge-node incidence matrix A of a bipartite graph G, a partition (I, L) of its column-set, and an half-integral vector b, we consider the following mixed-integer set:

$$X(G, b, I) = \{x \ : \ Ax \geq b, \ x \geq 0, \ x_i \text{ integer for all } i \in I\}. \tag{1}$$

In this paper we provide a *formulation* for the polyhedron $\operatorname{conv}(X(G, b, I))$, where a formulation for a polyhedron P is a description of P as the intersection of a finite number of half-spaces. So it consists of a finite set of inequalities $Cx \geq d$ such that $P = \{x \ : \ Cx \geq d\}$.

M. Fischetti and D.P. Williamson (Eds.): IPCO 2007, LNCS 4513, pp. 324–336, 2007.

An *extended formulation* of P is a formulation for a polyhedron P' in a higher dimensional space that includes the original space, so that P is the projection of P' onto the original space.

A general technique to describe an extended formulation for the set of solutions of a system $Ax \geq b$, when $A^\top$ is a network matrix and some of the variables are restricted to be integer, was introduced in [1]. In Section 2 we derive such an extended formulation for $\mathrm{conv}(X(G, b, I))$, while in Section 3 we describe a formulation in the original space by explicitly computing the projection of the polyhedron defined by the extended formulation. Finally, in Section 4, we give a polynomial-time algorithm to solve the separation problem for $\mathrm{conv}(X(G, b, I))$.

1.1 The Main Result

Given a bipartite graph $G = (U, V; E)$, a partition (I, L) of $U \cup V$, and an half-integral vector b, we say that a path P of G is an *I-path* if at least one endnode of P is in I, and no intermediate node of P is in I. We say that P is *odd* if P has an odd number of edges e such that $b_e = \frac{1}{2}$ mod 1. Whenever we have a vector v with entries indexed by some set S, given a subset T of S we denote $v(T) = \sum_{i \in T} v_i$. In this paper we show the following:

Theorem 1. *The polyhedron* $\mathrm{conv}(X(G, b, I))$ *is defined by the following inequalities:*

$$x_i + x_j \geq \quad b_{ij} \qquad ij \in E, \tag{2}$$
$$2x(V(P) \cap L) + x(V(P) \cap I) \geq b(P) + \tfrac{1}{2} \quad P \text{ odd } I\text{-path}, \tag{3}$$
$$x_i \geq \quad 0 \qquad i \in U \cup V. \tag{4}$$

Eisenbrand [4] conjectured that the inequalities in (2)-(4) are sufficient to characterize $\mathrm{conv}(X(G, b, I))$ when G is a path. Theorem 1 shows that this conjecture holds in a quite more general setting (and it certainly cannot be extended beyond that). Preliminary results for the path case were obtained by Skutella [11] and Eisenbrand [4].

1.2 First Chvátal Closure

The following observation allows us to describe $X(G, b, I)$ in terms of a pure integer set.

Observation 2. *Let* $\bar{x}$ *be a vertex of* $\mathrm{conv}(X(G, b, I))$. *Then* $2\bar{x}$ *is integral.*

Proof: If not, let U' and V' be the sets of nodes i in U and V, respectively, such that $2\bar{x}_i$ is not integer. Then, for ϵ small enough, the vectors $\bar{x} + \epsilon\chi^{U'} - \epsilon\chi^{V'}$ and $\bar{x} - \epsilon\chi^{U'} + \epsilon\chi^{V'}$ are both in $\mathrm{conv}(X(G, b, I))$, where we denote by χ^S the incidence vector of S for any $S \subseteq U \cup V$. $\qquad \square$

Let $b' = 2b$, A' be obtained form A by multiplying by 2 the columns corresponding to nodes in I. By Observation 2, the linear transformation $x'_i = x_i, i \in I$,

$x_i' = 2x_i, i \in L$, maps $X(G, b, I)$ into $\{x' : A'x' \geq b', x' \geq 0, x' \text{ integral}\}$, which is a pure integer set.

Let $P = v_1, \ldots v_n$ be an I-path. Notice that $b(P) = \frac{1}{2} \bmod 1$ is equivalent to $b'(P)$ odd. Then the inequality

$$\sum_{i \in V(P)} x_i' \geq \left\lceil \frac{b'(P)}{2} \right\rceil \tag{5}$$

is a Gomory-Chvátal inequality of $\{x' : A'x' \geq b', x' \geq 0\}$. Indeed, assume $v_1 \in I$. If $v_n \in I$, then (5) is obtained from

$$\frac{1}{2}(2x_{v_1}' + x_{v_2}' \geq b_{v_1 v_2}') + \sum_{i=2}^{n-2} \frac{1}{2}(x_{v_i}' + x_{v_{i+1}}' \geq b_{v_i v_{i+1}}') + \frac{1}{2}(x_{v_{n-1}}' + 2x_{v_n}' \geq b_{v_{n-1} v_n}')$$

by rounding up the right-hand-side. If $x_n \notin I$, then (5) is obtained from

$$\frac{1}{2}(2x_{v_1}' + x_{v_2}' \geq b_{v_1 v_2}') + \sum_{i=2}^{n-1} \frac{1}{2}(x_{v_i}' + x_{v_{i+1}}' \geq b_{v_i v_{i+1}}') + \frac{1}{2}(x_{v_n}' \geq 0)$$

by rounding up the right-hand-side.

Furthermore the inequalities in (5) correspond to the inequalities in (3). Therefore Theorem 1 implies that the polyhedron defined by $A'x' \geq b'$, $x' \geq 0$ has Chvátal rank 1. In the case where G is a path with no intermediate node in I, this last fact follows immediately from a theorem of Edmonds and Jonhnson [2,3], since in this case A' satisfies the condition that the sum of the absolute values of the entries of each column is at most 2.

1.3   The Motivation

A (general) mixed-integer set is a set of the form

$$\{x \mid Ax \geq b, \, x_i \text{ integer } i \in I\} \tag{6}$$

where I is a subset of the columns of A and b is a vector that may contain fractional components.

In [1], it is shown that the problem of deciding if the above set is nonempty is NP-complete, even if b is an half-integral vector and A is a network matrix. (We refer the reader to [7] or [10] for definitions and results related to network matrices and, more generally, totally unimodular matrices.)

However, it may be possible that, when A is the *transpose* of a network matrix, the associated mixed-integer programming problem is polynomially solvable. Indeed, let MIX^{2TU} be a mixed-integer set of the form (6) when $A^\top$ is a network matrix.

An extended formulation of the polyhedron $\text{conv}(MIX^{2TU})$ was described in [1]. The extended formulation involves an additional variable for each possible fractional part taken by the variables at any vertex of $\text{conv}(MIX^{2TU})$. If this number is polynomial in the size of (A, b), then such a formulation is compact, i.e. of polynomial size in the size of (A, b). Therefore the problem of optimizing a linear function over MIX^{2TU} can be efficiently solved in this case. However, it seems to be rather difficult to compute the projection in the original x-space. It follows from Observation 2 that if $\bar{x}$ is a vertex of $\text{conv}(X(G, b, I))$, then $\bar{x}_i - \lfloor \bar{x}_i \rfloor \in \{0, \frac{1}{2}\}$. Therefore the extended formulation for $\text{conv}(X(G, b, I))$ (which will be introduced in Section 2) is compact. The main contribution of this paper is the explicit description of the projection of the polyhedron defined by this extended formulation in the original x-space.

The mixed-integer set $X(G, b, I)$ is related to certain mixed-integer sets that arise in the context of production planning (see [9]). The case when G is a star with center node in L and leaves in I has been studied by Pochet and Wolsey in [8], where they gave a compact extended formulation for the convex hull of feasible solutions. Günlük and Pochet [5] projected this formulation onto the original space, thus showing that the family of *mixing inequalities* gives the formulation in the x-space.

Miller and Wolsey [6] extended the results in [8] to general bipartite graphs, with the restriction that the partition (I, L) coincides with the bipartition (U, V) of the graph. Their result shows that the mixing inequalities associated with every single star of G having center a node in L and leaf nodes all nodes in I give a formulation for this case.

2   The Extended Formulation

We use here a modeling technique introduced by Pochet and Wolsey [8] and extensively investigated in [1].

Observation 2 allows to express each variable x_i , $i \in L$, as

$$x_i = \mu_i + \frac{1}{2}\delta_i, \ \mu_i \geq 0, \ 0 \leq \delta_i \leq 1, \ \mu_i, \delta_i \text{ integer.} \tag{7}$$

For now, we assume $I = \emptyset$, that is, $L = (U \cup V)$.

Lemma 3. *Let* $ij \in E$, *and suppose* $x_i, x_j, \mu_i, \mu_j, \delta_i, \delta_j$ *satisfy (7).*
If $b_{ij} = \frac{1}{2} \bmod 1$, x_i, x_j *satisfy* $x_i + x_j \geq b_{ij}$ *if and only if*

$$\begin{aligned} \mu_i + \mu_j &\geq \lfloor b_{ij} \rfloor \\ \mu_i + \delta_i + \mu_j + \delta_j &\geq \lceil b_{ij} \rceil \, . \end{aligned} \tag{8}$$

If $b_{ij} = 0 \bmod 1$, x_i, x_j *satisfy* $x_i + x_j \geq b_{ij}$ *if and only if*

$$\begin{aligned} \mu_i + \delta_i + \mu_j &\geq b_{ij} \\ \mu_i + \mu_j + \delta_j &\geq b_{ij} \, . \end{aligned} \tag{9}$$

Proof: Assume $x_i, x_j, \mu_i, \mu_j, \delta_i, \delta_j$ satisfy (7). Then, if $b_{ij} = \frac{1}{2} \bmod 1$, constraint $x_i + x_j \geq b_{ij}$ is satisfied if and only if $\mu_i + \mu_j \geq \lfloor b_{ij} \rfloor$ and $\delta_i + \delta_j \geq 1$ whenever $\mu_i + \mu_j = \lfloor b_{ij} \rfloor$. If $b_{ij} = 0 \bmod 1$, the constraint is satisfied if and only if $\mu_i + \mu_j \geq b_{ij} - 1$ and $\delta_i = \delta_j = 1$ whenever $\mu_i + \mu_j = b_{ij} - 1$.
It is easy to see that these conditions are enforced by the above constraints. $\square$

Observation 4. *Given* $ij \in E$, *the constraints* (8) *and* (9) *belong to the first Chvátal closure of the polyhedron defined by*

$$\mu_i + \frac{1}{2}\delta_i + \mu_j + \frac{1}{2}\delta_j \geq b_{ij}$$
$$\mu_i, \mu_j \geq 0$$
$$\delta_i, \delta_j \leq 1$$
$$\delta_i, \delta_j \geq 0$$

whenever $b_{ij} = \frac{1}{2} \bmod 1$ *and* $b_{ij} = 0 \bmod 1$, *respectively.*

By applying the unimodular transformation $\mu_i^0 = \mu_i$, $\mu_i^1 = \mu_i + \delta_i$, the constraints $x_i = \mu_i + \frac{1}{2}\delta_i$, $\mu_i \geq 0$, $0 \leq \delta_i \leq 1$ become

$$x_i - \frac{1}{2}(\mu_i^0 + \mu_i^1) = 0 \tag{10}$$

$$\mu_i^0 \geq 0$$
$$0 \leq \mu_i^1 - \mu_i^0 \leq 1 \tag{11}$$

and constraints (8) and (9) become:

$$\mu_i^0 + \mu_j^0 \geq \lfloor b_{ij} \rfloor$$
$$\mu_i^1 + \mu_j^1 \geq \lceil b_{ij} \rceil \tag{12}$$

$$\mu_i^1 + \mu_j^0 \geq b_{ij}$$
$$\mu_i^0 + \mu_j^1 \geq b_{ij} \tag{13}$$

Theorem 5. *The projection onto the space of the* x *variables of the polyhedron* Q *defined on the space of the variables* (x, μ^0, μ^1) *by the inequalities*

$$\begin{array}{ll} (10), (11) & \text{for every } i \in U \cup V, \\ (12) & \text{for every } ij \in E \text{ s.t. } b_{ij} = \frac{1}{2} \bmod 1 \\ (13) & \text{for every } ij \in E \text{ s.t. } b_{ij} = 0 \bmod 1 \end{array}$$

is the polyhedron $\operatorname{conv}(X(G, b, \emptyset))$.

Proof: Since the variable x_i is determined by (10) for all $i \in U \cup V$, we only need to show that the polyhedron defined by inequalities (11) for every $i \in U \cup V$, (12) for every $ij \in E$ s.t. $b_{ij} = \frac{1}{2} \bmod 1$, and (13) for every $ij \in E$ s.t. $b_{ij} = 0 \bmod 1$, is integral. Let A_μ be the constraint matrix of the above system. Since G is a bipartite graph, then the matrix $\bar{A}$, obtained by multiplying by -1 the columns of A_μ relative to the variables μ_i^0, μ_i^1, $i \in V$, has at most a 1 and at most a -1 in each row. Therefore $\bar{A}$ is the transpose of a network matrix, so A_μ is totally unimodular (see [10]). Since the right-hand-sides of (11)-(13) are all integer, the statement follows from the theorem of Hoffman and Kruskal. $\qquad\square$

Observation 6. *For any* $i \in U \cup V$, x_i *is integer valued if and only if* $\delta_i = 0$. *Therefore, for a given* $I \subseteq (U \cup V)$, *the polyhedron* $\operatorname{conv}(X(G, b, I))$ *is the projection onto the space of the* x *variables of the face* Q_I *of* Q *defined by the equations* $\mu_i^1 - \mu_i^0 = 0$, $i \in I$ *(which correspond to* $\delta_i = 0$, $i \in I$).

3   The Formulation in the Original Space

In this section we prove Theorem 1 by projecting the polyhedron Q_I onto the space of the x variables.

Let $p_i = \frac{\mu_i^0 - \mu_i^1}{2}$. The $\mu_i^0 = x_i + p_i$ and $\mu_i^1 = x_i - p_i$. The inequalities (10)-(13), defining Q, become:

$$
\begin{aligned}
p_i + p_j &\geq \lfloor b_{ij} \rfloor - x_i - x_j, & ij \in E \text{ s.t. } b_{ij} = \tfrac{1}{2} \bmod 1, \\
-p_i - p_j &\geq \lceil b_{ij} \rceil - x_i - x_j, & ij \in E \text{ s.t. } b_{ij} = \tfrac{1}{2} \bmod 1, \\
p_i - p_j &\geq b_{ij} - x_i - x_j, & ij \in E \text{ s.t. } b_{ij} = 0 \bmod 1, \\
-p_i + p_j &\geq b_{ij} - x_i - x_j, & ij \in E \text{ s.t. } b_{ij} = 0 \bmod 1, \\
p_i &\geq -\tfrac{1}{2}, & i \in U \cup V, \\
-p_i &\geq 0, & i \in U \cup V, \\
p_i &\geq -x_i, & i \in U \cup V.
\end{aligned}
$$

By Observation 6, $\operatorname{conv}(X(G, B, I))$ is the projection onto the x-space of the polyhedron defined by the above inequalities and by $p_i = 0$ for every $i \in I$.

Associate multipliers to the above constraints as follows:

$$
\begin{aligned}
(u_{ij}^{++}) & \quad p_i + p_j \geq \lfloor b_{ij} \rfloor - x_i - x_j \\
(u_{ij}^{--}) & \quad -p_i - p_j \geq \lceil b_{ij} \rceil - x_i - x_j \\
(u_{ij}^{+-}) & \quad p_i - p_j \geq b_{ij} - x_i - x_j \\
(u_{ij}^{-+}) & \quad -p_i + p_j \geq b_{ij} - x_i - x_j \\
(u_i^{\frac{1}{2}}) & \quad p_i \geq -\tfrac{1}{2} \\
(u_i^0) & \quad -p_i \geq 0 \\
(u_i^x) & \quad p_i \geq -x_i
\end{aligned}
\tag{14}
$$

Any valid inequality for $\operatorname{conv}(X(G, b, I))$ has the form $\alpha_u x \geq \beta_u$, where

$$
\begin{aligned}
\alpha_u x = \sum_{b_{ij} = \frac{1}{2} \bmod 1} (u_{ij}^{++} + u_{ij}^{--})(x_i + x_j) + \\
\sum_{b_{ij} = 0 \bmod 1} (u_{ij}^{+-} + u_{ij}^{-+})(x_i + x_j) + \sum_{i \in U \cup V} u_i^x x_i
\end{aligned}
\tag{15}
$$

$$
\begin{aligned}
\beta_u = \sum_{b_{ij} = \frac{1}{2} \bmod 1} (u_{ij}^{--} \lceil b_{ij} \rceil + u_{ij}^{++} \lfloor b_{ij} \rfloor) + \\
\sum_{b_{ij} = 0 \bmod 1} (u_{ij}^{+-} + u_{ij}^{-+}) b_{ij} - \sum_{i \in L} \frac{1}{2} u_i^{\frac{1}{2}}
\end{aligned}
\tag{16}
$$

for some nonnegative vector $u = (u_{ij}^{++}, u_{ij}^{--}, u_{ij}^{+-}, u_{ij}^{-+}, u_i^{\frac{1}{2}}, u_i^0, u_i^x)$ such that $uP = 0$, where P is the column-submatrix of the above system (14) involving columns corresponding to variables p_i, $i \in L$ (see e.g. Theorem 4.10 in [7]). For instance the inequality $x_i + x_j \geq b_{ij}$, for $ij \in E$ with $b_{ij} = \frac{1}{2} \bmod 1$, is obtained by setting $u_{ij}^{++} = u_{ij}^{--} = \frac{1}{2}$, and all other entries of u to be 0.

We are interested in characterizing the nonnegative vectors u such that $uP = 0$ and $\alpha_u x \geq \beta_u$ is facet-defining for $\text{conv}(X(G, b, I))$, and such that the inequality $\alpha_u x \geq \beta_u$ is not of the form $x_i + x_j \geq b_{ij}$, for some $ij \in E$, or $x_i \geq 0$, for some $i \in U \cup V$. From now on we will assume, w.l.o.g., that the entires of u are integer and relatively prime.

We define an auxiliary graph $\Gamma_u = (L \cup \{d\}, F)$, where d is a dummy node not in $U \cup V$, and F is defined as follows.

- For every edge $ij \in E$ such that $i, j \in L$, there are $u_{ij}^{++} + u_{ij}^{--} + u_{ij}^{+-} + u_{ij}^{-+}$ parallel edges between i and j in F, each edge corresponding to a multiplier among $u_{ij}^{++}, u_{ij}^{--}, u_{ij}^{+-}, u_{ij}^{-+}$.

- For each node $i \in L$, there are $u_i^{\frac{1}{2}} + u_i^0 + u_i^x + \sum_{j \in I : ij \in E}(u_{ij}^{++} + u_{ij}^{--} + u_{ij}^{+-} + u_{ij}^{-+})$ parallel edges between d and i in F, each edge corresponding to a multiplier among $u_i^{\frac{1}{2}}, u_i^0, u_i^x$, or $u_{ij}^{++}, u_{ij}^{--}, u_{ij}^{+-}, u_{ij}^{-+}$, for some $j \in I$.

We impose a *bi-orientation* ω on Γ_u, that is, to each edge $e \in F$, and each endnode i of e that belongs to L, we associate the value $\omega(e, i) = tail$ if e corresponds to an inequality of (14) where p_i has coefficient -1, while we associate the value $\omega(e, i) = head$ if e corresponds to an inequality of (14) where p_i has coefficient $+1$. The dummy node d is neither a tail nor a head of any edge. Thus, each edge of Γ_u can have one head and one tail, two heads, two tails, or, if d is one of the two endnodes, only one head and no tail or only one tail and no head.

For each $i \in L$, we denote with $\delta_\omega^{in}(i)$ the number of edges in F of which i is a head, and with $\delta_w^{out}(i)$ the number of edges in F of which i is a tail.

We say that Γ_u is ω-eulerian if $\delta_\omega^{in}(i) = \delta_\omega^{out}(i)$ for every $i \in L$.

Observation 7. Γ_u is ω-eulerian if and only if $uP = 0$.

We define a *closed ω-eulerian walk* in Γ_u as a closed-walk in Γ_u,

$$v_0, e_0, v_1, e_1, \ldots, v_k, e_k, v_{k+1},$$

where $v_0 = v_{k+1}$, with the property that $\omega(e_{h-1}, v_h) \neq \omega(e_h, v_h)$ for every h such that v_h is in L, $h = 0, \ldots, k, k+1$, where the indices are taken modulo k. That is, if $v_h \in L$, then v_h is a head of e_{h-1} if and only if v_h is a tail of e_h.

Observation 8. Γ_u is ω-eulerian if and only if Γ_u is the disjoint union of closed ω-eulerian walks. In particular, every node in $L \cup \{d\}$ has even degree in Γ_u.

Observe that, if $v_0, e_0, \ldots, e_k, v_{k+1}$ is a closed ω-eulerian walk in Γ_u, then both graphs Γ', Γ'' on $L \cup \{d\}$ with edge-sets $F' = \{e_1, \ldots, e_k\}$ and $F'' = F \setminus F'$, respectively, are ω-eulerian. Suppose $F'' \neq \emptyset$. Then there are nonnegative integer vectors u' and u'', both different from zero, such that $u'P = 0$, $u''P = 0$, $\Gamma' = \Gamma_{u'}$ and $\Gamma'' = \Gamma_{u''}$, and $u = u' + u''$. By the fact that Γ' and Γ'' are ω-eulerian, and by the structure of the inequalities in (14), the vectors $(\alpha_{u'}, \beta_{u'})$ and $(\alpha_{u''}, \beta_{u''})$ are both non-zero. Furthermore $\alpha_u = \alpha_{u'} + \alpha_{u''}$ and $\beta_u = \beta_{u'} + \beta_{u''}$, contradicting the fact that $\alpha_u x \geq \beta_u$ is facet-defining and the entries of u are relatively prime.

Hence we have shown the following.

Observation 9. *Every closed ω-eulerian walk of Γ_u traverses all the edges in F. In particular, there exists a closed ω-eulerian walk $v_0, e_0, \ldots, e_k, v_{k+1}$ of Γ_u such that $F = \{e_h \mid h = 1, \ldots, k\}$.*

Suppose d has positive degree in Γ. Then we may assume, w.l.o.g., that $v_0 = v_{k+1} = d$. Suppose $d = v_h$ for some $h = 1, \ldots, k$. Then $v_0, e_0, v_1, \ldots, e_{h-1}v_h$ is a closed ω-eulerian walk, contradicting the previous observation. Hence we have the following.

Observation 10. *Node d has degree 0 or 2 in Γ_u.*

Next we show the following.

Lemma 11. *Every node in $L \cup \{d\}$ has degree 0 or 2 in Γ_u.*

Proof: We have already shown d has degree 0 or 2 in Γ_u. If d has degree 2, we assume $d = v_0 = v_{k+1}$, else v_0 is arbitrarily chosen. If there is a node in L with degree at least 4, then there exists distinct indices $s, t \in \{1, \ldots, k\}$ such that $v_s = v_t$. We choose s and t such that $t - s$ is positive and as small as possible. Therefore $C = v_s, e_s, \ldots, e_{t-1}, v_t$ is a cycle of Γ_u containing only nodes in L. Since G is a bipartite graph, C has even length, hence the edges in C can be partitioned into two matchings M_0, M_1 of cardinality $|C|/2$. We will denote with HH, TT, HT the sets of edges of F with, respectively, two heads, two tails, one head and one tail.

If v_s is the head of exactly one among e_s and e_{t-1}, then C is a closed ω-eulerian walk, contradicting Observation 9. Hence v_s is either a head of both e_s and e_{t-1} or a tail of both e_s and e_{t-1}. This shows that $|C \cap TT| = |C \cap HH| \pm 1$. Therefore there is an odd number of edges e in C such that $b_e = \frac{1}{2} \bmod 1$. By symmetry, we may assume $\sum_{e \in M_0} b_e \geq \sum_{e \in M_1} b_e + \frac{1}{2}$. Then the inequality

$$2 \sum_{i \in V(C)} x_i \geq \sum_{e \in C} b_e + \frac{1}{2} \tag{17}$$

is valid for $\mathrm{conv}(X(G, b, I))$, since it is implied by the valid inequalities $x_i + x_j \geq b_{ij}$, $ij \in M_0$, because

$$2 \sum_{i \in V(C)} x_i = 2 \sum_{ij \in M_0} (x_i + x_j) \geq 2 \sum_{ij \in M_0} b_{ij} \geq \sum_{e \in M_0} b_e + \sum_{e \in M_1} b_e + \frac{1}{2} = \sum_{e \in C} b_e + \frac{1}{2}.$$

Case 1: Node v_s is a tail of both e_s and e_{t-1}.

Then $|C \cap TT| = |C \cap HH| + 1$, hence

$$\sum_{e \in C \cap TT} \lfloor b_e \rfloor + \sum_{e \in C \cap HH} \lceil b_e \rceil + \sum_{e \in C \cap HT} b_e = \sum_{e \in C} b_e + \frac{1}{2}. \tag{18}$$

Let u' be the vector obtained from u as follows

$$\begin{cases} u'^{**}_{ij} = u^{**}_{ij} - 1 & \text{for every } ij \in C \\ u'^{0}_{v_s} = u^{0}_{v_s} + 2 \end{cases}$$

all other components of u' and u being identical, where u_{ij}^{**} is the variable among $u_{ij}^{++}, u_{ij}^{--}, u_{ij}^{+-}, u_{ij}^{-+}$ corresponding to edge ij of C.

Then one can easily see that $\Gamma_{u'}$ is the graph obtained from Γ_u by removing the edges $e_s, \ldots, e_t$, and adding two parallel edges $v_s d$ both with tail in v_s, hence $\Gamma_{u'}$ is ω-eulerian and $u'P = 0$. By (18)

$$\beta_{u'} = \beta_u - \sum_{e \in C} b_e - \frac{1}{2},$$

while by construction

$$\alpha_u x = \alpha_{u'} x + 2 \sum_{i \in V(C)} x_i.$$

Thus $\alpha_u x \geq \beta_u$ can be obtained by taking the sum of $\alpha_{u'} x \geq \beta_{u'}$ and (17), contradicting the assumption that $\alpha_u x \geq \beta_u$ is facet-defining.

Case 2: Node v_s is a head of both e_s and e_{t-1}.

Then $|C \cap TT| = |C \cap HH| - 1$, hence

$$\sum_{e \in C \cap TT} \lfloor b_e \rfloor + \sum_{e \in C \cap HH} \lceil b_e \rceil + \sum_{e \in C \cap HT} b_e = \sum_{e \in C} b_e - \frac{1}{2}. \qquad (19)$$

Let u' be the vector obtained from u as follows

$$\begin{cases} u_{ij}'^{**} = u_{ij}^{**} - 1 \text{ for every } ij \in C \\ u_{v_s}'^{\frac{1}{2}} = u_{v_s}^{\frac{1}{2}} + 2 \end{cases}$$

all other components of u' and u being identical.

Then one can easily see that $\Gamma_{u'}$ is the graph obtained from Γ_u by removing the edges $e_s, \ldots, e_t$, and adding two parallel edges $v_s d$ both with head in v_s, hence $u'P = 0$. By (19)

$$\beta_{u'} = \beta_u - \sum_{e \in C} b_e + \frac{1}{2} - 2\frac{1}{2},$$

while by construction

$$\alpha_u x = \alpha_{u'} x + 2 \sum_{i \in V(C)} x_i.$$

Thus $\alpha_u x \geq \beta_u$ can be obtained by taking the sum of $\alpha_{u'} x \geq \beta_{u'}$ and (17), contradicting the assumption that $\alpha_u x \geq \beta_u$ is facet-defining. $\qquad \square$

We are now ready to give the proof of the main theorem.

Proof of Theorem 1. We show that all facet-defining inequalities $\alpha_u x \geq \beta_u$, where u is nonnegative, integral, and with entries that are relatively prime, that are not inequalities in (2) or (4), are of the form (3).

First we show the following.

$$\sum_{ij\in E} u_{ij}^{--} > \sum_{ij\in E} u_{ij}^{++} + \sum_{i\in U\cup V} u_i^{\frac{1}{2}} \tag{20}$$

In fact, we can write the inequality

$$\alpha_u x \geq \sum_{b_{ij}=\frac{1}{2}\ \mathrm{mod}\ 1} (u_{ij}^{--} + u_{ij}^{++})b_{ij} + \sum_{b_{ij}=0\ \mathrm{mod}\ 1} (u_{ij}^{+-} + u_{ij}^{-+})b_{ij}$$

as nonnegative combination of inequalities of the form (2) or (4), therefore we must have

$$\beta_u > \sum_{b_{ij}=\frac{1}{2}\ \mathrm{mod}\ 1} (u_{ij}^{--} + u_{ij}^{++})b_{ij} + \sum_{b_{ij}=0\ \mathrm{mod}\ 1} (u_{ij}^{+-} + u_{ij}^{-+})b_{ij}.$$

Thus

$$0 < \beta_u - \sum_{b_{ij}=\frac{1}{2}\ \mathrm{mod}\ 1} (u_{ij}^{--} + u_{ij}^{++})b_{ij} - \sum_{b_{ij}=0\ \mathrm{mod}\ 1} (u_{ij}^{+-} + u_{ij}^{-+})b_{ij}$$

$$= \frac{1}{2}(\sum_{ij\in E} u_{ij}^{--} - \sum_{ij\in E} u_{ij}^{++} - \sum_{i\in U\cup V} u_i^{\frac{1}{2}})$$

which proves (20).

By Lemma (11) and Observation (9), Γ_u consists of an induced cycle C and isolated nodes, where every node in $V(C)\cap L$ is a head of exactly one edge and a tail of exactly one edge.

If d is an isolated node, then each edge ij of C corresponds to a variable of the form u_{ij}^{**}, and since the total number of heads in C equals the number of tails, then $\sum_{ij\in E} u_{ij}^{--} = \sum_{ij\in E} u_{ij}^{++}$ and $\sum_{i\in U\cup V} u_i^{\frac{1}{2}} = 0$, contradicting (20). Thus we may assume that $C = v_0, e_0, \ldots, e_k, v_{k+1}$ where $d = v_0 = v_{k+1}$.

Claim: The following are the only possible cases, up to symmetry.
1. Edges dv_1, dv_k of Γ_u correspond to variables $u_{v_1}^x$ and $u_{v_k}^x$, respectively;
2. dv_1 corresponds to variable $u_{wv_1}^{--}$ or $u_{wv_1}^{-+}$ for some $w \in I$, and dv_k corresponds to $u_{v_k}^x$;
3. dv_1 corresponds to variables $u_{wv_1}^{--}$ or $u_{wv_1}^{-+}$ for some $w \in I$, and dv_k corresponds to variable $u_{w'v_k}^{--}$ or $u_{w'v_k}^{-+}$ for some $w' \in I$.

Proof of claim. If v_1 is a head of e_0 and v_k is a head of e_k, then the number of edges among $e_1, \ldots, e_{k-1}$ with two tails is one plus the number of edges with two heads. Since the former correspond to variables of type u_{ij}^{--} for some $ij \in E$, and the latter correspond to to variables of type u_{ij}^{++} for some $ij \in E$, then by (20) dv_1 does not correspond to variable $u_{v_1}^{\frac{1}{2}}$ or to a variable $u_{wv_1}^{++}$ for any $w \in I$, and dv_k does not correspond to variable $u_{v_k}^{\frac{1}{2}}$ or to a variable $u_{wv_k}^{++}$ for any $w \in I$, thus one of the above three cases holds.

If v_1 is a tail of e_0 and v_k is a head of e_k, then the number of edges among $e_1, \ldots, e_{k-1}$ with two tails is equal the number of edges with two heads. By (20), dv_1 corresponds to variable $u^{--}_{wv_1}$ for some $w \in I$, and dv_k corresponds to either $u^x_{v_k}$ or to a variable $u^{-+}_{w'v_k}$ for some $w' \in I$, thus case 2 or 3 holds.

If v_1 is a tail of e_0 and v_k is a tail of e_k, then the number of edges among $e_1, \ldots, e_{k-1}$ with two tails is equal one minus the number of edges with two heads. By (20), dv_1 corresponds to variable $u^{--}_{wv_1}$ for some $w \in I$, and dv_k corresponds to a variable $u^{--}_{w'v_k}$ for some $w' \in I$, thus case 3 holds. This completes the proof of the claim.

Case 1: *Edges dv_1, dv_k of Γ_u correspond to variables $u^x_{v_1}$ and $u^x_{v_k}$, respectively.*

In this case the path $P = v_1, e_1, \ldots, e_{k-1}, v_k$ of Γ_u is also a path of G containing only nodes in L, and P contains an odd number of edges e such that $b_e = \frac{1}{2} \bmod 1$. The inequality $\alpha_u x \geq \beta_u$ is then $2x(V(P)) \geq b(P) + \frac{1}{2}$. The edges of P can be partitioned into two matchings M_0 and M_1, thus we may assume, w.l.o.g., $\sum_{e \in M_0} b_e \geq \sum_{e \in M_1} b_e + \frac{1}{2}$. Thus $2x(V(P)) \geq 2 \sum_{ij \in M_0} (x_i + x_j) \geq 2 \sum_{ij \in M_0} b_{ij} \geq \sum_{e \in M_0} b_e + \sum_{e \in M_1} b_e + \frac{1}{2} = b(P) + \frac{1}{2}$, hence $\alpha_u x \geq \beta_u$ is not facet-defining.

Case 2: *dv_1 corresponds to variable $u^{--}_{wv_1}$ or $u^{-+}_{wv_1}$ for some $w \in I$, and dv_k corresponds to $u^x_{v_k}$.*

In this case, $P = w, v_1, e_1, \ldots, e_{k-1}, v_k$ is an odd I-path of G between $w \in I$ and $v_k \in L$. The inequality $\alpha_u x \geq \beta_u$ is $2x(V(P) \cap L) + x_w \geq b(P) + \frac{1}{2}$, which is one of the inequalities in (3).

Case 3: *dv_1 corresponds to variables $u^{--}_{wv_1}$ or $u^{-+}_{wv_1}$ for some $w \in I$, and dv_k corresponds to variable $u^{--}_{w'v_k}$ or $u^{-+}_{w'v_k}$ for some $w' \in I$.*

If $w \neq w'$, then the path $P = w, v_1, e_1, \ldots, e_{k-1}, v_k, w'$ is an odd I-path of G between $w \in I$ and $w' \in I$. The inequality $\alpha_u x \geq \beta_u$ is $2x(V(P) \cap L) + x_w + x_{w'} \geq b(P) + \frac{1}{2}$, which is one of the inequalities in (3).

If $w = w'$, then we must have $v_1 \neq v_k$, since otherwise v_1 would be either the head or the tail of both edges of Γ_u incident to v_1. Thus $C' = w, v_1, \ldots, v_k, w$ is a cycle of G. Since G is a bipartite graph, C' has even length, hence the edges in C' can be partitioned into two matchings M_0, M_1 of cardinality $|C'|/2$. Since C' contains an odd number of edges e such that $b_w = \frac{1}{2} \bmod 1$, then we may assume, w.l.o.g., $\sum_{e \in M_0} b_e \geq \sum_{e \in M_1} b_e + \frac{1}{2}$. The inequality $\alpha_u x \geq \beta_u$ is $2x(V(C')) \geq b(C') + \frac{1}{2}$. But $2x(V(C')) = 2 \sum_{ij \in M_0} (x_i + x_j) \geq 2 \sum_{ij \in M_0} b_{ij} \geq \sum_{e \in M_0} b_e + \sum_{e \in M_1} b_e + \frac{1}{2} = b(C') + \frac{1}{2}$, hence $\alpha_u x \geq \beta_u$ is not facet-defining. $\qquad\square$

4   Separation

Theorem 5 and Observation 6 imply that the problem of minimizing a linear function over the set $X(G, b, I)$ is solvable in polynomial time, since it reduces to solving a linear programming problem over the set of feasible points for (10)-(13).

In this section we give a combinatorial polynomial-time algorithm for the separation problem for the set $\operatorname{conv}(X(G, b, I))$, thus giving an alternative proof that the problem of optimizing a linear function over such polyhedron, and thus over $X(G, b, I)$, is polynomial.

Clearly, given a nonnegative vector x^*, we can check in polynomial-time whether x^* satisfies (2) for every edge. Thus, by Theorem 1, we only need to describe a polynomial-time algorithm that, given a nonnegative vector x^* satisfying (2), either returns an inequality of type (3) violated by x^*, or proves that none exists.

For every $ij \in E$, let $s_{ij}^* = x_i^* + x_j^* - b_{ij}$. Since x^* satisfies (2), then s_e^* is nonnegative for every $e \in E$. Let $P = v_1, \ldots v_n$ be an odd I-path.

Claim. The vector x^* satisfies $2x^*(V(P) \cap L) + x^*(V(P) \cap I) \geq b(P) + \frac{1}{2}$ if and only if $s^*(P) + x^*(\{v_1, v_n\} \cap L) \geq \frac{1}{2}$.

Indeed, assume $v_1 \in I$. If $v_n \in I$ then

$$\sum_{i=1}^{n-1} s_{v_i v_{i+1}}^* = \sum_{i=1}^{n-1} (x_{v_i}^* + x_{v_{i+1}}^* - b_{v_i v_{i+1}})$$

gives the equality $s^*(P) = 2x^*(V(P) \cap L) + x^*(V(P) \cap I) - b(P)$, hence $2x^*(V(P) \cap L) + x^*(V(P) \cap I) \geq b(P) + \frac{1}{2}$ if and only if $s^*(P) \geq \frac{1}{2}$.

If $v_n \notin I$, then

$$\sum_{i=1}^{n-1} s_{v_i v_{i+1}}^* + x_{v_n}^* = \sum_{i=1}^{n-1} (x_{v_i}^* + x_{v_{i+1}}^* - b_{v_i v_{i+1}}) + x_{v_n}^*$$

gives the equality $s^*(P) + x_{v_n}^* = 2x^*(V(P) \cap L) + x^*(V(P) \cap I) - b(P)$, hence $2x^*(V(P) \cap L) + x^*(V(P) \cap I) \geq b(P) + \frac{1}{2}$ if and only if $s^*(P) + x_{v_n}^* \geq \frac{1}{2}$. This completes the proof of the Claim.

Therefore, if we assign length s_e^* to every $e \in E$, we need to give an algorithm that, for any two nodes r, t such that $r \in I$, either determines that the shortest odd I-path between r and t (if any) has length at least $\frac{1}{2} - x^*(\{t\} \cap L)$, or returns an odd I-path P for which $2x^*(V(P) \cap L) + x^*(V(P) \cap I) < b(P) + \frac{1}{2}$.

Observe that any walk W between r and t that contains an odd number of edges e such that $b_e = \frac{1}{2} \bmod 1$ either contains a sub-path P that is an odd I-path or it contains a cycle C that contains an odd number of edges e such that $b_e = \frac{1}{2} \bmod 1$. In the former case, either both endnodes of P are in I, or t is the only endnode of P in L. Hence, if $s^*(W) < \frac{1}{2} - x^*(\{t\} \cap L)$, then also $s^*(P) < \frac{1}{2} - x^*(\{t\} \cap L)$, hence $2x^*(V(P) \cap L) + x^*(V(P) \cap I) < b(P) + \frac{1}{2}$. In the second case, since G is bipartite, the edges of C can be partitioned into two matchings M_0 and M_1 such that $b(M_0) \geq b(M_1) + \frac{1}{2}$. Thus $s^*(C) = \sum_{ij \in C} (x_i^* + x_j^* - b_{ij}) = 2x^*(V(C)) - b(C) \geq 2(x^*(V(C)) - b(M_0)) + \frac{1}{2} = 2\sum_{ij \in M_0} (x_i^* + x_j^* - b_{ij}) + \frac{1}{2} \geq \frac{1}{2}$, hence $s^*(W) \geq \frac{1}{2}$.

Thus we only need to find, for every pair $r, t \in U \cup V$ with $r \in I$, the shortest walk W between r and t, w.r.t. the distance s^*, among all such walks containing an odd number of edges e such that $b_e = \frac{1}{2} \bmod 1$. If, for a given choice of r, t,

$s(W) < \frac{1}{2} - x^*(\{t\} \cap L)$, then by the above argument we can find in polynomial time a sub-path P of W such that P is an odd I-path and $2x^*(V(P) \cap L) + x^*(V(P) \cap I) < b(P) + \frac{1}{2}$, otherwise we can conclude that $x^* \in \text{conv}(X(G, b, I))$.

To conclude, we only need to show a polynomial-time algorithm that, given an undirected graph Γ with nonnegative lengths on the edges $\ell_e, e \in E(\Gamma)$, a subset $F \subseteq E(\Gamma)$, and a pair of nodes $r, t \in V(\Gamma)$, determines the walk W of minimum length between r and t such that $|E(W) \cap F|$ is odd, or determines that no such walk exists. The latter problem can be solved in polynomial time. Since, as far as we know, this fact is folklore, we briefly describe an algorithm.

We construct a new graph Γ' as follows. For every node $v \in V(\Gamma)$, there is a pair of nodes v, v' in $V(\Gamma')$. For every edge $uv \in E(\Gamma)$, $E(\Gamma')$ contains the edges uv' and $u'v$ if $uv \in F$, and the edges uv and $u'v'$ if $uv \notin F$, each with length ℓ_{uv}. One can verify that a walk W between r and t with an odd number of edges in F exists in Γ if and only if there exists a walk of the same length between r and t' in Γ'. Hence we only need to find a shortest path between r and t' in Γ', if any exists, and output the corresponding walk in Γ.

References

1. M. Conforti, M. Di Summa, F. Eisenbrand, L.A. Wolsey, Network formulations of mixed-integer programs, In preparation, 2006.
2. J. Edmonds and E.L. Johnson, Matching: a well-solved class of integer linear programs, *Combinatorial Structures and Their Applications* (R.K. Guy, et al., eds.), Gordon and Breach, New York, 1970, 89-92.
3. J. Edmonds and E.L. Johnson, Matching, Euler tours and the Chinese postman, *Mathematical Programming* **5** (1973), 88-124.
4. F. Eisenbrand, Mixed Integer Programming over TU systems, Manuscript, 2006.
5. O. Günlük and Y. Pochet, Mixing mixed integer inequalities, *Mathematical Programming* **90** (2001), 429-457 .
6. A. Miller and L.A. Wolsey, Tight formulations for some simple MIPs and convex objective IPs, *Mathematical Programming* B **98** (2003), 73–88.
7. G.L. Nemhauser, L.A. Wolsey, *Integer and Combinatorial Optimization*, Wiley Interscience, New York, 1988.
8. Y. Pochet and L.A. Wolsey, Polyhedra for lot-sizing with Wagner-Whitin costs, *Mathematical Programming* **67** (1994), 297–324.
9. Y. Pochet and L.A. Wolsey, *Production Planning by Mixed Integer Programming*, Springer Series in Operations Research and Financial Engineering, New York, 2006.
10. A. Schrijver, *Theory of Linear and Integer Programming*, Wiley, New York, 1986.
11. M. Skutella, Mixed Integer vertex cover on paths, Manuscript, 2005.

On the MIR Closure of Polyhedra

Sanjeeb Dash[1], Oktay Günlük[2], and Andrea Lodi[3]

[1] IBM T.J. Watson Research Center, P.O. Box 218, Yorktown Heights, NY 10598
sanjeebd@us.ibm.com
[2] IBM T.J. Watson Research Center, P.O. Box 218, Yorktown Heights, NY 10598
gunluk@us.ibm.com
[3] DEIS, University of Bologna, viale Risorgimento 2 - 40136 Bologna - Italy
alodi@deis.unibo.it

Abstract. We study the mixed-integer rounding (MIR) closure of polyhedra. The MIR closure of a polyhedron is equal to its split closure and
the associated separation problem is NP-hard. We describe a mixed-
integer programming (MIP) model with linear constraints and a nonlinear objective for separating an arbitrary point from the MIR closure
of a given mixed-integer set. We linearize the objective using additional
variables to produce a linear MIP model that solves the separation problem approximately, with an accuracy that depends on the number of
additional variables used. Our analysis yields a short proof of the result
of Cook, Kannan and Schrijver (1990) that the split closure of a polyhedron is again a polyhedron. We also present some computational results
with our approximate separation model.

1 Introduction

We study the mixed-integer rounding (MIR) closure of a given mixed-integer set

$$P = \{v \in R^{|J|},\ x \in Z^{|I|} \ :\ Cv + Ax \geq b,\ v, x \geq 0\}$$

where all numerical data is rational. In other words, we are interested in the set
of points that satisfy all MIR inequalities

$$(\lambda C)^+ v + (-\lambda)^+(Cv + Ax - b) + \min\{\lambda A - \lfloor \lambda A \rfloor, r\} x + r \lfloor \lambda A \rfloor x \geq r \lceil \lambda b \rceil$$

that can be generated by some λ of appropriate dimension. Here $r = \lambda b - \lfloor \lambda b \rfloor$,
$(\cdot)^+$ denotes $\max\{0, \cdot\}$ and all operators are applied to vectors component-wise.
In Section 2, we discuss in detail how these inequalities are derived and why
they are called MIR inequalities.

The term *mixed-integer rounding* was first used by Nemhauser and Wolsey
[18, pp.244] to denote valid inequalities that can be produced by what they
call the MIR procedure. These authors in [17] strengthen and redefine the MIR
procedure and the resulting inequality. The same term was later used to denote
seemingly simpler inequalities in Marchand and Wolsey [16], and Wolsey [20].

M. Fischetti and D.P. Williamson (Eds.): IPCO 2007, LNCS 4513, pp. 337–351, 2007.
© Springer-Verlag Berlin Heidelberg 2007

The definition of the MIR inequality we use in this paper is equivalent to the one in [17], though our presentation is based on [20].

Split cuts were defined by Cook, Kannan and Schrijver in [9], and are a special case of the disjunctive cuts introduced by Balas [2]. In [17], Nemhauser and Wolsey show that MIR cuts are equivalent to split cuts in the sense that, for a given polyhedron, the MIR closure is identical to the split closure. In [9], Cook, Kannan and Schrijver show that the split closure of a polyhedron is again a polyhedron. In this paper, we present a short proof of the same fact by analyzing MIR closure of polyhedra. This is not a new result but our proof is significantly easier to follow and present.

The problem of separating an arbitrary point from the MIR closure of a polyhedron is NP-hard as it was shown (using split cuts) by Caprara and Letchford [7]. The same also holds for the (Gomory-)Chvátal closure of a polyhedron as shown by Eisenbrand [13]. Gomory-Chvátal cuts are dominated by MIR cuts and therefore Chvátal closure is contained in the MIR closure, usually strictly.

In [15], Fischetti and Lodi show that, even though it is theoretically hard, in practice it is possible to separate points from the Chvátal closure in a reasonable time. Their approach involves formulating the separation problem as an MIP, and solving it with a black-box MIP solver. By repeatedly applying their separation algorithm to MIPLIB instances, they are able to approximately optimize over the Chvátal closure and obtain very tight bounds on the value of optimal solutions. Motivated by their work, we describe an MIP model for separating from the MIR closure of a polyhedron and present computational results on approximately optimizing over the MIR closure for problems in the MIPLIB 3.0.

Our work is also closely related with two recent papers written independently. The first one is a paper [4] by Balas and Saxena who experiment with a parametric MIP model to find violated split cuts. The second one is the paper by Vielma [19] which presents a proof of the fact that the split closure of a polyhedron is again a polyhedron.

The paper is organized as follows: In Section 2, we define MIR inequalities and their basic properties. In Section 3 we present a mixed-integer programming model that approximately separates an arbitrary point from the MIR closure of a given polyhedron. In Section 4, we present a simple proof that the MIR (or, split) closure of a polyhedron is again a polyhedron. In Section 5 we present a summary of the experiments with the approximate separation model.

2   Mixed-Integer Rounding Inequalities

In [20], Wolsey develops the MIR inequality as the only non-trivial facet of the following simple mixed-integer set:

$$Q^0 = \left\{ v \in R, \ x \in Z \ : \ v + x \ \geq \ b, \ v \geq 0 \right\}.$$

It is easy to see that

$$v \geq \hat{b}(\lceil b \rceil - x) \tag{1}$$

where $\hat{b} = b - \lfloor b \rfloor$ is valid and facet defining for Q^0. In [20] this inequality is called the *basic mixed-integer* inequality.

To apply this idea to more general sets defined by a single inequality, one needs to group variables in a way that resembles Q^0. More precisely, given a set

$$Q^1 = \left\{ v \in R^{|J|},\ x \in Z^{|I|} : \sum_{j \in J} c_j v_j + \sum_{i \in I} a_i x_i \geq b,\ v,\ x \geq 0 \right\}$$

the defining inequality is relaxed to obtain

$$\left(\sum_{j \in J} \max\{0, c_j\} v_j + \sum_{i \in I'} \hat{a}_i x_i \right) + \left(\sum_{i \in I \setminus I'} x_i + \sum_{i \in I} \lfloor a_i \rfloor x_i \right) \geq b$$

where $\hat{a}_i = a_i - \lfloor a_i \rfloor$ and $I' \subseteq I$. As the first part of the left hand side of this inequality is non-negative, and the second part is integral, the MIR inequality

$$\sum_{j \in J} \max\{0, c_j\} v_j + \sum_{i \in I'} \hat{a}_i x_i \geq \hat{b} \left(\lceil b \rceil - \sum_{i \in I \setminus I'} x_i - \sum_{i \in I} \lfloor a_i \rfloor x_i \right)$$

is valid for Q^1. Notice that $I' = \{i \in I : \hat{a}_i < \hat{b}\}$ gives the strongest inequality of this form and therefore the MIR inequality can also be written as

$$\sum_{j \in J} (c_j)^+ v_j + \sum_{i \in I} \min\{\hat{a}_i, \hat{b}\} x_i + \hat{b} \sum_{i \in I} \lfloor a_i \rfloor x_i \geq \hat{b} \lceil b \rceil. \tag{2}$$

To apply this idea to sets defined by $m > 1$ inequalities, the first step is to combine them to obtain a single *base* inequality and then apply inequality (2). Let

$$P = \left\{ v \in R^l,\ x \in Z^n : Cv + Ax \geq b,\ v, x \geq 0 \right\}$$

be a mixed-integer set where C, A and b are vectors of appropriate dimension. To obtain the base inequality, one possibility is to use a vector $\lambda \in R^m$, $\lambda \geq 0$ to combine the inequalities defining P. This approach leads to the base inequality $\lambda Cv + \lambda Ax \geq \lambda b$ and the corresponding MIR inequality

$$(\lambda C)^+ v + \min\{\lambda A - \lfloor \lambda A \rfloor, r\} x + r \lfloor \lambda A \rfloor x \geq r \lceil \lambda b \rceil, \tag{3}$$

where operators $(\cdot)^+$, $\lfloor \cdot \rfloor$ and $\min\{\cdot, \cdot\}$ are applied to vectors component-wise, and $r = \lambda b - \lfloor \lambda b \rfloor$.

Alternatively, it is also possible to first introduce slack variables to the set of inequalities defining P and combine them using a vector λ which is not necessarily non-negative. This gives the base inequality $\lambda Cv + \lambda Ax - \lambda s = \lambda b$ and the corresponding MIR inequality

$$(\lambda C)^+ v + (-\lambda)^+ s + \min\{\lambda A - \lfloor \lambda A \rfloor, r\} x + r \lfloor \lambda A \rfloor x \geq r \lceil \lambda b \rceil, \tag{4}$$

where s denotes the (non-negative) slack variables. Finally, substituting out the slack variables gives the following MIR inequality in the original space of P:

$$(\lambda C)^+ v + (-\lambda)^+ (Cv + Ax - b) + \min\{\lambda A - \lfloor \lambda A \rfloor, r\} x + r \lfloor \lambda A \rfloor x \geq r \lceil \lambda b \rceil. \tag{5}$$

These inequalities are what we call MIR inequalities in this paper.

Notice that when $\lambda \geq 0$, inequality (5) reduces to inequality (3). When $\lambda \ngeq 0$, however, there are inequalities (5) which cannot be written in the form (3). We present an example to emphasize this point (a similar one was independently developed in [5]).

Example 1. *Consider the simple mixed-integer set $T = \{v \in R, \ x \in Z \ : \ -v-4x \geq -4, \ -v+4x \geq 0, \ v, x \geq 0\}$ and the base inequality generated by $\lambda = [-1/8, 1/8] \ x + s_1/8 - s_2/8 \geq 1/2$ where s_1 and s_2 denote the slack variables for the first and second constraint, respectively. The corresponding MIR inequality is $1/2x + s_1/8 \geq 1/2$, which after substituting out s_1, becomes $-v/8 \geq 0$ or simply $v \leq 0$. This inequality defines the only non-trivial facet of T.*

Notice that it is not possible to generate this inequality using non-negative multipliers. Any base inequality generated by $\lambda_1, \lambda_2 \geq 0$ has the form $(-\lambda_1 - \lambda_2)v + (-4\lambda_1 + 4\lambda_2)x \geq -4\lambda_1$ where variable v has a negative coefficient. Therefore, the MIR inequality generated by this base inequality would have a coefficient of zero for the v variable, establishing that $v \leq 0$ cannot be generated as an MIR inequality (3).

2.1   Basic Properties of MIR Inequalities

Let P^{LP} denote the continuous relaxation of P. A linear inequality $hv + gx \geq d$ is called a *split cut* for P if it is valid for both $P^{LP} \cap \{\bar{\alpha}x \leq \bar{\beta}\}$ and $P^{LP} \cap \{\bar{\alpha}x \geq \bar{\beta}+1\}$, where $\bar{\alpha}$ and $\bar{\beta}$ are integral. Inequality $hv + gx \geq d$ is said to be derived from the *disjunction* $\bar{\alpha}x \leq \bar{\beta}$ and $\bar{\alpha}x \geq \bar{\beta}+1$. Obviously all points in P satisfy any split cut for P. Note that multiple split cuts can be derived from the same disjunction.

The basic MIR inequality (1) is a split cut for Q^0 with respect to x derived from the disjunction $x \leq \lfloor b \rfloor$ and $x \geq \lfloor b \rfloor + 1$. Therefore, the MIR inequality (5) is also a split cut for P derived from the disjunction $\bar{\alpha}x \leq \bar{\beta}$ and $\bar{\alpha}x \geq \bar{\beta}+1$ where $\bar{\beta} = \lfloor \lambda b \rfloor$ and

$$\bar{\alpha}_i = \begin{cases} \lceil (\lambda A)_i \rceil \text{ if } (\lambda A)_i - \lfloor (\lambda A)_i \rfloor \geq \lambda b - \lfloor \lambda b \rfloor \\ \lfloor (\lambda A)_i \rfloor \text{ otherwise.} \end{cases}$$

We note that this observation also implies that if a point $(v^*, x^*) \in P^{LP}$ violates the MIR inequality (5) then $\bar{\beta} + 1 > \bar{\alpha}x^* > \bar{\beta}$.

Furthermore, Nemhauser and Wolsey [17] showed that every split cut for P can be derived as an MIR cut for P. As what we call MIR inequalities in this paper are equivalent to the MIR inequalities defined in [17], the same observation holds for the MIR inequalities written in the form of inequality (5). We next formally define the MIR closure of polyhedra.

Definition 2. *The MIR closure of a polyhedron is the set of points satisfying all MIR inequality (5) that can be generated by some multiplier vector $\lambda \in R^m$.*

Thus, the split closure of a polyhedron is the same as its MIR closure. We next show that in certain cases, the closure of a polyhedron is invariant to reformulation.

3   The Separation Problem

In this section, we study the problem of separating an arbitrary point from the MIR closure of the polyhedron $P = \{v \in R^l, \; x \in Z^n \; : \; Cv + Ax \geq b, \; v, x \geq 0\}$. In other words, for a given point, we are interested in either finding violated inequalities or concluding that none exists. For convenience of notation, we first argue that without loss of generality we can assume P is given in equality form.

Consider the MIR inequality (4) for P,

$$(\lambda C)^+ v + (-\lambda)^+ s + \min\{\lambda A - \lfloor \lambda A \rfloor, r\} x + r \lfloor \lambda A \rfloor x \geq r \lceil \lambda b \rceil,$$

where s denotes the slack expression $(Cv + Ax - b)$. If we explicitly define the slack variables, by letting $\tilde{C} = (C, \; -I)$ and $\tilde{v} = (v, \; s)$, then the constraints defining P become $\tilde{C}\tilde{v} + Ax = b$, $\tilde{v} \geq 0$, $x \geq 0$, and the MIR inequality can be written as

$$(\lambda \tilde{C})^+ \tilde{v} + \min\{\lambda A - \lfloor \lambda A \rfloor, r\} x + r \lfloor \lambda A \rfloor x \geq r \lceil \lambda b \rceil. \tag{6}$$

In other words, all continuous variables, whether slack or structural, can be treated uniformly. In the remainder we assume that P is given in the equality form $P = \{v \in R^l, x \in Z^n : Cv + Ax = b, v, x \geq 0\}$, and we denote its continuous relaxation by P^{LP}.

3.1   Relaxed MIR Inequalities

Let

$$\Pi = \Big\{ (\lambda, c^+, \hat{\alpha}, \bar{\alpha}, \hat{\beta}, \bar{\beta}) \in R^m \times R^l \times R^n \times Z^n \times R \times Z \; :$$

$$c^+ \geq \lambda C, \hat{\alpha} + \bar{\alpha} \geq \lambda A, \hat{\beta} + \bar{\beta} \leq \lambda b, c^+ \geq 0, 1 \geq \hat{\alpha} \geq 0, 1 \geq \hat{\beta} \geq 0 \Big\}.$$

Note that for any $(\lambda, c^+, \hat{\alpha}, \bar{\alpha}, \hat{\beta}, \bar{\beta}) \in \Pi$,

$$c^+ v + (\hat{\alpha} + \bar{\alpha}) x \geq \hat{\beta} + \bar{\beta} \tag{7}$$

is valid for P^{LP} as it is a relaxation of $(\lambda C)v + (\lambda A)x = \lambda b$. Furthermore, using the basic mixed-integer inequality (1), we infer that

$$c^+ v + \hat{\alpha} x + \hat{\beta} \bar{\alpha} x \geq \hat{\beta}(\bar{\beta} + 1) \tag{8}$$

is a valid inequality for P. We call inequality (8) where $(\lambda, c^+, \hat{\alpha}, \bar{\alpha}, \hat{\beta}, \bar{\beta}) \in \Pi$ a *relaxed MIR inequality* derived using the *base inequality* (7). We next show some basic properties of relaxed MIR inequalities.

Lemma 3. *A relaxed MIR inequality (8) violated by* $(v^*, x^*) \in P^{LP}$ *satisfies* $(i) 1 > \hat{\beta} > 0$, $(ii) 1 > \Delta > 0$, (iii) *the violation of the inequality is at most* $\hat{\beta}(1 - \hat{\beta}) \leq 1/4$, *where* $\Delta = \bar{\beta} + 1 - \bar{\alpha} x^*$ *and violation is defined to be the right hand side of inequality (8) minus its left hand side.*

Proof: If $\hat{\beta} = 0$, then the relaxed MIR is trivially satisfied by all points in P^{LP}. Furthermore, if $\hat{\beta} = 1$, then inequality (8) is identical to its base inequality (7) which again is satisfied by all points in P^{LP}. Therefore, a non-trivial relaxed MIR cut satisfies $1 > \hat{\beta} > 0$.

For part (ii) of the Lemma, note that if $\bar{\alpha}x^* \geq \bar{\beta} + 1$ then inequality (8) is satisfied, as $c^+, \hat{\alpha}, \hat{\beta} \geq 0$ and $(v^*, x^*) \geq 0$. Furthermore, if (v^*, x^*) satisfies inequality (7) and $\bar{\alpha}x^* \leq \bar{\beta}$, then so is inequality (8) as $\hat{\beta} \leq 1$. Therefore, as the cut is violated, $1 > \Delta > 0$. It is also possible to show this by observing that inequality (8) is a split cut for P derived from the disjunction $\Delta \geq 1$ and $\Delta \leq 0$.

For the last part, let $w = c^+v^* + \hat{\alpha}x^*$ so that the base inequality (7) becomes $w \geq \hat{\beta} + \Delta - 1$ and the relaxed MIR inequality (8) becomes $w \geq \hat{\beta}\Delta$. Clearly $\hat{\beta}\Delta - w \leq \hat{\beta}(w + 1 - \hat{\beta}) - w = \hat{\beta}(1 - \hat{\beta}) - (1 - \hat{\beta})w \leq \hat{\beta}(1 - \hat{\beta})$, and the last inequality follows from the fact that $w \geq 0$ and $\hat{\beta} \leq 1$. □

Next, we relate MIR inequalities to relaxed MIR inequalities.

Lemma 4. *For any $\lambda \in R^m$, the MIR inequality (6) is a relaxed MIR inequality.*

Proof: For a given multiplier vector λ, define α to denote λA. Further, set $c^+ = (\lambda C)^+$, $\bar{\beta} = \lceil \lambda b \rceil$ and $\hat{\beta} = \lambda b - \lfloor \lambda b \rfloor$. Also, define $\hat{\alpha}$ and $\bar{\alpha}$ as follows:

$$\hat{\alpha}_i = \begin{cases} \alpha_i - \lfloor \alpha_i \rfloor \text{ if } \alpha_i - \lfloor \alpha_i \rfloor < \hat{\beta} \\ 0 \qquad \text{otherwise} \end{cases}, \qquad \bar{\alpha}_i = \begin{cases} \lfloor \alpha_i \rfloor \text{ if } \alpha_i - \lfloor \alpha_i \rfloor < \hat{\beta} \\ \lceil \alpha_i \rceil \text{ otherwise} \end{cases},$$

Clearly, $(\lambda, c^+, \hat{\alpha}, \bar{\alpha}, \hat{\beta}, \bar{\beta}) \in \Pi$ and the corresponding relaxed MIR inequality (8) is the same as the MIR inequality (6). □

Lemma 5. *MIR inequalities dominate relaxed MIR inequalities.*

Proof: Let $(v^*, x^*) \in P^{LP}$ violate a relaxed MIR inequality $\mathcal{I}$ which is obtained with $(\lambda, c^+, \hat{\alpha}, \bar{\alpha}, \hat{\beta}, \bar{\beta}) \in \Pi$. We will show that (v^*, x^*) also violates the MIR inequality (6).

Due to Lemma 3, we have $\bar{\beta} + 1 - \bar{\alpha}x^* > 0$ and therefore increasing $\hat{\beta}$ only increases the violation of the relaxed MIR inequality. Assuming $\mathcal{I}$ is the most violated relaxed MIR inequality, $\hat{\beta} = \min\{\lambda b - \bar{\beta}, 1\}$. By Lemma 3, we know that $\hat{\beta} < 1$, and therefore $\hat{\beta} = \lambda b - \bar{\beta}$ and $\bar{\beta} = \lfloor \lambda b \rfloor$.

In addition, due to the definition of Π we have $c^+ \geq (\lambda C)^+$ and $\hat{\alpha} + \hat{\beta}\bar{\alpha} \geq \min\{\lambda A - \lfloor \lambda A \rfloor, \hat{\beta}\} + \hat{\beta}\lfloor \lambda A \rfloor$. As $(v^*, x^*) \geq 0$, the violation of the MIR inequality is at least as much as the violation of $\mathcal{I}$. □

Combining Lemmas 4 and 5, we observe that a point in P^{LP} satisfies all MIR inequalities, if and only if it satisfies all relaxed MIR inequalities. Therefore, we can define the MIR closure of polyhedra using relaxed MIR inequalities and thus without using operators that take minimums, maximums or extract fractional parts of numbers. Let $\bar{\Pi}$ be the projection of Π in the space of $c^+, \hat{\alpha}, \bar{\alpha}, \hat{\beta}$ and $\bar{\beta}$ variables. In other words, $\bar{\Pi}$ is obtained by projecting out the λ variables. We now describe the MIR closure of P as:

$$P^{MIR} = \left\{ (v,x) \in P^{LP} : c^+v + \hat{\alpha}x + \hat{\beta}\bar{\alpha}x \geq \hat{\beta}(\bar{\beta}+1) \text{ for all } (c^+, \hat{\alpha}, \bar{\alpha}, \hat{\beta}, \bar{\beta}) \in \bar{\Pi} \right\}.$$

We would like to emphasize that $\bar{\Pi}$ is not the polar of P^{MIR} and therefore even though $\bar{\Pi}$ is a polyhedral set (with a finite number of extreme points and extreme directions), we have not yet shown that the polar of P^{MIR} is polyhedral. The polar of a polyhedral set is defined to be the set of points that yield valid inequalities for the original set. If the original set is defined in R^n, its polar is defined in R^{n+1} and the first n coordinates of any point in the polar give the coefficients of a valid inequality for the original set, and the last coordinate gives the right hand side of the valid inequality. Therefore, polar of P^{MIR} is the collection of points $(c^+, \hat{\alpha} + \hat{\beta}\bar{\alpha}, \hat{\beta}(\bar{\beta}+1)) \in R^{l+n+1}$ where $(c^+, \hat{\alpha}, \bar{\alpha}, \hat{\beta}, \bar{\beta}) \in \bar{\Pi}$. A set is polyhedral if and only if its polar is polyhedral.

For a given point $(v^*, x^*) \in P^{LP}$, testing if $(v^*, x^*) \in P^{MIR}$ can be achieved by solving the following non-linear integer program (MIR-SEP):

$$\max \left\{ \hat{\beta}(\bar{\beta}+1) - (c^+v^* + \hat{\alpha}x^* + \hat{\beta}\bar{\alpha}x^*) \; : \; (c^+, \hat{\alpha}, \bar{\alpha}, \hat{\beta}, \bar{\beta}) \in \bar{\Pi} \right\}.$$

If the optimal value of this program is non-positive, then $(v^*, x^*) \in P^{MIR}$. Otherwise, if the optimal value is positive, the optimal solution gives a most violated MIR inequality.

3.2 An Approximate Separation Model

We next (approximately) linearize the nonlinear terms that appear in the objective function of MIR-SEP. To this end, we first define a new variable Δ that stands for the term $(\bar{\beta}+1-\bar{\alpha}x)$. We then approximate $\hat{\beta}$ by a number $\tilde{\beta} \leq \hat{\beta}$ representable over some $\mathcal{E} = \{\epsilon_k : k \in K\}$. We say that a number δ is representable over $\mathcal{E}$ if $\delta = \sum_{k \in \bar{K}} \epsilon_k$ for some $\bar{K} \subseteq K$. We can therefore write $\tilde{\beta}$ as $\sum_{k \in K} \epsilon_k \pi_k$ using binary variables π_k and approximate $\hat{\beta}\Delta$ by $\tilde{\beta}\Delta$ which can now be written as $\sum_{k \in K} \epsilon_k \pi_k \Delta$. Finally, we linearize terms $\pi_k \Delta$ using standard techniques as π_k is binary and $\Delta \in (0,1)$ for any violated inequality.

An approximate MIP model APPX-MIR-SEP reads as follows:

$$\max \quad \sum_{k \in K} \epsilon_k \Delta_k - (c^+v^* + \hat{\alpha}x^*) \tag{9}$$

$$\text{s.t.} \quad (c^+, \hat{\alpha}, \bar{\alpha}, \hat{\beta}, \bar{\beta}) \in \bar{\Pi} \tag{10}$$

$$\hat{\beta} \geq \sum_{k \in K} \epsilon_k \pi_k \tag{11}$$

$$\Delta = (\bar{\beta}+1) - \bar{\alpha}x^* \tag{12}$$

$$\Delta_k \leq \Delta \qquad \forall k \in K \tag{13}$$

$$\Delta_k \leq \pi_k \qquad \forall k \in K \tag{14}$$

$$\lambda \in R^m, \quad \pi \in \{0,1\}^{|K|} \tag{15}$$

In our experiments, we use $\mathcal{E} = \{2^{-k} : k = 1, \ldots, \bar{k}\}$ for some small number $\bar{k}$. With this choice of $\mathcal{E}$, notice that for any $\hat{\beta}$ there exists a $\tilde{\beta}$ representable over $\mathcal{E}$ such that $2^{-\bar{k}} \geq \hat{\beta} - \tilde{\beta} \geq 0$. This observation is used to bound the error of the approximate model.

Theorem 6. *Let $\mathcal{E} = \{2^{-k} : k = 1, \ldots, \bar{k}\}$ for some positive integer $\bar{k}$ and denote the optimal values of MIR-SEP and APPX-MIR-SEP by z^{sep} and $z^{apx-sep}$, respectively. Then,*

$$z^{sep} \geq z^{apx-sep} > z^{sep} - 2^{-\bar{k}}. \qquad (16)$$

Proof: By (10), any (integral) feasible solution of APPX-MIR-SEP yields a feasible solution of MIR-SEP. Further, define $\tilde{\beta}$ to be $\sum_{k \in K} \epsilon_k \pi_k$. As $\Delta_k \leq \Delta \pi_k$ for all $k \in K$, $\sum_{k \in K} \epsilon_k \Delta_k \leq \sum_{k \in K} \epsilon_k \pi_k \Delta = \tilde{\beta} \Delta \leq \hat{\beta} \Delta$. Therefore, $z^{sep} \geq z^{apx-sep}$.

Note that $z^{apx-sep} \geq 0$ as we can get a feasible solution of APPX-MIR-SEP with objective 0 by setting Δ to 1, and the remaining variables to 0. Therefore the second inequality in (16) holds if $z^{sep} \leq 0$. Assume that $z^{sep} > 0$. Let $(c^+, \hat{\alpha}, \bar{\alpha}, \hat{\beta}, \bar{\beta}) \in \bar{\Pi}$ be an optimal solution of MIR-SEP. For the variables in APPX-MIR-SEP common with MIR-SEP, set their values to the above optimal solution of MIR-SEP. Let $\tilde{\beta}$ be the largest number representable over $\mathcal{E}$ less than or equal to $\hat{\beta}$. Clearly, $2^{-\bar{k}} \geq \hat{\beta} - \tilde{\beta} \geq 0$. Choose $\pi \in \{0, 1\}^{\bar{k}}$ such that $\tilde{\beta} = \sum_{k \in K} \epsilon_k \pi_k$. Set $\Delta = \bar{\beta} + 1 - \bar{\alpha} x^*$. Set $\Delta_k = 0$ if $\pi_k = 0$, and $\Delta_k = \Delta$ if $\pi_k = 1$. Then $\Delta_k = \pi_k \Delta$ for all $k \in K$, and $\tilde{\beta} \Delta = \sum_{k \in K} \epsilon_k \Delta_k$. Therefore, $2^{-\bar{k}} > 2^{-\bar{k}} \Delta \geq \hat{\beta} \Delta - \tilde{\beta} \Delta = \hat{\beta} \Delta - \sum_{k \in K} \epsilon_k \Delta_k$. The second inequality in (16) follows. $\qquad \square$

The previous result says that a solution of APPX-MIR-SEP with positive objective value yields a violated MIR cut, and if there is an MIR cut with a "large enough" violation, we will find some violated MIR cut by solving APPX-MIR-SEP.

In the next section (Theorem 12) we show that APPX-MIR-SEP becomes an exact model for finding violated MIR cuts when $\mathcal{E}$ is chosen as $\{\epsilon_k = 2^k / \Phi, \forall k = \{1, \ldots, \lceil \log \Phi \rceil\}\}$ where Φ is the least common multiple of all subdeterminants of $A|C|b$.

4   A Simple Proof That the MIR Closure Is a Polyhedron

In this section we give a short proof that the MIR closure of a polyhedron is a polyhedron. As MIR cuts are equivalent to split cuts, this result obviously follows from the work of Cook, Kannan and Schrijver [9] on split cuts. Andersen, Cornuéjols and Li [1], and Vielma [19] give alternative proofs that the split closure of a polyhedron is a polyhedron.

The main tool in the proof is a finite bound on the multipliers λ needed for non-redundant MIR cuts given in Lemma 9 . The bounds on λ can be tightened if the MIP is a pure integer program, and we give these tighter bounds in the next lemma whose proof is omitted due to the lack of space.

In this section we assume that the coefficients in $Cv + Ax = b$ are integers. Denote the ith equation of $Ax + Cv = b$ by $c_iv + a_ix = b_i$. An equation $c_iv + a_ix = b_i$ is a *pure integer* equation if $c_i = 0$.

Lemma 7. *If some MIR inequality is violated by the point (v^*, x^*), then there is another MIR inequality violated by (v^*, x^*) derived using $\lambda_i \in [0, 1)$ for every pure integer equation.*

Definition 8. *We define Ψ to be the largest absolute value of subdeterminants of C, and 1 if $C = 0$, where m is the number of rows in $Ax + Cv = b$.*

Lemma 9. *If there is an MIR inequality violated by the point (v^*, x^*), then there is another MIR inequality violated by (v^*, x^*) with $\lambda_i \in (-m\Psi, m\Psi)$, where m is the number of rows in $Ax + Cv = b$.*

Proof: Let the MIR cut $(\lambda C)^+v + \hat{\alpha}x + \hat{\beta}\bar{\alpha}x \geq \hat{\beta}(\bar{\beta}+1)$ be violated by (v^*, x^*). Then $(\lambda, (\lambda C)^+, \hat{\alpha}, \bar{\alpha}, \hat{\beta}, \bar{\beta}) \in \Pi$ with $0 < \hat{\beta} < 1$. Let C_j stand for the jth column of C. Let $S_1 = \{j : \lambda C_j > 0\}$ and $S_2 = \{j : \lambda C_j \leq 0\}$.

Consider the cone $\mathcal{C} = \{v \in R^m : vC_i \leq 0 \ \forall i \in S_1, \ vC_i \geq 0 \ \forall i \in S_2\}$. Obviously λ belongs to $\mathcal{C}$. We will find a vector λ' in $\mathcal{C}$, such that $\bar{\lambda} = \lambda - \lambda'$ is integral and belongs to $\mathcal{C}$. $\mathcal{C}$ is a polyhedral cone, and is generated by a finite set of vectors $\mu_1, \ldots, \mu_t$, for some $t > 0$. (Observe that if $C = 0$, then $\mathcal{C} = R^m$, and $\mu_1, \ldots, \mu_t$ can be chosen to be the unit vectors times ± 1.) We can assume these vectors are integral (by scaling); we can also assume the coefficients of $\mu_1, \ldots, \mu_t$ have absolute value at most Ψ. Further, we can assume that $\mu_1, \ldots, \mu_k$ (here $k \leq m$) are linearly independent vectors such that $\lambda = \sum_{j=1}^k v_j\mu_j$, with $v_j \in R$, $v_j > 0$. If $v_j < 1$ for $j = 1, \ldots, k$, then each coefficient of λ has absolute value less than $m\Psi$, and there is nothing to prove. If $v_j \geq 1$ for any $j \in \{1, \ldots, k\}$, then let $\lambda' = \sum_{j=1}^k \hat{v}_j\mu_j \Rightarrow \lambda - \lambda' = \sum_{j=1}^k \lfloor v_j \rfloor \mu_j$, where $\hat{v}_j = v_j - \lfloor v_j \rfloor$. Clearly λ' belongs to $\mathcal{C}$, and has coefficients with absolute value at most $m\Psi$. Also, $\lambda' \neq 0$ as $\lambda' = 0 \Rightarrow \lambda$ is integral $\Rightarrow \hat{\beta} = 0$. Let $\bar{\lambda} = \lambda - \lambda'$; obviously $\bar{\lambda}$ belongs to $\mathcal{C}$ and is integral. Further, $(\lambda C)^+ - (\lambda'C)^+ = (\bar{\lambda}C)^+$. Therefore $(\lambda', (\lambda'C)^+, \hat{\alpha}, \bar{\alpha} - \bar{\lambda}A, \hat{\beta}, \bar{\beta} - \bar{\lambda}b) \in \Pi$. It follows that the multipliers λ' lead to the MIR

$$(\lambda'C)^+v + \hat{\alpha}x + \hat{\beta}(\bar{\alpha} - \bar{\lambda}A)x \geq \hat{\beta}(\bar{\beta} - \bar{\lambda}b + 1). \tag{17}$$

The rhs of the old MIR minus the rhs of the new MIR equals

$$\hat{\beta}\bar{\lambda}b = \hat{\beta}\bar{\lambda}(Ax^* + Cv^*) = \hat{\beta}\bar{\lambda}Ax^* + \hat{\beta}\bar{\lambda}Cv^*$$
$$\leq \hat{\beta}\bar{\lambda}Ax^* + \hat{\beta}(\bar{\lambda}C)^+v^*. \tag{18}$$

The lhs of the old MIR (with v^*, x^* substituted) minus the lhs of the new MIR equals the last term in (18). Therefore the new MIR is violated by at least as much as the old MIR. □

Theorem 10. *If there is an MIR inequality violated by the point (v^*, x^*), then there is another MIR inequality violated by (v^*, x^*) for which $\hat{\beta}$ and the components of $\lambda, \hat{\alpha}$ are rational numbers with denominator equal to a subdeterminant of $A|C|b$, and each component of λ is contained in the interval $[-m\Psi, m\Psi]$.*

Proof: Let (v^*, x^*) be a point in P^{LP} which violates an MIR cut. Let this MIR cut be defined by $(\lambda_o, c_o^+, \hat{\alpha}_o, \bar{\alpha}_o, \hat{\beta}_o, \bar{\beta}_o) \in \Pi$. By Lemma 9 , we can assume each component of λ_o lies in the range $(-m\Psi, m\Psi)$. Define $\Delta_o = \bar{\beta}_o + 1 - \bar{\alpha}_o^T x^*$. Then $\hat{\beta}_o \Delta_o - c_o^+ v^* - \hat{\alpha}_o x^* > 0$. Consider the following LP:

$$\max \left\{ \hat{\beta}\Delta_o - c^+ v^* - \hat{\alpha} x^* \; : \; (\lambda, c^+, \hat{\alpha}, \bar{\alpha}_o, \hat{\beta}, \bar{\beta}_o) \in \Pi, -m\Psi \le \lambda_i \le m\Psi \right\}.$$

Note that the objective is a linear function as Δ_o is fixed. Further, we have fixed the variables $\bar{\alpha}$ and $\bar{\beta}$ in the constraints defining Π. The bounds on λ come from Lemma 9, except that we weaken them to non-strict inequalities. This LP has at least one solution for $(\lambda, c^+, \hat{\alpha}, \hat{\beta})$ with positive objective value, namely $(\lambda_o, c_o^+, \hat{\alpha}_o, \hat{\beta}_o)$. Therefore a basic optimal solution of this LP has positive objective value. Consider the MIR cut defined by an optimal solution along with $\bar{\alpha}_o$ and $\bar{\beta}_o$. It is obviously an MIR cut with violation at least the violation of the original MIR cut. Therefore, $0 < \hat{\beta} < 1$. Further, it is easy to see that the LP constraints (other than the bounds on the variables) can be written as

$$\begin{bmatrix} A^T & -I & & \\ C^T & & -I & \\ b^T & & & -1 \end{bmatrix} \begin{pmatrix} \lambda \\ \hat{\alpha} \\ c^+ \\ \hat{\beta} \end{pmatrix} \begin{matrix} \le \\ \le \\ \ge \end{matrix} \begin{pmatrix} \bar{\alpha}_o \\ 0 \\ \bar{\beta}_o \end{pmatrix}.$$

The theorem follows.   □

Corollary 11 *The MIR closure of a polyhedron P is a polyhedron.*

Proof: By Theorem 10, each non-redundant MIR inequality is defined by $\lambda = (\lambda_i)$ where λ_i is a rational number in $[-m\Psi, m\Psi]$ with a denominator equal to a subdeterminant of $A|C|b$. Thus, the number of non-redundant MIR inequalities is finite.   □

As the MIR closure equals the split closure, it follows that the split closure of a polyhedron is again a polyhedron. Let the split closure of P be denoted by P_S. For integral c and d, define $P_{(c,d)}$ by

$$P_{(c,d)} = conv\{(P \cap \{cx \le d\}) \cup (P \cap \{cx \ge d+1\})\} \Rightarrow P_S = \cap_{c \in Z^n, d \in Z} P_{(c,d)},$$

where x has n components. Lemma 9 gives a characterization of the useful disjunctions in the definition of the split closure. Define the vector $\mu \in R^m$ by

$$\mu_i = \begin{cases} m\Psi & \text{if } c_i \ne 0 \\ 1 & \text{if } c_i = 0 \end{cases}$$

Define $D = \{(c,d) \in Z^n \times Z : -\mu|A| \le c \le \mu|A|, \lfloor -\mu|b| \rfloor \le d \le \lfloor \mu|b| \rfloor\}$, where D is clearly a finite set, and $P_S = \cap_{c \in Z^n, d \in Z} P_{(c,d)} = \cap_{(c,d) \in D} P_{(c,d)}$. To see this, let x^* be a point in P but not in P_S. Then some split cut, which is also an MIR cut, is violated by x^*. By Lemma 9 , there is an MIR cut with $-\mu < \lambda < \mu$ which is violated by x^*. This MIR cut has the form $(\lambda C)^+ v + \hat{\alpha} x + \hat{\beta}\bar{\alpha} x \ge$

$\hat{\beta}(\bar{\beta}+1)$, where $(\bar{\alpha}, \bar{\beta}) \in D$. Thus x^* does not belong to $P_{(\bar{\alpha}, \bar{\beta})}$. This implies that $\cap_{(c,d) \in D} P_{(c,d)} \subseteq \cap_{c \in Z^n, d \in Z} P_{(c,d)}$, and the two sets in the expression above are equal as the reverse inclusion is true by definition.

Theorem 12. *Let Φ be the least common multiple of all subdeterminants of $A|C|b$, $K = \{1, \ldots, log\Phi\}$, and $\mathcal{E} = \{\epsilon_k = 2^k/\Phi, \forall k \in K\}$. Then APPX-MIR-SEP is an exact model for finding violated MIR cuts.*

Proof: By Theorem 10, $\hat{\beta}$ in a violated MIR cut can be assumed to be a rational number with a denominator equal to a subdeterminant of $A|C|b$ and therefore of Φ. But such a $\hat{\beta}$ is representable over $\mathcal{E}$.   □

5   Computational Experiments

In this section we briefly discuss our computational experience with the approximate separation model MIR-SEP . The goal is to approximately optimize over the MIR closure of a given MIP instance by repeatedly solving APPX-MIR-SEP to get violated MIR cuts. The general idea is to start off with the continuous relaxation of the given MIP. Then the following separation step is repeated. APPX-MIR-SEP is solved to find one or more MIR inequalities violated by the optimal solution of the current relaxation of the MIP, and the current relaxation is strengthened by adding these cuts. Even though this procedure is guaranteed to terminate after a finite number of iterations (for any fixed precision), in practice, there is no guarantee that we can actually optimize over the (approximate) MIR closure in a reasonable amount of time. Our approach, therefore, should be considered as a heuristic that tries to find good bounds in a reasonable amount of time.

We next sketch some practical issues and heuristic ideas to obtain good bounds faster.

1. **Numerical Issues.** A major issue is that the point (v^*, x^*) to be separated from the MIR closure of P is only approximately contained in P^{LP} if it is obtained using a practical LP solver. We deal with these numerical issues by modifying (v^*, x^*) and b to get a truly feasible solution of a different set of constraints. We let $v' = \max\{v^*, \mathbf{0}\}$, and $x' = \max\{x^*, \mathbf{0}\}$, and then define b' as $Cv' + Ax'$. We then use APPX-MIR-SEP to separate (v', x') from the MIR closure of $Cv + Ax = b', v, x \geq 0, x \in Z$. We use the multipliers λ in the solution of APPX-MIR-SEP to compute an MIR cut for P. Of course, this cut may not be violated by (v^*, x^*), but mostly is, as the point (v', x') is usually very close to (v^*, x^*).

2. **Reducing the size of the separation problem.** It is clear that in APPX-MIR-SEP, the variables $c_i^+, \hat{a}_j, \bar{a}_j$ corresponding to $v_i^* = 0$ and $x_j^* = 0$ do not contribute to the objective. For some of the problems in MIPLIB 3.0, this approach is quite crucial in allowing us to use APPX-MIR-SEP at all. For example, the MIP nw04 has 36 constraints, and over 87000 0-1 variables. If we let $\bar{k} = 5$, the first separation MIP has at most $36+5$ integer variables.

3. **Separation Heuristics.** To speed up this process we implemented several ideas which can essentially be seen as finding heuristic solutions to MIR-SEP.

   (a) **Preprocessing.** We take a subset S of integer variables, and for every x_i with $i \in S$, we solve LPs to maximize and minimize x_i for $x \in P^{LP}$.

   (b) **Gomory mixed-integer cuts.** Gomory mixed-integer cuts for the initial LP-relaxation of the MIP are known to be MIR inequalities [16] where the multipliers used to aggregate the rows of the formulation are obtained from the inverse of the optimal basis. Of course, we use these cuts only in the first iteration of the cutting plane algorithm to be sure that they have rank 1.

   (c) **Cuts based on the rows of the formulation.** Another heuristic considers rows of the formulation, one at a time, and obtains base inequalities by scaling them. Variables that have upper bounds are sometimes complemented using the bound constraints. The procedure is in the spirit of [12].

   (d) **Cuts based on pure integer base inequalities.** One way to generate effective MIR cuts is to concentrate on base inequalities that only contain integer variables. To obtain such base inequalities, the multiplier vector λ, used to aggregate the rows of the formulation, is required to satisfy $\lambda C \leq 0$ so that $(\lambda C)^+ = 0$. This can simply be achieved by fixing variables c^+ to zero in MIR-SEP thus obtaining a model called INT-SEP. This heuristic in a way mimics the procedure to generate the so-called projected Chvátal-Gomory (pro-CG) cuts [6] for mixed integer programs.

   (e) **Cuts generated by MIR-SEP.** The only parameter which must be specified for the definition and solution of MIR-SEP is the value of $\bar{k}$, i.e., the parameter responsible of the degree of approximation we use for $\hat{\beta}$. In all computational experiments, we do use $\bar{k} = 6$ which is a good compromise between computational efficiency and precision. In such a way, as proved by Theorem 6, our approximate model is guaranteed to find a cut violated by at least $1/64 = .015625$ which can be considered a reasonable threshold value to distinguish effective violated cuts.

4. **Piloting the black-box MIP solver.** A few tricks in the line of what already done in [6,15] can be used to force the black-box MIP solver, in our experiments ILOG-Cplex 10.0.1, to return good heuristic solutions of both INT-SEP and MIR-SEP. Indeed, it has to be stressed that we do not need to solve any of the separation problems to optimality in our cutting plane algorithm but, eventually, a final MIR-SEP so as to prove that no MIR inequality exists, i.e., the MIR closure has been computed.

The detailed computational results are reported in Tables 1 and 2 where the bounds we obtain with a time limit of 1 hour of CPU time on a standard PC are compared with those obtained in [4,6,15]. Our results confirm what other authors have already noticed, i.e., that those closures indeed provide a very tight approximation of the optimal solution of the problems in the MIPLIB. Most of the times we are able to compute bounds comparable with the ones already

Table 1. IPs of the MIPLIB 3.0

instance	$|I|$	# iter	# cuts	% gap closed	time MIR	% CG gap closed	time CG	% gap split	time split
air03	10,757	1	36	100.00	1	100.0	1	100.00	3
air04	8,904	5	294	9.18	3,600	30.4	43,200	91.23	864,360
air05	7,195	8	238	12.08	3,600	35.3	43,200	61.98	24,156
cap6000	6,000	120	334	50.55	3,600	22.5	43,200	65.17	1,260
fast0507	63,009	14	330	1.66	3,600	5.3	43,200	19.08	304,331
gt2	188	83	254	98.21	664	91.0	10,800	98.37	599
harp2	2,993	122	796	59.99	260	49.5	43,200	46.98	7,671
l152lav	1,989	57	214	12.66	3,600	59.6	10,800	95.20	496,652
lseu	89	103	306	92.28	3,600	93.3	175	93.75	32,281
mitre	10,724	12	158	100.00	380	16.2	10,800	100.00	5,330
mod008	319	41	173	100.00	11	100.0	12	99.98	85
mod010	2,655	1	39	100.00	0	100.0	1	100.00	264
nw04	87,482	100	301	95.16	3,600	100.0	509	100.00	996
p0033	33	27	115	87.42	2,179	85.3	16	87.42	429
p0201	201	394	1357	74.43	3,600	60.6	10,800	74.93	31,595
p0282	282	223	1474	99.60	3,600	99.9	10,800	99.99	58,052
p0548	548	255	1309	96.35	3,600	62.4	10,800	99.42	9,968
p2756	2,756	83	717	35.32	3,600	42.6	43,200	99.90	12,673
seymour	1,372	1	559	8.35	3,600	33.0	43,200	61.52	775,116
stein27	27	70	325	0.00	3,600	0.0	521	0.00	8,163
stein45	45	420	1930	0.00	3,600	0.0	10,800	0.00	27,624

Table 2. MILPs of the MIPLIB 3.0. For instance `arki001` we used an upper bound of value 7,580,813.0459.

instance	$|I|$	$|J|$	# iter	# cuts	% gap closed	time MIR	% CG gap closed	time CG	% gap split	time split
10teams	1,800	225	338	3341	100.00	3,600	57.14	1,200	100.00	90
arki001	538	850	14	124	33.93	3,600	28.04	1,200	83.05	193,536
bell3a	71	62	21	166	98.69	3,600	48.10	65	65.35	102
bell5	58	46	105	608	93.13	3,600	91.73	4	91.03	2,233
blend2	264	89	723	3991	32.18	3,600	36.40	1,200	46.52	552
dano3mip	552	13,321	1	124	0.10	3,600	0.00	1,200	0.22	73,835
danoint	56	465	501	2480	1.74	3,600	0.01	1,200	8.20	147,427
dcmulti	75	473	480	4527	98.53	3,600	47.25	1,200	100.00	2,154
egout	55	86	37	324	100.00	31	81.77	7	100.00	18,179
fiber	1,254	44	98	408	96.00	3,600	4.83	1,200	99.68	163,802
fixnet6	378	500	761	4927	94.47	3,600	67.51	43	99.75	19,577
flugpl	11	7	11	26	93.68	3,600	19.19	1,200	100.00	26
gen	150	720	11	127	100.00	16	86.60	1,200	100.00	46
gesa2	408	816	433	1594	99.81	3,600	94.84	1,200	99.02	22,808
gesa2_o	720	504	131	916	97.74	3,600	94.93	1,200	99.97	8,861
gesa3	384	768	464	1680	81.84	3,600	58.96	1,200	95.81	30,591
gesa3_o	672	480	344	1278	69.74	3,600	64.53	1,200	95.20	6,530
khb05250	24	1,326	65	521	100.00	113	4.70	3	100.00	33
markshare1	50	12	4781	90628	0.00	3,600	0.00	1,200	0.00	1,330
markshare2	60	14	4612	87613	0.00	3,600	0.00	1,200	0.00	3,277
mas74	150	1	1	12	6.68	0	0.00	0	14.02	1,661
mas76	150	1	1	11	6.45	0	0.00	0	26.52	4,172
misc03	159	1	143	727	33.65	450	34.92	1,200	51.70	18,359
misc06	112	1,696	112	1125	99.84	376	0.00	0	100.00	229
misc07	259	1	432	2135	11.03	3,600	3.86	1,200	20.11	41,453
mod011	96	10,862	253	1781	17.30	3,600	0.00	0	72.44	86,385
modglob	98	324	357	2645	60.77	254	0.00	0	92.18	1,594
mkc	5,323	2	112	2745	12.18	3,600	1.27	1,200	36.16	51,519
pk1	55	31	4229	22088	0.00	3,600	0.00	0	0.00	55
pp08a	64	176	246	1400	95.97	3,600	4.32	1,200	97.03	12,482
pp08aCUTS	64	176	143	687	62.99	3,600	0.68	1,200	95.81	5,666
qiu	48	792	847	2243	28.41	3,600	10.71	1,200	77.51	200,354
qnet1	1,417	124	182	805	64.60	3,600	7.32	1,200	100.00	21,498
qnet1_o	1,417	124	90	409	83.78	3,600	8.61	1,200	100.00	5,312
rentacar	55	9,502	92	281	23.41	3,600	0.00	5	0.00	—
rgn	100	80	114	666	99.81	1,200	0.00	0	100.00	222
rout	315	241	2225	17230	16.07	3,600	0.03	1,200	70.70	464,634
set1ch	240	472	156	694	63.39	3,600	51.41	34	89.74	10,768
swath	6,724	81	167	1421	33.96	3,600	7.68	1,200	28.51	2,420
vpm1	168	210	53	241	99.93	158	100.00	15	100.00	5,010
vpm2	168	210	74	314	71.48	224	62.86	1,022	81.05	6,012

reported in [4,6,15] in a much shorter computing time although sometimes a very large computational effort seems customary to obtain tight approximations. In a few cases, we have been able to improve over the best bound known so far. Of course, 1 hour of CPU time to strengthen the initial formulation can be by far too much, but as shown in [4,15] it might be the case that such a preprocessing allows the solution of hard unsolved problems and making it quicker has an intrinsic value.

Acknowledgments

Part of this research was carried out when the third author was Herman Goldstine Fellow of the IBM T.J. Watson Research Center, whose support is strongly acknowledged. The third author was also supported in part by the EU projects ADONET (contract n. MRTN-CT-2003-504438) and ARRIVAL (contract n. FP6-021235-2).

References

1. K. Andersen, G. Cornuejols, Y. Li, Split Closure and Intersection Cuts, *Mathematical Programming* **102** (2005), 457-493.
2. E. Balas, Disjunctive programming, *Annals of Discrete Mathematics* **5** (1979), 3-51.
3. E. Balas, S. Ceria, G. Cornuéjols, G. Natraj, Gomory cuts revisited, *Operations Research Letters* **19** (1996) 1-9.
4. E. Balas, A. Saxena, Optimizing over the Split Closure, 2005. *Mathematical Programming*, to appear.
5. P. Bonami, G. Cornuéjols, A note on the MIR Closure. *Manuscript*, 2006.
6. P. Bonami, G. Cornuéjols, S. Dash, M. Fischetti, A. Lodi, Projected Chvátal-Gomory cuts for mixed integer linear programs, 2005. *Mathematical Programming*, to appear.
7. A. Caprara, A. Letchford, On the separation of split cuts and related inequalities, *Mathematical Programming* **94** (2003), 279-294.
8. V. Chvátal, Edmonds polytopes and a hierarchy of combinatorial problems, *Discrete Mathematics* **4** (1973), 305-337.
9. W. J. Cook, R. Kannan, A. Schrijver, Chvátal closures for mixed integer programming problems, *Mathematical Programming* **47** (1990) 155-174.
10. G. Cornuéjols, Y. Li, Elementary closures for integer programs, *Operations Research Letters* **28** (2001), 1-8.
11. G. Cornuéjols, Y. Li, On the Rank of Mixed 0,1 Polyhedra, *Mathematical Programming* **91** (2002), 391-397.
12. S. Dash, O. Günlük, M. Goycoolea. Two step MIR inequalities for mixed-integer programs. *Manuscript*, 2005.
13. F. Eisenbrand, On the membership problem for the elementary closure of a polyhedron, *Combinatorica* **19** (1999), 297-300.
14. R. E. Gomory, An algorithm for the mixed integer problem, RM-2597, The Rand Corporation, 1960.

15. M. Fischetti, A. Lodi, Optimizing over the first Chvátal closure, Integer Programming and Combinatorial Optimization (M. Juenger and V. Kaibel eds.), Lecture Notes in Computer Science 3509, Springer-Verlag, Berlin, 12-22, 2005. *Mathematical Programming*, to appear.
16. H. Marchand, L. A. Wolsey, Aggregation and Mixed Integer Rounding to solve MIPs, *Operations Research* **49** (2001), 363–371.
17. G. Nemhauser, L. A. Wolsey, A recursive procedure to generate all cuts for 0-1 mixed integer programs, *Mathematical Programming* **46** (1990), 379–390.
18. G. Nemhauser, L. A. Wolsey, *Integer and Combinatorial Optimization*, Wiley, New York (1988).
19. J. P. Vielma, A Constructive Characterization of the Split Closure of a Mixed Integer Linear Program, 2005. *Operations Research Letters*, to appear.
20. L. A. Wolsey, *Integer Programming*, Wiley, New York (1998).

The Intersection of Continuous Mixing Polyhedra and the Continuous Mixing Polyhedron with Flows⋆

Michele Conforti[1], Marco Di Summa[1], and Laurence A. Wolsey[2]

[1] Dipartimento di Matematica Pura ed Applicata, Università degli Studi di Padova,
Via Trieste 63, 35121 Padova, Italy
{conforti,mdsumma}@math.unipd.it
[2] Center for Operations Research and Econometrics (CORE), Université catholique
de Louvain, 34, Voie du Roman Pays, 1348 Louvain-la-Neuve, Belgium
wolsey@core.ucl.ac.be

Abstract. In this paper we investigate two generalizations of the continuous mixing set studied by Miller and Wolsey [5] and Van Vyve [7]: the intersection set

$$X^{\mathrm{I}} = \{(\sigma, r, y) \in \mathbb{R}^n_+ \times \mathbb{R}^n_+ \times \mathbb{Z}^n_+ : \sigma_k + r_t + y_t \geq b_{kt}, \ 1 \leq k, t \leq n\}$$

and the continuous mixing set with flows

$$X^{\mathrm{CMF}} = \{(s, r, x, y) \in \mathbb{R}_+ \times \mathbb{R}^n_+ \times \mathbb{R}^n_+ \times \mathbb{Z}^n_+ :$$
$$s + r_t + x_t \geq b_t, \ x_t \leq y_t, \ 1 \leq t \leq n\} \ ,$$

which appears as a strong relaxation of some single-item lot-sizing problems. We give two extended formulations for the convex hull of each of these sets. In particular, for X^{CMF} the sizes of the extended formulations are polynomial in the size of the original description of the set, thus proving that the corresponding linear optimization problem can be solved in polynomial time.

Keywords: integer programming.

1 Introduction

In the last 5-10 years several mixed-integer sets have been studied that are interesting in their own right as well as providing strong relaxations of single-item lot-sizing sets. One in particular is the *continuous mixing set* X^{CM}:

$$s + r_t + y_t \geq b_t, \ 1 \leq t \leq n$$
$$s \in \mathbb{R}_+, \ r \in \mathbb{R}^n_+, \ y \in \mathbb{Z}^n_+ \ .$$

⋆ This work was partly carried out within the framework of ADONET, a European network in Algorithmic Discrete Optimization, contract no. MRTN-CT-2003-504438.

M. Fischetti and D.P. Williamson (Eds.): IPCO 2007, LNCS 4513, pp. 352–366, 2007.
© Springer-Verlag Berlin Heidelberg 2007

The *continuous mixing polyhedron* $\text{conv}(X^{\text{CM}})$, which is the convex hull of the above set, was introduced and studied by Miller and Wolsey in [5], where an extended formulation of $\text{conv}(X^{\text{CM}})$ with $\mathcal{O}(n^2)$ variables and $\mathcal{O}(n^2)$ constraints was given. Van Vyve [7] gave a more compact extended formulation of $\text{conv}(X^{\text{CM}})$ with $\mathcal{O}(n)$ variables and $\mathcal{O}(n^2)$ constraints and a formulation of $\text{conv}(X^{\text{CM}})$ in its original space.

We study here two generalizations of the continuous mixing set. First we consider the *intersection set* X^{I}, the intersection of several continuous mixing sets with distinct σ_k variables and common r and y variables:

$$\sigma_k + r_t + y_t \geq b_{kt}, \ 1 \leq k, t \leq n \tag{1}$$

$$\sigma \in \mathbb{R}^n_+, \ r \in \mathbb{R}^n_+, \ y \in \mathbb{Z}^n_+ \ . \tag{2}$$

Then we consider X^{CMF}, the "flow version" of the continuous mixing set:

$$s + r_t + x_t \geq b_t, \ 1 \leq t \leq n \tag{3}$$

$$x_t \leq y_t, \ 1 \leq t \leq n \tag{4}$$

$$s \in \mathbb{R}_+, \ r \in \mathbb{R}^n_+, \ x \in \mathbb{R}^n_+, \ y \in \mathbb{Z}^n_+ \ . \tag{5}$$

We now show two links between the continuous mixing set with flows X^{CMF} and lot-sizing. The first is to the single-item constant capacity lot-sizing problems with backlogging over n periods, which can be formulated (including redundant equations) as:

$$s_{k-1} + \sum_{u=k}^{t} w_u + r_t = \sum_{u=k}^{t} d_u + s_t + r_{k-1}, \ 1 \leq k \leq t \leq n$$

$$w_u \leq C z_u, 1 \leq u \leq n; \ s \in \mathbb{R}^{n+1}_+, \ r \in \mathbb{R}^{n+1}_+, \ w \in \mathbb{R}^n_+, \ z \in \{0,1\}^n \ .$$

Here d_u is the demand in period u, s_u and r_u are the stock and backlog at the end of period u, z_u takes value 1 if there is a set-up in period u allowing production to take place, w_u is the production in period u and C is the capacity (i.e. the maximum production). To see that this set has a relaxation as the intersection of n continuous mixing sets with flows, take $C = 1$ wlog, fix k, set $s = s_{k-1}$, $x_t = \sum_{u=k}^{t} w_u$, $y_t = \sum_{u=k}^{t} z_u$ and $b_t = \sum_{u=k}^{t} d_u$, giving a first relaxation:

$$s + x_t + r_t \geq b_t, \ k \leq t \leq n \tag{6}$$

$$0 \leq x_u - x_{u-1} \leq y_u - y_{u-1} \leq 1, \ k \leq u \leq n \tag{7}$$

$$s \in \mathbb{R}_+, \ r \in \mathbb{R}^{n-k+1}_+, \ x \in \mathbb{R}^{n-k+1}_+, \ y \in \mathbb{Z}^{n-k+1} \ . \tag{8}$$

Now summing (7) over $k \leq u \leq t$ (for each fixed $t = k, \ldots, n$) and dropping the upper bound on y_t, one obtains precisely the continuous mixing set with flows X^{CMF}.

The set X^{CMF} also provides an exact model for the two stage stochastic lot-sizing problem with constant capacities and backlogging. Specifically, at time 0 one must choose to produce a quantity s at a per unit cost of h. Then in period 1, n different outcomes are possible. For $1 \leq t \leq n$, the probability of event t is ϕ_t, the demand is b_t and the unit production cost is p_t, with production in batches of size up to C; there are also a fixed cost of q_t per batch and a possible bound k_t on the number of batches. As an alternative to production there is a linear backlog (recovery) cost e_t. Finally the goal is to satisfy all demands and minimize the total expected cost. The resulting problem is

$$\min \quad hs + \sum_{t=1}^n \phi_t(p_t x_t + q_t y_t + e_t r_t)$$

$$\text{s.t.} \quad s + r_t + x_t \geq b_t, \ 1 \leq t \leq n \tag{9}$$

$$x_t \leq C y_t, \ y_t \leq k_t, \ 1 \leq t \leq n \tag{10}$$

$$s \in \mathbb{R}_+, \ r \in \mathbb{R}_+^n, \ x \in \mathbb{R}_+^n, \ y \in \mathbb{Z}_+^n \ . \tag{11}$$

When $k_t = 1$ for all t, this is a standard lot-sizing problem, and in general (assuming $C = 1$ wlog) this is the set $X^{\mathrm{CMF}} \cap \{(s, r, x, y) : y_t \leq k_t, 1 \leq t \leq n\}$.

Now we describe the contents of this paper. Note that throughout, a *formulation* of a polyhedron $P \subseteq \mathbb{R}^n$ is an external description of P in its original space. It consists of a finite set of inequalities $Ax \leq d$ such that $P = \{x \in \mathbb{R}^n : Ax \leq d\}$. A formulation of P is *extended* whenever it gives an external description of P in a space $\mathbb{R}^{n+m}$ that includes the original space, so that, given $Q = \{(x, w) \in \mathbb{R}^{n+m} : A'x + B'w \leq d'\}$, P is the projection of Q onto the x-space. Given a mixed-integer set X, an extended formulation of $\mathrm{conv}(X)$ is *compact* if the size of the matrix $(A' \mid B' \mid d')$ is polynomial in the size of the original description of X.

In Sect. 2 we give two extended formulations for the polyhedron $\mathrm{conv}(X^{\mathrm{I}})$. In the first one, we split X^{I} into smaller sets, where the fractional parts of the σ variables are fixed. We then find an extended formulation for each of these sets and we use Balas' extended formulation for the convex hull of the union of polyhedra [1] to obtain an extended formulation of $\mathrm{conv}(X^{\mathrm{I}})$.

To construct the second extended formulation, we introduce extra variables to represent all possible fractional parts taken by the continuous variables at a vertex of $\mathrm{conv}(X^{\mathrm{I}})$. We then strengthen the original inequalities and show that the system thus obtained yields an extended formulation of $\mathrm{conv}(X^{\mathrm{I}})$.

When $b_{kt} = b_t - b_k$, $1 \leq t, k \leq n$, the intersection set is called a *difference set*, denoted X^{DIF}. For $\mathrm{conv}(X^{\mathrm{DIF}})$, we prove in Sect. 3 that our two extended formulations are compact. On the other hand, we show in Sect. 4 that the extended formulations of $\mathrm{conv}(X^{\mathrm{I}})$ are not compact when the values b_{kt} are arbitrary.

We then study the polyhedron $\mathrm{conv}(X^{\mathrm{CMF}})$. We show in Sect. 5 that there is an affine transformation which maps the polyhedron $\mathrm{conv}(X^{\mathrm{CMF}})$ into the intersection of a polyhedron $\mathrm{conv}(X^{\mathrm{DIF}})$ with a polyhedron that admits an easy external description. This yields two compact extended formulations for $\mathrm{conv}(X^{\mathrm{CMF}})$, showing in particular that one can optimize over X^{CMF} in polynomial time.

2   Two Extended Formulations for the Intersection Set

The *intersection set* X^{I} is the mixed-integer set defined by (1)–(2). Note that X^{I} is the intersection of n continuous mixing sets X_k^{CM}, each one associated with a distinct variable σ_k and having common variables r, y.

In order to obtain extended formulations for $\mathrm{conv}(X^I)$, we introduce two versions of the intersection set in which the fractional parts of the continuous variables σ_k, r_t are restricted in value.

In the following we call *fractional part* any number in $[0, 1)$. Also, for a number $a \in \mathbb{R}$, $f(a) = a - \lfloor a \rfloor$ denotes the fractional part of a, and for a vector $v = (v_1, \dots, v_q)$, $f(v)$ is the vector $(f(v_1), \dots, f(v_q))$.

In the first case, we consider a list $\mathcal{L}_\sigma = \{f^1, \dots, f^\ell\}$ of n-vectors whose components are fractional parts and a list $\mathcal{L}_r = \{g_1, \dots, g_m\}$ of fractional parts and define the set

$$X_1^I = \{(\sigma, r, y) \in X^I : f(\sigma) \in \mathcal{L}_\sigma,\ f(r_t) \in \mathcal{L}_r,\ 1 \le t \le n\} \ .$$

We say that the lists $\mathcal{L}_\sigma, \mathcal{L}_r$ are *complete* for X^I if for every vertex $(\bar{\sigma}, \bar{r}, \bar{y})$ of $\mathrm{conv}(X^I)$, $f(\bar{\sigma}) \in \mathcal{L}_\sigma$ and $f(\bar{r}_t) \in \mathcal{L}_r$, $1 \le t \le n$.

Remark 1. If $\mathcal{L}_\sigma, \mathcal{L}_r$ are complete lists for X^I then $\mathrm{conv}(X_1^I) = \mathrm{conv}(X^I)$.

In the second case, we consider a single list $\mathcal{L} = \{f_1, \dots, f_\ell\}$ of fractional parts and define the set

$$X_2^I = \{(\sigma, r, y) \in X^I : f(\sigma_k) \in \mathcal{L},\ f(r_t) \in \mathcal{L},\ 1 \le k, t \le n\} \ .$$

We say that the list $\mathcal{L}$ is *complete* for X^I if for every vertex $(\bar{\sigma}, \bar{r}, \bar{y})$ of $\mathrm{conv}(X^I)$ and for every $1 \le k, t \le n$, $f(\bar{\sigma}_k) \in \mathcal{L}$ and $f(\bar{r}_t) \in \mathcal{L}$.

Remark 2. If $\mathcal{L}$ is a complete list for X^I then $\mathrm{conv}(X_2^I) = \mathrm{conv}(X^I)$.

2.1 An Extended Formulation for $\mathrm{conv}(X_1^I)$

We give here an extended formulation of $\mathrm{conv}(X_1^I)$ with $\mathcal{O}(\ell m n)$ variables and $\mathcal{O}(\ell n(m + n))$ constraints.

For each fixed vector $f^i \in \mathcal{L}_\sigma$, let $X_{1,i}^I = \{(\sigma, r, y) \in X_1^I : f(\sigma) = f^i\}$. Notice that $X_1^I = \bigcup_{i=1}^\ell X_{1,i}^I$. First we find an extended formulation for each of the sets $\mathrm{conv}(X_{1,i}^I)$, $1 \le i \le \ell$, and then, since $\mathrm{conv}(X_1^I) = \mathrm{conv}\left(\bigcup_{i=1}^\ell \mathrm{conv}(X_{1,i}^I)\right)$, we use Balas' extended formulation for the convex hull of the union of polyhedra [1], in the fashion introduced in [3].

In the following we assume wlog $g_1 > g_2 > \cdots > g_m$. The set $X_{1,i}^I$ can be modeled as the following mixed-integer set:

$$\sigma_k = \mu_k + f_k^i,\ 1 \le k \le n$$
$$r_t = \nu_t + \sum_{j=1}^m g_j \delta_{tj},\ 1 \le t \le n$$
$$\mu_k + \nu_t + \sum_{j=1}^m g_j \delta_{tj} + y_t \ge b_{kt} - f_k^i,\ 1 \le k, t \le n$$
$$\sum_{j=1}^m \delta_{tj} = 1,\ 1 \le t \le n$$
$$\mu_k, \nu_t, y_t, \delta_{tj} \ge 0,\ 1 \le t, k \le n,\ 1 \le j \le m$$
$$\mu_k, \nu_t, y_t, \delta_{tj}\ \text{integer},\ 1 \le t, k \le n,\ 1 \le j \le m \ .$$

Using Chvátal-Gomory rounding, the above system can be tightened to

$$\sigma_k = \mu_k + f_k^i, \ 1 \le k \le n \tag{12}$$

$$r_t = \nu_t + \sum_{j=1}^{m} g_j \delta_{tj}, \ 1 \le t \le n \tag{13}$$

$$\mu_k + \nu_t + \sum_{j:g_j \ge f(b_{kt} - f_k^i)} \delta_{tj} + y_t \ge \lfloor b_{kt} - f_k^i \rfloor + 1, \ 1 \le k, t \le n \tag{14}$$

$$\sum_{j=1}^{m} \delta_{tj} = 1, \ 1 \le t \le n \tag{15}$$

$$\mu_k, \nu_t, y_t, \delta_{tj} \ge 0, \ 1 \le t, k \le n, 1 \le j \le m \tag{16}$$

$$\mu_k, \nu_t, y_t, \delta_{tj} \ \text{integer}, \ 1 \le t, k \le n, 1 \le j \le m \ . \tag{17}$$

Let A be the constraint matrix of (14)–(15). We show that A is a totally unimodular (TU) matrix.

Order the columns of A according to the following ordering of the variables:

$$\mu_1, \ldots, \mu_n; \ y_1, \nu_1, \delta_{11}, \ldots, \delta_{1m}; \ y_2, \nu_2, \delta_{21}, \ldots, \delta_{2m}; \ \ldots ; \ y_n, \nu_n, \delta_{n1}, \ldots, \delta_{nm} \ .$$

For each row of A, the 1's that appear in a block $[y_t, \nu_t, \delta_{t1}, \ldots, \delta_{tm}]$ are consecutive and start from the first position. Furthermore, for each row of A only one of these blocks contains nonzero elements.

Consider an arbitrary column submatrix of A. We give color red to all the μ_i (if any) and then, for each of the blocks $[y_t, \nu_t, \delta_{t1}, \ldots, \delta_{tm}]$, we give alternating colors, always starting with blue, to the columns of this block which appear in the submatrix. Since this is an equitable bicoloring, the theorem of Ghouila-Houri [4] shows that A is TU. Since the right-hand side of the constraints is integer, the theorem of Hoffman and Kruskal implies that (14)–(15) (along with the nonnegativity conditions) define an integral polyhedron.

Since (12)–(13) just define variables σ_k, r_t, we can remove the integrality constraints from (12)–(17), thus obtaining an extended formulation for $\text{conv}(X_{1,i}^I)$:

$$\text{conv}(X_{1,i}^I) = \{(\sigma, r, y) \text{ such that there exist } \delta, \mu \text{ satisfying } (12)–(16)\} \ .$$

This formulation involves $\mathcal{O}(mn)$ variables and $\mathcal{O}(n(m+n))$ constraints.

Using Balas' description for the union of polyhedra [1], we obtain:

Theorem 3. *The following linear system is an extended formulation of the polyhedron* $\text{conv}(X_1^I)$ *with* $\mathcal{O}(\ell mn)$ *variables and* $\mathcal{O}(\ell n(m+n))$ *constraints:*

$$\sigma_k = \sum_{i=1}^{\ell} \sigma_k^i, \ 1 \le k \le n$$

$$r_t = \sum_{i=1}^{\ell} r_t^i, \ 1 \le t \le n$$

$$y_t = \sum_{i=1}^{\ell} y_t^i, \ 1 \le t \le n$$

$$\sum_{i=1}^{\ell} \lambda^i = 1$$

$$\sigma_k^i = \mu_k^i + f_k^i \lambda^i, \ 1 \le k \le n, 1 \le i \le \ell$$

$$r_t^i = \nu_t^i + \sum_{j=1}^{m} g_j \delta_{tj}^i, \ 1 \le t \le n, 1 \le i \le \ell$$

$$\mu_k^i + \nu_t^i + \sum_{j:g_j \ge f(b_{kt} - f_k^i)} \delta_{tj}^i + y_t^i \ge (\lfloor b_{kt} - f_k^i \rfloor + 1)\lambda^i, \ 1 \le k, t \le n, 1 \le i \le \ell$$

$$\sum_{j=1}^{m} \delta_{tj}^i = \lambda^i, \ 1 \le t \le n, 1 \le i \le \ell$$

$$\mu_k^i, \nu_t^i, y_t^i, \delta_{tj}^i, \lambda^i \ge 0, \ 1 \le k, t \le n, 1 \le j \le m, 1 \le i \le \ell \ .$$

By Remark 1 we then obtain:

Corollary 4. *If the lists $\mathcal{L}_\sigma, \mathcal{L}_r$ are complete for X^I then the linear system given in Theorem 3 is an extended formulation of* $\mathrm{conv}(X^I)$.

2.2   An Extended Formulation for $\mathrm{conv}(X_2^I)$

We give an extended formulation for $\mathrm{conv}(X_2^I)$ with $\mathcal{O}(\ell n)$ variables and $\mathcal{O}(\ell n^2)$ constraints. We include zero in the list $\mathcal{L}$. Also, for technical reasons we define $f_0 = 1$. Wlog we assume $1 = f_0 > f_1 > \cdots > f_\ell = 0$.

The set X_2^I can be modeled as the following mixed-integer set:

$$\sigma_k = \mu^k + \sum_{j=1}^{\ell} f_j \delta_j^k, \ 1 \leq k \leq n \tag{18}$$

$$r_t = \nu^t + \sum_{j=1}^{\ell} f_j \beta_j^t, \ 1 \leq t \leq n \tag{19}$$

$$\sigma_k + r_t + y_t \geq b_{kt}, \ 1 \leq k, t \leq n \tag{20}$$

$$\sum_{j=1}^{\ell} \delta_j^k = 1, \ 1 \leq k \leq n \tag{21}$$

$$\sum_{j=1}^{\ell} \beta_j^t = 1, \ 1 \leq t \leq n \tag{22}$$

$$\sigma_k \geq 0, \ r_t \geq 0, \ y_t \geq 0, \ 1 \leq k, t \leq n$$

$$\delta_j^k, \beta_j^t \geq 0, \ 1 \leq k, t \leq n, \ 1 \leq j \leq \ell$$

$$\mu^k, \nu^t, y_t, \delta_j^k, \beta_j^t \text{ integer}, \ 1 \leq k, t \leq n, \ 1 \leq j \leq \ell \ .$$

Now define the unimodular transformation

$$\mu_0^k = \mu^k, \ \mu_j^k = \mu^k + \sum_{h=1}^{j} \delta_h^k, \ 1 \leq k \leq n, \ 1 \leq j \leq \ell$$

$$\nu_0^t = \nu^t + y_t, \ \nu_j^t = \nu^t + y_t + \sum_{h=1}^{j} \beta_h^t, \ 1 \leq t \leq n, \ 1 \leq j \leq \ell \ .$$

Then (18) and (19) become

$$\sigma_k = \sum_{j=0}^{\ell-1} (f_j - f_{j+1}) \mu_j^k, \ 1 \leq k \leq n$$

$$r_t = -y_t + \sum_{j=0}^{\ell-1} (f_j - f_{j+1}) \nu_j^t, \ 1 \leq t \leq n \ ,$$

while (21)–(22) become $\mu_\ell^k - \mu_0^k = 1$, $1 \leq k \leq n$, and $\nu_\ell^t - \nu_0^t = 1$, $1 \leq t \leq n$.

Constraints $\delta_j^k \geq 0$, $1 \leq k \leq n, 1 \leq j \leq \ell$, can be modeled as $\mu_j^k - \mu_{j-1}^k \geq 0$. Similarly $\beta_j^t \geq 0$, $1 \leq t \leq n, 1 \leq j \leq \ell$, can be modeled as $\nu_j^t - \nu_{j-1}^t \geq 0$.

Inequalities $\sigma_k \geq 0$, $1 \leq k \leq n$, become $\mu_0^k \geq 0$, while $r_t \geq 0$, $1 \leq t \leq n$, become $\nu_0^t - y_t \geq 0$.

We now model (20). Define $\ell_{kt} = \max\{\tau : f_\tau \geq f(b_{kt})\}$. Also, for an index $0 \leq j \leq \ell_{kt} - 1$, define $h_{kt}^j = \max\{\tau : f_\tau \geq 1 + f(b_{kt}) - f_{j+1}\}$ and for an index $\ell_{kt} \leq j \leq \ell - 1$, define $h_{kt}^j = \max\{\tau : f_\tau \geq f(b_{kt}) - f_{j+1}\}$.

Lemma 5. *Assume that a point (σ, r, y) satisfies (18), (19), (21) and (22). Then (σ, r, y) satisfies (20) if and only if the following inequalities are valid for (σ, r, y):*

$$\mu_{h_{kt}^j}^k + \nu_j^t \geq \lfloor b_{kt} \rfloor, \ 0 \leq j \leq \ell_{kt} - 1 \tag{23}$$

$$\mu_{h_{kt}^j}^k + \nu_j^t \geq \lfloor b_{kt} \rfloor + 1, \ \ell_{kt} \leq j \leq \ell - 1 \ . \tag{24}$$

Proof. We first assume that (σ, r, y) satisfies (18)–(22). Suppose $0 \leq j \leq \ell_{kt} - 1$. Constraint (20) can be written as $\mu^k + \nu^t + y_t + \sum_{i=1}^{\ell} f_i \delta_i^k + \sum_{i=1}^{\ell} f_i \beta_i^t \geq (\lfloor b_{kt} \rfloor - 1) + 1 + f(b_{kt})$. Since the δ_i^k's (resp. β_i^t's) are binary variables such that $\sum_{i=1}^{\ell} \delta_i^k = 1$ (resp. $\sum_{i=1}^{\ell} \beta_i^t = 1$), this implies $\mu^k + \nu^t + y_t + \sum_{i=1}^{h_{kt}^j} f_i \delta_i^k + f_{h_{kt}^j + 1} + \sum_{i=1}^{j} f_i \beta_i^t + f_{j+1} \geq (\lfloor b_{kt} \rfloor - 1) + 1 + f(b_{kt})$, thus $\mu_{h_{kt}^j}^k + \nu_j^t \geq (\lfloor b_{kt} \rfloor - 1) + 1 + f(b_{kt}) - f_{h_{kt}^j + 1} - f_{j+1}$. As $1 + f(b_{kt}) - f_{h_{kt}^j + 1} - f_{j+1} > 0$ for $0 \leq j \leq \ell_{kt} - 1$ and as $\mu_{h_{kt}^j}^k + \nu_j^t$ is an integer, (23) is valid.

Suppose now $\ell_{kt} \leq j \leq \ell - 1$. Constraint (20) can be written as $\mu^k + \nu^t + y_t + \sum_{i=1}^{\ell} f_i \delta_i^k + \sum_{i=1}^{\ell} f_i \beta_i^t \geq \lfloor b_{kt} \rfloor + f(b_{kt})$. Similarly as before, this implies $\mu_{h_{kt}^j}^k + \nu_j^t \geq \lfloor b_{kt} \rfloor + f(b_{kt}) - f_{h_{kt}^j + 1} - f_{j+1}$. As $f(b_{kt}) - f_{h_{kt}^j + 1} - f_{j+1} > 0$ for $\ell_{kt} \leq j \leq \ell - 1$ and as $\mu_{h_{kt}^j}^k + \nu_j^t$ is an integer, (24) is valid.

Now assume that (σ, r, y) satisfies (18), (19), (21) and (22), along with (23)–(24). Specifically, assume $\sigma_k = \mu^k + f_i$ and $r_t = \nu^t + f_l$.

Suppose $l \leq \ell_{kt}$. Inequality (23) for $j = l - 1$ is $\mu_{h_{kt}^{l-1}}^k + \nu^t + y^t \geq \lfloor b_{kt} \rfloor$. If $i \leq h_{kt}^{l-1}$, the inequality is $\mu^k + \nu^t + y_t \geq \lfloor b_{kt} \rfloor - 1$, thus $\sigma_k + r_t + y_t \geq \lfloor b_{kt} \rfloor - 1 + f_i + f_l \geq \lfloor b_{kt} \rfloor + f(b_{kt}) = b_{kt}$. And if $i > h_{kt}^{l-1}$, the inequality is $\mu^k + \nu^t + y_t \geq \lfloor b_{kt} \rfloor$, thus $\sigma_k + r_t + y_t \geq \lfloor b_{kt} \rfloor + f_l \geq \lfloor b_{kt} \rfloor + f(b_{kt}) = b_{kt}$. Thus (20) is satisfied when $l \leq \ell_{kt}$. The case $l > \ell_{kt}$ is similar. $\quad\square$

Thus we obtain the following result.

Theorem 6. *The following linear system is an extended formulation of the polyhedron* $\mathrm{conv}(X_2^{\mathrm{I}})$ *with* $\mathcal{O}(\ell n)$ *variables and* $\mathcal{O}(\ell n^2)$ *constraints:*

$$\sigma_k = \sum_{j=0}^{\ell-1} (f_j - f_{j+1}) \mu_j^k, \quad 1 \leq k \leq n \tag{25}$$

$$r_t = -y_t + \sum_{j=0}^{\ell-1} (f_j - f_{j+1}) \nu_j^t, \quad 1 \leq t \leq n \tag{26}$$

$$\mu_{h_{kt}^j}^k + \nu_j^t \geq \lfloor b_{kt} \rfloor, \quad 1 \leq k, t \leq n, \ 0 \leq j \leq \ell_{kt} - 1 \tag{27}$$

$$\mu_{h_{kt}^j}^k + \nu_j^t \geq \lfloor b_{kt} \rfloor + 1, \quad 1 \leq k, t \leq n, \ \ell_{kt} \leq j \leq \ell - 1 \tag{28}$$

$$\mu_\ell^k - \mu_0^k = 1, \ \nu_\ell^t - \nu_0^t = 1, \quad 1 \leq k, t \leq n \tag{29}$$

$$\mu_j^k - \mu_{j-1}^k \geq 0, \ \nu_j^t - \nu_{j-1}^t \geq 0, \quad 1 \leq k, t \leq n, \ 1 \leq j \leq \ell \tag{30}$$

$$\mu_0^k \geq 0, \ \nu_0^t - y_t \geq 0, \ y_t \geq 0, \quad 1 \leq k, t \leq n \ . \tag{31}$$

Proof. X_2^{I} is the set of points (σ, r, y) such that there exist *integral* vectors δ, μ satisfying (25)–(31). Changing the sign of the ν_j^t and y_t variables, the constraint matrix of (27)–(31) is a *dual network matrix* (that is, the transpose of a network flow matrix), in particular it is TU. Since the right-hand side is an integer vector and since (25)–(26) just define variables σ_k, r_t,

$$\mathrm{conv}(X_2^{\mathrm{I}}) = \{(\sigma, r, y) \text{ such that there exist } \delta, \mu \text{ satisfying } (25)\text{–}(31)\} \ . \quad\square$$

By Remark 2 we then obtain:

Corollary 7. *If the list $\mathcal{L}$ is complete for X^I then the linear system given in Theorem 6 is an extended formulation of* $\operatorname{conv}(X^I)$.

3   The Difference Set

The following set is the *difference set* X^{DIF}:

$$\sigma_k + r_t + y_t \geq b_t - b_k, \ 0 \leq k < t \leq n$$
$$\sigma \in \mathbb{R}_+^{n+1}, \ r \in \mathbb{R}_+^n, \ y \in \mathbb{Z}_+^n \ ,$$

where $0 = b_0 \leq b_1 \leq \ldots \leq b_n$. Note that X^{DIF} is an intersection set where $b_{kt} = b_t - b_k$, as for $k \geq t$ the constraint $\sigma_k + r_t + y_t \geq b_t - b_k$ is redundant.

Here we prove that the extended formulations given in Sect. 2 are compact for a set of the type X^{DIF}. This will be useful in Sect. 5, where we study X^{CMF}.

Theorem 8. *Let (σ^*, r^*, y^*) be a vertex of* $\operatorname{conv}(X^{\mathrm{DIF}})$. *Then there exists an index $h \in \{0, \ldots, n\}$ such that $\sigma_k^* > 0$ for $k < h$ and $\sigma_k^* = 0$ for $k \geq h$. Furthermore there is an index $\ell \geq h$ such that $f(\sigma_k^*) = f(b_\ell - b_k)$ for $0 \leq k < h$.*

Proof. Let (σ^*, r^*, y^*) be a vertex of $\operatorname{conv}(X^{\mathrm{DIF}})$, let $\alpha = \max_{1 \leq t \leq n} \{b_t - r_t^* - y_t^*\}$ and let $T_\alpha \subseteq \{1, \ldots, n\}$ be the subset of indices for which this maximum is achieved.

CLAIM 1: *For each $1 \leq k \leq n$, $\sigma_k^* = \max\{0, \alpha - b_k\}$.*
PROOF. The inequalities that define X^{DIF} show that $\sigma_k^* \geq \max\{0, \alpha - b_k\}$. If $\sigma_k^* > \max\{0, \alpha - b_k\}$, then there is an $\varepsilon > 0$ such that $(\sigma^*, r^*, y^*) \pm \varepsilon(e_k, \mathbf{0}, \mathbf{0})$ are both in $\operatorname{conv}(X^{\mathrm{DIF}})$, a contradiction to the fact that (σ^*, r^*, y^*) is a vertex. This concludes the proof of the claim.

Let $h = \min\{k : \alpha - b_k \leq 0\}$. (This minimum is well defined: since the only inequality involving σ_n is $\sigma_n \geq 0$, certainly $\sigma_n^* = 0$; then, by Claim 1, $\alpha - b_n \leq 0$.) Since $0 = b_0 \leq b_1 \leq \cdots \leq b_n$, Claim 1 shows that $\sigma_k^* > 0$ for $k < h$ and $\sigma_k^* = 0$ for $k \geq h$ and this proves the first part of the theorem. Furthermore $\sigma_k^* + r_t^* + y_t^* = b_t - b_k$ for all $k < h$ and $t \in T_\alpha$.
CLAIM 2: *Either $r_t^* = 0$ for some $t \in T_\alpha$ or $f(r_t) = f(b_t - b_h)$ for every $t \in T_\alpha$.*
PROOF. We use the fact that (σ^*, r^*) is a vertex of the polyhedron:

$$Q = \{(\sigma, r) \in \mathbb{R}_+^{n+1} \times \mathbb{R}_+^n : \sigma_k + r_t \geq b_t - b_k - y_t^*, 0 \leq k < t \leq n\} \ .$$

We now consider the following two cases:
CASE 1: $\alpha - b_h < 0$.
For $k \geq h$, the only inequality that is tight for (σ^*, r^*) and contains σ_k in its support is $\sigma_k \geq 0$. For $k < h$, the only inequalities that are tight for (σ^*, r^*) and contain σ_k in their support are $\sigma_k + r_t \geq b_t - b_k - y_t^*$, $t \in T_\alpha$.

Let e_H be the $(n+1)$-vector having the first h components equal to 1 and the others to 0, let e_{T_α} be the incidence vector of T_α and assume that $r_t^* > 0$ for all

$t \in T_\alpha$. Then the vectors $(\sigma^*, r^*) \pm \varepsilon(e_H, -e_{T_\alpha})$ for some $\varepsilon > 0$ are both in Q, contradicting the fact that (σ^*, r^*) is a vertex of Q. So $r_t^* = 0$ for some $t \in T_\alpha$.

CASE 2: $\alpha - b_h = 0$.

Then (σ^*, r^*, y^*) satisfies $\sigma_h^* + r_t^* + y_t^* = b_t - b_h$ for all $t \in T_\alpha$. Since $\sigma_h^* = 0$ and y_t^* is integer, then $f(r_t^*) = f(b_t - b_h)$ for all $t \in T_\alpha$ and this completes the proof of Claim 2.

Assume $r_t^* = 0$ for some $t \in T_\alpha$. Since $\sigma_k^* + r_t^* + y_t^* = b_t - b_k$ for all $k < h$ and y_t^* is an integer, then $f(\sigma_k^*) = f(b_t - b_k)$ for all $k < h$.

If $f(r_t^*) = f(b_t - b_h)$ for all $t \in T_\alpha$, since $\sigma_k^* + r_t^* + y_t^* = b_t - b_k$ for all $t \in T_\alpha$ and for all $k < h$ and since y^* is an integer vector, then $f(\sigma_k^*) = f(b_h - b_k)$ for all $k < h$. □

Corollary 9. *If (σ^*, r^*, y^*) is a vertex of* $\mathrm{conv}(X^{\mathrm{DIF}})$, *then $f(r_t^*) \in \{f(b_t - b_k), 1 \le k \le n\}$ for $1 \le t \le n$.*

Proof. The result follows from Theorem 8 and the observation that at a vertex of $\mathrm{conv}(X^{\mathrm{DIF}})$ either $r_t^* = 0$ or $\sigma_k^* + r_t^* + y_t^* = b_t - b_k$ for some k. □

We then obtain the following result.

Theorem 10. *The polyhedron* $\mathrm{conv}(X^{\mathrm{DIF}})$ *admits an extended formulation of the type given in Theorem 3 with $\mathcal{O}(n^5)$ variables and constraints and an extended formulation of the type given in Theorem 6 with $\mathcal{O}(n^3)$ variables and $\mathcal{O}(n^4)$ constraints.*

Proof. Recall that X^{DIF} is an intersection set. Define $\mathcal{L}_\sigma$ as the set of all possible $(n+1)$-vectors of fractional parts taken by σ at a vertex of $\mathrm{conv}(X^{\mathrm{DIF}})$ and $\mathcal{L}_r$ as the set of all possible fractional parts taken by the variables r_t at a vertex of $\mathrm{conv}(X^{\mathrm{DIF}})$. Since these lists are complete for X^{DIF}, Corollary 4 implies that the linear system given in Theorem 3 is an extended formulation of $\mathrm{conv}(X^{\mathrm{DIF}})$. By Theorem 8, $\ell = |\mathcal{L}_\sigma| = \mathcal{O}(n^2)$ and by Corollary 9, $m = |\mathcal{L}_r| = \mathcal{O}(n^2)$, therefore this formulation has $\mathcal{O}(n^5)$ variables and $\mathcal{O}(n^5)$ constraints.

Now define $\mathcal{L}$ as the set of all possible fractional parts taken by the variables σ_k, r_t at a vertex of $\mathrm{conv}(X^{\mathrm{DIF}})$. Since this list is complete for X^{DIF}, by Corollary 7 the system given in Theorem 6 is an extended formulation of $\mathrm{conv}(X^{\mathrm{DIF}})$. Since $\ell = |\mathcal{L}| = \mathcal{O}(n^2)$ (see Theorem 8 and Corollary 9), this formulation has $\mathcal{O}(n^3)$ variables and $\mathcal{O}(n^4)$ constraints. □

We point out that the result of the above theorem can be improved as follows.

Consider the first formulation. If for each set $X_{1,i}^{\mathrm{I}}$ we define a different list of fractional parts for the variables r_t, say $\mathcal{L}_r^i$, then we can easily choose such lists so that $|\mathcal{L}_r^i| = \mathcal{O}(n)$. In this case the first extended formulation for $\mathrm{conv}(X^{\mathrm{DIF}})$ involves $\mathcal{O}(n^4)$ variables and constraints.

Consider now the second formulation. Instead of defining a unique list for all variables, we can define a list for each variable, say $\mathcal{L}_{\sigma_k}$ and $\mathcal{L}_{r_t}$, $1 \le k, t \le n$. It is not difficult to verify that the construction of the extended formulation can be carried out with straightforward modifications. Since in this case $|\mathcal{L}_{\sigma_k}| = \mathcal{O}(n)$ (by Theorem 8) and $|\mathcal{L}_{r_t}| = \mathcal{O}(n)$ (by Corollary 9), the second extended formulation involves $\mathcal{O}(n^2)$ variables and $\mathcal{O}(n^3)$ constraints.

Theorem 11. *The polyhedron* $\mathrm{conv}(X^{\mathrm{CMF}})$ *admits an extended formulation with* $\mathcal{O}(n^2)$ *variables and* $\mathcal{O}(n^3)$ *constraints.*

4 Intersection Sets with an Exponential Number of Fractional Parts

In this section we show that the extended formulations derived in Sect. 2 are not compact in general. Specifically, we prove here the following result:

Theorem 12. *In the set of vertices of the polyhedron defined by*

$$\sigma_k + r_t \geq \frac{3^{(t-1)n+k}}{3^{n^2+1}}, \quad 1 \leq k, t \leq n \tag{32}$$

$$\sigma \in \mathbb{R}_+^n, \, r \in \mathbb{R}_+^n \tag{33}$$

the number of distinct fractional parts taken by variable σ_n *is exponential in* n.

Remark 13. Since the vertices of the above polyhedron are the vertices on the face defined by $y = 0$ of the polyhedron $\mathrm{conv}(X^I)$ with the same right-hand side, Theorem 12 shows that any extended formulation that explicitly takes into account a list of all possible fractional parts taken at a vertex by the continuous variables (such as those introduced to model $\mathrm{conv}(X_1^I)$ and $\mathrm{conv}(X_2^I)$) will not be compact in general.

Now let b_{kt} be as in the theorem, i.e. $b_{kt} = \frac{3^{(t-1)n+k}}{3^{n^2+1}}$, $1 \leq k, t \leq n$.

Remark 14. $b_{kt} < b_{k't'}$ if and only if $(t,k) \prec (t',k')$, where $\prec$ denotes the lexicographic order. Thus $b_{11} < b_{21} < \cdots < b_{n1} < b_{12} < \cdots < b_{nn}$.

Lemma 15. *The following properties hold.*

1. *Suppose that* $\alpha \in \mathbb{Z}_+^q$ *with* $\alpha_j < \alpha_{j+1}$ *for* $1 \leq j \leq q-1$, *and define* $\Phi(\alpha) = \sum_{j=1}^q (-1)^{q-j} 3^{\alpha_j}$. *Then* $\frac{1}{2} 3^{\alpha_q} < \Phi(\alpha) < \frac{2}{3} 3^{\alpha_q}$.
2. *Suppose that* α *is as above and* $\beta \in \mathbb{Z}_+^{q'}$ *is defined similarly. Then* $\Phi(\alpha) = \Phi(\beta)$ *if and only if* $\alpha = \beta$.

Proof. 1. $\sum_{j=0}^{\alpha_q - 1} 3^j = \frac{3^{\alpha_q}-1}{3-1} < \frac{1}{2} 3^{\alpha_q}$. Now $\Phi(\alpha) \geq 3^{\alpha_q} - \sum_{j=1}^{\alpha_q - 1} 3^j > 3^{\alpha_q} - \frac{1}{2} 3^{\alpha_q} = \frac{1}{2} 3^{\alpha_q}$, and $\Phi(\alpha) \leq 3^{\alpha_q} + \sum_{j=1}^{\alpha_q - 1} 3^j < 3^{\alpha_q} + \frac{1}{2} 3^{\alpha_q} = \frac{3}{2} 3^{\alpha_q}$.
   2. Suppose $\alpha \neq \beta$. Wlog we assume $q \geq q'$. Assume first $(\alpha_{q-q'+1}, \ldots, \alpha_q) = \beta$. Then $q > q'$ (otherwise $\alpha = \beta$) and, after defining $\bar{\alpha} = (\alpha_1, \ldots, \alpha_{q-q'})$, we have $\Phi(\alpha) - \Phi(\beta) = \Phi(\bar{\alpha}) > 0$ by 1. Now assume $(\alpha_{q-q'+1}, \ldots, \alpha_q) \neq \beta$. Define $h = \min\{\tau : \alpha_{q-\tau} \neq \beta_{q'-\tau}\}$ and suppose $\alpha_{q-h} > \beta_{q'-h}$ (the other case is similar). If we define the vectors $\bar{\alpha} = (\alpha_1, \ldots, \alpha_{q-h})$ and $\bar{\beta} = (\beta_1, \ldots, \beta_{q'-h})$, 1. gives $\Phi(\alpha) - \Phi(\beta) = \Phi(\bar{\alpha}) - \Phi(\bar{\beta}) > \frac{1}{2} 3^{\alpha_{q-h}} - \frac{3}{2} 3^{\beta_{q'-h}} \geq 0$, as $\alpha_{q-h} > \beta_{q'-h}$. $\square$

We now give a construction of an exponential family of vertices of (32)–(33) such that at each vertex variable σ_n takes a distinct fractional part. Therefore this construction proves Theorem 12.

Let $(k_1, \ldots, k_m)$ and $(t_1, \ldots, t_{m-1})$ be two increasing subsets of $\{1, \ldots, n\}$ with $k_1 = 1$ and $k_m = n$. For $1 \le k, t \le n$, let $p(k) = \max\{j : k_j \le k\}$ and $q(t) = \max\{j : t_j \le t\}$, with $q(t) = 0$ if $t < t_1$.

Consider the following system of equations:

$$\sigma_{k_1} = 0$$
$$\sigma_{k_j} + r_{t_j} = b_{k_j t_j}, \qquad 1 \le j \le m-1,$$
$$\sigma_{k_{j+1}} + r_{t_j} = b_{k_{j+1} t_j}, \quad 1 \le j \le m-1,$$
$$\sigma_{k_{q(t)+1}} + r_t = b_{k_{q(t)+1} t}, \quad t \notin \{t_1, \ldots, t_{m-1}\}$$
$$\sigma_k + r_{t_{p(k)}} = b_{k t_{p(k)}}, \quad k \notin \{k_1, \ldots, k_m\} .$$

The unique solution of this system is:

$$\sigma_{k_1} = 0$$
$$\sigma_{k_j} = \sum_{\ell=1}^{j-1} b_{k_{\ell+1} t_\ell} - \sum_{\ell=1}^{j-1} b_{k_\ell t_\ell}, \quad 2 \le j \le m$$
$$r_{t_j} = \sum_{\ell=1}^{j} b_{k_\ell t_\ell} - \sum_{\ell=1}^{j-1} b_{k_{\ell+1} t_\ell}, \quad 1 \le j \le m-1$$
$$\sigma_k = b_{k t_{p(k)}} - r_{t_{p(k)}}, \qquad k \notin \{k_1, \ldots, k_m\}$$
$$r_t = b_{k_{q(t)+1} t} - \sigma_{k_{q(t)+1}}, \qquad t \notin \{t_1, \ldots, t_{m-1}\} .$$

As each of these variables σ_k, r_t takes a value of the form $\Phi(\alpha)/3^{n^2+1}$, by Lemma 15 (i) we have that $\sigma_{k_j} > \frac{1}{2} b_{k_j t_{j-1}} > 0$ for $2 \le j \le m$, $r_{t_j} > \frac{1}{2} b_{k_j t_j} > 0$ for $1 \le j \le m-1$, $\sigma_k > \frac{1}{2} b_{k t_{p(k)}} > 0$ for $k \notin \{k_1, \ldots, k_m\}$ and $r_t > \frac{1}{2} b_{k_{q(t)+1} t} > 0$ for $t \notin \{t_1, \ldots, t_{m-1}\}$. Therefore the nonnegativity constraints are satisfied.

Now we show that the other constraints are satisfied. Consider the k, t constraint with $t \notin \{t_1, \ldots, t_{m-1}\}$. We distinguish some cases.

1. $p(k) \le q(t)$. Then $\sigma_k + r_t \ge r_t > \frac{1}{2} b_{k_{q(t)+1} t} \ge \frac{1}{2} b_{k_{p(k)+1} t} \ge \frac{3}{2} b_{kt} > b_{kt}$.
2. $p(k) > q(t)$ and $k \notin \{k_1, \ldots, k_m\}$. Then $\sigma_k + r_t \ge \sigma_k > \frac{1}{2} b_{k t_{p(k)}} \ge \frac{1}{2} b_{k t_{q(t)+1}} \ge \frac{3^n}{2} b_{kt} > b_{kt}$.
3. $p(k) = q(t) + 1$ and $k = k_j$ for some $1 \le j \le m$ (thus $p(k) = j = q(t) + 1$). In this case the k, t constraints is satisfied at equality by construction.
4. $p(k) > q(t) + 1$ and $k = k_j$ for some $1 \le j \le m$ (thus $p(k) = j > q(t) + 1$). Then $\sigma_k + r_t \ge \sigma_k > \frac{1}{2} b_{k t_{j-1}} \ge \frac{1}{2} b_{k t_{q(t)+1}} \ge \frac{3^n}{2} b_{kt} > b_{kt}$.

The argument with $k \notin \{k_1, \ldots, k_m\}$ is similar.

Finally suppose that $k = k_j$ and $t = t_h$ with $h \notin \{j-1, j\}$. If $h > j$, $\sigma_k + r_t \ge r_t > \frac{1}{2} b_{k_h t_h} \ge \frac{3}{2} b_{k_j t_h} > b_{kt}$. If $h < j-1$, $\sigma_k + r_t \ge \sigma_k > \frac{1}{2} b_{k_j t_{j-1}} \ge \frac{3^n}{2} b_{k_j t_h} > b_{kt}$.

This shows that the solution is feasible and as it is unique, it defines a vertex of (32)–(33).

Now let $a_{kt} = (t-1)n + k$, so that $b_{kt} = 3^{a_{kt}}/3^{n^2+1}$ and take

$$\alpha = (a_{k_1 t_1}, a_{k_2 t_1}, a_{k_2 t_2}, a_{k_3 t_2}, \ldots, a_{k_m t_{m-1}}) .$$

As $\sigma_n = \Phi(\alpha)/3^{n^2+1}$, Lemma 15 (ii) implies that in any two vertices constructed as above by different sequences $(k_1, \ldots, k_m)$, $(t_1, \ldots, t_{m-1})$ and $(k'_1, \ldots, k'_{m'})$, $(t'_1, \ldots, t'_{m'-1})$, the values of σ_n are distinct numbers in the interval $(0, 1)$. As the number of such sequences is exponential in n, this proves Theorem 12.

5   An Extended Formulation for $\mathrm{conv}(X^{\mathrm{CMF}})$

Now we address the question of showing that the linear optimization problem over the continuous mixing set with flows (3)–(5) is solvable in polynomial time. Specifically we derive compact extended formulations for $\mathrm{conv}(X^{\mathrm{CMF}})$.

We assume that $0 < b_1 \leq \cdots \leq b_n$. Consider the set Z:

$$s + r_t + y_t \geq b_t, \ 1 \leq t \leq n \tag{34}$$

$$s + r_k + x_k + r_t + y_t \geq b_t, \ 1 \leq k < t \leq n \tag{35}$$

$$s + r_t + x_t \geq b_t, \ 1 \leq t \leq n \tag{36}$$

$$s \in \mathbb{R}_+, \ r \in \mathbb{R}_+^n, \ x \in \mathbb{R}^n, \ y \in \mathbb{Z}_+^n \ . \tag{37}$$

Note that x is unrestricted in Z.

Proposition 16. *Let X^{CMF} and Z be defined on the same vector b. Then $X^{\mathrm{CMF}} \subseteq Z$ and $X^{\mathrm{CMF}} = Z \cap \{(s, r, x, y) : 0 \leq x \leq y\}$.*

Proof. Clearly (34)–(37) are valid for the points in X^{CMF}. The only inequalities that define X^{CMF} but do not appear in the definition of Z are $0 \leq x \leq y$.   □

Lemma 17. *The $3n+1$ extreme rays of $\mathrm{conv}(X^{\mathrm{CMF}})$ are the vectors $(1, 0, 0, 0)$, $(0, e_i, 0, 0)$, $(0, 0, 0, e_i)$, $(0, 0, e_i, e_i)$. The $3n+1$ extreme rays of $\mathrm{conv}(Z)$ are the vectors $(1, 0, -1, 0)$, $(0, e_i, -e_i, 0)$, $(0, 0, e_i, 0)$, $(0, 0, 0, e_i)$. Therefore both recession cones of $\mathrm{conv}(X^{\mathrm{CMF}})$ and $\mathrm{conv}(Z)$ are full-dimensional simplicial cones, thus showing that $\mathrm{conv}(X^{\mathrm{CMF}})$ and $\mathrm{conv}(Z)$ are full-dimensional polyhedra.*

Proof. The first part is obvious. We characterize the extreme rays of $\mathrm{conv}(Z)$. The recession cone C of $\mathrm{conv}(Z)$ is defined by

$$s + r_k + x_k + r_t + y_t \geq 0, \ 1 \leq k < t \leq n$$

$$s + r_t + x_t \geq 0, \ 1 \leq t \leq n$$

$$s \in \mathbb{R}_+, \ r \in \mathbb{R}_+^n, \ x \in \mathbb{R}^n, \ y \in \mathbb{R}_+^n \ .$$

One can verify that the vectors $\rho = (1, 0, -1, 0)$, $u_i = (0, e_i, -e_i, 0)$, $v_i = (0, 0, e_i, 0)$, $z_i = (0, 0, 0, e_i)$ are extreme rays of $\mathrm{conv}(Z)$ by checking that each of them satisfies at equality $3n$ linearly independent inequalities defining C (including nonnegativity constraints).

Thus we only have to show that every vector in C can be expressed as conic combination of the above rays. Let $(\bar{s}, \bar{r}, \bar{x}, \bar{y})$ be in C. Notice that $(\bar{s}, \bar{r}, \bar{x}, \bar{y}) = \bar{s}\rho + \sum_{i=1}^n \bar{r}_i u_i + \sum_{i=1}^n (\bar{s} + \bar{r}_i + \bar{x}_i)v_i + \sum_{i=1}^n \bar{y}_i w_i$. Since $(\bar{s}, \bar{r}, \bar{x}, \bar{y}) \in C$, all the coefficients appearing in the above combination are nonnegative.

It can also be checked that the above rays are linearly independent.   □

Lemma 18. *Let* (s^*, r^*, x^*, y^*) *be a vertex of* $\mathrm{conv}(Z)$. *Then*

$$s^* = \max\{0; b_t - r_t^* - y_t^*, 1 \le t \le n\},$$
$$x_k^* = \max\{b_k - s^* - r_k^*; b_t - s^* - r_k^* - r_t^* - y_t^*, 1 \le k < t \le n\}.$$

Proof. Assume $s^* > 0$ and $s^* + r_t^* + y_t^* > b_t$, $1 \le t \le n$. Then, there is an $\varepsilon \ne 0$ such that $(s^*, r^*, x^*, y^*) \pm \varepsilon(1, 0, -1, 0)$ belong to $\mathrm{conv}(Z)$, a contradiction. This proves the first statement. The second one is obvious. $\qquad\square$

Proposition 19. *Let* (s^*, r^*, x^*, y^*) *be a vertex of* $\mathrm{conv}(Z)$. *Then* $\mathbf{0} \le x^* \le y^*$.

Proof. Assume that $\{t : x_t^* < 0\} \ne \emptyset$ and let $h = \min\{t : x_t^* < 0\}$. Then $s^* + r_h^* > b_h > 0$ and together with $y_h^* \ge 0$, this implies $s^* + r_h^* + y_h^* > b_h$.
CLAIM: $r_h^* > 0$.
PROOF. Assume $r_h^* = 0$. Then $s^* > b_h > 0$. By Lemma 18, $s^* + r_t^* + y_t^* = b_t$ for some index t. It follows that $s^* \le b_t$, thus $t > h$ (as $b_h < s^* \le b_t$). Equation $s^* + r_t^* + y_t^* = b_t$, together with $s^* + r_h^* + x_h^* + r_t^* + y_t^* \ge b_t$, gives $r_h^* + x_h^* \ge 0$, thus $r_h^* > 0$, as $x_h^* < 0$, and this concludes the proof of the claim.
The inequalities $s^* + r_h^* + y_h^* > b_h$ and $r_k^* + x_k^* \ge 0$, $1 \le k < h$, imply $s^* + r_k^* + x_k^* + r_h^* + y_h^* > b_h$, $1 \le k < h$.
All these observations show the existence of an $\varepsilon \ne 0$ such that both points $(s^*, r^*, x^*, y^*) \pm \varepsilon(0, e_h, -e_h, 0)$ belong to $\mathrm{conv}(Z)$, a contradiction to the fact that the point (s^*, r^*, x^*, y^*) is a vertex of $\mathrm{conv}(Z)$. Thus $x^* \ge \mathbf{0}$.
Suppose now that there exists h such that $x_h^* > y_h^*$. Then constraint $s + r_h + y_h \ge b_h$ gives $s^* + r_h^* + x_h^* > b_h$. Lemma 18 then implies that $s^* + r_h^* + x_h^* + r_t^* + y_t^* = b_t$ for some $t > h$. This is not possible, as inequalities $x_h^* > y_h^* \ge 0$, $r_h^* \ge 0$ and $s^* + r_t^* + y_t^* \ge b_t$ imply $s^* + r_h^* + x_h^* + r_t^* + y_t^* > b_t$. Thus $x^* \le y^*$. $\qquad\square$

For the main theorem of this section we present a lemma whose proof is given in [2].

For a polyhedron P in $\mathbb{R}^n$ and a vector $a \in \mathbb{R}^n$, let $\mu_P(a)$ be the value $\min\{ax, x \in P\}$ and $M_P(a)$ be the face $\{x \in P : ax = \mu_P(a)\}$, where $M_P(a) = \emptyset$ whenever $\mu_P(a) = -\infty$.

Lemma 20. *Let* $P \subseteq Q$ *be two pointed polyhedra in* $\mathbb{R}^n$, *with the property that every vertex of* Q *belongs to* P. *Let* $Cx \ge d$ *be a system of inequalities that are valid for* P *such that for every inequality* $cx \ge \delta$ *of the system,* $P \not\subseteq \{x \in \mathbb{R}^n : cx = \delta\}$. *If for every* $a \in \mathbb{R}^n$ *such that* $\mu_P(a)$ *is finite but* $\mu_Q(a) = -\infty$, $Cx \ge d$ *contains an inequality* $cx \ge \delta$ *such that* $M_P(a) \subseteq \{x \in \mathbb{R}^n : cx = \delta\}$, *then* $P = Q \cap \{x \in \mathbb{R}^n : Cx \ge d\}$.

Proof. See [2].

Theorem 21. *Let* X^{CMF} *and* Z *be defined on the the same vector* b. *Then* $\mathrm{conv}(X^{\mathrm{CMF}}) = \mathrm{conv}(Z) \cap \{(s, r, x, y) : \mathbf{0} \le x \le y\}$.

Proof. By Proposition 16, $\mathrm{conv}(X^{\mathrm{CMF}}) \subseteq \mathrm{conv}(Z)$. By Propositions 19 and 16, every vertex of $\mathrm{conv}(Z)$ belongs to $\mathrm{conv}(X^{\mathrm{CMF}})$.

Let $a = (h, d, p, q)$, $h \in \mathbb{R}^1$, $d \in \mathbb{R}^n$, $p \in \mathbb{R}^n$, $q \in \mathbb{R}^n$, be such that $\mu_{\text{conv}(X^{\text{CMF}})}(a)$ is finite and $\mu_{\text{conv}(Z)}(a) = -\infty$. Since by Lemma 17, the extreme rays of $\text{conv}(Z)$ that are not rays of $\text{conv}(X^{\text{CMF}})$ are the vectors $(0, \mathbf{0}, e_i, \mathbf{0})$, $(0, e_i, -e_i, \mathbf{0})$ and $(1, \mathbf{0}, -\mathbf{1}, \mathbf{0})$, then either $p_i < 0$ for some index i or $d_i < p_i$ for some index i or $h < \sum_{t=1}^n p_t$.

If $p_i < 0$, then $M_{\text{conv}(X^{\text{CMF}})}(a) \subseteq \{(s, r, x, y) : x_i = y_i\}$.

If $d_i < p_i$, then $M_{\text{conv}(X^{\text{CMF}})}(a) \subseteq \{(s, r, x, y) : x_i = 0\}$, otherwise, given an optimal solution with $x_i > 0$, we could increase r_i by a small $\varepsilon > 0$ and decrease x_i by ε, thus obtaining a feasible point with lower objective value.

If $h < \sum_{t=1}^n p_t$, let $N^+ = \{j : p_j > 0\}$ and $k = \min\{j : j \in N^+\}$: we show that $M_{\text{conv}(X^{\text{CMF}})}(a) \subseteq \{(s, r, x, y) : x_k = 0\}$. Suppose that $x_k > 0$ in some optimal solution. As the solution is optimal and $p_k > 0$, we cannot just decrease x_k and remain feasible. Thus $s + r_k + x_k = b_k$, which implies that $s < b_k$. Then for all $j \in N^+$ we have $r_j + x_j \geq b_j - s > b_j - b_k \geq 0$, as $j \geq k$. Since we can assume $d_t \geq p_t$ for every t (otherwise we are in the previous case), $r_t = 0$ for every t: if not, chosen an index t such that $r_t > 0$, one can decrease r_t by a small $\varepsilon > 0$ and increase x_t by ε, thus obtaining a feasible point with lower objective value, a contradiction. So $r_t = 0$ for every t and thus, since $r_j + x_j > 0$ for all $j \in N^+$, we have $x_j > 0$ for all $j \in N^+$. Then we can increase s by a small $\varepsilon > 0$ and decrease x_j by ε for all $j \in N^+$. The new point is feasible in X^{CMF} and has lower objective value, a contradiction.

We have shown that for every vector a such that $\mu_{\text{conv}(X^{\text{CMF}})}(a)$ is finite and $\mu_{\text{conv}(Z)}(a) = -\infty$, the system $\mathbf{0} \leq x \leq y$ contains an inequality which is tight for the points in $M_{\text{conv}(X^{\text{CMF}})}(a)$. To complete the proof, since $\text{conv}(X^{\text{CMF}})$ is full-dimensional (Lemma 17), the system $\mathbf{0} \leq x \leq y$ does not contain an improper face of $\text{conv}(X^{\text{CMF}})$. So we can now apply Lemma 20 to $\text{conv}(X^{\text{CMF}})$, $\text{conv}(Z)$ and the system $\mathbf{0} \leq x \leq y$. □

Therefore, if we have a compact extended formulation of $\text{conv}(Z)$, then this will immediately yield a compact extended formulation of $\text{conv}(X^{\text{CMF}})$. Such a formulation exists, as Z is equivalent to a difference set:

Theorem 22. *Let X^{DIF} be a difference set and X^{CMF} be defined on the same vector b. The affine transformation $\sigma_0 = s$, $\sigma_t = s + r_t + x_t - b_t$, $1 \leq t \leq n$, maps $\text{conv}(X^{\text{CMF}})$ into $\text{conv}(X^{\text{DIF}}) \cap \{(\sigma, r, y) : 0 \leq \sigma_k - \sigma_0 - r_k + b_k \leq y_k, 1 \leq k \leq n\}$.*

Proof. Let Z be defined on the same vector b. It is straightforward to check that the affine transformation $\sigma_0 = s$, $\sigma_t = s + r_t + x_t - b_t$, $1 \leq t \leq n$, maps $\text{conv}(Z)$ into $\text{conv}(X^{\text{DIF}})$. By Theorem 21, $\text{conv}(X^{\text{CMF}}) = \text{conv}(Z) \cap \{(s, r, x, y) : \mathbf{0} \leq x \leq y\}$ and the result follows. □

Then the extended formulations of $\text{conv}(X^{\text{DIF}})$ described in Sects. 2–3 give extended formulations of $\text{conv}(X^{\text{CMF}})$ which are compact. By Theorem 11 we have:

Theorem 23. *The polyhedron $\text{conv}(X^{\text{CMF}})$ admits an extended formulation with $\mathcal{O}(n^2)$ variables and $\mathcal{O}(n^3)$ constraints. It follows that the linear optimization problem over X^{CMF} can be solved in polynomial time.*

5.1 An Extended Formulation for the Two Stage Stochastic Lot-Sizing Problem with Constant Capacities and Backlogging

We briefly consider the set $X^{\mathrm{CMF}} \cap W$, where

$$W = \{(s, r, x, y) : l_j \leq y_j \leq u_j,\ l_{jk} \leq y_j - y_k \leq u_{jk},\ 1 \leq j, k \leq n\}\ ,$$

with $l_j, u_j, l_{jk}, u_{jk} \in \mathbb{Z} \cup \{+\infty, -\infty\}$, $1 \leq j, k \leq n$. We assume that for each $1 \leq i \leq n$, W contains a point satisfying $y_i \geq 1$.

In the following we show that an extended formulation of $\mathrm{conv}(X^{\mathrm{CMF}} \cap W)$ is obtained by adding the inequalities defining W to one of the extended formulations of $\mathrm{conv}(X^{\mathrm{CMF}})$ derived above. The proof uses the same technique as in Sect. 5, where Z (resp. X^{CMF}) has to be replaced with $Z \cap W$ (resp. $X^{\mathrm{CMF}} \cap W$). We only point out the main differences.

To see that the proof of Theorem 21 is still valid, note that the extreme rays of $\mathrm{conv}(Z \cap W)$ are of the following types:

1. $(1, \mathbf{0}, -\mathbf{1}, \mathbf{0})$, $(0, e_i, -e_i, \mathbf{0})$, $(0, \mathbf{0}, e_i, \mathbf{0})$;
2. $(0, \mathbf{0}, \mathbf{0}, y)$ for suitable vectors $y \in \mathbb{Z}^n$.

However, the rays of type 2 are also rays of $\mathrm{conv}(X^{\mathrm{CMF}} \cap W)$. Also, the condition that for every index i, W contains a vector with $y_i > 0$, shows that none of the inequalities $0 \leq x_i \leq y_i$ defines an improper face of $\mathrm{conv}(X^{\mathrm{CMF}} \cap W)$ and Lemma 20 can still be applied. Thus the proof of Theorem 21 is still valid.

The rest of the proof is a straightforward adaptation of Theorem 22.

Since (9)–(11) define a set of the type $X^{\mathrm{CMF}} \cap W$ (assuming $C = 1$ wlog), the above result yields an extended formulation for the feasible region of the two stage stochastic lot-sizing problem with constant capacities and backlogging.

References

1. Balas, E.: Disjunctive programming: properties of the convex hull of feasible points. Invited paper, Discrete Appl. Math. **89** (1988) 1–44
2. Conforti, M., Di Summa, M., Wolsey, L. A.: The mixing set with flows. CORE DP 2005/92, Université catholique de Louvain. SIAM J. Discret. Math. (to appear)
3. Conforti, M., Wolsey, L. A.: Compact formulations as a union of polyhedra. CORE DP 2005/62, Université catholique de Louvain. Math. Program. (to appear)
4. Ghouila-Houri, A.: Caractérisations des matrices totalement unimodulaires. C. R. Acad. Sci. Paris **254** (1968) 155–163
5. Miller, A. J., Wolsey, L. A.: Tight formulations for some simple mixed integer programs and convex objective integer programs. Math. Program. B **98** (2003) 73–88
6. Van Vyve, M.: A solution approach of production planning problems based on compact formulations for single-item lot-sizing models. Ph. D. thesis, Faculté des Sciences Appliquées, Université catholique de Louvain (2003)
7. Van Vyve, M.: The continuous mixing polyhedron. Math. Oper. Res. **30** (2005) 441–452

Simple Explicit Formula for Counting Lattice Points of Polyhedra

Jean B. Lasserre[1] and Eduardo S. Zeron[2]

[1] LAAS-CNRS and Institute of Mathematics,
LAAS, 7 Av. du Colonel Roche, 31077 Toulouse, France
lasserre@laas.fr
http://www.laas.fr/~lasserre
[2] Depto. Matemáticas, Apdo. Postal 14-740
Cinvestav-IPN, Mexico D.F. 07000, Mexico
eszeron@math.cinvestav.mx
http://www.cinvestav.mx/SantillanSp.htm

Abstract. Given $z \in \mathbb{C}^n$ and $A \in \mathbb{Z}^{m \times n}$, we provide an explicit expression and an algorithm for evaluating the counting function $h(y; z) := \sum \{ z^x \mid x \in \mathbb{Z}^n; Ax = y, x \geq 0 \}$. The algorithm only involves simple (but possibly numerous) calculations. In addition, we exhibit *finitely many* fixed convex cones of $\mathbb{R}^n$ explicitly and exclusively defined by A, such that for *any* $y \in \mathbb{Z}^m$, $h(y; z)$ is obtained by a simple formula that evaluates $\sum z^x$ over the integral points of those cones only. At last, we also provide an alternative (and different) formula from a decomposition of the generating function into simpler rational fractions, easy to invert.

1 Introduction

Consider the (not necessarily compact) polyhedron

$$\Omega(y) := \{x \in \mathbb{R}^n \mid Ax = y; \ x \geq 0\}, \tag{1}$$

with $y \in \mathbb{Z}^m$ and $A \in \mathbb{Z}^{m \times n}$ of maximal rank for $n \geq m$; besides, given $z \in \mathbb{C}^n$, let $h : \mathbb{Z}^m \to \mathbb{C}$ be the *counting* function

$$y \mapsto h(y; z) := \sum_{x \in \Omega(y) \cap \mathbb{Z}^n} z^x \tag{2}$$

(where z^x stands for $\prod_k z_k^{x_k}$). The complex vector $z \in \mathbb{C}^n$ may be chosen close enough to zero in order to ensure that $h(y; z)$ is well defined even when $\Omega(y)$ is not compact. If $\Omega(y)$ is compact, then $y \mapsto h(y; z)$ provides us with the exact number of points in the set $\Omega(y) \cap \mathbb{Z}^n$ by either evaluating $h(y, 1)$, or even rounding $h(y; z)$ up to the nearest integer when all the entries of z are close enough to one.

Computation of h has attracted a lot of attention in recent years, from both theoretical and practical computation viewpoints. Barvinok and Pommersheim [4], Brion and Vergne [8], have provided nice exact (theoretical) formulas

M. Fischetti and D.P. Williamson (Eds.): IPCO 2007, LNCS 4513, pp. 367–381, 2007.
© Springer-Verlag Berlin Heidelberg 2007

for $h(y; z)$; see also Szenes and Vergne [15]. For instance, Barvinok considers $z \mapsto h(y; z)$ as the generating function (evaluated at $z := e^c \in \mathbb{C}^n$) of the indicator function $x \mapsto I_{\Omega(y) \cap \mathbb{Z}^n}(x)$ of the set $\Omega(y) \cap \mathbb{Z}^n$ and provides a decomposition into a sum of simpler generating functions associated with supporting cones (themselves having a signed decomposition into unimodular cones). We call this a *primal* approach because y is *fixed*, and one works in the primal space $\mathbb{R}^n$ in which $\Omega(y)$ is defined. Remarkably, Barvinok's counting algorithm which is implemented in the software LattE (see De Loera et al. [10]) runs in time polynomial in the problem size when the dimension n is fixed. The software developed by Verdoolaege [17] extends the LattE software to handle *parametric polytopes*. On the other hand, Brion and Vergne [8] consider the generating function $H : \mathbb{C}^m \to \mathbb{C}$ of $y \mapsto h(y; z)$, that is,

$$w \mapsto H(w) := \sum_{y \in \mathbb{Z}^m} h(y; z) w^y = \prod_{k=1}^{n} \frac{1}{1 - z_k w^{A_k}}. \tag{3}$$

They provide a generalized residue formula, and so obtain $h(y; z)$ in closed form by *inversion*. We call this latter approach *dual* because z is fixed, and one works in the space $\mathbb{C}^m$ of variables w associated with the m constraints $Ax = y$.

As a result of both primal and dual approaches, $h(y; z)$ is finally expressed as a weighted sum over the vertices of $\Omega(y)$. Similarly, Beck [5], and Beck, Diaz and Robins [6] provided a complete analysis based on residue techniques for the case of a tetrahedron ($m = 1$). Despite its theoretical interest, Brion and Vergne's formula is not directly *tractable* because it contains many products with complex coefficients (roots of unity) which makes the formula difficult to evaluate numerically. However, in some cases, this formula can be exploited to yield an efficient algorithm as e.g. in [2] for flow polytopes, in [7] for transportation polytopes, and more generally when the matrix A is totally unimodular as in [9]. Finally, in [12,13], we have provided two algorithms based on Cauchy residue techniques to invert H in (3), and an alternative algebraic technique based on partial fraction expansion of H. A nice feature of the latter technique of [13] is to avoid computing residues.

Contribution: Our contribution is twofold as it is concerned with both primal and dual approaches. On the primal side, we provide an explicit expression of $h(y; z)$ and an algorithm which involves only elementary operations. It uses Brion's identity along with an explicit description of the supporting cones at the vertices of $\Omega(y)$. It also has a simple equivalent formulation as a (finite) *group problem*. Finally, we exhibit *finitely many* fixed convex *cones* of $\mathbb{R}^n$, explicitly and exclusively defined from A, such that for *any* $y \in \mathbb{Z}^m$, the sum $h(y; z)$ is obtained by a simple formula which evaluates $\sum z^x$ over the integral points of those cones only.

On the dual side, we analyze the *counting* function h, via its generating function H in (3). Inverting H is difficult in general, except if an appropriate expansion of H into simple fractions is available, as in e.g. [13]. In their landmark paper [8], Brion and Vergne provided a *generalized residue* formula which yields the generic expansion

$$H(w) = \sum_{\sigma \in \mathbb{J}_A} \sum_{g \in G_\sigma} \widehat{Q}_{g,\sigma} \prod_{k \in \sigma} \frac{[z_k w^{A_k}]^{\delta_{k,\sigma}}}{1 - \rho_q^{g_k}[z_k w^{A_k}]^{1/q}}. \tag{4}$$

Here, $\sigma \in \mathbb{J}_A$ whenever A_σ is invertible, q is the smallest common multiple of all $|\det A_\sigma| \neq 0$, $\rho_q = e^{2\pi i/q}$ is the q-root of unity, $\delta_{k,\sigma} \in \{0, 1/q\}$, and $\widehat{Q}_{g,\sigma} \in \mathbb{C}$. The finite group G_σ has q^m elements. The coefficients $\widehat{Q}_{g,\sigma}$ are difficult to evaluate. Our contribution is to expand H in (3) in the form

$$H(w) = \sum_{\sigma \in \mathbb{J}_A} \left[\prod_{j \in \sigma} \frac{1}{1 - z_j w^{A_j}} \right] \times \frac{1}{R_2(\sigma; z)} \sum_{u_{\bar{\sigma}} \in \mathbb{Z}_{\mu_\sigma}^{n-m}} z^{\eta[\sigma, u_{\bar{\sigma}}]} w^{A\eta[\sigma, u_{\bar{\sigma}}]}, \tag{5}$$

where: $\mathbb{Z}_{\mu_\sigma} = \{0, 1, \ldots, \mu_\sigma - 1\}$, $\mu_\sigma = |\det A_\sigma|$, each $\eta[\sigma, u_{\bar{\sigma}}] \in \mathbb{Z}^n$ and:

$$z \mapsto R_2(\sigma; z) := \prod_{k \notin \sigma} \left[1 - \left(z_k z_\sigma^{-A_\sigma^{-1} A_k} \right)^{\mu_\sigma} \right]. \tag{6}$$

Identity (5) is a nontrivial simplification of the residue formula (4) because the $\eta[\sigma, u_{\bar{\sigma}}]$'s are given *explicitly*. And so the coefficients of the rational fraction (5) in w are very simple to evaluate with no root of unity involved (it can also be done symbolically); however this task can be tedious as for each $\sigma \in \mathbb{J}_A$ one has $|\det A_\sigma|^{n-m}$ terms $\eta[\sigma, u_{\bar{\sigma}}]$ to determine. But once determined, (5) is easy to invert and provides $h(y; z)$ for *any* $y \in \mathbb{Z}^m$.

2   Brion's Decomposition

2.1   Notation and Definitions

The notation $\mathbb{C}$, $\mathbb{R}$ and $\mathbb{Z}$ stand for the usual sets of complex, real and integer numbers, respectively. Moreover, the set of natural numbers $\{0, 1, 2, \ldots\}$ is denoted by $\mathbb{N}$, and for every natural number $\mu \in \mathbb{N}$, the finite set $\{0, 1, \ldots, \mu - 1\}$ of cardinality μ is denoted by $\mathbb{Z}_\mu$. The notation B' stands for the transpose of a matrix (or vector) $B \in \mathbb{R}^{s \times t}$; and the kth column of the matrix B is denoted by $B_k := (B_{1,k}, \ldots, B_{s,k})'$. When $y = 0$, the cone $\mathbf{\Omega}(0)$ in (1) is convex, and its *dual* cone is given by,

$$\mathbf{\Omega}(0)^* := \{b \in \mathbb{R}^n \,|\, b'x \geq 0 \text{ for every } x \in \mathbf{\Omega}(0)\}. \tag{7}$$

Notice that $\mathbf{\Omega}(0)^* \equiv \mathbb{R}^n$ if $\mathbf{\Omega}(0) = \{0\}$, which is the case if $\mathbf{\Omega}(y)$ is compact.

Definition 1. Let $A \in \mathbb{Z}^{m \times n}$ be of maximal rank. An ordered set $\sigma = \{\sigma_1, \ldots, \sigma_m\}$ of natural numbers is said to be a *basis* if it has cardinality $|\sigma| = m$, the sequence of inequalities $1 \leq \sigma_1 < \sigma_2 < \cdots < \sigma_m \leq n$ holds, and the square $[m \times m]$ submatrix :

$$A_\sigma := [A_{\sigma_1} | A_{\sigma_2} | \cdots | A_{\sigma_m}] \quad \text{is invertible.} \tag{8}$$

We denote the set of all bases σ by $\mathbb{J}_A$.

Definition 2. Given a maximal rank matrix $A \in \mathbb{Z}^{m \times n}$, and any basis $\sigma \in \mathbb{J}_A$, the complementary matrices $A_\sigma \in \mathbb{Z}^{m \times n}$ and $A_{\sigma} \in \mathbb{Z}^{m \times (n-m)}$ stand for $[A_k]_{k \in \sigma}$ and $[A_k]_{k \notin \sigma}$, respectively. Similarly, given $z \in \mathbb{C}^n$, the complementary vectors $z_\sigma \in \mathbb{C}^m$ and $z_{\sigma} \in \mathbb{C}^{n-m}$ stand for $(z_k)_{k \in \sigma}$ and $(z_k)_{k \notin \sigma}$, respectively.

For each basis $\sigma \in \mathbb{J}_A$ with associated matrix $A_\sigma \in \mathbb{Z}^{m \times m}$, introduce the *indicator* function $\delta_\sigma : \mathbb{Z}^m \to \mathbb{N}$ defined by :

$$y \mapsto \delta_\sigma(y) := \begin{cases} 1 \text{ if } A_\sigma^{-1} y \in \mathbb{Z}^m, \\ 0 \text{ otherwise.} \end{cases} \tag{9}$$

Notice that δ_σ is a multi-periodic function with periods A_σ and $\mu_\sigma := |\det A_\sigma|$, meaning that $\delta_\sigma(y + A_\sigma q) = \delta_\sigma(y + \mu_\sigma q) = \delta_\sigma(y)$ for all $y, q \in \mathbb{Z}^m$. Finally, given a triplet $(z, x, u) \in \mathbb{C}^n \times \mathbb{Z}^n \times \mathbb{R}^n$, introduce the notation :

$$\begin{aligned} z^x &:= z_1^{x_1} z_2^{x_2} \cdots z_s^{x_n}, \\ \|z\| &:= \max\{|z_1|, |z_2|, \ldots, |z_n|\}, \\ \ln\langle z \rangle &:= (\ln(z_1), \ln(z_2), \ldots, \ln(z_n)). \end{aligned} \tag{10}$$

Notice that $z^x = z_\sigma^{x_\sigma} z_{\sigma}^{x_{\sigma}}$, for all bases $\sigma \in \mathbb{J}_A$ and all $z \in \mathbb{C}^n$, $x \in \mathbb{Z}^n$.

2.2 Brion's Decomposition

Let $\Omega(y)$ be the convex polyhedron in (1) with $y \in \mathbb{Z}^m, A \in \mathbb{Z}^{m \times n}$ being of maximal rank, and let $h : \mathbb{Z}^m \to \mathbb{C}$ be the counting function in (2), with $\|z\| < 1$.

Obviously $h(y; z) = 0$ whenever the equation $Ax = y$ has no solution $x \in \mathbb{N}^n$. The main idea is to decompose the function h following Brion's ideas. Given any convex rational polyhedron $P \subset \mathbb{R}^n$, let $[P] : \mathbb{R}^n \to \{0, 1\}$ be its characteristic function, and $f[P] : \mathbb{C} \to \mathbb{C}$ its associated rational function, such that

$$z \mapsto f[P, z] := \sum_{x \in P \cap \mathbb{Z}^n} z^x, \tag{11}$$

holds whenever the sum converges absolutely. For every vertex V of P, define $\mathrm{Co}(P, V) \subset \mathbb{R}^n$ to be the supporting cone of P at V. Then, Brion's formula yields the decomposition :

$$[P] = \sum_{\text{vertices } V} [\mathrm{Co}(P, V)], \tag{12}$$

modulo the group generated by the characteristic functions of convex polyhedra which contain affine lines. And so,

$$f[P, z] = \sum_{\text{vertices } V} f[\mathrm{Co}(P, V), z]. \tag{13}$$

The above summation is *formal* because in general there is no $z \in \mathbb{C}^n$ for which the series

$$\sum\{z^x \mid x \in P \cap \mathbb{Z}^n\} \quad \text{and} \quad \sum\{z^x \mid x \in \mathrm{Co}(P, V) \cap \mathbb{Z}^n\}$$

converge absolutely for all vertices V. The notation $\sum E$ stands for the sum of all elements of a countable set $E \subset \mathbb{C}$. It is a complex number whenever the resulting series converges absolutely; otherwise it stands for a formal series.

Example: Let $P := [0,1] \subset \mathbb{R}$ so that $\mathrm{Co}(P, \{0\}) = [0, +\infty)$ and $\mathrm{Co}(P, \{1\}) = (-\infty, 1]$. Simple enumeration yields $f[P, z] = z^0 + z = 1 + z$, but one also has:

$$f[P, z] = f[\mathrm{Co}(P, \{0\}), z] + f[(P, \{1\}), z] = 1/(1 - z) + z^2/(z - 1) = 1 + z.$$

3 Computing $h(y; z)$: A Primal Approach

Let $C(\mathbb{J}_A) := \{Ax \mid x \in \mathbb{N}^n\} \subset \mathbb{R}^m$ be the cone generated by the columns of A, and for any basis $\sigma \in \mathbb{J}_A$, let $C(\sigma) \subset \mathbb{R}^m$ be the cone generated by the columns A_k with $k \in \sigma$. As A has maximal rank, $C(\mathbb{J}_A)$ is the union of all $C(\sigma)$, $\sigma \in \mathbb{J}_A$. With any $y \in C(\mathbb{J}_A)$ associate the intersection of all cones $C(\sigma)$ that contain y. This defines a subdivision of $C(\mathbb{J}_A)$ into polyhedral cones. The interiors of the maximal subdivisions are called *chambers*. In each chamber γ, the polyhedron $\Omega(y)$ is *simple*, i.e. $A_\sigma^{-1} y > 0$ for all $\sigma \in \mathbb{J}_A$ such that $A_\sigma^{-1} y \geq 0$.

For any chamber γ, define,

$$\mathcal{B}(\mathbb{J}_A, \gamma) := \{\sigma \in \mathbb{J}_A \mid \gamma \subset C(\sigma)\}. \tag{14}$$

The intersection of all $C(\sigma)$ with $\sigma \in \mathcal{B}(\mathbb{J}_A, \gamma)$ is the closure $\overline{\gamma}$ of γ.

Back to our original problem, and setting $P := \Omega(y)$, the rational function $f[P, z]$ is equal to $h(y; z)$ in (2) whenever $\|z\| < 1$. We next provide an explicit description of the rational function $f[\mathrm{Co}(P, V), z]$ for every vertex V of P.

Let δ_σ be the function defined in (9), and let $\mathbb{Z}_{\mu_\sigma} := \{0, 1, \dots, \mu_\sigma - 1\}$ with $\mu_\sigma := |\det A_\sigma|$. A vector $V \in \mathbb{R}^n$ is a vertex of $P = \Omega(y)$ if and only if there exists a basis $\sigma \in \mathbb{J}_A$ such that :

$$V_\sigma = A_\sigma^{-1} y \geq 0 \quad \text{and} \quad V_{\not\sigma} = 0, \tag{15}$$

where V_σ and $V_{\not\sigma}$ are given in Definition 2. Moreover, the supporting cone of P at the vertex V is described by :

$$\mathrm{Co}(\Omega(y), V) := \{x \in \mathbb{R}^n \mid Ax = y; \ x_k \geq 0 \text{ if } V_k = 0\}. \tag{16}$$

Let us now define the larger set

$$C(\Omega(y), \sigma) := \{x \in \mathbb{R}^n \mid A_\sigma x_\sigma + A_{\not\sigma} x_{\not\sigma} = y; \ x_{\not\sigma} \geq 0\}, \tag{17}$$

so that $\mathrm{Co}(\Omega(y), V)$ is a subcone of $C(\Omega(y), \sigma)$ for all bases $\sigma \in \mathbb{J}_A$ and vertex V of $\Omega(y)$ which satisfy $V_{\not\sigma} = 0$ (recall (15)). Besides, when $V_{\not\sigma} = 0$ and $y \in \gamma$ for some chamber γ, then $C(\Omega(y), \sigma)$ and $\mathrm{Co}(\Omega(y), V)$ are identical because $\Omega(y)$ is a simple polytope, and so $A_\sigma^{-1} y > 0$ for all $\sigma \in \mathbb{J}_A$.

Recall that $A_\sigma \in \mathbb{Z}^{m \times n}$ and $A_{\not\sigma} \in \mathbb{Z}^{m \times (n-m)}$ stand for $[A_k]_{k \in \sigma}$ and $[A_k]_{k \notin \sigma}$, respectively. Similarly, given a vector $x \in \mathbb{Z}^n$, the vectors x_σ and $x_{\not\sigma}$ stand for $(x_k)_{k \in \sigma}$ and $(x_k)_{k \notin \sigma}$ respectively. The following result is from [8, p. 818].

Proposition 1. Let $y \in \mathbb{R}^m$ and let $\boldsymbol{\Omega}(y)$ be as in (1), and let $y \in \overline{\gamma}$ for some chamber γ. Then,

$$[\boldsymbol{\Omega}(y)] = \sum_{\sigma \in \mathcal{B}(\mathbb{J}_A, \gamma)} [C(\boldsymbol{\Omega}(y), \sigma)], \tag{18}$$

modulo the group generated by the characteristic functions of convex polyhedra which contain affine lines.

Proof. Using notation of [8, p. 817], define the linear mapping $p : \mathbb{R}^n \to \mathbb{R}^m$ with $p(x) = Ax$, so that the polyhedra $P_\Delta(y)$ and $\boldsymbol{\Omega}(y)$ are identical. Moreover, for every basis $\sigma \in \mathcal{B}(\mathbb{J}_A, \gamma)$, $v_\sigma : \mathbb{R}^m \to \mathbb{R}^n$ is the linear mapping:

$$y \mapsto \quad [v_\sigma(y)]_\sigma = A_\sigma^{-1} y \quad \text{and} \quad [v_\sigma(y)]_{\sigma\!\!\!/} = 0, \quad y \in \mathbb{R}^m.$$

Finally, for every $\hat{x} \in \mathbb{R}^n$ with $\hat{x} \geq 0$, $\rho_\sigma(\hat{x}) := \hat{x} - v_\sigma(A\hat{x})$ satisfies,

$$[\rho_\sigma(\hat{x})]_\sigma = -A_\sigma^{-1} A_{\sigma\!\!\!/} \hat{x}_{\sigma\!\!\!/} \quad \text{and} \quad [\rho_\sigma(\hat{x})]_{\sigma\!\!\!/} = \hat{x}_{\sigma\!\!\!/}.$$

Therefore, the cone $[v_\sigma(y) + \rho_\sigma(C)]$ in [8] is the set of points $x \in \mathbb{R}^m$ such that $x_{\sigma\!\!\!/} \geq 0$ and $x_\sigma = A_\sigma^{-1}(y - A_{\sigma\!\!\!/} x_{\sigma\!\!\!/})$; and so this cone is just $[C(\boldsymbol{\Omega}(y), \sigma)]$ in (17). Therefore a direct application of (3.2.1) in [8, p. 818] yields (18). $\qed$

Theorem 1. Let $y \in \mathbb{Z}^m$, $z \in \mathbb{C}^n$ with $\|z\| < 1$, and let $y \in \overline{\gamma}$ for some chamber γ. Recall the set of bases $\mathcal{B}(\mathbb{J}_A, \gamma)$ defined in (14). With $P := \boldsymbol{\Omega}(y)$, the rational function h defined in (2) can be written:

$$h(y; z) = \sum_{\sigma \in \mathcal{B}(\mathbb{J}_A, \gamma)} f[C(\boldsymbol{\Omega}(y), \sigma), z] = \sum_{\sigma \in \mathcal{B}(\mathbb{J}_A, \gamma)} \frac{R_1(y, \sigma; z)}{R_2(\sigma; z)}, \tag{19}$$

$$\text{with} \quad z \mapsto R_1(y, \sigma; z) := z_\sigma^{A_\sigma^{-1} y} \sum_{u \in \mathbb{Z}_{\mu_\sigma}^{n-m}} \frac{\delta_\sigma(y - A_{\sigma\!\!\!/} u) z_{\sigma\!\!\!/}^u}{z_\sigma^{A_\sigma^{-1} A_{\sigma\!\!\!/} u}}, \tag{20}$$

$$\text{and} \quad z \mapsto R_2(\sigma; z) := \prod_{k \notin \sigma} \left[1 - \left(z_k z_\sigma^{-A_\sigma^{-1} A_k} \right)^{\mu_\sigma} \right]. \tag{21}$$

The pair $\{R_1, R_2\}$ is well defined whenever $z \in \mathbb{C}^n$ satisfies $z_k \neq 0$ and $z_k \neq z_\sigma^{A_\sigma^{-1} A_k}$ for every basis $\sigma \in \mathbb{J}_A$ which does not contain the index $k \notin \sigma$.

Proof. By a direct application of Brion's theorem to the sum (18), the associated rational functions $f[\boldsymbol{\Omega}(y), z]$ and $f[C(\boldsymbol{\Omega}(y), \sigma), z]$ satisfy:

$$h(y, z) = f[\Omega(y), z] = \sum_{\sigma \in \mathcal{B}(\mathbb{J}_A, \gamma)} f[C(\boldsymbol{\Omega}(y), \sigma), z]. \tag{22}$$

Therefore, in order to show (19), one only needs to prove that the rational function $\frac{R_1(y, \sigma; z)}{R_2(\sigma; z)}$ is equal to $f[C(\boldsymbol{\Omega}(y), \sigma), z]$, i.e.,

$$\frac{R_1(y, \sigma; z)}{R_2(\sigma; z)} = \sum \{ z^x \mid x \in C(\Omega(y), \sigma) \cap \mathbb{Z}^n \}, \tag{23}$$

on the domain $D_\sigma = \{z \in \mathbb{C}^n \,|\, 1 > |z_k z_\sigma^{-A_\sigma^{-1} A_k}|$ for each $k \notin \sigma\}$. Notice that

$$\frac{1}{R_2(\sigma; z)} = \prod_{k \notin \sigma} \frac{1}{1 - \left(z_k z_\sigma^{-A_\sigma^{-1} A_k}\right)^{\mu_\sigma}} =$$

$$= \prod_{k \notin \sigma} \sum_{v_k \in \mathbb{N}} \left[\frac{z_k}{z_\sigma^{A_\sigma^{-1} A_k}}\right]^{\mu_\sigma v_k} = \sum_{v \in \mathbb{N}^{n-m}} \frac{z_{\not\sigma}^{\mu_\sigma v}}{z_\sigma^{\mu_\sigma A_\sigma^{-1} A_{\not\sigma} v}},$$

on D_σ. On the other hand, according to (17), the integer vector $x \in \mathbb{Z}^n$ lies inside the cone $C(P, V_\sigma)$ if and only if :

$$x_\sigma = A_\sigma^{-1}(y - A_{\not\sigma} x_{\not\sigma}), \quad \delta_\sigma(y - A_{\not\sigma} x_{\not\sigma}) = 1 \quad \text{and}$$
$$x_{\not\sigma} = u + \mu_\sigma v, \quad \text{with} \quad u \in \mathbb{Z}_{\mu_\sigma}^{n-m} \quad \text{and} \quad v \in \mathbb{N}^{n-m}.$$

From the definition (20) of $R_1(y, \sigma; z)$ and $z^x = z_{\not\sigma}^{x_{\not\sigma}} z_\sigma^{x_\sigma} = z_{\not\sigma}^{x_{\not\sigma}} z_\sigma^{A_\sigma^{-1}(y - A_{\not\sigma} x_{\not\sigma})}$,

$$\frac{R_1(y, \sigma; z)}{R_2(\sigma; z)} = z_\sigma^{A_\sigma^{-1} y} \sum_{u \in \mathbb{Z}_{\mu_\sigma}^{n-m}} \sum_{v \in \mathbb{N}^{n-m}} \frac{\delta_\sigma(y - A_{\not\sigma} u) \, z_{\not\sigma}^{x_{\not\sigma}}}{z_\sigma^{A_\sigma^{-1} A_{\not\sigma} x_{\not\sigma}}}, \qquad (24)$$

$$= \sum \{z^x \,|\, x \in C(\mathbf{\Omega}(y), \sigma) \cap \mathbb{Z}^n\} = f[C(\mathbf{\Omega}(y)\sigma), z],$$

which is exactly (23). Notice that $x_{\not\sigma} = u + \mu_\sigma v$, and so $\delta_\sigma(y - A_{\not\sigma} u) = \delta_\sigma(y - A_{\not\sigma} x_{\not\sigma})$ because of the definition (9) of δ_σ. Finally, using (24) in (22) yields that (19) holds whenever $\|z\| < 1$ and $R_1(y, \sigma; z)$ and $R_2(\sigma; z)$ are all well defined.

Notice that R_2 is constant with respect to y, and from the definition (9) of δ_σ, R_1 is *quasiperiodic* with periods A_σ and μ_σ, meaning that

$$R_1(y + A_\sigma q, \sigma; z) = R_1(y, \sigma; z) \, z_\sigma^q \quad \text{and}$$
$$R_1(y + \mu_\sigma q, \sigma; z) = R_1(y, \sigma; z) \left(z_\sigma^{A_\sigma^{-1} q}\right)^{\mu_\sigma} \qquad (25)$$

hold for all $y, q \in \mathbb{Z}^m$. Obviously, the more expensive part in calculating $R_2(\cdot)$ in (21) is to compute the determinant $\mu_\sigma = |\det A_\sigma|$. On the other hand, computing $R_1(\cdot)$ in (20) may become quite expensive when μ_σ is large, as one must evaluate μ_σ^{n-m} terms, the cardinality of $\mathbb{Z}_{\mu_\sigma}^{n-m}$. However, as detailed below, a more careful analysis of (20) yields some simplifications.

3.1 Simplifications Via Group Theory

From the proof of Theorem 1, the closed forms (20)–(21) for $R_1(\cdot)$ and $R_2(\cdot)$ are deduced from (24), i.e.,

$$\frac{R_1(y, \sigma; z)}{R_2(\sigma; z)} = z_\sigma^{A_\sigma^{-1} y} \sum_{x_{\not\sigma} \in \mathbb{Z}^{n-m}} \frac{\delta_\sigma(y - A_{\not\sigma} x_{\not\sigma}) \, z_{\not\sigma}^{x_{\not\sigma}}}{z_\sigma^{A_\sigma^{-1} A_{\not\sigma} x_{\not\sigma}}},$$

after setting $x_\mathscr{J} = u + \mu_\sigma v$ and recalling that $\delta_\sigma(y)$ is a periodic function, i.e., $\delta_\sigma(y + \mu_\sigma q) = \delta_\sigma(y)$ for all $y, q \in \mathbb{Z}^m$. However, we have not used yet that $\delta_\sigma(y + A_\sigma q) = \delta_\sigma(y)$ as well. For every $\sigma \in \mathbb{J}_A$, consider the lattice :

$$\Lambda_\sigma := \bigoplus_{j \in \sigma} A_j \mathbb{Z} \subset \mathbb{Z}^m, \tag{26}$$

generated by the columns A_j, $j \in \sigma$. The following quotient group

$$\mathcal{G}_\sigma := \mathbb{Z}^m / \Lambda_\sigma = \mathbb{Z}^m \Big/ \bigoplus_{j \in \sigma} A_j \mathbb{Z} \tag{27}$$

$$= \{\mathrm{Ec}[0, \sigma], \mathrm{Ec}[2, \sigma], \dots, \mathrm{Ec}[\mu_\sigma - 1, \sigma]\}$$

is commutative, with $\mu_\sigma = |\det A_\sigma|$ elements (or, equivalence classes) $\mathrm{Ec}[j, \sigma]$, and so, $\mathcal{G}_\sigma$ is isomorphic to a finite Cartesian product of cyclic groups $\mathbb{Z}_{\eta_k}$, i.e.,

$$\mathcal{G}_\sigma \cong \mathbb{Z}_{\eta_1} \times \mathbb{Z}_{\eta_2} \times \cdots \times \mathbb{Z}_{\eta_s}.$$

Obviously, $\mu_\sigma = \eta_1 \eta_2 \cdots \eta_s$, and so, $\mathcal{G}_\sigma$ is isomorphic to the cyclic group $\mathbb{Z}_{\mu_\sigma}$ whenever μ_σ is a prime number. Actually, $\mathcal{G}_\sigma = \{0\}$ whenever $\mu_\sigma = 1$. Notice that the Cartesian product $\mathbb{Z}_{\eta_1} \times \cdots \times \mathbb{Z}_{\eta_s}$ can be seen as the integer space $\mathbb{Z}^s$ modulo the vector $\eta := (\eta_1, \eta_2, \cdots, \eta_s)' \in \mathbb{N}^s$.

Hence, for every finite commutative group $\mathcal{G}_\sigma$, there exist a positive integer $s_\sigma \geq 1$, a vector $\eta_\sigma \in \mathbb{N}^{s_\sigma}$ with positive entries, and a group isomorphism,

$$g_\sigma : \mathcal{G}_\sigma \to \mathbb{Z}^{s_\sigma} \bmod \eta_\sigma, \tag{28}$$

where $g_\sigma(\xi) \bmod \eta_\sigma$ means evaluating $[g_\sigma(\xi)]_k \bmod [\eta_\sigma]_k$, for all indices $1 \leq k \leq s_\sigma$. For every $y \in \mathbb{Z}^m$, there exists a unique equivalence class $\mathrm{Ec}[j_y, \sigma]$ which contains y, and so we can define the following group epimorphism,

$$\hat{h}_\sigma : \mathbb{Z}^m \to \mathbb{Z}^{s_\sigma} \bmod \eta_\sigma, \tag{29}$$

$$y \mapsto \hat{h}_\sigma(y) := g_\sigma(\mathrm{Ec}[j_y, \sigma]).$$

On the other hand, the unit element of $\mathcal{G}_\sigma$ is the equivalence class $\mathrm{Ec}[0, \sigma]$ which contains the origin, that is, $\mathrm{Ec}[0, \sigma] = \{A_\sigma q \mid q \in \mathbb{Z}^m\}$.

Hence, $\hat{h}_\sigma(y) = 0$ if and only if there exists $q \in \mathbb{Z}^m$ such that $y = A_\sigma q$. We can then redefine the function δ_σ as follows,

$$y \mapsto \delta_\sigma(y) := \begin{cases} 1 \text{ if } \hat{h}_\sigma(y) = 0, \\ 0 \text{ otherwise,} \end{cases} \tag{30}$$

One also needs the following additional notation; given any matrix $B \in \mathbb{Z}^{m \times t}$,

$$\hat{h}_\sigma(B) := [\hat{h}_\sigma(B_1) | \hat{h}_\sigma(B_2) | \cdots | \hat{h}_\sigma(B_t)] \in \mathbb{Z}^{s_\sigma \times t}. \tag{31}$$

And so, from (20), $\hat{h}_\sigma(y - A_\mathscr{J} u) \equiv \hat{h}_\sigma(y) - \hat{h}_\sigma(A_\mathscr{J}) u \bmod \eta_\sigma$. Finally, using (30) in (20), one obtains a simplified version of $R_1(\cdot)$ in the form:

$$R_1(y, \sigma; z) = \sum \left\{ \frac{z_\sigma^{A_\sigma^{-1} y} z_\mathscr{J}^u}{z_\sigma^{A_\sigma^{-1} A_\mathscr{J} u}} \;\middle|\; \begin{array}{l} u \in \mathbb{Z}_{\mu_\sigma}^{n-m}; \\ \hat{h}_\sigma(y) \equiv \hat{h}_\sigma(A_\mathscr{J}) u \bmod \eta_\sigma \end{array} \right\}. \tag{32}$$

Next, with $q \in \mathbb{Z}^m$ fixed, $\nu_q A_\sigma^{-1} q \in \mathbb{Z}^m$ for some integer ν_q, if and only if $\nu_q \hat{h}_\sigma(q) = 0 \bmod \eta_\sigma$. If we set $\nu_q = \mu_\sigma$, then $\mu_\sigma A_\sigma^{-1} q \in \mathbb{Z}^m$, and $\mu_\sigma \hat{h}_\sigma(q) = 0 \bmod \eta_\sigma$, because $\mathcal{G}_\sigma$ has $\mu_\sigma = |\det A_\sigma|$ elements. Nevertheless, μ_σ may not be the smallest positive integer with that property. So, given $\sigma \in \mathbb{J}_A$ and $k \notin \sigma$, define $\nu_{k,\sigma} \geq 1$ to be $order$ of $\hat{h}_\sigma(A_k)$. That is, $\nu_{k,\sigma}$ is the smallest positive integer such that $\nu_{k,\sigma} \hat{h}_\sigma(A_k) = 0 \bmod \eta_\sigma$, or equivalently :

$$\nu_{k,\sigma} A_\sigma^{-1} A_k \in \mathbb{Z}^m. \tag{33}$$

Obviously $\nu_{k,\sigma} \leq \mu_\sigma$. Moreover, μ_σ is a multiple of $\nu_{k,\sigma}$ for it is the order of an element in $\mathcal{G}_\sigma$. For example, the group $\mathbb{Z}^2$ modulo $\eta = \binom{2}{7}$ has 14 elements; and the elements $b_1 = \binom{1}{0}$, $b_2 = \binom{0}{1}$ and $b_3 = \binom{1}{1}$ have respective orders : 2, 7 and 14. Notice that, $2b_1 \equiv 7b_2 \equiv 14b_3 \equiv 0 \bmod \eta$. But, $2b_3 \equiv 2b_2 \not\equiv 0$ and $7b_3 \equiv b_1 \not\equiv 0 \bmod \eta$.

The important observation is that $\delta_\sigma(y - \nu_{k,\sigma} A_k q) = \delta_\sigma(y)$ for all $q \in \mathbb{Z}^m$ and $k \notin \sigma$, which follows from (33) and (9). Thus, following step by step the proof of Theorem 1, we obtain:

Corollary 1. Let $y \in \mathbb{Z}^m$, $z \in \mathbb{C}^n$ with $\|z\| < 1$, and let $y \in \bar{\gamma}$ for some chamber γ. Recall the set of bases $\mathcal{B}(\mathbb{J}_A, \gamma)$ defined in (14). With $\sigma \in \mathcal{B}(\mathbb{J}_A, \gamma)$, let R_1 and R_2 be as in Theorem 1. Then

$$\frac{R_1(y, \sigma; z)}{R_2(y; z)} = \frac{R_1^*(y, \sigma; z)}{R_2^*(y; z)}, \tag{34}$$

$$where : \qquad R_2^*(\sigma; z) := \prod_{k \notin \sigma} \left[1 - \left(z_k z_\sigma^{-A_\sigma^{-1} A_k} \right)^{\nu_{k,\sigma}} \right], \tag{35}$$

$$R_1^*(y, \sigma; z) := z_\sigma^{A_\sigma^{-1} y} \sum_{u_\sigma \in U_\sigma} \frac{\delta_\sigma(y - A_\sigma u_\sigma) z_\sigma^{u_\sigma}}{z_\sigma^{A_\sigma^{-1} A_\sigma u_\sigma}} = \tag{36}$$

$$= \sum \left\{ \frac{z_\sigma^{A_\sigma^{-1} y} z_\sigma^{u_\sigma}}{z_\sigma^{A_\sigma^{-1} A_\sigma u_\sigma}} \;\middle|\; \begin{array}{l} u_\sigma \in U_\sigma; \\ \hat{h}_\sigma(y) \equiv \hat{h}_\sigma(A_\sigma) u_\sigma \bmod \eta_\sigma \end{array} \right\},$$

with $U_\sigma := \{ u_\sigma \in \mathbb{N}^{n-m} \mid 0 \leq u_k \leq \nu_{k,\sigma} - 1; \ k \notin \sigma \}$.

One can also obtain (34) by noticing that:

$$\frac{R_1(y, \sigma; z)}{R_1^*(y, \sigma; z)} = \frac{R_2(\sigma; z)}{R_2^*(\sigma; z)} = \prod_{k \notin \sigma} \left(1 + \beta^{\nu_{k,\sigma}} + \cdots + \beta^{\mu_\sigma - \nu_{k,\sigma}} \right),$$

where $\beta_{k,\sigma} = z_k z_\sigma^{-A_\sigma^{-1} A_k}$, and μ_σ is a multiple of $\nu_{k,\sigma}$.

3.2   Simplifications Via Finite Number of Generators

Decompose $\mathbb{Z}^m$ into $\mu_\sigma := |\det A_\sigma|$ disjoint equivalent classes, where $y, \xi \in \mathbb{Z}^m$ are equivalent if and only if $\delta_\sigma(y - \xi) = 1$. For every basis $\sigma \in \mathbb{J}_A$, let $\mathcal{G}_\sigma$ be the quotient group defined in (27), that is,

$$\mathcal{G}_\sigma = \mathbb{Z}^m \Big/ \bigoplus_{j \in \sigma} A_j \mathbb{Z} = \{\mathrm{Ec}[0, \sigma], \dots, \mathrm{Ec}[\mu_\sigma - 1, \sigma]\}.$$

Notice that $y, \xi \in \mathbb{Z}^n$ belong to $\mathrm{Ec}[j, \sigma]$ if and only if $A_\sigma^{-1}(y - \xi) \in \mathbb{Z}^n$, and that $\mathbb{Z}^m$ is equal to the disjoint union of all classes $\mathrm{Ec}[j, \sigma]$.

Next, pick up a *minimal* representative element of every class, i.e., fix

$$\xi[j, \sigma] \in \mathrm{Ec}[j, \sigma] \quad \text{such that} \quad A_\sigma^{-1} y \geq A_\sigma^{-1} \xi[j, \sigma] \geq 0, \tag{37}$$

for every $y \in \mathrm{Ec}[j, \sigma]$ with $A_\sigma^{-1} y \geq 0$. The minimal representative elements $\xi[j, \sigma]$ in (37) can be computed as follows: Let $d \in \mathrm{Ec}[j, \sigma]$, arbitrary, and let $d^* \in \mathbb{Z}^m$ be such that his k-entry d_k^* is the smallest integer greater than or equal to the k-entry of $-A_\sigma^{-1} d$. The vector $\xi[j, \sigma]$ defined by $d + A_\sigma d^*$ satisfies (37).

Notice that $d^* + A_\sigma^{-1} d \geq 0$. Besides, let $d, y \in \mathrm{Ec}[j, \sigma]$ with $A_\sigma^{-1} y \geq 0$. There exists $q \in \mathbb{Z}^m$ such that $y = d + A_\sigma q$. Hence $q \geq -A_\sigma^{-1} d$; in addition, $q \geq d^*$ follows from the above definition of d^*, and so $A_\sigma^{-1} y \geq d^* + A_\sigma^{-1} d \geq 0$.

Therefore, the vector $\xi[j, \sigma] := d + A_\sigma d^*$ satisfies (37). In particular, if $\mathrm{Ec}[0, \sigma]$ is the class which contains the origin of $\mathbb{Z}^m$, then $\xi[0, \sigma] = 0$. Notice that for every integer vector $y \in \mathbb{Z}^m$, there exists a unique $\xi[j, \sigma]$ such that :

$$y = \xi[j, \sigma] + A_\sigma q, \quad \text{for} \quad q \in \mathbb{Z}^m.$$

Moreover, the extra condition $A_\sigma^{-1} y \geq 0$ holds if and only if:

$$y = \xi[j, \sigma] + A_\sigma q \quad \text{with} \quad q \in \mathbb{N}^m. \tag{38}$$

We obtain a compact form of $h(y; z)$ when $y \in \mathbb{Z}^m \cap \overline{\gamma}$, for some chamber γ.

Theorem 2. *Let h and $\xi[j, \sigma]$ be as in (2) and (37), respectively. Let $y \in \mathbb{Z}^m \cap \overline{\gamma}$, for some chamber γ. Recall the set of bases $\mathcal{B}(\mathbb{J}_A, \gamma)$ defined in (14). For every basis $\sigma \in \mathcal{B}(\Delta, \gamma)$ there is a unique index $0 \leq \jmath[y, \sigma] < \mu_\sigma$ such that y is contained in the equivalence class $\mathrm{Ec}[\jmath[y, \sigma], \sigma]$ defined in (27), and so:*

$$h(y; z) = \sum_{\sigma \in \mathcal{B}(\Delta, \gamma)} \frac{R_1(\xi[\jmath[y, \sigma], \sigma], \sigma; z)}{R_2(\sigma; z)} z_\sigma^{\lfloor A_\sigma^{-1} y \rfloor}, \tag{39}$$

where $\lfloor A_\sigma^{-1} y \rfloor \in \mathbb{Z}^m$ is such that his k-entry is the largest integer less than or equal to the k-entry of $A_\sigma^{-1} y$.

Proof. Recall that if $y \in \mathbb{Z}^m \cap \overline{\gamma}$

$$h(y; z) = \sum_{\sigma \in \mathcal{B}(\Delta, \gamma)} \frac{R_1(y, \sigma; z)}{R_2(y; z)}$$

Next, recalling the definition (14) of $\mathcal{B}(\mathbb{J}_A, \gamma)$, $A_\sigma^{-1} y \geq 0$ for every basis $\sigma \in \mathcal{B}(\mathbb{J}_A, \gamma)$ with $y \in \overline{\gamma}$. Recall that there is a unique index $\jmath[y, \sigma] < \mu_\sigma$ such that $y = \xi[\jmath[y, \sigma], \sigma] + A_\sigma q$ with $q \in \mathbb{N}^m$; see (38) and the comment just before.

To obtain the vector $q \in \mathbb{N}^m$, recall that the minimal representative element $\xi[\jmath[y, \sigma], \sigma]$ in (37) is the sum $y + A_\sigma y^*$ where $y^* \in \mathbb{Z}^m$ is such that his k-entry y_k^* is the smallest integer greater than or equal to $-A_\sigma^{-1} y$, for we only need to fix $d = y$ in the paragraph that follows (37). In particular, $\lfloor A_\sigma^{-1} y \rfloor = -y^*$, and $\xi[\jmath[y, \sigma], \sigma] = y - A_\sigma \lfloor A_\sigma^{-1} y \rfloor$, which when used in (20) and (25), yields,

$$R_1(y, \sigma; z) = R_1(\xi[\jmath[y, \sigma], \sigma], \sigma; z) \, z_\sigma^{\lfloor A_\sigma^{-1} y \rfloor}.$$

And so (19) implies (39).

Theorem 2 explicitly shows that it suffices to compute $R_1(v, \sigma; z)$ for finitely many values $v = \xi[j, \sigma]$, with $\sigma \in \mathcal{B}(\Delta, \gamma)$ and $0 \leq j < \mu_\sigma$, in order to calculate $h(y; z)$ for arbitrary values $y \in \mathbb{Z}^m \cap \overline{\gamma}$, via (39).

In other words, in the closure $\overline{\gamma}$ of a chamber γ, one only needs to consider *finitely many* fixed convex cones $C(\Omega(\xi[j, \sigma]), \sigma) \subset \mathbb{R}^n$, where $\sigma \in \mathcal{B}(\Delta, \gamma)$ and $0 \leq j < \mu_\sigma$, and compute their associated rational function (39). The counting function $h(y; z)$ is then obtained as follows.

Input: $y \in \mathbb{Z}^m \cap \overline{\gamma}$, $z \in \mathbb{C}^n$.
Output $\rho = h(y; z)$.
Set $\rho := 0$. For every $\sigma \in \mathcal{B}(\Delta, \gamma)$:
- Compute $\xi[\jmath[y, \sigma], \sigma] := y - A_\sigma \lfloor A_\sigma^{-1} y \rfloor \in \mathbb{Z}^m$.
- Read the value $R_1(\xi[\jmath[y, \sigma], \sigma], \sigma; z)/R_2(\sigma; z)$, and update ρ by:

$$\rho := \rho + \frac{R_1(\xi[\jmath[y, \sigma], \sigma], \sigma; z)}{R_2(\sigma; z)} \, z_\sigma^{\lfloor A_\sigma^{-1} y \rfloor}.$$

For the whole space $\mathbb{Z}^m$ it suffices to consider *all* chambers γ and all cones $C(\Omega(\xi[j, \sigma]), \sigma) \subset \mathbb{R}^n$, where $\sigma \in \mathcal{B}(\Delta, \gamma)$ and $0 \leq j < \mu_\sigma$.

Finally, in view of (20)-(21), the above algorithm can be symbolic, i.e., $z \in \mathbb{C}^m$ can be treated symbolically, and ρ becomes a rational fraction of z.

4   Generating Function

An appropriate tool for computing the exact value of $h(y; z)$ in (2) is the formal generating function $H : \mathbb{C}^m \to \mathbb{C}$,

$$s \mapsto H(s) := \sum_{y \in \mathbb{Z}^m} h(y; z) \, s^y = \prod_{k=1}^n \frac{1}{1 - z_k s^{A_k}}, \qquad (40)$$

where s^y is defined in (10) and the sum is understood as a formal power series, so that we need not consider conditions for convergence. This generating function was already considered in Brion and Vergne [8] with $\lambda = \ln\langle s \rangle$.

Following notation of [8, p. 805], let $0 \le \hat{x} \in \mathbb{R}^n$ be a *regular* vector, i.e., no entry $[A_\sigma^{-1} A\hat{x}]_j$ vanishes for any basis $\sigma \in \mathbb{J}_A$ or index $1 \le j \le m$. Define :

$$\varepsilon_{j,\sigma} := \begin{cases} 1 & \text{if } [A_\sigma^{-1} A\hat{x}]_j > 0, \\ -1 & \text{if } [A_\sigma^{-1} A\hat{x}]_j < 0. \end{cases} \tag{41}$$

Next, for every basis $\sigma \in \mathbb{J}_A$, index $j \in \sigma$ and vector $u_{\not\sigma} \in \mathbb{Z}^{n-m}$, fix :

$$\theta[j, \sigma, u_{\not\sigma}] \in \mathbb{Z} : \quad \text{the smallest integer greater than or equal to} \quad -\varepsilon_{j,\sigma} [A_\sigma^{-1} A_{\not\sigma} u_{\not\sigma}]_j. \tag{42}$$

Define also the vector $\eta[\sigma, u_{\not\sigma}] \in \mathbb{Z}^n$ by :

$$\eta[\sigma, u_{\not\sigma}]_j = \begin{cases} u_j & \text{if } j \notin \sigma; \\ \theta[j, \sigma, u_{\not\sigma}] & \text{if } j \in \sigma, \ \varepsilon_{j,\sigma} = 1; \\ 1 - \theta[j, \sigma, u_{\not\sigma}] & \text{if } j \in \sigma, \ \varepsilon_{j,\sigma} = -1. \end{cases} \tag{43}$$

The following expansion can be deduced from [8].

Theorem 3. *Let $0 \le \hat{x} \in \mathbb{R}^n$ be regular and consider the vectors $\eta[\sigma, u_{\not\sigma}] \in \mathbb{Z}^n$ defined in (43) for $\sigma \in \mathbb{J}_A$ and $u_{\not\sigma} \in \mathbb{Z}^{n-m}$. The following expansion holds:*

$$\prod_{k=1}^{n} \frac{1}{1 - z_k s^{A_k}} = \sum_{\sigma \in \mathbb{J}_A} \left[\prod_{j \in \sigma} \frac{1}{1 - z_j s^{A_j}} \right] \times \tag{44}$$

$$\times \frac{1}{R_2(\sigma; z)} \sum_{u_{\not\sigma} \in \mathbb{Z}_{\mu_\sigma}^{n-m}} z^{\eta[\sigma, u_{\not\sigma}]} s^{A\eta[\sigma, u_{\not\sigma}]},$$

where $\mathbb{Z}_{\mu_\sigma} = \{0, 1, \ldots, \mu_\sigma - 1\}$, $\mu_\sigma = |\det A_\sigma|$ and:

$$z \mapsto R_2(\sigma; z) := \prod_{k \notin \sigma} \left[1 - \left(z_k z_\sigma^{-A_\sigma^{-1} A_k} \right)^{\mu_\sigma} \right]. \tag{45}$$

Proof. From Brion and Vergne's identity [8, p. 813],

$$\prod_{j=1}^{n} \frac{1}{1 - e^{w_k}} = \sum_{\sigma \in \mathbb{J}_A} \left[\prod_{j \in \sigma} \varepsilon_{j,\sigma} \right] F(C_{\hat{x}}^\sigma + \rho_\sigma(C), L), \tag{46}$$

where $F(C_{\hat{x}}^\sigma + \rho_\sigma(C), L)$ is the formal power series $\sum_l e^l$ added over all elements l in the intersection of the cone $C_{\hat{x}}^\sigma + \rho_\sigma(C)$ with the integer lattice $L = \mathbb{Z}[w_1, \ldots, w_n]$. Moreover, the coefficients $\varepsilon_{j,\sigma}$ are defined in (41) and the cone $C_{\hat{x}}^\sigma$ is defined by the following formula [8, p. 805],

$$C_{\hat{x}}^\sigma = \left\{ \sum_{j \in \sigma} \varepsilon_{j,\sigma} x_j w_j \ \middle|\ x_\sigma \in \mathbb{R}^m, \ x_\sigma \ge 0 \right\}. \tag{47}$$

Finally, given the real vector space $W = \mathbb{R}[w_1, \ldots, w_n]$, every $\rho_\sigma : W \to W$ is a linear mapping defined by its action on each basis element w_k of W,

$$\rho_\sigma(w_k) := w_k - \sum_{j \in \sigma} [A_\sigma^{-1} A_k]_j w_j. \tag{48}$$

Hence, $\rho_\sigma(w_j) = 0$ for every $j \in \sigma$, and the cone $\rho_\sigma(C)$ is given by

$$\rho_\sigma(C) = \left\{ \sum_{k \notin \sigma} x_k w_k - \sum_{j \in \sigma} [A_\sigma^{-1} A_{\sigma'} x_{\sigma'}]_j w_j \;\middle|\; \begin{matrix} x_{\sigma'} \in \mathbb{R}^{n-m}, \\ x_{\sigma'} \ge 0 \end{matrix} \right\}; \tag{49}$$

see [8, p.805]. Thus, every element in the intersection of the cone $C_{\hat{x}}^\sigma + \rho_\sigma(C)$ with the lattice $\mathbb{Z}[w_1, \ldots, w_n]$ must be of the form :

$$\sum_{k \notin \sigma} x_k w_k + \sum_{j \in \sigma} \varepsilon_{j,\sigma} \, \xi_j \, w_j, \quad \text{with} \quad x_{\sigma'} \in \mathbb{N}^{n-m}, \tag{50}$$

$$\xi_\sigma \in \mathbb{Z}^m \quad \text{and} \quad \xi_j \ge -\varepsilon_{j,\sigma}[A_\sigma^{-1} A_{\sigma'} x_{\sigma'}]_j.$$

On the other hand, for every basis σ, define $\mu_\sigma = |\det A_\sigma|$ and :

$$x_{\sigma'} = u_{\sigma'} + \mu_\sigma v_{\sigma'}, \quad \text{with} \quad u_{\sigma'} \in \mathbb{Z}_{\mu_\sigma}^{n-m} \quad \text{and} \quad v_{\sigma'} \in \mathbb{N}^{n-m}. \tag{51}$$

Moreover, as in (42), fix $\theta[j, \sigma, u_{\sigma'}] \in \mathbb{Z}$ to be the smallest integer greater than or equal to $-\varepsilon_{j,\sigma}[A_\sigma^{-1} A_{\sigma'} u_{\sigma'}]_j$. Thus, we can rewrite (50) so that the intersection of the cone $C_{\hat{x}}^\sigma + \rho_\sigma(C)$ with the lattice $\mathbb{Z}[w_1, \ldots, w_n]$ must be of the form :

$$\sum_{k \notin \sigma} [u_k w_k + v_k \mu_\sigma \rho(w_k)] + \sum_{j \in \sigma} \varepsilon_{j,\sigma} w_j [\theta[j, \sigma, u_{\sigma'}]_j + q_j], \tag{52}$$

$$\text{with} \quad u_{\sigma'} \in \mathbb{Z}_{\mu_\sigma}^{n-m}, \quad v_{\sigma'} \in \mathbb{N}^{n-m} \quad \text{and} \quad q_\sigma \in \mathbb{N}^m.$$

We can deduce (52) from (50) by recalling the definition (48) of $\rho_\sigma(w_k)$ and letting :

$$\xi_j := \theta[j, \sigma, u_{\sigma'}] + q_j - \varepsilon_{j,\sigma} \mu_\sigma [A_\sigma^{-1} A_{\sigma'} v_{\sigma'}]_j.$$

Since $F(C_{\hat{x}}^\sigma + \rho_\sigma(C), L)$ is the formal power series $\sum_l e^l$ with summation over all elements l in (52), one obtains

$$F(C_{\hat{x}}^\sigma + \rho_\sigma(C), L) = \tag{53}$$

$$\sum_{u_{\sigma'} \in \mathbb{Z}_{\mu_\sigma}^{n-m}} \left[\prod_{j \in \sigma} \frac{e^{\varepsilon_{j,\sigma} \theta[j, \sigma, u_{\sigma'}] w_j}}{1 - e^{\varepsilon_{j,\sigma} w_j}} \right] \left[\prod_{k \notin \sigma} \frac{e^{u_k w_k}}{1 - e^{\mu_\sigma \rho_\sigma(w_k)}} \right].$$

With $\eta[\sigma, u_{\sigma'}] \in \mathbb{Z}^n$ being as in (43), using (53) into (46) yields the expansion

$$\prod_{j=1}^n \frac{1}{1 - e^{w_k}} = \sum_{\sigma \in \mathbb{J}_A} \sum_{u_{\sigma'} \in \mathbb{Z}_{\mu_\sigma}^{n-m}} \left[\prod_{j \in \sigma} \frac{1}{1 - e^{w_j}} \right] \times \tag{54}$$

$$\times \left[\prod_{j=1}^n e^{\eta[\sigma, u_{\sigma'}]_j w_j} \right] \left[\prod_{k \notin \sigma} \frac{1}{1 - e^{\mu_\sigma \rho_\sigma(w_k)}} \right].$$

Finally, we defined $w_k := \ln(z_k) + \ln\langle s \rangle A_k$ for every index $1 \le k \le n$, where the vectors $s, z \in \mathbb{C}^n$ have all their entries different from zero and $\ln\langle s \rangle$ is the

$[1 \times n]$ matrix defined in (10). So $e^{w_k} = z_k s^{A_k}$. Moreover, recalling the definition (48) of $\rho_\sigma(w_k)$, the following identities hold for all $1 \leq k \leq n$,

$$\rho_\sigma(w_k) = \ln(z_k) - \sum_{j \in \sigma} \ln(z_j)[A_\sigma^{-1} A_k]_j. \tag{55}$$

Notice $\sum_{j \in \sigma} A_j [A_\sigma^{-1} A_k]_j = A_k$. A direct application of (55) and the identities $e^{w_k} = z_k s^{A_k}$ yields (44), i.e.:

$$\prod_{k=1}^n \frac{1}{1 - z_k s^{A_k}} = \sum_{\sigma \in \mathbb{J}_A} \sum_{u_\sigma \in \mathbb{Z}_{\mu_\sigma}^{n-m}} \frac{z^{\eta[\sigma, u_\sigma]} s^{A\eta[\sigma, u_\sigma]}}{R_2(\sigma; z)} \prod_{j \in \sigma} \frac{1}{1 - z_j s^{A_j}},$$

$$\text{with} \quad R_2(\sigma; z) = \prod_{k \notin \sigma} \left[1 - \left(z_k z_\sigma^{-A_\sigma^{-1} A_k}\right)^{\mu_\sigma}\right].$$

A direct expansion of (44) yields the following:

Theorem 4. *Let $0 \leq \hat{x} \in \mathbb{R}^n$ be regular, and let h and η be as in (2) and (43), respectively. Let $\mathbb{J}_A$ be the set of bases associated with A. Then for every pair of $(y, z) \in \mathbb{Z}^m \times \mathbb{C}^n$ with $\|z\| < 1$:*

$$h(y; z) = \sum_{\sigma \in \mathbb{J}_A, \, A_\sigma^{-1} y \geq 0} \frac{z_\sigma^{A_\sigma^{-1} y}}{R_2(\sigma; z)} \sum_{u \in \mathbb{Z}_{\mu_\sigma}^{n-m}} \frac{z_\sigma^u}{z_\sigma^{A_\sigma^{-1} A_\sigma u}} \times \tag{56}$$

$$\times \begin{cases} 1 \text{ if } A_\sigma^{-1}(y - A\eta[\sigma, u]) \in \mathbb{N}^m, \\ 0 \text{ otherwise}, \end{cases}$$

where:

$$\mathbb{Z}_{\mu_\sigma} = \{0, 1, \ldots, \mu_\sigma - 1\}, \, \mu_\sigma = |\det A_\sigma|,$$

$$0 \leq [A_\sigma^{-1} A\eta[\sigma, u]]_j \leq 1 \quad \text{for each} \quad j \in \sigma \tag{57}$$

$$\text{and} \quad R_2(\sigma; z) := \prod_{k \notin \sigma} \left[1 - \left(z_k z_\sigma^{-A_\sigma^{-1} A_k}\right)^{\mu_\sigma}\right]. \tag{58}$$

The proof is based on arguments similar to those developed in [13].

Observe that (56) is different from (19) or (34) because in (19) and (34) the summation is over bases σ in the subset $\mathcal{B}(\Delta, \gamma) \subset \{\mathbb{J}_A; A_\sigma^{-1} y \geq 0\}$.

References

1. Baldoni-Silva, W., Vergne, M.: Residues formulae for volumes and Ehrhart polynomials of convex polytopes. arXiv:math.CO/0103097 v1, 2001.
2. Baldoni-Silva, W., De Loera, J.A., Vergne, M.: Counting integer flows in networks. Found. Comput. Math. **4** (2004), 277–314.
3. Barvinok, A.I.: Computing the volume, counting integral points and exponentials sums. Discr. Comp. Geom. **10** (1993), 123–141.

4. Barvinok, A.I., Pommersheim, J.E.: An algorithmic theory of lattice points in polyhedral. in: *New Perspectives in Algebraic Combinatorics*, MSRI Publications **38** (1999), 91–147.

5. Beck, M.: Counting Lattice Points by means of the Residue Theorem. Ramanujan Journal **4** (2000), 399–310.

6. Beck, M., Diaz, R., Robins, S.: The Frobenius problem, rational polytopes, and Fourier-Dedekind sums. J. Numb. Theor. **96** (2002), 1–21.

7. Beck, M., Pixton, D.: The Ehrhart polynomial of the Birkhoff polytope. Discr. Comp. Math. **30** (2003), 623–637.

8. Brion, M., Vergne, M.: Residue formulae, vector partition functions and lattice points in rational polytopes. J. Amer. Math. Soc. **10** (1997), 797–833.

9. Cochet, C.: Réduction des graphes de Goretsky-Kottwitz-MacPherson; nombres de Kostka et coefficients de Littlewodd-Richardson. Thèse de Doctorat: Mathématiques, Université Paris 7, Paris, Décembre 2003.

10. De Loera, J.A., R. Hemmecke, R., Tauzer, J., Yoshida, R.: Effective lattice point counting in rational convex polytopes. J. of Symb. Comp., to appear.

11. Pukhlikov, A.V., Khovanskii, A.G.: A Riemann-Roch theorem for integrals and sums of quasipolynomials over virtual polytopes. St. Petersburg Math. J. **4** (1993), 789–812.

12. Lasserre, J.B., E.S. Zeron, E.S.: On counting integral points in a convex rational polytope. Math. Oper. Res. **28** (2003), 853–870.

13. Lasserre, J.B., Zeron, E.S.: An alternative algorithm for counting lattice points in a convex polytope. Math. Oper. Res. **30** (2005), 597–614.

14. Schrijver, A.: Theory of Linear and Integer Programming. John Wiley & Sons, Chichester, 1986.

15. Szenes, A., Vergne, M.: Residue formulae for vector partitions and Euler-MacLaurin sums. Adv. in Appl. Math.**30** (2003), 295–342.

16. Szenes, A.: Residue theorem for rational trigonometric sums and Verlinde's formula. *Duke Math. J.* **118** (2003), 189–227.

17. Verdoolaege, S., Beyls, K., Bruynooghe, M., Seghir, R., Loechner, V.: Analytical Computation of Ehrhart Polynomials and its Applications for Embedded Systems. Technical report # 376, Computer Science Department, KUL University, Leuwen, Belgium.

Characterizations of Total Dual Integrality[*]

Edwin O'Shea[1] and András Sebő[2]

[1] Department of Mathematics, University of Kentucky, Lexington,
KY 40506-0027, USA
oshea@ms.uky.edu
[2] CNRS, Laboratoire G-SCOP, 46, Avenue Félix Viallet, 38000 Grenoble 38031
Grenoble, Cedex 1, France
Andras.Sebo@g-scop.inpg.fr

Abstract. In this paper we provide new characterizing properties of TDI systems. A corollary is Sturmfels' theorem relating toric initial ideals generated by square-free monomials to unimodular triangulations. A reformulation of these test-sets to polynomial ideals actually generalizes the existence of square-free monomials to arbitrary TDI systems, providing new relations between integer programming and Gröbner bases of toric ideals. We finally show that stable set polytopes of perfect graphs are characterized by a refined fan that is a triangulation consisting only of unimodular cones, a fact that endows the Weak Perfect Graph Theorem with a computationally advantageous geometric feature. Three ways of implementing the results are described and some experience about one of these is reported.

1 Introduction

Let $A = [\mathbf{a}_1\, \mathbf{a}_2 \cdots \mathbf{a}_n] \in \mathbb{Z}^{d \times n}$ and assume that A has rank d. With an abuse of notation the ordered vector configuration consisting of the columns of A will also be denoted by A. For every $\sigma \subseteq [n] := \{1, \ldots, n\}$ we have the $d \times |\sigma|$ matrix A_σ given by the columns of A indexed by σ. Let $\mathrm{cone}(A)$, $\mathbb{Z}A$ and $\mathbb{N}A$ denote the non-negative real, integer and non-negative integer span of A respectively and assume that $\mathbb{Z}A = \mathbb{Z}^d$.

Fixing $\mathbf{c} \in \mathbb{R}^n$, for each $\mathbf{b} \in \mathbb{R}^d$ the *linear program* (or *primal program*) $\mathrm{LP}_{A,\mathbf{c}}(\mathbf{b})$ and its *dual program* $\mathrm{DP}_{A,\mathbf{c}}(\mathbf{b})$ are defined by

$$\mathrm{LP}_{A,\mathbf{c}}(\mathbf{b}) := \text{minimize}\,\{\,\mathbf{c} \cdot \mathbf{x}\,:\,A\mathbf{x} = \mathbf{b},\,\mathbf{x} \geq \mathbf{0}\,\}$$

and $\mathrm{DP}_{A,\mathbf{c}}(\mathbf{b}) := \text{maximize}\,\{\,\mathbf{y} \cdot \mathbf{b}\,:\,\mathbf{y}A \leq \mathbf{c}\,\}$. Let $P_\mathbf{b}$ and $Q_\mathbf{c}$ denote the feasible regions of $\mathrm{LP}_{A,\mathbf{c}}(\mathbf{b})$ and $\mathrm{DP}_{A,\mathbf{c}}(\mathbf{b})$ respectively. Note that the linear program $\mathrm{LP}_{A,\mathbf{c}}(\mathbf{b})$ is feasible if and only if $\mathbf{b} \in \mathrm{cone}(A)$. We refer to Schrijver [21] for basic terminology and facts about linear programming.

[*] The first author was supported by a Fulbright grant and by NSF grants DMS-9983797 and DMS-0401047. The research of the second author was supported by the "Marie Curie Training Network" ADONET of the European Community.

M. Fischetti and D.P. Williamson (Eds.): IPCO 2007, LNCS 4513, pp. 382–396, 2007.

The *integer program* is defined as

$$\mathrm{IP}_{A,\mathbf{c}}(\mathbf{b}) := \text{minimize} \{ \mathbf{c} \cdot \mathbf{x} : A\mathbf{x} = \mathbf{b}, \mathbf{x} \in \mathbb{N}^n \}.$$

We say that $\mathbf{c} \in \mathbb{R}^n$ is *generic* for A if the integer program $\mathrm{IP}_{A,\mathbf{c}}(\mathbf{b})$ has a unique optimal solution for all $\mathbf{b} \in \mathbb{N}A$. In this case, each linear program $\mathrm{LP}_{A,\mathbf{c}}(\mathbf{b})$ also has a unique optimal solution for all $\mathbf{b} \in \mathrm{cone}(A)$ but the converse is not true in general. (However, for TDI systems the two are equivalent.)

The system $\mathbf{y}A \leq \mathbf{c}$ is *totally dual integral (TDI)* if $\mathrm{LP}_{A,\mathbf{c}}(\mathbf{b})$ has an integer optimal solution $\mathbf{x} \in \mathbb{N}^n$ for each $\mathbf{b} \in \mathrm{cone}(A) \cap \mathbb{Z}^d$. In other words, the system $\mathbf{y}A \leq \mathbf{c}$ is TDI exactly if the optima of $\mathrm{LP}_{A,\mathbf{c}}(\mathbf{b})$ and of $\mathrm{IP}_{A,\mathbf{c}}(\mathbf{b})$ coincide for all $\mathbf{b} \in \mathrm{cone}(A) \cap \mathbb{Z}^d$. This is a slight twist of notation when compared to habits in combinatorial optimization: we defined the TDI property for the dual problem. We do this in order to be in accordance with notations in computational algebra.

Totally dual integral (TDI) systems of linear inequalities play a central role in combinatorial optimization. The recognition of TDI systems and the task of efficiently solving integer linear programs constrained by TDI systems of inequalities and their duals are among the main challenges of the field. This problem is open even for generic systems (Problem 1). Recent graph theory results of Chudnovsky, Cornuéjols, Xinming and Vušković [7] allows one to recognize TDI systems with $0-1$ coefficient matrices A and right hand sides $\mathbf{b}$. However, solving the corresponding dual pair of integer linear programs (including the coloration of perfect graphs) in polynomial time with combinatorial algorithms remains open even in this special case.

In Section 2, new characterizing properties of TDI systems are provided. These properties involve tools from both combinatorial optimization and computational algebra. Section 3 specializes these results to integral set packing polytopes. Finally, Section 4 will exhibit the utility of the computational algebraic tools in recognizing TDI systems.

If A is a matrix whose first $d \times (n-d)$ submatrix is a $0-1$ matrix and whose last $d \times d$ submatrix is $-I_d$, and $\mathbf{c}$ is all 1 except for the last d coordinates which are 0, then $\mathrm{DP}_{A,\mathbf{c}}(\mathbf{b})$ is called a *set packing problem*, and $Q_{\mathbf{c}}$ a *set packing polytope*. We will show that if the set packing polytope is integral then the lexicographic perturbation technique of linear programming can be used to make the set packing polytope *non-degenerate* while keeping TDI-ness. This means that the normal fan of the set packing polytope has a refinement which is a unimodular triangulation, and this does not hold for TDI systems in general.

The remainder of this introduction is devoted to providing some background.

A collection of subsets $\{\sigma_1, \ldots, \sigma_t\}$ of $[n]$ will be called a *regular subdivision* of A if there exists $\mathbf{c} \in \mathbb{R}^n$, and $\mathbf{z}_1, \ldots, \mathbf{z}_t \in \mathbb{R}^d$, such that $\mathbf{z}_i \cdot \mathbf{a}_j = c_j$ for all $j \in \sigma_i$ and $\mathbf{z}_i \cdot \mathbf{a}_j < c_j$ for all $j \notin \sigma_i$. The sets $\sigma_1, \ldots, \sigma_t$ are called the *cells* of the regular subdivision and the regular subdivision is denoted by $\Delta_{\mathbf{c}}(A) = \{\sigma_1, \ldots, \sigma_t\}$ or simply $\Delta_{\mathbf{c}}$ when A is unambiguous.

Equivalently, regular subdivisions are simply capturing *complementary slackness* from linear programming. Namely, a feasible solution to $\mathrm{LP}_{A,\mathbf{c}}(\mathbf{b})$ is optimal if and only if the support of the feasible solution is a subset of some cell of $\Delta_{\mathbf{c}}$. Geometrically, $\Delta_{\mathbf{c}}$ can be thought of as a partition of cone(A) by the inclusion-wise maximal ones among the cones cone(A_{σ_1}), . . . , cone(A_{σ_t}); each such cone is generated by the normal vectors of defining inequalities of faces of $Q_{\mathbf{c}}$, each maximal cell indexes the set of normal vectors of defining inequalities of a vertex (or minimal face) of $Q_{\mathbf{c}}$. So the regular subdivision $\Delta_{\mathbf{c}}$ is geometrically realized as the *normal fan* of $Q_{\mathbf{c}}$.

A regular subdivision of A is called a *triangulation* if the columns of each A_{σ_i} are linearly independent for all $i = 1, . . . , t$. Note that a regular subdivision $\Delta_{\mathbf{c}}$ is a triangulation if and only if every vertex is contained in exactly d facets; that is, the polyhedron $Q_{\mathbf{c}}$ is *simple*, or, *non-degenerate*. A triangulation $\Delta_{\mathbf{c}}$ is called *unimodular* if $\det(\sigma_i) = \pm 1$ for each maximal cell of $\Delta_{\mathbf{c}}$. The *refinement* of a subdivision $\Delta_{\mathbf{c}}$ of A is another subdivision $\Delta_{\mathbf{c}'}$ of A so that each cell of $\Delta_{\mathbf{c}'}$ is contained in some cell of $\Delta_{\mathbf{c}}$. A vector configuration $B \subset \mathbb{Z}^d$ is a *Hilbert basis* if $\mathbb{N}B = \text{cone}(B) \cap \mathbb{Z}^d$. Note that if for some $\mathbf{c} \in \mathbb{R}^n$ $\Delta_{\mathbf{c}}$ is a unimodular triangulation of A then Cramer's rule implies that A itself is a Hilbert basis.

A simple but helpful characterization of the TDI property in terms of the Hilbert basis property of regular subdivisions has been provided by Schrijver [21]. We prove another elementary characterization in Section 2 in terms of test-sets:

Let $\text{IP}_{A,\mathbf{c}} := \{\text{IP}_{A,\mathbf{c}}(\mathbf{b}) : \mathbf{b} \in \mathbb{N}A\}$ denote the family of integer programs $\text{IP}_{A,\mathbf{c}}(\mathbf{b})$ having a feasible solution. Informally, a *test set* for the family of integer programs $\text{IP}_{A,\mathbf{c}}$ is a finite collection of integer vectors, called *test vectors*, with the property that any non-optimal feasible solution can be improved (in objective value) by subtracting a test vector from it. Test sets for the family of integer programs $\text{IP}_{A,\mathbf{c}}$ were first introduced by Graver [13].

Theorem 1 (one of the equivalences). *A system of linear inequalities is TDI if and only if its coefficient vectors form a Hilbert basis, and there exists a test set for $\text{IP}_{A,\mathbf{c}}$ where all test vectors have positive entries equal to 1, and a linearly independent positive support.*[1]

This simple result has the virtue of presenting a not too big test-set: there is at most one test-vector for each at most d element subset of $\{1, . . . , n\}$, so the number of test-vectors is $O(n^d)$. This will allow to deduce shortly Cook, Lovász and Schrijver's result on testing for TDI in fix dimension, providing a short proof for this result.

It also has the other virtue that it has a nice and useful reformulation to polynomial ideals. This reformulation generalizes a well-known algebraic result proved by Sturmfels [26, Corollary 8.9] relating toric initial ideals to unimodular triangulations. The basic connections between integer programming and

---

[1] In oral and electronic communication the condition on test-sets was replaced by the following still equivalent condition: "A system of linear inequalities is TDI if and only if the coefficient vectors form a Hilbert basis, and there exists an integer dual solution *for objective functions that are sums of linearly independent coefficient vectors*", implying TDI test in fix dimension [5], in practically all interesting cases. This is just another wording of Applegate, Cook and McCormick's Theorem 2 (Operations Research Letters 10 (1991) 37–41), as we learnt from several colleagues.

computational algebra was initiated by Conti and Traverso [3] and studied by Sturmfels and Thomas, Weismantel and Ziegler and further explained from various viewpoints in [26], [25], [27] and [28]. Knowledge of this algebraic viewpoint will not be assumed and a useful part will be described in Section 2.

In Section 3 we show that the converse of the following fact (explained at the end of Section 2) holds for normal fans of integral set packing polytopes: *if* $\mathbf{c}, \mathbf{c}' \in \mathbb{R}^n$ *are such that* $\Delta_{\mathbf{c}'}$ *is a refinement of* $\Delta_{\mathbf{c}}$, *where* $\Delta_{\mathbf{c}'}$ *is a unimodular triangulation, then* $\mathbf{y}A \leq \mathbf{c}$ *is TDI.* In general, the converse does not hold. Thus Schrijver's above mentioned result cannot necessarily be strengthened by asserting a unimodular refinement of A. In general, the most that is known in this direction is the existence of just one full dimensional subset of the columns of A which is unimodular [11]. Not even a "unimodular covering" of a Hilbert basis may be possible [1]. However, the converse does hold for normal fans of integral set packing polytopes. More precisely, the main result of Section 3 is the following:

Theorem 2. *Given a set-packing problem defined by A and $\mathbf{c}$, $Q_{\mathbf{c}}$ has integer vertices if and only if there exists $\mathbf{c}'$ such that $\Delta_{\mathbf{c}'}$ is a refinement of the normal fan $\Delta_{\mathbf{c}}$ of $Q_{\mathbf{c}}$, where $\Delta_{\mathbf{c}'}$ is a unimodular triangulation.*

The proof relies on the basic idea of Fulkerson's famous "pluperfect graph theorem" [12] stating that the integrality of such polyhedra implies their total dual integrality in a very simple "greedy" way. Chandrasekaran and Tamir [2] and Cook, Fonlupt and Schrijver [4] exploited Fulkerson's method by pointing out its lexicographic or advantageous Caratheodory feature. In [23, §4] it is noticed with the same method that the active rows of the dual of integral set packing polyhedra (the cells of their normal fan) have a unimodular subdivision, which can be rephrased as follows: *the normal fan of integral set packing polyhedra has a unimodular refinement.* However, the proof of the regularity of such a refinement appears for the first time in the present work.

These results offer three methods for recognizing TDI systems, explained and illustrated in Section 4.

2   TDI Systems

In this section we provide some new characterizations of TDI systems. We show the equivalence of five properties, three polyhedral (one of them is the TDI property) and two concern polynomial ideals. A third property is also equivalent to these in the generic case.

While the proofs of the equivalences of the three polyhedral properties use merely polyhedral arguments, the last among them – (iii) – has an appealing reformulation into the language of polynomial ideals. Therefore, we start this section by introducing the necessary background on polynomial ideals; namely, toric ideals, their initial ideals and Gröbner bases. The characterizations of TDI systems involving polynomial ideals are useful generalizations of known results in computational algebra. See [8] and [26] for further background.

An *ideal* I in a polynomial ring $R := \mathbf{k}[x_1, \ldots, x_n]$ is an R-vector subspace with the property that $I \cdot R = I$. It was proven by Hilbert that every ideal is finitely generated. That is, given an ideal I there exists a finite set of polynomials $f_1, \ldots, f_t \in I$ such that for every $f \in I$ there exists $h_1, \ldots, h_t \in R$ with $f = h_1 f_1 + \cdots + h_t f_t$. We call such a collection $f_1, \ldots, f_t \in I$ a *generating set* for the ideal I and denote this by $I = \langle f_1, \ldots, f_t \rangle$. For the monomials in R we write $\mathbf{x}^{\mathbf{u}} = x_1^{u_1} \cdots x_n^{u_n}$ for the sake of brevity. We call $\mathbf{u}$ the *exponent vector* of $\mathbf{x}^{\mathbf{u}}$. A monomial $\mathbf{x}^{\mathbf{u}}$ is said to be *square-free* if $\mathbf{u} \in \{0,1\}^n$. An ideal is called a *monomial ideal* if it has a generating set consisting only of monomials. For any ideal J of R, mono(J) denotes the largest monomial ideal in R contained in J. Alternatively, mono(J) is the ideal generated by all monomials in J. There is an algorithm [20, Algorithm 4.4.2] for computing the generators of the monomial ideal mono(J).

Every weight vector $\mathbf{c} \in \mathbb{R}^n$ induces a partial order $\succeq$ on the monomials in R via $\mathbf{x}^{\mathbf{u}} \succeq \mathbf{x}^{\mathbf{v}}$ if $\mathbf{c} \cdot \mathbf{u} \geq \mathbf{c} \cdot \mathbf{v}$. If $\mathbf{c} \in \mathbb{R}^n$ where 1 is the monomial of minimum $\mathbf{c}$-cost (that is, $\mathbf{c} \cdot \mathbf{u} \geq 0$ for every monomial $\mathbf{x}^{\mathbf{u}}$), then we can define initial terms and initial ideals. Given a polynomial $f = \sum_{\mathbf{u} \in \mathbb{N}^n} r_{\mathbf{u}} \mathbf{x}^{\mathbf{u}} \in I$ the *initial term* of f with respect to $\mathbf{c}$, is denoted by $\text{in}_{\mathbf{c}}(f)$, and equals the sum of all $r_{\mathbf{u}} \mathbf{x}^{\mathbf{u}}$ of f, where $\mathbf{c} \cdot \mathbf{u}$ is maximum. The *initial ideal* of I with respect to $\mathbf{c}$ is defined as the ideal in R generated by the initial terms of the polynomials in I: $\text{in}_{\mathbf{c}}(I) := \langle \text{in}_{\mathbf{c}}(f) : f \in I \rangle$. A *Gröbner basis* of an ideal I *with respect to* $\mathbf{c}$, is a finite collection of elements $g_1, \ldots, g_s$ in I such that $\text{in}_{\mathbf{c}}(I) = \langle \text{in}_{\mathbf{c}}(g_1), \text{in}_{\mathbf{c}}(g_2), \ldots, \text{in}_{\mathbf{c}}(g_s) \rangle$. Every Gröbner basis is a generating set for the ideal I.

If $\text{in}_{\mathbf{c}}(I)$ is a monomial ideal then a Gröbner basis is *reduced* if for every $i \neq j$, no term of g_i is divisible by $\text{in}_{\mathbf{c}}(g_j)$. The reduced Gröbner basis is unique. In this case, the set of monomials in $\text{in}_{\mathbf{c}}(I)$ equal $\{\mathbf{x}^{\mathbf{u}} : \mathbf{u} \in U\}$ with $U := D + \mathbb{N}^n$ where D is the set of exponent vectors of the monomials $\text{in}_{\mathbf{c}}(g_1), \text{in}_{\mathbf{c}}(g_2), \ldots, \text{in}_{\mathbf{c}}(g_s)$. Dickson's lemma states that sets of the form $D + \mathbb{N}^n$, where D is arbitrary have only a finite number of minimal elements (with respect to coordinate wise inequalities). This is an alternative proof to Hilbert's result that every polynomial ideal is finitely generated. In this case, the Gröbner basis also provides a generalization of the Euclidean algorithm for polynomial rings with two or more variables called Buchberger's algorithm (see [8]). This algorithm solves the *ideal membership problem*: decide if a given polynomial is in an ideal or not. However, a Gröbner basis for an ideal can have many elements (relative to a minimal generating set for the ideal).

The *toric ideal* of A is the ideal $I_A = \langle \mathbf{x}^{\mathbf{u}} - \mathbf{x}^{\mathbf{v}} : A\mathbf{u} = A\mathbf{v}, \mathbf{u}, \mathbf{v} \in \mathbb{N}^n \rangle$ and is called a *binomial ideal* since it is generated by polynomials having at most terms. Every reduced Gröbner basis of a toric ideal consists of binomials. A *toric initial ideal* is any initial ideal of a toric ideal. The following lemma is a natural connection between integer programming and toric initial ideals.

Lemma 1. *[20, Lemma 4.4.7] Let $A \in \mathbb{Z}^{d \times n}$ and $\mathbf{c} \in \mathbb{R}^n$. Then the monomial ideal* $\mathrm{mono}(\mathrm{in}_{\mathbf{c}}(I_A))$ *is equal to*

$$\langle\, \mathbf{x}^{\omega} \,:\, \omega \in \mathbb{N}^n \text{ is non-optimal solution for } \mathrm{IP}_{A,\mathbf{c}}(A\omega)\,\rangle.$$

One direction of the proof of Lemma 1 is straightforward: let ω be a non-optimal solution, and ω' an optimal solution to $\mathrm{IP}_{A,\mathbf{c}}(A\omega)$. Then $\mathbf{x}^{\omega} - \mathbf{x}^{\omega'} \in I_A$ is a binomial with $\mathbf{x}^{\omega}$ as its initial term with respect to $\mathbf{c}$ and $\mathbf{x}^{\omega}$ is a monomial in $\mathrm{mono}(\mathrm{in}_{\mathbf{c}}(I_A))$. Our proof of the converse made essential use of Gröbner bases, and was longer, it is intuitive enough to be used without proof with the reference [20, Lemma 4.4.7] in the background.

A *test set* for the family of integer programs $\mathrm{IP}_{A,\mathbf{c}}$ is a collection of integer vectors $\{\mathbf{v}_i^+ - \mathbf{v}_i^- \,:\, A\mathbf{v}_i^+ = A\mathbf{v}_i^-, \ \mathbf{v}_i^+, \mathbf{v}_i^- \in \mathbb{N}^n, i = 1, \dots, s\}$ with the property that $\mathbf{u}$ is a feasible, non-optimal solution to $\mathrm{IP}_{A,\mathbf{c}}(\mathbf{b})$ if and only if there exists an $i, 1 \leq i \leq s$, such that $\mathbf{u} - (\mathbf{v}_i^+ - \mathbf{v}_i^-) \geq \mathbf{0}$. We can now state our characterizations:

Theorem 1. *Fix $A \in \mathbb{Z}^{d \times n}$ and $\mathbf{c} \in \mathbb{R}^n$, where A is a Hilbert basis. The following statements are equivalent:*

(i) *The system $\mathbf{y}A \leq \mathbf{c}$ is TDI.*

(ii) *The subconfiguration A_{σ} of A is a Hilbert basis for every cell σ in $\Delta_{\mathbf{c}}$.*

(iii) *There exists a test-set for $\mathrm{IP}_{A,\mathbf{c}}$ where all the positive coordinates are equal to 1, the positive support consists of linearly independent columns, (and the negative support is a subset of a cell of $\Delta_{\mathbf{c}}$).*

(iv) *The monomial ideal $\langle\, \mathbf{x}^{\omega} : \omega \in \mathbb{N}^n$ is not an optimal solution for $\mathrm{IP}_{A,\mathbf{c}}(A\omega)\rangle$ has a square-free generating set.*

(v) *The monomial ideal generated by the set of monomials in $\mathrm{in}_{\mathbf{c}}(I_A)$ has a square-free generating set, that is, $\mathrm{mono}(\mathrm{in}_{\mathbf{c}}(I_A))$ has a square-free generating set.*

Proof. **(i) is equivalent to (ii)** : This is well-known from Schrijver's work, (see for instance [21]), but we provide the (very simple) proof here for the sake of completeness: Suppose the system $\mathbf{y}A \leq \mathbf{c}$ is TDI, and let $\sigma \in \Delta_c$. We show that A_{σ} is a Hilbert basis. Let $\mathbf{b} \in \mathrm{cone}(A_{\sigma})$. Since the optimal solutions for $\mathrm{LP}_{A,\mathbf{c}}(\mathbf{b})$ are exactly the non-negative combinations of the columns of A_{σ} with result $\mathbf{b}$, the TDI property means exactly that $\mathbf{b}$ can also be written as a non-negative integer combination of columns in A_{σ}, as claimed.

(ii) implies (iii) : Suppose (ii) holds true for $\Delta_{\mathbf{c}}$ of A. For every $\tau \subseteq [n]$ with τ not contained in any cell of $\Delta_{\mathbf{c}}$, let $\mathbf{b}_{\tau} := \sum_{i \in \tau} a_i = A(\sum_{i \in \tau} \mathbf{e}_i)$. Since τ is not contained in any cell of $\Delta_{\mathbf{c}}$, there exists an optimal solution β_{τ} to $\mathrm{LP}_{A,\mathbf{c}}(\mathbf{b}_{\tau})$ with $\mathbf{c} \cdot \beta_{\tau} < \mathbf{c} \cdot \sum_{i \in \tau} \mathbf{e}_i$. By the optimality of β_{τ} we must have $\mathrm{supp}(\beta_{\tau}) \subseteq \sigma$ for some σ a cell of $\Delta_{\mathbf{c}}$. Since (ii) holds A_{σ} is a Hilbert basis for every cell of $\Delta_{\mathbf{c}}$ and therefore β_{τ} can be chosen to be an integral vector. Let

$$\mathcal{T}_{A,\mathbf{c}} := \{\, \sum_{i \in \tau} \mathbf{e}_i - \beta_{\tau} \,:\, \tau \text{ not contained in any cell of } \Delta_{\mathbf{c}} \,\}.$$

We claim that $\mathcal{T}_{A,\mathbf{c}}$ is a test set for $\text{IP}_{A,\mathbf{c}}$. Suppose $\mathbf{b} \in \mathbb{Z}^d$ and $\omega \in \mathbb{N}^n$ satisfies $A\omega = \mathbf{b}$. That is, ω is a feasible solution to $\text{LP}_{A,\mathbf{c}}(\mathbf{b})$.

If ω is an optimal solution then $\text{supp}(\omega)$ is contained in a cell in $\Delta_\mathbf{c}$. Thus no vector in $\mathcal{T}_{A,\mathbf{c}}$ can be subtracted from it and remain in $\mathbb{N}^n$. Conversely, if ω is not an optimal solution to $\text{LP}_{A,\mathbf{c}}(\mathbf{b})$ then $\text{supp}(\omega) \subseteq [n]$ is not contained in any cell σ of $\Delta_\mathbf{c}$ and so by basic linear programming there exists $\tau \subseteq \text{supp}(\omega)$, A_τ is linearly independent which is also not contained in any cell. $\omega - (\sum_{i \in \tau} \mathbf{e}_i - \beta_\tau) \geq \mathbf{0}$. Note that this integer vector is cheaper than ω with respect to $\mathbf{c}$.

(iii) implies (i): Suppose (iii) is true but for some $\mathbf{b} \in \text{cone}(A)$ the linear program $\text{LP}_{A,\mathbf{c}}(\mathbf{b})$ does not have an integer optimal solution. Let $\omega \in \mathbb{N}^n$ be the optimal solution to the integer program $\text{IP}_{A,\mathbf{c}}(\mathbf{b})$ and let α/D be the optimal solution to $\text{LP}_{A,\mathbf{c}}(\mathbf{b})$ where $\alpha \in \mathbb{N}^n$, and D is a positive integer. Since $\text{LP}_{A,\mathbf{c}}(\mathbf{b})$ does not have an integer optimal solution, we have $\mathbf{c} \cdot \alpha/D < \mathbf{c} \cdot \omega$. This also implies that $D\omega$ is not an optimal solution to $\text{IP}_{A,\mathbf{c}}(D\mathbf{b})$.

By (iii) there exists a test set for solving the integer program $\text{IP}_{A,\mathbf{c}}(D\mathbf{b})$ and so there exists a $\gamma^+ - \gamma^-$ with $\gamma^+ \in \{0,1\}^n$ and $\gamma^- \in \mathbb{N}^n$ such that $\mathbf{c} \cdot (\gamma^+ - \gamma^-) > 0$ and with $D\omega - (\gamma^+ - \gamma^-) \in \mathbb{N}^n$. Hence, $\text{supp}(\gamma^+) \subseteq \text{supp}(D\omega) = \text{supp}(\omega)$. Since the value of all elements in γ^+ is 0 or 1 then we also have $\omega \geq \gamma^+$, so $\omega - (\gamma^+ - \gamma^-) \in \mathbb{N}^n$ is also a feasible solution to $\text{IP}_{A,\mathbf{c}}(\mathbf{b})$ with $\mathbf{c} \cdot (\omega - (\gamma^+ - \gamma^-)) < \mathbf{c} \cdot \omega$, in contradiction to the optimality of ω.

(iii) is equivalent to (iv): Both (iii) and (iv) can be reformulated as follows: If $\omega \in \mathbb{N}^n$ is not an optimal solution to $\text{LP}_{A,\mathbf{c}}(A\omega)$ then the vector ω' defined as $\omega'_i := 1$ if $i \in \text{supp}(\omega)$ and 0 otherwise is also a non-optimal solution to $\text{LP}_{A,\mathbf{c}}(A\omega')$.

(iv) is equivalent to (v): This is a special case of Lemma 1. $\square$

Recall that we defined $\mathbf{c} \in \mathbb{R}^n$ to be *generic* with the first of the following conditions, but the others are also equivalent to the definition [28]:

- The integer program $\text{IP}_{A,\mathbf{c}}(\mathbf{b})$ has a unique optimal solution for all $\mathbf{b} \in \mathbb{N}A$.
- The toric initial ideal $\text{in}_\mathbf{c}(I_A)$ is a monomial ideal.
- There exists a Gröbner basis $\{\mathbf{x}^{\mathbf{u}_1^+} - \mathbf{x}^{\mathbf{u}_1^-}, \ldots, \mathbf{x}^{\mathbf{u}_s^+} - \mathbf{x}^{\mathbf{u}_s^-}\}$ of I_A with $\mathbf{c} \cdot \mathbf{u}_i^+ > \mathbf{c} \cdot \mathbf{u}_i^-$ for each $i = 1, \ldots, s$.

In the generic case, by Cramer's rule, (ii) is equivalent to $\Delta_\mathbf{c}$ being a unimodular triangulation which gives the following corollary.

Corollary 1. (Sturmfels) [26, Corollary 8.9] *Let $A \in \mathbb{Z}^{d \times n}$ and let $c \in \mathbb{R}^n$ be generic with respect to A. Then $\Delta_\mathbf{c}$ is a unimodular triangulation if and only if the toric initial ideal $\text{in}_\mathbf{c}(I_A)$ is generated by square-free monomials.*

Still concerning generic $\mathbf{c}$ it is worth to note the following result of Conti and Traverso which provides another connection between integer linear programming and Gröbner bases. Here we think of an element $\mathbf{x}^{\mathbf{v}^+} - \mathbf{x}^{\mathbf{v}^-}$ in the reduced Gröbner basis as a vector $\mathbf{v}^+ - \mathbf{v}^-$.

Proposition 1. (Conti-Traverso) [2] – see [29, Lemma 3] *If* $IP_{A,c}(b)$ *has a unique optimal solution for every* $b \in \mathbb{N}A$ *then the reduced Gröbner basis is a minimal test set for the family of integer programs* $IP_{A,c}$.

This proposition means for us that in the generic case the following (vi) can be added to Theorem 1:

(vi) *The initial terms in the reduced Gröbner basis are square-free.*

In particular, in the generic case of condition (iii) of Theorem 1 the unique inclusion wise minimal test set is defined by the reduced Gröbner basis, which, by (vi) has only square-free terms initial terms.

As is typically the case in combinatorial optimization, the cost vector **c** is not generic for A. Theorem 1 was found by a desire to generalize Sturmfels' theorem. In the rest of this section we study the limits of profiting from the advantages of the generic case by refinement. ¿From the implication "(ii) implies (i)" we immediately get the following:

Proposition 2. *If* $c, c' \in \mathbb{R}^n$ *are such that* $\Delta_{c'}$ *of* A *is a refinement of* Δ_c *of* A, *where* $\Delta_{c'}$ *is a unimodular triangulation of* A, *then* $yA \leq c$ *is TDI*.

Clearly, the unimodular triangulation does not even need to be regular – a unimodular cover of the cells is actually enough as well for verifying – by Cramer's rule – that A_σ is a Hilbert basis, and therefore (ii) holds. We are interested in the converse of Proposition 2, that is, the existence of such a **c'** for every TDI system. In general such a converse does not hold. It is not even true that a Hilbert basis has a unimodular partition or a unimodular covering [1]. This counterexample [1] inspires two important remarks. First, it cannot be expected that the equivalence of (i) and (v) can be reduced to Sturmfels' generic case, even though square-free generating sets exist for general TDI systems as well. Secondly, it should be appreciated that the converse of this remark does hold in the important set packing special case, as we will see in the next Section 3.

3 Set Packing

Let a set packing problem be defined with a matrix A and vector **c**, and recall $c := (1, 0) \in \mathbb{R}^n$, where the last d entries of **c** are 0. If the set packing polytope Q_c has integer vertices then the matrix A and the polytope Q_c are said to be *perfect*. (We will not use the well-known equivalence of this definition with the integer values of optima: this will follow.) Lovász' (weak) perfect graph theorem [16] is equivalent to: *the matrix* A *defining a set packing polytope is perfect if and only if its first* $(n - d)$ *columns form the incidence vectors (indexed by the vertices) of the inclusion wise maximal complete subgraphs of a perfect graph.*

A polyhedral proof of the perfect graph theorem can be split into two parts: Lovász' *replication lemma* [16] and Fulkerson's *pluperfect graph theorem* [12]. The latter states roughly that a set packing polytope with integer vertices is described by a TDI system of linear inequalities. In this section we restate Fulkerson's result in a sharper form: there is a unimodular regular triangulation that refines the normal fan of any integral set packing polytope. We essentially repeat Fulkerson's proof, completing it with a part that shows unimodularity along the lines of the proof of [23, Theorem 3.1]. The following theorem contains the weak perfect graph theorem and endows it with an additional geometric feature. Denote the common optimal value of $LP_{A,\mathbf{c}}(\mathbf{b})$ and $DP_{A,\mathbf{c}}(\mathbf{b})$ by $\gamma_{\mathbf{c}}(\mathbf{b})$. Note that $\gamma_{\mathbf{c}}$ is a monotone increasing function in all of the coordinates.

Theorem 2. *Let $Q_{\mathbf{c}}$ be a set packing polytope defined by A and $\mathbf{c}$. Then there exists a vector $\varepsilon \in \mathbb{R}^n$ such that $\mathbf{c}' := (\mathbf{1}, \mathbf{0}) + \varepsilon$ defines a regular triangulation $\Delta_{\mathbf{c}'}$ refining $\Delta_{\mathbf{c}}$, and this triangulation is unimodular, if and only if $Q_{\mathbf{c}}$ is perfect.*

We do not claim that the following proof of this theorem is novel. All essential ingredients except unimodularity are already included in the proof of Fulkerson's pluperfect graph theorem [12]. Cook, Fonlupt and Schrijver [4] and Chandrasekaran, Tamir [2] both exploited the fact that the greedy way of taking active rows leads to integer basic solutions in this case. The latter paper extensively used *lexicographically best* solutions, which is an important tool in linear programming theory, and this was used in observing the existence of a unimodular refinement of the normal fan in [23]. This same lexicographic perturbation is accounted for by the vector ε of Theorem 2, showing that the unimodular refinement is regular. This motivated the following problem, thus containing perfectness test:

Problem 1. [24] *Given a $d \times n$ integer matrix A and an n dimensional integer vector c, decide in polynomial time whether the normal fan of $Q_{\mathbf{c}}$ consists only of unimodular cones. Equivalently, can it be decided in polynomial time that $Q_{\mathbf{c}}$ is non-degenerate, and the determinant of A_σ is ± 1 for all $\sigma \in \Delta_{\mathbf{c}}$.*

Theorem 2 is a last step in a sharpening series of observations all having essentially the same proof. We begin similarly, with the proof of Fulkerson's pluperfect graph theorem which will indicate what the $\mathbf{c}'$ of Theorem 2 should be, and then finish by showing that $\Delta_{\mathbf{c}'}$ is a unimodular triangulation.

Assume that A is a perfect matrix for the remainder of this section and that $\mathbf{c} = (\mathbf{1}, \mathbf{0})$ as before. For all $\mathbf{b} \in \mathbb{Z}^d$ and column index $i \in \{1, \ldots, n\}$ let

$$\lambda_{\mathbf{c},i}(\mathbf{b}) := \max\{x_i : \mathbf{x} \text{ is an optimal solution of } LP_{A,\mathbf{c}}(\mathbf{b})\}.$$

That is, $\lambda_{\mathbf{c},i}(\mathbf{b})$ is the largest value of x_i such that $\mathbf{c} \cdot \mathbf{x}$ is minimum under $\mathbf{x} \in P_{\mathbf{b}}$.

An additional remark: if σ is the minimal cell of $\Delta_{\mathbf{c}}$ such $\mathbf{b} \in \mathrm{cone}(A_\sigma)$, then $\mathbf{b} - \lambda_{\mathbf{c},i}(\mathbf{b})\mathbf{a}_i \in \mathrm{cone}(A_{\sigma'})$ where $\sigma' \in \Delta_{\mathbf{c}}$, $\sigma' \subseteq \sigma$ and the dimension of $\mathrm{cone}(A_{\sigma'})$ is strictly smaller than that of $\mathrm{cone}(A_\sigma)$. Furthermore, $\mathbf{b} - \lambda\mathbf{a}_i \notin \mathrm{cone}(A_\sigma)$ if $\lambda > \lambda_{\mathbf{c},i}(\mathbf{b})$.

For all $\mathbf{b} \in \mathbb{Z}^d$ we show that $\lambda_{\mathbf{c},i}(\mathbf{b})$ is an integer for every $i = 1, \ldots, n$. This is the heart of Fulkerson's pluperfect graph theorem [12, Theorem 4.1]. We state it here in a way that is most useful for our needs:

Lemma 2. *Suppose $\gamma_{\mathbf{c}}(b) \in \mathbb{Z}$ for all $\mathbf{b} \in \mathbb{Z}^d$. If $\mathbf{x}$ is an optimal solution to $\mathrm{LP}_{A,\mathbf{c}}(\mathbf{b})$ with $x_l \neq 0$ for some $1 \leq l \leq n$, then there exists $\mathbf{x}^*$ also optimal for the same $\mathbf{b}$, such that $x_l^* \geq 1$.*

Note that this lemma implies the integrality of $\lambda := \lambda_{\mathbf{c},l}(\mathbf{b})$ for all $l = 1, \ldots, n$: if λ were not an integer then setting $\mathbf{b}' := \mathbf{b} - \lfloor \lambda \rfloor \mathbf{a}_l$ we have $\lambda_{\mathbf{c},l}(\mathbf{b}') = \{\lambda\}$ where $0 \leq \{\lambda\} := \lambda - \lfloor \lambda \rfloor < 1$, contradicting Lemma 2.

Proof. Suppose $\mathbf{x} \in P_{\mathbf{b}}$ with $\mathbf{c} \cdot \mathbf{x} = \gamma(\mathbf{b})$ and $x_l > 0$ for some $1 \leq l \leq n$. We have two cases: either $1 \leq l \leq n - d$ or $n - d + 1 \leq l \leq n$.

If $n - d + 1 \leq l \leq n$ then $\mathbf{a}_l = -\mathbf{e}_{l-(n-d)} \in \mathbb{R}^d$ and $c_l = 0$. In this case, we have $\gamma_{\mathbf{c}}(\mathbf{b}) = \gamma_{\mathbf{c}}(\mathbf{b} + x_l \mathbf{e}_{l-(n-d)})$ because replacing x_l by 0 in $\mathbf{x}$ we get a solution of the same objective value for the right hand side $\mathbf{b} + x_l \mathbf{e}_{l-(n-d)}$ which gives $\gamma_{\mathbf{c}}(\mathbf{b}) \geq \gamma_{\mathbf{c}}(\mathbf{b} + x_l \mathbf{e}_{l-(n-d)})$. The reverse inequality follows from the (coordinate-wise) monotonicity of $\gamma_{\mathbf{c}}$. But then

$$\gamma_{\mathbf{c}}(\mathbf{b} + \mathbf{e}_{l-(n-d)}) \leq \gamma_{\mathbf{c}}(\mathbf{b} + x_l \mathbf{e}_{l-(n-d)}) + 1 - x_l = \gamma_{\mathbf{c}}(\mathbf{b}) + 1 - x_l,$$

and since $\gamma_{\mathbf{c}}(\mathbf{b} + \mathbf{e}_{l-(n-d)})$ is integer and $1 - x_l < 1$, we conclude that $\gamma_{\mathbf{c}}(\mathbf{b} + \mathbf{e}_{l-(n-d)}) = \gamma_{\mathbf{c}}(\mathbf{b})$.

So for any optimal $\mathbf{x}' \in P_{\mathbf{b}+\mathbf{e}_{l-(n-d)}}$ where $\mathbf{c} \cdot \mathbf{x}' = \gamma_{\mathbf{c}}(\mathbf{b})$, letting $\mathbf{x}^* := \mathbf{x}' + \mathbf{e}_{l-(n-d)} \in P_{\mathbf{b}}$ we have $\mathbf{c} \cdot \mathbf{x}^* \leq \gamma_{\mathbf{c}}(\mathbf{b})$ and so $\mathbf{x}^*$ is optimal and $x_l^* \geq 1$.

On the other hand, suppose $1 \leq l \leq n - d$. By the monotonicity of $\gamma_{\mathbf{c}}$, and noting that replacing x_l in $\mathbf{x}$ by 0 we get a point in $P_{\mathbf{b} - x_l \mathbf{a}_l}$. This point has objective value $\mathbf{c} \cdot \mathbf{x} - x_l < \mathbf{c} \cdot \mathbf{x} = \gamma_{\mathbf{c}}(\mathbf{b})$, and so we have

$$\gamma(\mathbf{b} - \mathbf{a}_l) \leq \gamma(\mathbf{b} - x_l \mathbf{a}_l) < \gamma(\mathbf{b}).$$

Since the left and right hand sides are both integer values then $\gamma(\mathbf{b} - \mathbf{a}_l) \leq \gamma(\mathbf{b}) - 1$. In other words, for any optimal $\mathbf{x}' \in P_{\mathbf{b} - \mathbf{a}_l}$ we have $\mathbf{c} \cdot \mathbf{x}' \leq \gamma_{\mathbf{c}}(\mathbf{b}) - 1$. Letting $\mathbf{x}^* := \mathbf{x}' + \mathbf{e}_l \in P_{\mathbf{b}}$ we get $\mathbf{c} \cdot \mathbf{x}^* \leq \gamma_{\mathbf{c}}(\mathbf{b}) - 1 + 1 = \gamma_{\mathbf{c}}(\mathbf{b})$ with $x_l^* \geq 1$. $\square$

Let us know define the appropriate $\mathbf{c}'$ for the theorem, depending only on $\mathbf{c}$. Define $\mathbf{c}' := \mathbf{c} + \varepsilon \in \mathbb{R}^n$ where $\varepsilon_i := -(1/n^{n+2})^i$ for each $i = 1, \ldots, n$. Note that the absolute value of the determinant of a $\{-1, 0, 1\}$-matrix cannot exceed n^n. It follows, by Cramer's rule, that the coefficients of linear dependencies between the columns of A are at most n^n in absolute value, and then the sum of absolute values of the coefficients between two solutions of an equation $A\mathbf{x} = \mathbf{b}$ for any $b \in \mathbb{R}^n$ can differ by at most a factor of n^{n+2}. After this observation two facts can be immediately checked (this is well-known from courses of linear programming):

(i) Any optimal solution to $\mathrm{LP}_{A,\mathbf{c}'}(\mathbf{b})$ is also optimal for $\mathrm{LP}_{A,\mathbf{c}}(\mathbf{b})$.
(ii) If $\mathbf{x}'$ and $\mathbf{x}''$ are both optimal solutions to $\mathrm{LP}_{A,\mathbf{c}}(\mathbf{b})$ then $\mathbf{x}'$ is *lexicographically bigger* than $\mathbf{x}''$ (that is, the first non-zero coordinate of $\mathbf{x}' - \mathbf{x}''$ is positive) if and only if $\mathbf{c}' \cdot \mathbf{x}' < \mathbf{c}' \cdot \mathbf{x}''$.

Fact (i) means that $\Delta_{\mathbf{c}'}$ refines $\Delta_{\mathbf{c}}$, and (ii) means that an optimal solution to $\mathrm{LP}_{A,\mathbf{c}'}(\mathbf{b})$ is constructed by defining $\mathbf{b}^0 := \mathbf{b}$ and recursively

$$x_i := \lambda_{\mathbf{c},i}(\mathbf{b}^{i-1}), \quad \mathbf{b}^i := \mathbf{b}^{i-1} - x_i \mathbf{a}_i.$$

Furthermore, this optimum is unique and it follows that $\Delta_{\mathbf{c}'}$ is a triangulation. We are now ready to prove Theorem 2.

Proof of Theorem 2. The necessity of the condition is straightforward: each vertex $y \in Q_{\mathbf{c}}$ satisfies the linear equation of the form $yA_{\sigma'} = 1$, where σ' is a cell of $\Delta_{\mathbf{c}'}$, $b \in \text{cone}(A_{\sigma'}) \subseteq \text{cone}(A_\sigma)$, $\sigma \in \Delta_{\mathbf{c}}$. Since the determinant of A_σ is ± 1, by Cramer's rule, y is integer.

Conversely, we will prove the assertion supposing only that $\gamma_{\mathbf{c}}(\mathbf{b})$ is integer for all $\mathbf{b} \in \mathbb{Z}^d$. (Note that then by the already proven easy direction we will have proved from this weaker statement that $Q_{\mathbf{c}}$ is perfect, as promised at the definition of perfectness.)

Without loss of generality, suppose that $\mathbf{b}$ cannot be generated by less than d columns of A, that is, the minimal cell σ of $\Delta_{\mathbf{c}}$ such that $\mathbf{b} \in \text{cone}(A_\sigma)$ is a maximal cell of $\Delta_{\mathbf{c}}$. That is, $\text{cone}(A_\sigma)$ is d-dimensional. Because of fact (i), an optimal solution to $\text{LP}_{A,\mathbf{c}}(\mathbf{b})$ will have support in σ and fact (ii) implies that such an optimal solution is constructed as follows:

Let $s_1 := \min\{i : i \in \sigma\}$ and $x_{s_1} := \lambda_{\mathbf{c},s_1}(\mathbf{b})$. Recursively, for $j = 2, \ldots, d$ let s_j be the smallest element in σ indexing a column of A on the minimal face of $\text{cone}(A_\sigma)$ containing

$$\mathbf{b} - \sum_{i=1}^{j-1} x_{s_i} \mathbf{a}_{s_i}.$$

Since b is in the interior of $\text{cone}(A_\sigma)$ then $x_{s_i} > 0$ for each $i = 1, ..., d$, and by Lemma 2, these d x_{s_i}'s are integer. Moreover, since the dimension of $\text{cone}(A_{\sigma \setminus \{s_1, ..., s_i\}})$ is strictly decreasing as $i = 2, \ldots, d$ progresses then

$$\mathbf{b} - \sum_{i=1}^{d} x_{s_i} \mathbf{a}_{s_i} = 0$$

and, setting $U := \{s_1, \ldots, s_d\} \subseteq \sigma$, we have the columns of A_U are linearly independent. Note that U is a cell of $\Delta_{\mathbf{c}'}$ and every maximal cell of $\Delta_{\mathbf{c}'}$ arises in this fashion. We show that the matrix A_U has determinant ± 1.

Suppose not. Then the inverse of the matrix A_U is non-integer, and from the matrix equation $(A_U)^{-1} A_U = \text{id}$ we see that there exists a unit vector $\mathbf{e}_j \in \mathbb{R}^d$ which is a noninteger combination of columns in A_U:

$$\sum_{i=1}^{d} x_{s_i} \mathbf{a}_{s_i} = \mathbf{e}_j.$$

For $\alpha \in \mathbb{R}$ let $\{\alpha\} := \alpha - \lfloor \alpha \rfloor$, and define:

$$\sum_{i=1}^{d} \{x_{s_i}\} \mathbf{a}_{s_i} =: \mathbf{z}$$

Clearly, $\mathbf{z} \in \text{cone}(A_U)$ and furthermore $\mathbf{z} \in \mathbb{Z}^d$ since it differs from $\mathbf{e}_j$ by an integer combination of the columns of A_U. So Lemma 2 can be applied to $\mathbf{b} := \mathbf{z}$:

letting $l := \min\{i : \{x_{s_i}\} \neq 0\}$ we see that $\lambda_{\mathbf{c},s_l}(\mathbf{z}) < 1$ contradicting Lemma 2. Hence both A_U and $(A_U)^{-1}$ are integer, their determinant is ± 1; since A_U was an arbitrary maximal cell of $\Delta_{\mathbf{c'}}$, we conclude that $\Delta_{\mathbf{c'}}$ is unimodular.   □

The argument concerning the inverse matrix replaces the use of parallelepipeds (compare with [23, proof of Theorem 3.1]) that we wanted to avoid here to stay in elementary terms.

Note that all the numbers in the definition of $\mathbf{c'}$ are at most n^{n^2}, so they have a polynomial number of digits: the perturbed problem has polynomial size in terms of the original one, reducing perfectness test to Problem 1.

4   Computation

In this section we wish to give an idea of how the results presented in this work lead to practical algorithms. There are three essentially different approaches.

A first, general, elementary algorithm can be based on Theorem 1, or more precisely on the proof of its Corollary ??. Indeed, the procedure described in this corollary is a general algorithm for testing the TDI property in $O(n^d)$ time. If d is fixed, it is a polynomial algorithm. This is very recent and has not yet been implemented.

The equivalences of (i) and (v) in Theorem 1 along with an algorithm [20, Algorithm 4.4.2] for computing the generators of the monomial ideal $\mathrm{mono}(\mathrm{in}_{\mathbf{c}}(I_A))$ permit us to detect TDI using algebraic methods: the generators are square-free if and only if the system $\mathbf{y}A \leq \mathbf{c}$ is TDI.

This algorithm works for all cost vectors, be they generic or non-generic, but it is not yet implemented and our suspicion is that $\mathrm{mono}(\mathrm{in}_{\mathbf{c}}(I_A))$ could be rather difficult to compute in the non-generic case. However, in the generic case, $\mathrm{in}_{\mathbf{c}}(I_A)$ is already a monomial ideal and can be attained in practice. In addition, even if $\mathbf{c}$ is non-generic, it may have a generic perturbation yielding a unimodular triangulation and then the toric initial ideals can be studied with respect to the perturbed vector. Computing the toric initial ideal may be far easier than investigating the unimodularity of the corresponding triangulation.

Let us have a look at one example of an A and $\mathbf{c}$ coming from a set packing problem. A more efficient way of treating the data is at hand in the generic case. Then we can use the computationally well studied reduced Gröbner bases according to Proposition 1.

The perfect graph in Figure 1 with 21 maximal cliques on 20 vertices was constructed by Padberg in [18]. The matrix A is a (20×41)-matrix and the toric ideal I_A lives in the polynomial ring $\mathbf{k}[a, \ldots, u, v_1, \ldots, v_{20}]$ where $a, \ldots, u$ correspond to the maximal cliques of G (the first 21 columns of A) and where $v_1, \ldots, v_{20}$ correspond to the vertices of G (the ordered columns of $-I_{20}$, the last 20 columns of A) as before.

The toric initial ideal with an appropriate perturbation has 61 elements, all of which are square-free. The computation was carried out in Macaulay 2 [14] (in less than 1 second) and its implementation can be seen in [17, Appendix A].

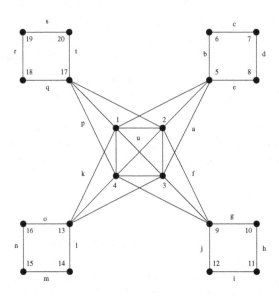

Fig. 1. Padberg's graph G with 21 maximal cliques on 20 vertices

However, we could (equivalently) have asked if a well-defined triangulation refining $\Delta_\mathbf{c}$, was a unimodular triangulation. This is a far more exhausting task than computing the monomial toric initial ideal. Because of the bijection between the cells of $\Delta_\mathbf{c}$ and the vertices of $Q_\mathbf{c}$, using PORTA [9] we computed that $Q_\mathbf{c}$ had precisely 5901 vertices. Next, using TOPCOM [19] a number of these 5901 cells are each refined into many pieces by the refinement. To confirm TDI, the determinant indexed by each of the many refined cells would have to be computed.

Acknowledgments

The authors wish to thank Rekha Thomas for her valuable input and suggestions. Some work related to the results of this article, including the computational experimentation, can be found in the first author's Ph.D. dissertation at the University of Washington. Thanks are also due to our colleagues who developed the computational packages Macaulay 2 and TOPCOM.

References

1. W. BRUNS, J. GUBELADZE, Normality and Covering Properties of Affine Semigroups, manuscript.
2. R. CHANDRASEKARAN, A. TAMIR , On the integrality of an extreme solution to pluperfect graph and balanced systems, *Oper. Res. Let.*, **3**, (1984), 215–218.
3. P. CONTI, C. TRAVERSO, Buchberger algorithm and integer programming, *Applied algebra, algebraic algorithms and error-correcting codes*, Lecture Notes in Comput. Sci., **539**, Springer, Berlin, 1991

4. W. COOK, J. FONLUPT, A. SCHRIJVER, An integer analogue of Carathéodory's theorem. *J. Combin. Theory*(B), **40**, (1986), 63–70.

5. W. COOK, L. LOVÁSZ, A. SCHRIJVER, A polynomial-time test for total dual integrality in fixed dimension, *Mathematical Programming Study* **22**, (1984), 64–69.

6. G. CORNUÉJOLS, *Combinatorial optimization: packing and covering*, CBMS-NSF regional conference series in applied mathematics, **74** SIAM, 2001.

7. M. CHUDNOVSKY, G. CORNUÉJOLS, L. XINMING, K. VUŠKOVIĆ, Recognizing Berge graphs. *Combinatorica*, **25**, (2005), no. 2, 143–186.

8. D. COX, J. LITTLE, D. O'SHEA, *Ideals, varieties and algorithms*, 2nd edition, Springer-Verlag, NY, 1996.

9. T. CHRISTOF, A. LÖBEL, *PORTA (POlyhedron Representation Transformation Algorithm)*, available from http://www.zib.de/Optimization/Software/Porta/.

10. V. CHVÁTAL, On certain polytopes associated with graphs *Journal of Comb. Theory*,(B), **18**, (1975), 138–154.

11. A. GERARDS, A. SEBŐ, Total dual integrality implies local strong unimodularity, Mathematical Programming, 38 (1987), 69-73.

12. D. R. FULKERSON, Anti-blocking polyhedra, *Journal of Comb. Theory*,(B), **12**, (1972), 50–71.

13. J. GRAVER, On the foundations of linear and integer programming I, *Math. Programming* **8**, (1975), 207–226.

14. D. GRAYSON, M. STILLMAN, *Macaulay 2, a software system for research in algebraic geometry*, available from http://www.math.uiuc.edu/Macaulay2/.

15. S. HOŞTEN, R.R. THOMAS, Gomory integer programs, *Mathematical Programming*(B), **96**, (2003), 271–292.

16. L. LOVÁSZ, Normal hypergraphs and the perfect graph conjecture, *Discrete Mathematics*, **2**, (1972), 253–267.

17. E. O'SHEA, *Toric algebra and the weak perfect graph theorem*, Ph.D. dissertation, University of Washington, 2006.

18. M. PADBERG, Perfect zero-one matrices, *Math. Programming* **6** (1974), 180–196.

19. J. RAMBAU, *TOPCOM (Triangulations Of Point Configurations and Oriented Matroids)*, available from http://www.uni-bayreuth.de/departments/wirtschaftsmathematik/rambau/TOPCOM/.

20. M. SAITO, B. STURMFELS, N. TAKAYAMA, *Gröbner deformations of hypergeometric differential equations*, Algorithms and Computation in Mathematics, **6** Springer-Verlag, Berlin, 2000.

21. A. SCHRIJVER, *Theory of linear and integer programming*, Wiley, 1986.

22. A. SCHRIJVER, *Combinatorial optimization: polyhedra and efficiency*, Algorithms and Combinatorics **24**, Springer, 2003.

23. A. SEBŐ, Hilbert bases, Caratheodory's theorem and combinatorial optimization, *Integer Programming and Combinatorial Optimization* (eds: R. Kannan, W. Pulleyblank) Mathematical Programming Society, University of Waterloo Press, Waterloo, 1990.

24. A. SEBŐ, Problem A.6, "TDI Matrices", in *Open Problems* (Workshop on "The Perfect Graph Conjecture", (2002), http://www.aimath.org/pastworkshops/perfectgraph.html.

25. B. STURMFELS, R. WEISMANTEL, G. M. ZIEGLER, Gröbner bases of lattices, corner polyhedra, and integer programming, *Beiträge Algebra Geom.*, **36**, (1995), 281–298.

26. B. STURMFELS, *Gröbner bases and convex polytopes*, University Lecture Series **8**, American Mathematical Society, Providence, RI, 1996.

27. B. STURMFELS, Algebraic recipes for integer programming, *AMS Shortcourse: Trends in Optimization* (eds: S. Hoşten, J. Lee, R.R. Thomas) Proceedings of Symposia in Applied Mathematics, **61**, American Mathematical Society, Providence, RI, 2004.

28. B. STURMFELS AND R. R. THOMAS, Variations of cost functions in integer programming, *Math. Programming* **77**, (1997), 357–387.

29. R. R. THOMAS, Algebraic methods in integer programming, *Encyclopedia of Optimization* (eds: C. Floudas and P. Pardalos), Kluwer Academic Publishers, Dordrecht, 2001

Sign-Solvable Linear Complementarity Problems

Naonori Kakimura

Department of Mathematical Informatics,
Graduate School of Information Science and Technology,
The University of Tokyo, Tokyo 113-8656, Japan
`naonori_kakimura@mist.i.u-tokyo.ac.jp`

Abstract. This paper presents a connection between qualitative matrix theory and linear complementarity problems (LCPs). An LCP is said to be *sign-solvable* if the set of the sign patterns of the solutions is uniquely determined by the sign patterns of the given coefficients. We provide a characterization for sign-solvable LCPs such that the coefficient matrix has nonzero diagonals, which can be tested in polynomial time. This characterization leads to an efficient combinatorial algorithm to find the sign pattern of a solution for these LCPs. The algorithm runs in $O(\gamma)$ time, where γ is the number of the nonzero coefficients.

Keywords: Linear Complementarity Problems, Combinatorial Matrix Theory.

1   Introduction

This paper deals with linear complementarity problems (LCPs) in the following form:

$$\text{LCP}(A, b): \text{ find } (w, z)$$
$$\text{s.t. } w = Az + b,$$
$$w^{\mathrm{T}} z = 0,$$
$$w \geq 0, \ z \geq 0,$$

where A is a real square matrix, and b is a real vector. The LCP, introduced by Cottle [4], Cottle and Dantzig [5], and Lemke [16], is one of the most widely studied mathematical programming problems, which contains linear programming and convex quadratic programming. Solving $\text{LCP}(A, b)$ for an arbitrary matrix A is NP-complete [3], while there are several classes of matrices A for which the associated LCPs can be solved efficiently. For details of the theory of LCPs, see the books of Cottle, Pang, and Stone [6] and Murty [20].

The *sign pattern* of a real matrix A is the $\{+, 0, -\}$-matrix obtained from A by replacing each entry by its sign. When we develop an LCP model in practice, the entries of A and b are subject to many sources of uncertainty including errors of measurement and absence of information. On the other hand, the sign patterns of A and b are structural properties independent of such uncertainty. This motivates us to provide a combinatorial method that exploits the sign patterns before using numerical information.

M. Fischetti and D.P. Williamson (Eds.): IPCO 2007, LNCS 4513, pp. 397–409, 2007.
© Springer-Verlag Berlin Heidelberg 2007

Sign pattern analysis for matrices and linear systems, called *qualitative matrix theory*, was originated in economics by Samuelson [24]. Various results about qualitative matrix theory are compiled in the book of Brualdi and Shader [1]. For a matrix A, we denote by $\mathcal{Q}(A)$ the set of all matrices having the same sign pattern as A, called the *qualitative class* of A. The qualitative class of a vector is defined similarly. A square matrix A is said to be *sign-nonsingular* if $\tilde{A}$ is nonsingular for any $\tilde{A} \in \mathcal{Q}(A)$. The problem of recognizing sign-nonsingular matrices has many equivalent problems in combinatorics [17,21,25,27], while its time complexity had been open for a long time. In 1999, Robertson, Seymour, and Thomas [22] presented a polynomial-time algorithm for solving this problem (cf. McCuaig [18,19]).

For linear programming, Iwata and Kakimura [11] proposed sign-solvability in terms of qualitative matrix theory. A linear program $\max\{cx \mid Ax = b, x \geq 0\}$, denoted by $\mathrm{LP}(A, b, c)$, is *sign-solvable* if the set of the sign patterns of the optimal solutions of $\mathrm{LP}(\tilde{A}, \tilde{b}, \tilde{c})$ is the same as that of $\mathrm{LP}(A, b, c)$ for any $\tilde{A} \in \mathcal{Q}(A)$, $\tilde{b} \in \mathcal{Q}(b)$, and $\tilde{c} \in \mathcal{Q}(c)$. They showed that recognizing sign-solvability of a given LP is NP-hard, and gave a sufficient condition for sign-solvable linear programs, which can be tested in polynomial time. Moreover, they devised a polynomial-time algorithm to obtain the sign pattern of an optimal solution for linear programs satisfying this sufficient condition.

In this paper, we introduce sign-solvability for linear complementarity problems. We say that $\mathrm{LCP}(A, b)$ is *sign-solvable* if the set of the sign patterns of the solutions of $\mathrm{LCP}(\tilde{A}, \tilde{b})$ coincides with that of $\mathrm{LCP}(A, b)$ for any $\tilde{A} \in \mathcal{Q}(A)$ and $\tilde{b} \in \mathcal{Q}(b)$. An $\mathrm{LCP}(A, b)$ such that all diagonal entries of A are nonzero is said to have *nonzero diagonals*. The class of LCPs with nonzero diagonals includes LCPs associated with positive definite matrices, P-matrices, and nondegenerate matrices, which are all of theoretical importance in the context of LCPs (e.g. [6, Chapter 3]). LCPs with P-matrices are related to a variety of applications such as circuit equations with piecewise linear resistances [8] and linear systems of interval linear equations [23]. We present a characterization for a sign-solvable $\mathrm{LCP}(A, b)$ with nonzero diagonals, and describe a polynomial-time algorithm to solve them from the sign patterns of A and b.

We first provide a sufficient condition for sign-solvable LCPs with nonzero diagonals. A square matrix A is *term-nonsingular* if the determinant of A contains at least one nonvanishing expansion term. A square matrix A is *term-singular* if it is not term-nonsingular. A matrix A is term-singular if and only if $\tilde{A}$ is singular for any $\tilde{A} \in \mathcal{Q}(A)$. An $m \times n$ matrix with $m \leq n$ is said to be *totally sign-nonsingular* if all submatrices of order m are either sign-nonsingular or term-singular, namely, if the nonsingularity of each submatrix of order m is determined uniquely by the sign pattern of the matrix. Totally sign-nonsingular matrices were investigated in the context of sign-solvability of linear systems [1,12,13,26] (the terms "matrices with signed mth compound" and "matrices with signed null space" are used instead). Recognizing totally sign-nonsingular matrices can be done in polynomial time by testing sign-non singularity of related square matrices [11]. We show that, if the matrix $M = (A\ b)$ is totally sign-nonsingular and A has nonzero diagonals, then $\text{LCP}(A, b)$ is sign-solvable.

We then present a characterization of sign-solvable LCPs with nonzero diagonals. A row of a matrix is called *mixed* if it has both positive and negative entries. A matrix is *row-mixed* if every row is mixed. For an $\text{LCP}(A, b)$ with nonzero diagonals, we introduce the *residual row-mixed* matrix, which is the special submatrix of $M = (A\ b)$ defined in Sect. 3. Then $\text{LCP}(A, b)$ with nonzero diagonals is sign-solvable if and only if its residual row-mixed matrix M' satisfies one of followings: M' does not contain the subvector of b, M' has no rows, or M' is totally sign-nonsingular. The residual row-mixed matrix can be obtained in polynomial time. Thus the sign-solvability of a given $\text{LCP}(A, b)$ with nonzero diagonals can be recognized in polynomial time.

This characterization leads to an efficient combinatorial algorithm to solve a given $\text{LCP}(A, b)$ with nonzero diagonals from the sign patterns of A and b. The algorithm tests the sign-solvability, and finds the sign pattern of a solution if it is a sign-solvable LCP with solutions. In this algorithm, we obtain a solution of $\text{LCP}(\tilde{A}, \tilde{b})$ for some $\tilde{A} \in \mathcal{Q}(A)$ and $\tilde{b} \in \mathcal{Q}(b)$. If $\text{LCP}(A, b)$ is sign-solvable, then $\text{LCP}(A, b)$ has a solution with the same sign pattern as the obtained one. The time complexity is $\text{O}(\gamma)$, where γ is the number of nonzero entries in A and b. We note that the obtained sign pattern easily derives a solution of the given LCP by Gaussian elimination. Thus a sign-solvable LCP with nonzero diagonals is a class of LCPs which can be solved in polynomial time.

Before closing this section, we give some notations and definitions used in the following sections.

For a matrix A, the row and column sets are denoted by U and V. If A is a square matrix, suppose that U and V are both identical with N. We denote by a_{ij} the (i, j)-entry in A. Let $A[I, J]$ be the submatrix in A with row subset I and column subset J, where the orderings of the elements of I and J are compatible with those of U and V. The submatrix $A[J, J]$ is abbreviated as $A[J]$. The *support* of a row subset I, denoted by $\Gamma(I)$, is the set of columns having nonzero entries in the submatrix $A[I, V]$, that is, $\Gamma(I) = \{j \in V \mid \exists i \in I, a_{ij} \neq 0\}$. For a vector b, the jth entry of b is denoted by b_j. The vector $b[J]$ means the subvector with index subset J. The *support* of a vector b is the column index subset $\{j \mid b_j \neq 0\}$.

For a square matrix A, let π be a bijection from the row set N to the column set N. We denote by $p(A|_\pi) = \text{sgn}\,\pi \prod_{i \in N} a_{i\pi(i)}$ the expansion term of $\det A$ corresponding to π. Then a matrix A is term-nonsingular if and only if there exists a bijection $\pi : N \to N$ with $p(A|_\pi) \neq 0$. A square matrix A is sign-nonsingular if and only if A is term-nonsingular and every nonvanishing expansion term of $\det A$ has the same sign [1, Theorem 1.2.5]. Thus, if A is sign-nonsingular, the determinant of every matrix in $\mathcal{Q}(A)$ has the same sign. It is also shown in [1, Theorem 2.1.1] that, if a square matrix A is sign-nonsingular, then A is not row-mixed.

This paper is organized as follows. In Sect. 2, we provide a sufficient condition using totally sign-nonsingular matrices. Section 3 gives a characterization for sign-solvable LCPs with nonzero diagonals. In Sect. 4, we describe a polynomial-time algorithm to solve sign-solvable LCPs with nonzero diagonals from the sign patterns of the given coefficients.

2 Totally Sign-Nonsingular Matrices

In this section, we give a sufficient condition for sign-solvable LCPs using totally sign-nonsingular matrices. For that purpose, we define *sign-nondegenerate* matrices. A square matrix A is *nondegenerate* if every principal minor is nonzero. A matrix A is nondegenerate if and only if $\mathrm{LCP}(A, b)$ has a finite number of solutions for any vector b [6]. Recognizing nondegenerate matrices is co-NP-complete [2,20]. A square matrix A is said to be *sign-nondegenerate* if $\tilde{A}$ is nondegenerate for any $\tilde{A} \in \mathcal{Q}(A)$. Then the following lemma holds, which implies that sign-nondegeneracy can be tested in polynomial time.

Lemma 2.1. *A square matrix A is sign-nondegenerate if and only if A is a sign-nonsingular matrix with nonzero diagonals.*

Proof. To see the necessity, suppose that A is sign-nondegenerate. Let $\tilde{A}$ be a matrix in $\mathcal{Q}(A)$. Since all principal minors in $\tilde{A}$ are nonzero, all diagonal entries are nonzero. Moreover, $\det \tilde{A}$ is nonzero, which implies that A is sign-nonsingular. Thus A is a sign-nonsingular matrix with nonzero diagonals.

To see the sufficiency, suppose that A is a sign-nonsingular matrix with nonzero diagonals. Let $J \subseteq N$ be an index subset. Since the principal submatrix $A[J]$ has nonzero diagonals, $A[J]$ is term-nonsingular. Let σ_1 and σ_2 be bijections from J to J such that $p(A[J]|_{\sigma_1}) \neq 0$ and $p(A[J]|_{\sigma_2}) \neq 0$. Define bijections $\pi_k : N \to N$ to be $\pi_k(j) = j$ if $j \in N \setminus J$ and $\pi_k(j) = \sigma_k(j)$ if $j \in J$ for $k = 1, 2$. Since A has nonzero diagonals, $p(A|_{\pi_1})$ and $p(A|_{\pi_2})$ are both nonzero. By $p(A|_{\pi_k}) = p(A[J]|_{\sigma_k}) \prod_{i \in N \setminus J} a_{ii}$ for $k = 1, 2$, it follows from sign-nonsingularity of A that the two nonzero terms $p(A[J]|_{\sigma_1})$ and $p(A[J]|_{\sigma_2})$ have the same sign. Thus $A[J]$ is sign-nonsingular, which implies that A is sign-nondegenerate. $\square$

We now obtain the following theorem. For $\mathrm{LCP}(A, b)$, let M be the matrix in the form of $M = (A \ b)$, where the column set is indexed by $N \cup \{g\}$.

Theorem 2.2. *For a linear complementarity problem $\mathrm{LCP}(A, b)$ with nonzero diagonals, if the matrix $M = (A \ b)$ is totally sign-nonsingular, then $\mathrm{LCP}(A, b)$ is sign-solvable.*

Proof. First assume that $\mathrm{LCP}(A, b)$ has a solution (w, z). Let J be the support of z. Then we have $A_J \begin{pmatrix} w[N \setminus J] \\ z[J] \end{pmatrix} + b = 0$, where A_J is the matrix in the form of

$$A_J = \begin{pmatrix} O & A[J] \\ -I & A[N \setminus J, J] \end{pmatrix}.$$

Since A is sign-nondegenerate by Lemma 2.1, each principal submatrix is sign-nonsingular, and hence A_J is also sign-nonsingular by $\det A_J = \pm \det A[J]$. Then it holds by Cramer's rule that

$$z_j = \begin{cases} -\det A_J^j / \det A_J, & \text{if } j \in J, \\ 0, & \text{if } j \in N \setminus J, \end{cases} \tag{1}$$

$$w_j = \begin{cases} 0, & \text{if } j \in J, \\ -\det A_J^j / \det A_J, & \text{if } j \in N \setminus J, \end{cases} \tag{2}$$

where A_J^j is the matrix obtained from A_J by replacing the jth column vector of A_J with b. The determinant of A_J^j is represented by

$$\det A_J^j = \begin{cases} \pm \det M[J, J - j + g], & \text{if } j \in J, \\ \pm \det M[J + j, J + g], & \text{if } j \in N \setminus J, \end{cases} \tag{3}$$

where $J - j + g$ means $J \setminus \{j\} \cup \{g\}$ with g being put at the position of j in J, the set $J + j$ coincides with $J \cup \{j\}$, and $J + g$ means $J \cup \{g\}$ in which g is put at the same position as that of j in $J + j$.

We show that A_J^j is either term-singular or sign-nonsingular for any $J \subseteq N$ and $j \in N$. Assume that there exists $j \in N$ such that A_J^j is term-nonsingular, but not sign-nonsingular. First suppose that $j \in J$. By (3), the submatrix $M[J, J - j + g]$ is term-nonsingular, but not sign-nonsingular. Then there exist two bijections σ_1 and σ_2 from J to $J - j + g$ such that $p(M[J, J - j + g]|_{\sigma_1})$ and $p(M[J, J - j + g]|_{\sigma_2})$ are both nonzero, and have the opposite signs. Define two bijections $\pi_k : N \to N - j + g$ to be $\pi_k(i) = i$ if $i \in N \setminus J$ and $\pi_k(i) = \sigma_k(i)$ if $i \in J$ for $k = 1, 2$. By $p(M[N, N - j + g]|_{\pi_k}) = p(M[J, J - j + g]|_{\sigma_k}) \prod_{i \in N \setminus J} a_{ii}$ for $k = 1, 2$, the two nonzero terms $p(M[N, N - j + g]|_{\pi_1})$ and $p(M[N, N - j + g]|_{\pi_2})$ are both nonzero, and have the opposite signs. This contradicts the total sign-nonsingularity of M. Next suppose that $j \in N \setminus J$. Then, by (3), $M[J + j, J + g]$ is term-nonsingular, but not sign-nonsingular. Let σ_1 and σ_2 be bijections from $J + j$ to $J + g$ such that $p(M[J + j, J + g]|_{\sigma_1})$ and $p(M[J + j, J + g]|_{\sigma_2})$ are both nonzero, and have the opposite signs. Define two bijections $\pi_k : N \to N - j + g$ for $k = 1, 2$ to be $\pi_k(i) = i$ if $i \in N \setminus (J \cup \{j\})$ and $\pi_k(i) = \sigma_k(i)$ if $i \in J \cup \{j\}$. Then the two nonzero terms $p(M[N, N - j + g]|_{\pi_1})$ and $p(M[N, N - j + g]|_{\pi_2})$ have the opposite signs, which contradicts the total sign-nonsingularity of M.

Thus A_J^j is either term-singular or sign-nonsingular for any index j. The matrix A_J is sign-nonsingular. Therefore, it follows from (1) that the sign pattern of (w, z) is independent of the magnitudes of A and b. Hence $\text{LCP}(\tilde{A}, \tilde{b})$ has a solution with the same sign pattern as that of (w, z) for any $\tilde{A} \in \mathcal{Q}(A)$ and $\tilde{b} \in \mathcal{Q}(b)$. Thus $\text{LCP}(A, b)$ is sign-solvable.

Next assume that $\text{LCP}(A, b)$ has no solutions. Note that $\text{LCP}(A, b)$ has no solutions if and only if $A_J x + b = 0$ has no nonnegative solutions for any $J \subseteq N$, that is, there exists $j \in N$ such that $(A_J^{-1} b)_j < 0$ for any $J \subseteq N$. It follows from Cramer's rule that we have $(A_J^{-1} b)_j = -\det A_J^j / \det A_J < 0$. Since $\det A_J^j \neq 0$, the matrix A_J^j is sign-nonsingular. Hence it holds that $-\det \tilde{A}_J^j / \det \tilde{A}_J < 0$ for any $\tilde{A} \in \mathcal{Q}(A)$ and $\tilde{b} \in \mathcal{Q}(b)$. Thus $\text{LCP}(\tilde{A}, \tilde{b})$ has no solutions for any $\tilde{A} \in \mathcal{Q}(A)$ and $\tilde{b} \in \mathcal{Q}(b)$, which means that $\text{LCP}(A, b)$ is sign-solvable.   □

Sign-solvable LCPs do not necessarily satisfy this sufficient condition. Indeed, consider $\text{LCP}(A, b)$, where A and b are defined to be

$$A = \begin{pmatrix} -p_1 & -p_2 \\ +p_3 & +p_4 \end{pmatrix} \text{ and } b = \begin{pmatrix} 0 \\ +p_5 \end{pmatrix}$$

for positive constants $p_1, \ldots, p_5 > 0$. Then $\text{LCP}(A, b)$ has a unique solution $w = (0 \ p_5)^\text{T}$ and $z = 0$, and hence $\text{LCP}(A, b)$ is sign-solvable. However, this does not satisfy the condition of Theorem 2.2, as A is not sign-nonsingular.

We conclude this section with sign-solvability of LCPs associated with another class of matrices. A square matrix A is a *P-matrix* if every principal minor is positive. A P-matrix is clearly nondegenerate. It is known that A is a P-matrix if and only if $\text{LCP}(A, b)$ has a unique solution for any vector b. Recognizing P-matrices is co-NP-complete [7]. A matrix A is a *sign-P-matrix* if all matrices in $\mathcal{Q}(A)$ are P-matrices. Then similar statements to Lemma 2.1 and Theorem 2.2 hold for sign-P-matrices.

Corollary 2.3. *A square matrix A is a sign-P-matrix if and only if A is a sign-nonsingular matrix with positive diagonals.*

Corollary 2.4. *For a linear complementarity problem $\text{LCP}(A, b)$ with positive diagonals, if the matrix $M = (A \ b)$ is totally sign-nonsingular, then $\text{LCP}(\tilde{A}, \tilde{b})$ has a unique solution with the same sign pattern as that of $\text{LCP}(A, b)$.*

3   Sign-Solvable LCPs with Nonzero Diagonals

In this section, we describe a characterization for a sign-solvable $\text{LCP}(A, b)$ with nonzero diagonals. Recall that M is the matrix in the form of $M = (A \ b)$, where the column set is indexed by $N \cup \{g\}$.

3.1   The Residual Row-Mixed Matrix

We first introduce the *residual row-mixed* matrix of $\text{LCP}(A, b)$ with nonzero diagonals.

For each row index i, the ith equation of $\text{LCP}(A, b)$ is represented by

$$w_i = \sum_{j \in \Gamma(\{i\})} a_{ij} z_j + b_i. \tag{4}$$

First assume that M has a nonpositive row i, that is, $b_i \leq 0$ and $a_{ij} \leq 0$ for all $j \in N$. Suppose that $b_i < 0$. Since any solution of $\text{LCP}(A, b)$ is nonnegative, the ith row implies that $\text{LCP}(A, b)$ has no solutions. Next suppose that $b_i = 0$. Then, if $\text{LCP}(A, b)$ has a solution (w, z), the solution (w, z) must satisfy that $z_j = 0$ for any $j \in \Gamma(\{i\})$.

Next assume that M has a nonnegative row i, that is, $b_i \geq 0$ and $a_{ij} \geq 0$ for all $j \in N$. Let (w, z) be a solution of $\mathrm{LCP}(A, b)$. If $w_i > 0$, then the complementarity implies $z_i = 0$. Suppose that $w_i = 0$. Since any solution is nonnegative, (w, z) must satisfy $z_j = 0$ for any $j \in \Gamma(\{i\})$, and hence $z_i = 0$ by $a_{ii} \neq 0$. Thus, if $\mathrm{LCP}(A, b)$ has a solution and M has a nonnegative row i, any solution of $\mathrm{LCP}(A, b)$ must satisfy that $z_i = 0$. Note that there exists $j \in \Gamma(\{i\})$ with $z_j > 0$ if and only if the left-hand side of (4) is positive, i.e., $w_i > 0$.

Therefore, if M has a nonnegative or nonpositive row, then we know that some entries of any solution must be zero. We can repeat this process as follows. Set $M^{(1)} = M$. For a positive integer ν and a matrix $M^{(\nu)}$, let $I_-^{(\nu)}$ be the set of nonpositive rows in $M^{(\nu)}$, and $I_+^{(\nu)}$ be the set of nonnegative rows that have a nonzero entry in $M^{(\nu)}$. If $\Gamma(I_-^{(\nu)})$ contains the index g, then the LCP has no solutions. Define $I^{(\nu)} = I_+^{(\nu)} \cup I_-^{(\nu)}$ and $J^{(\nu)} = I_+^{(\nu)} \cup \Gamma(I_-^{(\nu)})$. Then any solution (w, z) of $\mathrm{LCP}(A, b)$ satisfies $z_j = 0$ for any $j \in J^{(\nu)}$. Let $M^{(\nu+1)}$ be the matrix obtained from $M^{(\nu)}$ by deleting the rows indexed by $I^{(\nu)}$ and the columns indexed by $J^{(\nu)}$. Repeat this for $\nu = 1, 2, \dots$ until $I^{(\nu)} = J^{(\nu)} = \emptyset$, that is, until either $M^{(\nu)}$ is row-mixed or $M^{(\nu)}$ has no rows.

We call the remaining row-mixed submatrix M' the *residual row-mixed* matrix of $\mathrm{LCP}(A, b)$. Note that, if $\mathrm{LCP}(A, b)$ has solutions, the column index g is not deleted in each iteration.

Assume that the column set of M' contains the index g. Let M' be in the forms of $M' = (A' \; b')$, where b' is the subvector of b and A' is the submatrix of A with row set U' and column set V'. We denote $\bar{U}' = N \setminus U'$ and $\bar{V}' = N \setminus V'$. Since A has nonzero diagonals, $\bar{U}' \subseteq \bar{V}'$ holds, and hence we have $V' \subseteq U'$. Suppose that M' has no rows. Then $\bar{V}' = N$ holds, which means that any solution (w, z) of $\mathrm{LCP}(A, b)$ must satisfy $z = 0$. Since g is not deleted in each iteration, the vector b is nonnegative. Thus $(b, 0)$ is a unique solution of $\mathrm{LCP}(A, b)$. Next suppose that M' is row-mixed. Consider the following system:

$$
\begin{aligned}
& w = A'z + b', \\
& w_i^{\mathrm{T}} z_i = 0, \text{ for any } i \in V', \\
& w \geq 0, \; z \geq 0.
\end{aligned}
\tag{5}
$$

We claim that there exists a one-to-one correspondence between solutions of $\mathrm{LCP}(A, b)$ and (5). For a solution (w, z) of $\mathrm{LCP}(A, b)$, the pair $(w[U'], z[V'])$ is a solution of (5). Conversely, let (w', z') be a solution of (5). Define (w, z) to be $z[V'] = z'$, $z[\bar{V}'] = 0$, and $w = Az + b$. Then $w[U'] = A'z' + b' = w' \geq 0$ holds. Moreover, since each row in $A[\bar{U}', V']$ is nonnegative, we have $w[\bar{U}'] = A[\bar{U}', V']z' + b[\bar{U}'] \geq 0$. By $V' \subseteq U'$, the pair (w, z) satisfies the complementarity $w^{\mathrm{T}} z = 0$. Thus (w, z) is a solution of $\mathrm{LCP}(A, b)$.

3.2   Characterization

Using the residual row-mixed matrix M' of $\mathrm{LCP}(A, b)$, we have the following theorem.

Theorem 3.1. *Let* LCP(A, b) *be a linear complementarity problem with nonzero diagonals, and M' be the residual row-mixed matrix. Then* LCP(A, b) *is sign-solvable if and only if one of the followings holds:*

- *The column set of M' does not contain the index g.*
- *The residual row-mixed matrix M' has no rows.*
- *The residual row-mixed matrix M' is totally sign-nonsingular.*

In order to prove this theorem, we give some definitions. A linear system $Ax = b$ has *signed nonnegative solutions* if the set of the sign patterns of nonnegative solutions of $\tilde{A}x = \tilde{b}$ is the same as that of nonnegative solutions of $Ax = b$ for any $\tilde{A} \in \mathcal{Q}(A)$ and $\tilde{b} \in \mathcal{Q}(b)$. A matrix A is said to have *signed nonnegative null space* if $Ax = 0$ has signed nonnegative solutions. Matrices with signed nonnegative null space were examined by Fisher, Morris, and Shapiro [9]. They showed that a row-mixed matrix has signed nonnegative null space if and only if it is the matrix called *mixed dominating*, which is defined to be a row-mixed matrix which does not contain a square row-mixed submatrix. By the result of mixed dominating matrices, the following two lemmas hold.

Lemma 3.2 (Fischer and Shapiro [10]). *If a row-mixed matrix A has signed nonnegative null space, then the rows of A are linearly independent.*

A matrix A is said to have *row-full term-rank* if A has a term-nonsingular submatrix with row size. A matrix A has *column-full term-rank* if A^{T} has row-full term-rank.

Lemma 3.3 (Fischer, Morris, and Shapiro [9]). *An $n \times (n + 1)$ row-mixed matrix has signed nonnegative null space if and only if it is a totally sign-nonsingular matrix with row-full term-rank.*

We have the following lemmas.

Lemma 3.4. *Suppose that the matrix $(A\ b)$ is row-mixed. If the linear system $Ax + b = 0$ has signed nonnegative solutions, then it has a solution all of whose entries are positive.*

Proof. Since $(A\ b)$ is row-mixed, there exist $\tilde{A} \in \mathcal{Q}(A)$ and $\tilde{b} \in \mathcal{Q}(b)$ such that the sum of the columns of $\tilde{A}$ and $\tilde{b}$ is zero, that is, $\tilde{A}\mathbf{1} + \tilde{b} = 0$, where $\mathbf{1}$ is the column vector whose entries are all one. This implies that $\tilde{A}x = \tilde{b}$ has a solution all of whose entries are positive for any $\tilde{A} \in \mathcal{Q}(A)$ and $\tilde{b} \in \mathcal{Q}(b)$.   □

Lemma 3.5. *Suppose that $M = (A\ b)$ is row-mixed. The linear system $Ax + b = 0$ has signed nonnegative solutions if and only if M has signed nonnegative null space.*

Proof. Suppose that the matrix M has signed nonnegative null space. Since $\{x \mid Ax + b = 0, x \geq 0\} = \{x \mid (A\ b)\binom{x}{1} = 0, x \geq 0\}$ is contained in the set of nonnegative vectors in the null space of M, the linear system $Ax + b = 0$ has signed nonnegative solutions.

Next suppose that $Ax + b = 0$ has signed nonnegative solutions, and that M does not have signed nonnegative null space. Then M is not mixed dominating, which means that there exists a row-mixed square submatrix in M. Note that a row-mixed square submatrix which does not contain any row-mixed square proper submatrix is term-nonsingular. Choose a row-mixed term-nonsingular submatrix $M[I, J]$ such that $|J|$ is maximum. Since M is row-mixed, the maximality implies that each row of $M[N \setminus I, J]$ is mixed or zero.

We define B to be $B = M[N, J \setminus \{g\}]$ if $g \in J$, and $B = M[N, J]$ otherwise. Then $(B\ b)$ does not have signed nonnegative null space. The set of the nonnegative vectors in the null space of $(B\ b)$ consists of the union of $\{x \mid Bx = 0, x \geq 0\}$ and $\{x \mid (B\ b)\binom{x}{x_g} = 0, x \geq 0, x_g > 0\}$. Since the set of sign patterns in the second one coincides with that of $\{x \mid Bx + b = 0, x \geq 0\}$ and $Bx + b = 0$ has signed nonnegative solutions, we may assume that B does not have signed nonnegative null space. Let $\tilde{B} \in \mathcal{Q}(B)$ be a matrix such that $\tilde{B}$ has column-full rank. Then the null space of $\tilde{B}$ is empty, and $\tilde{B}x + b = 0$ has a unique solution all of whose entries are positive by Lemma 3.4. By the assumption, there exists $\hat{B} \in \mathcal{Q}(B)$ such that $\hat{B}x = 0$ has a nonnegative, nonzero solution x^*. Lemma 3.4 implies that $\hat{B}x + b = 0$ has a solution x^0 all of whose entries are positive. Then $x^0 - \mu x^*$, where $\mu = \min_{i \in N} x_i^0/x_i^*$, is also a nonnegative solution of $\hat{B}x + b = 0$. Thus the linear system $Bx + b = 0$ does not have signed nonnegative solutions, which contradicts that $Ax + b = 0$ has signed nonnegative solutions.   □

We are now ready to prove Theorem 3.1.

Proof of Theorem 3.1. To show the necessity, suppose that $\text{LCP}(A, b)$ is sign-solvable. Assume that M' has a row and that M' is in the form of $M' = (A'\ b')$, where b' is the subvector of b indexed by g. Let x be a nonnegative vector with $A'x + b' = 0$. Since there exists a one-to-one correspondence between solutions of $\text{LCP}(A, b)$ and (5), $(0, x)$ is a solution of (5). Hence the sign-solvability of $\text{LCP}(A, b)$ implies that the linear system $A'x + b' = 0$ has signed nonnegative solutions. It follows from Lemma 3.5 that $M' = (A'\ b')$ has signed nonnegative null space. By Lemma 3.2 and $V' \subseteq U'$, it holds that $U' = V'$, i.e., A' is square. Therefore, Lemma 3.3 implies that M' is totally sign-nonsingular.

We next show the sufficiency. If the column set of M' does not contain the index g, then clearly $\text{LCP}(A, b)$ is a sign-solvable LCP with no solutions. Suppose that M' is in the forms of $M' = (A'\ b')$. If M' has no rows, then $(b, 0)$ is a unique solution of $\text{LCP}(A, b)$, which means that $\text{LCP}(A, b)$ is sign-solvable. Next suppose that $M' = (A'\ b')$ is totally sign-nonsingular. By $V' \subseteq U'$, it holds that $|U'| = |V'|$ or $|U'| = |V'| + 1$. If $|U'| = |V'|$, then M' is sign-nonsingular, which contradicts that M' is row-mixed. Hence we have $|U'| = |V'| + 1$. Since A' has nonzero diagonals, (5) forms the linear complementarity problem with nonzero diagonals. By Theorem 2.2, $\text{LCP}(A', b')$ is sign-solvable, and hence so is $\text{LCP}(A, b)$.   □

Note that $\text{LCP}(A, b)$ is a sign-solvable LCP with no solutions if and only if the column set of M' does not contain g.

If M is row-mixed, then the residual row-mixed matrix is M itself. Hence Theorem 3.1 implies the following corollary.

Corollary 3.6. *Let A have nonzero diagonals, and $M = (A\ b)$ be a row-mixed matrix. Then $\mathrm{LCP}(A, b)$ is sign-solvable if and only if the matrix M is totally sign-nonsingular.*

We close this section with an example of sign-solvable LCPs with nonzero diagonals. Consider $\mathrm{LCP}(A, b)$, where A and b have the sign patterns, respectively,

$$
\begin{pmatrix}
+ & + & 0 & 0 & 0 \\
- & + & + & 0 & + \\
+ & - & + & - & 0 \\
- & 0 & - & - & + \\
0 & - & + & 0 & +
\end{pmatrix}
\quad \text{and} \quad
\begin{pmatrix}
0 \\
+ \\
0 \\
0 \\
-
\end{pmatrix}.
$$

The residual row-mixed matrix is

$$
\begin{pmatrix}
+ & - & 0 & 0 \\
- & - & + & 0 \\
+ & 0 & + & -
\end{pmatrix},
$$

which is obtained from the matrix $(A\ b)$ by deleting the first two rows and the first two columns. This residual row-mixed matrix is totally sign-nonsingular, and hence $\mathrm{LCP}(A, b)$ is sign-solvable.

4 Algorithm for Sign-Solvable LCPs with Nonzero Diagonals

In this section, we describe an algorithm for a given $\mathrm{LCP}(A, b)$ with nonzero diagonals. The algorithm tests sign-solvability of $\mathrm{LCP}(A, b)$, and finds the sign pattern of a solution of $\mathrm{LCP}(A, b)$ if it is sign-solvable.

The algorithm starts with finding the residual row-mixed matrix M' as described in the previous section. If the column set of M' does not contain the index g, then $\mathrm{LCP}(A, b)$ is sign-solvable and has no solutions. Let M' be in the forms of $M' = (A'\ b')$, where b' is the subvector of b and A' is the submatrix of A with row set U' and column set V'. We denote $\bar{U}' = N \setminus U'$ and $\bar{V}' = N \setminus V'$. Note that $V' \subseteq U'$ holds. If M' has a row and M' is not totally sign-nonsingular, then return that $\mathrm{LCP}(A, b)$ is not sign-solvable by Theorem 3.1.

Assume that M' has no rows. Then $\mathrm{LCP}(A, b)$ is sign-solvable, and $(b, 0)$ is a unique solution of $\mathrm{LCP}(A, b)$. Next assume that M' has a row and $M' = (A'\ b')$ is totally sign-nonsingular. Then $\mathrm{LCP}(A, b)$ is sign-solvable by Theorem 3.1. Since M' is row-mixed, there exists $\tilde{M} = (\tilde{A}\ \tilde{b}) \in \mathcal{Q}(M)$ such that the sum of the columns of $\tilde{M}' \in \mathcal{Q}(M')$ is zero. Hence it follows from (5) that the pair (w, z), defined to be $z[\bar{V}'] = 0$, $z[V'] = +1$, and $w = \tilde{A}z + \tilde{b}$, is a solution of $\mathrm{LCP}(\tilde{A}, \tilde{b})$. This means that the vector w satisfies that $w_j > 0$ if $j \in \bar{U}'$ and $A[\{j\}, V']$ has nonzero entries, and $w_j = 0$ otherwise. Since $\mathrm{LCP}(A, b)$ is sign-solvable, (w, z) is the sign pattern of a solution of $\mathrm{LCP}(A, b)$.

We now summarize the algorithm description.

Algorithm: An algorithm for LCPs with nonzero diagonals.

Input: A linear complementarity problem $LCP(A, b)$ with nonzero diagonals.

Output: The sign pattern of a solution if $LCP(A, b)$ is sign-solvable.

Step 1: Set $M^{(1)} = M$ and $\nu = 1$. Repeat the following until $I^{(\nu)} = J^{(\nu)} = \emptyset$.

    1-1: Find $I_-^{(\nu)}$ and $I_+^{(\nu)}$, where $I_-^{(\nu)}$ is the set of nonpositive rows in $M^{(\nu)}$, and $I_+^{(\nu)}$ is the set of nonnegative rows that have a nonzero entry in $M^{(\nu)}$.

    1-2: If $g \in \Gamma(I_-^{(\nu)})$, then return that $LCP(A, b)$ is sign-solvable and has no solutions.

    1-3: Let $I^{(\nu)} = I_+^{(\nu)} \cup I_-^{(\nu)}$ and $J^{(\nu)} = I_+^{(\nu)} \cup \Gamma(I_-^{(\nu)})$. Define $M^{(\nu+1)}$ to be the matrix obtained by deleting the rows indexed by $I^{(\nu)}$ and the columns indexed by $J^{(\nu)}$ from $M^{(\nu)}$.

    1-4: Set $\nu = \nu + 1$ and go back to Step 1.

Step 2: Let $M' = (A' \; b')$ be the remaining submatrix, and U', V' be the row and column sets of A', respectively. If M' has a row and M' is not totally sign-nonsingular, then return that $LCP(A, b)$ is not sign-solvable. Otherwise go to Step 3.

Step 3: Return that $LCP(A, b)$ is sign-solvable and do the following.

    3-1: If U' is empty, then return the sign pattern of a solution $(w, z) = (b, 0)$.

    3-2: Otherwise, return the sign pattern of (w, z) defined to be

$$\operatorname{sgn} z_j = \begin{cases} +, & \text{if } j \in V' \\ 0, & \text{otherwise} \end{cases} \quad \text{and} \quad \operatorname{sgn} w_j = \begin{cases} +, & \text{if } j \in K \\ 0, & \text{otherwise} \end{cases} \tag{6}$$

    where K is the set of rows which have nonzero entries in $A[\bar{U}', V']$, that is, $K = \{j \in \bar{U}' \mid \Gamma(\{j\}) \cap V' \neq \emptyset\}$.

Applying this algorithm to the example at the end of Sect. 3, we obtain the sign pattern of a solution, $w = (\; 0 \; + \; 0 \; 0 \; 0)^{\mathrm{T}}$ and $z = (\; 0 \; 0 \; + \; + \; +)^{\mathrm{T}}$.

Based on this algorithm, we can compute a solution of a sign-solvable LCP as well as the sign pattern of a solution. Suppose that M' has a row. The solution (w, z) with the obtained sign pattern satisfies that $A'z[V'] + b' = 0$, $z[\bar{V}'] = 0$. Since A' is nonsingular by total sign-nonsingularity of M', we can compute a solution of $LCP(A, b)$ by performing Gaussian elimination.

The running time bound of the algorithm is now given as follows. Note that an $n \times (n + 1)$ row-mixed matrix A is a totally sign-nonsingular matrix with row-full term-rank if and only if all square submatrices of order n are sign-nonsingular [1, Theorem 5.3.3]. Such matrix is called an *S-matrix* in [1,15], which can be recognized in $O(n^2)$ time [14].

Theorem 4.1. *For a linear complementarity problem $LCP(A, b)$ with nonzero diagonals, let n be the matrix size of A, and γ the number of nonzero entries in A and b. Then the algorithm tests sign-solvability in $O(n^2)$ time, and, if $LCP(A, b)$ is sign-solvable, the algorithm finds the sign pattern of a solution in $O(\gamma)$ time.*

Proof. In the νth iteration in Step 1, it requires $O(\gamma_\nu)$ time to find $I^{(\nu)}$ and $J^{(\nu)}$, where γ_ν is the number of nonzero entries in the columns deleted in the νth iteration. Since each column is deleted at most once, Step 1 takes $O(\gamma)$ time in total. In Step 2, if the residual row-mixed matrix M' is totally sign-nonsingular, M' has row-full term-rank and the column size is one larger than the row size. Hence testing total sign-nonsingularity of M' is equivalent to recognizing S-matrices. Thus it requires $O(n^2)$ time to test sign-solvability in Step 2. Step 3 requires $O(\gamma)$ time. Thus this statement holds. □

Acknowledgements

The author is very obliged to Satoru Iwata for his suggestions and reading the draft of the paper. This work is supported by the 21st Century COE Program on Information Science and Technology Strategic Core at the University of Tokyo from the Ministry of Education, Culture, Sports, Science and Technology of Japan.

References

1. Brualdi, R.A., Shader, B.L.: Matrices of Sign-solvable Linear Systems. Cambridge University Press, Cambridge (1995)
2. Chandrasekaran, R., Kabadi, S.N., Murty, K.G.: Some NP-complete problems in linear programming. Operations Research Letters **1** (1982) 101–104
3. Chung, S.J.: NP-completeness of the linear complementarity problem. Journal of Optimization Theory and Applications **60** (1989) 393–399
4. Cottle, R.W.: The principal pivoting method of quadratic programming. In Dantzig, G.B., Veinott, A.F., eds.: Mathematics of Decision Sciences, Part 1. American Mathematical Society, Providence R. I. (1968) 142–162
5. Cottle, R.W., Dantzig, G.B.: Complementary pivot theory of mathematical programming. Linear Algebra and Its Applications **1** (1968) 103–125
6. Cottle, R.W., Pang, J.S., Stone, R.E.: The Linear Complementarity Problem. Academic Press (1992)
7. Coxson, G.E.: The P-matrix problem is co-NP-complete. Mathematical Programming **64** (1994) 173–178
8. Eijndhoven, J.T.J.V.: Solving the linear complementarity problem in circuit simulation. SIAM Journal on Control and Optimization **24** (1986) 1050–1062
9. Fischer, K.G., Morris, W., Shapiro, J.: Mixed dominating matrices. Linear Algebra and Its Applications **270** (1998) 191–214
10. Fischer, K.G., Shapiro, J.: Mixed matrices and binomial ideals. Journal of Pure and Applied Algebra **113** (1996) 39–54
11. Iwata, S., Kakimura, N.: Solving linear programs from sign patterns. Mathematical Programming, to appear.
12. Kim, S.J., Shader, B.L.: Linear systems with signed solutions. Linear Algebra and Its Applications **313** (2000) 21–40
13. Kim, S.J., Shader, B.L.: On matrices which have signed null-spaces. Linear Algebra and Its Applications **353** (2002) 245–255

14. Klee, V.: Recursive structure of S-matrices and $O(m^2)$ algorithm for recognizing strong sign-solvability. Linear Algebra and Its Applications **96** (1987) 233–247

15. Klee, V., Ladner, R., Manber, R.: Sign-solvability revisited. Linear Algebra and Its Applications **59** (1984) 131–158

16. Lemke, C.E.: Bimatrix equilibrium points and mathematical programming. Management Science **11** (1965) 681–689

17. Lovász, L., Plummer, M.D.: Matching Theory. Volume 29 of Annals of Discrete Mathematics. North-Holland, Amsterdam (1986)

18. McCuaig, W.: Brace decomposition. Journal of Graph Theory **38** (2001) 124–169

19. McCuaig, W.: Pólya's permanent problem. The Electronic Journal of Combinatorics **11, R79** (2004)

20. Murty, K.G.: Linear Complementarity, Linear and Nonlinear Programming. Internet Edition (1997)

21. Pólya, G.: Aufgabe 424. Archiv der Mathematik und Physik **20** (1913) 271

22. Robertson, N., Seymour, P.D., Thomas, R.: Permanents, Pfaffian orientations, and even directed circuits. Annals of Mathematics **150** (1999) 929–975

23. Rohn, J.: Systems of linear interval equations. Linear Algebra and Its Applications **126** (1989) 39–78

24. Samuelson, P.A.: Foundations of Economics Analysis. Harvard University Press, 1947; Atheneum, New York, 1971.

25. Seymour, P., Thomassen, C.: Characterization of even directed graphs. Journal of Combinatorial Theory, Series B **42** (1987) 36–45

26. Shao, J.Y., Ren, L.Z.: Some properties of matrices with signed null spaces. Discrete Mathematics **279** (2004) 423–435

27. Vazirani, V.V., Yannakakis, M.: Pfaffian orientations, 0-1 permanents, and even cycles in directed graphs. Discrete Applied Mathematics **25** (1989) 179–190

An Integer Programming Approach for Linear Programs with Probabilistic Constraints

James Luedtke, Shabbir Ahmed, and George Nemhauser

H. Milton Stewart School of Industrial and Systems Engineering
Georgia Institute of Technology
Atlanta, GA, USA
{jluedtke,sahmed,gnemhaus}@isye.gatech.edu

Abstract. Linear programs with joint probabilistic constraints (PCLP) are known to be highly intractable due to the non-convexity of the feasible region. We consider a special case of PCLP in which only the right-hand side is random and this random vector has a finite distribution. We present a mixed integer programming formulation and study the relaxation corresponding to a single row of the probabilistic constraint, yielding two strengthened formulations. As a byproduct of this analysis, we obtain new results for the previously studied mixing set, subject to an additional knapsack inequality. We present computational results that indicate that by using our strengthened formulations, large scale instances can be solved to optimality.

Keywords: Integer programming, probabilistic constraints, stochastic programming.

1 Introduction

Consider a linear program with a probabilistic or chance constraint

$$(PCLP) \quad \min\left\{cx : x \in X,\ P\{\tilde{T}x \geq \xi\} \geq 1 - \epsilon\right\} \tag{1}$$

where $X = \{x \in \mathbb{R}_+^d : Ax = b\}$ is a polyhedron, $c \in \mathbb{R}^d$, $\tilde{T}$ is an $m \times d$ random matrix, ξ is a random vector taking values in $\mathbb{R}^m$, and ϵ is a confidence parameter chosen by the decision maker, typically near zero, e.g., $\epsilon = 0.01$ or $\epsilon = 0.05$. Note that in (1) we enforce a single probabilistic constraint over *all* rows, rather than requiring that each row independently be satisfied with high probability. Such a constraint is known as a *joint probabilistic constraint*, and is appropriate in a context in which it is important to have all constraints satisfied simultaneously and there may be dependence between random variables in different rows.

Problems with joint probabilistic constraints have been extensively studied; see [1] for background and an extensive list of references. Probabilistic constraints have been used in various applications including supply chain management [2], production planning [3], optimization of chemical processes [4,5] and surface water quality management [6]. Unfortunately, linear programs with probabilistic

M. Fischetti and D.P. Williamson (Eds.): IPCO 2007, LNCS 4513, pp. 410–423, 2007.

constraints are still largely intractable except for a few very special cases. There are two primary reasons for this intractability. First, in general, for a given $x \in X$, the quantity $\phi(x) := P\{\tilde{T}x \geq \xi\}$ is hard to compute, as it requires multidimensional integration. Second, the feasible region defined by a probabilistic constraint is generally not convex.

Recently, several approaches have been proposed which can find highly reliable feasible solutions to probabilistic programs. Examples of these conservative approximations include scenario approximation [7,8], Bernstein approximation [9] and robust optimization, e.g., [10,11,12]. These methods are attractive when high reliability is most important and solution cost is a secondary objective. However, when very high reliability is not crucial, for example if the probabilistic constraint represents a service level constraint, a decision maker may be interested in exploring the trade-off between solution cost and system reliability, and would be interested in obtaining solutions which are on or near the efficient frontier of these competing objectives. The aforementioned conservative approximations generally do not yield bounds on the optimal solution cost at a given reliability level ϵ, and hence cannot distinguish whether the produced solutions are close to the efficient frontier. This latter context is the motivation for using integer programming to solve PCLP so that we can obtain solutions that are *provably* optimal or near optimal.

In this work, we demonstrate that by using integer programming techniques, PCLP can be solved efficiently under the following two simplifying assumptions:

(**A1**) Only the right-hand side vector ξ is random; the matrix $\tilde{T} = T$ is deterministic.

(**A2**) The random vector ξ has a finite distribution.

Despite its restrictiveness, the special case given by assumption A1 has received a lot of attention in the literature, see, e.g., [1,13,14]. A notable result for this case is that if the distribution of the right-hand side is log-concave, then the feasible region defined by the joint probabilistic constraint is convex [15]. This allows problems with small dimension of the random vector to be solved to optimality, but higher dimensional problems are still intractable due to the previously mentioned difficulty in checking feasibility of the probabilistic constraint. Specialized methods have been developed in [14] for the case in which assumption A1 holds and the random vector has discrete but not necessarily finite distribution. However, these methods also do not scale well with the dimension of the random vector. Assumption A2 may also seem very restrictive. However, if the possible values for ξ are generated by taking Monte Carlo samples from a general distribution, we can think of the resulting problem as an approximation of the problem with this distribution. Under some reasonable assumptions we can show that the optimal solution of the sampled problem converges exponentially fast to the optimal solution of the original problem as the number of scenarios increases. Also, the optimal objective of the sampled problem can be used to develop statistical lower bounds on the optimal objective of the original problem. See [16,17,18] for some related results. It seems that the reason such a sampling approach has not been seriously considered for PCLP in the past is that the resulting sampled problem has a non-convex feasible region, and thus is still generally intractable. Our contribution is to demonstrate that, at least under assumption A1, it is nonetheless possible to solve the sampled problem in practice.

Under assumption A2 it is possible to write a mixed integer programming formulation for PCLP, as has been done, for example, in [19]. In the general case, such a formulation requires the introduction of "big-M" type constraints, and hence is difficult to solve. However, the particular case of assumption A1 has not been studied from an integer programming perspective; by doing so, we are able to develop strong mixed integer programming formulations. Our approach in developing these formulations is to consider the relaxation obtained from a single row in the probabilistic constraint. It turns out that this yields a system similar to the *mixing set* introduced by Günlük and Pochet [20], subject to an additional knapsack inequality. We are able to derive strong valid inequalities for this system by first using the knapsack inequality to "pre-process" the mixing set, then applying the mixing inequalities of [20], see also [21,22]. We also derive an extended formulation, equivalent to one given by Miller and Wolsey in [23]. Making further use of the knapsack inequality, we are able to derive more general classes of valid inequalities, for both the original and extended formulations. If all scenarios are equally likely, the knapsack inequality reduces to a cardinality restriction. In this case, we are able to characterize the convex hull of feasible solutions to the extended formulation for the single row case. Although these results are motivated by the application to PCLP, they can be used in any problem in which a mixing set appears along with a knapsack constraint.

2   The MIP Formulation

We now consider a probabilistically constrained linear programming problem, with random right-hand side given by

$$
\begin{aligned}
(PCLPR) \quad \min \ & cx \\
\text{s.t. } & Ax = b \\
& P\{Tx \geq \xi\} \geq 1 - \epsilon \\
& x \geq 0 \ .
\end{aligned}
\tag{2}
$$

Here A is an $r \times d$ matrix, $b \in \mathbb{R}^r$, T is an $m \times d$ matrix, ξ is a random vector in $\mathbb{R}^m$, $\epsilon \in (0,1)$ (typically small) and $c \in \mathbb{R}^d$. We assume that ξ has finite support, that is there exist vectors, $\xi_i, i = 1, \ldots, n$ such that $P\{\xi = \xi_i\} = \pi_i$ for each i where $\pi_i \geq 0$ and $\sum_{i=1}^n \pi_i = 1$. We will refer to the possible outcomes as scenarios. We assume without loss of generality that $\xi_i \geq 0$ and $\pi_i \leq \epsilon$ for each i. We also define the set $N = \{1, \ldots, n\}$.

Before proceeding, we note that PCLPR is NP-hard even under assumptions A1 and A2.

Theorem 1. *PCLPR is NP-hard, even in the special case in which $\pi_i = 1/n$ for all $i \in N$, the constraints $Ax = b$ are not present, T is the $m \times m$ identity matrix, and $c = (1, \ldots, 1) \in \mathbb{R}^m$.*

We now formulate PCLPR as a mixed integer program [19]. To do so, we introduce for each $i \in N$, a binary variable z_i, where $z_i = 0$ will guarantee that $Tx \geq \xi_i$. Observe that because $\epsilon < 1$ we must have $Tx \geq \xi_i$ for at least one $i \in N$, and because $\xi_i \geq 0$ for all i, this implies $Tx \geq 0$ in any feasible solution of PCLPR. Then, introducing variables $v \in \mathbb{R}^m$ to summarize Tx, we obtain the MIP formulation of PCLPR given by

$$(PMIP) \quad \min cx$$

$$\text{s.t. } Ax = b, \ Tx - v = 0 \tag{3}$$

$$v + \xi_i z_i \geq \xi_i \quad i = 1, \ldots, n \tag{4}$$

$$\sum_{i=1}^{n} \pi_i z_i \leq \epsilon \tag{5}$$

$$x \geq 0, \ z \in \{0,1\}^n \ .$$

3   Strengthening the Formulation

Our approach is to strengthen PMIP by ignoring (3) and finding strong formulations for the set

$$F := \left\{ (v,z) \in \mathbb{R}_+^m \times \{0,1\}^n : (4), (5) \right\} \ . \tag{6}$$

Note that

$$F = \bigcap_{j=1}^{m} \{(v,z) : (v_j, z) \in G_j\} \ ,$$

where for $j = 1, \ldots, m$

$$G_j = \{(v_j, z) \in \mathbb{R}_+ \times \{0,1\}^n : (5), \ v_j + \xi_{ij} z_i \geq \xi_{ij} \quad i = 1, \ldots, n\} \ .$$

Thus, a natural first step in developing a strong formulation for F is to develop a strong formulation for each G_j. In particular, note that if an inequality is facet-defining for $\mathrm{conv}(G_j)$, then it is also facet-defining for $\mathrm{conv}(F)$. This follows because if an inequality valid for G_j is supported by $n + 1$ affinely independent points in $\mathbb{R}^{n+1}$, then because this inequality will not have coefficients on v_i for any $i \neq j$, the set of supporting points can trivially be extended to a set of $n + m$ affinely independent supporting points in $\mathbb{R}^{n+m}$ by appropriately setting the v_i values for each $i \neq j$.

The above discussion leads us to consider the generic set

$$G = \{(y, z) \in \mathbb{R}_+ \times \{0,1\}^n : (5), \ y + h_i z_i \geq h_i \quad i = 1, \ldots, n\} \tag{7}$$

obtained by dropping the index j in the set G_j and setting $y = v_j$ and $h_i = \xi_{ij}$ for each i. For convenience, we assume that the h_i are ordered so that $h_1 \geq h_2 \geq \cdots \geq h_n$. The *mixing set*

$$P = \{(y, z) \in \mathbb{R}_+ \times \{0,1\}^n : y + h_i z_i \geq h_i \quad i = 1, \ldots, n\}$$

has been extensively studied, in varying degrees of generality, by Atamtürk et. al [21], Günlük and Pochet [20], Guan et. al [22] and Miller and Wolsey [23]. If we ignore the knapsack constraint in G, we can apply these results to obtain the set of valid inequalities

$$y + \sum_{j=1}^{l}(h_{t_j} - h_{t_{j+1}})z_{t_j} \geq h_{t_1} \quad \forall T = \{t_1, \ldots, t_l\} \subseteq N , \tag{8}$$

where $t_1 < t_2 < \cdots < t_l$ and $h_{t_{l+1}} := 0$. Following [21], we call (8) the *star inequalities*. In addition, these inequalities can be separated in polynomial time [20,21,22]. It has been shown in these same works that these inequalities define the convex hull of P and are facet defining if and only if $t_1 = 1$. We can do considerably better, however, by using the knapsack constraint in G to first strengthen the inequalities, and then derive the star inequalities. In particular, let $p := \max\left\{k : \sum_{i=1}^{k} \pi_i \leq \epsilon\right\}$. Then, due to the knapsack constraint, we cannot have $z_i = 1$ for all $i = 1, \ldots, p+1$ and thus we have $y \geq h_{p+1}$. This also implies that the mixed integer constraints in G are redundant for $i = p+1, \ldots, n$. Thus, we can write a tighter formulation of G as

$$G = \{(y, z) \in \mathbb{R}_+ \times \{0,1\}^n : (5), \quad y + (h_i - h_{p+1})z_i \geq h_i \quad i = 1, \ldots, p\} . \tag{9}$$

Remark 1. In addition to yielding a tighter relaxation, this description of G is also more compact. In typical applications, ϵ will be near 0, suggesting $p << n$. When applied for each j in the set F, this will yield a formulation with $mp <<$ mn rows.

If we now apply the star inequalities to the improved formulation of G, we obtain the following result, which can be obtained by applying results in [20],[21] or [22].

Theorem 2. *The inequalities*

$$y + \sum_{j=1}^{l}(h_{t_j} - h_{t_{j+1}})z_{t_j} \geq h_{t_1} \quad \forall T = \{t_1, \ldots, t_l\} \subseteq \{1, \ldots, p\} \tag{10}$$

with $t_1 < \ldots < t_l$ *and* $h_{t_{l+1}} := h_{p+1}$, *are valid for* G. *Moreover,* (10) *is facet-defining for* conv(G) *if and only if* $h_{t_1} = h_1$.

We refer to the inequalities (10) as the *strengthened star inequalities*.

Remark 2. The difference between the star inequalities (8) and strengthened star inequalities (10) is that in (10) we have $h_{t_{l+1}} := h_{p+1}$ whereas in (8) we have $h_{t_{l+1}} := 0$, corresponding to the fact that our lower bound on y was shifted from 0 to h_{p+1} by using the knapsack inequality.

Remark 3. The strengthened star inequalities are not sufficient to characterize the convex hull of G, even in the special case in which all probabilities are equal, that is $\pi_i = 1/n$ for all i.

We now consider the special case in which $\pi_i = 1/n$ for all $i \in N$. Note that in this case we have $p := \max\left\{k : \sum_{i=1}^{k} 1/n \leq \epsilon\right\} = \lfloor n\epsilon \rfloor$ and the knapsack constraint (5) becomes

$$\sum_{i=1}^{n} z_i \leq n\epsilon$$

which, by integrality on z_i can be strengthened to the simple cardinality restriction

$$\sum_{i=1}^{n} z_i \leq p \ . \tag{11}$$

Thus, the feasible region for our single row formulation becomes

$$G' = \{(y, z) \in \mathbb{R}_+ \times \{0,1\}^n : (11), \quad y + (h_i - h_{p+1})z_i \geq h_i \quad i = 1, \ldots, p\} \ .$$

Now, observe that for any $(\gamma, \alpha) \in \mathbb{R}^{n+1}$, the problem

$$\min\{\gamma y + \alpha z : (y, z) \in G'\}$$

is easy. Indeed, if $\gamma < 0$, then the problem is unbounded, so we can assume $\gamma \geq 0$. Then, one need only consider setting y to h_k for $k = 1, \ldots, p+1$, and setting the z_i accordingly. That is, if $y = h_k$ for $k \in \{1, \ldots, p+1\}$, then we must set $z_i = 1$ for $i = 1, \ldots, k-1$. The remaining z_i can be set to 0 or 1 as long as $\sum_{i=k}^{n} z_i \leq p-k+1$. Hence, we set $z_i = 1$ if and only if $i \in S_k^*$ where

$$S_k^* \in \underset{S \subseteq \{k,\ldots,n\}}{\arg\min} \left\{\sum_{i \in S} \alpha_i : |S| \leq p-k+1\right\} \ .$$

Since we can optimize over G' efficiently, we know that we can separate over $\mathrm{conv}(G')$ efficiently. Indeed, given (y^*, z^*) we can write an explicit polynomial size linear program for separation over $\mathrm{conv}(G')$. Although this would yield a theoretically efficient way to separate over $\mathrm{conv}(G')$, it still may be too expensive to solve a linear program to generate cuts. We would therefore prefer to have an explicit characterization of a class or classes of valid inequalities for G' with an associated combinatorial algorithm for separation. The following theorem gives an example of one such class.

Theorem 3. *Let* $m \in \{1, \ldots, p-1\}$, $T = \{t_1, \ldots, t_l\} \subseteq \{1, \ldots, m\}$ *and* $Q = \{q_1, \ldots, q_{p-m}\} \subseteq \{p+1, \ldots, n\}$. *Define* $\Delta_1^m = h_{m+1} - h_{m+2}$ *and*

$$\Delta_i^m = \max\left\{\Delta_{i-1}^m, h_{m+1} - h_{m+i+1} - \sum_{j=1}^{i-1} \Delta_j^m\right\} \quad \text{for } i = 2, \ldots, p-m \ .$$

Then, with $h_{t_{l+1}} := h_{m+1}$,

$$y + \sum_{j=1}^{l} (h_{t_j} - h_{t_{j+1}})z_{t_j} + \sum_{j=1}^{p-m} \Delta_j^m(1 - z_{q_j}) \geq h_{t_1} \tag{12}$$

is valid for G' *and facet-defining for* $\mathrm{conv}(G')$ *if and only if* $h_{t_1} = h_1$.

Example 1. Let $n = 10$ and $\epsilon = 0.4$ so that $p = 4$ and suppose $h_{1-5} = \{20, 18, 14, 11, 6\}$. Let $m = 2$, $T = \{1, 2\}$ and $Q = \{5, 6\}$. Then, $\Delta_1^2 = 3$ and $\Delta_2^2 = \max\{3, 8 - 3\} = 5$ so that (12) yields

$$y + 2z_1 + 4z_3 + 3(1 - z_5) + 5(1 - z_6) \geq 20 .$$

Separation of inequalities (12) can be accomplished by a simple modification to the routine for separating the strengthened star inequalities. We have identified other classes of valid inequalities, but have so far failed to find a general class that characterizes the convex hull of G'.

4   A Strong Extended Formulation

Let

$$FS = \{(y, z) \in \mathbb{R}_+ \times [0, 1]^n : (5), (10)\} .$$

FS represents the polyhedral relaxation of G, augmented with the strengthened star inequalities. Note that the inequalities $y + (h_i - h_{p+1})z_i \geq h_i$ are included in FS by taking $T = \{i\}$, so that enforcing integrality in FS would yield a valid single row formulation for the set G. Our aim is to develop a reasonably compact extended formulation which is equivalent to FS. To do so, we introduce variables $w_1, \ldots, w_p$ and let

$$EG = \left\{(y, z, w) \in \mathbb{R}_+ \times \{0, 1\}^{n+p} : (13) - (16)\right\}$$

where

$$w_i - w_{i+1} \geq 0 \quad i = 1, \ldots, p \tag{13}$$

$$z_i - w_i \geq 0 \quad i = 1, \ldots, p \tag{14}$$

$$y + \sum_{i=1}^{p}(h_i - h_{i+1})w_i \geq h_1 \tag{15}$$

$$\sum_{i=1}^{n} \pi_i z_i \leq \epsilon . \tag{16}$$

and $w_{p+1} := 0$. The variables w_i can be interpreted as deciding whether or not scenario i is satisfied for the single row under consideration, and because they are specific to this single row, the inequalities (13) can be safely added. The inequalities (14) then ensure that if a scenario is infeasible for this row, then it is infeasible overall, and the lower bound on y is now given by the single inequality (15). We let EF be the polyhedron obtained by relaxing integrality in EG.

Theorem 4. $Proj_{(y,z)}(EG) = G$, *that is, EG is a valid formulation for G.*

An interesting result is that the linear relaxation of this extended formulation is as strong as having all strengthened star inequalities in the original formulation. A similar result has been proved in [23].

Theorem 5. $Proj_{(y,z)}(EF) = FS$.

Because of the equivalence between EF and FS, Remark 3 holds for this formulation as well, that is, even in the special case in which all probabilities are equal, this formulation does not characterize the convex hull of feasible solutions of G. We therefore investigate what other valid inequalities exist for this formulation. We first introduce the notation

$$f_k := \sum_{i=1}^{k} \pi_i, \quad k = 0, 1, \ldots, p .$$

Theorem 6. Let $k \in \{1, \ldots, p\}$ and let $S \subseteq \{k, \ldots, n\}$ be such that $\sum_{i \in S} \pi_i \leq \epsilon - f_{k-1}$. Then,

$$\sum_{i \in S} \pi_i z_i + \sum_{i \in \{k \ldots, p\} \setminus S} \pi_i w_i \leq \epsilon - f_{k-1} \tag{17}$$

is valid for EG.

Now, consider the special case in which $\pi_i = 1/n$ for $i = 1, \ldots, n$. Then the extended formulation becomes

$$EG' = \left\{ (y, z, w) \in \mathbb{R}_+ \times \{0, 1\}^{n+p} : (11) \text{ and } (13) - (15) \right\} .$$

Letting $\mathcal{S}_k = \{S \subseteq \{k, \ldots, n\} : |S| \leq p-k+1\}$ for $k = 1, \ldots, p$, the inequalities (17) become

$$\sum_{i \in S} z_i + \sum_{i \in \{k, \ldots, p\} \setminus S} w_i \leq p-k+1 \qquad \forall S \in \mathcal{S}_k, \ k = 1, \ldots, p . \tag{18}$$

Example 2. Let $n = 10$ and $\epsilon = 0.4$ so that $p = 4$. Let $k = 2$ and $S = \{4, 5, 6\}$. Then (18) becomes

$$z_4 + z_5 + z_6 + w_2 + w_3 \leq 3 .$$

Now, let

$$EH' = \{(y, z, w) \in \mathbb{R}_+ \times [0, 1]^{n+p} : (11), (13) - (15) \text{ and } (18)\}$$

be the linear relaxation of the extended formulation, augmented with this set of valid inequalities.

Theorem 7. *The convex hull of the extended formulation EG' is given by the inequalities defining EG' and the inequalities (18); that is, $EH' = \text{conv}(EG')$.*

We close this section by noting that inequalities (18) can be separated in polynomial time. Indeed, suppose we wish to separate the point (z^*, w^*). Then separation can be accomplished by solving

$$\max_{S \in \mathcal{S}_k} \left\{ \sum_{i \in S} z_i^* + \sum_{i \in \{k, \ldots, p\} \setminus S} w_i^* \right\} = \max_{S \in \mathcal{S}_k} \left\{ \sum_{i \in S} \theta_i^* \right\} + \sum_{i=k}^{p} w_i^*$$

for $k = 1, \ldots, p$ where

$$\theta_i^* = \begin{cases} z_i^* - w_i^* & i = 1, \ldots, p \\ z_i^* & i = p+1, \ldots, n \end{cases} .$$

Hence, a trivial separation algorithm is to first sort the values θ_i^* in non-increasing order, then for each k, find the maximizing set $S \in \mathcal{S}_k$ by searching this list. This yields an algorithm with complexity $O(n \log n + p^2) = O(n^2)$. However, by considering the values of k in the order $p, \ldots, 1$ and updating an ordered list of *eligible* indices $\mathcal{S}_k$ for each k, it is possible to improve the complexity to $O(n \log n)$. For the more general inequalities (17), (heuristic) separation can be accomplished by (heuristically) solving p knapsack problems.

5   Computational Experience

We performed computational tests on a probabilistic version of the classical transportation problem. We have a set of suppliers S and a set of customers D with $|D| = m$. The suppliers have limited capacity M_i for $i \in S$. There is a per-unit transportation cost c_{ij} for (producing and) shipping a unit of product from supplier $i \in S$ to customer $j \in D$. The customer demands are random and are represented by a random vector $\tilde{d} \in \mathbb{R}_+^m$. We assume we must choose the shipment quantities before the customer demands are known. We enforce the following probabilistic constraint:

$$P\{\sum_{i \in S} x_{ij} \geq \tilde{d}_j, j = 1, \ldots, m\} \geq 1 - \epsilon . \tag{19}$$

The objective is to minimize distribution costs subject to (19), non-negativity on the flow variables x_{ij}, and the supply capacity constraints

$$\sum_{j \in D} x_{ij} \leq M_i, \quad \forall i \in S .$$

We randomly generated instances with the number of suppliers fixed at 40 and varying numbers of customers and scenarios. The supply capacities and cost coefficients were generated using normal and uniform distributions respectively. For the random demands, we experimented with independent normal, dependent normal and independent Poisson distributions. We found qualitatively similar results in all cases, but the independent normal case yielded the most challenging instances, so for our experiments we focus on this case. For each instance, we first randomly generated the mean and variance of each customer demand. We then generated the number n of scenarios required, independently across scenarios and across customer locations, as Monte Carlo samples with these fixed parameters. In most instances we assumed all scenarios occur with probability $1/n$, but we also did some tests in which the scenarios have general probabilities, which were also randomly generated. CPLEX 9.0 was used as the MIP solver and all experiments were done on a computer with two 2.4 Ghz processors (although no parallelism is used) and 2.0 Gigabytes of memory. We set a time limit of one hour. For each problem size we generated 5 random instances and, unless otherwise specified, the computational results reported are averages over the 5 instances.

5.1   Comparison of Formulations

In Table 1 we compare the results obtained by solving our instances using

1. formulation PMIP given by (3) - (5),
2. formulation PMIP with strengthened star inequalities (10), and
3. the extended formulation of Sect. 4, but without (17) or (18).

When the strengthened star inequalities are not used, we still used the improved formulation of G corresponding to (9). Recall that the strengthened star inequalities subsume the rows defining the formulation PMIP; therefore, when we using these inequalities we initially add only a small subset of the mp inequalities in the formulation. Subsequently separating the strengthened star inequalities as needed guarantees the formulation remains valid. For formulation PMIP without strengthened star inequalities, we report the average optimality gap that remained after the hour time limit was reached. For the other two formulations, which we refer to as the strong formulations, we report the geometric average of the time to solve the instances to optimality. We used $\epsilon = 0.05$ and $\epsilon = 0.1$, reflecting the natural assumption that we want to meet demand with high probability.

The first observation from Table 1 is that formulation PMIP without the strengthened star inequalities fails to solve these instances within an hour, often leaving large optimality gaps, whereas the instances are solved efficiently using the strong formulations. The number of nodes required to solve the instances for the strong formulations is very small. The instances with equi-probable scenarios were usually solved at the root, and even when branching was required, the root relaxation usually gave an exact lower bound. Branching in this case was only required to find an integer solution which achieved this bound. The instances with general probabilities required slightly more branching, but generally not more than 100 nodes. Observe that the number of strengthened star inequalities that were added is small relative to the number of rows in the formulation PMIP itself. For example, for $\epsilon = 0.1$, $m = 200$ and $n = 3,000$, the number of rows in PMIP would be $mp = 60,000$, but on average, only $5,541$ strengthened star inequalities were added. Next we observe that in most cases the computation time using the extended formulation is significantly less than the formulation with strengthened star inequalities. Finally, we observe that the instances with general probabilities take somewhat longer to solve than those with equi-probable scenarios but can still be solved efficiently.

5.2   Testing Inequalities (18)

With small ϵ the root relaxation given by the extended formulation is extremely tight, so that adding the inequalities (18) is unlikely to have a positive impact on

Table 1. Average solution times for different formulations

Probabilities	ϵ	m	n	PMIP Gap	Cuts	PMIP+Star Time(s)	Extended Time(s)
Equal	0.05	100	1000	0.18%	734.8	7.7	1.1
		100	2000	1.29%	1414.2	31.8	4.6
		200	2000	1.02%	1848.4	61.4	12.1
		200	3000	2.56%	2644.0	108.6	12.4
	0.10	100	1000	2.19%	1553.2	34.6	12.7
		100	2000	4.87%	2970.2	211.3	41.1
		200	2000	4.48%	3854.0	268.5	662.2
		200	3000	5.84%	5540.8	812.7	490.4
General	0.05	100	1000	0.20%	931.8	9.0	3.9
		100	2000	1.04%	1806.6	55.2	13.2
	0.10	100	1000	1.76%	1866.0	28.7	52.5
		100	2000	4.02%	3686.2	348.5	99.2

computation time. However, for larger ϵ, we have seen that the extended formulation may have a substantial optimality gap. We therefore investigated whether using inequalities (18) can improve solution time in this case. In Table 3 we present results comparing solution times and node counts with and without inequalities (18) for instances with larger ϵ. We performed these tests on smaller instances since these instances are already hard for these values of ϵ. We observe that adding inequalities (18) at the root can decrease the root optimality gap significantly. For the instances that could be solved in one hour, this leads to a significant reduction in the number of nodes explored, and a moderate reduction in solution time. For the instances which were not solved in one hour, the remaining optimality gap was usually, but not always, lower when the inequalities (18) were used. These results indicate that when ϵ is somewhat larger, inequalities (18) may be helpful on smaller instances. However, they also reinforce the difficulty of the instances with larger ϵ, since even with these inequalities, only the smallest of these smaller instances could be solved to optimality within an hour.

5.3   The Effect of Increasing ϵ

The results of Table 1 indicate that the strong formulations can solve large instances to optimality when ϵ is small, which is the typical case. However, it is still an interesting question to investigate how well this approach works for larger ϵ. Note first that we should expect solution times to grow with ϵ if only because the formulation sizes grow with ϵ. However, we observe from Table 2 that the situation is much worse than this. This table shows the root LP solve times and optimality gaps achieved after an hour of computation time for an example instance with equi-probable scenarios, $m = 50$ rows and $n = 1,000$ scenarios at increasing levels of ϵ, using the formulation PMIP with strengthened star inequalities. Root LP solve time here refers to the time until no further strengthened star inequalities could be separated. We see that the time to solve

Table 2. Effects of increasing ϵ on an instance with $m = 50$ and $n = 1000$

ϵ	0.10	0.20	0.30	0.40	0.50	0.60	0.70	0.80	0.90
Root LP Time (s)	21.7	37.1	82.7	144.3	227.8	327.6	505.6	792.6	1142.6
Optimality Gap	0.0%	0.0%	2.2%	5.8%	10.5%	16.2%	28.7%	35.1%	44.4%

the root linear programs does indeed grow with ϵ as expected, but the optimality gaps achieved after an hour of computation time deteriorate even more drastically with growing ϵ. This is explained by the increased time to solve the linear programming relaxations *combined with* a weakening of the relaxation bound as ϵ increases.

Table 3. Results with and without inequalities (18)

m	ϵ	n	Root Gap		Nodes		Time(s) or Gap	
			Ext	+(18)	Ext	+(18)	Ext	+(18)
25	0.3	250	1.18%	0.67%	276.9	69.0	121.2	93.9
	0.3	500	1.51%	0.58%	455.0	165.8	750.6	641.3
	0.35	250	2.19%	1.50%	1259.4	409.0	563.2	408.4
	0.35	500	2.55%	1.61%	2297.6	968.8	0.22%	0.06%
50	0.3	500	2.32%	2.00%	991.8	238.6	1.37%	1.41%
	0.3	1000	2.32%	1.75%	28.3	8.5	1.98%	1.66%
	0.35	500	4.10%	3.31%	650.4	92.9	3.03%	2.66%
	0.35	1000	4.01%	3.23%	22.7	6.2	3.58%	3.17%

6 Concluding Remarks

We have presented strong integer programming formulations for linear programs with probabilistic constraints in which the right-hand side is random with finite distribution. In the process we made use of existing results on mixing sets, and have introduced new results for the case in which the mixing set additionally has a knapsack restriction. Computational results indicate that these formulations are extremely effective on instances in which reasonably high reliability is enforced, which is the typical case. However, instances in which the desired reliability level is lower remain difficult to solve, partly due to increased size of the formulations, but more significantly due to the weakening of the formulation bounds. Moreover, these instances remain difficult even when using the inequalities which characterize the single row relaxation convex hull. This suggests that relaxations which consider multiple rows simultaneously need to be studied to yield valid inequalities which significantly improve the relaxation bounds for these instances.

Future work in this area should focus on addressing the two assumptions we made at the beginning of this paper. The finite distribution assumption can be addressed by using the results about the statistical relationship between a problem with probabilistic constraints and its Monte Carlo sample approximation to establish methods for generating bounds on the optimal value of the original problem. Computational studies will need to be performed to establish the practicality of this approach. We expect that relaxing the assumption that only the right-hand side is random will be more challenging. A natural first step in this direction will be to extend results from the *generalized* mixing set [23,24] to the case in which an additional knapsack constraint is present.

Acknowledgments. This research has been supported in part by the National Science Foundation under grants DMI-0121495 and DMI-0522485.

References

1. Prékopa, A.: Probabilistic programming. In Ruszczyński, A., Shapiro, A., eds.: Stochastic Programming. Volume 10 of Handbooks in Operations Research and Management Science. Elsevier (2003)
2. Lejeune, M.A., Ruszczyński, A.: An efficient trajectory method for probabilistic inventory-production-distribution problems. Operations Research (Forthcoming, 2007)
3. Murr, M.R., Prékopa, A.: Solution of a product substitution problem using stochastic programming. In Uryasev, S.P., ed.: Probabilistic Constrained Optimization: Methodology and Applications. Kluwer Academic Publishers (2000) 252–271
4. Henrion, R., Li, P., Möller, A., Steinbach, M.C., Wendt, M., Wozny, G.: Stochastic optimization for operating chemical processes under uncertainty. In Grötschel, M., Krunke, S., Rambau, J., eds.: Online Optimization of Large Scale Systems. (2001) 457–478
5. Henrion, R., Möller, A.: Optimization of a continuous distillation process under random inflow rate. Computers and Mathematics with Applications **45** (2003) 247–262
6. Takyi, A.K., Lence, B.J.: Surface water quality management using a multiple-realization chance constraint method. Water Resources Research **35** (1999) 1657–1670
7. Calafiore, G., Campi, M.: Uncertain convex programs: randomized solutions and confidence levels. Mathematical Programming, Ser. A **102** (2005) 25–46
8. Nemirovki, A., Shapiro, A.: Scenario approximation of chance constraints. Preprint available at www.optimization-online.org (2004)
9. Nemirovski, A., Shapiro, A.: Convex approximations of chance constrained programs. Preprint available at www.optimization-online.org (2004)
10. Ben-Tal, A., Nemirovski, A.: Robust convex optimization. Mathematics of Operations Research **23** (1998) 769–805
11. Bertsimas, D., Sim, M.: The price of robustness. Operations Research **52** (2004) 35–53
12. Ghaoui, L.E., Lebret, H.: Robust solutions to least-squares problems with uncertain data. SIAM Journal on Matrix Analysis and Applications **18** (1997) 1035–1064
13. Cheon, M.S., Ahmed, S., Al-Khayyal, F.: A branch-reduce-cut algorithm for the global optimization of probabilistically constrained linear programs. Mathematical Programming, Ser B **108** (2006) 617–634
14. Dentcheva, D., Prékopa, A., Ruszczyński, A.: Concavity and efficient points of discrete distributions in probabilistic programming. Mathematical Programming, Ser A **89** (2000) 55–77

15. Prékopa, A.: On probabilistic constrained programmming. In Kuhn, H.W., ed.: Proceedings of the Princeton Symposium on Mathematical Programming, Princeton, N.J., Princeton University Press (1970) 113–138
16. Ahmed, S., Shapiro, A.: The sample average approximation method for stochastic programs with integer recourse. Preprint available at www.optimization-online.org (2002)
17. Atlason, J., Epelman, M.A., Henderson, S.G.: Call center staffing with simulation and cutting plane methods. Annals of Operations Research **127** (2004) 333–358
18. Shapiro, A., Homem-de-Mello, T.: On the rate of convergence of optimal solutions of monte carlo approximations of stochastic programs. SIAM Journal of Optimization **11** (2000) 70–86
19. Ruszczyński, A.: Probabilistic programming with discrete distributions and precedence constrained knapsack polyhedra. Mathematical Programming, Ser A **93** (2002) 195–215
20. Günlük, O., Pochet, Y.: Mixing mixed-integer inequalities. Mathematical Programming **90** (2001) 429–457
21. Atamtürk, A., Nemhauser, G.L., Savelsbergh, M.W.P.: The mixed vertex packing problem. Mathematical Programming **89** (2000) 35–53
22. Guan, Y., Ahmed, S., Nemhauser, G.L.: Sequential pairing of mixed integer inequalities. Discrete Optimization (Forthcoming, 2007)
23. Miller, A.J., Wolsey, L.A.: Tight formulations for some simple mixed integer programs and convex objective integer programs. Mathematical Programming **98** (2003) 73–88
24. Van Vyve, M.: The continuous mixing polyhedron. Mathematics of Operations Research **30** (2005) 441–452

Infrastructure Leasing Problems[*]

Barbara M. Anthony[1] and Anupam Gupta[2]

[1] Dept. of Mathematical Sciences, Carnegie Mellon University, Pittsburgh PA 15213
banthony@andrew.cmu.edu
[2] Computer Science Department, Carnegie Mellon University, Pittsburgh PA 15213
anupamg@cs.cmu.edu

Abstract. Consider the following Steiner Tree leasing problem. Given a graph $G = (V, E)$ with root r, and a sequence of terminal sets $D_t \subseteq V$ for each day $t \in [T]$. A feasible solution to the problem is a set of edges E_t for each t connecting D_t to r. Instead of obtaining edges for a single day at a time, or for infinitely long (both of which give Steiner tree problems), we *lease* edges for say, { *a day, a week, a month, a year* }. Naturally, leasing an edge for a longer period costs less per unit of time. What is a good leasing strategy? In this paper, we give a general approach to solving a wide class of such problems by showing a close connection between deterministic leasing problems and problems in multistage stochastic optimization. All our results are in the offline setting.

Keywords: Approximation algorithms, graph and network algorithms, stochastic combinatorial optimization, randomized algorithms.

1 Introduction

Traditional network design problems require us to make decisions about how to send data, and how to provision bandwidth on various links of the network. A standard feature in most models for network design that have been considered, and in the algorithms that have been developed, has been the *permanence* of the bandwidth allocation—and this has been true even in cases where demands arrive online: once some amount of bandwidth is allocated on an edge, this bandwidth can be used *at any time in the future* (perhaps by paying some additional incremental "routing cost" per unit of flow). Some works have also considered the question of buying versus renting, but the simplifying assumption again has been that buying gives permanent access to the commodity. *But what if we are allowed only to leaseↄ bandwidth on the links of the network for fixed lengths of time: which leases on which network links should we obtain over time to satisfy our demands?*

Given a situation with multiple lease lengths, it is natural to assume that a longer lease is a cheaper one (per day), and that we pay more dearly for the

---

[*] This research is partly supported by an NSF CAREER award CCF-0448095, and by an Alfred P. Sloan Fellowship.

flexibility afforded by the short-term leases.[1] Hence, if our traffic consists of some stable parts and other bursty parts, we can use long-term leases to satisfy the stable traffic, and the short-term leases to handle the more volatile demands: a clever leasing strategy can reduce costs substantially over a naïve one. Note that solving this problem requires us to simultaneously perform clustering over space (in order to figure out which edges to allocate bandwidth on) and over time (to figure out which traffic is stable and requires longer leases, and which is bursty and is best served by shorter leases).

The question of finding good leasing strategies is relevant in the context of other problems as well: in planning for demands arriving over multiple periods in classical facility location problems, one might want to lease warehouses/plants for varying lengths of time. Moreover, the idea that leases of varying lengths are available is fairly natural: even in situations where there is a standard lease length (say plants are usually leased for a year), the presence of a secondary market for reselling or sub-letting might naturally give rise to the situation with multiple lease lengths we consider in this paper.

In this paper, we initiate a systematic study of *Leasing problems*, and give algorithms for several classic infrastructure design problems in the presence of finite-duration leases. To illustrate our general model, we will use the STEINER TREE LEASING problem as our running example.

We are given a graph $G = (V, E)$ with a root r. For each day t, we are given a set of terminals D_t and a set K of permissible lease lengths, where the cost of leasing any edge e for length $\ell \in K$ is $c(\ell)$: we ensure that for any lengths $\ell_1 < \ell_2$ in K, $c(\ell_2) \leq c(\ell_1) \times \frac{\ell_2}{\ell_1}$. Note that an edge leased on day t for duration ℓ can be used on any of the days $t, t+1, \ldots, t+\ell-1$, and is said to be *active* on all these days. Define $X_t(\ell) \subseteq E$ to be the set of edges leased for duration ℓ on day t, and $F_t = \cup_{\ell \in K} \cup_{j \in [t-\ell+1,t]} X_j(\ell)$ to be the set of active edges on day t. A solution (given by edge sets $X_t(\ell)$ for all t and ℓ) is feasible if on each day t, the induced active edge sets F_t connect the demand set D_t to the root r. The goal is to find a feasible solution of minimum cost $\sum_{t,\ell}[c(\ell) \times |X_t(\ell)|]$.

One can follow this general idea and define other infrastructure design problems: in FACILITY LOCATION LEASING, we are given demand sets D_t for each day, and may want to lease different facilities for different periods of time, with the goal of minimizing the resulting cumulative facility opening costs plus the connection costs for the clients on their respective days. (In this case, one may even imagine a "non-uniform" scenario where the different facilities have different lease cost functions.) And an even more general problem is that of SET COVER LEASING, where we are given sets $D_t \subseteq U$ of elements to cover on the tth day, and want to lease sets such that the active sets at time t form a feasible cover of the set D_t.

While such problems of finite-period leases are related to the substantial body of work on *perishable commodities* [29,13] in inventory theory, we are not aware

---

[1] More formally, we assume leasing for length ℓ costs no more than two leases of length $\ell/2$. This sub-additive cost structure also allows amortization of one-time costs.

of work that directly addresses the questions under consideration in this paper. Loosely speaking, given supply of a perishable good—e.g., cartons of milk with a lifetime of ℓ days—and demands over time, research on perishable commodities has considered questions pertaining to inventory positions (in deterministic vs stochastic settings, with several classes of customers, etc.), and to pricing such perishable goods. At a high level, our leasing problems can be viewed as solving multiple perishable goods problems to solve a global network design problem.

1.1 Our Results and Techniques

The main result of this paper is the following, showing a close connection between leasing problems as described above, and stochastic optimization problems.

Theorem 1 (General Leasing Theorem). *The offline leasing version of a subadditive combinatorial optimization problem Π with $|K| = k$ lease lengths can be reduced to the stochastic optimization version of Π in the model of k-stage stochastic optimization with recourse.*

We feel this theorem is somewhat surprising: even though the leasing version of the problem Π can be *completely deterministic* with a given input and no stochastic component, this theorem shows that an algorithm to solve the (multistage) stochastic version of the problem suffices to solve the (non-stochastic) leasing problem. The proof of this theorem turns out to be fairly clean, and appears in Section 4.1. Given this main theorem, we can use recently-developed algorithms for multistage stochastic combinatorial optimization [34,37] to infer:

Corollary 1 (Optimal Algorithms for Leasing). *There exist $O(1)$-approximation algorithms for the Facility Location Leasing and Vertex Cover Leasing problems, and an $O(\log n)$-approximation for the Set Cover Leasing problem.*

All these results are asymptotically optimal (up to constants). For the Steiner Tree Leasing problem we were using as our running example, we get the following result by combining Theorem 1 with known results [17,19].

Theorem 2 (Steiner Tree Leasing). *There is an $O(\min\{k, \log n\})$-approximation algorithm for offline Steiner Tree Leasing with $|K| = k$ lease lengths.*

It seems improving the approximation to $o(k)$ requires techniques that also improve results for the Stochastic Steiner Problem, which remains an open question.

New Algorithms for Network Problems: We go on to study other network leasing problems that generalize the Steiner Tree Leasing problem. In these problems, instead of just connecting up the terminals, we are now required to allocate "sufficient" bandwidth on the connecting edges as well. However, the cost of allocating bandwidth is itself a concave function $g(b)$ of the amount of bandwidth b allocated on the edge: these are commonly known as *buy-at-bulk* problems. In the leasing framework, this translates into problems where the cost of leasing b units of bandwidth for a period of length ℓ is $c(\ell) \times g(b)$.

Theorem 3 (Buy-at-Bulk Theorems). *There is an $O(k)$ approximation for the k-stage Stochastic versions of the single-sink Rent-or-Buy, and the single-sink Buy-at-Bulk problems. Moreover, the Stochastic Buy-at-Bulk problem with multiple sinks has an $O(k \log n)$ approximation algorithm.*

By Theorem 1, we get the same approximation ratios for the corresponding network leasing versions of these problems as well.

Related Work. There has been a tremendous amount of work on network design where the the cost of bandwidth obeys natural economies of scale (often called "buy-at-bulk" network design). It is beyond the scope of this paper to survey this body of work, so we point the reader to [25,26,4,32,2,14,12,38,1,10] and the many references therein. This line of work is related to our work both in spirit, as well as in some of the technical methodology. In this paper, we also show how we can extend some of the current algorithms for these "buy-at-bulk" problems to the case when the bandwidth is leased and not bought permanently.

As mentioned above, leasing for finite periods is related to a large body of work on perishable commodities [29,13] in inventory theory; however, to the best of our knowledge, such problems have not been directly considered in the literature.

The Steiner Tree Leasing problem was first explored in a paper on the "parking permit problem" [27]. The paper noted that dynamic programming could be used to solve the Steiner Tree Leasing problem when the graph was a single edge (or to obtain an approximation scheme if the numbers are large), and gave an $O(\log k)$ competitive algorithm in the *online* case where the terminal set D_t is revealed only on day t. These results can be extended naturally to general graphs using standard tree-approximation techniques [5,11] by losing an extra $O(\log n)$ factor. However, it does not seem clear how to improve their techniques directly in the offline case to avoid this loss of $O(\log n)$ and obtain an approximation dependent only on k, or to extend them to the other problems we consider here.

In this paper, we show a concrete connection of network leasing to multistage stochastic optimization problems. While the history of stochastic optimization begins in the 1950s, this work is directly related to recent work on approximation algorithms for stochastic combinatorial problems [9,20,31,16,19,33,8,7]. We draw most directly from the results of [19,17,34] on the multistage stochastic optimization problems, and on the results in [16,17] to convert algorithms for the non-stochastic versions of problems to their multistage stochastic counterparts.

A standard tool in algorithms design today is the tree approximation technique of [5,11], as well as the general techniques for solving covering problems from, e.g., [30,35,36,23]. These techniques will allow us to get some simple approximation bounds; one of the goals of this paper is to develop algorithms that beat these naïve bounds by making use of the combinatorics of the problems, and to explore connections to problems in multistage stochastic optimization.

As an aside, let us note that a problem called the "Network Leasing" problem has been previously studied in the literature [3]; since that problem has come to be better known as the "Rent-or-Buy" problem, we have taken the liberty of claiming the term "leasing" to refer to an orthogonal concept in this paper.

2   Models and Notation

Consider a general subadditive optimization problem Π with k lease lengths. Formally, we are given a set U of potential *clients* or *demands*, such that on each day $t \in \{1, 2, \ldots, T\}$, some subset $D_t \subseteq U$ of these clients actually appear and demand service. (We will soon discuss how these sets D_t are given to us.) We also have a set of *elements* X that we can use to build solutions: for each subset of clients $D \subseteq U$, we are given some set of *solutions* $\mathsf{Sols}(D) \subseteq 2^X$ to the client set D. On day t, we would like to own a set of elements $F_t \in \mathsf{Sols}(D_t)$.

If each element could only be leased for a single day at a time, then this would just require us to solve T instances of the problem Π; on the other hand, if elements could only be leased indefinitely (i.e., "bought"), we would just solve the problem on $\cup_t D_t$. The "leasing" aspect of the problem is reflected in the fact that each of these elements $e \in X$ can be leased for several periods: i.e., on any day t, given any duration $\ell \in K$, one can obtain a lease of length ℓ on element $e \in X$ for cost $c_e(\ell)$ and use it on days $t, t+1, \ldots, t+\ell-1$. Formally, let $X_t(\ell)$ be the elements for which leases of length ℓ were obtained on day t, and $F_t = \cup_{\ell \in K} \cup_{t'=t-\ell+1}^{t} X_t(\ell)$, then a *feasible leasing strategy* is a sequence of sets $X_t(\ell)$ which results in $F_t \in \mathsf{Sols}(D_t)$ for each day t.

Definition 1 (Uniform vs. Non-Uniform). *A leasing problem is called uniform if the cost functions $c_e(\cdot)$ for all elements $e \in X$ are identical (here we will drop the subscript and refer to it as $c(\cdot)$), and is called non-uniform otherwise.*

As may be expected, we will be able to obtain better results for uniform problems in some cases. One immediate advantage of uniform network design problems will be the applicability of tree-approximation techniques (see Lemma 3); see also Section 4.1 for other advantages of uniformity.

Stochastic Optimization. The relevant stochastic model is k-stage stochastic optimization with recourse: the demand set D is revealed on day-k drawn from some known distribution π, but on each of days $1, 2, \ldots, k-1$, we are given additional information about the set D. (One can view this process as having a joint distribution over "signals" $s_1, s_2, \ldots, s_{k-1}, s_k$ received on the various days, with the actual demand set some known function of this signals.) One can see, e.g., [34,17] for more details about the model. The costs of elements change over time (usually getting more expensive over time): the *uniform inflation* model assumes the cost $cost_t(e)$ of element $e \in X$ on day-t (or *stage-t*) to be $\sigma_i \times cost_{t-1}(e)$ (and hence $cost_1(e) \prod_{1 < j \leq i} \sigma_j$). Note that the σ_i's are uniform, and independent of the element e. In the more powerful *non-uniform* model, the costs of different elements can change differently as time progresses.

We use the Boosted Sampling framework to develop new algorithms for some network design problems: these will require us to use terminology about cost shares, which can be found in Appendix A.

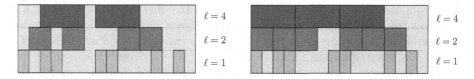

Fig. 1. A solution, and the corresponding nested version (right), as in Lemma 2

3   Observations and Reductions

Before we give the main results of this paper, we give some observations which will be helpful in the rest of the paper. We investigate how solutions can be assumed to have a simple structure, what results tree-approximations can give for Steiner Tree Leasing (giving us a baseline to compare to), and what tree-approximation techniques can give for more complex network leasing problems.

Structure of Solutions. The following two lemmas allow us to impose a simple structure on the instances we solve and solutions we seek. They are fairly standard (e.g., [27, Thms 2.1 & 2.2]) and are given for completeness. Recall that the set of permissible lease lengths is $K = \{\ell_1, \ell_2, \dots, \ell_k\}$ with $\ell_1 < \ell_2 < \dots < \ell_k$.

Lemma 1. *Given any instance $\mathcal{I}$ of a leasing problem, we can convert it into an instance $\mathcal{I}'$ in which the lengths of leases exactly divide each other (i.e., $\ell_i | \ell_j$ for $i < j$), and where the costs satisfy $c(\ell_j) < c(\ell_i) \times (\ell_j / \ell_i)$. Moreover, there is an optimal solution to $\mathcal{I}'$ which has cost at most 2 times the optimal cost for $\mathcal{I}$.*

The above lemma can be proved, e.g., by rounding all the lease lengths down to the closest powers of 2, and by discarding leases that do not satisfy the subadditivity property. The following lemma shows that we can focus our attention only on "nested" solutions; i.e., solutions where we never have a short-term lease still active when a longer-term lease begins or ends.

Lemma 2 (Nested Solutions). *Given an instance $\mathcal{I}$ of a leasing problem, there is a solution which has cost at most 2 times the optimum, where a lease of length ℓ is obtained only for intervals of the form $[t, t + \ell)$ with t a multiple of ℓ.*

See Fig. 1 for an example of a non-nested solution on the left, and a nested solution whose cost is at most twice the cost of the former.

Reduction to Trees/Single-Edges. Given a graph $G = (V, E)$, a theorem of Fakcharoenphol et al. [11] (see also [5]) says that there is a distribution $\mathcal{D}$ over dominating trees T (i.e., $d_G(u, v) \le d_T(u, v)$ for any T in the support of $\mathcal{D}$) such that the expected stretch $\frac{E_{T \leftarrow \mathcal{D}}[d_T(u,v)]}{d_G(u,v)} \le O(\log n)$. The following use of this result is fairly standard by now (see [2]).

Lemma 3 (Reduction to Trees/Edges). *Given an instance of Steiner Tree Leasing which is* uniform *(where the cost functions $c_e(\cdot)$ are the same for all edges), an α-approximation for the single-edge case gives an α approximation for trees, and an $O(\alpha \log n)$ approximation for the general graph case.*

The proof uses the fact that the reduction to a tree instance loses an $O(\log n)$; once on a tree, the paths to be chosen are unique, and hence it suffices to run the single-edge algorithm on each edge to determine when to lease it. (The simple details are deferred to the final version of the paper.) Since we can solve the leasing problem on a single edge exactly, we get an $O(\log n)$-approximation for the Steiner Tree Leasing problem.

General (Uniform) Leasing Strategies and CIPs. Consider a much more general network design problem where at each time step t we are given a traffic matrix D_t, and want to allocate enough bandwidth to route D_t. We are now given a set $\mathcal{L} = \{L_j = (I_j, b_j, p_j)\}_j$ of possible leases, where each lease L_j in $\mathcal{L}$ is specified by a time interval I_j during which this lease is active, an amount b_j of bandwidth and a price p_j for it. Moreover, for any lease L_j, we may have an upper bound u_j on the number of copies of this lease we can buy per edge. This is a much more general model than the one we have been looking at, since we allow "one-time-only" offers (a special deal valid only for some days at a special price, limit one only), etc: this captures Buy-at-Bulk Leasing, and much more.

However, as long as the problem is *uniform* (i.e., each edge e has the same set $\mathcal{L}$ of potential leases), we can use a reduction akin to Lemma 3 to randomly reduce the problem to a tree and hence to a single edge, where it can be solved using general theorems on CIPs, covering integer problems techniques (e.g., see [30,35,6,36,23]). Applying these techniques to our problems give us approximation ratios that typically depend on $\log \ell_{max}$, and $\log b_{max}$, where b_{max} is the maximum bandwidth requirement. (See the full version for precise details.) In this paper, we attempt to give algorithms that are better—i.e., independent of ℓ_{max}; it is easy to see that $\log \ell_{max} \geq k$, and we think of $\log \ell_{max} \gg k$.

4   Algorithms for Leasing Problems

In this section, we will prove the main result: that Leasing Problems can be cast as Stochastic Optimization problems. This will allow us to get approximation algorithms for a variety of leasing problems from the corresponding algorithms for stochastic optimization. While we use many stochastic algorithms already in the literature, we will give new algorithms for some problems like Stochastic Rent-or-Buy and Stochastic Buy-at-Bulk, and hence for their leasing versions.

4.1   Reduction to Multistage Stochastic Optimization

Let us assume, without loss of generality, that $\ell_1 = 1$, and denote the maximum lease length by ℓ_{max}. By Lemma 2 we can assume that our solutions are nested.

Theorem 4. *[Reduction to k-stage Stochastic Optimization] Any (non-uniform) offline problem Π in the above framework with $|K| = k$ lease lengths can be reduced to the standard k-stage stochastic optimization version of Π.*

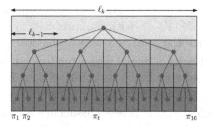 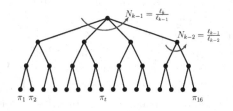

Fig. 2. A nested leasing instance ($k = 4$), and the resulting stochastic tree $\mathfrak{T}$

Proof. As in the proof of Theorem 2, let us consider tiling time by intervals of length ℓ_k, each of which are divided into $N_{k-1} = \frac{\ell_k}{\ell_{k-1}}$ consecutive intervals of length ℓ_{k-1}, each of which are further subdivided into $N_{k-2} = \frac{\ell_{k-1}}{\ell_{k-2}}$ intervals of length ℓ_{k-2}, and so on. Note that this gives a different representation of time: we can describe time $t = \sum_{p=1}^{k} j_p \, \ell_p$ as a k-tuple of the form $(j_k, j_{k-1}, \ldots, j_1)$—and we will denote this tuple by $\bar{\tau}(t)$. (Note that j_p is simply $\lfloor t/\ell_p \rfloor \pmod{\ell_{p+1}}$, where we assume $\ell_{k+1} = \infty$). Corresponding to this notation, we will refer to the set $X_t(\ell_i)$ also as $X_{(j_k, j_{k-1}, \ldots, j_1)}(\ell_i)$, where t, ℓ_i and the j_k's are as above.

Recall that we are looking for nested solutions, and hence each lease of length ℓ_i will be obtained at the beginning of some interval of length ℓ_i; hence $X_t(\ell_i) = \emptyset$ for $t \not\equiv 0 \pmod{\ell_i}$. Moreover, since the longest interval is of length ℓ_k, *all* permits will have to be purchased afresh at the end of each length ℓ_k interval, and hence we can focus on the time interval from 0 to $T = \ell_k - 1$. Using these facts, consider a leasing solution that for each $t \in [T]$ and $p \in [k]$, buys leases of length ℓ_p on the elements in $X_t(\ell_p)$ at time t. The (expected) cost of this solution is[2]

$$\mathbf{E}\left[\sum_{e \in X_0(\ell_k)} c_e(\ell_k) + \sum_{t:\ell_{k-1}|t} \sum_{e \in X_t(\ell_{k-1})} c_e(\ell_{k-1}) + \sum_{t:\ell_{k-2}|t} \sum_{e \in X_t(\ell_{k-2})} c_e(\ell_{k-2}) + \ldots\right]. \quad (1)$$

We now define an instance of the k-stage stochastic optimization problem $\mathsf{Stoc}_k(\Pi)$ with the same optimal value as (1), and hence an α-approximation to the stochastic problem gives an α-approximation to our network leasing problem.

The Stochastic Instance. Consider the tree $\mathfrak{T}$ in Fig. 2 where the root has $N_{k-1} = \frac{\ell_k}{\ell_{k-1}}$ children, each node at depth 1 has $N_{k-2} = \frac{\ell_{k-1}}{\ell_{k-2}}$ children, and so on. This gives rise to ℓ_k leaves associated with the distributions $\pi_1, \pi_2, \ldots, \pi_{\ell_t}$ from left to right. The k-stage stochastic problem now involves k stages of decision-making. In the first stage, a particle is placed at the root, and we buy a set $Y_1 \subseteq X$, where element $e \in X$ costs $c_e(\ell_k)$. After this, the particle moves to one of the children of the root at random; after we learn the identity of this vertex of $\mathfrak{T}$, we can buy a "stage-2" set $Y_2 \subseteq X$, but the cost of e now becomes

---

[2] We allow randomized leasing policies, and so expectation is over the coin tosses of our algorithm, as well as over randomness in the choice of the sets S_t in case we are working in the *stochastic* offline model where S_t is drawn from the distribution π_t.

$c_e(\ell_{k-1}) \times N_{k-1}$. In this way, after t steps, the particle reaches some node at depth t, whence we buy some "stage-$t+1$" set $Y_{t+1} \subseteq X$ with the costs $c_e(\ell_{k-t}) \times \prod_{1 \leq p \leq t} N_{k-p} = c_e(\ell_{k-t}) \times \frac{\ell_k}{\ell_{k-t}}$. Finally, when the particle reaches some leaf v_k (at depth $k-1$, say it is the tth leaf), the algorithm finally gets a random set of clients $S_t \in_R \pi_t$, and must output a set Y_k such that $Y_1 \cup Y_2 \cup \ldots \cup Y_k \in \mathsf{Sols}(S_t)$; as above, the costs are now $c_e(\ell_1) \times \prod_{1 \leq p \leq k} N_{k-p} = c_e(\ell_1) \times \frac{\ell_k}{\ell_1}$.

The Correspondence. Note that a solution to this process associates a (potentially) random set $Y(v)$ with each vertex of tree $\mathfrak{T}$; the expected cost is

$$\mathbf{E}\left[\sum_{e \in Y(\text{root})} c_e(\ell_k) + \sum_{p=1}^{k-1} \sum_{v \text{ at depth } p} \Pr[\text{reach } v] \sum_{e \in Y(v)} c_e(\ell_{k-p}) \times \frac{\ell_k}{\ell_{k-p}} \right] \quad (2)$$

Finally, we place the nodes at level p of $\mathfrak{T}$ in correspondence with integers t such that $\ell_{k-p}|t$, associate $Y(v)$ with $X_t(\ell_{k-p})$, and observe the probability of reaching any fixed node at level p is $\frac{\ell_{k-p}}{\ell_k}$ to get that (2) and (1) are identical.

Costs and Inflations. The instances of $\mathsf{Stoc}_k(\Pi)$ created by the reduction above have the property that when we go *from stage $p-1$ to stage p* of the stochastic problem, the cost of each element e increases by an *inflation factor* of

$$\sigma_{e,p} \doteq \frac{c_e(\ell_{k-p+1}) \times N_{k-p+1}}{c_e(\ell_{k-p+2})}, \quad (3)$$

which by our assumptions is at least 1. If the leasing problem was uniform (the functions $c_e(\cdot)$ were the same for all $e \in X$), this inflation parameter depends only on the stage p but not on the element e (the *uniform inflation* case). But, if the leasing problem was non-uniform, we get a *non-uniform* inflation stochastic problem. This distinction will be useful, since depending on the problem Π, different approximation guarantees exist for uniform and non-uniform versions.

4.2 Leasing Algorithms from Existing Stochastic Algorithms

There has been much recent work on designing algorithms for multistage stochastic optimization with provable guarantees; see [34,17,19]; some are in the uniform inflation model, whereas others are more general. Using Theorem 4, we get:

Problem	Inflation type	Approximation Ratio for Leasing problem	Stochastic Citation
Steiner Tree	uniform	$8k$	$2k$ [17,19]
Facility Location	non-uniform	9.4	2.36 [37]
Vertex Cover	non-uniform	8	2 [28,37]
Set Cover	non-uniform	$4 \ln n$	$\ln n$ [37]

We note that as presented, the algorithms for the k-stage stochastic problems specify which elements to buy in an "online-like" fashion; given the observations of what has happened in the past, the stochastic algorithms prescribe the elements to buy at the current time instant. In particular, they do not give an *explicit representation* of the sets $Y(v)$ of elements to buy for each node v of the distribution tree $\mathcal{T}$. However, the above algorithms can easily be altered to give all these sets; the details are deferred to the final version of the paper.

5   New Stochastic/Leasing Approximations

In this paper, we give new results for k-stage stochastic optimization (and hence for Network Leasing) on a group of network design problems, all of which lie under the umbrella of "buy-at-bulk"-type problems. In these problems, the demand D_t for day t is not just a set of clients that have to be connected (as in Steiner Tree), but instead is a traffic matrix specifying how much traffic flows between various pairs of nodes in the network. In addition to the lease-cost function $c : K \rightarrow \mathbb{R}_+$ given earlier, we are also given a "bandwidth-cost" function $g : \mathbb{R}_+ \rightarrow \mathbb{R}_+$. The cost of leasing b bandwidth on an edge for ℓ length of time is now $\mathsf{Cost}(b, \ell) = g(b) \times c(\ell)$. (We consider these problems only in the *uniform model*, and hence both the functions $c(\cdot)$ and $g(\cdot)$ are the same for each edge.)

We will give the following results for some buy-at-bulk type problems, using the Boosted Sampling approach and defining "strict" cost-shares to prove these results; a quick overview is provided in Appendix A.

Problem	Inflation	Approximation Ratio	Citation
Buy-at-Bulk	uniform	$O(k \log n)$	Theorem 5
Single-Sink Rent-or-Buy	uniform	$O(k)$	Theorem 6
Single-Sink Buy at Bulk	uniform	$O(k)$	Theorem 7

5.1   Multiple-Sink Buy-at-Bulk

There are many ways to specify the Buy-at-Bulk problem which are all equivalent to within a factor of 2 (see, e.g., [38]), so let us fix one. We are given a demand matrix $D \in \mathbb{R}^{n \times n}$ where D_{ij} gives the traffic from v_i to v_j. We have a monotone subadditive cost function $g(\cdot)$, where the cost of bandwidth b is $g(b)$. By well-known properties of subadditive functions, we can find a *concave* cost function $h(\cdot)$ such that $g(b) \le h(b) \le 2g(b)$ for all $b \neq 0$. We assume that the cost of bandwidth allocation is $h(b)$ for all non-zero values of b; this only changes the problem by a factor of 2.

The best-known algorithm for the Buy-at-Bulk problem is by Awerbuch and Azar [2]. We approximate the graph by a random tree (as in Lemma 3), and given the Buy-at-Bulk problem on the tree, we can solve it on an edge-by-edge basis. We now show how to get an algorithm for the stochastic version.

Theorem 5. *The k-stage stochastic version of the Buy-at-Bulk problem on the tree has an $O(k)$ approximation, and hence Buy-at-Bulk on general graphs has an $O(k \log n)$ approximation.*

Proof. Let us give an algorithm for a single edge in the tree that separates V into A and $V \setminus A$: we can calculate the traffic crossing this edge e as $D_e = \sum_{ij \in \partial A} D_{ij}$. For this edge, we allocate capacity D_e and divide the cost $h(D_e)$ equally among each of the D_e units of demand. Clearly the cost shares are cross-monotone: if more demand passes through the edge, the cost only decreases because h is concave. Moreover, the algorithm is a 1-approximation with respect to these cost-shares, since we share the exact cost of the algorithm amongst the players.

Moreover, we can check that these cost shares are 1-c-strict (as defined in (5)): indeed, if we divided the traffic D_e into two parts S and T, and allocated S units of bandwidth first, then the cost shares $\xi(X/\mathcal{A}(S), T, T) = h(S+T) - h(S)$ would be at most the cost-shares $\xi(X, S \cup T, T) = h(S+T) \times \frac{T}{S+T}$ ascribed to T when both S and T were in the fray; this follows from the concavity of h.

Given that we have 1-c-strict and cross-monotone cost shares ξ and a 1-approximation algorithm $\mathcal{A}$ with respect to ξ, we can apply Theorem 8 to infer a k-approximation (with respect to the cost function h), and hence a $2k$-approximation with respect to the original cost function g. Finally, since we moved to a random tree, we lose another $O(\log n)$ in translating the solution back to the original graph G. This concludes the proof.

5.2   Single-Sink Buy-at-Bulk Problems

In the Single-Sink Rent-or-Buy problem (a special case of the Buy-at-Bulk problem), we are given a graph $G = (V, E)$ with a distinguished *root* vertex r. Each vertex j wants to send d_j amount of traffic to r. The bandwidth cost function is $g(b) = \min\{b, M\}$ for some parameter M. We show the following result:

Theorem 6. *The Single-Sink Rent-or-Buy problem has an $O(1)$-approximation algorithm with respect to 1-c-strict cost sharing functions; moreover, these cost-shares are cross-monotone.*

Combined with Theorem 8, this gives an $O(k)$-approximation for stochastic Single-Sink Rent-or-Buy, and hence an $O(k)$-approximation for the Single-Sink Rent-or-Buy Leasing problem, where buying b bandwidth for ℓ costs $g(b) \cdot c(\ell)$.

Proof. The algorithm $\mathcal{A}$ is the SimpleCFL algorithm from [15]. This algorithm starts off with $F = \{r\}$, and add each vertex j to F independently with probability d_j/M. It then builds an approximate Steiner tree on F using the MST heuristic, and allocates unlimited capacity on its edges (hence paying M on each such edge). It then sends d_j units of flow from j to its closest vertex in F (which may be j itself, in case $j \in F$); for this it pays cost 1 per unit of flow.

Define the cost-share for node j as $\xi_{RoB}(v) = \mathbf{E}[M \, \xi_{MST}(v)] + \mathbf{E}[d_j \, l(v, F)]$. (Here ξ_{MST} is a cross-monotonic cost-sharing function ξ_{MST} for the minimum spanning tree problem—e.g., given in [22,21], and $l(v, F)$ is the distance from v to the nearest vertex in F.) It is known that ξ_{RoB} is cross-monotone, and moreover that $\mathcal{A}$ is a 4-approximation for Single-Sink Rent-or-Buy with respect to these cost-shares ξ_{RoB} [24,18].

We claim ξ_{RoB} is 1-c-strict with respect to $\mathcal{A}$. By the definition of 1-c-strict-ness, we want to show that given $S, T \subseteq V$, $\xi(G, S \cup T, T) \geq \mathbf{E}[\xi(G/\mathcal{A}(S), T, T)]$; here the expectation on the right hand side is over the coins flipped by $\mathcal{A}(S)$.[3] Of course, to compute both the cost shares ξ's, we also have to take expectations. Since the expressions on the left and the right both involve flipping an independent coin for each of the nodes in $S \cup T$, let us couple the two random processes in the natural way by making the same set of coin tosses in both expressions.

Consider a particular choice of coin flips for $S \cup T$, which chooses $F_S \subseteq S$ and $F_T \subseteq T$; set $F = \{r\} \cup F_S \cup F_T$. The cost-shares on the right involve paying for the MST on F_T (in the graph $G/\mathcal{A}(S)$), and paying for connections from each $j \in T \setminus F_T$ to F. Charging for the latter is easy, since we pay for the distance from j to F in the left expression too. To pay for the former, we look at the primal-dual process that generates ξ_{MST}. In the run on $G/\mathcal{A}(S)$ with terminals F_T, a node j in F_T obtains cost-shares as long as its moat does not contain the root of the graph $G/\mathcal{A}(S)$. Since all nodes in F_S are contracted to the root in $G/\mathcal{A}(S)$, in the process for the left hand side the moat of j must not have hit any moat of $F_S \cup \{r\}$, and hence must get at least as much cost-share. This implies that for any particular set of coin flips, the cost-share on the right is bounded above by the cost-share on the left, and hence this holds in expectation as well.

This can be extended to give the following theorem:

Theorem 7. *The Single-Sink Buy-at-Bulk problem has an $O(1)$-approximation algorithm with respect to 1-c-strict cross-monotone cost sharing functions.*

The proof of Theorem 7 extends the proof of Theorem 6. While we defer it until the final version of the paper, we sketch it here: the algorithm is essentially the SimpleSSBB algorithm from [15], which uses the above SimpleCFL algorithm repeatedly to collect the traffic, which is then aggregated at some randomly chosen locations. Each time the aggregation is done using cables of larger capacity, and results in fewer and fewer locations, until finally all the traffic is at one location, whence it is sent to the root. Since we repeatedly use the algorithm SimpleCFL, the cost-share of a node u is just the expected cost-share of u accumulated over the various runs of SimpleCFL (where its cost-share is zero when there is no more traffic at u). The proof of strictness again proceeds by coupling the run on $S \cup T$ to the run where we build a solution on S, and then augment it to T.

6 Conclusions

In this paper, we defined several natural "Leasing" problems, in which an optimization problem is solved repeatedly over time (each time with a different set of clients), and the elements chosen to serve the clients can be leased for extended periods of time to take advantage of temporal trends in the sets of clients. The costs of these leases satisfy standard economies of scale, and hence longer leases cost less per unit of time. We study leasing problems in an offline

---

[3] The added expectation sign over the definition (5) is required since $\mathcal{A}$ is randomized.

setting, and give approximation algorithms for them via a connection with multistage stochastic optimization. We also give new algorithms for some network design problems in the multistage stochastic framework.

Many future directions of research suggest themselves: an important one is to extend the results to *online* or *stochastic* versions of leasing problems. In this paper, the demands D_t were given up front, but one can also consider cases where the demands D_t appear only on day t, chosen adversarially (i.e., the online model) or from some probability distribution (i.e., the stochastic model). While some of these problems can be solved by solving associated LPs and rounding them online (as in [27]), obtaining general results for these online problems is a direction we are exploring in ongoing work. There seem to be interesting questions involved in pricing these leases as well. It would be good to extend the "buy-at-bulk" results to cases where the cost function is not separable $g(b)f(\ell)$. Finally, getting $o(k)$-approximations for the Steiner Tree Leasing problem is an intriguing question—it seems that the ideas for such an improvement would be useful for the multistage stochastic versions as well.

References

1. Andrews, M., Zhang, L.: Wavelength assignment in optical networks with fixed fiber capacity. In: 31st ICALP. Volume 3142 of LNCS. (2004) 134–145
2. Awerbuch, B., Azar, Y.: Buy-at-bulk network design. In: 38th FOCS. (1997) 542–547
3. Awerbuch, B., Azar, Y., Bartal, Y.: On-line generalized Steiner problem. Theoret. Comput. Sci. **324**(2-3) (2004) 313–324
4. Balakrishnan, A., Magnanti, T.L., Mirchandani, P.: Network design. In Dell'Amico, M., Maffioli, F., Martello, S., eds.: Annotated bibliographies in combinatorial optimization. John Wiley & Sons Ltd., Chichester (1997) 311–334
5. Bartal, Y.: Probabilistic approximations of metric spaces and its algorithmic applications. In: 37th FOCS. (1996) 184–193
6. Carr, R., Fleischer, L., Leung, V., Phillips, C.: Strengthening integrality gaps for capacitated network design and covering problems. In: 11th SODA. (2000) 106–115
7. Charikar, M., Chekuri, C., Pál, M.: Sampling bounds for stochastic optimization. In: 9th APPROX. Volume 3624 of LNCS. Springer, Berlin (2005) 257–269
8. Dhamdhere, K., Ravi, R., Singh, M.: On two-stage stochastic minimum spanning trees. In: IPCO. Volume 3509 of LNCS. Springer, Berlin (2005) 321–334
9. Dye, S., Stougie, L., Tomasgard, A.: The stochastic single resource service-provision problem. Naval Research Logistics **50**(8) (2003) 869–887
10. Eisenbrand, F., Grandoni, F., Oriolo, G., Skutella, M.: New approaches for virtual private network design. In: 32nd ICALP. Volume 3580 of LNCS. (2005) 1151–1162
11. Fakcharoenphol, J., Rao, S., Talwar, K.: A tight bound on approximating arbitrary metrics by tree metrics. J. Comput. System Sci. **69**(3) (2004) 485–497
12. Garg, N., Khandekar, R., Konjevod, G., Ravi, R., Salman, F.S., Sinha, A.: On the integrality gap of a natural formulation of the single-sink buy-at-bulk network design formulation. In: 8th IPCO. Volume 2081 of LNCS. (2001) 170–184
13. Goyal, S., Giri, B.C.: Recent trends in modeling of deteriorating inventory. European Journal of Operational Research **134**(1) (2001) 1–16

14. Guha, S., Meyerson, A., Munagala, K.: Hierarchical placement and network design problems. In: 41th FOCS. (2000) 603–612
15. Gupta, A., Kumar, A., Roughgarden, T.: Simpler and better approximation algorithms for network design. In: 35th STOC. (2003) 365–372
16. Gupta, A., Pál, M., Ravi, R., Sinha, A.: Boosted sampling: Approximation algorithms for stochastic optimization problems. In: 36th STOC. (2004) 417–426
17. Gupta, A., Pál, M., Ravi, R., Sinha, A.: What about Wednesday? approximation algorithms for multistage stochastic optimization. In: 8th APPROX. (2005) 86–98
18. Gupta, A., Srinivasan, A., Tardos, É.: Cost-sharing mechanisms for network design. In: 7th APPROX. Volume 3122 of LNCS. (2004) 139–150
19. Hayrapetyan, A., Swamy, C., Tardos, E.: Network design for information networks. In: 16th SODA. (2005) 933–942
20. Immorlica, N., Karger, D., Minkoff, M., Mirrokni, V.: On the costs and benefits of procrastination: Approximation algorithms for stochastic combinatorial optimization problems. In: 15th SODA. (2004) 684–693
21. Jain, K., Vazirani, V.V.: Equitable cost allocations via primal-dual-type algorithms. In: 34th STOC, ACM Press (2002) 313–321
22. Kent, K.J., Skorin-Kapov, D.: Population monotonic cost allocations on MSTs. In: Proceedings of the 6th International Conference on Operational Research (Rovinj, 1996), Croatian Oper. Res. Soc., Zagreb (1996) 43–48
23. Kolliopoulos, S.G., Young, N.E.: Tight approximation results for general covering integer programs. In: 42nd FOCS. (2001) 522–528
24. Leonardi, S., Schäfer, G.: Cross-monotonic cost sharing methods for connected facility location games. Theoret. Comput. Sci. **326**(1-3) (2004) 431–442
25. Magnanti, T.L., Wong, R.T.: Network design and transportation planning: Models and algorithms. Transportation Science **18** (1984) 1–55
26. Magnanti, T.L., Mirchandani, P., Vachani, R.: Modeling and solving the two-facility capacitated network loading problem. Oper. Res. **43**(1) (1995) 142–157
27. Meyerson, A.: The parking permit problem. In: 46th FOCS. (2005) 274–284
28. Munagala, K. personal communication
29. Nahmias, S.: Perishable inventory theory: A review. Operations Research **30**(4) (1982) 680–708
30. Raghavan, P., Thompson, C.D.: Randomized rounding: a technique for provably good algorithms and algorithmic proofs. Combinatorica **7**(4) (1987) 365–374
31. Ravi, R., Sinha, A.: Hedging uncertainty: Approximation algorithms for stochastic optimization problems. In: 10th IPCO. (2004) 101–115
32. Salman, F.S., Cheriyan, J., Ravi, R., Subramanian, S.: Approximating the single-sink link-installation problem in network design. SIAM J. Optimization **11**(3) (2000) 595–610
33. Shmoys, D., Swamy, C.: Stochastic optimization is (almost) as easy as deterministic optimization. In: 45th FOCS. (2004) 228–237
34. Shmoys, D., Swamy, C.: Sampling-based approximation algorithms for multi-stage stochastic optimization. In: 46th FOCS. (2005) 357–366
35. Srinivasan, A.: Improved approximation guarantees for packing and covering integer programs. SIAM J. Comput. **29**(2) (1999) 648–670
36. Srinivasan, A.: New approaches to covering and packing problems. In: 12th SODA. (2001) 567–576
37. Srinivasan, A.: Approximation algorithms for stochastic and risk-averse optimization. In: 18th SODA. (2007) 1305–1313
38. Talwar, K.: Single-sink buy-at-bulk LP has constant integrality gap. In: 9th IPCO. Volume 2337 of LNCS. (2002) 475–486

A   Cost Shares and Stochastic Algorithms

We will draw on some techniques developed in recent work on converting approximation algorithms for standard (non-stochastic) versions of optimization problems into those for the stochastic versions of the problems [16,17]. In particular, we use the following theorem.

Theorem 8 ([17]). *Given a problem Π, if $\mathcal{A}$ is an α-approximation algorithm w.r.t. a 1-c-strict cost-sharing function ξ, and if ξ is cross-monotone, then there is an αk-approximation algorithm for the k-stage stochastic version of Π.*

Let us briefly discuss the basics of the cost-sharing concepts we will use in this paper; we refer the reader to [17] for a detailed discussion of the concepts. Loosely, a *cost-sharing* function ξ divides the cost of a solution among the client set S. We use the notation $\xi(G, S, j)$ to denote the share of the client j when the input is the graph G and the set of clients is S. (By convention, we will assume that $\xi(G, S, j) > 0 \implies j \in S$.) The function ξ is *cross-monotone* if for every pair of client sets $S \subseteq T$ and a client j such that $j \in S$, we have $\xi(G, T, j) \leq \xi(G, S, j)$. (I.e., if more clients join the system, the share of any individual client does not increase.)

Competitiveness. We will try to relate algorithms $\mathcal{A}$ to cost-sharing functions ξ, and hence ξ will conceptually behave like a "dual". Hence a crucial property is that ξ give a lower bound on the cost of the optimal solution: A cost-sharing function ξ is *competitive* if for every client set S, it holds that

$$\sum_{j \in S} \xi(G, S, j) \leq \mathsf{OPT}(X, S). \tag{4}$$

We will focus only on competitive cost-sharing functions. (Henceforth, we will use the notation $\xi(G, S, S')$ to denote $\sum_{j \in S'} \xi(G, S, j)$.)

Strictness. Let $S, T \subseteq V$ be sets of users. Suppose G is the original graph, and $G/\mathcal{A}(G, S)$ is the graph after the client set S has already been served by running the algorithm $\mathcal{A}$ on it. Then the cost-sharing function ξ is β-c-strict if

$$\xi(G/\mathcal{A}(G, S), T, T) \leq \beta \times \xi(G, S \cup T, T). \tag{5}$$

In other words, the total cost shares for the set T of users in the reduced instance $G/\mathcal{A}(G, S)$ is at most β times the cost-shares for T if the users in S were present as well. In this paper, we will deal only with the case when $\beta = 1$; i.e., cases where the cost shares for T when it appears with S are at least as much as when S is served earlier, and then T has to be served by itself.

Finally, we call $\mathcal{A}$ an *α-approximation algorithm with respect to the cost-sharing function ξ*

$$c(\mathcal{A}(G, S)) \leq \alpha \, \xi(G, S, S). \tag{6}$$

Note that chaining the inequalities (6) and (4) implies that $\mathcal{A}$ is an α-approximation algorithm in the conventional sense as well.

Robust Combinatorial Optimization with Exponential Scenarios

Uriel Feige[1], Kamal Jain[1], Mohammad Mahdian[2], and Vahab Mirrokni[1]

[1] Microsoft Research
{urifeige,kjain,mirrokni}@microsoft.com
[2] Yahoo! Research
mahdian@yahoo-inc.com

Abstract. Following the well-studied two-stage optimization framework for stochastic optimization [15,18], we study approximation algorithms for robust two-stage optimization problems with an exponential number of scenarios. Prior to this work, Dhamdhere et al. [8] introduced approximation algorithms for two-stage robust optimization problems with explicitly given scenarios. In this paper, we assume the set of possible scenarios is given implicitly, for example by an upper bound on the number of active clients. In two-stage robust optimization, we need to pre-purchase some resources in the first stage before the adversary's action. In the second stage, after the adversary chooses the clients that need to be covered, we need to complement our solution by purchasing additional resources at an inflated price. The goal is to minimize the cost in the worst-case scenario. We give a general approach for solving such problems using LP rounding. Our approach uncovers an interesting connection between robust optimization and online competitive algorithms. We use this approach, together with known online algorithms, to develop approximation algorithms for several robust covering problems, such as set cover, vertex cover, and edge cover. We also study a simple *buy-at-once* algorithm that either covers all items in the first stage or does nothing in the first stage and waits to build the complete solution in the second stage. We show that this algorithm gives tight approximation factors for unweighted variants of these covering problems, but performs poorly for general weighted problems.

1   Introduction

In many combinatorial optimization problems, the objective is to minimize the cost of building an installation to serve a number of clients. In classical optimization problems, it is often assumed that the parameters of the system are known in advance. However, in reality, it is almost always impossible or costly to obtain accurate data about various parameters of the optimization problem at the time of planning. For example, the cost of acquiring a resource or the set of clients that need to be serviced might be unknown. The goal of the fields of *stochastic optimization* and *robust optimization* is to provide algorithms for minimizing the cost in presence of uncertainty.

M. Fischetti and D.P. Williamson (Eds.): IPCO 2007, LNCS 4513, pp. 439–453, 2007.

In *stochastic optimization* [6,7], it is assumed that we have information about the probability distribution governing the data. Given this information, the goal is to plan ahead to minimize the expected cost. In particular, in *two-stage stochastic optimization*, a solution is built in two stages: in the first stage, we need to decide which resources to purchase given only distributional information about the instance. In the second stage, the exact information about the data is revealed and we are allowed to complement the solution built in the first stage by purchasing extra resources at an inflated cost.

Robust optimization [4,5,17,3] can be considered the worst-case analogue of the stochastic optimization. In a robust optimization problem, we are given bounds on various parameters of the system, and the goal is to find a solution that minimizes the cost in a worst-case scenario (or be feasible in a worst-case scenario). A *two-stage robust optimization* problem (introduced in [3,8]) is similar to a two-stage stochastic problem except instead of a distribution, we have a set of possible scenarios (given either explicitly, or implicitly by giving bounds on various parameters), and instead of expectation, we would like to minimize the maximum cost of the solution, where maximum is taken over the set of all possible scenarios.

During the past few years, stochastic optimization (and in particular, two-stage stochastic optimization) has received considerable attention from the perspective of approximation algorithms. Efficient approximation algorithms are given for a wide class of optimization problems, both for cases where the distribution is given explicitly by listing the set of all possible scenarios and the corresponding probabilities [18,13], and in the more general case where the distribution is given implicitly, as the product of a number of independent trials, or by an oracle [15,20,12,8].

For robust optimization, Ben-tal et al [3] initiated the study of two-stage robust optimization problems. Dhamdhere et al. [8] introduced the first approximation algorithms for two-stage robust covering problems when the set of scenarios is given *explicitly*. In this paper, we take on the task of studying approximation algorithms for two-stage robust optimization problems, where the set of possible scenarios is given *implicitly*. In particular, we focus on the case where the set of possible scenarios is given by an upper bound on the number of active clients, and give approximation algorithms for the robust version of several classical covering problems such as set cover, vertex cover, and edge cover.

1.1   The Model

In this section we give a formal definition of the robust optimization model that will be studied in this paper. This model is a generalization of the model introduced by Dhamdhere et al. [8] (in the case of explicitly listed scenarios), and is motivated by similar models for two-stage stochastic optimization [15,13].

In a *covering problem*, we have a set $\mathcal{C}$ of potential *clients*, and a set $\mathcal{R}$ of *resources*. Each resource $r \in R$ can be *purchased* at a cost c_r. In order to serve a set of clients, a set of resources must be purchased. The collection of all sets of resources which can serve the set $S \subset \mathcal{C}$ of clients is denoted by $sol(S)$. In covering problems the collection $sol(S)$ is an upper ideal, i.e., if a set of resources can serve S, so can any superset of this set. An *unweighted* covering problem is a covering problem in which all c_r's are equal to 1.

Generally, the collection $sol(S)$ is given implicitly by specifying a set of constraints. Three examples that we will focus on in this paper are set cover, vertex cover, and edge cover. In the *set cover* problem, each resource $r \in \mathcal{R}$ corresponds to a set of clients, and the collection $sol(S)$ consists of all sets of resources whose union covers S. In the *vertex cover* problem, the set of clients and the set of resources are the edge set and the vertex set of a given graph, respectively, and a set $S \subset \mathcal{C}$ can be served by any set of vertices that contain at least one of the endpoints of each edge in S. In the *edge cover* problem, each resource is an edge and each client is a vertex of a given graph, and a set S of clients can be covered by any set of edges that has at least one edge adjacent to any vertex in S.

In a *two-stage robust covering problem*, we have a collection $\mathcal{S}$ of scenarios, each given by a set of *active* clients (i.e., clients that need to be covered). The objective is to purchase a set of resources in the first stage to minimize the cost of these resources plus a given inflation factor λ times the maximum over scenarios S in $\mathcal{S}$, of the cost of completing the solution for scenario S. In other words, after we purchase a set of resources in the first stage, an adversary decides in which scenario we are. After that, we need to complete the solution by purchasing more resources at costs inflated by a factor λ (or more generally, by an inflation factors λ_r^S which depends on the resource $r \in \mathcal{R}$ and the scenario $S \in \mathcal{S}$).

The robust optimization problem can be studied in several different models, depending on how the list of scenarios $\mathcal{S}$ is given to the algorithm. One model, studied by Dhamdhere et al. [8] and Golovin et al [11], is to assume that the list of all possible scenarios is given explicitly. This model is suitable for situations where the number of possible scenarios is not very large. An alternative model, motivated by the independent trials model of stochastic optimization, is to assume that the list of scenarios is given implicitly by an upper bound on the maximum number of active clients. More formally, in this model an integer k is given and $\mathcal{S}$ is defined as $\{S \subseteq \mathcal{C} : |S| \le k\}$.[1] Finally, motivated by the oracle model in stochastic optimization, we define an oracle model for robust optimization where the list of possible scenarios is given by an oracle which, given the set of resources purchased in the first stage, outputs the worst-case (or approximately the worst-case) scenario for the second stage.

An important distinction between our oracle model and the oracle model for stochastic optimization is that in our model, the problem the oracle needs to solve is often computationally intractable. For example, if the set of scenarios in a robust set cover problem is given by an upper bound on the number of active clients, the oracle needs to find a subset of k clients whose minimum cost of covering is maximized. We call this problem the *max-min set cover problem*,

---

[1] More generally, we can consider a model where the set of clients is partitioned into subsets $\mathcal{C}_1, \ldots, \mathcal{C}_t$, and the set of scenarios is the collection of all sets that have at most k_i clients from the set $\mathcal{C}_i$. Although the results in Sections 2 and 3 work for this more general model, for clarity of exposition we restrict ourselves to the simpler model.

and will observe that it is computationally hard. Considering the hardness of the oracle problem, the algorithms designed for the oracle model need to be able to work with an *approximate oracle* as well. Furthermore, in order to solve a robust optimization problem in the model where the scenarios are given by an upper bound on the number of active clients, in addition to designing an algorithm for the oracle model, we need to give an approximation algorithm for the max-min version of the problem.

1.2   Our Contribution

In this paper, we mostly focus on the model where the scenarios are given implicitly by an upper bound on the number of active scenarios. This is motivated by real-world situations where a good estimate of the total number of clients who will show up is available, but we do not exactly know where they will appear. We will also give a general LP-based algorithm for the oracle model, assuming that the oracle gives a good approximation of the worst-case scenario with respect to the fractional solution.

A naive idea to solve the robust optimization problems is a *buy-at-once* algorithm: either cover all items in the first stage in which case nothing needs to be done in the second stage. Or do nothing in the first stage and construct a solution in the second stage, after the adversary makes its choices. The choice of which of the two options to use is based on a polynomial-time test that is problem specific. We study this algorithm in Section 4 and prove that when the inflation factor is the same for all scenarios, the approximation ratio of this algorithm for robust unweighted set cover, vertex cover, and edge cover problems are $\max(\log m, \log n)$, 2, and 2, respectively. However, the following example shows that for the weighted version of robust vertex cover, any buy-at-once algorithm (even with unbounded computing power) performs poorly. Consider a clique on n vertices, with $k = 1$ and $\lambda = \sqrt{n}$. All vertices have weight 1, except for two vertices that have weight $w = \sqrt{n}$. The buy-at-once algorithm will either pay at least n in the first stage, or at least $\lambda w = n$ in the second stage. However, an optimal algorithm can choose only one of the heavy vertices in the first stage, and then pay at most $w + \lambda k = 2\sqrt{n}$. Hence the approximation ratio of the buy-at-once algorithm for weighted robust vertex cover is no better than $\Omega(\sqrt{n})$. This example indicates the need for more sophisticated approximation algorithms for robust two-stage optimization problems.

In Section 2, we give a general LP-based framework for solving robust covering problems given access to an oracle that solves the fractional *max-min problem* (or the adversary's problem) and another oracle that rounds the LP solution for the classical (i.e., non-robust) optimization problem. For example, for the robust set cover problem, we need an oracle that given an integer k and a collection of subsets $S_1, S_2, \ldots, S_m$ of a universe F each with a cost $c(S_i)$, finds a subset $T \subseteq F$ of size at most k for which the cost of fractional set cover is maximized, and another oracle that rounds a fractional set cover to an integral one. In Section 3, we show how an online algorithm can be used to solve the max-min problem when the set of feasible scenarios are given by an upper bound on the number of active clients. We use this to give an $O(\log m)$-approximation algorithm for max-min fractional set cover. We also show that the max-min fractional set cover problem is not approximable within a factor better than $\Omega(\frac{\log m}{\log \log m})$ under reasonable complexity assumptions. As a result of this framework, we get an $O(\log n \log m)$-approximation for the robust set cover problem. Following similar ideas, we design constant-factor approximation algorithms for robust vertex cover and edge cover problems. This framework can be extended easily to more general settings in which the scenarios are given implicitly in more general ways. The main step for these extensions is to design good approximation algorithms for the max-min problems.

Finally, in Section 5 we show that our algorithms for max-min fractional set cover and edge cover achieve essentially the best possible approximation factors, assuming reasonable complexity assumptions.

2   An LP-Rounding Approach for Robust Set Cover

In this section, we give an LP-based approach for robust set cover. Our techniques work for a more general covering problem where each resource $r \in \mathcal{R}$ can be picked an integer number of times x_r, and a client is covered if a corresponding inequality of the form $\sum_r a_{ir} x_r \geq 1$ (where a_{ir} are given non-negative coefficients) is satisfied. The details of this generalization are omitted here.

We start by giving an LP formulation of two-stage robust set cover.[2]

$$\text{minimize} \quad Z + \sum_{r \in \mathcal{R}} c_r y_r^0 \tag{1}$$

$$\text{subject to} \ \forall S \in \mathcal{S}, \forall i \in S : \ \sum_{r:\ i \in r} (y_r^0 + y_r^S) \geq 1 \tag{2}$$

$$\forall S \in \mathcal{S} : \ \sum_{r \in \mathcal{R}} \lambda c_r y_r^S \leq Z. \tag{3}$$

The variable y_r^0 in the above LP indicates whether the resource r is purchased in the first stage. Similarly, the variable y_r^S indicates whether this resource is purchased in the second stage, if the adversary selects the set S as the set of active clients. The variable Z indicates the maximum cost of the second stage, where the maximum is taken over all possible scenarios. Clearly, if the variables y_r^0 and y_r^S are restricted to be integers, the above integer program captures the robust set cover problem precisely. Therefore, relaxing the integrality condition gives us a linear program whose solution is a lower bound on the cost of the optimal solution to the robust set cover problem.

The main difficulty with this LP formulation is that it contains an exponential number of constraints *and* an exponential number of variables, and therefore cannot be solved directly using the ellipsoid method. We can deal with this problem

---

[2] We present this LP in the case that the inflation factor λ does not depend on the resource r or the scenario S. However, it is easy to see that all proofs in this section apply to the more general case.

using a technique developed by Shmoys and Swamy [20] for stochastic optimization: we consider the projection of the above LP onto the space corresponding to the variables y_r^0's and Z, and then give a separation oracle for the reduced LP. The projection of the above LP corresponds to the following program.

$$\text{minimize} \quad Z + \sum_{r \in \mathcal{R}} c_r y_r^0 \qquad (P)$$

$$\text{subject to} \quad \forall S \in \mathcal{S} : \ Z \geq \text{cost}_2(S, y^0)$$

Here $\text{cost}_2(S, y^0)$ denotes the cost of the optimal fractional solution for the second stage when the set of active clients is S, given that resource r is already purchased to the extent of y_r^0 in the first stage.

The separation oracle for this LP corresponds to an algorithm that computes the optimal strategy for the adversary of the robust fractional set cover problem. We call this *the max-min fractional set cover problem*. More precisely, the max-min fractional set cover problem is the following: given a fractional first-stage solution (i.e., y_r^0's), select a scenario (in the example we will focus on in this paper, a set of at most k clients) so that the cost of a fractional solution for the second stage is maximized. The following lemma, proved using a simple application of the ellipsoid method, shows that given an approximation algorithm for the max-min fractional problem, we can compute an approximate solution of the above LP in polynomial time.

Lemma 1. *Assume we have a polynomial time γ-approximation algorithm for the max-min fractional problem. Then, we can compute a γ-approximation to the solution of the linear program (P) in polynomial time.*

The proof, which is omitted here, is based on the ellipsoid algorithm and the techniques used by Shmoys and Swamy [20] in the context of stochastic optimization.

The above lemma requires us to be able to solve the max-min fractional set cover problem given a *fractional* first-stage solution. In other words, for each client i we are given a fractional value θ_i, so that if the adversary chooses i in the set of active clients, we will have to cover i to the extent of θ_i. In the following lemma, we show that it is enough to be able to solve the max-min problem given that θ_i's are zero or one. In other words, given a subset C' of the clients (corresponding to those with $\theta_i = 1$), we need to be able to find a set of at most k clients in C' whose minimum fractional covering cost is maximized. We call this problem the max-min fractional set cover problem with integer requirements.

Lemma 2. *Assume we have a polynomial time γ-approximation algorithm A for the max-min fractional set cover problem with integer requirements. Then, we can compute a $(\gamma+1)$-approximation to the solution of the linear program (P) in polynomial time.*

Proof. We iteratively run the ellipsoid algorithm to check whether (P) has a solution with an objective function value of at most R, and use binary search to find the smallest value of R for which the ellipsoid algorithm declares that there is such a solution. For the separation oracle, we do the following: let $C' = \{i \in C : \sum_{r:i \in r} y_r^0 < 1/(\gamma + 1)\}$ denote the set of clients covered by the fractional first-stage solution to an extent less than $1/(\gamma + 1)$. We run algorithm A to find the max-min fractional set cover among clients in C'. Let T denote the cost of the solution returned by A. This means that there is at least one scenario $S^* \in \mathcal{S}$ such that the cost of minimum fractional cover for $S^* \cap C'$ is at least T, and for every scenario $S \in \mathcal{S}$, the cost of the minimum fractional cover for $S \cap C'$ is at most γT. Our separation oracle accepts (y^0) if the cost of the first stage solution y^0 plus $\frac{\gamma}{\gamma+1}T$ is at most R; otherwise it rejects (i.e., declares that there is no solution with a first stage solution of y^0 of total cost at most R).

First, we show that if the above separation oracle rejects y^0, then an exact separation oracle would do the same. This is because with a first stage solution of y^0, clients in C' need to be covered to the extent of at least $\frac{\gamma}{\gamma+1}$ in the second stage, and therefore the cost of the second stage in scenario S^* cannot be less than $\frac{\gamma}{\gamma+1}$ times the cost of the minimum fractional cover for $S^* \cap C'$, which, by definition, is at least T.

Next, we prove that if our separation oracle accepts y^0, then we can build a feasible solution for (P) of cost at most $(\gamma + 1)R$. This is done by multiplying y^0 by $(\gamma + 1)$. The set of clients not covered by this inflated first stage solution is a subset of C'. Therefore, the cost of the second stage is at most γT. The overall cost of this solution is at most $(\gamma + 1) \sum_{r \in \mathcal{R}} c_r y_r^0 + \gamma T \leq (\gamma + 1)R$, where the latter inequality follows from the fact that our separation oracle accepts y^0.

Now, let R^* be the smallest value of R for which our algorithm decides that the linear program has a solution. This means that for $R < R^*$, our separation oracle never accepts any first stage solution y^0. By our first observation, the ellipsoid algorithm with an exact separation oracle would return the same answer. Hence, R^* is a lower bound on the solution of (P). Since the ellipsoid algorithm for $R = R^*$ finds a y^0 which our separation oracle accepts, our second observation implies that there is a solution of value $(\gamma + 1)R^*$ for (P). This is a $(\gamma + 1)$-approximate solution for the program (P). $\qquad\square$

The solution obtained by solving the linear program (P) can be rounded into an integral solution using an LP-based algorithm that solves the (non-robust) optimization problem. Combining this with Lemma 2, we obtain the following.

Theorem 1. *Assume there is an α-approximation algorithm A_1 for the max-min fractional set cover problem with integer requirements, and an algorithm A_2 that given a subset S of clients, finds an integral solution that covers the clients in S and whose cost is at most β times the minimum cost of fractionally covering S. Then there is a $(\alpha + 1)\beta$-approximation algorithm for the robust set cover problem.*

Proof. We run the algorithm described in the proof of Lemma 2 to compute an $(\alpha + 1)$-approximate solution to the LP (P). The solution that this algorithm finds corresponds to $(\alpha + 1)y^0$, where y^0 is a first stage solution accepted by our separation oracle. As in the separation oracle, we define $C' = \{i \in C : \sum_{r:i \in r} y_r^0 < 1/(\alpha + 1)\}$. Now, we run the algorithm A_2 to find an integral set cover solution that covers clients in C'. The cost of this first stage solution is at most $(\alpha + 1)\beta$ times the cost of y^0. Also, for every scenario in the second stage, we use A_2 to find a β-approximation to the optimal fractional second stage cost. This defines a $(\alpha + 1)\beta$ approximation algorithm for robust set cover.   □

By the above theorem, the main ingredient in solving a robust optimization problem with implicitly given scenarios is the algorithm for the max-min problem. In the next section, we show how an online algorithm for the underlying optimization problem can be used to approximately solve the max-min problem.

3   The Max-Min Problems

The results of the previous section show that in order to solve the LP relaxation of the robust set cover problem, we need to consider the max-min problem. In this section, we design an $O(\log m)$-approximation algorithm for max-min fractional set cover problem. In fact, we present a general framework to design an approximation algorithms for a max-min problem using online competitive algorithms for the underlying optimization problem. Note that the max-min problems that we need to solve for approximating the robust covering problems are the fractional variants of the problems.

Given a universe F of clients and a subset $T \subseteq F$, let $\mathrm{opt}(T)$ be the cost of an optimal (fractional) solution to cover all clients in T. Let A be an α-competitive online algorithm for a covering problem. Namely, upon the arrival of any client a_k to an existing set of clients $a_1, a_2, \ldots, a_{k-1}$, A augments the current solution to a feasible solution for $a_1, \ldots, a_{k-1}, a_k$. The algorithm is α-competitive if for every sequence of clients $a_1, \ldots, a_k$ the cost of the online solution produced by A is at most α times the cost of the optimal (offline) solution for $a_1, a_2, \ldots, a_k$. Let $A(b|a_1, a_2, \ldots, a_k)$ denote the *marginal increase* in the cost of the solution constructed by algorithm A when we add a new element b to an existing sequence of clients $(a_1, \ldots, a_k)$.

Consider two solutions w and w' for a fractional covering problem. Solution w' dominates solution w if for each set S the fractional value given to its respective variable in w' is at least as large as that given in w. We say that the covering problem satisfies the *monotonicity* property, if for any two given solutions w and w' such that w' dominates w and any element a, the optimal marginal increase in expanding w' to cover a is not more than the optimal marginal increase in expanding w to cover a. It is not hard to prove that the set cover problem and its special cases satisfy this property.

The following theorem presents a relation between competitive online algorithms and approximation algorithms for the max-min problem.

Theorem 2. *Let A be an α-competitive online algorithm for a covering problem. If the covering problem satisfies the monotonicity property then the corresponding max-min problem admits an $(\frac{e}{e-1})\alpha$-approximation algorithm.*

Proof. Given the online algorithm A for the covering problem, we prove that the following algorithm B is a $(1 - \frac{1}{e})\alpha$-approximation algorithm for the max-min problem:

1. $T = \emptyset$.
2. for $i = 1, \ldots, k$ do
   (a) Find a client a_i that maximizes $\mathcal{A}(a_i | a_1, a_2, \ldots, a_{i-1})$ and add it to T.

Let the optimal solution to the max-min problem be the set $\{b_1, \ldots, b_k\}$ of clients. Let OPT^* be the optimal cost of covering $\{b_1, \ldots, b_k\}$. Let W_i be the cost of the solution of the online algorithm after i elements $a_1, \ldots, a_i$ have arrived and $L_i = \max[0, \mathrm{OPT}^* - W_i]$. We prove that $L_i \leq (1 - \frac{1}{k}) L_{i-1}$. Consider expanding the solution of the online algorithm for $\{a_1, \ldots, a_{i-1}\}$ in the optimal way so that it covers $\{b_1, b_2, \ldots, b_k\}$. The cost of this new solution is at least OPT^*. Hence there is some item b_j (with $1 \leq j \leq k$) such that there is difference of at least $\frac{\mathrm{OPT}^* - W_{i-1}}{k}$ in cost between the case in which the clients $b_1, \ldots, b_{j-1}$ are added (or no clients at all, if $j = 1$) and the case in which the clients $b_1, \ldots, b_j$ are added. Since the covering problem satisfies the monotonicity property, adding b_j alone to $\{a_1, \ldots, a_{i-1}\}$ requires an increase in cost of at least $\frac{\mathrm{OPT}^* - W_{i-1}}{k}$ compared to the cost of $\mathcal{A}$ covering $\{a_1, \ldots, a_{i-1}\}$. Hence $W_{i-1} + (\frac{\mathrm{OPT}^* - W_{i-1}}{k})$ is a lower bound on the cost of $\mathcal{A}$ for covering $\{a_1, \ldots, a_{i-1}, b_j\}$. Since algorithm $\mathcal{B}$ chooses in the ith step the a_i that maximizes the marginal increase in the cost of $\mathcal{A}$, we will indeed have that $W_i \geq W_{i-1} + \frac{\mathrm{OPT}^* - W_{i-1}}{k}$, and thus, $L_i \leq L_{i-1}(1 - \frac{1}{k})$. Thus $L_i \leq (1 - \frac{1}{k})^i L_0$. Therefore, $L_k \leq (1 - \frac{1}{k})^k \mathrm{OPT}^* \leq \frac{1}{e} \mathrm{OPT}^*$. This shows that $W_k \geq (1 - \frac{1}{e}) \mathrm{OPT}^*$. Since $\mathcal{A}$ is an α-competitive algorithm, the true cost of covering $\{a_1, \ldots, a_k\}$ is at least W_k / α, and algorithm $\mathcal{B}$ is a $(\frac{e}{e-1}) \alpha$-approximation algorithm for the max-min problem. $\square$

Using Theorem 2 and known online algorithms, we can design approximation algorithms for the max-min problems. For example, an $O(\log m)$-competitive algorithm for the online fractional set cover problem by Alon et al. [1] and Theorem 2 implies an $O(\log m)$-approximation algorithm for the max-min fractional set cover problem. In Section 5, we show that this result is nearly best possible (assuming certain complexity theoretic assumptions). This algorithm, together with the $O(\log n)$ randomized rounding algorithm for set cover and Theorem 1, imply the following.

Theorem 3. *There exists an $O(\log m \log n)$-approximation algorithm for the robust two-stage set cover problem.*

Using the ideas of the 2-approximation algorithm for vertex cover by Bar-Yehuda and Even [2], we can design a 2-competitive online algorithm for vertex cover problem as follows. In the online algorithm, we keep track of a value $r(u)$ for each vertex u of the graph. We initialize these values to $r(u) = w(u)$. Upon the arrival of a new edge $e = uv$, the online algorithm sets $r_u = r_u - \min(r_u, r_v)$ and $r_v = r_v - \min(r_u, r_v)$. Observe that after this update either $r(u) = 0$ or $r(v) = 0$. At any moment, the fractional vertex cover solution is to pick $1 - \frac{r(u)}{w(u)}$ fraction of each vertex u. This means that we fully pick u or v for edge $e = uv$ and this solution is a feasible fractional vertex cover. Similar to the proof of Bar-Yehuda and Even [2], we can prove that this algorithm is a 2-competitive online algorithm. Using Theorem 2, this 2-competitive online algorithm implies a $(\frac{2e}{e-1})$-approximation algorithm for the max-min fractional vertex cover problem. Applying the above results and the 2-approximate rounding procedure for the vertex cover problem, we get a $2(\frac{2e}{e-1}+1)$-approximation algorithm for the robust (weighted) vertex cover problem. The details are omitted.

For the edge cover problem, a simple 2-competitive online algorithm is to cover every arriving vertex with the cheapest edge incident to it. This, together with Theorems 2 and 1, give a constant-factor approximation algorithm for the robust edge cover problem. We do not optimize the constants of the approximation ratio for the weighted problems. However, in Section 4 we show a tight buy-at-once 2-approximation for unweighted vertex cover and edge cover.

4   Improved Algorithms for Unweighted Problems

In this section, we give buy-at-once approximation algorithms for unweighted variants of robust set cover, vertex cover, and edge cover.

4.1   Robust Unweighted Set Cover

In the robust unweighted set cover problem, all sets have unit cost. The input of the problem consists of n items, a collection of m sets, a parameter k (for number of items to be chosen by adversary), and an inflation factor $\lambda > 1$. To simplify notation, we assume here that parameters such as k, m and n are sufficiently large, and hence we shall ignore effects such as rounding $\ln m$ to the nearest integer. They affect the approximation ratio only by low order terms. The buy-at-once approximation algorithm for robust set cover is as follows:

1. Compute a minimum fractional set cover that covers all potential clients and let t be its size.
2. If $t < \frac{\lambda k}{\ln n}$, use the greedy algorithm to find a set cover. It will be of size at most $t \ln n$. Nothing needs to be done in the second stage.
3. If $t \geq \frac{\lambda k}{\ln n}$, do nothing in first stage. In the second stage, use a greedy algorithm to cover the items chosen by the adversary.

Theorem 4. *The above buy-at-once algorithm achieves an approximation ratio no worse than* $\max(\ln n, \ln m)$ *(up to low order terms) for unweighted robust set cover.*

Proof. Observe that by duality, t is the size of the maximum fractional packing. Let αt be the number of sets chosen by the optimal solution (to the robust set cover problem) in the first stage. Removing all items covered by these sets, the remaining set cover instance still has a fractional packing of value at least $(1 - \alpha)t$. (We may assume that $\alpha \leq 1$, as otherwise the analysis becomes even simpler.) Pick a set T of items, where each item is selected into T independently, with probability equal to its fractional value in the maximum fractional packing. The expected size of T is at least $(1 - \alpha)t$. In fact, known bounds by Siegel [19] imply that $|T| \geq \lfloor(1 - \alpha)t\rfloor$ with probability at least $1/2$. Moreover, every set is expected to contain at most one item from T, and Chernoff bound implies that with probability at least $1/2$, no set will contain more than $\ln 2m$ items. For simplicity of notation, we shall assume that T contains exactly $(1 - \alpha)t$ items, and no set contains more than $\ln m$ items from T. (This assumption affects only low order terms in the approximation ratio.) Hence in the second stage opt will pay at least $\min((1 - \alpha)t, k)\frac{\lambda}{\ln m}$, and in the two stages combined opt pays at least $\alpha t + \min((1 - \alpha)t, k)\frac{\lambda}{\ln m}$. This is a piecewise linear function in α, and its minimum is achieved when α is either 0 or 1, or when $(1 - \alpha)t = k$. It follows that opt pays at least $\min(t, \frac{k\lambda}{\ln m})$.

Now we can analyze the approximation ratio of our algorithm. When $t < \frac{\lambda k}{\ln n}$, the algorithm pays at most $t \ln n \leq \lambda k$, which is a factor of $\ln n$ larger than t, and at most a factor of $\ln m$ larger than $\frac{k\lambda}{\ln m}$. Hence the approximation ratio in this case is at most $\max(\ln m, \ln n)$.

When $t \geq \frac{\lambda k}{\ln n}$, the algorithm pays nothing in the first stage, and at most $\lambda k \leq t \ln n$ in the second stage. Again, the approximation ratio can be seen to be at most $\max(\ln m, \ln n)$. $\qquad\square$

The following example shows that the above analysis for the buy-at-once algorithm is tight up to a $\log \log m$ factor: consider an instance of the two-stage robust set cover where the ground set consists of $n + n^{1/4}$ elements and $k = n^{1/4}$ and $\lambda = n^{1/4}$. The family of subsets in the set cover instance is the family of all subsets of size $n^{1/4}$ of set $\{1, 2, \ldots, n\}$ and all singleton sets $\{n+1\}, \{n+2\}, \ldots, \{n+n^{1/4}\}$. The optimal solution is to buy all singleton sets in advance and wait for the scenario. Since the adversary should choose a set of size $n^{1/4}$ of $\{1..n\}$ and this set is in the family of sets in the set cover instance, we can cover any scenario by buying one set at cost $\lambda = n^{1/4}$ later. Thus, the cost of the optimal solution is at most $2n^{1/4}$. If we do not buy any set in advance, the adversary selects $\{n+1, n+2, \ldots, n+n^{1/4}\}$ and we should pay $\lambda k = n^{1/2}$ in the second stage. On the other hand, in order to cover all elements in the first stage, we need to buy at least $n^{3/4}$ sets. Both of these cases are more than a factor of $n^{1/4} = \Omega(\frac{\log m}{\log \log m})$ larger than the optimal solution.

Also, observe that the term $\ln n$ in the approximation ratio cannot be improved by any polynomial-time algorithm (e.g., when $\lambda = \infty$), due to the hardness of approximating minimum set cover [10]. It is not clear whether the term $\ln m$ can be improved.

4.2   Robust Unweighted Vertex Cover

Consider the following buy-at-once algorithm for the robust unweighted vertex cover problem: compute a maximum matching M in G, and let $|M|$ denote its size. If $|M| < \lambda k$, then we pick a vertex cover of size no larger than $2|M|$ in the first stage (for example, by picking both endpoints of every edge in M) and nothing needs to be done in the second stage. If $|M| \geq \lambda k$, we do nothing in the first stage and in the second stage, for each edge that is present in the realized scenario, we pick one of its endpoints arbitrarily.

Theorem 5. *The above algorithm achieves an approximation ratio of 2 for the unweighted robust vertex cover problem.*

Proof. Let OPT denote an optimal algorithm, and let x denote the number of vertices purchased by this algorithm in the first stage. Thus, at least $|M| - x$ of the edges in M are not covered by OPT in the first stage. Consider the scenario where the adversary picks $\min(k, |M| - x)$ of these edges in the second stage. The overall cost of OPT in this scenario is $T := x + \lambda \min(k, |M| - x)$. Since $\lambda \geq 1$, we have $T \geq \min(\lambda k, |M|)$.

Now, we show that the cost of our algorithm is always at most $2T$. We consider two cases: if $|M| < \lambda k$, our algorithm buys a vertex cover of cost at most $2|M|$ in the first stage. Therefore, the cost of our algorithm is at most $2|M| = 2 \min(\lambda k, |M|) \leq 2T$. If $|M| \geq \lambda k$, our algorithm incurs a cost of $\lambda k = \min(\lambda k, |M|) \leq T$. Therefore, in this case our algorithm is actually optimal.   □

Note that any algorithm that approximates unweighted robust vertex cover within a ratio better than 2 can be used to approximate the minimum vertex cover problem within a ratio better than 2 (e.g., by setting $\lambda = \infty$), and achieving this would resolve a long standing open problem.

4.3   Robust Unweighted Edge Cover

The input of the unweighted edge cover problem is a graph with n vertices, m edges, a parameter k (for number of vertices to be chosen by adversary), and an inflation factor $\lambda > 1$. All edges have unit cost. Observe that the number of edges needed to cover ℓ vertices is always between $\ell/2$ and ℓ. This fact serves as a basis for a tight 2-approximation for the robust unweighted edge cover problem. The algorithm and the proof are left to the full version of the paper. Moreover, we can prove that if P≠NP, then the max-min variant and the robust two-stage variant of the edge cover problem cannot be approximated within a factor better than 2.

5   Hardness of Max-Min Problems

In this section, we give a strong inapproximability result for the max-min (fractional) set cover problem. First, we show a hardness result for the max-min (fractional) edge cover problem.

Theorem 6. *For every $\epsilon > 0$, it is NP-hard to approximate the max-min unweighted edge cover problem within a ratio better than $2 - \epsilon$.*

Proof. The proof is by reduction from the maximum independent set problem. As shown in [14], for every sufficiently small $\epsilon > 0$, it is NP-hard to distinguish between the following two classes of graphs:

Yes instances: graphs on n vertices that contain an independent set of size ϵn.

No instances: graphs on n vertices with no independent set of size $\epsilon^5 n$.

In order to distinguish whether a given graph G is a yes instance or a no instance, we construct an instance of the max-min edge cover problem that consists of G and $k = \epsilon n$. On yes instances, one can select k vertices that form an independent set in G, and then k edges are needed in order to cover them. On no instances, whenever there are more than $\epsilon^5 k$ vertices, two of them share an edge in G. It follows that any selection of k vertices can be covered by $\epsilon^5 k + (1-\epsilon^5)k/2 < k/(2-\epsilon)$ edges. Therefore, any algorithm that approximates the max-min unweighted edge cover problem within a factor better than $2 - \epsilon$ can be used to distinguish between these classes.   □

The proof of Theorem 6 can be adopted easily for the max-min fractional edge cover problem. This implies that for any $\epsilon > 0$, it is NP-hard to approximate the max-min fractional edge cover problem within a factor better than $2 - \epsilon$. This hardness ratio can be strengthened to nearly logarithmic factors for the fractional set cover problem, but proving this using current techniques seems to require assumptions stronger than $P \neq NP$. Picking $p(n) = \sqrt{n}$ in Theorem 7 shows that the max-min (fractional) set cover problem cannot be approximated within a ratio better than $\Omega(\frac{\log N}{\log \log N})$ (on instances of size N) unless 3SAT can be solved in time $2^{O(\sqrt{n})}$ (on instances of size n).

Theorem 7. *For every $0 < \delta < 1$ and $p(n) = n^\delta$, the max-min fractional set cover problem cannot be approximated within a ratio better than $\Omega(\frac{p(n)}{\log p(n)})$ on instances of size $N = 2^{O(p(n))}$ (in time polynomial in N), unless NP problems (say 3SAT) can be solved in time $2^{O(p(n))}$.*

Proof. The proof is presented for the integral set cover problem, but the approximation hardness applies also to the max-min fractional set cover problem, because in the *yes* instance the cover is disjoint. The proof is based on the structure of instances of set cover that are generated by the reduction described in [10], and specifically, on the parameters given in Section 6 in [10]. Here we only sketch the proof.

Recall that in [10], the hardness of approximation result is based on a certain multiple-prover proof system. We shall need the number of provers (denoted in [10] by k) to be $p(n)$. (Hence one cannot use here the earlier [16] instead of [10].) In [10] it suffices that the number of parallel repetitions ℓ is logarithmic in the number of provers, hence we can have $\ell = O(\log(p(n)))$. (Remark: later work [9] used a version of a multilayered PCP which is somewhat simpler than the multiple prover system of [10]. This simpler version requires ℓ to grow at a faster rate than $p(n)$, and would result in weaker conclusions if used in the proof of Theorem 7.) This results in a set cover instance with $2^{O(p(n))}$ clients and sets.

Each subset in [10] would be an item in the max-min set cover problem. Each item in [10] would be a set in the max-min set cover problem. Note that in [10] all sets are of the same size, and there is a disjoint set cover for *yes* instances, say, by t sets. We shall set k for the max-min set cover problem to be equal to this t. Hence *yes* instances of [10] correspond to *yes* instances of max-min set cover for which k clients can be selected that require k sets in order to be covered.

The property of *no* instances of [10] that we shall use is the following: for every $q < p(n)$, for every collection of $tq/p(n)$ sets, there is some item that belongs to $O(p(n)/q)$ of the sets. Extensions of the analysis in [10] can be used in order to prove this property, but this is omitted from the current paper.

The property above implies that for *no* instances in [10], for every collection of t sets there are $O(t\frac{\log(p(n))}{p(n)})$ clients that hit all the sets. This implies that in *no* instances of the max-min set cover problem, the optimum solution has value $O(t\frac{\log(p(n))}{p(n)})$. $\qquad\square$

Acknowledgements

This work benefited greatly from insightful comments that anonymous reviewers provided on previous versions of this manuscript.

References

1. N. Alon, B. Awerbuch, Y. Azar, N. Buchbinder, and J. Naor. A general approach to online network optimization problems. In *SODA*, pages 577–586, 2004.
2. R. Bar-Yehuda and S. Even. A linear time approximation algorithm for the weighted vertex cover problem. *Journal of Algorithms*, 2:198–203, 1981.
3. A. Ben-Tal, A. Goryashko, E. Guslitzer, A. Nemirovski. Adjustable robust solutions of uncertain linear programs. 351-376 *Mathematical Programming*, 99:2:351-376, 2004.
4. D. Bertsimas and M. Sim. The price of robustness. *Operation Research*, 52:35–53.
5. D. Bertsimas and M. Sim. Robust discrete optimization and network flows. *Mathematical Programming Series B*, 98:49–71.
6. J. Birge and F. Louveaux. Introduction to stochastic programming. *Springer*, Berlin, 1997.
7. G. B. Dantzig. Linear programming under uncertainty. *Management Science*, 1:197–206, 1955.
8. K. Dhamdhere, V. Goyal, R. Ravi, and M. Singh. How to pay, come what may: Approximation algorithms for demand-robust covering problems. *FOCS*, 2005.
9. I. Dinur, V. Guruswami, S. Khot, and O. Regev. New multilayered pcp and the hardness of hypergraph vertex cover. *SIAM Journal of Computing*, 34(5):1129–1146, 2005.
10. U. Feige. A threshold of ln n for approximating set cover. *JACM*, 45(4):634–652, 1998.
11. D. Golovin, V. Goyal, and R. Ravi. Pay today for a rainy day: Improved approximation algorithms for demand-robust min-cut and shortest path problems. *STACS*, 2006.
12. A. Gupta, M. Pal, R. Ravi, and A. Sinha. Boosted sampling: Approximation algorithms for stochastic optimization. In *STOC*, pages 170–178, 2004.
13. A. Gupta, R. Ravi, and A. Sinha. An edge in time saves nine: Lp rounding approximation algorithms for stochastic network design. *FOCS*, 45, 2004.
14. J. Hastad. Clique is hard to approximate. In *FOCS*, pages 627–636, 1996.
15. N. Immorlica, D. Karger, M. Minkoff, and V. S. Mirrokni. On the costs and benefits of procrastination: Approximation algorithms for stochastic combinatorial optimization problems. In *SODA*, 2004.

16. C. Lund and M. Yannakakis. On the hardness of approximating minimization problems. *JACM*, 41(5):960–981, 1994.

17. Y. Nikulin. Robustness in combinatorial optimization and scheduling theory: An annotated bibliography. *Technical Report SOR-91-13, Statistics and Operation Research*, `http://www.optimization-online.org/DB_FILE/2004/11/995.pdf`, 2004.

18. R. Ravi and A. Sinha. Hedging uncertainty: Approximation algorithms for stochastic optimization problems. *IPCO*, pages 101–115, 2004.

19. A. Siegel, Median Bounds and Their Application. J. Algorithms 38(1): 184-236 (2001).

20. D. Shmoys and S. Swamy. Stochastic optimization is (almost) as easy as deterministic optimization. In *FOCS*, 2004.

Approximation Algorithms for the Multi-item Capacitated Lot-Sizing Problem Via Flow-Cover Inequalities

Retsef Levi[1], Andrea Lodi[2], and Maxim Sviridenko[3]

[1] Sloan School of Management, MIT, Cambridge, MA, 02139
retsef@mit.edu
[2] DEIS, University of Bologna, viale Risorgimento 2 - 40136 Bologna - Italy
alodi@deis.unibo.it
[3] IBM T.J. Watson Research Center, P.O. Box 218, Yorktown Heights, NY 10598
sviri@us.ibm.com

Abstract. We study the classical *multi-item capacitated lot-sizing problem* with hard capacities. There are N items, each of which has specified sequence of demands over a finite planning horizon of discrete T periods; the demands are known in advance but can vary from period to period. All demands must be satisfied on time. Each order incurs a time-dependent *fixed ordering cost* regardless of the combination of items or the number of units ordered, but the total number of units ordered cannot exceed a given capacity C. On the other hand, carrying inventory from period to period incurs *holding costs*. The goal is to find a feasible solution with minimum overall ordering and holding costs.

We show that the problem is strongly NP-Hard, and then propose a novel facility location type LP relaxation that is based on an exponentially large subset of the well-known *flow-cover inequalities*; the proposed LP can be solved to optimality in polynomial time via an efficient separation procedure for this subset of inequalities. Moreover, the optimal solution of the LP can be rounded to a feasible integer solution with cost that is at most twice the optimal cost; this provides a 2-Approximation algorithm, being the first constant approximation algorithm for the problem. We also describe an interesting *on-the-fly* variant of the algorithm that does not require to solve the LP a-priori with all the flow-cover inequalities. As a by-product we obtain the first theoretical proof regarding the strength of flow-cover inequalities in capacitated inventory models. We believe that some of the novel algorithmic ideas proposed in this paper have a promising potential in constructing strong LP relaxations and LP-based approximation algorithms for other inventory models, and for the capacitated facility location problem.

Keywords: approximation algorithms, integer programming, polyhedral combinatorics, randomized algorithms, scheduling theory and algorithms.

1 Introduction

The issue of capacity constrains arises in many practical and theoretical inventory management problems as well as in problems in other application domains, such as facility location problems. In most practical inventory systems there exist capacity constrains

M. Fischetti and D.P. Williamson (Eds.): IPCO 2007, LNCS 4513, pp. 454–468, 2007.

that limit the quantities that one can order, ship or produce. Unfortunately, it is often the case that models with capacity constrains are computationally far more challenging than their counterpart models with no capacity constrains. In particular, in many problems with capacity constrains computing optimal policies and sometimes even feasible policies is a very challenging task.

In recent years there has been an immense work to develop integer programming methods for solving hard, large-scale deterministic inventory management problems. (We refer the reader to the recent book of Pochet and Wolsey [18].) A major part of this work has been focused on constructing strong formulations for the corresponding inventory models. In fact, it is essential to have an integer programming formulation with a strong linear programming relaxation. Stronger formulations are achieved by identifying *valid inequalities* that are satisfied by all feasible integral solutions and cut off fractional solutions. Another key aspect within an integer programming framework is the ability to construct good feasible integer solutions to the corresponding model. This has been known to have a huge impact on decreasing the computational effort involved. In models with capacity constrains, finding good feasible solutions can be very challenging.

In this paper, we study the classical *multi-item capacitated lot-sizing problem*, which is an extension of the *single-item economic lot-sizing problem* [18]. We propose a novel facility location type linear programming (LP), and show how to round its optimal solution to a feasible integral solution with cost that is guaranteed to be at most twice the optimal cost. This is called a *2-Approximation algorithm*, that is, the cost of the solution constructed by the algorithm is guaranteed to be at most twice the optimal cost. (This is the first constant approximation algorithm for this problem.) The LP relaxation is based on a variant of a well-known class of valid inequalities called *flow-cover inequalities*. These inequalities have been introduced over two decades ago [16] and have been shown empirically to be very effective in solving several inventory and facility location problems with capacity constrains [1,18]. (In Section 2 below, we discuss the relevant literature on flow-cover inequalities in more details.) Our results have several significant contributions: (i) To the best of our knowledge, this is the first theoretical evidence for the strength of flow-cover inequalities applied to capacitated inventory models. (All the previous theoretical results have been obtained for fixed-charge single-node problems. See Section 2 below for details.) (ii) Our approach provides a conceptually simple way to generate provably good feasible solutions, and can be easily implemented within an integer programming framework. (iii) Several of the newly proposed algorithmic ideas in this paper have a promising potential of applicability in other inventory models with capacity constrains. Moreover, we believe that they can be used to develop strong LP relaxations and LP-based approximation algorithms for the capacitated facility location problem.

The model. The details of the inventory model discussed in this paper are as follows. There are N items indexed by $i = 1, \ldots, N$, each of which has a specified sequence of demands over a finite planning horizon of T discrete periods indexed by $t = 1, \ldots, T$. The demand of item i in period t is denoted by d_{it}. The demands are known in advance but can vary from period to period. Moreover, all of the demands must be fully satisfied on time, that is, d_{it} must be fully ordered by time period t. At the beginning of each period $s = 1, \ldots, T$, it is possible to place an order for any combination of items, and this incurs a *fixed ordering cost* K_s regardless of the combination of items or the number of units ordered. However, the overall quantity of units ordered in period s cannot exceed a certain capacity limit $C_s \geq 0$. These are usually called *hard* capacity constrains in contrast to *soft* capacity constrains, where in each period s, the order is placed in *batches*, each of which has capacity C_s and incurs an additional fixed ordering cost K_s. We consider the special case with *uniform capacities*, i.e., $C_s = C$, for each $s = 1, \ldots, T$.

The units ordered in period s are assumed to arrive instantaneously, and can be used to satisfy demands in that period and subsequent periods. The fixed ordering cost is balanced with a cost to maintain physical inventory that is called *holding cost*. In most of the existing literature the holding costs are linear and additive. Specifically, for each item i and period t, there is a holding cost parameter $h_{it} \geq 0$ that denotes the per-unit cost to carry one unit of item i in inventory from period t to period $t + 1$. Following Levi, Roundy and Shmoys [12], we model the holding cost in a more general way. For each demand point (i, t) and a potential order $s \leq t$, let $h_{st}^i \geq 0$ be the per-unit cost of holding one unit of item i in inventory from period s to period t. The only assumption is that, for a fixed (i, t), the parameters h_{st}^i are non-increasing in s. (This implies that if d_{it} is ordered from a closer period to t the resulting holding cost is not bigger.) The way we model the holding cost is more general, and can capture several important phenomena such as perishable goods. We also note that we can incorporate a per-unit ordering cost into the holding cost parameters. The goal is to find a feasible policy that satisfies all of the demands on time and has minimum overall ordering and holding cost.

Literature review. As we already mentioned, this is a classical model in inventory theory that has been studied by several researchers throughout the years. The special case with a single-item ($N = 1$) and uniform capacities is polynomially solvable both with hard capacities [11] and soft capacities [17]. (This is usually called *single-item capacitated economic lot-sizing problem.*) Moreover, there are known *extended LPs*, that is, LPs with integrality property that provide an exact description of the set of feasible solutions. The single item problem with non-uniform capacities is known to be weakly NP-Hard [11], but there is a fully polynomial time approximation scheme (FPTAS) [19]. For results on other variants of single-item models, we refer the reader to [6,18].

Federgruen, Meisner and Tzur [10] have studied the model discussed in this paper with traditional holding cost, but with additional *fixed item ordering costs* that are incurred in each period, in which item i is ordered. Under the assumption that all of the demands and the cost parameters are uniformly bounded by constants, they have proposed a dynamic-programming-based algorithm, and shown that it is asymptotically optimal as the number of periods increases to infinity. In a subsequent paper [9], they provide a probabilistic analysis of the algorithm. Another dynamic-programming-based algorithm for a special case of the model discussed in this paper has been proposed by Anily and Tzur [3]. (They have studied a model with traditional holding costs and stationary cost parameters, i.e., $h_{it} = h$ and $K_t = K$, for each i and t.) However, the running time of their algorithm grows exponentially fast in the number of items, and thus, it is not practical unless there are few items. In a recent paper Anily, Tzur and Wolsey [4] have considered the same model with time-dependent cost parameters, but with the additional *monotonicity assumption* on the holding parameters. In particular, the assumption is that the items are ordered, such that each item has higher holding costs than all previously ordered items, uniformly for all periods. Specifically, $h_{1t} \leq h_{2t} \leq \cdots \leq h_{Nt}$, for all periods $t = 1, \ldots, T$. For this problem, they have proposed an extended linear programming formulation with $O(NT^2)$ constrains and variables that solves the problem to optimality. (This implies that this special case is polynomially solvable.)

Our results and techniques. Our first result shows that the multi-item capacitated lot-sizing problem with hard or soft capacities is strongly NP-Hard. (This is done by reduction from 3-PARTITION Problem, the details are omitted due lack of space.) This implies that the monotonicity assumption of Anily, Tzur and Wolsey [4] is somewhat essential to get a polynomial time optimization algorithm. We propose a novel facility location type LP relaxation for the problem that is very different than the one used by Anily, Tzur and Wolsey [4]. Our LP is based on the family of flow-cover inequalities in the same spirit as the LP proposed by Aardal, Pochet and Wolsey for the capacitated facility location problem [2]. However, it incorporates only a subset of the class of flow-cover inequalities: there are exponentially many inequalities in this subset, but we show that they can be separated in polynomial time. Thus, the LP can be solved optimally in polynomial time, using the Ellipsoid method. We then use an extremely simple rounding algorithm. The optimal solution of the LP relaxation is scaled by a suitably chosen factor, and the scaled solution is used to execute a randomized rounding procedure that outputs the sequence of periods, in which orders are placed. Given the output of the first phase, demands are assigned to orders by solving the induced transportation problem, and this minimizes the resulting holding costs. The main challenge in the worst-case analysis is to show that the first phase of the algorithm opens capacity that is sufficient to serve all of the demands, and that the resulting solution is of low cost. This is done by exploiting the structure of the flow-cover inequalities. In particular, we show that together with the scaling at the first phase of the algorithm, they guarantee that the resulting transportation problem has a low cost feasible solution. This provides a randomized 2-Approximation algorithm. The randomized procedure can be derandomized to provide a deterministic 2-Approximation algorithm. As a by-product, we obtain the first theoretical proof of the strength of flow-cover inequalities in capacitated inventory models. (As already mentioned, all previous results are restricted to fixed-charge single-node problems, see Section 2 below.)

Finally, the insights from the worst-case analysis are used to construct an *on-the-fly variant of the algorithm.* Instead of solving the LP a-priori with all the corresponding flow-cover inequalities, we propose an iterative procedure. In each iteration, a well designed rounding procedure is applied to the optimal fractional solution of the LP relaxation. If this procedure comes to an end successfully, it can be shown that the resulting integral solution is feasible and has cost that is at most twice the optimal cost. On the other hand, if the procedure is terminated in the middle, it is guaranteed to identify a violated flow-cover inequality. The corresponding inequality is added to the LP, which is then solved again. The on-the-fly algorithm can be viewed as running the Ellipsoid method until termination or until the first time the rounding procedure is 'stuck', whereas then we are guaranteed to have a good feasible integral solution. We believe that the on-the-fly algorithm might be computationally more efficient, since it does not require solving the LP a-priori with all the flow-cover inequalities. (This algorithmic approach is similar in spirit to what is discussed in Carr at al. [8] in the context of a single-node fixed charge problem.)

The rest of the paper is organized as follows. In Section 2, we describe the LP relaxation and discuss the flow-cover inequalities. In Section 3 we describe the rounding algorithms and the worst-case analysis.

2 A Flow-Cover-Inequality-Based LP Relaxation

A natural Mixed Integer linear Programming (MIP) formulation of the multi-item capacitated lot-sizing problem can be obtained by using two sets of variables:

- For each $s = 1, \ldots, T$, let y_s be a binary variable that is equal to 1 if an order is placed in period s and 0 otherwise.
- For each $i = 1, \ldots, N$, $t = 1, \ldots, T$ and $s = 1, \ldots, t$, let x^i_{st} be the fraction of the demand d_{it} satisfied by an order placed in period s.
- For each $i = 1, \ldots, N$, $t = 1, \ldots, T$ and $s = 1, \ldots, t$, let $H^i_{st} = h^i_{st} d_{it}$ be the cost of holding the all demand d_{it} if ordered in period s.

The corresponding MIP formulation is as follows:

$$\min \sum_{s=1}^{T} K_s y_s \; + \; \sum_{i=1}^{N} \sum_{s=1}^{T} \sum_{t=s}^{T} H^i_{st} x^i_{st} \tag{1}$$

$$\sum_{s \leq t} x^i_{st} = 1 \qquad i = 1, \ldots, N, \; t = 1, \ldots, T, \; d_{it} > 0, \tag{2}$$

$$x^i_{st} \leq y_s \qquad i = 1, \ldots, N, \; s = 1, \ldots, T, \; t \geq s, \tag{3}$$

$$\sum_{i=1}^{N} \sum_{t \geq s} d_{it} x^i_{st} \leq C y_s \qquad s = 1, \ldots, T, \tag{4}$$

$$x^i_{st} \geq 0 \qquad i = 1, \ldots, N, \tag{5}$$

$$y_s \in \{0, 1\} \qquad s = 1, \ldots, T. \tag{6}$$

If we relax the integrality constrains to $0 \leq y_s \leq 1$, we get an LP relaxation that provides a lower bound on the cost of the optimal solution. However, this LP relaxation is weak in that the gap between its optimal value and the value of the optimal integral solution can be arbitrarily high. Thus, there is no hope to use the LP to construct constant approximation algorithms. For example, consider an instance with a single-item and 2 periods, no holding costs, fixed ordering costs $K_1 = 0$ and $K_2 = 1$, and demands $d_1 = 0$ and $d_2 = C + 1$. The optimal policy must open two orders incurring a cost of 1. The optimal fractional solution can achieve a cost of $1/C$ by setting $y_1 = 1$, $y_s = 1/C$, $x_{12} = C/(C+1)$ and $x_{22} = 1/(C+1)$.

2.1 Flow-Cover Inequalities

In this section, we introduce the class of flow-cover inequalities that we use to strengthen the LP induced by (1)-(6). Flow-cover inequalities have been introduced by Padberg, Van Roy and Wolsey [16] over two decades ago in the context of the *fixed charge single-node problem*. In this problem there is a single-node of demand D and a collection of T capacitated arcs. The goal is to open arcs and send a flow of D units to the demand node. Opening arc s incurs a fixed cost K_s, and sending flow over arc s incurs a per-unit cost h_s, for each unit of flow. Padberg, Van Roy and Wolsey [16] have used flow-cover inequalities to construct an extended LP for this problem with uniform arc capacities. They have also shown that these flow-cover inequalities can be separated in polynomial time. Carr et al. [8] have shown that another variant of flow-cover inequalities can be used to construct an LP relaxation for the fixed-charge single-node problem with nonuniform capacities, whose optimal solution can be rounded to a feasible solution with cost that is at most twice the optimal cost. Carnes and Shmoys [7] have used the same LP to construct a prima-dual algorithm with the same worst-case performance guarantee. Aardal, Pochet and Wolsey [2] have used aggregation of constrains to leverage the flow-cover inequalities to multi-location problems, specifically, hard capacitated facility location problems. They have reported that flow-cover inequalities seem to be effective in narrowing the integrality gap and enhance integer programming solution procedures. However, to the best of our knowledge there has been no theoretical analysis regarding the strength of flow-cover inequalities in facility location or inventory models.

In the spirit of [2], we next introduce flow-cover inequalities for the multi-item capacitated lot-sizing problem. Given a subset A of demand points, i.e., a collections of pairs (i, t), $i = 1, \ldots, N$, $t = 1, \ldots, T$, let $D(A) = \sum_{(i,t) \in A} d_{it}$ denote the cumulative demand of the set A; $\ell_A = \lceil \frac{D(A)}{C} \rceil$ be the *cover number* of A, i.e., the minimum number of orders required to satisfy the demands in A; $\lambda_A = \ell_A C - D(A)$; $R_A = C - \lambda_A$ be the *residual capacity* of A, i.e., the capacity required to satisfy the demands in A after $\ell_A - 1$ orders are fully used; and $r_A = R_A / C (= \frac{D(A)}{C} - \lfloor \frac{D(A)}{C} \rfloor)$ be the *fraction of the residual capacity*. Observe that by definition $0 < R_A \leq C$ and $0 < r_A \leq 1$. Moreover, a subset F of orders (i.e., $F \subseteq \{1, \ldots, T\}$) is called a *cover of* A if $|F| \geq \ell_A$.

Then, we claim that the following inequalities are valid:

$$\sum_{(i,t) \in A} \sum_{s \in F} d_{it} x_{st}^i - R_A \sum_{s \in F} y_s \leq D(A) - \ell_A R_A. \tag{7}$$

The validity of inequalities (7) in the multi-item capacitated lot-sizing problem can be obtained as a special case of the general *mixed integer rounding inequalities*, or in short MIR inequalities (see, e.g., Nemhauser and Wolsey [15]). An MIR inequality is defined with respect to the simple mixed-integer set $Q = \{x \in R, y \in Z \ : \ x + y \geq b, x \geq 0\}$, for which it is easy to prove the validity of the inequality $x + \hat{b} y \geq \hat{b} \lceil b \rceil$, where $\lceil b \rceil$ is equal to b rounded up to the next integer, and $\hat{b} = \lceil b \rceil - b$. This can be generalized to more complicated sets that involve more variables, as long as the variables can be split into an integral part and a continuous nonnegative part. In particular, we apply an MIR derivation to:

$$\frac{1}{C} \sum_{(i,t)\in A} \sum_{s\notin F} d_{it}x_{st}^i + \sum_{s\in F} y_s \geq \frac{D(A)}{C}. \tag{8}$$

It is easy to see that Inequality (8) is valid for the system (1)–(6). Specifically, Constraint (2) implies that $D(A) = \sum_{(i,t)\in A} \sum_{s\in F} d_{it}x_{st}^i + \sum_{(i,t)\in A} \sum_{s\notin F} d_{it}x_{st}^i$; then replace the first term in the right hand side of the equality by an upper bound $C\sum_{s\in F} y_s$ (see Constraint (4)) and divide by C to get the desired Inequality (8).

Thus, by applying an MIR derivation to (8) one obtains:

$$\frac{1}{C} \sum_{(i,t)\in A} \sum_{s\notin F} d_{it}x_{st}^i + r_A \sum_{s\in F} y_s \geq r_A \ell_A, \tag{9}$$

which coincides with (7) after splitting $D(A)$ as done before and dividing by C.

Separation. To the best of our knowledge the complexity of separating flow-cover inequalities is unknown. Aardal [1] has shown that flow-cover inequalities can be separated in polynomial time for a fixed set of demand points. (There is a simple greedy procedure.)

Next we consider a fixed subset of orders $\bar{F} \subseteq \{1,\dots,T\}$, and describe a polynomial time algorithm to separate flow-cover inequalities that correspond to the subset of orders $\bar{F}$. For the description of the algorithms, it will be useful to rewrite flow-cover inequalities that correspond to the set $\bar{F}$ as

$$\sum_{(i,t)\in A} \frac{d_{it}}{C}\left(1 - \sum_{s\in \bar{F}} x_{st}^i\right) \geq r_A\left(\ell_A - \sum_{s\in \bar{F}} y_s\right). \tag{10}$$

Observe that (10) above may still contain exponentially many constrains, one for each subset A of demand points that can be covered by $\bar{F}$. However, this is similar to the *residual capacity inequalities* introduced by Magnanti, Marchandani and Vachani [14] for the mixed-integer set called *splittable flow arc set* $\mathcal{X} = \{(x,y) : \sum_{i=1}^n a_i x_i \leq a_0 + y, x \in [0,1]^n, y \in \{0,1\}\}$. It has been shown that residual capacity inequalities are sufficient to characterize $conv(\mathcal{X})$ [14,5]. Moreover, Atamtürk and Rajan [5] have described an $O(n)$ time separation algorithm .

Building on the results of Atamtürk and Rajan [5], we can obtain the following theorem. (The proof is omitted due to lack of space.)

Theorem 1. *Consider a subset of orders $\bar{F}$. Then there exists a polynomial time separation algorithm for the inequalities in (10). The algorithm runs in $O(NT^2)$ time.*

An LP. Next we describe an LP based on (1)-(5), relaxation of the integrality constraint of (6) and a subset of the flow-cover inequalities defined in (7) above.

Let $\mathcal{F} := \{[s,t] : 1 \leq s \leq t \leq T\}$ be the collection of all subsets of orders defined by intervals $[s,t]$. Consider the LP defined by (1)-(5), the relaxation of (6) and only the inequalities in (7) that correspond to subsets of orders $F \in \mathcal{F}$. Recall that in Theorem 1 we have shown that flow-cover inequalities that correspond to a fixed subset of orders can be separated in polynomial time. Since the cardinality of the set $\mathcal{F}$ is $O(T^2)$, it follows that the above LP can be solved to optimality in polynomial time by using the Ellipsoid method. Let $(\hat{x}, \hat{y})$ be the optimal solution of that LP and V_{LP} be the respective optimal value. Since the LP is a relaxation of the problem, it follows that V_{LP} is a lower bound on the optimal cost denoted by V_{OPT}.

3   The Random-Shift Algorithm with Median Demand Assignment

In this section, we describe an approximation algorithm for the multi-item capacitated lot-sizing problem with hard capacities that is based on the linear programming relaxation defined above by (1)-(5), relaxation of the integrality constraint of (6) and the flow-cover inequalities in (7) that correspond to the collection of subsets $\mathcal{F}$ defined above.

We shall first show how to round the optimal fractional solution $(\hat{x}, \hat{y})$ of this LP to a feasible integer solution with cost that is at most twice V_{LP}. Since V_{LP} is a lower bound on the optimal cost, this implies that the algorithm is a 2-Approximation. In addition, we shall describe an *on-the-fly* variant of the algorithm that does not require to add all the respective flow-cover inequalities a-priori, but instead adds violated constrains on-the-fly until a (good) integer solution is obtained.

First, we present a randomized rounding procedure that we call *Random-Shift with Median Assignment*. This procedure rounds the fractional optimal solution $(\hat{x}, \hat{y})$ to a feasible integer solution $(\tilde{x}, \tilde{y})$ with expected cost that is at most twice the optimal cost V_{OPT}. We then discuss how to derandomize the algorithm, and get a deterministic 2−Approximation algorithm.

The rounding algorithm runs in two phases. In the first phase of the algorithms we determine in which periods to place orders. Based on the outcome of the first phase of the algorithm, we decide how to assign demand points to orders.

3.1   Phase I: The Random-Shift Procedure

We first describe Phase I of the algorithm which we call the *Random-Shift procedure*. (This is similar in spirit to the work of Levi, Roundy, Shmoys and Sviridenko [13] on the single-warehouse and multi-retailer problem.) In this phase we decide, in which periods to place orders. This simple randomized procedure is based on the values $\hat{y}_1, \dots, \hat{y}_T$. For each $s = 1, \dots, T$, let $\bar{y}_s = \min\{2\hat{y}_s, 1\}$, i.e., we double the original value of each variable $\hat{y}_s \leq 0.5$ and make it equal to 1 if $\hat{y}_s > 0.5$. We call $\hat{y}_s$ and $\bar{y}_s$ the *fractional order* and *scaled fractional order* in period $s = 1, \dots, T$, respectively. Next we shall use the values $\bar{y}_1, \dots, \bar{y}_T$ to determine the periods in which orders are placed.

For the description of the Random-Shift procedure, consider the interval $(0, \sum_{s=1}^{T} \bar{y}_s]$, which corresponds to the total weight of scaled fractional orders. Each period $r = 1, \dots, T$ is then associated with the corresponding interval $(\sum_{s=1}^{r-1} \bar{y}_s, \sum_{s=1}^{r} \bar{y}_s]$, which is of length $\bar{y}_r$. In particular, some periods can correspond to empty intervals of length 0 (if $\hat{y}_t = \bar{y}_t = 0$). The input for this procedure is a *shift-parameter* α that is chosen uniformly at random in $(0, 1]$. Let W be the smallest integer that is greater than or equal to $\sum_{s=1}^{T} \bar{y}_s$. Specifically, W is the upper ceiling of the total cumulative weight of the scaled fractional orders; that is, $W = \lceil \sum_{s=1}^{T} \bar{y}_s \rceil$. Note that the interval $(0, \sum_{s=1}^{T} \bar{y}_s]$ is contained in the interval $[0, W]$. Within the interval $[0, W]$ focus on the sequence of points $0, 1, \ldots, W - 1$. The shift-parameter α induces a sequence of what we call *shift-points*. Specifically, the set of shift-points is defined as $\{\alpha + w : w = 0, \ldots, W - 1\}$. This set is constructed through a *shift* of random length α to the right of the points $0, 1, \ldots, W - 1$. Thus, there are W shift-points that are all located within the interval $[0, W]$. Observe that the sequence of shift-points is a-priori random and is realized with the shift-parameter α.

The shift-points determine the periods, in which orders are placed. For each period $r = 1, \ldots, T$, we place an order in that period if there is a shift-point within the interval $(\sum_{s=1}^{r-1} \bar{y}_s, \sum_{s=1}^{r} \bar{y}_s]$ that is associated with period r. That is, we place an order in period r, if for some integer $0 \leq w \leq W - 1$ there exists a shift- point $\alpha + w$ that falls within the interval $(\sum_{s=1}^{r-1} \bar{y}_s, \sum_{s=1}^{r} \bar{y}_s]$. Let $\mathcal{T} := \{r_1 < r_2 < \ldots < r_Q\}$ be the set of periods of the orders as determined in the first phase of the algorithm using the random-shift procedure. We set $\tilde{y}_{r_m} = 1$, for each $m = 1, \ldots, Q$, and call $r_1, \ldots, r_Q$ the *opened orders*.

Next we bound the expected ordering cost incurred by the random shift procedure. (The proof is omitted due lack of space.)

Lemma 2. *Consider the Random-Shift procedure described above. Then, for each period* $r = 1, \ldots, T$, *the probability to place an order in period* r *is at most* $\bar{y}_r \leq 2\hat{y}_r$. *Thus, the total expected ordering cost of the Random-Shift procedure, denoted by* $\mathcal{K}$ *is at most twice the total ordering costs in the optimal LP solution. That is,* $\mathcal{K} \leq \sum_{s=1}^{T} \bar{y}_s K_s \leq 2 \sum_{s=1}^{T} \hat{y}_s K_s$.

Given the opened orders $r_1, \ldots, r_Q$, we can compute the cheapest assignments of demand points to opened orders by solving the corresponding transportation problem. The solution of the transportation problem will determine the values of $\tilde{x}_{st}^i$, for each (i, t) and $s \leq t$. However, it is not clear a-priori that the induced transportation problem has a feasible solution, and even if it has one, there is a question regarding the cost of this solution. Next we shall show that the induced transportation problem indeed has a feasible solution with cost denoted by $\mathcal{H}$ that is at most twice the holding cost incurred by the optimal fractional solution $(\hat{x}, \hat{y})$. That is, the holding cost incurred by the algorithm is $\mathcal{H} \leq 2 \sum_{i=1}^{N} \sum_{t=1}^{T} \sum_{s=1}^{t} H_{st}^i x_{st}^i$.

3.2   The Median Assignment

Next we describe a constructive procedure, called the *Median Assignment*, that assigns all the demand points to the opened orders $r_1, \ldots, r_Q$, and incurs an holding cost that is at most twice the holding cost incurred by the optimal fractional solution $(\hat{x}, \hat{y})$. Observe that the optimal solution to the transportation problems induced by the opened orders $r_1, \ldots, r_Q$ incurs even lower holding cost. To describe the procedure we introduce the notion of *flow-requirements* of demand point (i, t). Focus on a specific demand point (i, t), and let $s_1 < s_2 < \cdots < s_G$ be the fractional orders that fractionally serve this demand point in the optimal LP solution $(\hat{x}, \hat{y})$. In particular, $\hat{x}^i_{s_g,t} > 0$, for each $g = 1\ldots, G$, and $\sum_{g=1}^{G} \hat{x}^i_{s_g,t} = 1$. Let s_M be the *median order* of (i,t), i.e., the latest point in time such that at least half of the demand d_{it} is satisfied from orders within $[s_M, t]$. That is, $M = \max\{m : \sum_{g=m}^{G} \hat{x}^i_{s_g,t} \geq 0.5\}$. For each $g = 1, \ldots, G$, let $z^i_{s_g,t}$ be the *flow-requirement of* (i,t) *that is due* s_g. Specifically, for each $g = 1, .., M-1$, we define $z^i_{s_g,t} = 2\hat{x}^i_{s_g,t}d_{it}$; for $g = M$ we define $z^i_{s_M,t} = 2(\sum_{q=1}^{M} \hat{x}^i_{s_q,t} - 0.5)d_{it}$; and for each $g = M+1, \ldots, G$, we define $z^i_{s_g,t} = 0$.

Note that the flow-requirements defined above do not necessarily provide a feasible assignment of demands to orders. Intuitively, we consider the median order that splits the assignment of demand point (i,t) in the optimal fractional solution $(\hat{x}, \hat{y})$ into two equal halves. We then ignore the upper (later) half and scale the lower (earlier) half by 2. However, we shall use the flow-requirements $z^i_{s_1,t}, \ldots, z^i_{s_G,t}$ to construct a feasible assignment of demands with relatively low holding costs. First, observe that $\sum_{g=1}^{G} z^i_{s_g,t} = d_{it}$. We wish to construct a feasible assignment that, for each demand point (i,t) and an order s_g, satisfies at least $z^i_{s_g,t}$ units of d_{it} from orders within the interval $[s_g, t]$. That is, the flow-requirement $z^i_{s_g,t}$ is satisfied either from s_g or from orders later in time. We will say that such an assignment satisfies all the flow-requirements. (Recall that s_g and $z^i_{s_g,t}$ are specific to demand point (i,t) based on the optimal fractional solution $(\hat{x}, \hat{y})$.)

Consider any assignment of demands that satisfies all the flow-requirements of all demands. Since the assignment satisfies $z^i_{s_g,t}$ units of d_{it} either from s_g or even from orders later in time, we conclude that the holding cost incurred by each demand point (i,t) is at most $\sum_{g=1}^{G} z^i_{s_g,t}h^i_{s_g,t}$. However, by the definition of the flow-requirements, we have

$$\sum_{g=1}^{G} z^i_{s_g,t}h^i_{s_g,t} \leq 2\sum_{g=1}^{G} \hat{x}^i_{s_g,t}d_{it}h^i_{s_g,t} = 2\sum_{g=1}^{G} \hat{x}^i_{s_g,t}H^i_{s_g,t}.$$

That is, the holding cost incurred is at most twice the holding costs incurred by (i,t) in $(\hat{x}, \hat{y})$. In light of Lemma 2 above, if such an assignment exists, the resulting feasible solution $(\tilde{x}, \tilde{y})$ has cost that is at most twice the optimal values of the LP V_{LP}. Since V_{LP} is a lower bound on the optimal cost, it follows that the cost of the solution is at most twice the optimal cost. It is then left to show that such an assignment does exist. Next we shall describe the details of the Median Assignment procedure.

We construct the Median Assignment in stages indexed by $\tau = T, \ldots, 1$. In each stage τ, we consider the set of positive flow-requirements due within τ, i.e., the set $\mathcal{B}_\tau = \{z^i_{\tau t} > 0 : i = 1, \ldots, N, t = \tau, \ldots, T\}$. These are the flow-requirements that need to be satisfied from orders within $[\tau, T]$. Partition the set $\mathcal{B}_\tau$ into sets $\mathcal{B}_{\tau t}$, for $t = \tau, \ldots, T$, where $\mathcal{B}_{\tau t} = \{z^i_{\tau t} > 0 : i = 1, \ldots, N\}$. We then consider the sets $\mathcal{B}_{\tau t}$ in decreasing order $t = T, \ldots, \tau$. For each flow-requirement $z^i_{\tau t} \in \mathcal{B}_{\tau t}$, we consider the opened orders (decided upon in Phase I) within $[\tau, t]$ in decreasing order from latest to earliest. The flow-requirement $z^i_{\tau t}$ is then assigned to these orders greedily according to the current available capacity. More rigorously, consider a specific flow-requirement $z^i_{\tau t}$ and let $\mathcal{T}_{[\tau, t]} = \mathcal{T} \cap [\tau, t] = \{e_1 < e_2 < \cdots < e_V\}$ be the set of opened orders within the interval $[\tau, t]$. Let $\delta_V = \min\{z^i_{\tau t}, (C - \sum_{j=1}^{N} \sum_{u=e_V}^{T} \tilde{x}^j_{e_V,u}d_{ju})^+\}$, and for each

$v = V - 1, \ldots, 1$, let $\delta_v = \min\{(z_{\tau t}^i - \sum_{q=v+1}^V \delta_q)^+, (C - \sum_{j=1}^N \sum_{u=e_v}^T \tilde{x}_{e_v,u}^j d_{ju})^+\}$.
(Recall that $(x)^+ = \max\{x, 0\}$.) We then update $\tilde{x}_{e_v,t}^i = \tilde{x}_{e_v,t}^i + \delta_v/d_{it}$, for each $v = 1, \ldots, V$. By construction it follows that if completed successfully, the Median Assignment described above satisfies the flow-requirements of all demand points. Thus, to establish a bound on the holding cost incurred by the algorithm, it is sufficient to show that the Median Assignment can be completed successfully.

Before we prove that, we would like to state a technical lemma that draws a connection between the cumulative fractional orders opened by the fractional solution $(\hat{x}, \hat{y})$, and the corresponding number of integral orders opened in Phase I of the algorithm. (The proof is omitted due lack of space.)

Lemma 3. *Consider the interval* $[s, t]$ *for some* $s \leq t$, *and suppose that the cumulative fractional orders opened by the LP optimal solution* $(\hat{x}, \hat{y})$ *is equal* $L + \beta$, *where* L *is a non-negative integer and* β *is between 0 and 1. That is,* $\sum_{u=s}^t \hat{y}_u = L + \beta$. *Then if* $\beta \geq 0.5$, *the number of orders placed in Phase I of the algorithm over the interval* $[s, t]$ *is at least* $L + 1$. *That is,* $\sum_{u=s}^t \tilde{y}_u \geq L + 1$.

Lemma 4. *The Median Assignment can be completed successfully.*

Proof : Assume by contradiction that the Median Assignment cannot be completed at some stage τ due lack of capacity to satisfy the flow-requirement $z_{\tau, \bar{t}}^i$ of some demand point $(i, \bar{t})$. It follows that all the opened orders within the interval $[\tau, \bar{t}]$ are currently fully used by the integer partial solution $(\tilde{x}, \tilde{y})$. That is, for each $r \in \mathcal{T} \cap [\tau, \bar{t}]$, we have $\sum_{i=1}^N \sum_{u=r}^T \tilde{x}_{ru}^i d_{iu} = C$. Now let $\bar{r}$ be the earliest opened order within $(\bar{t}, T] \cap \mathcal{T}$ that still has free capacity or $T + 1$ if no such order exists. That is,

$$\bar{r} = \min\{\operatorname{argmin}\{r \in \mathcal{T} \cap (\bar{t}, T] : \sum_{i=1}^N \sum_{u=r}^T \tilde{x}_{ru}^i d_{iu} < C\}, T + 1\}.$$

Let $F = [\tau, \bar{r})$ be the corresponding interval of orders.

Next we focus on the set of demand points (i, u) with positive flow-requirements that are due within $[\tau, \bar{r})$, i.e., the set $A = \{(i, t) : t \in [\tau, \bar{r}) \text{ and } \sum_{s=\tau}^t z_{st}^i > 0\}$. Using the notation in Section 2 we write $D(A) = \sum_{(i,t) \in A} d_{it} = (\ell_A - 1)C + R_A$, where $\ell_A \geq 1$ is an integer and $0 < R_A \leq C$. Consider again the integer partial solution $(\tilde{x}, \tilde{y})$ at the moment the Median Assignment terminated due to lack of capacity. Recall that $\bar{r} = T + 1$ or $\bar{r} \in \mathcal{T}$ is an opened order with free capacity. By the construction of the Median Assignment it follows that no demand point outside the interval $[\tau, \bar{r})$ is being served by the partial solution $(\tilde{x}, \tilde{y})$ from orders within the interval. That is, $\tilde{x}_{st}^i = 0$ for each (i, t) with $t \geq \bar{r}$ and $s \in [\tau, \bar{r})$. This implies that all the available capacity of the opened orders within the interval $[\tau, \bar{r})$ is *fully used* by the integer partial solution $(\tilde{x}, \tilde{y})$ to serve demand points in A. Moreover, since the Median Assignment could have not been completed, it follows that the flow-requirements of demand points in A that are due within the interval F exceed the total opened capacity over F. That is,

$$\sum_{(i,t) \in A} \sum_{u \in F} z_{ut}^i > \sum_{u \in F} \tilde{y}_u C. \qquad (11)$$

Now consider the set of fractional orders in the optimal LP solution $(\hat{x}, \hat{y})$ over F. Let $\sum_{u \in F} \hat{y}_u C = (L-1)C + R$, where $L \geq 1$ is a nonnegative integer and $0 < R \leq C$. The rest of the proof is based on comparing ℓ_A and R_A to L and R, respectively, and deriving a contradiction.

We first claim that $L \leq \ell_A$. We have already seen in the proof of Lemma 3 that if $\sum_{u \in F} \hat{y}_u \geq L - 1$, then the Random-Shift procedure will open at least $L - 1$ orders over the interval F, i.e., $\sum_{u \in F} \tilde{y}_u C \geq (L - 1)C$. However, Inequality (11) implies that $\sum_{(i,t) \in A} \sum_{u \in F} z_{ut}^i > \tilde{y}_u C \geq (L - 1)C$. Finally, observe that the overall flow-requirements of demand points in A cannot exceed $D(A)$, which is at most $\ell_A C$. The claim then follows.

Next we claim that $R/C < 0.5$. Assume otherwise. It follows that $\sum_{u \in F} \hat{y}_u \geq (L-1) + 0.5$, and by Lemma 3 we conclude that there are at least L opened orders over F. That is, $\sum_{u \in F} \tilde{y}_u C \geq LC$. However, the flow-requirements are always bounded by the original flow in the fractional optimal solution $(\hat{x}, \hat{y})$. That is,

$$\sum_{u \in F} \sum_{(i,t) \in A} z_{ut}^i \leq \sum_{u \in F} \sum_{(i,t) \in A} \hat{x}_{ut}^i d_{it} \leq \sum_{u \in F} \hat{y}_u C = (L-1)C + R,$$

where the last inequality follows from (4). It follows that capacity of LC units is sufficient to satisfy all the flow-requirements that are due within F of all demand points in A, which leads to contradiction. (There are at least L opened orders over F, all of which are used to satisfy flow-requirement of demand points in A.)

Next we claim that $L = \ell_A$. Assume otherwise, i.e., $\ell_A > L$. Since each demand point $(i, t) \in A$ has positive flow-requirements over F, it follows that $\sum_{u \in F} \hat{x}_{ut}^i > 0.5$. However, by the construction of the flow-requirements this implies that its total flow-requirements over F can be expressed as

$$\sum_{u \in F} z_{ut}^i = 2 d_{it} \left(\sum_{u \in F} \hat{x}_{ut}^i - 0.5 \right).$$

Thus, the total flow-requirements of demand points in A over F can be expressed as

$$\sum_{(i,t) \in A} \sum_{u \in F} z_{ut}^i = 2 \left(\sum_{(i,t) \in A} \sum_{u \in F} \hat{x}_{ut}^i d_{it} - 0.5 D(A) \right)$$

$$\leq 2 \left(\sum_{u \in F} \hat{y}_u C - 0.5 D(A) \right) = 2(L-1)C + 2R - (\ell_A - 1)C - R_A$$

$$\leq (L-1)C.$$

The last inequality follows from the assumptions that $\ell_A - 1 \geq L$ and that $2R < C$. Moreover, this implies that capacity of $(L-1)C$ units is sufficient to satisfy all the flow-requirements that are due within F of the demand points $(i, t) \in A$. However, we have already seen that there are at least $L - 1$ opened orders over F, i.e., $\sum_{u \in F} \tilde{y}_u \geq L - 1$. Since all of them are fully used to satisfy flow-requirements of demand points in A, this again leads to contradiction.

Suppose now that $R/C < 0.5$ and $\ell_A = L$. This implies that the set of orders F is a cover of the set of demands A. Moreover, $F \in \mathcal{F}$, which implies that the solution $(\hat{x}, \hat{y})$

satisfies the flow-cover inequality that corresponds to F and A. It follows that

$$D(A) - \sum_{(i,t)\in A}\sum_{u\in F} \hat{x}^i_{ut}d_{it} \geq R_A(\ell_A - \sum_{u\in F}\hat{y}_u) = R_A\left(\ell_A - (L-1) - R/C\right)$$

$$= R_A(1 - R/C) \geq 0.5R_A.$$

The first inequality follows from the flow-cover inequality with respect to F and A. The first equality follows from the fact that $\sum_{u\in F}\hat{y}_u = L-1+R/C$. The last inequality follows from the fact that $R/C < 0.5$. We conclude that $2(D(A)-\sum_{(i,t)\in A}\sum_{u\in F}\hat{x}^i_{ut}d_{it})$ $\geq R_A$. However, $D(A) - \sum_{(i,t)\in A}\sum_{u\in F}\hat{x}^i_{ut}d_{it}$ is exactly the portion of $D(A)$ that is being served in the optimal fractional solution $(\hat{x}, \hat{y})$ from *outside* F. Moreover, we have already seen that, for each demand point $(i, t) \in A$, more than half of the demand d_{it} is served by $(\hat{x}, \hat{y})$ from within F, i.e., $\sum_{u\in F}\hat{x}^i_{ut} > 0.5$. By the construction of the flow-requirements this implies that the total flow-requirements of demand points $(i, t) \in A$ that are due *outside* F is exactly $2(D(A) - \sum_{(i,t)\in A}\sum_{u\in F}\hat{x}^i_{ut}d_{it}) \geq R_A$. In turn, this implies that the total flow-requirements that are due within F is at most $(\ell_A - 1)C = (L-1)C$. However, as we have already seen, this leads to contradiction since there are at least $(L-1)$ opened orders over F. We conclude that the Median Assignment can be completed successfully. ∎

Corollary 5. *The overall holding cost incurred by the algorithm is at most* $2\sum_{i=1}^{N}\sum_{t=1}^{T}\sum_{s=1}^{t}\hat{x}^i_{st}H^i_{st}.$

Lemma 2 and Corollary 5 imply the following theorem.

Theorem 6. *The Random-Shift algorithm is a randomized 2-Approximation algorithm for the multi-item capacitated lot-sizing problem with hard capacities.*

We note that the same analysis holds in the presence of soft capacities. Finally, we describe how to derandomize the algorithms and get a deterministic 2-Approximation algorithm. We have already mentioned that once the periods, in which orders are placed are determined, the problem is reduced to solving a transportation problem that minimizes the holding costs. The worst-case analysis implies that for *any* outcome of Phase I, the induced transportation problem has a low-cost feasible solution with cost that is at most twice the holding costs incurred by the optimal fractional solution. Thus, it is sufficient to derandomize Phase I, and this can be done by enumerating over all the values of α that yield a different set of orders in Phase I. (There are only $O(T)$ such values.)

Theorem 7. *There exists a deterministic 2-Approximation algorithm for the multi-item capacitated lot-sizing problem with hard capacity constrains.*

3.3   On-The-Fly Algorithm

In this section, we shall describe an *on-the-fly* variant of the algorithm described above. The underlying idea is similar to what discussed by Carr et al. [8] in the context of the fixed-charge single-node problem.

This variant does not require to solve the LP a-priori with all the flow-cover inequalities defined by the collection of subsets $\mathcal{F}$. Instead, we shall have an iterative procedure that is based on an oracle that, in each iteration, either finds a violated flow-cover inequality or generates a feasible solution with cost that is at most twice the optimal cost.

Having an efficient oracle that can separate the respective flow-cover inequalities enables us to run the Ellipsoid method and solve the corresponding LP, and then use the rounding algorithm described in Sections 3.1 and 3.2 above. The resulting integer solution has cost that is at most twice the optimal cost.

However, instead of using an oracle that can separate all the corresponding flow-cover inequalities, the on-the-fly algorithm will use the Median Assignment procedure. If the Median Assignment procedure is stuck, then we can easily identify a violated flow-cover inequality that corresponds to the set of demand points A and the set of orders F as defined in Section 3.2 above. As long as this is the case we execute the Ellipsoid method. On the other hand, if the Median Assignment procedure is completed successfully and the Ellipsoid method is stuck, then we have constructed a solution with cost that is at most twice the optimal cost. The main observation is that the our rounding algorithm can be applied to any feasible fractional solution, where it either ends up with a violated flow-cover inequality or in turn, provides a feasible integer solution with cost that is at most twice the cost of the fractional solution.

We note that in practice the on-the-fly algorithm can be implemented using the Simplex method. Since in each iteration we add a constraint to the primal LP, the Dual-Simplex method might be very attractive to find the new optimal solution of the LP.

Acknowledgements

Part of this research has been carried out when the first and the second authors were Herman Goldstine Postdoctoral Fellows in the Department of Mathematical Sciences of the IBM T.J. Watson Research Center, whose support is strongly acknowledged. The second author was also supported in part by the EU project ADONET (contract n. MRTN-CT-2003-504438). We warmly thank Oktay Günlük who brought to our attention reference [5], and David Shmoys for numerous fruitful discussions.

References

1. K. Aardal. Capacitated facility location: separation algorithms and computational experience. *Mathematical Programming*, 81:149–175, 1998.
2. K. Aardal, Y. Pochet, and L. A. Wolsey. Capacitated facility location: valid inequalities and facets. *Mathematics of Operations Research*, 20:562–582, 1995.
3. S. Anily and M. Tzur. Shipping multiple-items by capacitated vehicles - an optimal dynamic programming approach. *Transportation Science*, 39:233–248, 2005.
4. S. Anily, M. Tzur, and L. A. Wolsey. Multi-item lot-sizing with joint set-up cost. Technical Report 2005/70, CORE, 2005. Working paper.
5. A. Atamtürk and D. Rajan. On splittable and unsplittable flow capacity network design arc-set polyhedra. *Mathematical Programming*, 92:315–333, 2002.
6. G. R. Bitran and H. H. Yanasee. Computational complexity of the capacitated lot-size problem. *Management Science*, 28:1174–1186, 1982.

7. T. Carnes and D. B. Shmoys. A primal-dual 2-approximation algorithm the single-demand fixed-charge minimum-cost flow problem. Working paper, 2006.

8. R. D. Carr, L. K. Fleischer, V. J. Leung, and C. A. Phillips. Strengthening integrality gaps for capacitated network design and covering problems. In *Proceedings of the 11th ACM/SIAM Symposium on Discrete Algorithms (SODA)*, 2000.

9. A. Federgruen and J. Meissner. Probabilistic analysis of multi-item capacitated lot sizing problems. Working paper, 2004.

10. A. Federgruen, J. Meissner, and M. Tzur. Progressive interval heuristics for the multi-item capacitated lot sizing problem. To appear in *Operations Research*, 2003.

11. M. Florian, J. K. Lenstra, and A. H. G. Rinooy Kan. Deterministic production planning: Algorithms and complexity. *Management Science*, 26:669–679, 1980.

12. R. Levi, R. O. Roundy, and D. B. Shmoys. Primal-dual algorithms for deterministic inventory problems. *Mathematics of Operations Research*, 31:267–284, 2006.

13. R. Levi, R. O. Roundy, D. B. Shmoys, and M. Sviridenko. First constant approximation algorithm for the single-warehouse multi-retailer problem. Under revision in *Management Science*, 2004.

14. T. L. Magnanti, P. Mirchandani, and R. Vachani. The convex hull of two core capacitated network design problems. *Mathematical Programming*, 60:233–250, 1993.

15. G. Nemhauser and L. A. Wolsey. *Integer Programming and Combinatorial Optimization*. Wiley, 1990.

16. M. W. Padberg, T. J. V. Roy, and L. A. Wolsey. Valid inequalities for fixed charge problems. *Operations Research*, 33:842–861, 1985.

17. Y. Pochet and L. A. Wolsey. Lot-sizing with constant batches: *Formulation* and valid inequalities. *Mathematics of Operations Research*, 18:767–785, 1993.

18. Y. Pochet and L. A. Wolsey. *Production Planning by Mixed Integer Programming*. Springer Verlag, 2006.

19. C. P. M. van Hoesel and A. P. M. Wagelmans. Fully polynomial approximation schemes for single-item capacitated economic lot-sizing problems. *Mathematics of Operations Research*, 26:339–357, 2001.

Optimal Efficiency Guarantees for Network Design Mechanisms*

Tim Roughgarden** and Mukund Sundararajan***

Department of Computer Science, Stanford University,
353 Serra Mall, Stanford, CA 94305

Abstract. A cost-sharing problem is defined by a set of players vying to receive some good or service, and a cost function describing the cost incurred by the auctioneer as a function of the set of winners. A cost-sharing mechanism is a protocol that decides which players win the auction and at what prices. Three desirable but provably mutually incompatible properties of a cost-sharing mechanism are: incentive-compatibility, meaning that players are motivated to bid their true private value for receiving the good; budget-balance, meaning that the mechanism recovers its incurred cost with the prices charged; and efficiency, meaning that the cost incurred and the value to the players served are traded off in an optimal way.

Our work is motivated by the following fundamental question: for which cost-sharing problems are incentive-compatible mechanisms with good approximate budget-balance and efficiency possible? We focus on cost functions defined implicitly by NP-hard combinatorial optimization problems, including the metric uncapacitated facility location problem, the Steiner tree problem, and rent-or-buy network design problems. For facility location and rent-or-buy network design, we establish for the first time that approximate budget-balance and efficiency are simultaneously possible. For the Steiner tree problem, where such a guarantee was previously known, we prove a new, optimal lower bound on the approximate efficiency achievable by the wide and natural class of "Moulin mechanisms". This lower bound exposes a latent approximation hierarchy among different cost-sharing problems.

1 Introduction

Mechanism Design. In the past decade, there has been a proliferation of large systems used and operated by independent agents with competing objectives (most

* Preliminary versions of most of these results appear in a technical report [32].
** Supported in part by ONR grant N00014-04-1-0725, an NSF CAREER Award, and an Alfred P. Sloan Fellowship.
*** Supported in part by OSD/ONR CIP/SW URI "Software Quality and Infrastructure Protection for Diffuse Computing" through ONR Grant N00014-01-1-0795 and by OSD/ONR CIP/SW URI "Trustworthy Infrastructure, Mechanisms, and Experimentation for Diffuse Computing" through ONR Grant N00014-04-1-0725.

M. Fischetti and D.P. Williamson (Eds.): IPCO 2007, LNCS 4513, pp. 469–483, 2007.

notably the Internet). Motivated by such applications, an increasing amount of algorithm design research studies optimization problems that involve self-interested entities. Naturally, game theory and economics are important for modeling and solving such problems. *Mechanism design* is a classical area of microeconomics that has been particularly influential. The field of mechanism design studies how to solve optimization problems in which part of the problem data is known only to self-interested players. It has numerous applications to, for example, auction design, pricing problems, and network protocol design [8,15,24,27].

Selling a single good to one of n potential buyers is a paradigmatic problem in mechanism design. Each bidder i has a *valuation* v_i, expressing its maximum willingness to pay for the good. We assume that this value is known only to the bidder, and not to the auctioneer. A *mechanism* (or *auction*) for selling a single good is a protocol that determines the winner and the selling price. Each bidder i is "selfish" in the sense that it wants to maximize its "net gain" $(v_i - p)x_i$ from the auction, where p is the price, and x_i is 1 (0) if the bidder wins (loses).

What optimization problem underlies a single-good auction? One natural goal is *economic efficiency*, which in this context demands that the good is sold to the bidder with the highest valuation. This goal is trivial to accomplish if the valuations are known a priori. Can it be achieved when the valuations are private?

Vickrey [34] provided an elegant solution. First, each player submits a sealed bid b_i to the seller, which is a proxy for its true valuation v_i. Second, the seller awards the good to the highest bidder. This achieves the efficient allocation *if* we can be sure that players bid their true valuations—if $b_i = v_i$ for every i. To encourage players to bid truthfully, we must charge the winner a non-zero price. (Otherwise, all players will bid gargantuan amounts in an effort to be the highest.) On the other hand, if we charge the winning player its bid, it encourages players to underbid. (Bidding your maximum willingness to pay ensures a net gain of zero, win or lose.) Vickrey [34] suggested charging the winner the value of the *second-highest* bid, and proved that this price transforms truthful bidding into an optimal strategy for each bidder, independent of the bids of the other players. In turn, the Vickrey auction is guaranteed to produce an efficient allocation of the good, provided all players bid in the obvious, optimal way.

Cost-Sharing Mechanisms. Economic efficiency is not the only important objective in mechanism design. *Revenue* is a second obvious concern, especially in settings where the mechanism designer incurs a non-trivial cost. This cost can represent production costs, or more generally some revenue target.

A *cost-sharing problem* is defined by a set U of players vying to receive some good or service, and a cost function $C : 2^U \to \mathcal{R}^+$ describing the cost incurred by the mechanism as a function of the auction outcome—the set S of winners. We assume that $C(S)$ is nonnegative for every set $S \subseteq U$, that $C(\emptyset) = 0$, and that C is nondecreasing ($S \subseteq T$ implies $C(S) \le C(T)$). Note that there is no explicit limit on the number of auction winners, although a large number of winners might result in extremely large costs. With outcome-dependent costs, the *efficient allocation* is the one that maximizes the *social welfare* $W(S) =$

$\sum_{i \in S} v_i - C(S)$—the outcome that trades offs the valuations of the winners and the cost incurred in an optimal way. The problem of selling a single good can be viewed as the special case in which $C(S) = 0$ if $|S| \leq 1$ and $C(S) = +\infty$ otherwise.

In this paper, we focus on cost functions that are defined implicitly by an instance of a combinatorial optimization problem. For example, U could represent a set of potential clients, located in an undirected graph with fixed edge costs, that want connectivity to a server r [7,17]. In this application, $C(S)$ denotes the cost of connecting the terminals in S to r—the cost of the minimum-cost Steiner tree that spans $S \cup \{r\}$.

A *cost-sharing mechanism*, given a set U and a function C, is a protocol that decides which players win the auction and at what prices. Typically, such a mechanism is also (perhaps approximately) *budget-balanced*, meaning that the cost incurred is passed on to the auction's winners. Budget-balanced cost-sharing mechanisms provide control over the revenue generated, relative to the cost incurred by the mechanism designer.

Summarizing, we have identified three natural goals in auction and mechanism design: (1) *incentive-compatibility*, meaning that every player's optimal strategy is to bid its true private value v_i for receiving the service; (2) *budget-balance*, meaning that the mechanism recovers its incurred cost with the prices charged; and (3) *efficiency*, meaning that the cost and valuations are traded off in an optimal way.

Unfortunately, properties (1)–(3) cannot be simultaneously achieved, even in very simple settings [10,30]. This impossibility result motivates relaxing at least one of the these properties. Until recently, nearly all work in cost-sharing mechanism design completely ignored either budget-balance or efficiency. If the budget balance constraint is discarded, then there is an extremely powerful and flexible mechanism that is incentive-compatible and efficient: the *VCG mechanism* (see e.g. [26]). This mechanism specializes to the Vickrey auction in the case of selling a single good, but is far more general. Since the VCG mechanism is typically not approximately budget-balanced for any reasonable approximation factor (see e.g. [6]), it is not suitable for many applications.

The second approach is to insist on incentive-compatibility and budget-balance, while regarding efficiency as a secondary objective. The only general technique for designing mechanisms of this type is due to Moulin [25]. Over the past five years, researchers have developed approximately budget-balanced Moulin mechanisms for cost-sharing problems arising from numerous different combinatorial optimization problems, including fixed-tree multicast [1,6,7]; the more general submodular cost-sharing problem [25,26]; Steiner tree [17,18,20]; Steiner forest [21,22]; facility location [23,29]; rent-or-buy network design [14,29]; and various covering problems [5,16]. Most of these mechanisms are based on novel primal-dual approximation algorithms for the corresponding optimization problem. With one exception discussed below, none of these works provided any guarantees on the efficiency achieved by the proposed mechanisms.

Approximately Efficient Cost-Sharing Mechanisms. Impossibility results are, of course, common in optimization. From conditional impossibility results like Cook's Theorem to information-theoretic lower bounds in restricted models of computation, as with online and streaming algorithms, algorithm designers are accustomed to devising heuristics and proving worst-case guarantees about them using approximation measures. This approach can be applied equally well to cost-sharing mechanism design, and allows us to quantify the inevitable efficiency loss in incentive-compatible, budget-balanced cost-sharing mechanisms. As worst-case approximation measures are rarely used in economics, this research direction has only recently been pursued.

Moulin and Shenker [26] were the first to propose quantifying the efficiency loss in budget-balanced Moulin mechanisms. They studied an additive notion of efficiency loss for submodular cost functions. This notion is useful for ranking different mechanisms according to their worst-case efficiency loss, but does not imply bounds on the quality of a mechanism's outcome relative to that of an optimal outcome. A more recent paper [31] provides an analytical framework for proving approximation guarantees on the efficiency attained by Moulin mechanisms. The present paper builds on this framework. (See [4,11] for other very recent applications.)

Several definitions of approximate efficiency are possible, and the choice of definition is important for quantifying the inefficiency of Moulin mechanisms. Feigenbaum et al. [6] showed that, even for extremely simple cost functions, budget-balance and social welfare cannot be simultaneously approximated to within any non-trivial factor. This negative approximation result is characteristic of mixed-sign objective functions such as welfare.

An alternative formulation of exact efficiency is to choose a subset minimizing the *social cost*, where the social cost $\pi(S)$ of a set S is the sum of the incurred service cost and the excluded valuations: $C(S) + \sum_{i \notin S} v_i$. Since $\pi(S) = -W(S) + \sum_{i \in U} v_i$ for every set S, where U denotes the set of all players, a subset maximizes the social welfare if and only if it minimizes the social cost. The two functions are not, of course, equivalent from the viewpoint of approximation. Similar transformations have been used for "prize-collecting" problems in combinatorial optimization (see e.g. [3]). We call a cost-sharing mechanism α-*approximate* if it always produces an outcome with social cost at most an α factor times that of an optimal outcome. Also, a mechanism is β-budget-balanced if the sum of prices charged is always at most the cost incurred and at least a $1/\beta$ fraction of this cost.

Previous work [31] demonstrated that $O(\text{polylog}(k))$-approximate, $O(1)$-budget-balanced Moulin mechanisms exist for two important types of cost-sharing problems: submodular cost functions, and Steiner tree cost functions. (Here k denotes the number of players.) This was the first evidence that properties (1)–(3) above can be approximately simultaneously satisfied, and motivates the following fundamental question: *which cost-sharing problems admit incentive-compatible mechanisms that are approximately budget-balanced and approximately efficient?*

Our Results. This paper presents three contributions. We first consider metric uncapacitated facility location (UFL) cost-sharing problems, where the input is a UFL instance, the players U are the demands of this instance, and the cost $C(S)$ is defined as the cost of an optimal solution to the UFL sub-instance induced by S. The only known $O(1)$-budget-balanced Moulin mechanism for this problem is due to Pál and Tardos [29] (the *PT mechanism*). The PT mechanism is 3-budget-balanced [29], and no Moulin mechanism for the problem has better budget balance [16]. We provide the first efficiency guarantee for the PT mechanism by proving that it is $O(\log k)$-approximate, where k is the number of players. Simple examples show that every $O(1)$-budget-balanced Moulin mechanism for UFL is $\Omega(\log k)$-approximate. Thus the PT mechanism simultaneously optimizes both budget balance and efficiency over the class of Moulin mechanisms for UFL.

Second, we design and analyze Moulin mechanisms for rent-or-buy network design cost-sharing problems. For example, the single-sink rent-or-buy (SSRoB) problem is a generalization of the Steiner tree problem in which several source vertices of a network (corresponding to the players U) want to simultaneously send one unit of flow each to a common root vertex. For a subset $S \subseteq U$ of players, the cost $C(S)$ is defined as the minimum-cost way of installing sufficient capacity for the players of S to simultaneously send flow to the root. Capacity on an edge can be rented on a per-unit basis, or an infinite amount of capacity can be bought for M times the per-unit renting cost, where $M \geq 1$ is a parameter. (Steiner tree is the special case where $M = 1$.) Thus the SSRoB problem is a simple model of capacity installation in which costs obey economies of scale. The multicommodity rent-or-buy (MRoB) problem is the generalization of SSRoB in which each player corresponds to a source-sink vertex pair, and different players can have different sink vertices.

Gupta, Srinivasan, and Tardos [14] and Leonardi and Schäfer [23] independently showed how to combine the SSRoB algorithm of [13] with the Jain-Vazirani Steiner tree mechanism [17] to obtain an $O(1)$-budget-balanced SSRoB mechanism. (Earlier, Pál and Tardos [29] designed an $O(1)$-budget-balanced SSRoB mechanism, but it was more complicated and its budget balance factor was larger.) We note that the mechanism design ideas in [14,23], in conjunction with the recent 2-budget-balanced Steiner forest mechanism due to Könemann, Leonardi, and Schäfer [21], lead to an $O(1)$-budget-balanced MRoB mechanism. Much more importantly, we prove that this SSRoB mechanism and a variant of this MRoB mechanism are $O(\log^2 k)$-approximate, the first efficiency guarantees for any approximately budget-balanced mechanisms for these problems. Our third result below implies that these are the best-achievable efficiency guarantees for $O(1)$-budget-balanced Moulin mechanisms for these problems.

Third, we prove a new lower bound that exposes a non-trivial, latent hierarchy among different cost-sharing problems. Specifically, we prove that every $O(1)$-budget-balanced Moulin mechanism for Steiner tree cost functions is $\Omega(\log^2 k)$-approximate. This lower bound trivially also applies to Steiner forest, SSRoB, and MRoB cost functions.

This lower bound establishes a previously unobservable separation between submodular and facility location cost-sharing problems on the one hand, and the above network design cost-sharing problems on the other. All admit $O(1)$-budget-balanced Moulin mechanisms, but the worst-case efficiency loss of Moulin mechanisms is provably larger in the second class of problems than in the first one.

All previous lower bounds on the efficiency of Moulin mechanisms were derived from either budget-balance lower bounds or, as for the problems considered in this paper, from a trivial example equivalent to a cost-sharing problem in a single-link network [31]. This type of example cannot prove a lower bound larger than the kth Harmonic number $\mathcal{H}_k = \Theta(\log k)$ on the approximate efficiency of a Moulin mechanism. We obtain the stronger bound of $\Omega(\log^2 k)$ by a significantly more intricate construction that exploits the complexity of Steiner tree cost functions.

2 Preliminaries

Cost-Sharing Mechanisms. We consider a cost function C that assigns a cost $C(S)$ to every subset S of a universe U of players. We assume that C is nonnegative and nondecreasing (i.e., $S \subseteq T$ implies $C(S) \le C(T)$). We sometimes refer to $C(S)$ as the *service cost*, to distinguish it from the social cost (defined below). We also assume that every player $i \in U$ has a private, nonnegative *valuation* v_i.

A *mechanism* collects a nonnegative bid b_i from each player $i \in U$, selects a set $S \subseteq U$ of players, and charges every player i a price p_i. In this paper, we focus on cost functions that are defined implicitly as the optimal solution of an instance of a (NP-hard) combinatorial optimization problem. The mechanisms we consider also produce a feasible solution to the optimization problem induced by the served set S, which has cost $C'(S)$ that in general is larger than the optimal cost $C(S)$. We also impose the following standard restrictions and assumptions. We only allow mechanisms that are "individually rational" in the sense that $p_i = 0$ for players $i \notin S$ and $p_i \le b_i$ for players $i \in S$. We require that all prices are nonnegative ("no positive transfers"). Finally, we assume that players have *quasilinear* utilities, meaning that each player i aims to maximize $u_i(S, p_i) = v_i x_i - p_i$, where $x_i = 1$ if $i \in S$ and $x_i = 0$ if $i \notin S$.

Our incentive-compatibility constraint is the well-known strategyproofness condition, which intuitively requires that a player cannot gain from misreporting its bid. Formally, a mechanism is *strategyproof (SP)* if for every player i, every bid vector b with $b_i = v_i$, and every bid vector b' with $b_j = b'_j$ for all $j \ne i$, $u_i(S, p_i) \ge u_i(S', p'_i)$, where (S, p) and (S', p') denote the outputs of the mechanism for the bid vectors b and b', respectively.

For a parameter $\beta \ge 1$, a mechanism is *β-budget balanced* if $C'(S)/\beta \le \sum_{i \in S} p_i \le C(S)$ for every outcome (set S, prices p, feasible solution with service cost $C'(S)$) of the mechanism. In particular, this requirement implies that the feasible solution produced by the mechanism has cost at most β times that of optimal.

As discussed in the Introduction, a cost-sharing mechanism is α-*approximate* if, assuming truthful bids, it always produces a solution with social cost at most an α factor times that of an optimal solution. Here, the social cost incurred by the mechanism is defined as the service cost $C'(S)$ of the feasible solution it produces for the instance corresponding to S, plus the sum $\sum_{i \notin S} v_i$ of the excluded valuations. The optimal social cost is $\min_{S \subseteq U}[C(S) + \sum_{i \notin S} v_i]$. A mechanism thus has two sources of inefficiency: first, it might choose a suboptimal set S of players to serve; second, it might produce a suboptimal solution to the optimization problem induced by S.

Moulin Mechanisms and Cross-Monotonic Cost-Sharing Methods. Next we review *Moulin mechanisms*, the preeminent class of SP, approximately budget-balanced mechanisms. Such mechanisms are based on *cost sharing methods*, defined next.

A cost-sharing method χ is a function that assigns a non-negative *cost share* $\chi(i, S)$ for every subset $S \subseteq U$ of players and every player $i \in S$. We consider cost-sharing methods that, given a set S, produce both the cost shares $\chi(i, S)$ for all $i \in S$ and also a feasible solution for the optimization problem induced by S. A cost-sharing method is β-*budget balanced* for a cost function C and a parameter $\beta \geq 1$ if it always recovers a $1/\beta$ fraction of the cost: $C'(S)/\beta \leq \sum_{i \in S} \chi(i, S) \leq C(S)$, where $C'(S)$ is the cost of the produced feasible solution. A cost-sharing method is *cross-monotonic* if the cost share of a player only increases as other players are removed: for all $S \subseteq T \subseteq U$ and $i \in S$, $\chi(i, S) \geq \chi(i, T)$.

A cost-sharing method χ for C defines the following *Moulin mechanism M_χ* for C. First, collect a bid b_i for each player i. Initialize the set S to all of U and invoke the cost-sharing method χ to define a feasible solution to the optimization problem induced by S and a price $p_i = \chi(i, S)$ for each player $i \in S$. If $p_i \leq b_i$ for all $i \in S$, then halt, output the set S, the corresponding feasible solution, and charge prices p. If $p_i > b_i$ for some player $i \in S$, then remove an arbitrary such player from the set S and iterate. A Moulin mechanism based on a cost-sharing method thus simulates an iterative auction, with the method χ suggesting prices for the remaining players at each iteration. The cross-monotonicity constraint ensures that the simulated auction is ascending, in the sense that the prices that are compared to a player's bid are only increasing with time. Note that if χ produces a feasible solution in polynomial time, then so does M_χ. Also, M_χ clearly inherits the budget-balance factor of χ. Finally, Moulin [25] proved the following.

Theorem 1 ([25]). *If χ is a cross-monotonic cost-sharing method, then the corresponding Moulin mechanism M_χ is strategyproof.*[1]

Theorem 1 reduces the problem of designing an SP, β-budget-balanced cost-sharing mechanism to that of designing a cross-monotonic, β-budget-balanced cost-sharing method.

---

[1] Moulin mechanisms also satisfy a stronger notion of incentive compatibility called groupstrategyproofness (GSP), which is a form of collusion resistance [26].

Summability and Approximate Efficiency. Roughgarden and Sundararajan [31] showed that the approximate efficiency of a Moulin mechanism is completely controlled by its budget-balance and one additional parameter of its underlying cost-sharing method. We define this parameter and the precise guarantee next.

Definition 1 (Summability [31]). *Let C and χ be a cost function and a cost-sharing method, respectively, defined on a common universe U of players. The method χ is α-summable for C if*

$$\sum_{\ell=1}^{|S|} \chi(i_\ell, S_\ell) \leq \alpha \cdot C(S)$$

for every ordering σ of U and every set $S \subseteq U$, where S_ℓ and i_ℓ denote the set of the first ℓ players of S and the ℓth player of S (with respect to σ), respectively.

Theorem 2 ([31]). *Let U be a universe of players and C a nondecreasing cost function on U with $C(\emptyset) = 0$. Let M be a Moulin mechanism for C with underlying cost-sharing method χ. Let $\alpha \geq 0$ and $\beta \geq 1$ be the smallest numbers such that χ is α-summable and β-budget-balanced. Then the mechanism M is $(\alpha + \beta)$-approximate and no better than $\max\{\alpha, \beta\}$-approximate.*

In particular, an $O(1)$-budget-balanced Moulin mechanism is $\Theta(\alpha)$-approximate if and only if the underlying cost-sharing method is $\Theta(\alpha)$-summable. Analyzing the summability of a cost-sharing method, while non-trivial, is a tractable problem in many important cases. Because summability is defined as the accrued cost over a worst-case "insertion order" of the players, summability bounds are often reminiscent of performance analyses of online algorithms.

3   An Optimal Facility Location Cost-Sharing Mechanism

In this section we consider the metric uncapacitated facility location (UFL) problem.[2] The input is given by a set U of demands (the players), a set F of facilities, an opening cost f_q for each facility $q \in F$, and a metric c defined on $U \cup F$. The cost $C(S)$ of a subset $S \subseteq U$ of players is then defined as the cost of an optimal solution to the UFL problem induced by S. In other words, $C(S) = \min_{\emptyset \neq F^* \subseteq F} [\sum_{q \in F^*} f_q + \sum_{i \in S} \min_{q \in F^*} c(q, i)]$. We seek an $O(1)$-budget-balanced Moulin mechanism for UFL with the best-possible approximate efficiency. Theorems 1 and 2 reduce this goal to the problem of designing an $O(1)$-budget-balanced cross-monotonic cost-sharing method with the smallest-possible summability.

We begin with a simple lower bound, similar to that given in [31] for submodular cost-sharing problems.

Proposition 1 (Lower Bound on UFL Approximate Efficiency). *For every $k \geq 1$, there is a k-player UFL cost function C with the following property: for every $\beta \geq 1$ and every β-budget-balanced Moulin mechanism M for C, M is no better than $\mathcal{H}_k / \beta$-approximate.*

---

[2] Due to space constraints, we omit all proofs. Details are in [32].

Pál and Tardos [29] showed that every UFL cost function admits a 3-budget-balanced cross-monotonic cost-sharing method χ_{PT}. We call this the *PT method*, and the induced Moulin mechanism the *PT mechanism*. (See [29] or [32] for details.) Our main result in this section shows that the PT mechanism matches the lower bound in Proposition 1, up to a constant factor.

Theorem 3 (Upper Bound on PT Summability). *Let C be a k-player UFL cost function and χ_{PT} the corresponding PT method. Then χ_{PT} is $\mathcal{H}_k$-summable for C.*

Applying Theorem 2 yields an efficiency guarantee for the PT mechanism.

Corollary 1 (Upper Bound on PT Approximate Efficiency). *Let C be a k-player UFL cost function and M_{PT} the corresponding PT mechanism. Then M_{PT} is $(\mathcal{H}_k + 3)$-approximate.*

Theorem 3 follows from two lemmas. The first states that single-facility instances supply worst-case examples for the summability of the PT method.

Lemma 1. *For every $k \geq 1$, the summability of PT methods for k-player UFL cost functions is maximized by the cost functions that correspond to single-facility instances.*

Lemma 1 is based on a monotonicity property that we prove for the PT method: increasing the distance between a demand and a facility can only increase cost shares. This monotonicity property allows us to argue that in worst-case UFL instances, players are partitioned into non-interacting groups, each clustered around one facility. We complete the proof of Lemma 1 by arguing that the summability of the PT method for one of these single-facility clusters in at least that in the original facility location instance.

Our second lemma bounds the summability of PT methods in single-facility instances.

Lemma 2. *Let C be a k-player UFL cost function corresponding to a single-facility instance. If χ_{PT} is the corresponding PT method, then χ_{PT} is $\mathcal{H}_k$-summable for C.*

4   Optimal Rent-or-Buy Cost-Sharing Mechanisms

Single-Sink Rent-or-Buy: Next we consider single-sink rent-or-buy (SSRoB) cost-sharing problems. The input is given by a graph $G = (V, E)$ with edge costs that satisfy the Triangle Inequality, a root vertex t, a set U of demands (the players), each of which is located at a vertex of G, and a parameter $M \geq 1$. A feasible solution to the SSRoB problem induced by S is a way of installing sufficient capacity on the edges of G so that every player in S can simultaneously route one unit of flow to t. Installing x units of capacity on an edge e costs $c_e \cdot \min\{x, M\}$; the parameter M can be interpreted as the ratio between the cost of "buying" infinite capacity for a flat fee and the cost of "renting" a single unit of capacity. The cost $C(S)$ of a subset $S \subseteq U$ of players is then defined as the cost of an optimal solution to the SSRoB problem induced by S. We sometimes abuse notation and use $i \in U$ to denote both a player and the vertex of G that hosts the player.

Gupta, Srinivasan, and Tardos [14] and Leonardi and Schäfer [23] independently designed the following $O(1)$-budget-balanced cross-monotonic cost-sharing method for SSRoB, which we call the *GST method*. Given an SSRoB cost function and a set $S \subseteq U$ of players, we use the randomized algorithm of [13] to produce a feasible solution. This algorithm first chooses a random subset $D \subseteq S$ by adding each player $i \in S$ to D independently with probability $1/M$. Second, it computes an approximate Steiner tree spanning $D \cup \{t\}$ using, for example, the 2-approximate MST heuristic [33], and buys infinite capacity on all of the edges of this tree. Third, for each player $i \notin D$, it rents one unit of capacity for exclusive use by i on a shortest path from its vertex to the closest vertex in $D \cup \{t\}$. This defines a feasible solution with probability 1, and the expected cost of this solution at most 4 times that of an optimal solution to the SSRoB instance induced by S [13].

The GST cost share $\chi_{GST}(i, S)$ is defined as the expectation of the following random variable X_i, over the random choice of the set D in the above algorithm: if $i \notin D$, then X_i equals one quarter of the length of the shortest path used to connect i to a vertex in $D \cup \{t\}$; if $i \in D$, then X_i equals $M/2$ times the Jain-Vazirani cost share $\chi_{JV}(i, D)$ of i with respect to the Steiner tree instance defined by G, c, t, and the players D (see [17] for the details of χ_{JV}). These cost shares are 4-budget-balanced with respect to the optimal cost of the SSRoB instance induced by S, as well as the expected cost of the above randomized algorithm that produces a feasible solution to this instance. We prove the following result.

Theorem 4. *For every k-player SSRoB cost function, the corresponding GST mechanism is $O(\log^2 k)$-approximate.*

Theorem 6 below implies that this is the best efficiency guarantee possible for an $O(1)$-budget-balanced SSRoB Moulin mechanism.

With an eye toward extending Theorem 4 to the MRoB problem, we summarize very briefly the main steps in the proof (details are in [32]). First, we decompose each GST cost share $\chi_{GST}(i, S)$ into two terms, a term $\chi_{buy}(i, S)$ for the contributions of samples $D \subseteq S$ in which $i \in D$, and a term $\chi_{rent}(i, S)$ for the contributions of the remaining samples. Proving Theorem 4 reduces to proving that both χ_{buy} and χ_{rent} are $O(\log^2 k)$-summable. Second, we use the $O(\log^2 k)$-summability of χ_{JV} [31] together with a counting argument inspired by [13,19] to prove that χ_{buy} is $O(\log^2 k)$-summable. Third, we prove that the cost-sharing method χ_{JV} is $O(1)$-*strict* in the sense of [12]. This roughly means that whenever a player i is included in the random sample D, then the cost share $\chi_{JV}(i, D)$ is at least a constant factor times the cost share it would have received had it not been included.[3] We leverage the strictness of χ_{JV} to prove that the summability of χ_{rent} is at most a constant times that of χ_{buy}.

Multicommodity Rent-or-Buy. We next extend Theorem 4 to the MRoB problem, where each player i corresponds to a vertex pair (s_i, t_i). (All other aspects of the problem are the same.) The high-level approach is similar, but the technical challenges are much more formidable. In the proof of Theorem 4, the Jain-Vazirani cost-sharing method χ_{JV} played a heroic role: it is cross-monotonic, which is necessary for the GST cost-sharing method to be cross-monotonic; it is $O(\log^2 k)$-summable, which is necessary for χ_{buy} to be $O(\log^2 k)$-summable; and it is $O(1)$-strict in the sense of [12] with respect to the MST heuristic for Steiner tree, which is necessary for χ_{rent} to be $O(\log^2 k)$-summable. Is there a comparably all-purpose cost-sharing method for the *Steiner Forest* problem—the problem of finding the min-cost subgraph of a given graph that includes a path between every given vertex pair (s_i, t_i)? The only known cross-monotonic cost-sharing method χ_{KLS} for Steiner Forest cost-sharing problems was recently given by Könemann, Leonardi, and Schäfer [21]. This method is defined by a primal-dual algorithm; the cost shares are a natural byproduct of a dual growth process, and the primal is a 2-approximate feasible solution to the given Steiner Forest instance. Using the ideas in [9,12,14,23], these facts suffice to define an $O(1)$-budget-balanced Moulin mechanism for MRoB cost-sharing problems. Moreover, the KLS method was very recently shown to be $O(\log^2 k)$-summable [4]; thus, the corresponding cost-sharing method χ_{buy} is $O(\log^2 k)$-summable. Unfortunately, the KLS cost-sharing method is $\Omega(k)$-strict with respect to the corresponding primal solution [12], which precludes bounding the summability of χ_{rent} in terms of χ_{buy}. While several strict cost-sharing methods are known for different Steiner Forest approximation algorithms [2,9,12,28], none of these are cross-monotonic methods.

Our high-level approach is to modify the above composition of the KLS method with the mechanism design techniques of [14,23] in a way that achieves $O(1)$-strictness while sacrificing only a small constant factor in the budget balance. Similar ideas have been used previously to obtain strictness guarantees for other Steiner forest algorithms [2,12,28].

Theorem 5. *Every k-player MRoB cost function admits an $O(1)$-budget-balanced, $O(\log^2 k)$-approximate Moulin mechanism.*

5   An $\Omega(\log^2 k)$ Lower Bound for Steiner Tree Problems

An instance of the *Steiner tree cost-sharing problem* [17] is given by an undirected graph $G = (V, E)$ with a root vertex t and nonnegative edge costs, with each player of U located at some vertex of G. For a subset $S \subseteq U$, the cost $C(S)$ is defined as that of a minimum-cost subgraph of G that spans all of the

---

[3] Formally, strictness of a cost-sharing method is defined with respect to some primal algorithm; see [12] for a precise definition.

players of S as well as the root t. There are $O(1)$-budget-balanced, $O(\log^2 k)$-approximate Moulin mechanisms for such problems [4,17,21,31]. The main result of this section is a matching lower bound on the approximate efficiency of every $O(1)$-budget-balanced Moulin mechanism.

Theorem 6. *There is a constant $c > 0$ such that for every constant $\beta \geq 1$, every β-budget-balanced Moulin mechanism for Steiner tree cost-sharing problems is at least $(\beta^{-1}c\log^2 k)$-approximate, where k is the number of players served in an optimal outcome.*

Theorem 6 implies that Steiner tree cost-sharing problems and their generalizations are fundamentally more difficult for Moulin mechanisms than facility location (Theorem 3) and submodular cost-sharing problems (see [31]).

We now outline the proof of Theorem 6. Fix values for the parameters $k \geq 2$ and $\beta \geq 1$. We construct a sequence of networks, culminating in G. The network G_0 consists of a set V_0 of two nodes connected by an edge of cost 1. One of these is the root t. The player set U_0 is $\sqrt{k}$ players that are co-located at the non-root node. (Assume for simplicity that k is a power of 4.) For $j > 0$, we obtain the network G_j from G_{j-1} by replacing each edge (v, w) of G_{j-1} with m internally disjoint two-hop paths between v and w, where m is a sufficiently large function of k of β. (We will choose $m \geq 8\beta\sqrt{k} \cdot (2\beta)^{\sqrt{k}}$.) See Figure 1. The cost of each of these $2m$ edges is half of the cost of the edge (v, w). Thus every edge in G_j has cost 2^{-j}.

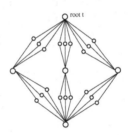

Fig. 1. Network G_2 in the proof of Theorem 6, with $m = 3$. All edges have length 1/4

Let V_j denote the vertices of G_j that are not also present in G_{j-1}. We augment the universe by placing $\sqrt{k}$ new co-located players at each vertex of V_j; denote these new players by U_j. The final network G is then G_p, where $p = (\log k)/2$. Let $V = V_0 \cup \cdots \cup V_p$ and $U = U_0 \cup \cdots \cup U_p$ denote the corresponding vertex and player sets. Let C denote the corresponding Steiner tree cost function.

Now fix $\beta \geq 1$ and an arbitrary cross-monotonic, β-budget balanced Steiner tree cost-sharing method χ. By Theorem 2, we can prove Theorem 6 by exhibiting a subset $S \subseteq U$ of size k and an ordering σ of the players of S such that $\sum_{\ell=1}^{k} \chi(i_\ell, S_\ell) \geq (c\log^2 k/\beta) \cdot C(S)$, where i_ℓ and S_ℓ denote the ℓth player and the first ℓ players with respect to σ.

We construct the set S iteratively. For $j = 0, 1, \ldots, p$, we will identify a subset $S_j \subseteq U_j$ of players; the set S will then be $S_0 \cup \cdots \cup S_p$. Recall that U_j consists of groups of $\sqrt{k}$ players, each co-located at a vertex of V_j, with m such groups for each edge of G_{j-1}. The set S_j will consist of zero or one such group of $\sqrt{k}$ players for each edge of G_{j-1}.

The set S_0 is defined to be U_0. For $j > 0$, suppose that we have already defined $S_0, \ldots, S_{j-1}$. Call a vertex $v \in V_0 \cup \cdots \cup V_{j-1}$ *active* if v is the root t or if the $\sqrt{k}$ players co-located at v were included in the set $S_0 \cup \cdots \cup S_{j-1}$. Call an edge (v, w) of G_{j-1} *active* if both of its endpoints are active and *inactive* otherwise.

To define S_j, we consider each edge (v, w) of G_{j-1} in an arbitrary order. Each such edge gives rise to m groups of $\sqrt{k}$ co-located players in G_j. If (v, w) is inactive in G_{j-1}, then none of these $m\sqrt{k}$ players are included in S_j. If (v, w) is active in G_{j-1}, then we will choose precisely one of the m groups of players, and will include these $\sqrt{k}$ co-located players in S_j. We first state two lemmas that hold independently of how this choice is made; we then elaborate on our criteria for choosing groups of players.

Lemma 3. *For every $j \in 1, 2, \ldots, p$, $|S_j| = 2^{j-1}\sqrt{k}$. Also, $|S_0| = \sqrt{k}$.*

Lemma 3 implies that $|S| = \sqrt{k}(1 + \sum_{j=0}^{p-1} 2^j) = k$. The next lemma states that our construction maintains the invariant that the players selected in the first j iterations lie "on a straight line" in G.

Lemma 4. *For every $j \in 0, 1, \ldots, p$, $C(S_0 \cup \cdots \cup S_j) = 1$.*

Lemmas 3 and 4 both follow from straightforward inductions on j.

We now explain how to choose one out of the m groups of co-located players that arise from an active edge. Fix an iteration $j > 0$ and let $\hat{S}$ denote the set of players selected in previous iterations $(S_0, \ldots, S_{j-1})$ and previously in the current iteration. Let (v, w) be the active edge of G_{j-1} under consideration and $A_1, \ldots, A_m \subseteq U_j$ the corresponding groups of co-located players. We call the group A_r *good* if the $\sqrt{k}$ players of A_r can be ordered $i_1, i_2, \ldots, i_{\sqrt{k}}$ so that

$$\chi(i_\ell, \hat{S} \cup \{i_1, \ldots, i_\ell\}) \geq \frac{1}{4\beta} \cdot \frac{2^{-j}}{\ell} \tag{1}$$

for every $\ell \in \{1, 2, \ldots, \sqrt{k}\}$. We then include an arbitrary good group A_r in the set S_j. See [32] for a proof of the following lemma.

Lemma 5. *Provided m is a sufficiently large function of k and β, for every $j \in \{1, \ldots, p\}$, every ordering of the active edges of G_{j-1}, and every edge (v, w) in this ordering, at least one of the m groups of players of U_j that corresponds to (v, w) is good. Also, the group S_0 is good.*

We conclude by using the lemma to finish the proof of Theorem 6.

We have already defined the subset $S \subseteq U$ of players. We define the ordering σ of the players in S as follows. First, for all $j \in \{1, \ldots, p\}$, all players of S_{j-1}

precede all players of S_j in σ. Second, for each $j \in \{1, \ldots, p\}$, the players of S_j are ordered according to groups, with the $\sqrt{k}$ players of a group appearing consecutively in σ. The ordering of the different groups of players of S_j is the same as the corresponding ordering of the active edges of G_{j-1} that was used to define these groups. Third, each (good) group of $\sqrt{k}$ co-located players is ordered so that (1) holds.

Now consider the sum $\sum_{\ell=1}^{k} \chi(i_\ell, S_\ell)$, where i_ℓ and S_ℓ denote the ℓth player and the first ℓ players of S with respect to σ, respectively. Since (1) holds for every group of players, for every $j \in \{0, 1, \ldots, p\}$, every group of players in S_j contributes at least

$$\sum_{\ell=1}^{\sqrt{k}} \frac{1}{4\beta} \cdot \frac{2^{-j}}{\ell} = \frac{2^{-j}\mathcal{H}_{\sqrt{k}}}{4\beta}$$

to this sum. By Lemma 3, for each $j \in \{1, \ldots, p\}$, there are 2^{j-1} such groups. There is also the group S_0. Thus the sum $\sum_{\ell=1}^{k} \chi(i_\ell, S_\ell)$ is at least

$$\frac{\mathcal{H}_{\sqrt{k}}}{4\beta} \left(1 + \sum_{j=1}^{(\log k)/2} 2^{j-1} \cdot 2^{-j}\right) \geq \frac{c}{\beta} \log^2 k = \left(\frac{c}{\beta} \log^2 k\right) \cdot C(S)$$

for some constant $c > 0$ that is independent of k and β. This completes the proof of Theorem 6.

References

1. A. Archer, J. Feigenbaum, A. Krishnamurthy, R. Sami, and S. Shenker. Approximation and collusion in multicast cost sharing. *Games and Economic Behavior*, 47(1):36–71, 2004.
2. L. Becchetti, J. Könemann, S. Leonardi, and M. Pál. Sharing the cost more efficiently: Improved approximation for multicommodity rent-or-buy. In *SODA '05*, pages 375–384.
3. D. Bienstock, M. X. Goemans, D. Simchi-Levi, and D. P. Williamson. A note on the prize-collecting traveling salesman problem. *Mathematical Programming*, 59(3):413–420, 1993.
4. S. Chawla, T. Roughgarden, and M. Sundararajan. Optimal cost-sharing mechanisms for network design. In *WINE '06*.
5. N. R. Devanur, M. Mihail, and V. V. Vazirani. Strategyproof cost-sharing mechanisms for set cover and facility location games. In *EC '03*, pages 108–114.
6. J. Feigenbaum, A. Krishnamurthy, R. Sami, and S. Shenker. Hardness results for multicast cost sharing. *Theoretical Computer Science*, 304:215–236, 2003.
7. J. Feigenbaum, C. Papadimitriou, and S. Shenker. Sharing the cost of multicast transmissions. *Journal of Computer and System Sciences*, 63(1):21–41, 2001.
8. J. Feigenbaum and S. J. Shenker. Distributed algorithmic mechanism design: Recent results and future directions. In *DIAL M '02*, pages 1–13.
9. L. Fleischer, J. Könemann, S. Leonardi, and G. Schäfer. Simple cost-sharing schemes for multicommodity rent-or-buy and stochastic Steiner tree. In *STOC '06*, pages 663–670.

10. J. Green, E. Kohlberg, and J. J. Laffont. Partial equilibrium approach to the free rider problem. *Journal of Public Economics*, 6:375–394, 1976.
11. A. Gupta, J. Könemann, S. Leonardi, R. Ravi, and G. Schäfer. An efficient cost-sharing mechanism for the prize-collecting Steiner forest problem. In *SODA '07*.
12. A. Gupta, A. Kumar, M. Pál, and T. Roughgarden. Approximation via cost-sharing: A simple approximation algorithm for the multicommodity rent-or-buy problem. In *FOCS '03*, pages 606–615.
13. A. Gupta, A. Kumar, and T. Roughgarden. Simpler and better approximation algorithms for network design. In *STOC '03*, pages 365–372.
14. A. Gupta, A. Srinivasan, and É. Tardos. Cost-sharing mechanisms for network design. In *APPROX '04*, pages 139–150.
15. J. D. Hartline. *Optimization in the Private Value Model: Competitive Analysis Applied to Auction Design*. PhD thesis, University of Washington, 2003.
16. N. Immorlica, M. Mahdian, and V. S. Mirrokni. Limitations of cross-monotonic cost-sharing schemes. In *SODA '05*, pages 602–611.
17. K. Jain and V. Vazirani. Applications of approximation algorithms to cooperative games. In *STOC '01*, pages 364–372.
18. K. Jain and V. Vazirani. Equitable cost allocations via primal-dual-type algorithms. In *STOC '02*, pages 313–321.
19. D. R. Karger and M. Minkoff. Building Steiner trees with incomplete global knowledge. In *FOCS '00*, pages 613–623.
20. K. Kent and D. Skorin-Kapov. Population monotonic cost allocation on MST's. In *Operational Research Proceedings KOI*, pages 43–48, 1996.
21. J. Könemann, S. Leonardi, and G. Schäfer. A group-strategyproof mechanism for Steiner forests. In *SODA '05*, pages 612–619.
22. J. Könemann, S. Leonardi, G. Schäfer, and S. van Zwam. From primal-dual to cost shares and back: A stronger LP relaxation for the steiner forest problem. In *ICALP '05*, pages 1051–1063.
23. S. Leonardi and G. Schäfer. Cross-monotonic cost-sharing methods for connected facility location. In *EC '04*, pages 242–243.
24. A. Mas-Colell, M. D. Whinston, and J. R. Green. *Microeconomic Theory*. Oxford University Press, 1995.
25. H. Moulin. Incremental cost sharing: Characterization by coalition strategy-proofness. *Social Choice and Welfare*, 16:279–320, 1999.
26. H. Moulin and S. Shenker. Strategyproof sharing of submodular costs: Budget balance versus efficiency. *Economic Theory*, 18:511–533, 2001.
27. M. J. Osborne and A. Rubinstein. *A Course in Game Theory*. MIT Press, 1994.
28. M. Pál. *Cost Sharing and Approximation*. PhD thesis, Cornell University, 2005.
29. M. Pál and É. Tardos. Group strategyproof mechanisms via primal-dual algorithms. In *FOCS '03*, pages 584–593.
30. K. Roberts. The characterization of implementable choice rules. In J. J. Laffont, editor, *Aggregation and Revelation of Preferences*. North-Holland, 1979.
31. T. Roughgarden and M. Sundararajan. New trade-offs in cost-sharing mechanisms. In *STOC '06*, pages 79–88.
32. T. Roughgarden and M. Sundararajan. Approximately efficient cost-sharing mechanisms. Technical Report cs.GT/0606127, arXiv, 2006.
33. V. V. Vazirani. *Approximation Algorithms*. Springer, 2001.
34. W. Vickrey. Counterspeculation, auctions, and competitive sealed tenders. *Journal of Finance*, 16(1):8–37, 1961.

The Set Connector Problem in Graphs

Takuro Fukunaga and Hiroshi Nagamochi

Department of Applied Mathematics and Physics,
Graduate School of Informatics, Kyoto University, Japan
{takuro,nag}@amp.i.kyoto-u.ac.jp

Abstract. Given a graph $G = (V, E)$ with an edge cost and families $\mathcal{V}_i \subseteq 2^V$, $i = 1, 2, \ldots, m$ of disjoint subsets, an edge subset $F \subseteq E$ is called a set connector if, for each $\mathcal{V}_i$, the graph $(V, F)/\mathcal{V}_i$ obtained from (V, F) by contracting each $X \in \mathcal{V}_i$ into a single vertex x has a property that every two contracted vertices x and x' are connected in $(V, F)/\mathcal{V}_i$. In this paper, we introduce a problem of finding a minimum cost set connector, which contains several important network design problems such as the Steiner forest problem, the group Steiner tree problem, and the NA-connectivity augmentation problem as its special cases. We derive an approximate integer decomposition property from a fractional packing theorem of set connectors, and present a strongly polynomial 2α-approximation algorithm for the set connector problem, where $\alpha = \max_{1 \leq i \leq m}(\sum_{X \in \mathcal{V}_i} |X|) - 1$.

1 Introduction

Let $G = (V, E)$ be an undirected graph with vertex set V and edge set E. For a family $\mathcal{V} \subseteq 2^V$ of disjoint vertex subsets, we let $G/\mathcal{V}$ stand for the graph obtained from G by contracting each $X \in \mathcal{V}$ into a single vertex x, which is called a $\mathcal{V}$-terminal. As a general concept of the edge connectivity between two vertices, we define the edge-connectivity $\lambda(\mathcal{V}; G)$ for $\mathcal{V} \subseteq 2^V$ as the minimum edge-connectivity of two $\mathcal{V}$-terminals in $G/\mathcal{V}$. If $\mathcal{V}$ consists of two singletons $\{u\}$ and $\{v\}$, then $\lambda(\mathcal{V}; G)$ is equivalent to the edge-connectivity between two vertices u and v.

Let $X \subseteq V$ be a vertex subset. We define $\delta(X)$ as the set of edges in E that join a vertex in X and another in $V - X$, where we let $\delta(V) = \emptyset$ for convenience. We say that X *separates* $\mathcal{V}$ if either $Y \subseteq X$ or $Y \subseteq V - X$ holds for each $Y \in \mathcal{V}$, and $Y \subseteq X \subseteq V - Y'$ for some $Y, Y' \in \mathcal{V}$. We note that $\lambda(\mathcal{V}; G)$ is also defined as $\min\{|\delta(X)| \mid X \subset V \text{ separates } \mathcal{V}\}$.

In this paper, we consider the *set connector problem*, which is defined as follows.

Set connector problem
Given a simple undirected graph $G = (V, E)$, an edge cost $c : E \to \mathbb{Q}_+$, and families $\mathcal{V}_1, \ldots, \mathcal{V}_m \subseteq 2^V$ of disjoint vertex subsets, find a minimum cost edge subset $F \subseteq E$ such that $\lambda(\mathcal{V}_i; G_F) \geq 1$ for $1 \leq i \leq m$, where G_F denotes the graph (V, F).

M. Fischetti and D.P. Williamson (Eds.): IPCO 2007, LNCS 4513, pp. 484–498, 2007.

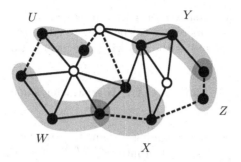

Fig. 1. An instance of the set connector problem with $\mathcal{V}_1 = \{U, W, Z\}$ and $\mathcal{V}_2 = \{U, X, Y\}$, where a set connector F consists of the edges depicted by dashed lines

We call a feasible solution for the set connector problem a *set connector*. Notice that a minimal set connector is a forest. Figure 1 shows an instance $(G, \mathcal{V}_1 = \{U, W, Z\}, \mathcal{V}_2 = \{U, X, Y\})$ of the set connector problem, where the subsets $U, W, X, Y, Z \subseteq V$ are respectively depicted by gray areas, and a set connector F is given by the edges represented by dashed lines.

The set connector problem contains many fundamental problems. For example, it is equivalent to the Steiner forest problem when each $\mathcal{V}_i$ consists of singletons. Besides this, it contains the group Steiner tree problem, which is another generalization of the Steiner tree problem. As will be stated in Section 5, the group Steiner tree problem contains the set cover problem, the tree cover problem, and the terminal Steiner tree problem as its special cases.

Our main contribution of this paper is to present a 2α-approximation algorithm for the set connector problem, where $\alpha = \max_{1 \le i \le m}(\sum_{X \in \mathcal{V}_i} |X|) - 1$. To the best of our knowledge, this is the first approximation algorithm that approximates the Steiner forest problem and the group Steiner tree problem simultaneously. The approximation ratio of our algorithm to the set connector problem matches with the best approximation ratios of several special cases such as the Steiner forest problem, as will be discussed in Section 5. Our algorithm is based on the *approximate integer decomposition property* [5]. A polyhedron P has an f-approximate integer decomposition property for a real $f > 0$ if, for every rational vector $x \in$ P and every integer k such that kx is an integer vector, there exist k integer vectors $x_1, \ldots, x_k \in$ P such that $fkx \ge x_1 + \cdots + x_k$ holds. This property implies that the integrality gap of polyhedron P is at most f for any non-negative cost vector c, since an integer vector x_j attaining $\min\{c^T x_i \mid i = 1, \ldots, k\}$ satisfies $fc^T x \ge c^T x_j$. C. Chekuri and F. B. Shepherd [5] showed the 2-approximate integer decomposition property of an LP relaxation for the Steiner forest problem via the following Steiner packing theorem, which generalizes a well-known spanning tree packing theorem due to Gusfield [12] in Eulerian graphs.

Theorem 1. [5] *Let G be an Eulerian multigraph. Then G contains k edge-disjoint forests $F_1, \ldots, F_k$ such that $\lambda(u, v; G_{F_i}) \ge 1$, $1 \le i \le k$ holds for every two vertices u and v that belong to the same $2k$-edge-connected component in G.* □

For an $|E|$-dimensional real vector x and an edge subset $F \subseteq E$, we let $x(F)$ denote the sum of elements of x corresponding to edges in F. The set connector problem can be formulated as the following integer programming.

minimize $c^T x$
subject to $x(\delta(X)) \geq 1$ for every $X \subset V$ separating some $\mathcal{V}_i \in \{\mathcal{V}_1, \ldots, \mathcal{V}_m\}$
$x \in \{0, 1\}^E$.

Let $\mathrm{LP_{sc}}$ be the linear programming obtained by relaxing the integrality constraint $x \in \{0, 1\}^E$ of this problem into $x \in \mathbb{R}_+^E$, and let $\mathrm{P_{sc}}$ denote its feasible region. For obtaining the 2α-approximate integer decomposition property of $\mathrm{P_{sc}}$, it suffices to show the following set connector packing theorem, which is a generalization of Theorem 1.

Theorem 2. *Let G be an Eulerian multigraph, $\mathcal{V}_1, \ldots, \mathcal{V}_m$ be families of disjoint vertex subsets, and $\alpha = \max_{1 \leq i \leq m}(\sum_{X \in \mathcal{V}_i} |X|) - 1$. If $\lambda(\mathcal{V}_i; G) \geq 2\alpha k$ for $1 \leq i \leq m$, then G contains k edge-disjoint set connectors.* $\square$

The approximate integer decomposition property depends on the fact that $x \in \mathrm{P}$ is a rational vector. Hence we actually prove the following fractional packing theorem instead of Theorem 2. The proof of the theorem can be easily modified to imply Theorem 2.

Theorem 3. *Let $x \in \mathrm{P_{sc}}$ and $\alpha = \max_{1 \leq i \leq m}(\sum_{X \in \mathcal{V}_i} |X|) - 1$ for a simple undirected graph $G = (V, E)$ and families $\mathcal{V}_1, \ldots, \mathcal{V}_m$ of disjoint vertex subsets. Then there exist set connectors $C_1, \ldots, C_k$ with positive weights $w_1, \ldots, w_k$ such that $2\alpha x \geq \sum_{i=1}^k w_i \mathcal{X}_{C_i}$ and $\sum_{i=1}^k w_i = 1$, where $\mathcal{X}_{C_i} \in \{0, 1\}^E$ denotes the incidence vector of C_i.* $\square$

This paper is organized as follows. Section 2 introduces notations and induction techniques. Section 3 provides a proof of Theorem 3, and Section 4 describes a 2α-approximation algorithm for the set connector problem. Section 5 shows applications of the set connector problem. Section 6 concludes this paper with some remarks.

2   Preliminaries

2.1   Notations

Let $\mathbb{R}_+$, $\mathbb{Q}_+$ and $\mathbb{Z}_+$ stand for the sets of non-negative reals, rationals, and integers, respectively. Let $G = (V, E)$ be an undirected graph and $x \in \mathbb{R}_+^E$. For an edge $e \in E$, $x(e)$ denotes the element of x corresponding to e. Let E_x represent the support for x, i.e., $E_x = \{e \in E \mid x(e) > 0\}$. For an edge subset $F \subseteq E$, let $x(F) = \sum_{e \in F} x(e)$, where we define $x(\emptyset) = 0$ for convenience. Let $G_F = (V, F)$, $x_F \in \mathbb{R}_+^F$ denote the projection of x onto F, and $\mathcal{X}_F \in \{0, 1\}^E$ denote the incidence vector of F.

For a vertex subset $U \subseteq V$, $E[U]$ denotes the set of edges whose both end vertices are in U, and $G[U]$ denotes the subgraph $(U, E[U])$ of G induced by U. Moreover $\delta(U)$ represents the set of edges in G that join a vertex in U and another in $V - U$. A singleton set $U = \{u\}$ may be written as u. For a partition $\mathcal{P} = \{V_1, \ldots, V_p\}$ of V into non-empty subsets, $\delta(\mathcal{P})$ denotes $\cup_{i=1}^{p} \delta(V_i)$. For a family $\mathcal{V} \subseteq 2^V$ of disjoint vertex subsets, $G/\mathcal{V}$ denotes a graph obtained by contracting each $X \in \mathcal{V}$ into a single vertex.

In this paper, we often discuss the edge-connectivity of a simple graph $G = (V, E)$ whose edges are weighted by a vector $x \in \mathbb{R}_+^E$. In this case, we assume without loss of generality that G is the complete graph on V by augmenting E with edges $e \in \binom{V}{2} - E$, where we let $x(e) = 0$, $e \in \binom{V}{2} - E$. We denote such an edge-weighted graph by (V, x). We define the edge-connectivity $\lambda(\mathcal{V}; V, x)$ of a family $\mathcal{V} \subseteq 2^V$ of disjoint vertex subsets in (V, x) as $\min\{x(\delta(X)) \mid X \subset V \text{ separates } \mathcal{V}\}$. If $\mathcal{V}$ consists of two elements X and Y, we may denote $\lambda(\mathcal{V}; V, x)$ by $\lambda(X, Y; V, x)$. A k-edge-connected component of (V, x) is an inclusion-wise maximal subset $U \subseteq V$ that satisfies $\lambda(u, v; V, x) \geq k$ for all $u, v \in U$.

For $F \subseteq \binom{V}{2}$ and a positive real w, we let (F, w) stand for a subgraph (V, F) weighted by w. A set of weighted subgraphs $(F_1, w_1), (F_2, w_2), \ldots, (F_k, w_k)$ is called a *fractional forest packing* of an edge-weighted graph (V, x) if F_i is a forest, $1 \leq i \leq k$, $x \geq \sum_{1 \leq i \leq k} w_i \mathcal{X}_{F_i}$, and $\sum_{1 \leq i \leq k} w_i = 1$. Notice that $F_i \subseteq E_x$ holds for $1 \leq i \leq k$ here. If each of $F_1, \ldots, F_k$ is a spanning tree on V (resp., set connector), we especially call it *fractional spanning tree packing* (resp., *fractional set connector packing*). We may simply say that a set of edge subsets $F_1, F_2, \ldots, F_k$ is a fractional forest packing of (V, x) if there are weights $w_1, w_2, \ldots, w_k$ such that (F_i, w_i), $i = 1, 2, \ldots, k$ is a fractional forest packing of (V, x).

2.2 Induction Techniques

In this subsection, we review graph operations called *contraction* and *splitting*. In this paper, we use these operations in order to prove some claims inductively.

First, let us see the contraction. *Contracting* a vertex set $S \subseteq V$ into a single vertex s means that S is replaced by s, resultant loops are deleted, and one end vertex of every edge in $\delta(S)$ is changed from a vertex in S to s. Let V' denote the vertex set obtained by the contraction, i.e., $V' = (V - S) \cup s$. If we execute the contraction in (V, x), then x is modified into $x' \in \mathbb{R}_+^{\binom{V'}{2}}$ so that $x'(e) = x(e)$ for each $e \in \binom{V'}{2} - \delta(s)$, and $x'(e) = \sum_{u \in S} x(uv)$ for each $e = sv \in \delta(s)$.

Lemma 1. *Let (V', x') be an edge-weighted undirected graph obtained from (V, x) by contracting $S \subseteq V$ into a single vertex s. If there exists a fractional forest packing $\mathcal{C}'$ of (V', x'), then we can obtain a fractional forest packing $\mathcal{C}$ of (V, x), every forest in which consists of edges in $E_x - \binom{S}{2}$. Every two vertices in $V - S$ connected by all forests in $\mathcal{C}'$ are also connected by the union of every forest in $\mathcal{C}$ and every spanning tree on S.* □

Next, let us see the splitting. *Splitting* a pair $\{sa, sb\}$ of edges by $\epsilon > 0$ is an operation that decreases $x(sa)$ and $x(sb)$ by ϵ and increases $x(ab)$ by ϵ, where possibly $a = b$ and ϵ is supposed to be at most $\min\{x(su), x(sv)\}$. Historically, this operation was introduced by L. Lovász to study the edge-connectivity of multigraphs G (i.e., $x \in \mathbb{Z}_+^E$ and $\epsilon \in \mathbb{Z}_+$). W. Mader [21] showed that if $|\delta(s; G)| \neq 3$, there always exists a pair of edges incident to s such that splitting them by 1 preserves the edge-connectivity between every two vertices in $V - s$. Furthermore, A. Frank [7] showed that for any edge incident to s, there always exists such a pair that contains the edge if $|\delta(s; G)|$ is even. A proof of his theorem uses the fact that splitting $\{sa, sb\}$ by a real $\epsilon \in \mathbb{R}_+$ preserves the edge-connectivity between every two vertices in $V - s$ if and only if

$$\epsilon \leq \frac{1}{2} \min\{x(\delta(X)) - \lambda(u, v; V, x) \mid a, b, u \in X \subseteq V - (s \cup v), s \neq v\}. \tag{1}$$

This fact can be derived from the observation that splitting $\{sa, sb\}$ by ϵ decreases $x(\delta(X))$ by 2ϵ if $a, b \in X \subset V - s$, and does not change $x(\delta(X))$ otherwise. We let $q_x(a, b)$ denote the right hand side of inequality (1), and $\epsilon_{a,b} = \min\{x(sa), x(sb), q_x(a, b)\}$. Notice that $\epsilon_{a,b}$ can be regarded as the maximum value such that splitting $\{sa, sb\}$ by $\epsilon_{a,b}$ preserves the edge-connectivity between every two vertices in $V - s$.

In this paper, we use the splitting in order to isolate a vertex $s \in V$ in (V, x), i.e., $x(sv) = 0$ for every $v \in V - s$. A *complete splitting* at s denotes an operation that isolates s by repeating splitting edges incident to s. The following theorem tells that it always can be executed in strongly polynomial time while preserving the edge-connectivity between every two vertices in $V - s$.

Theorem 4. *Let s be an arbitrary vertex in (V, x). There exists a complete splitting at s such that $\lambda(u, v; V, x) = \lambda(u, v; V - s, x')$ holds for every $u, v \in V - s$, where $x' \in \mathbb{R}_+^{\binom{V-s}{2}}$ is the resulting edge weight from the complete splitting. Such a complete splitting can be found in strongly polynomial time.* □

We note that splitting every pair $\{sa, sb\}$ of edges incident to s by $\epsilon_{a,b}$ gives the complete splitting in the theorem. The strong polynomiality of the complete splitting comes from the fact that $\epsilon_{a,b}$ can be computed in strongly polynomial time. We use the splitting for the induction as described below.

Lemma 2. *Let $x' \in \mathbb{R}_+^{\binom{V-s}{2}}$ be the edge-weight obtained from $x \in \mathbb{R}_+^{\binom{V}{2}}$ by a complete splitting at s. If there exists a fractional forest packing $\mathcal{C}'$ of (V, x'), then we can construct a fractional forest packing $\mathcal{C}$ of (V, x). Every two vertices in $V - s$ connected by all forests in $\mathcal{C}'$ are also connected by all in $\mathcal{C}$.* □

3   Proof of the Fractional Set Connector Packing Theorem

In this section, we give a proof of Theorem 3. First of all, let us review a fractional version of Tutte's tree packing theorem [25].

Theorem 5. [25] *Let $G = (V, x)$ be an edge-weighted undirected graph. Then there exits a fractional spanning tree packing of G if and only if*

$$x(\delta(\mathcal{P})) \geq |\mathcal{P}| - 1 \text{ for every partition } \mathcal{P} \text{ of } V \text{ into nonempty classes.} \quad (2)$$

$\square$

We can derive the following lemma from the above theorem.

Lemma 3. *Let $G = (V, x)$ be an edge-weighted undirected graph, and $K \subset V$ be an inclusion-wise minimal subset such that $x(\delta(K)) < 2$. Then there exists a fractional spanning tree packing of $(K, x_{\binom{K}{2}})$.*

Proof. We show that (2) holds for graph $(K, x_{\binom{K}{2}})$. Let $\mathcal{P}$ be a partition of K into nonempty classes. Then for any $X \in \mathcal{P}$ (i.e., $X \subset K$), it holds that $x(\delta(X)) \geq 2$ by the minimality of K. Therefore $x_{\binom{K}{2}}(\delta(\mathcal{P})) = (\sum_{X \in \mathcal{P}} x(\delta(X)) - x(\delta(K)))/2 > |\mathcal{P}| - 1$ holds. Then by applying Theorem 5 to $(K, x_{\binom{K}{2}})$, we can obtain a fractional spanning tree packing of $(K, x_{\binom{K}{2}})$. $\square$

To prove Theorem 3, we use a result on the Steiner forest packing due to C. Chekuri and F. B. Shepherd [5]. Here we state a fractional packing version of Theorem 1. The proof is based on that of C. Chekuri and F. B. Shepherd [5].

Theorem 6. *Let $G = (V, x)$ be an edge-weighted undirected graph. Then there exists a fractional forest packing $\mathcal{C}$ of G such that $\lambda(u, v; G_F) \geq 1$ for every $F \in \mathcal{C}$ and $u, v \in V$ with $\lambda(u, v; V, x) \geq 2$.*

Proof. We prove this theorem by an induction on the number N of 2-edge-connected components in (V, x). First, let us consider the case of $N = 1$. Then for any nonempty $X \subset V$, it holds that $x(\delta(X)) \geq 2$, which implies that (2) holds for x because $x(\delta(\mathcal{P})) = \sum_{X \in \mathcal{P}} x(\delta(X))/2 \geq |\mathcal{P}|$. Therefore, we can obtain a required fractional forest packing by Theorem 5.

Next, consider the case of $N \geq 2$. Let $K \subset V$ be an inclusion-wise minimal subset such that $x(\delta(K)) < 2$ (such K exists since the edge-connectivity between two vertices in different components is less than 2). Then K is the union of some 2-edge-connected components. By Lemma 3, there exists a fractional spanning tree packing $\{(T_i, \beta_i) \mid 1 \leq i \leq p\}$ of $(K, x_{\binom{K}{2}})$. Let $G' = (V' = (V - K), x' \in \binom{V'}{2})$ be the graph obtained by contracting K into a single vertex v_K, executing the complete splitting at v_K, and removing isolated v_K. Note that any two vertices $u, v \in V'$ that belong to the same 2-edge-connected component in (V, x) remains 2-edge-connected in (V', x').

By the inductive hypothesis, (V', x') has a fractional forest packing $\{(H_i, \theta_i) \mid 1 \leq i \leq q\}$ such that each of $H_1, \ldots, H_q$ connects every two vertices $u, v \in V'$ with $\lambda(u, v; V', x') \geq 2$ (and hence $\lambda(u, v; V, x) \geq 2$). Let $\{(H_i', \theta_i') \mid i = 1, \ldots, q'\}$ be the fractional forest decomposition of (V, x) obtained from $\{(H_i, \theta_i) \mid i = 1, \ldots, q\}$ by applying Lemmas 1 and 2. Then clearly $\{(T_i \cup H_j', \beta_i \theta_j') \mid 1 \leq i \leq p, 1 \leq j \leq q'\}$ is a required fractional forest packing. $\square$

Lemma 4. *Let $G = (V, x)$ be an edge-weighted undirected graph, and $\mathcal{V}_1, \ldots, \mathcal{V}_m \subseteq 2^V$ be families of disjoint vertex subsets such that $\lambda(\mathcal{V}_i; V, x) \geq 2(\sum_{X \in \mathcal{V}_i} |X|) - 2$ for every $i = 1, \ldots, m$. If $F \subseteq \binom{V}{2}$ satisfies $\lambda(u, v; G_F) \geq 1$ for all $u, v \in V$ with $\lambda(u, v; V, x) \geq 2$, then F is a set connector for $\mathcal{V}_1, \ldots, \mathcal{V}_m$.*

Proof. Consider a family $\mathcal{V}_i \in \{\mathcal{V}_1, \ldots, \mathcal{V}_m\}$, and let $\{\{X_1, \ldots, X_q\}, \{Y_1, \ldots, Y_r\}\}$ be a partition of $\mathcal{V}_i$ into two classes. We denote $\cup_{j=1}^q X_j$ by X and $\cup_{j=1}^r Y_j$ by Y. In the following, we show that there exists two vertices $u \in X$ and $v \in Y$ with $\lambda(u, v; G, x) \geq 2$. This implies the lemma since an edge set that connects such vertices is a set connector in this case.

Now we suppose conversely that $\lambda(u, v; V, x) < 2$ holds for every $u \in X$ and $v \in Y$. We construct a partition $\mathcal{P}$ of V and a family $\mathcal{Q} \subseteq 2^V$ of vertex subsets as follows. First we set $\mathcal{P} = \{V\}$ and $\mathcal{Q} = \emptyset$. Let us consider the moment at which some two vertices $u \in X$ and $v \in Y$ belong to the same class of $\mathcal{P}$. Then choose $W \subset V$ such that $u \in W$, $v \in V - W$ and $x(\delta(W)) < 2$ (such W exists since $\lambda(u, v; G, x) < 2$) and update $\mathcal{P} := \cup_{Z \in \mathcal{P}}\{Z \cap W, Z - W\}$ and $\mathcal{Q} := \mathcal{Q} \cup \{W\}$. Repeat this procedure until every two vertices in X and in Y belong to different classes of $\mathcal{P}$.

We can see that the number of the repetitions is at most $|X| + |Y| - 1$ by the induction on $|X| + |Y|$ as follows. Let W be chosen as a member of $\mathcal{Q}$ after running the procedure once. For separating vertices in $W \cap X$ from those in $W \cap Y$, at most $|W \cap (X \cup Y)| - 1$ repetitions are enough by the inductive hypothesis. Similarly, at most $|(X \cup Y) - W| - 1$ repetitions separates vertices in $X - W$ from those in $Y - W$. Since W separates vertices in W from those in $(X \cup Y) - W$, the number of the repetitions is at most $1 + (|W \cap (X \cup Y)| - 1) + (|(X \cup Y) - W| - 1) = |X| + |Y| - 1$.

From this fact, $|\mathcal{Q}| \leq |X| + |Y| - 1 = (\sum_{Z \in \mathcal{V}_i} |Z|) - 1$ holds. Moreover, we can see that $\delta(\mathcal{P}) \subseteq \cup_{W \in \mathcal{Q}} \delta(W)$. Now let $U = \cup_{j=1}^p V_j$, where $V_1, \ldots, V_p$ be the classes of $\mathcal{P}$ that contain vertices in X. Notice that U separates $\mathcal{V}_i$. Since $x(\cup_{W \in \mathcal{Q}} \delta(W)) < 2|\mathcal{Q}| \leq 2(\sum_{Z \in \mathcal{V}_i} |Z|) - 2$, it holds that $x(\delta(U)) \leq x(\cup_{j=1}^p \delta(V_j)) \leq x(\delta(\mathcal{P})) < 2(\sum_{Z \in \mathcal{V}_i} |Z|) - 2$. These facts imply that $\lambda(\mathcal{V}_i; V, x) < 2(\sum_{Z \in \mathcal{V}_i} |Z|) - 2$, a contradiction. $\square$

Now we are ready to prove Theorem 3. In the proof, we show the following observation together with Theorem 3.

Observation 1. *Set connectors in Theorem 3 can be given as forests connecting all vertices in each 2-edge-connected component of $(V, 2\alpha x)$.* $\square$

Proof (**Proof of Theorem 3 and Observation 1**). Since $x \in P_{sc}$, we see that $\lambda(\mathcal{V}_i; V, x) = \min\{x(\delta(X)) \mid X \text{ separates } \mathcal{V}_i\} \geq 1$ holds for every $1 \leq i \leq m$. Therefore, $\lambda(\mathcal{V}_i; G, 2\alpha x) = 2\alpha\lambda(\mathcal{V}_i; G, x) \geq 2\alpha \geq 2(\sum_{X \in \mathcal{V}_i} |X| - 1)$ holds for $1 \leq i \leq m$. By Lemma 4, at least one pair $\{u, v\}$ of vertices $u \in \cup_{i=1}^q X_i$ and $v \in \cup_{i=1}^r Y_i$ is contained in the same 2-edge-connected component of $(V, 2\alpha x)$ for any partition $\{\{X_1, \ldots, X_q\}, \{Y_1, \ldots, Y_r\}\}$ of $\mathcal{V}_i$ into two classes. Hence every forest that connects all vertices in each 2-edge-connected component of $(V, 2\alpha x)$ is a set connector.

By Theorem 6, there exist a fractional forest packing $\{F_1, \ldots, F_k\}$ of $(V, 2\alpha x)$ such that every two vertices $u, v \in V$ with $\lambda(u, v; V, 2\alpha x)$ are connected by each of $F_1, \ldots, F_k$. By the above observation, this is a desired factional set connector packing.   □

As a corollary of Theorem 3, we can see that the integrality gap of LP_{sc} is at most 2α.

Corollary 1. *For any vectors $x \in P_{\text{sc}}$ and $c \in \mathbb{Q}_+^E$, there always exists a set connector $F \subseteq E$ such that $2\alpha c^T x \geq c(F)$. Such F can be given as a forest connecting all vertices in each 2-edge-connected component of $(V, 2\alpha x)$.*   □

This gap is tight in the following instance. Given an integer $d \geq 1$, let $G = (V, E)$ be the complete graph on a vertex set V of cardinality $n > 2d$, and $c(e) = 1$ for all $e \in E$. Moreover specify a vertex $s \in V$ and define $\mathcal{V}_1, \ldots, \mathcal{V}_m$ as the families $\{\{s\}, U\}$ for all subsets $U \subseteq V - s$ with $|U| = \alpha$, where $m = \binom{|V|-1}{\alpha}$. In this instance, $\alpha = \max_{1 \leq i \leq m}(\sum_{X \in \mathcal{V}_i} |X|) - 1$ holds.

Define a rational vector $x \in \mathbb{Q}_+^E$ as $x(e) = 1/(n-1)$ if e is incident to s, and $x(e) = 1/(a(n-1))$ otherwise. Then we can verify that $x \in P_{\text{sc}}$ holds. Hence the optimal cost of rational solutions is at most $c^T x = (n-1)/(n-1) + \binom{n-1}{2}/(\alpha(n-1)) = (n + 2\alpha - 2)/(2\alpha)$. On the other hand, let us consider an optimal integral solution $F \subseteq E$. Consider the connected component S that contains s in G_F. If $|S| < n - \alpha + 1$, i.e., $|V - S| \geq \alpha$, then $0 = \delta(S; G_F) \geq \lambda(\mathcal{V}_i; G_F)$ would hold for some $\mathcal{V}_i = \{\{s\}, U\}$ with a set $U \subseteq V - S$. Hence $|S| \geq n - \alpha + 1$. By this, $|F| \geq |S| - 1 \geq n - \alpha + 1 - 1 = n - \alpha$. Therefore the integrality gap of this instance is

$$\frac{\text{The optimal cost of integer solutions}}{\text{The optimal cost of rational solutions}} \geq \frac{c(F)}{c^T x} \geq \frac{n - \alpha}{(n + 2\alpha - 2)/(2\alpha)}.$$

We can see that the most right term approaches 2α as n gets larger.

4   Approximation Algorithm

In Corollary 1, we saw that any vector $x \in P_{\text{sc}}$ can be rounded to a set connector F with $c(F) \leq 2\alpha c^T x$, and that such F can be given as a forest connecting all vertices in each 2-edge-connected component of $(V, 2\alpha x)$ (i.e., $1/\alpha$-edge-connected component of (V, x)). Hence by applying a ρ-approximation algorithm of the Steiner forest problem to constructing such a forest in G, we have a $2\alpha\rho$-approximation algorithm for the set connector problem, where currently $\rho \leq 2$ is known [11]. However, the arguments in Section 3 indicate a 2α-approximation algorithm for the set connector problem. In this section, we describe this.

In the first step, our algorithm computes an optimal solution x of LP_{sc} for the given instance consisting of $G = (V, E)$, $c \in \mathbb{Q}_+^E$, and $\mathcal{V}_1, \ldots, \mathcal{V}_m \subseteq 2^V$. We then augment x into $\mathbb{R}_+^{\binom{V}{2}}$ by adding 0's, and c into $\mathbb{Q}_+^{\binom{V}{2}}$ by adding $+\infty$'s if G is not complete. Then our algorithm constructs a forest $F \subseteq E_x \subseteq E$ that connects all vertices in each $1/\alpha$-edge-connected component of (V, x) as follows.

Let $K \subseteq V$ be an inclusion-wise minimal vertex set such that $x(\delta(K)) < 1/\alpha$. Recall that the proof of Theorem 6 computes a fractional tree packing of $(K, x_{\binom{K}{2}})$ by applying Lemma 3. Instead of this, our algorithm computes a minimum cost tree $T_K \subseteq E_x \cap E[K]$ spanning K. Then we contract K into a single vertex v_K, and execute a complete splitting at v_K. When our algorithm executes contraction or splitting, it modifies the edge cost simultaneously. After this, it recursively computes a sequence of trees in the resulting edge-weighted graph and edge cost until the vertex set becomes a singleton. As reverse operations of contraction and splitting, our algorithm modifies the forest and output the sum of T_K and the modified forest as a solution. Below, we describe how to modify the edge cost and how to modify the forest in the reverse operations of contraction and splitting.

First, let us consider the contraction. Let $x' \in \mathbb{R}_+^{\binom{V'}{2}}$ be the vector obtained from x by the contracting K into v_K, where $V' = (V - K) \cup v_K$. Together with this contraction, our algorithm modifies edge cost c into $c' \in \mathbb{Q}_+^{\binom{V'}{2}}$ so that $c'(uv_K) = \min\{c(us) \mid s \in K, x(us) > 0\}$ for each $u \in V - K$ and $c'(uv) = c(uv)$ for each $u, v \in V - K$. Suppose our algorithm has computed a forest $F' \subseteq E_{x'}$ for (V', x') and c'. Then it constructs a forest $F \subseteq E_x - \binom{K}{2}$ for (V, x) and c from F' in the reverse operation of the contraction as follows. If F' contains no edge in $\delta(v_K)$, we set F to F'. Otherwise, prepare an edge uv such that $c(uv) = c'(uv_K)$ for each $uv_K \in F' \cap \delta(v_K)$, and let F'' be the set of those edges. Then F is defined as $(F' - \delta(v_K)) \cup F''$. Notice that $c(F) = c'(F')$ holds. Moreover, $F' \cup T_K$ connects every two vertices connected by F.

Next, let us consider the splitting. Let $x' \in \mathbb{R}_+^{\binom{V'}{2}}$ be the vector from $x \in \mathbb{R}_+^{\binom{V'}{2}}$ by splitting a pair $\{v_K a, v_K b\}$ of edges by $\epsilon_{v_K a, v_K b} > 0$ in the complete splitting at v_K. Together with this splitting, our algorithm modifies the edge cost c into a new cost $c' \in \mathbb{Q}_+^{\binom{V'}{2}}$ so that $c'(ab) = \min\{c(ab), c(v_K a) + c(v_K b)\}$ if $x(ab) > 0$ and $c'(ab) = c(v_K a) + c(v_K b)$ otherwise while $c'(e) = c(e)$ for $e \in \binom{V'}{2} - ab$. Suppose our algorithm has computed a forest $F' \subseteq E_{x'}$ for (V', x') and c'. Then it constructs a forest $F \subseteq E_x$ for (V, x) and c from F' in the reverse operation of the splitting as follows. If $c'(ab) = c(v_K a) + c(v_K b)$, then F is set to $(F' - ab) \cup \{v_K a, v_K b\}$ Otherwise, F is set to F'. Notice that $c(F) = c'(F')$ holds in both cases.

We note that the reverse operation of contraction and splitting described above can be easily executed by maintaining $p(e)$ for each $e \in \binom{V}{2}$. At the beginning of our algorithm, $p(e)$ is set to $\{e\}$. Our algorithm then updates $p(uv_K) := p(uv)$ when a set K containing v is contracted into v_K and $c'(uv_K)$ is defined as $c(uv)$, and $p(ab) := p(v_K a) \cup p(v_K b)$ when a pair $\{v_K a, v_K b\}$ is split and $c'(ab)$ is updated to $c(v_K a) + c(v_K b)$. Observe that $\cup_{e \in F'} p(e)$ represents the edge set constructed from a forest F' in both reverse operations.

Now we are ready to see the entire algorithm. The following describes how to compute a solution after an optimal solution x of $\mathrm{LP_{sc}}$ is given.

Algorithm SETCONNECT

Input: A vertex set V, a vector $x \in \mathbb{R}_+^{\binom{V}{2}}$, and an edge cost $c \in \mathbb{Q}_+^{\binom{V}{2}}$

Output: A forest $F \subseteq \binom{V}{2}$

1: $K :=$ an inclusion-wise minimal $X \subseteq V$ with $x(\delta(X)) < 1/\alpha$;
2: Compute a minimum cost tree $T_K \subseteq E_x \cap \binom{K}{2}$ spanning K; # possibly $|K| = 1$ or $K = V$
3: **if** $|V| - 1 \le |K| \le |V|$ **then**
4:     Return $F := T_K$ as a solution and halt
5: **end if**;
       # contract K into v_K
6:     $c' := c$;
7:     For each $e \in \binom{V}{2}$, define $p(e) := \{e\}$;
8:     $V' := (V - K) \cup v_K$; $x'_{\binom{V-K}{2}} := x_{\binom{V-K}{2}}$;
9: **for** $u \in V - K$ **do**
10:     $x'(uv_K) := \sum_{v \in K} x(uv)$;
11:     **if** $x'(uv_K) > 0$ **then**
12:         $e :=$ an edge attaining $\min\{c(uv) \mid v \in K, x(uv) > 0\}$;
13:         $c'(uv_K) := c(e)$; $p(uv_K) := p(e)$
14:     **end if**
15: **end for**;
       # complete splitting at v_K
16: **for** distinct $a, b \in V' - v_K$ **do**
17:     Compute $\epsilon_{a,b}$ in (V', x');
18:     **if** $\epsilon_{a,b} > 0$ **and** $x'(ab) = 0$ or $c'(ab) > c'(v_K a) + c'(v_K b)$ **then**
19:         $c'(ab) := c'(v_K a) + c'(v_K b)$; $p(ab) := p(v_K a) \cup p(v_K b)$
20:     **end if**;
21:     $x'(v_K a) := x'(v_K a) - \epsilon_{a,b}$; $x'(v_K b) := x'(v_K b) - \epsilon_{a,b}$; $x'(ab) := x'(ab) + \epsilon_{a,b}$
22: **end for**;
23: $V' := V' - v_K$;
24: $F' :=$ A solution output by SETCONNECT applied to V', $x'_{\binom{V'}{2}}$ and c';
25: Return $F := T_K \cup_{e \in F'} p(e)$ as a solution;

Theorem 7. *The set connector problem can be approximated within factor of 2α by applying algorithm SETCONNECT to an optimal solution x of* $\mathrm{LP_{sc}}$.

Proof. connecting all vertices in a $1/\alpha$-edge-connected component of (V, x) by the induction on $|V|$, the combination of which and Lemma 4 implies that F is a set connector for G and $\mathcal{V}_1, \ldots, \mathcal{V}_m$.

By the choice of T_K, it holds that $T_K \subseteq E_x$. By the induction hypothesis, $F' \subseteq E_{x'}$, and then $\cup_{e \in F'} p(e) \subseteq E_x$. Since $F = T_K \cup_{e \in F'} p(e)$, $F \subseteq E_x$ holds. On the other hand, let u and v be two vertices in V such that $\lambda(u, v; V, x) \ge 1/\alpha$. Then these are contained either in K or in $V - K$ during the algorithm. If

$u, v \in K$, these are connected by F since F contains a tree T_K spanning K. In what follows, we suppose that $u, v \in V - K$. Let x' represent the vector maintained in the end of the algorithm. Since contracting K into v_K and the complete splitting at v_K does not decrease the edge-connectivity between u and v, it follows that $\lambda(u, v; V', x'_{\binom{V'}{2}}) \geq 1/\alpha$. By the inductive hypothesis, F' connects u and v, and thereby $F = T_K \cup_{e \in F'} p(e)$ connects such u and v.

Next, let $\{(C_i, w_i) \mid i = 1, \ldots, k\}$ be a fractional set connector packing of $(V, 2\alpha x)$ and $\mathcal{V}_1, \ldots, \mathcal{V}_m$ appeared in Theorem 3. In the following, we show that $c(F) \leq c(C_i)$ for every $i = 1, \ldots, k$ by the induction on $|V|$ again. This implies that F is a 2α-approximate solution for the set connector problem.

Recall that the proof of Theorem 3 constructs C_i as the union of T and $\cup_{e \in H} p(e)$, where $T \subseteq E_x \cap \binom{K}{2}$ is a spanning tree on K and $H \subseteq E_{x'} \cap \binom{V-K}{2}$ is a forest in a fractional forest packing of (V', x'). By the choice of T_K, obviously $c(T_K) \leq c(T)$ holds. On the other hand, $c'(F') \leq c'(H)$ by the inductive hypothesis. As observed in the above, it holds that $c'(F') = c(\cup_{e \in F'} p(e))$ and $c'(H) = c(\cup_{e \in H} p(e))$. Since $F = T_K \cup_{e \in F'} p(e)$ and $C_i = T \cup_{e \in H} p(e)$, we have obtained $c(F) \leq c(C_i)$. $\qquad\square$

We note that running time of algorithm SETCONNECT is strongly polynomial, where we use Tardos' algorithm [27] to solve LP_{sc}. All steps of algorithm SETCONNECT except solving LP_{sc} are combinatorial.

5   Applications

In this section, we review some problems related to the set connector problem.

5.1   NA-Connectivity

Here we mention the prior works on the *node to area connectivity* (*NA-connectivity*). H. Ito [16] considered the edge-connectivity $\lambda(v, X)$ between a vertex $v \in V$ and a vertex subset $X \subseteq V$, and called it NA-connectivity. Then augmentation-type problem of NA-connectivity was considered by some researchers [15,17,22]. For example, the following problem was shown to be NP-hard by H. Miwa and H. Ito [22].

1-NA-connectivity augmentation problem
Given an undirected graph $G = (V, E)$ and a family $\mathcal{V} \subseteq 2^V$, find an edge set $F \subseteq \binom{V}{2} - E$ of minimum cardinality such that $\lambda(v; X; G_{E \cup F}) \geq 1$ holds for all $X \in \mathcal{V}$ and $v \in V - X$.

By using an algorithm due to Z. Nutov [23], this problem can be approximated within $7/4$.

The edge-connectivity for a family of vertex subsets we defined in this paper generalizes the NA-connectivity since $\lambda(v, X; G) = \lambda(\mathcal{V}_X; G)$ holds if we set $\mathcal{V}_X = \{\{v\}, X\}$ for $X \in \mathcal{V}$. Hence the above augmentation problem is contained in the set connector problem even if it is generalized so that an edge cost $c : \binom{V}{2} - E \to \mathbb{Q}_+$ is also given and $c(F)$ is minimized.

Theorem 8. *The 1-NA-connectivity augmentation problem with an edge cost can be approximated within a factor of* $2 \max_{X \in \mathcal{V}} |X|$. □

5.2 Steiner Forest Problem

The Steiner forest problem is formulated as follows.

Steiner forest problem

Given an undirected graph $G = (V, E)$ and disjoint vertex subsets $X_1, \ldots, X_\ell \subseteq V$, find a minimum cost edge set $F \subseteq E$ that connects every two vertices in X_i for every $i = 1, \ldots, \ell$.

The Steiner forest problem can be formulated as the set connector problem by setting each family $\mathcal{V}_i$ of vertex subsets as $\{\{u\}, \{v\}\}$, where $u, v \in X_j, j = 1, \ldots, \ell$. Our algorithm to the set connector problem attains the approximation factor of $2\alpha = 2$, which coincides with the prior best result on the Steiner forest problem [11].

5.3 Group Steiner Tree Problem

The group Steiner tree problem is a generalization of the Steiner tree problem. It is formulated as follows.

Group Steiner tree problem

Given an undirected graph $G = (V, E)$, an edge cost $c : E \to \mathbb{Q}_+$, and a family $\mathcal{U} \subseteq 2^V$ of vertex subsets, find a minimum cost tree $T \subseteq E$ which spans at least one vertex in every $X \in \mathcal{U}$.

The group Steiner tree problem was introduced by G. Reich and P. Widmayer [24]. Their motivation came from the wire routing with multi-port terminals in VLSI design. After their work, it turned out that this problem has a close relationship with the Steiner tree problem both in undirected graphs and in directed graphs [13,28]. In addition to the Steiner tree problem, the problem is known to generalize several important other problems such as the tree cover problem [1,8,9,18], the terminal (full) Steiner tree problem [6,19,20], and the set cover problem. Especially a reduction from the set cover problem implies that the group Steiner tree problem does not admit any approximation factor of $(1 - o(1)) \ln m$ unless NP $\subseteq$ DTIME$(n^{\log \log n})$, where $m = |\mathcal{U}|$ and $n = |\cup_{X \in \mathcal{U}} X|$. Besides this, E. Halperin and R. Krauthgamer [14] proved that the group Steiner tree problem is hard to approximate within a factor better than $\Omega(\log^{2-\epsilon} m)$ for every $\epsilon > 0$ unless NP problems have quasi-polynomial time Las-Vegas algorithms. On the other hand, a $(1 + \ln m/2)\sqrt{m}$-approximation algorithm was proposed by C. D. Bateman et. al. [2]. Currently the best approximation factors are $O(\log m \log |V| \log N)$ due to [3,4,10], and $2N(1 - 1/n)$ due to P. Slavík [26], where $N = \max_{X \in \mathcal{U}} |X|$.

Although the set connector problem resembles the group Steiner tree problem, they are different in the fact that the set connectors may be forests. However, the group Steiner tree problem can be reduced to the set connector problem as follows. Pick up a designated subset $S \in \mathcal{U}$. For each $s \in S$, run the algorithm of the set connector problem for the instance with G, c, and $\mathcal{V}_U = \{s, U\}$, $U \in \mathcal{U} - S$. Then this provides the approximation factor of $2\alpha = 2N$. This approximation factor almost coincides with Slavík's result [26].

Theorem 9. *The group Steiner tree problem can be approximated within a factor of* $2 \max_{X \in \mathcal{V}} |X|$. $\qquad\qquad\qquad\qquad\qquad\qquad\qquad\qquad\qquad\qquad\qquad\quad$ $\square$

6   Concluding Remarks

In this paper, we have introduced the set connector problem as an important generalization of previously known fundamental problems such as the Steiner forest problem, and have presented a 2α-approximation algorithm to the problem, where $\alpha = \max_{1 \le i \le m}(\sum_{X \in \mathcal{V}_i} |X|) - 1$. Our algorithm is based on the 2α-approximate integer decomposition property, which is proven via the set connector decomposition theorem.

Some problems remain open yet. One is whether the set connector problem admits the approximation factor better than 2α. In the example presented in Section 4 for the tightness of the integrality gap, $\mathcal{V}_i$ consists of two vertex subsets one of which is always singleton. Hence this does not deny the possibility of a better approximation factor than $2 \max_{1 \le i \le m} \max_{X \in \mathcal{V}_i} |X|$. Constructing combinatorial approximation algorithms for the set connector problem is also an interesting and important issue.

Acknowledgement

This research was partially supported by the Scientific Grant-in-Aid from Ministry of Education, Culture, Sports, Science and Technology of Japan.

References

1. E. M. Arkin, M. M. Halldórsson, and R. Hassin, *Approximating the tree and tour covers of a graph*, Information Processing Letters, 47 (1993), pp. 275–282.
2. C. D. Bateman, C. S. Helvig, G. Robins, and A. Zelikovsky, *Provably good routing tree construction with multi-port terminals*, in Proceedings of the 1997 International Symposium on Physical Design, 1997, pp. 96–102.
3. M. Charikar, C. Chekuri, A. Goel, and S. Guha, *Rounding via trees: deterministic approximation algorithms for group Steiner trees and k-median*, in Proceedings of the thirtieth Annual ACM Symposium on Theory of Computing, 1998, pp. 114–123.
4. C. Chekuri, G. Even, and G. Kortsarzc, *A greedy approximation algorithm for the group Steiner problem*, Discrete Applied Mathematics, 154 (2006), pp. 15–34.
5. C. Chekuri and F. B. Shepherd, *Approximate integer decompositions for undirected network design problems*. Manuscript, 2004.

6. D. E. Drake and S. Hougardy, *On approximation algorithms for the terminal Steiner tree problem*, Information Processing Letters, 89 (2004), pp. 15–18.

7. A. Frank, *On a theorem of Mader*, Discrete Mathematics, 191 (1992), pp. 49–57.

8. T. Fujito, *On approximability of the independent/connected edge dominating set problems*, Information Processing Letters, 79 (2001), pp. 261–266.

9. T. Fujito, *How to trim an MST: A 2-approximation algorithm for minimum cost tree cover*, in Proceedings of Automata, Languages and Programming, 33rd International vol. 4051 of Lecture Notes in Computer Science, Venice, Italy, 2006, pp. 431–442.

10. N. Garg, G. Konjevod, and R. Ravi, *A polylogarithmic approximation algorithm for the group Steiner tree problem*, Journal of Algorithms, 37 (2000), pp. 66–84.

11. M. X. Goemans and D. P. Williamson, *A general approximation technique for constrained forest problems*, SIAM Journal on Computing, 24 (1995), pp. 296–317.

12. D. Gusfield, *Connectivity and edge-disjoint spanning trees*, Information Processing Letters, 16 (1983), pp. 87–89.

13. E. Halperin, G. Kortsarz, R. Krauthgamer, A. Srinivasan, and N. Wang, *Integrality ratio for group Steiner trees and directed steiner trees*, in Proceedings of the Fourteenth Annual ACM-SIAM Symposium on Discrete Algorithms, 2003, pp. 275–284.

14. E. Halperin and R. Krauthgamer, *Polylogarithmic inapproximability*, in Proceedings of the Thirty-Fifth Annual ACM Symposium on Theory of Computing, 2003, pp. 585–594.

15. T. Ishii and M. Hagiwara, *Augmenting local edge-connectivity between vertices and vertex subsets in undirected graphs*, in Proceedings of 28th International Symposium on Mathematical Foundations of Computer Science, vol. 2747 of Lecture Notes in Computer Science, 2003, pp. 490–499.

16. H. Ito, *Node-to-area connectivity of graphs*, Transactions of the Institute of Electrical Engineers of Japan, 11C (1994), pp. 463–469.

17. H. Ito and M. Yokoyama, *Edge connectivity between nodes and node-subsets*, Networks, 31 (1998), pp. 157–164.

18. J. Könemann, G. Konjevod, O. Parekh, and A. Sinha, *Improved approximations for tour and tree covers*, Algorithmica, 38 (2004), pp. 441–449.

19. G. Lin and G. Xue, *On the terminal Steiner tree problem*, Information Processing Letters, 84 (2002), pp. 103–107.

20. C. L. Lu, C. Y. Tang, and R. C.-T. Lee, *The full Steiner tree problem*, Theoretical Computer Science, 306 (2003), pp. 55–67.

21. W. Mader, *A reduction method for edge-connectivity in graphs*, Annals of Discrete Mathematics, 3 (1978), pp. 145–164.

22. H. Miwa and H. Ito, *Edge augmenting problems for increasing connectivity between vertices and vertex subsets*, in 1999 Technical Report of IPSJ, vol. 99-AL-66, 1999, pp. 17–24.

23. Z. Nutov, *Approximating connectivity augmentation problems*, in Proceedings of the 16th Annual ACM-SIAM Symposium on Discrete Algorithms, 2005, pp. 176–185.

24. G. Reich and P. Widmayer, *Beyond Steiner's problem: a VLSI oriented generalization*, in Proceedings of Graph-Theoretic Concepts in Computer Science, vol. 411 of Lecture Notes in Computer Science, 1990, pp. 196–210.

25. A. Schrijver, *Combinatorial Optimization: Polyhedra and Efficiency*, Springer, 2003.

26. P. Slavík, *Approximation algorithms for set cover and related problems*, PhD thesis, University of New York, 1998.

27. E. Tardos, *A strongly polynomial algorithm to solve combinatorial linear programs*, Operations Research, 34 (1986), pp. 250–256.
28. L. Zosin and S. Khuller, *On directed Steiner trees*, in Proceedings of the thirteenth annual ACM-SIAM symposium on Discrete algorithms, 2002, pp. 59–63.

Author Index

Lecture Notes in Computer Science

For information about Vols. 1–4432

please contact your bookseller or Springer